Movement is arguably the most fundamental and important function of the nervous system, since the organism cannot exist without it. Purposive movement requires the coordination of actions within many areas of the cerebral cortex, cerebellum, basal ganglia, and spinal cord, as well as in peripheral nerves and sensory receptors — in other words, most of the nervous system. Together, these neural structures must control a highly complex and sometimes capricious biomechanical apparatus made up of the skeleton and muscles. It should not be surprising, therefore, that we are only beginning to understand how coordinated movement can take place.

Although the specific topics in movement control addressed in this volume are not exhaustive, they were chosen to provide a broad view of this area of neuroscience. The topics are presented in a hierarchical order, beginning at the level of biomechanics and spinal reflexes and proceeding to brain structures in the cerebellum, brainstem, and cerebral cortex. Each chapter highlights important issues that need to be addressed by researchers to further our understanding of movement and how it is produced by the nervous system.

One unique feature is the critical treatment of each chapter by 10–20 commentaries. The authors of the commentaries were selected to provide a balanced treatment of the chapters by experts in a variety of areas related to movement, including behavior, physiology, robotics, and mathematics. This interactive format is used in the scientific journal *Behavioral and Brain Sciences*, and the contents of this volume are reproduced in their entirety from an issue of that journal dedicated specifically to the conference "Controversies in Neuroscience."

Movement control

Movement control

Edited by

PAUL CORDO
Robert S. Dow Neurological Center
Good Samaritan Hospital and
Medical Center

STEVAN HARNAD
Princeton University

Published by the Press Syndicate of the University of Cambridge
The Pitt Building, Trumpington Street, Cambridge CB2 1RP
40 West 20th Street, New York, NY 10011-4211, USA
10 Stamford Road, Oakleigh, Melbourne 3166, Australia

First published 1994

Library of Congress Cataloging-in-Publication Data
Movement control / edited by Paul Cordo, Stevan Harnad.
    p.   cm.
"Originally published as a special issue of the interdisciplinary
journal Behavioral and Brain Sciences"–Pref.
Includes bibliographical references and index.
ISBN 0–521–45241–4. – ISBN 0–521–45607–X (pbk.)
1. Human locomotion.   2. Muscles – Innervation.   3. Motor learning.
I. Harnad, Stevan R.
QP301.M695   1994
612′.04–dc20   93–51093
                                                            CIP

A catalog record for this book is available from the British Library.

ISBN 0–521–45241–4 hardback
ISBN 0–521–45607–X paperback

Transferred to digital printing 2004

Contents

Preface   *page* ix

1   **E. Bizzi, N. Hogan, F. A. Mussa-Ivaldi, and S. Giszter**
     Does the nervous system use equilibrium-point control to guide single and multiple joint
     movements?   1
2   **S. C. Gandevia and D. Burke**
     Does the nervous system depend on kinesthetic information to control natural limb
     movements?   12
3   **D. A. McCrea**
     Can sense be made of spinal interneuron circuits?   31
4   **D. A. Robinson**
     Implications of neural networks for how we think about brain function   42
5   **G. E. Alexander, M. R. DeLong, and M. D. Crutcher**
     Do cortical and basal ganglionic motor areas use "motor programs" to control movement? 54
6   **J. R. Bloedel**
     Functional heterogeneity with structural homogeneity: How does the cerebellum operate? 64
7   **E. E. Fetz**
     Are movement parameters recognizably coded in the activity of single neurons?   77
8   **J. F. Stein**
     The representation of egocentric space in the posterior parietal cortex   89

Open Peer Commentary and Authors' Responses

Table of Commentators   100

Open Peer Commentary
     Adamovich, S. V. – How does the nervous system control the equilibrium trajectory?   102
     Agarwal, G. C. – Movement control hypotheses: A lesson from history   103
     Alexander, G. E. – Neurophysiology of motor systems: Coming to grips with connectionism 104
     Andersen, R. A. and Brotchie, P. R. – Spatial maps versus distributed representations 105
     Barmack, N. H., Errico, P. and Fagerson, M. – Microzones, topographic maps and cerebellar
     "operations"   107
     Berkinblit, M. B., Sidorova, V. Y., Smetanin, B.N. and Tkach, T. V. – Afferent influence on central
     generators and the integration of proprioceptive input with afferent input from other modalities 107
     Beuter, A. – Modulation of kinesthetic information can be explored with nonlinear dynamics 109
     Bischof, H. and Pinz, A. J. – Artificial versus real neural networks   110
     Borrett, D. S., Yeap, T. H. and Kwan, H. C. – The nonlinear dynamics of connectionist networks:
     The basis of motor control   110
     Bossut, D. F. – Implication of neural networks for how we think about brain function 112
     Bower, J. M. – Is the cerebellum a motor control device?   112
     Braitenberg, V. and Preissl, H. – Why is the output of the cerebellum inhibitory?   113
     Bridgeman, B. – Taking distributed coding seriously   115
     Bullock, D. and Contreras-Vidal, J. L. – Adaptive behavioral phenotypes enabled by spinal interneuron
     circuits: Making sense the Darwinian way   115
     Burgess, P. R. – Equilibrium points and sensory templates   118
     Burke, D. – Movement programs in the spinal cord   120

Contents

Carey, D. P. and Servos, P. – Is "attention" necessary for visuomotor transformations? 121

Cavallari, P. – From neuron to hypothesis 121

Cavanagh, P. R., Simoneau, G. G. and Ulbrecht, J. S. – Posture and gait in diabetic distal symmetrical polyneuropathy 122

Clark, F. J. – How accurately can we perceive the positions of our limbs? 123

Clarke, T. L. – Mathematics is a useful guide to brain function 124

Colby, C. L., Duhamel, J-R. and Goldberg, M. E. – Posterior parietal cortex and retinocentric space 125

Connolly, C. I. – A robotics perspective on motor programs and path planning 126

Cordo, P. J. and Bevan, L. – Successive approximation in targeted movement: An alternative hypothesis 127

Dawson, M. R. W. – FINSTs, tag-assignment and the parietal gazetteer 128

Dean, J. – Is equilibrium-point control all there is to coding movement and do insects do it, too? 129

Dietz, V. – Control of natural movements: Interaction of various neuronal mechanisms 130

Duysens, J. and Gielen, C. C. A. M. – Spinal integration: From reflexes to perception 131

Eagleson, R. and Carey, D. P. – Connectionist networks do not model brain function 132

Feldman, A. G. – Fundamentals of motor control, kinesthesia and spinal neurons: In search of a theory 133

Flanders, M. and Soechting, J. F. – Network magic 136

Frolov, A. A. and Biryukova, E. V. – Adaptive neural networks organize muscular activity to generate equilibrium trajectories 137

Fuchs, A. F., Ling, L., Kaneko, C. R. S. and Robinson, F. R. – Network simulations and single-neuron behavior: The case for keeping the bath water 138

Fuster, J. M. – Brain systems have a way of reconciling "opposite" views of neural processing; the motor system is no exception 139

Gandevia, S. C. – How complex is a simple arm movement? 141

Gilbert, P. F. C. and Yeo, C. H. – Cerebellar function: On-line control *and* learning 141

Giszter, S. – Spinal movement primitives and motor programs: A necessary concept for motor control 142

Gnadt, J. W. – Area LIP: Three-dimensional space and visual to oculomotor transformation 143

Goodale, M. A. and Jakobson, L. S. – Action systems in the posterior parietal cortex 145

Gordon, A. M. and Inhoff, A. W. – Intermittent use of feedback during movement phase transitions and during the updating of internal models 146

Gottlieb, G. L. – Kinematics is only a (good) start 147

Graziano, M. S. and Gross, C. G. – Somatotopically organized maps of near visual space exist 148

Grobstein, P. – Information processing styles and strategies: Directed movement, neural networks, space and individuality 148

Gutman, S. R. and Gottlieb, G. L. – Virtual trajectory as a solution of the inverse dynamic problem 150

Hallett, M. – Operations of the motor system 152

Hamm, T. M. and McCurdy, M. L. – Making sense of recurrent inhibition: Comparisons of circuit organization with function 154

Hasan, Z. – Is stiffness the mainspring of posture and movement? 154

Heuer, H. – Computations, neural networks and the limits of human understanding 156

Horak, F. B., Shupert, C. and Burleigh, A. – Implications for human motor control 156

Iansek, R. – Converging approaches to the problem of single-cell recording 158

Ingle, D. – Spatial short-term memory: Evolutionary perspectives and discoveries from split-brain studies 158

Ioffe, M. E. – The necessity of a complex approach in studying brain mechanisms of movement 160

Ito, M. – Function versus synapse: Still a missing link? 161

Jaeger, D. – Toward an integration of neurophysiology, performance analysis, connectionism and compartmental modeling 161

Kalaska, J. F. and Crammond, D. J. – Neurophysiological mechanisms for the planning of movement and for spatial representations 162

Kirkwood, P. A. – The identification of corticomotoneuronal connections 164

Kuo, A. D. and Zajac, F. E. – What is the nature of the feedforward component in motor control? 165

Kupfermann, I. – Neural networks: They do not have to be complex to be complex 165

Lacquaniti, F. – Reflex control of mechanical interaction in man 166

Lan, N. and Crago, P. E. – Equilibrium-point hypothesis, minimum effort control strategy and the triphasic muscle activation pattern 167

Latash, M. L. – Are we able to preserve a motor command in the changing environment? 169

Lemon, R. – The meaning for movement of activity in single cortical output neurons 171

Levine, D. S. – Toward a genuine theoretical neuroscience of motor control 172

Loeb, G. E. – Past the equilibrium point 172

Lundberg, A. – To what extent are brain commands for movements mediated by spinal interneurones? 173

MacKay, W. A. and Riehle, A. – The single neuron is not for hiding — 174

Masson, G. and Pailhous, J. – Locomotion, oscillating dynamic systems and stiffness regulation by the basal ganglia — 176

McCollum, G. – Global organizations: Movement and spinal — 177

Morasso, P. and Sanguineti, V. – Equilibrium point and self-organization — 179

Neilson, P. D. and Neilson, M.D. – Adaptive model theory — 180

Nichols, T. R. – Stiffness regulation revisited — 181

Ostry, D. J. and Flanagan, J. R. – Aspects of the equilibrium-point hypothesis (λ model) for multijoint movements — 182

Paillard, J. – Between perception and reflex: A role for contextual kinaesthetic information — 184

Phillips, J. G., Jones, D. L., Bradshaw, J. L. and Iansek, R. – Levels of explanation and other available clinical models for motor theory — 185

Pouget, A. & Sejnowski, T. J. – A distributed common reference frame for egocentric space in the posterior parietal cortex — 185

Pratt, C. A. and Macpherson, J. M. – The many disguises of "sense": The need for multitask studies of multiarticular movements — 186

Prochazka, A. – A vital clue: Kinesthetic input is greatly enhanced in sensorimotor "vigilance" — 187

Proctor, R. W. and Franz, E. A. – Is the posterior parietal cortex the site for sensorimotor transformation? Cross-validation from studies of stimulus-response compatibility — 188

Quinlan, P. – Real space in the head? — 189

Rager, J. E. – There is much information in neural network unit activations — 190

Ross, H. E. – Command signals and the perception of force, weight and mass — 191

Rudomin, P. – Presynaptic inhibition and information transmission in neuronal populations — 191

Schieppati, M. – Selection of task-related motor output through spinal interneurones — 192

Schwarz, G. and Pouget, A. – Signals, brains and explanation — 193

Seltzer, B. – An anatomy of parallel distributed processing — 194

Smeets, J. B. J. – What do fast goal-directed movements teach us about equilibrium-point control? — 194

Smith, A. M. – Can the inferior olive both excite and inhibit Purkinje cells? — 195

Stein, J. F. – The role of the cerebellum in calibrating feedforward control — 196

Stein, R. B. – Varying the invariants of movement — 197

Summers, J. J. – The demise of the motor program — 198

Tanji, J. – Cortical area-specific activity not yet found? — 198

Thompson, R. F. – The cerebellum and memory — 199

Tsuda, I. – Nonlinear dynamical systems theory and engineering neural network: Can each afford plausible interpretation of "how" and "what"? — 200

Van Gisbergen, J. A.M. and Duysens, J. – Coordinate transformations in sensorimotor control: Persisting issues — 201

Van Ingen Schenau, G. J., Beek, P. J. and Bootsma, R. J. – Is position information alone sufficient for the control of external forces? — 202

Winters, J. M. and Mullins, P. – Synthesized neural/biochemical models used for realistic 3-D tasks are more likely to provide answers — 203

Authors' Responses

Bizzi, E., Hogan, N., Mussa-Ivaldi, F.A. and Giszter, S. – The equilibrium-point framework: A point of departure — 206

Gandevia, S. C. and Burke, D. – Afferent feedback, central programming and motor commands — 213

McCrea, D. A. – Spinal interneuronal connections: Out of the dark comes a ray of hope — 217

Robinson, D. A. – How far into brain function can neural networks take us? — 221

Alexander, G. E., DeLong, M. R. and Crutcher, M. D. – Naturalizing motor control theory: Isn't it time for a new paradigm? — 226

Bloedel, J. R. – Concepts of cerebellar integration: Still more questions than answers — 231

Fetz, E. E. – Saving the baby: Toward a meaningful reincarnation of single-unit data — 236

Stein, J. F. – Real spatial maps? — 240

References — 243
Index — 275

Preface

Producing purposive movement is one of the most fundamental functions of the nervous sytem, yet it is arguably one of the most complex. Other functions of the brain, such as memory, vision, and the automatic control of homeostatic systems, have close analogues in engineering and computer science, but the production of robots that move "naturally" continues to elude us. This lack of success in replicating animallike movement is surely due to the complexity of the motor apparatus: skeletal, muscular, and neural. The human skeleton consists of 206 bones, roughly 100 articulations, and more than 600 muscles. This mechanical system is far more complex than any current robotic device. A large proportion of the nervous system – including the peripheral nerves, much of the spinal grey and white matter, and large portions of the brainstem, cerebellum, basal ganglia, and cerebral cortex – is involved in the production of coordinated movements. And yet we move almost effortlessly and without having to "think" about it.

Given this complexity, it should not be surprising that there exist in neuroscience research a large number of controversies concerning how the nervous sytem actually controls purposive movement. The articles and commentaries in this volume originated at the first of a series of conferences entitled "Controversies in Neuroscience." The purpose of these conferences is to address the most controversial areas in neuroscience, bringing together diverse groups of investigators who work in the same area but focus on different levels of the nervous system and adopt different approaches in their research.

These conferences are sponsored by the Robert S. Dow Neurological Sciences Institute, Portland, Oregon, U.S.A., which is affiliated with Good Samaritan Hospital and Medical Center, a private community hospital with a strong commitment to scientific research. Robert S. Dow is a neurologist and well-known researcher in the area of cerebellar neuroanatomy and neuropathology. The institute grew out of his research program in the 1960s and currently consists of about 20 independent research laboratories focusing on a wide range of neuroscience, from molecular biology to human psychophysics. This institute is an ideal environment for the Controversies conferences. The diversity of its research programs allows us to draw from a wide range of expertise to organize each conference in the series. Since the first of these conferences, on movement control (1990), we have held others on neural transplantation (1991), G protein-receptors (1992), and synaptic plasticity in the cerebellum (1993), and we are currently planning conferences on persistent pain (1994) and the vestibular system (1995).

The papers and commentaries in this volume were originally published as a special issue of the interdisciplinary journal *Behavioral and Brain Sciences (BBS)* and

appear here with virtually no changes. *BBS*'s "open peer commentary" format was a natural one for recapturing the interactive flavor of the presentations and discussion, each accorded equal time during the conference. *BBS* also made it possible to include many investigators who were not present at the conference and to have all contributions refereed and revised for publication. We plan to publish the papers from each of the conferences in *BBS* and, when appropriate, as a book.

It is clearly impossible to address all of the controversial issues in any area of neuroscience in one conference, but we try to cover a wide range of topics. Thus the topics addressed in this volume run the gamut of movement control, from biomechanics to behavior. Chapter 1, by Bizzi, Hogan, Mussa-Ivaldi, and Giszter, addresses a topic related directly to biomechanics and indirectly to the structure of motor commands. In a reassessment of the so-called equilibrium point hypothesis, the authors evaluate the extent to which the biomechanics of muscles and their attachments to the skeleton constrain (or simplify) the organization of motor commands. Chapter 2, by Gandevia and Burke, explores the evidence linking kinesthetic input from sensory receptors in muscles, tendons, skin, and joint capsules to the control of movement. Whereas the question of *whether* kinesthetic information influences movement has become somewhat academic, the question of *how* this influence comes about remains controversial. Moving to the topic of the spinal cord in Chapter 3, McCrea attempts to make sense of the myriad interconnections of spinal cord circuitry with peripheral, spinal, and descending input. He uses the "wiring diagram" approach to understand this circuitry, arguing that the functions of different connections can be understood, not as rigid reflexes, but as a flexible, state-dependent substrate for processing multimodal information. In Chapter 4, Robinson uses eye movements, which are controlled principally by the brainstem, to attack the idea that the nervous system can be understood on a neuron-by-neuron basis. He argues that we must instead try to understand the processing of movement information at higher levels, such as that of entire networks.

Chapter 5, by Alexander, DeLong, and Crutcher, focuses on the basal ganglia–motor cortex loop as a substrate for motor processing. In contrast to more traditional views of serial processing of information by localized neural networks, the authors' position is that much of the processing through the basal ganglia occurs in parallel loops through distributed networks. In Chapter 6, Bloedel examines the long-debated question of what function is served by the cerebellum in controlling movement. Based on anatomical and physiological evidence, he hypothesizes that climbing fiber input to the cerebellum modulates the influence of mossy fiber input on Purkinje cells to integrate spatial features of external and internal space in targeted movements. In Chapter 7, Fetz reexamines the rationale of chronic unit recording studies in which the discharge of single motor cortical neurons has been used to infer coding of movement parameters. He argues that functional interpretations of discharge properties may be misleading, and that understanding the processing of motor output will require the examination of whole networks of neurons. In Chapter 8, Stein reviews a number of basic and clinical studies of the posterior parietal cortex (PPC) to understand better how this structure participates in the control of movement. He argues that attention plays an important role in the sensory transformations occurring in the PPC, which involve visual, auditory, somaesthetic, and vestibular sensory input.

Following these eight chapters are roughly 100 commentaries by leading researchers in neuroscience and related fields, who critically evaluate the position papers in the preceding chapters.

Most of the position papers from the Controversies conference present models of various operations or structures to illustrate the authors' positions on controversial issues. The purpose of these models differs from article to article, so it will be worthwhile to conclude this preface by addressing the proper use of modeling.

Nicolai Bernstein wrote about the modeling of motor function and organization in 1958. He pointed out that it is difficult to compare different models of the same system – to determine which model is "correct." The mere fact that one model can predict the behavior of a system under more conditions than another does not mean it is correct. Bernstein reasoned that the only strong inference one can draw from a model is that it is wrong. Modeling involves the use of a construct, often mathematical, to describe a highly complex system. It must be kept in mind that a model is not a realistic description of the actual system being modeled but a way to help us think about it and develop hypotheses to test our ideas. If a particular system were already understood, modeling it would serve little practical purpose. The very presentation of a model is therefore evidence that the operation or structure being modeled is not understood.

The value of modeling is in provoking new lines of thinking about a particular problem, not in explaining how things actually work. As modelers of the nervous system and movement, we sometimes find it difficult to accept Bernstein's observations. We want our models to be right, yet it seems inevitable that most will be wrong, at least in their specifics. Nevertheless, modeling is a necessary step to understanding. Models should not produce stress; that is, they should conform to the system they are attempting to describe. Humans and other animals cannot be described by the same rules as robots. The nervous system, in particular that of humans, has been built up through evolution into a multilayered and redundant structure. A complete description of how the nervous system produces coordinated movement will have to include not one nor even a few but many different models: for different types of movements, using different parts of the motor apparatus, in different environments. Attempts to stretch models beyond reasonable limits to predict the behavior of a system accurately can be costly and counterproductive. We must learn to accept the limits of our models.

This volume provides useful information for all students of movement, including neuroscientists (physiologists, anatomists, and behaviorists), neurologists, physical therapists, and physical educators, as well as those who have yet to determine the ultimate direction of their research careers. Its interactive format can also provide a useful new kind of text for graduate level courses in movement. The uniqueness of this volume is in the critical evaluation of existing research, in both the questions being asked and the techniques brought to bear on these questions. Whereas most textbooks provide a somewhat fixed and perhaps complacent view of a topic such as movement control, this volume was designed to point out what we do not know and, by inference, what we need to know about this area of neuroscience. We hope the chapters and discussion contained herein will provoke novice and experienced researchers alike to explore certain areas in movement control that need a more critical scientific evaluation. From a historic perspective, the volume also provides a snapshot of what we know and do not know about movement at the beginning of the last decade of the millenium.

In closing, we wish to acknowledge the contributions of the many individuals who contributed either directly or indirectly to this volume. First and foremost, we are grateful to the organizers of the first Controversies conference: Neal Barmack, Curt Bell, Fay Horak, Jane Macpherson, Carol Pratt, and all scientists at the R.S. Dow Neurological Sciences Institute (RSD-NSI). Also important in making the conference possible were Dr. Robert S. Dow, Scientist Emeritus, and Dr. Tom Morrow, founding member of Bioject Inc. and Chairman of the RSD-NSI Development Committee. Funding for the conference was generously contributed by the National Science Foundation and the Good Samaritan Foundation. Of course, the biggest contribution came from the authors of the eight chapters in this volume and the many individuals who contributed commentaries. We also acknowledge the efforts of Jim Alexander, Julia Hough, and Robin Smith of Cambridge University Press, who made possible the publication of the *BBS* issue and this book. We look forward to collaborating on future volumes on controversies in other areas of neuroscience.

Paul Cordo
Stevan Harnad

1

Does the nervous system use equilibrium-point control to guide single and multiple joint movements?

**E. Bizzi, N. Hogan,[a] F. A. Mussa-Ivaldi
and S. Giszter**

Department of Brain and Cognitive Sciences and [a]Department of
Mechanical Engineering, Massachusetts Institute of Technology,
Cambridge, MA 02139
Electronic mail: emilio@wheaties.ai.mit.edu

Abstract: The hypothesis that the central nervous system (CNS) generates movement as a shift of the limb's equilibrium posture has been corroborated experimentally in studies involving single- and multijoint motions. Posture may be controlled through the choice of muscle length-tension curve that set agonist-antagonist torque-angle curves determining an equilibrium position for the limb and the stiffness about the joints. Arm trajectories seem to be generated through a control signal defining a series of equilibrium postures. The equilibrium-point hypothesis drastically simplifies the requisite computations for multijoint movements and mechanical interactions with complex dynamic objects in the environment. Because the neuromuscular system is springlike, the instantaneous difference between the arm's actual position and the equilibrium position specified by the neural activity can generate the requisite torques, avoiding the complex "inverse dynamic" problem of computing the torques at the joints. The hypothesis provides a simple, unified description of posture and movement as well as contact control task performance, in which the limb must exert force stably and do work on objects in the environment. The latter is a surprisingly difficult problem, as robotic experience has shown. The prior evidence for the hypothesis came mainly from psychophysical and behavioral experiments. Our recent work has shown that microstimulation of the frog spinal cord's premotoneural network produces leg movements to various positions in the frog's motor space. The hypothesis can now be investigated in the neurophysiological machinery of the spinal cord.

Keywords: contact tasks; equilibrium point; force field; inverse dynamics; microstimulation; motor control; multijoint coordination; robotics; spinal cord

1. Introduction

The purpose of this target article is to present a critical evaluation of the equilibrium-point hypothesis. Before discussing its strengths and weaknesses, we would like to make clear why this hypothesis was proposed. To this end, we summarize briefly the transformations that are thought to occur when a sensory stimulus (such as an object to be reached) appears in the environment. The first step in carrying out a reaching task involves a transformation performed by cortical parietal cells. These cells receive visual, orbital, and neck afferent information. The integration of the information from these different sources generates a neural code representing the location of an object with respect to the body and the head (Andersen et al. 1985b). The second step involves the planning of the direction of hand motion and presumably its velocity and amplitude. Psychophysical observations by Morasso (1981) have suggested that this planning stage is carried out in extrinsic coordinates that represent the motion of the hand in space. In the same vein, recordings

from single cells in cortical and subcortical areas have shown a correlation between their firing pattern and the direction of hand motion (Georgopoulos et al. 1982; 1983). Whether such a correlation reflects an encoding of spatial coordinates or of muscle synergies is still an object of debate (Caminiti et al. 1990; Georgopoulos 1991; Mussa-Ivaldi 1988); it appears evident, however, that some high center of the brain such as the motor cortex must represent motor behavior in terms of extrinsic spatial coordinates. Subsequent representation in other coordinates (e.g., joint angles or muscle lengths) may also occur as part of the process of implementing the motor plan. This observation was first made in 1935 by Bernstein, who noted that our ability to control movements is independent of movement scale or location (Bernstein 1967).

If the spatial features of a hand movement are planned and represented by some structure of the CNS then there must be another set of neural processes devoted to transforming the desired hand trajectory into muscle activations. A third step in carrying out a reaching task therefore consists in the conversion by the CNS of the desired direction, amplitude, and velocity of movement into signals that control the mechanical action of the muscles. The equilibrium-point hypothesis is related to this third step and the communication between the processes of movement planning and movement execution.

Investigators of motor control have become increasingly aware of the computational complexities in the production of muscle forces. Some have proposed that the CNS derives a motion of the joints from the desired path of the end point (inverse kinematics) and that it then derives the forces to be delivered to the muscles (inverse dynamics; Hollerbach & Atkeson 1987). The idea that the CNS performs these inverse computations implies that it can somehow estimate precisely limb inertias, center of mass, and the moment arm of muscles. Small errors in the estimation of these parameters can result in inappropriate movements. Robotic experience with similar approaches has shown that inertial parameter errors as small as 5% can result in instability (Slotine 1985). Most motor control investigations regard this feedforward computation as rather unrealistic. As an alternative, we and others have proposed a different solution to the inverse dynamics problem: the equilibrium-point hypothesis.

2. Definition of the equilibrium-point hypothesis

The hypothesis was first proposed by Feldman (1966b), who viewed joint posture as an equilibrium resulting from the length-dependent forces generated by agonist-antagonist muscles. A key feature of the equilibrium-point hypothesis is that muscles have springlike behavior. Experimental evidence has indicated that muscles behave like tunable springs in the sense that the force they generate is a function of their length and neural activation level (Matthews 1972; Rack & Westbury 1974). The force-length relationship in individual muscle fibers was studied by Gordon et al. (1966), who related the development of tension at different muscle lengths to the degree of overlap between actin and myosin filaments. This overlap limits the formation of cross-bridges. The increase in muscular stiffness observed when the motoneuronal drive increases is considered a direct consequence of the generation of new cross-bridges.

In 1966, Feldman put forward the idea that the CNS may execute a movement by generating CNS signals that change the relative activation of agonist and antagonist muscles. This change in activation generates joint torques; the resulting joint motion will depend upon the muscle torques and the external loads.

There are at least two variants of the equilibrium-point hypothesis. Feldman (1986) called them the "alpha" and "lambda" models. We would like to stress that the alpha model, which has been attributed to our group, reflects our views only in part. The following discussion describes our interpretation of the two models.

2.1. The alpha model. A central postulate of the alpha model is that the CNS generates a temporal sequence of signals that specify, at all times, an equilibrium position of a limb and the stiffness of the muscles acting on the limb. Although the terminology of the equilibrium-point hypothesis is by now firmly rooted in the literature, the term

equilibrium position is a source of some confusion. We use the term in the following sense: It is the location at which the limb would rest if the centrally generated commands were "frozen" at any given value and the limb were free to move in the absence of external loads or forces. In the presence of static external loads or forces, the actual equilibrium position of the limb will in general differ from this position. Hence we introduced the term *virtual position* to distinguish the two. A time sequence of central commands gives rise to a time sequence of virtual positions, which is called a *virtual trajectory*.

The experimental evidence supporting this view derives from three sets of experiments performed in monkeys (Bizzi et al. 1984). The movements used in these experiments were visually evoked single-joint flexion and extension of the elbow, which lasted approximately 700 msec for a 60-degree amplitude.

The first set of experiments was performed both with intact monkeys and monkeys deprived of sensory feedback. The monkey's forearm was briefly held in its initial position after a target light that indicated final position had been presented. It was found that movements to the target after the forearm was released were consistently faster than control movements in the absence of a holding action.

Figure 1 shows a plot of the initial accelerative transients against the durations of the holding period in the same animal before and after interruption of the nerves conveying sensory information. The time course of the increase in the amplitude of the initial accelerative transient was virtually identical in the two conditions.

The initial acceleration after the release of the forearm increased gradually with the duration of the holding period, reaching a steady-state value no sooner than 400 msec after muscles' activation. These results demonstrated that the CNS has programmed a slow, gradual shift of the equilibrium position instead of a sudden, discontinuous transition to the final position.

The same conclusions were supported by a second set

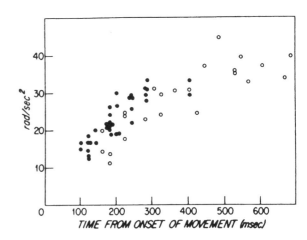

Figure 1. The forearm of intact and deafferented animals was held in its initial position while the animal attempted to move toward a target light. Then the forearm was released at various times. This figure is a plot of acceleration (immediately following release) versus holding time. The abscissa shows time in milliseconds; the ordinate shows radians per second squared. Solid circles: intact animal; open circles: deafferented animal. (From Bizzi et al. 1984.)

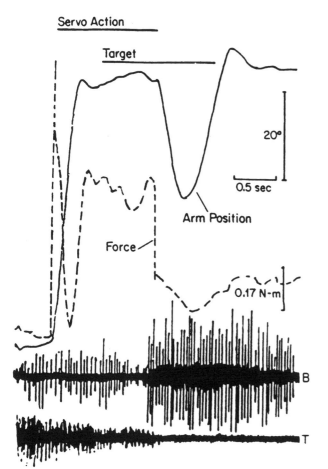

Figure 2. Forearm movements of deafferented monkeys with a holding action in the final position. While the target light remained off, the servo moved the arm to the target position. Then the target light was activated and the servo was turned off. The arm returned to a position intermediate between the initial and target positions before moving back to the target position. Similar results were obtained in many trials in two monkeys. The upper bar indicates duration of servo action. The lower bar indicates onset of the target light. The broad trace shows arm position; the dashed trace shows torque. B: flexor (biceps); T: extensor (triceps). (From Bizzi et al. 1984.)

of experiments (Bizzi et al. 1984) based on forcing the forearm to a target position through an assisting torque pulse applied at the beginning of a visually triggered forearm movement. The goal of the experiments was to move the limb ahead of the postulated equilibrium position with an externally imposed displacement in the direction of the target. It was found that the forearm, after being forced by the assisting pulse to the target position, returned to a point between the initial and final position before moving to the end point. This outcome results from a restoring force that is generated by the springlike muscle properties. If muscles merely generated force, or if the springlike properties were negligible, this return motion of the limb would not have been observed. Again, these experiments were performed in both intact and deafferented monkeys.

In the third set of experiments, performed in monkeys deprived of proprioceptive feedback, the forearm was first driven passively from the initial position to a new position in the absence of a target light and held there for a variable amount of time (1 to 3 sec), after which a target light at the new position was activated (Figure 2).

A cover prevented the animal from seeing its arm. After the reaction time to the presentation of the target light, the monkey activated its muscles (flexors in the case of Figure 2) to reach the target position even though the elbow was already there. At this point, usually shortly after the onset of muscle activity, the servo that held the arm was deactivated. Once released, the arm moved toward the original starting position to a point intermediate between the initial position and the target position before moving back to the target position. Note that during the return movement requiring extension, evident flexor activity was present. The amplitude of the return movement depended on the duration of the holding action. If enough time elapsed between the activation of the target light and the deactivation of the servo, the arm remained in the target position upon release. These observations provide further support for the view that the CNS specifies a series of equilibrium positions throughout the movement.

The idea of a moving equilibrium point (termed a virtual trajectory) is a direct consequence of two known facts: (1) that a limb is at static equilibrium in the absence of external loads when all the torques generated by opposing muscles cancel out, and (2) that the neural input to each muscle has the effect of selecting a length-tension curve. If the net stiffness due to muscle action on the limb is positive (see below), it follows that at all times the neural activities directed to all the muscles acting on a limb can be "translated" into a corresponding equilibrium angle, which is given by the balance of the springlike torques that keep the limb at rest (Hogan 1988a).

During the execution of a movement, these equilibria or virtual positions act as centers of attraction. The difference between actual and virtual position generates a springlike force directed toward a virtual position. The course of the movement is then determined by the interaction of the springlike force with limb inertia and viscosity and the velocity-based modulation of the muscle length-tension relationship. On this basis, a computer simulation developed by Hogan (1984) successfully reproduced all of the above experimental observations.

It should be stressed that a comparison of forearm movements in intact and deafferented monkeys revealed some quantitative, but no qualitative, differences. Insofar as the deafferented animal could execute movement, sensory feedback was not essential for movement. The major role of feedback in the experiments of Bizzi et al. (1984) may have been to augment the inherent properties of muscles such as stiffness, a role for which a considerable body of experimental evidence (Hoffer & Andreassen 1981) has been presented. It follows that in these highly trained and deafferented animals, the simple task of moving only one joint was executed primarily by a movement program of central origin. It should be noted that Bizzi et al. (1984) studied large arm movements with undemanding requirements on final position accuracy. In addition, these movements were performed at moderate speed. It is conceivable that under these circumstances feedback plays a minor role. In contrast, during the execution of motor tasks demanding greater accuracy, feedback may well play a much more important role, as shown by Sanes and Evarts (1983a) and by Day and

Marsden (1982). It is known that muscle and cutaneous receptors are most sensitive to signals of small amplitude (Matthews & Stein 1969).

In summary, the alpha equilibrium-point model rests upon studies in which fairly large single-joint movements were performed by deafferented animals at moderate speeds. We hasten to add that whereas our animals performed adequately in our restricted condition, there is no question that the performance of deafferented monkeys and humans is far from normal during the execution of multijoint movements. Clearly, sensory input must play a role. The lambda model and its predecessor, Merton's (1953) reflex servo control model, are directed at including sensory input into a motor-control scheme.

2.2. The lambda model. Alpha and lambda models have many common features. For example, both posit a unified description of posture and movement, and both attribute springlike properties to muscles and view movements as being generated by a shift in the equilibrium position of the limb. The main difference between the alpha and lambda models lies in the importance accorded the distinction between feedforward and feedback mechanisms generating an equilibrium position. In the alpha model the torque, T, produced by the muscles about a joint, is expressed as a function of the joint equilibrium angle, θ, and the centrally generated commands. In the deafferented animal, those commands are reflected in the muscle active states, presumably determined by the set of α-motoneuron activities $\{\alpha\}$:

$$T = \psi(\theta,\{\alpha\}). \qquad (1)$$

In the absence of external loads or forces, the joint equilibrium angle θ_0 may accordingly be expressed as a function[1] of $\{\alpha\}$, $\theta_0 = g(\{\alpha\})$, provided the stiffness is nonzero. In subsequent work, the same formalism was applied with success (Flash 1987) in a study simulating multijoint arm trajectories. In our view, it is the apparent mechanical behavior of the neuromuscular system, whether due to reflex action or intrinsic muscle behavior, that determines the stability and dynamic behavior of the limbs and how they interact with objects in the environment. Thus, in our formulation we deliberately make no attempt to distinguish between central and reflex effects on the α-activity. In contrast, Feldman (1986) has proposed that the net effect of the central commands impinging upon the α-motoneurons is to modulate the threshold of the stretch reflex. More precisely, in Feldman's lambda model the activity of the α-motoneurons can be expressed as a function

$$\alpha = \alpha(x - \lambda) \qquad (2)$$

of a muscle's length, x, and a centrally determined control parameter, λ. This parameter has the same dimension of muscle length and $\alpha(x - \lambda)$ is a threshold function ($\alpha = 0$ if $x \leq \lambda$ and $\alpha > 0$ otherwise). Thus, each muscle force, F, is expressed as a function of the difference between muscle length, x, and a control parameter, λ, that is $F = F(x - \lambda)$. Note that for $x > \lambda$, the function follows an invariant characteristic, and for $x \leq \lambda$ the active force is zero. Feldman and coworkers have suggested using the lambda model to account for the formation of known electromyographic (EMG) patterns (Berkinblit et al. 1986;

Feldman 1986). They suggested that this important goal could be achieved by relating the parameter λ on one hand to the equilibrium position, θ_0, and on the other to the net motoneuron activation, α.

Our main critique of the lambda model is related to its striking resemblance to Merton's servo reflex control hypothesis. Like the lambda model, Merton's (1953) hypothesis postulated that voluntary movements are initiated and controlled by the CNS as a central modulation of feedback. More precisely, the mismatch between extra- and intrafusal muscle fiber lengths generates an input signal to the alpha motoneurons via the monosynaptic pathways. Merton proposed considering this input as a length-error signal that the CNS specifies via the gamma system. Therefore, according to this hypothesis, movements are initiated and controlled by a specification of gamma fusimotor activity.

It is well known that the experimental evidence does not support Merton's hypothesis for the following reasons: (1) There is no gamma lead with respect to alpha activity, (2) the gain of the stretch reflex during movement is too small, and (3) deafferented animals can perform goal-directed movements, albeit in a clumsy way. The lambda model avoids the first criticism by postulating an (unspecified) relation between the centrally determined control parameter, λ, and the α- and γ-motoneuron activities. In this respect the λ model resembles the modification of Merton's hypothesis known as the servo-assistance hypothesis (Granit 1970; Stein 1974). The lambda model avoids the second criticism on the grounds that a significant stretch-reflex gain is not necessary for postural stability. However, in that case the relative contribution of reflexes to the expression of movement would presumably also be insignificant. The third criticism still applies to the lambda model.

Like Feldman, we believe that reflex activity may modify supraspinal commands. We part company, however, on the question of the relative contribution of reflexes to the expression of movement. The experimental work of Bizzi et al. (1978) investigated the contribution of reflex mechanisms in generating the forces produced by the neck muscles when loads were unexpectedly applied during centrally programmed head movements in monkeys. The results indicated that the compensating torque of reflex origin was less than 10% to 30% of that required for perfect compensation. Similar results were obtained by Vallbo (1973a), Grillner (1972), and Allum (1975). The conclusion from these experiments is that the reflex apparatus contributes in a modest way to force generation. These observations may be consistent with the fact that feedback-loop delays may cause instability if reflex gains are too high (Hogan et al. 1987).

Regarding the ability of deafferented animals to perform goal-directed movements, Feldman has suggested that in such a pathological condition the CNS replaces the lambda control with a coactivation strategy (Feldman 1986). This could be the case. We would like to point out, however, that even such a cocontraction strategy would require the remarkable ability of the CNS to control and coordinate directly a complex pattern of alpha activations. The available evidence suggests that these activations are correctly mapped by the deafferented animal into a desired equilibrium location, as suggested by the alpha model.

2.3. The alpha/lambda debate. Feldman (1986) has suggested that the alpha and lambda models are mutually exclusive.[2] In contrast, we believe that these models are mutually compatible and also that one, the lambda model, is a subset (or constrained version) of the other, the alpha model.

To illustrate this point let us consider a single-joint system. The torque-angle function of this system is given by Equation 1 in the alpha model. In the lambda model, at a single joint, each muscle's length is unique demonstrated by the joint angle and each muscle makes a specific contribution to the joint angle and a set of command parameters $\{\lambda\}$:

$$T = \phi(\theta, \{\lambda\}) \tag{3}$$

As was the case with the alpha model, provided the joint stiffness is nonzero, the equilibrium condition, $\phi(\theta_0, \{\lambda\}) = 0$, defines a map from $\{\lambda\}$ to the equilibrium position, θ_0, $\theta_0 = f(\{\lambda\})$.

However, the specific form assumed for the dependence on λ, a threshold function, means that for a single muscle there exists a range of joint angles (corresponding to the muscle lengths $x \leq \lambda$) for which that muscle stiffness is nearly zero. The required nonzero joint stiffness may be achieved by cocontraction of opposing muscles across the joint, just as in the alpha model.

The lambda model is related to the alpha model by a set of equations (Equation 2) that establish a dependence of the alpha signal upon the muscle length, x, and the centrally generated command, λ. Given the relation between muscle length and joint angle, we may rewrite these equations as

$$\{\alpha\} = \{\alpha(\theta - \lambda)\} \tag{4}$$

The equilibrium condition ($T = 0$) applied to the alpha model provides a map g from the set $\{\alpha\}$ to the equilibrium angle θ_0. The same equilibrium condition applied to the lambda model provides a different map f from the set $\{\lambda\}$ to the equilibrium angle. In set-theoretic notation, the first map is a set of ordered pairs, $g = \{(\theta_0, \{\alpha\})\}$, and the second is a set of pairs, $f = \{(\theta_0, \{\lambda\})\}$. It is easy to show[3] that under Feldman's conditions, f is a subset of g. The converse is not true. For example, unlike the lambda model, the alpha model does not require an invariant characteristic for the form of the torque-angle relation. In this case, there are elements of g that do not belong to f.

In summary, we believe that the alpha and lambda models are by no means mutually exclusive. Each model tends to direct attention toward a different aspect of motor control. The lambda model has been primarily applied to the explanation of EMG patterns. In contrast, we have been more concerned with the mapping of different motor behaviors – such as posture, movement, and contact – into the corresponding sets of equilibrium positions. In this regard, the fact that reflexes may induce a restriction of the alpha model may be of significant value.

We would like to reiterate that we feel uncomfortable being cast in the role of advocating a pure alpha model which we have never proposed. The experimental evidence reported in Bizzi et al. (1976; 1982; 1984), Polit and Bizzi (1978), and in Hogan (1982; 1984) was directed at establishing whether the CNS adopts a final-position control-strategy in order to generate arm trajectories in animals performing highly practiced forearm move-

ments. The results indicate that the transition from the initial to the final position was implemented by a gradual shift in the control signal establishing both a trajectory and a final equilibrium condition. To make the observation we used deafferented animals. However, our experiments were not intended to ascertain the relative contribution in the intact animal of feedforward commands versus feedback signals.

3. Multijoint posture and movement

The study of multijoint arm movement presents problems radically different from those of the single joint. In a multijoint situation, if a displacement is externally imposed on the hand, the amount of stretch experienced by the muscle depends not only upon the amplitude of the perturbation, but also upon its direction. Then, a single number is no longer sufficient to describe the force-displacement relation. This relation is expressed by a matrix whose elements characterize the ratio of each component of the restoring force vector to each component of the applied displacement vector.

To deal with the more complex situation of multijoint arm movement, a specific experimental approach to the study of posture was developed (see Mussa-Ivaldi et al. 1985). This approach was based on measuring the net springlike behavior of the multijoint arm by displacing the hand in several directions from rest and measuring the restoring forces, $F = (F_x, F_y)$. Because the displacements of the hand, $\delta r = (\delta x, \delta y)$, were small enough to justify neglecting higher order effects, a linear relation of the following form was assumed:

$$F_x = K_{xx}\, \delta x + K_{xy}\, \delta y$$

$$F_y = K_{yx}\, dx + K_{yy}\, \delta y. \tag{5}$$

Then, by measuring forces and displacements in different directions, it was possible to estimate the K coefficients from a linear regression applied independently to both of the above expressions. These coefficients could be represented by a single entity – a table, or matrix, expressing the multidimensional stiffness of the hand:

$$K = \begin{bmatrix} K_{xx} & K_{xy} \\ K_{yx} & K_{yy} \end{bmatrix} \tag{6}$$

With this notation, Equation 5 assumes a more compact form, $F = K\, \delta r$.

We first determined whether the behavior of the multijoint arm was in fact springlike. The curl of the force field $F = F(\delta r)$ must be zero for a springlike system (Hogan 1984). In terms of the stiffness matrix, the off-diagonal coefficients K_{xy} and K_{yx} (which were measured independently) would have to be identical. Our measurements showed that this was the case.

Because the curl of the force field was zero, the hand stiffness matrix K was symmetric and could be represented as an ellipse characterized by three parameters: magnitude (the total area derived from the determinant of the stiffness matrix); orientation (the direction of maximum stiffness); and shape (the ratio between maximum and minimum stiffness). The hand stiffness was actually estimated by Mussa-Ivaldi et al. (1985). Subjects were asked to maintain the hand at a set of workspace locations. At each location, K was derived from measured force and displacement vectors as outlined in Equation 1. The corresponding stiffness ellipses captured the main geometrical features of the springlike-force field associated with a given hand posture and provided an understanding of how the arm interacts with the environment.

To sum up, the experimental evidence indicates that an equilibrium position of the hand is established by the coordinated interaction of spring like forces generated by the arm muscles (Mussa-Ivaldi et al. 1985). According to the virtual-trajectory hypothesis, which was tested first in the context of single-joint movements (Bizzi et al. 1984), the multijoint arm trajectories are achieved as the CNS gradually shifts the centrally determined virtual position between the initial and final positions. In this control scheme the hand tracks its virtual equilibrium point and torque need not be computed explicitly.

Evidence supporting this equilibrium-point hypothesis in the context of multijoint hand movements was obtained by combining observations of hand movements with computer simulation studies. A model developed by Flash (1987) has successfully captured the kinematic features of measured planar-arm trajectories. In the simulation, Flash made the assumption that the hand's virtual trajectories (but not necessarily the actual trajectories) are invariantly straight. In addition, she assumed that each virtual trajectory has a unimodal velocity profile, regardless of the target locations in the workspace.

The arm dynamics were simulated by obtaining torques derived from the difference between actual and virtual positions multiplied by the stiffness (Flash 1987). It must be stressed that the stiffness parameters used in the simulation of movements were derived from experimentally measured postural stiffness values. The results of the simulation showed that with straight virtual trajectories, the actual movements were slightly curved. Moreover, the direction of curvature, in different workspace locations and with different movement directions, was in good agreement with the experimentally observed movements. This result suggests that during movement planning, the CNS ignores the inertial and viscous properties of the arm and directly translates the desired trajectory into a sequence of equilibrium positions. Therefore, when the movement is executed, the inertial and viscous forces act as perturbations, causing deviations of the actual path with respect to the planned path.

The success of the simulation in capturing the kinematic details of measured arm movements is important as a step toward providing us with a framework for understanding the CNS's trajectory formation in the multijoint context. This work indicates a planning strategy whereby the motor controller may avoid complex computations such as the solution of the inverse dynamics problem. Recent findings by Flanagan et al. (1990) agree partially with Flash's simulations. Flanagan's results support the notion that multijoint movements are planned in endpoint coordinates. In contrast with Flash, their results suggest that the equilibrium position is shifted at a constant velocity.

It should be noted that the simulations described by Flash were for relatively slow movements. The question whether fast movements could be achieved in the same way is dealt with in section 7.2. It is also important to note that the good agreement between simulated and experimental trajectories was contingent upon using stiffness

fields whose shape and orientation were identical with those recorded under static conditions by Flash; any change in these two parameters led to substantially different simulated trajectories. Hence, a question left unanswered by Flash's experiments was whether the neural signals to the muscles involved in the execution of natural movements could significantly alter the shape and orientation of the stiffness field, thus undermining the significance of the simulation.

An answer to the latter question was provided by Mussa-Ivaldi et al. (1987; see also Bizzi & Mussa-Ivaldi 1990). Mussa-Ivaldi et al. (1987) found that the shape and orientation of the stiffness field did not change when externally imposed disturbances acting in different directions were applied to the hand. These disturbances generated large shifts in the EMG activation of arm muscles but failed to modify these two parameters. The results of Mussa-Ivaldi et al. provided the evidence for assuming that these parameters may not change when the hand moves through the locations at which the field was measured.

Flash's simulation showed that her model can generate multijoint arm trajectories. The experiment of McKeon et al. (1984), which complement Flash's results (1987), provided qualitative evidence supporting the equilibrium-point hypothesis in the context of human multijoint movements. McKeon et al. (1984) asked subjects to perform pointing movements between two targets while gripping the handle of a two-link manipulandum. A clutch mounted on the inner joint of the manipulandum was used to brake the inner link under computer control. Because the clutch was activated randomly at the onset of a movement, the hand trajectory was restricted to a circular path with a radius equal to the length of the outer link of the manipulandum. While the clutch was engaged, the force exerted on the handle was always strongly oriented so as to restore the hand to the unconstrained path and not to the end point of the path. This result is in accordance with the equilibrium-point trajectory: The muscle's springlike properties and the proprioceptive reflexes generate forces attracting the hand toward the original path.

4. Control of contact tasks

The equilibrium-point hypothesis also provides a simple but highly effective way to solve the much more complex problem of contact control tasks in which the limb must exert force stably and do work on objects in the environment. In general, the manipulated object can have its own dynamic behavior, which may be arbitrarily complex. Because of the mechanical interaction, that dynamic behavior is added to the already complex dynamic behavior of the limb.

The ability to control contact with objects is clearly a fundamental prerequisite for the use of tools, one of the distinctive features of human behavior. [See also Parker & Gibson: "A Developmental Model for the Evolution of Language and Intelligence in Early Hominids" *BBS* 1 (3) 1979; Chevalier-Skolnikoff: "Spontaneous Tool Use and Sensorimotor Intelligence in *Cebus* Compared With Other Monkeys and Apes" *BBS* 12 (3) 1989; MacNeilage et al.: "Primate Handedness Reconsidered" *BBS* 10 (2) 1987; and Greenfield: "Language, Tools and Brain" *BBS*

14 (4) 1991.] The subtlety and difficulty of this problem is disguised by the ease with which humans manipulate objects. Experience with robots has repeatedly shown that even the apparently trivial problem of controlling the force exerted on a surface has proven surprisingly difficult. In robotics, an "obvious" approach is to measure the force of contact and send that information to the controlling computer so that it can adjust or regulate the force exerted. Unfortunately, this approach has been plagued by a phenomenon called *contact instability*. Robotics researchers in numerous laboratories have reported that a robot capable of executing unrestrained motions stably and accurately will break into a pathologically uncontrollable chattering instability upon contact with a rigid surface, bouncing off the surface and bumping it repeatedly. This problem has been identified as one of the prominent challenges of robotics (Paul 1987). Yet biological systems clearly have little difficulty contacting and manipulating objects.

The necessary and sufficient condition for a manipulator to remain stable when coupled to an arbitrarily complex passive object has recently been derived mathematically and verified experimentally. Details are provided in Colgate (1988; Colgate & Hogan 1988; Hogan 1988b). The essence of the result is that an arbitrary collection of passive objects such as springs and masses can temporarily store energy, but cannot supply it indefinitely. Consequently, if there is nonzero dissipation associated with the motion of the system it will converge on a stable state of minimum energy, a result first proved by Lord Kelvin.

In contrast, a typical actuator (e.g., a robot motor or a muscle) can supply energy indefinitely (or at least over time scales that are long compared to the characteristic dynamic behavior of the system they act upon; i.e., muscle may continuously supply energy for far longer than the duration of a typical voluntary movement). If that energy supply is improperly controlled (e.g., so that the energy supplied exceeds that dissipated), unstable behavior may result.

However, if the actuator control system is designed so that the apparent behavior of the actuator is that of an object that can only temporarily store energy (e.g., a spring), then Lord Kelvin's result is recovered: That actuator can be connected to an arbitrary collection of springs and masses and the combined system will be stable. This is precisely what is achieved by the springlike behavior of single muscles or agonist-antagonist muscle groups about a single joint.

In the multijoint case, new factors arise because of the possibility of complex interactions between joints. If the off-diagonal terms in the stiffness matrix were not identical, $K_{xy} \neq K_{yx}$, it would imply that energy could be supplied indefinitely by perturbing the hand so as to make small circular motions about the equilibrium point. It is therefore highly significant that our psychophysical experiments on human subjects have established that the entire multijoint upper limb mimics the behavior of a passive, multijoint spring (Mussa-Ivaldi et al. 1985) even though that requires finely balanced interjoint feedback (Hogan 1985a).

The relation between the properties of the hand stiffness and the equilibrium-point hypothesis may be summarized as follows: In the single-joint case, in order to define a map relating central commands to a virtual position the stiffness must be nonzero. In the multijoint case, the corresponding requirement is that the stiffness matrix, K, must be nonsingular (determinant $(K) \neq 0$). For static stability about that equilibrium point, the stiffness matrix must be positive definite, a stronger requirement. Passive springlike behavior adds a further requirement: The stiffness matrix must be symmetric.

To establish fully stability or passivity of the arm, its dynamic behavior must be considered in addition to its stiffness; measurements of the arm's dynamic response to perturbation (its mechanical impedance) are required. In the absence of that information we conclude that our observation of multijoint springlike arm behavior is consistent with the theoretical requirements to preserve stability on contact with passive objects, although a stronger conclusion would require a more thorough analysis of dynamic behavior.

The way equilibrium-point control may be used in contact tasks is illustrated conceptually in Figure 3. Although the real-life situation may be more complex, the basic mechanics are sufficiently similar to illustrate the concept. Figure 3A depicts a hand being moved downward to contact a surface and push on it. The hand is assumed to be controlled so that a simple relation between its force and its position is maintained. For simplicity, the force, F, is assumed to be proportional to the separation of the virtual position, X_v, and the actual position, X, of the hand:

$$F = K(X_v - X) \qquad (7)$$

where K is a constant, the stiffness at the hand.

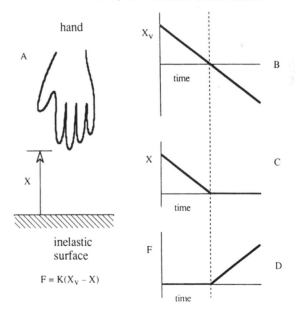

Figure 3. Diagram illustrating how virtual position may be used to control contact tasks. A: The hand is controlled by a relation between force exerted, F, and the difference between a virtual position, X_v, and the actual hand position, X. B: A trajectory of the virtual position, X_v, that could be used to move to a surface and exert force on it. C: While the hand is not in contact, its actual position, X, tends to follow the virtual position. On contact, the position of the hand is constrained by the surface. D: While the hand is not in contact, the force exerted by the hand, F, is constrained to zero. On contact, the force is proportional to the distance the virtual position moves "into" the surface.

Next assume that the virtual position, X_v, is moved slowly and steadily downward, as shown in Figure 3B. As long as the hand is clear of the surface, the force exerted is zero and the actual position of the hand will approximately track the equilibrium position, as depicted in Figure 3C. As soon as the hand contacts the surface, the actual position can no longer follow the virtual one because of the physical constraint imposed by the surface. But once the hand is in contact with the surface, the problem is to control the force exerted. This is easily accomplished by controlling the depth the equilibrium position "penetrates" into the surface. The fact that it may penetrate into a surface is the main reason the equilibrium position defined by the action of neuromuscular system is called *virtual*. Figure 3D shows that as the virtual position moves deeper into the surface, the force rises in the proportion required by the relation between force and position.

This simple example illustrates that *the same process* that can simplify the computational burden of controlling unrestrained motions can also be an effective way to control the force exerted on an object. The idea is a generalization of the unified description of posture and movement outlined above. If the response of the limb to perturbations is appropriate, the equilibrium position, which may simplify motion control, is also an effective way to control interaction. During interactive tasks, the behavior of the limb is dominated by the dynamic response of the skeletal and neuromuscular system to perturbation. Functional interactive behavior such as tool use can therefore be controlled by modulating that dynamic response, the mechanical impedance of the neuromuscular system (Hogan 1985b).

5. Neurophysiological basis of the equilibrium-point hypothesis

Until recently, the equilibrium-point hypothesis rested mostly on data derived from psychophysical and behavioral experiments. Recently, Mussa-Ivaldi et al. (1990), Giszter et al. (1991c; 1992b) and Bizzi et al. (1991) addressed directly the problem of providing a neurophysiological underpinning for the equilibrium-point hypothesis. To this end, they investigated the characteristics of the spinal circuitry involved in producing coordinated multijoint leg movements and postures. For these studies they used the spinal frog.

The spinal frog is a simplified preparation that retains significant multijoint motor abilities (Berkinblit et al. 1986; Fukson et al. 1980; Giszter et al. 1989). It is well known, for example, that the spinal frog is capable of generating a coordinated sequence of multijoint hindlimb movements to remove a noxious stimulus from the skin. This "wiping reflex" requires complex information processing. Thus, the spinal cord must contain enough circuitry to coordinate the motion of multiple limb segments.

One possible approach to understanding the motor behavior of a spinalized frog consists of postulating that a noxious stimulus on the skin triggers some form of an "inverse-dynamics" computation within the spinal cord. This computation must ultimately generate a coordinated pattern of joint torques in the hindlimb. In contrast,

according to the equilibrium-point hypothesis, the motion of the hindlimb is generated by the development of neural patterns that specify a sequence of equilibrium points with the limb's workspace. In support of the latter view, evidence for extensive cocontraction during flexion and wiping has indeed been found (Schotland et al. 1989).

We have addressed these different hypotheses in experiments in which we microstimulated the gray matter of spinalized frogs. According to the view that favors inverse dynamics, the activation of a region in the spinal gray matter is expected to generate a timed pattern of joint torques. These torques need not define an equilibrium point within the workspace.

Alternatively, the equilibrium-point hypothesis predicts that we should be able to induce a stable equilibrium of the leg within its range of action by activating the spinal gray matter. The equilibrium-point hypothesis also implies that the development of neural patterns corresponds to a movement of the equilibrium point.

Mussa-Ivaldi et al. (1990), Giszter et al. (1992a; 1992b), and Bizzi et al. (1991) elicited motor responses by microstimulating the spinal gray matter in a region located from the base of the dorsal horn to the upper ventral horn. An important methodological feature of these experiments involved measuring the x and y force components at the ankle with a 6-axis force transducer. The x-y plane corresponded approximately to the horizontal plane, where most of the wiping movements tended to occur.

In these studies, the electrical stimulation of the spinal gray matter at threshold levels for movement always coactivated groups of muscles. In order to record the forces generated by the leg, a two-part procedure in a single recording session was followed: First, the frog's ankle was placed at one location in the leg's horizontal workspace (i.e., that region of the horizontal plane that can be reached by the ankle). Second, the direction and amplitude of the force at the ankle elicited by stimulating a site in the spinal cord were recorded. While stimulating the same site, the investigators in these studies repeated this procedure with the ankle placed at each of 14 locations covering the whole range of the workspace. The force vectors that were recorded varied in direction and amplitude as the experimenters placed the leg at different workspace locations. The measured force vectors were used to estimate the force field in a large region of the ankle's workspace. Remarkably, in most instances, the spatial distribution of these vectors resulted in a field characterized by a single equilibrium point (i.e., a point at which the amplitude of the Fx, Fy force components was zero). The fields, with their associated equilibria, were found to be distributed in several locations throughout the leg's workspace.

Note that the type of field shown in Figure 4A does not result merely from the mechanical properties of the musculoskeletal system. A radically different pattern of forces was found when the stimulating electrode was placed in the ventral roots or within gray matter regions populated by the motoneurons (Figure 4B). In this case, the structure of the field was often characterized by forces that were parallel or divergent, with no detectable equilibrium point. The striking differences obtained after stimulation in regions consisting predominantly of interneurons compared to those with predominantly motoneurons indicated that the stimulation and activation of

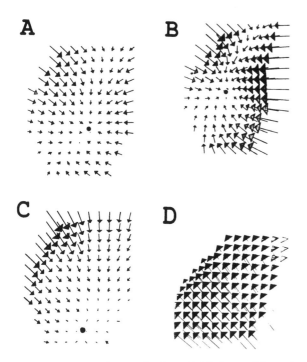

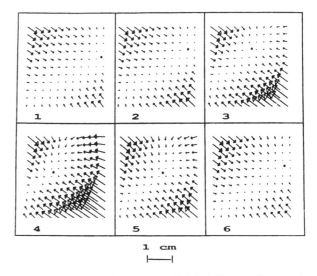

Figure 5. Temporal sequence of fields following the stimulation of a spinal site. The six frames are ordered by increasing latency from the stimulus and are separated by intervals of 86 msec. The filled circle indicates the equilibrium point.

Figure 4. Force fields. *A:* The force field obtained from the quiescent frog before the stimulation of the cord. The equilibrium of the force field is indicated by the filled circle. *B:* The alteration of the force field as a result of stimulating a site in the spinal cord grey matter in the lateral interneuron region. *C:* A force field resulting from stimulating a different spinal cord site (1 mm rostral to the site that generated the field shown in *B*) but still in the lateral interneuron region. Clearly, the equilibria lie in different locations. *D:* A force field resulting from stimulating motor fibers (8th ventral root) with the same currents used in the stimulation of the lateral interneuron region. Clearly, no equilibrium is present. We obtained similar results when the electrode was placed among the motoneurons.

the spinal cord's premotoneuronal network caused a balanced recruitment of motoneurons that imposed a structure on the forces generated by the limb muscles.

6. Temporal evolution of the force fields

After the delivery of a stimulus to the spinal gray matter, each measured force component changed with time. Consequently, the force field as a whole changed with time. The dependence of the force field on time is captured graphically by a sequence of "frames" (Figure 5). Each frame shows the force field measured at a given latency from the onset of the stimulus.

The first frame (latency = 0) shows the resting force field, that is, the field as it was before the stimulus had produced any mechanical effect. The subsequent frames are separated by intervals of 86 msec. They show the effect of the stimulus as a smooth change in the overall pattern of forces. In several instances we have observed the following sequence of events (as indicated in Figure 5):

(1) After a brief delay from the onset of the stimulus (about 50 msec), the pattern of forces began to change and the equilibrium position started to "move" in a given direction (Figure 5, frames 1 to 3).

(2) Then (Figure 5, frame 4), the equilibrium position reached a point of maximum displacement within the workspace. This point was maintained for a time interval that depended on the stimulation parameters (current, train duration, etc.). At the same time, the field forces reached a maximum amplitude around the equilibrium position, corresponding to a maximum in end-point stiffness.

(3) Finally (Figure 5, frames 5 & 6), the force vectors started to decrease in amplitude and rotate toward their original directions. At the same time, the equilibrium point moved back to the resting location. This sequence of static equilibria is by definition an "equilibrium trajectory": As the neuromuscular activity changes gradually in time, the equilibrium undergoes a gradual shift. Furthermore, after the EMG activities have returned to their resting value, the equilibrium returns to the resting location.

Summing up, these results show that the stimulation of the upper and middle layer of the spinal cord in conjunction with the positioning of the leg in different workspace locations produces a force field with a single equilibrium point. The equilibrium point represents a locus at which the leg would be at steady state. Further experiments are in progress to relate natural activation patterns to the data obtained from microstimulation of spinal-cord circuits. Basically, the neural signals originating from the spinal cord specify directional forces converging onto a location in the leg's workspace. Not surprisingly, perhaps, these results imply that natural activation of a group of spinal-cord premotor circuits also specifies the limb's final position and configuration.

7. Conclusions: Strengths and weaknesses of the equilibrium-point hypothesis

7.1. Strengths. The equilibrium-point hypothesis is strongly rooted in the biomechanics of muscles and in particular in their tunable springlike behavior (Hogan

9

1984; Rack & Westbury 1969): The isometric force generated by a muscle depends on the level of neuromuscular activity as well as the length of the muscle. In other words, the state of activation of a muscle does not determine tension alone but a whole length-tension curve.

It is significant that the springlike behavior of muscles is in conflict with the traditional engineering notion of an "ideal actuator." According to this notion, the output of an ideal actuator – for example, the torque produced by a torque motor – should be independent of the operating state (e.g., position and velocity). This requirement is analogous to the specification of an ideal voltage source in an electrical circuit. We believe there is a deep relationship between the characteristics of the actuators and the control and computational problems that have been central to research in robotics.

A clear example of the relation between actuator characteristics and control is the problem of contact instability: Measurements to date indicate that the dynamic behavior of the neuromuscular system has precisely those properties that are required to guarantee that contact with passive objects in the environment cannot induce instability. Thus, the apparent characteristics of the actuators circumvent a problem that has plagued the application of robots to contact tasks.

An example of the relation between actuator characteristics and computation is the inverse-dynamics problem: To make this computation, one finds the torque that must be applied to the joints to produce a desired motion when the inertial parameters of a manipulator are given. The formulation of this problem implicitly assumes the availability of either an ideal force generator or the computational machinery for translating the output of the inverse dynamics computation (a torque) into a motor command. From this perspective, the biological muscle would be a poor actuator.

The above argument can be reversed, however: Because the biological actuators are springlike, the inverse-dynamics problem does not need to be solved. In fact, according to the equilibrium-point hypothesis, the CNS can express the desired trajectory of a limb directly as a sequence of equilibrium positions. Then the muscles' springlike properties transform the difference between the actual and the desired position of the limb into a springlike restoring force. The actual motions that result are inexact but are produced without computing any dynamics. Consequently, there is no need to postulate neural structures to perform these complex computations.

Of course, the equilibrium-point hypothesis does not eliminate all computational problems; a pattern of neural activity may define a virtual trajectory, but there remains the formidable problem of how to select an appropriate pattern of neural activation to produce a desired virtual trajectory. Nevertheless, because it is based only on the static characteristics of muscles and their reflex connections and requires no knowledge of the dynamic parameters of the limbs (e.g., the inertias), this problem is significantly simpler than the direct computation of muscle forces or joint torques.

7.2. Weaknesses. Flash's simulation work indicates that the equilibrium-point hypothesis is indeed adequate for generating movements at moderate speeds (Flash 1987).

To move faster along the same path, either of two approaches may be taken: Increase the stiffness of the muscles or modify the equilibrium-point trajectory. With respect to the first option, note that the stiffness must be scaled with the square of the speed. That is, to move twice as fast, the muscle system must be four times as stiff. The question, then, is whether such levels of stiffness are biologically plausible.

The stiffness of the human arm has been estimated by measuring the response to disturbances applied during the movement. Stiffness values in the range of 4.0 to 36.0 newton-meters/radians have been recorded. These values are much lower than the theoretical value necessary to produce very fast forearm movements. The work reported by Bennett et al. (1989; Bennett 1990), Xu et al. (1989; 1990a; 1990b; 1991) and Lanman et al. (1978) identifies muscle stiffness during movement. The results indicate that the stiffness of the arm drops with the start of the movement and rises just before reaching the final position. Clearly, these findings do not support the idea that faster movements are achieved by increasing muscle stiffness.

It is possible that the speed of movement could be increased by another method – changing the equilibrium-point trajectory. Simulations of fast movements indicate that the equilibrium-point trajectory must lead initially and then lag behind the actual desired position during the course of the movement (see, e.g., Hogan 1984). The equilibrium position may actually overshoot the final desired position. Although this formulation of the model increases the efficiency of the system, it also increases the computational complexity of the problem. The equilibrium-point trajectory required to produce the movement is no longer simply a copy of the desired movement. The attractive computational simplicity of equilibrium-point control is therefore lost in the latter approach.

In summary, neither an increase in stiffness nor a modification of the equilibrium-point trajectory seems to be a biologically plausible mechanism for generating fast movements. Recently, a third option has been explored by McIntyre (1988; 1990; McIntyre & Bizzi 1992) who has developed a model competent to deal with the generation of faster movements in the context of the equilibrium-point hypothesis. McIntyre's model is basically a simple combination of the alpha and lambda hypotheses. The alpha command to the muscles is obtained from the sum of three elements: (1) a centrally defined signal, representing a desired equilibrium-point trajectory (position and velocity), (2) a position feedback signal, and (3) a velocity feedback signal. Consistent with the alpha model, the desired position signal is used as a feedforward component of the alpha activity. Consistent with the lambda model and the alpha-gamma coactivation, the same desired position cue is used as a reference signal in the position feedback loop. Furthermore, the desired velocity is used as a reference signal for the velocity feedback loop.

In equilibrium-point terms, the combination of feedback and feedforward signals can be regarded as follows: The feedforward signal (1) specifies a sequence of equilibrium points that corresponds to the desired trajectory. In the absence of feedback, this signal would be sufficient to drive the limb toward the final desired location, albeit in a sluggish way because of the limit on the achievable limb stiffness. Fortunately, feedback signals (2) and (3) come to the rescue by modifying effectively the equilibrium trajectory generated by the alpha activity. The addition of these feedback signals implements a position and a derivative controller that serve to improve the performance of the system. Remarkably, the suppression of feedback would impair but not abolish motor performance, a fact that has been observed in the deafferented preparation and not accounted for by the original lambda model. At the same time, this feedback signal allows the motor system to produce faster movements at a given level of stiffness while retaining the simplicity of an equilibrium trajectory specification at the input. Computer simulations testing such a control scheme have shown that the system is stable and produces fast movements at stiffness levels below those required by the equilibrium-point hypothesis. The movement's speed and velocity profiles are comparable to those achieved by a human subject asked to move "as fast as possible."

One major weakness of the equilibrium-point hypothesis is that it is difficult to test. The central concept is that posture and movement are subserved by the same processes. Static stability is arguably one of the defining requirements of posture; consequently, the equilibrium-point hypothesis makes the assumption that during movement as well as posture the limbs exhibit stability. Note that this is not a requirement for the motion of a mechanical system. Nor is it a fundamental requirement for a biological system, although it is physiologically plausible, given the known springlike behavior of muscles and their reflex connections.

The essence of the equilibrium-point hypothesis is that centrally planned motor intentions are expressed and transmitted to the periphery using the virtual trajectory. Evidence in support of this hypothesis may be derived from observations of simple patterns in the virtual trajectories underlying observed behavior (e.g., Flash's [1987] observation that the same straight virtual trajectory could give rise to a wide range of different reaching movements). The major drawbacks of this approach are the difficulty of arriving at a concise definition of "simplicity" and the difficulty of measuring limb stiffness under relevant conditions.

Much of the difficulty of arriving at a confirmation (or disconfirmation) of the equilibrium-point hypothesis stems from the problem of defining a (perhaps artificial) boundary between central and peripheral processes. Where should one look for a neural expression of the virtual trajectory?

Except in the deafferented case, it is of little value to define the boundary at the level of alpha motoneurons. Defining it at the level of commands descending into the spinal motoneuron pools seems more reasonable, and our recent studies of spinal frogs support the idea of a virtual trajectory expressed in the collective activity of spinal motoneuron pools. That activity, however, is modified by long loop reflexes.

The theory that motor intentions are expressed and transmitted to the periphery using the virtual trajectory has direct implications for studies of cell discharge in the brain. The important point is that according to the theory, neither the forces generated by the muscles nor the actual motions of the limbs are explicitly computed; they arise from the interplay between the virtual trajectory and the neuromuscular mechanics. Hence, neither the forces nor the motions need be explicitly represented in the brain. If this theory is correct, then cell discharge studies (e.g., Cheney & Fetz 1980; Evarts et al. 1983; Georgopoulos et al. 1983; 1982; Kalaska et al. 1983) might be better interpreted in terms of virtual trajectories and neuromuscular stiffness (or, more generally, impedance) than in terms of forces or motions (see especially Humphrey & Reed 1983).

ACKNOWLEDGMENTS
Research supported by Office of Naval Research Grant N00014/88/K/0372, National Institutes of Health Research Grants NS09343, AR26710, and AR40029 and National Science Foundation Grant 8914032-BCS.

NOTES
1. This is a consequence of a fundamental theorem on implicit functions. According to this theorem the equilibrium condition, $\psi(\theta_0, \{\alpha\}) = 0$, defines a unique map from the set $\{\alpha\}$ to θ_0, $\theta_0 = g(\{\alpha\})$, provided that the joint stiffness $\delta\psi/\delta\theta$ is different from zero. If the equilibrium position is also to be stable, it is necessary (though not sufficient) that the stiffness be positive. In this respect the equilibrium-point hypothesis can be regarded as a way to represent a high-dimensional control variable, the set $\{\alpha\}$, by means of another one, θ_0, which has the same dimension as the variable, θ, which describes the actual movement.

2. Part of Feldman's critique is addressed to a view of the alpha model that we can hardly share. For example, Feldman (1986) attributes to the alpha model the notion (1) that the muscle activation only modulates the muscle's stiffness but not its rest-length, and (2) that in an intact preparation, the level of alpha activity does not depend on feedback variables. Clearly, both statements fly in the face of physiological common sense. Endorsing them would merely transform the alpha model into a straw-man hypothesis whose refutation could not add anything to our knowledge of motor control.

3. The proof is as follows. Let $(\theta_0^*, \{\lambda^*\}) \in f$ be a pair satisfying the lambda-equilibrium condition $\phi(\theta_0^*, \{\lambda^*\}) = 0$; then the pair $(\theta_0^*, \{\alpha^*\})$ with $\{\alpha\} = \{\alpha(\theta_0^* - \lambda^*)\}$ satisfies the alpha-equilibrium condition $\psi(\theta_0^*, \{\alpha^*\}) = 0$. That is $(\theta_0^*, \{\alpha^*\}) \in g$.

2

Does the nervous system depend on kinesthetic information to control natural limb movements?

S. C. Gandevia and David Burke[1]

Department of Clinical Neurophysiology, The Prince Henry and Prince of Wales Hospitals and Prince of Wales Medical Research Institute, University of New South Wales, Sydney 2036, Australia

Abstract: This target article draws together two groups of experimental studies on the control of human movement through peripheral feedback and centrally generated signals of motor commands. First, during natural movement, feedback from muscle, joint, and cutaneous afferents changes; in human subjects these changes have reflex and kinesthetic consequences. Recent psychophysical and microneurographic evidence suggests that joint and even cutaneous afferents may have a proprioceptive role. Second, the role of centrally generated motor commands in the control of normal movements and movements following acute and chronic deafferentation is reviewed. There is increasing evidence that subjects can perceive their motor commands under various conditions, but that this is inadequate for normal movement; deficits in motor performance arise when the reliance on proprioceptive feedback is abolished either experimentally or because of pathology. During natural movement, the CNS appears to have access to functionally useful input from a range of peripheral receptors as well as from internally generated command signals. The unanswered questions that remain suggest a number of avenues for further research.

Keywords: deafferentation; kinesthesia; motor commands; motor control; muscle, joint, and cutaneous afferents

Movements are produced under a wide range of circumstances: They may be performed rapidly, with minimal time for feedback compensation (i.e., "ballistic movements"), or they may be performed slowly; they may be guided in part by visual or auditory cues, or even driven by metabolic demand (e.g., respiration). Furthermore, the exact trajectories of repeated movements vary so that both the neural drive that generates them and the resultant feedback differ (e.g., Abend et al. 1982; Bernstein 1967). Throughout the literature on movement control, there have been studies of feedback mechanisms (including their spinal circuitry) and central output mechanisms (ranging from the motor cortex to motoneurons), with less investigation of the dependence of the motor output on available feedback. The historical reasons why few studies considered natural movements are not clear. Perhaps there has been so much to establish about the properties of the central organization of motor output and motor feedback that their task-dependent interaction during movement has received less attention. Furthermore, kinesthetic control can operate at an apparently unconscious but not necessarily unimportant level.

This target article was formulated as part of a public debate about whether or not afferent feedback is necessary for movement control. Our objective is to show that human cutaneous, joint, and muscle afferents respond to movements, influence perception, and evoke reflex changes during movement. Because some motor tasks proceed well with limited afferent input we have reviewed the role of perceived motor commands in movement control and described some of the deficits ascribed to proprioceptive loss. A less restricted review has recently been published by Prochazka (1989).

The first part of this target article will consider (1) the discharge of skin, joint, and muscle receptors during movement, (2) evidence that input from such receptors can reach cortical levels and evoke sensations about movement, and (3) evidence that peripheral inputs exert reflex effects during movements. We will argue that the peripheral sensory apparatus can both provide information about and influence normal movement. In Part two, the role of central factors in movement control will be considered. This includes (1) evidence that the central nervous system has perceptual access to signals of central motor commands and (2) data on the deficits in motor control that arise when feedback is removed and reliance is placed on central commands. We will argue that, with some limitations, central motor programs alone may be sufficient to control simple learned movements but feedback is essential when movements involve small precise contractions, when disturbances occur, or during the learning process.

1. Sensory feedback and its effects during movement

1.1. Discharge of skin, joint, and muscle receptors during movement

Specialized receptors in the skin, joints, and muscles of a limb respond to appropriate movement whether it is active or passive. The technique of microneurography

Table 1. *Behavior of cutaneous, joint, and muscle afferents innervating receptors in the human hand*

Receptor class	Estimated frequency*** in sample	Background discharge	Response to movement		Response at both ends of angular range		Response to ≥2 axes of passive movement
			Active	Passive	Active	Passive	
Slowly adapting cutaneous receptors	27% SA I 20% SA II	0% SA I 30% SA II	81%	72%	50%	33%	24%
Rapidly adapting cutaneous receptors	24% RA (FA I) 7% PC (FA II)	~0%	71%	61%	87%	82%	27%
Joint receptors	9%	32%	*	90%	**	47%	69%
Muscle spindle endings	13%	31%	**	100%	*	0%	73%^

*Not measured, likely to be low (~10%).
**Not measured, likely to be high (~90%).
***Crude estimate from the two published studies; muscle afferent population is under-represented.
^The high response to ≥2 axes of movement reflects the predominance of interossei in the sample.
Abbreviations: SA I, slowly adapted type I (presumed Merkel cell complex); SA II, slowly adapting type II (presumed Ruffini ending); RA (or FA I) rapidly adapting (presumed Meissner's corpuscle); PC (or FA II) rapidly adapting (presumed Pacinian/paciniform corpuscle).
Source: Data are derived from Hulliger et al. (1979) for the responses of cutaneous afferents to active movement and from Burke et al. (1988) for the responses of muscle, joint, and cutaneous afferents to passive movements. Data for background discharges are derived from both studies in which more than 100 units were sampled. Afferents were recorded from the median and ulnar nerves.

introduced by Vallbo and Hagbarth (1968) has allowed documentation of the properties of afferent fibers in cooperative human subjects. Not only can an afferent's peripheral properties then be determined, but also the central responses to microstimulation in the region of the dominant axon (see sect. 1.2). The development of these techniques in human subjects has proceeded contemporaneously with methods to record from single peripheral units in freely moving animals (e.g., Loeb et al. 1977; Prochazka et al. 1976). Although microneurography does not yet allow the freedom to record in walking human subjects, it does permit recordings during a wide range of forces and movements confined to one joint.

Microneurography has been used to study the responses of cutaneous, joint, and muscle afferents innervating the human hand during active and passive movement (Burke et al. 1988; Edin & Abbs 1991; Hulliger et al. 1979). Results are summarized in Table 1 for afferents in the median and ulnar nerves. The majority of afferents from all classes of cutaneous receptors alter their discharge during both active isotonic movements and passive movements. With the hand in a rest position, however, there is little background discharge in any of the cutaneous receptors except the SA II mechanoreceptors, of which approximately one third have a background discharge. Responses during movement are much more common than static responses to different maintained positions. Responses to voluntary movement are similar in intensity to those associated with cutaneous stimuli applied to their receptive fields. Cutaneous afferents sometimes discharge only toward one end of a movement range (i.e., unidirectional response; Fig. 1A), but usually discharge in *both* flexion and extension. Marked changes in discharge rate occur toward extreme angular positions for both active (Hulliger et al. 1979) and passive (Burke et al. 1988) movements. Static responsiveness to changes in

joint angle has been documented for SA II receptors (Fig. 1A), with the discharge rate usually increasing as the end of an angular range is approached (see also Knibestöl 1975). Receptors responding to movement are usually located in the digit moved, whereas receptors in the palm may discharge with movement of one or two nearby digits. Cutaneous afferents innervating the dorsum of the hand, proximal to the metacarpophalangeal joints, discharge during movements of the digits (Edin & Abbs 1991): This emphasizes the potential role of "remote" cutaneous afferents in movement control. The proprioceptive function of cutaneous afferents innervating the hand may show considerable "regional" specialization.

Thus the overall input from cutaneous afferents should provide information on the occurrence of movement, the region of the hand involved in the movement, and the crude rates of oscillatory movements (particularly from PC [FA II] and RA [FA I] afferents), but less information on static joint angles (see Table 1 for definitions of abbreviations). Receptors in the densely innervated finger pulp maintain minimal background discharge rates but respond dramatically on first contact with grasped objects. Provided that their receptive fields contact the surface of an object, RA, PC, and SA I afferents may generate near-synchronous bursts at up to 300 Hz in response to imperceptible slippage between surface and skin (Johansson & Westling 1987; 1990; Kunesch et al. 1989; see also Westling & Johansson 1987). The reflex responses evoked by such tactile inputs will be considered later (sect. 1.3).

Dispute has long surrounded the properties of putative joint afferents in the cat (e.g., Boyd & Roberts 1953; Burgess & Clarke 1969; Clark 1975; Ferrell 1980; Skoglund 1956; Tracey 1979; for review see Proske et al. 1988). If data are pooled from two large microneurographic surveys in human subjects, they suggest that digital joint afferents probably make up less than 10%

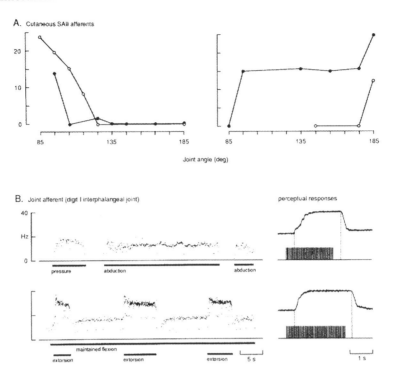

Figure 1A. Data from four SA II units with receptive fields in the hand. The static discharge rate during voluntary "staircase" movements is plotted against the angle of the finger joint. Units with decreasing relationships are on the left and those with increasing relationships are on the right. Many other slowly adapting cutaneous afferents discharge toward both movement extremes (e.g., Burke et al. 1988). Data redrawn from Hulliger et al. (1979).

1B. *Left:* Responses of a single joint afferent associated with the interphalangeal joint of the thumb to sustained pressure and passive joint movement in three axes of rotation: abduction (lateral stress to the distal phalanx), full flexion, and extorsion (external rotation of the distal phalanx). *Right:* Perceptual matching during microstimulation of the same joint afferent. The subject recorded the perceived movements by flexing the interphalangeal joint of the contralateral thumb: Changes in joint angle were registered by a potentiometer (upper trace in both panels). Stimulus pulses delivered to the intrafascicular recording site are represented in the lower traces (0.35 V, 0.1 msec, 20 Hz trains). In both examples, the latency from the onset of stimulation to the onset of perceived joint rotation was 350 msec. Reproduced from Macefield et al. (1990).

of the large-diameter afferent fibers in the digital nerves. Nearly all joint afferents innervating the metacarpophalangeal or interphalangeal joints of the hand were recruited or accelerated their discharge during passive movements (Burke et al. 1988). The discharges were sustained, with minimal adaptation, at the extremes of the physiological range (Fig. 1B). The overall capacity of the 19 joint receptors to signal static angular position was limited. Of the four receptors that altered their discharge across the angular range, the responses of two were unidirectional (i.e., monotonically increasing with joint angle). The majority discharged in more than one axis of joint rotation (e.g., flexion-extension; abduction-adduction), some discharging to nonphysiological bending and torsional stresses on the joint. Thus, only a minority of the slowly adapting joint afferents provide detailed information about joint angle, whereas nearly all can signal that movement is near to, or has exceeded,

usual physiological limits. Overall, broad agreement exists between the types of movement responses displayed by joint receptors in humans and those in experimental animals, but no comparable data for single units are yet available for the digital joints in other primates.

Each slowly adapting joint receptor can also be activated by strong focal pressure to a region of the joint, but the level of pressure required suggests that such a response occurs rarely during common manipulative tasks. Comparable data have been noted for the majority of articular afferents in the cat (Aloisi et al. 1988). The responses of joint receptors innervating the metacarpophalangeal and the interphalangeal joints were not qualitatively different. This observation is relevant because some midrange "units" in the cat hindlimb are actually muscle spindle afferents projecting in a joint nerve (Clark & Burgess 1975; McIntyre et al. 1978; Rossi & Grigg 1982): The focal receptive fields for the human digital joint afferents are not consistent with a muscle afferent origin. Joint receptors in experimental animals may respond to muscle contraction (McIntyre et al. 1978; Millar 1973), although the forces required are relatively large and must be exerted near the movement extreme which excites the receptor (Grigg & Greenspan 1977). Responses of human joint afferents to graded voluntary contractions have not yet been measured, but those in our sample did not respond to twitch contractions of intrinsic muscles of the hand (Burke et al. 1988). Paciniform corpuscles outnumber Ruffini endings by 2–3 to 1 in primate digital joints (Sathian & Devanandan 1983) but the behavior of the former is difficult to study because they have wide receptive fields involving much of the hand.

Muscle spindle endings are abundant in intrinsic muscles of the primate hand compared with other limb muscles (Devanandan et al. 1983). Whereas it is easy to equate the high number of specialized cutaneous receptors in the finger pulp with the high tactile acuity of this region, the high density of muscle spindle endings in distal muscles (expressed in terms of muscle weight) should not be equated automatically with high proprioceptive acuity. So far, several studies have failed to document increasing acuity for position and movement detection of joints in a proximo-distal direction (e. g., Clark et al. 1985; 1986; Goldscheider 1889; Hall & Mc-Closkey 1983; Taylor & McCloskey 1990a). The sensitivity of muscle spindle endings depends on their location within the muscle, their indirect coupling to the tendon (e.g., Griffiths 1987; Hoffer et al. 1989; Meyer-Lohmann et al. 1974) and their in-series coupling to adjacent muscle fibers (Burke et al. 1987). The data in Fig. 2A for the cat are redrawn from Meyer-Lohmann et al. (1974): Variability in the responsiveness of muscle spindle endings may be accounted for by differential lengthening of spindles located in different parts of the muscle. This problem may be greater in humans, given the larger size of their muscles, although it has not yet been formally studied. Certainly, there is a wide range of sensitivity to ramp stretch in human spindle endings under passive conditions.

The functionally important calculation of muscle spindle "numbers" remains a puzzle. It may not be their density or absolute number in a muscle that is critical but the number of muscle spindles per motor unit. When calculated in this way values for the intrinsic muscles of the hand are high but closer to values for other muscles. This calculation is relevant as individual spindles can monitor precisely the discharge of nearby motor units and reflexly affect the discharge of these motor units as suggested by Binder and Stuart (1980; see also Cameron et al. 1980; McKeon & Burke 1983; for review see Windhorst et al. 1989). However, if spindle numbers are expressed relative to the number of muscle *fibers* (arguably a more relevant transformation given the distributed territory of a motor unit), then the intrinsic hand muscles with motor units containing few muscles fibers are richly endowed with proprioceptors.

The discharge rate of muscle spindle endings increases monotonically with increasing passive stretch (Edin & Vallbo 1990a; Vallbo 1974a). During a voluntary contraction at different muscle lengths against a standard load, however, the spindle discharge rates commonly remain constant (Hulliger et al. 1982) so that they no longer provide a simple index of static muscle length. Directional information is available from the spindle discharge during slow isotonic movements of the metacarpophalangeal joints (2–5°/sec) generating relatively low forces (<30% maximum; Hulliger et al. 1985; see also Burke et al. 1978a), but again, position "sensitivity" is reduced compared with that during passive stretch. Most muscle spindle endings innervating finger extensor muscles (17/25) discharged with a pure stretch response (no response with shortening) during voluntary and passive movements about the metacarpophalangeal joint of

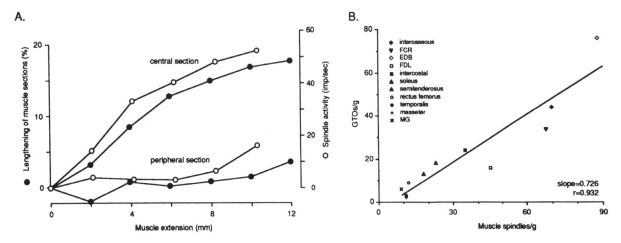

Figure 2A. Data from the extensor digitorum longus of the cat show the relationship between muscle extension and the static discharge of typical muscle spindle endings (open circles) in a peripheral section of the muscle (lower part of graph) and a central section of the muscle (upper part). The actual lengthening in different sections of the muscle was measured (filled circles) and the data have been superimposed. Thus the intramuscular location of a muscle spindle ending is an important determinant of its response to lengthening. Data redrawn from Meyer-Lohmann et al. (1974).

2B. Relationship between the density of muscle spindles (expressed as number per g) and Golgi tendon organs for a range of muscles in the cat. FCR: flexor carpi radialis; EDB: extensor digitorum brevis; FDL: flexor digitorum longus; MG: medial gastrocnemius. Data from Botterman et al. (1978) and Richmond & Stuart (1985).

15°/sec (Al-Falahe et al. 1990), even when the muscle contracted against loads equivalent to about 15% maximum. Spindle endings with a range of dynamic responsiveness behaved in this way. Thus for some relatively slow unloaded shortening movements, fusimotor drive to the agonist may be insufficient to offset the effect of spindle unloading. The discharge rates were more variable and were usually higher with increased loading (Vallbo 1974b). During isotonic contractions, the ability of muscle spindle endings to signal angular velocity improves as the velocity of movement increases (Burke et al. 1978b). This is consistent with data obtained for the hindlimb of the cat during natural movements: With velocities of active shortening above about 0.2 resting muscle lengths/s spindle discharge decreased, whereas this may be offset at lower shortening velocities by fusimotor activity (Prochazka et al. 1979; see also Loeb 1984; Prochazka 1986).

Individual muscle spindle endings in the interossei respond to stretch in the flexion-extension plane across the metacarpophalangeal and interphalangeal joints as well as in the abduction-adduction plane, whereas spindle endings in the lumbricals respond to movements in the former, but not the latter plane. Given the electromyographic silence of the lumbricals during many natural flexion movements of the fingers produced by the extrinsic muscles (Long & Brown 1962), the lumbrical muscle spindle endings would provide simple proprioceptive signals of muscle length uncontaminated by the complex effects of extrafusal and intrafusal activation. If so, this observation would support the suggestion that the high density of muscle spindles in short muscles (like the lumbricals) represents a useful proprioceptive strategy (Peck et al. 1984). Because the lumbrical acts over three joints of the finger, reference to the discharge of other receptors would be needed to decompose the signal from its spindle endings into the relative angular movement of each joint. Further specialization of the sensory input from the lumbrical is indicated by its lack of Golgi tendon organs (Devanandan et al. 1983). As shown in Figure 2, this is in contrast to most limb muscles in which the number of muscle spindles is positively correlated with the number of tendon organs.

Although there have been advances in the physiological classification of human muscle afferents (Burke et al. 1987; Edin & Vallbo 1990b), Golgi tendon organs have been encountered less frequently and studied less systematically in human recordings than muscle spindles. Human tendon organs appear to have the same properties as those described in animals (e.g., for review see Jami 1988; Matthews 1972; Proske 1981), including a dynamic response to small increments in a voluntary force (e.g., Al-Falahe et al. 1990; Edin & Vallbo 1990c; Vallbo 1974a). In the last decade, the conventional view has been that input from a population of Golgi tendon organs will accurately reflect overall muscle force, while an individual tendon organ will encode focal details of force production by a group of motor units dependent upon competing in-series and in-parallel effects (Proske 1981; Stauffer & Stephens 1977). Reassessment of this view may be necessary, because when several motor units affecting a Golgi tendon organ afferent are stimulated, the output of the receptor preserves the dynamic component of the partially fused twitches at the expense of the average force

(Horcholle-Bossavit et al. 1990; cf. Crago et al. 1982). The dynamic component may be less obvious when the muscle generates large forces (Stauffer & Stephens 1977), although this is hard to assess under natural conditions.

The behavior of tendon organs is broadly consistent with the view developed above for muscle spindle endings, namely, that within their ensemble input they preserve information about the focal dynamics of a portion of a whole muscle. The discharge of Golgi tendon organs in the cat hindlimb during walking is strongly coupled with the EMG to abrupt increases in discharge, presumably due to the recruitment of additional motor units, occurring only during slow changes in muscle force (Appenteng & Prochazka 1984; Prochazka & Wand 1980).

We conclude that specialized joint, cutaneous, and intramuscular receptors discharge during natural movements although most joint receptors will not discharge unless the joint is under moderate stress. Kinesthetic coding by cutaneous receptors remote and adjacent to joints requires further study, as do the mechanisms underlying the variability in coding by individual muscle spindle afferents. Another way to determine the role of spindle and tendon organ feedback would be to compare the population inputs from muscle afferents in a variety of muscles including intrinsic and extrinsic hand muscles during natural movements.

1.2. Sensations of limb "movement" originating in peripheral receptors

Hughlings Jackson (1931) introduced the concept that movements could be graded on a scale from the "most automatic" to the "least automatic," with the implication that the latter relied on the most delicate use of inputs and outputs involving the sensory, motor, and other cortical areas. Perception of afferent signals generated before and during movement is not an absolute requirement for movement. Indeed, some hypotheses about movement control have placed no such requirement on perception; for example, the original servo theory required muscle spindle feedback simply to drive reflex circuits (Merton 1953). However, while not theoretically necessary for control of movement, perceptual access for proprioceptive information evoked in skin, joint, and muscle receptors should provide the CNS with greater flexibility in control, particularly for small slow movements and isometric contractions, for the start and end of rapid movements, and also in the transition phases in a movement sequence (Cordo 1990). Such access to perception is presumably also critical when learning a new skill.

As shown above (sect 1.1), receptors in the skin, joint, and muscles respond to both active and passive movements and are all, in the Sherringtonian sense, providing proprioceptive information (Sherrington 1906). Below are some of the arguments that all three classes of receptor provide perceivable information about movement.

Pioneering studies by Goodwin et al. (1972) and Eklund (1972) established that perceived signals of joint position and movement derived from muscle spindle endings. The initial evidence relied on two classes of argument. First, vibration applied to muscles and their tendons produced illusory "movements." These illusions contained two components: a continual movement and a static change in position. Direct recordings from human spindle efferents (Burke et al. 1976a; 1976b; Roll & Vedel 1982) suggest that these illusions depend upon signals from primary and secondary spindle endings. The perceptual role of secondary spindle endings is unclear: Changes in the parameters of vibration implicate secondaries in the distortions of position (McCloskey 1973), whereas studies with weak electrical stimulation suggest that both illusions could be derived centrally from the signal of primary spindle endings (Gandevia 1985). No recordings of secondary spindle afferent discharge in human subjects have revealed a specific stimulus that would preferentially excite them. Second, proprioceptive acuity for passive movements was retained when joints (and relevant areas of skin) were anesthetized or when joints were removed. However, acuity can be diminished with skin and joint anesthesia, particularly with joints in the hand. The possibility that there were small residual deficits following joint removal, or that the joint signals were redundant, was initially put to one side. Establishment of a perceptual role for muscle spindle endings followed, and proceeded with studies of their cortical projection to the primate sensory and motor cortex (for reviews see Phillips & Porter 1977; Wiesendanger & Miles 1982). Some cells within the primate motor cortex can derive signals of velocity and position from muscle spindle signals (Hore et al. 1976). In human subjects, the techniques of intrafascicular stimulation and intramuscular microstimulation were used to reveal the cortical projection from individual proximal, distal, and truncal muscles (e.g., Burke et al. 1982; Gandevia & Burke 1988; Gandevia & Macefield 1989; Gandevia et al. 1984), and although assessed indirectly, the amplitude of the short-latency projections was similar for trunk and limb muscles. Currently both psychophysical (e.g., Soechting & Flanders 1989) and electrophysiological techniques (e.g., Georgopoulos et al. 1988; Gurfinkel et al. 1988) are revealing the processes whereby proprioceptive signals are transformed into a three-dimensional coordinate system for perception and movement. As many reviews have covered the material that established a major role for muscle spindle endings in the detection of movement and position (Burgess et al. 1982; Matthews 1982; 1988; McCloskey 1978), we concentrate here on some relevant points of argument and dispute.

1.2.1. Coding and detection during movement.
There are some clues about the way in which position and movement are coded, based upon the pattern of muscle spindle input. An increase in the frequency of vibration or of electrical stimulation (below motor threshold) increases the velocity of illusory movements, an effect presumably attributable to group Ia afferents (e.g., Burke et al. 1976a; 1976b; Gandevia 1985; Roll & Vedel 1982). An example is given for the illusory movements of the hand produced by electrical stimulation of the ulnar nerve in Figure 3. Spatial recruitment also alters the illusions: Velocity of illusory motion increases with stimulus intensity (Gandevia 1985). The perceptual effects of increasing the amplitude of vibration have been less well studied but must be more complicated because vibration is then a less selective stimulus (contrast Burke et al. 1976a with Roll & Vedel 1982).

Little information exists about detection thresholds for these illusory phenomena when they are evoked during

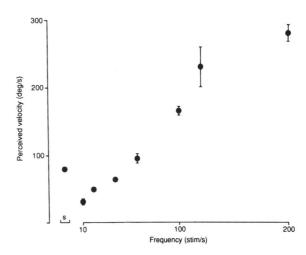

Figure 3. Relationship between stimulus frequency and the perceived angular velocity for the illusory movement produced by ulnar nerve stimulation at the wrist. Mean ± SEM shown for one subject. Stimuli were delivered through a needle electrode at below threshold for motor axons, and these stimuli did not produce cutaneous paresthesias. The illusory movement of the ring finger consisted of flexion of the interphalangeal joints and extension of the metacarpophalangeal joint; this corresponds to a perceived elongation of the interosseous and lumbrical muscles. The velocity of the illusory movement (at the metacarpophalangeal joint) was determined from a matching movement on the contralateral side. Note the "flick" movement experienced with a single stimulus (labelled "s"). Reproduced from Gandevia (1985).

natural movement. Other evidence implies that proprioceptive acuity for applied movement is enhanced markedly during a steady voluntary contraction (Colebatch & McCloskey 1987; Gandevia & McCloskey 1976; see also Paillard & Brouchon 1968). Several factors could contribute to the enhanced proprioceptive acuity during muscle contraction: First, tendon slack is eliminated and, second, the population discharge of muscle spindle endings (and tendon organ afferents) is elevated. This proprioceptive enhancement contrasts with the diminution in cutaneous sensibility: The detection and perceived intensity of cutaneous signals are reduced by both active and passive movement (e.g., Angel & Malenka 1982; Milne et al. 1988; Schady & Torebjörk 1983). This gating of cutaneous signals is due to primarily central processes *before* movement, while there is also a peripheral factor *during* movement probably mediated by cutaneous PC afferents (Ferrington et al. 1977). The perceptual gating of cutaneous inputs is much more prominent during movement than during an isometric contraction (e.g., Milne et al. 1988; see also Dimitrov et al. 1989; Rushton et al. 1981). However, reflex responses to unexpected cutaneous inputs from the fingertips are not present when muscles are relaxed and are largest for the first of a sequence of repeated movements (Johansson & Westling 1987; see also Fig. 4). Clearly, the mechanisms controlling reflexes and perception may be set differently during the same movement, and may also differ during the movement's isometric and isotonic conditions.

Under passive conditions, a subject responds with a normal reaction time to the muscle afferent stimuli which

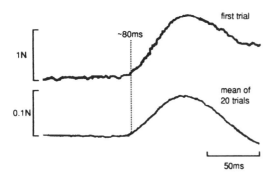

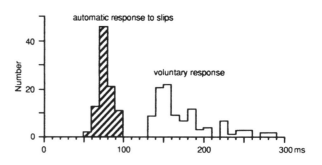

Figure 4. Data highlight the rapid and powerful responses to an unexpected electrical stimulus to the skin during a grip. *Left:* Change in grip force produced by a single electrical stimulus to the pulp of the thumb and one finger at just above perceptual threshold as a 400 g object was grasped and then lifted. The prominent force increment began at about 80 msec in the first trial. The average increase in force for trials 10–30 had the same latency but was of smaller amplitude as the subject had adapted to the irrelevant cutaneous stimulus.

    Right: Distribution of latencies between small "microslips" of the test object (detected by an accelerometer) and the onset of the force increments (automatic response to slips) and voluntary reaction times to a strong cutaneous stimulus. Mean latencies to the liminal stimuli and microslips were 74 msec and 78 msec respectively, much less than the mean reaction time of 176 msec. Data are redrawn from Johansson & Westling (1987).

evoke illusory movements and may even detect a single stimulus (Gandevia 1985). The CNS appears able to interpret correctly whether a change in spindle discharge is due to an external perturbation or to volitionally-generated activity presumably by reference to signals of motor command or corollary discharge (McCloskey 1973; 1981; see also sect. 2.1), so that the same proprioceptive input may evoke different sensations dependent upon the initial "set" of the subject. All proprioceptive signals must be interpreted by the CNS against a background input from other sources such as the visual, auditory, axial muscle, and vestibular systems (e.g., Biguer et al. 1988; Gurfinkel et al. 1988). Thus, for example, vision of the limb can attenuate or even abolish the illusory phenomena attributed to muscle spindle endings (e.g., Gandevia 1985; Goodwin et al. 1972), but when vision is prevented, proprioceptive cues can affect apparent visual localization (e.g., Levine & Lackner 1979). The degree of perceptual "trade-off" between competing signals may be complete (as with visual abolition of vibration-induced illusory limb movements) or more subtle (shown below for stimulation of single joint afferents). Furthermore, perception of overall limb position may be more acute than that of a constituent joint within it (Soechting 1982).

1.2.2. Muscle spindle aftereffects and nonlinearities.
Muscle spindle output is powerfully affected by the movement- and contraction-history of the parent muscle. Specific "aftereffects" in muscle spindle responsiveness after contraction at different muscle lengths have been exploited to provide further evidence for a muscle spindle role in position sense (Gregory et al. 1988). Such effects highlight the lability in the relationship between muscle length and spindle discharge (Baumann et al. 1982). Further variability is introduced by receptor adaptation and hysteresis across the angular range (e.g., Burgess et al. 1982). Sensation of limb position is systematically disturbed for several minutes after prolonged muscle vibration (Rogers et al. 1985), consistent with a perceived "shortening" of the muscle, possibly due to a postvibratory depression of muscle spindle receptors or their

central projections (Macefield & Burke 1990). Joint and cutaneous inputs could help "realign" any perceptual maps derived from muscle spindle endings.

1.2.3. How many muscles are required? Most joints are crossed by more than one muscle so that signals from muscle spindle endings will be available from more than one source. The relationship between spindle discharge and joint angle differs for different muscles crossing a joint (Burgess et al. 1982). For joints with movements in more than one axis, unique determination of joint "position" will require analysis of the input from more than one agonist-antagonist pair or from extramuscular receptors. For example, differences in the signals from muscle spindles in the interossei and lumbricals acting on one finger would provide information about the position of the metacarpophalangeal joint in the plane of abduction/adduction. Observations have confirmed that both the antagonist and agonist can contribute to movement detection for proximal (e.g., Gilhodes et al. 1986; Roll & Vedel 1982) and distal joints (Gandevia et al. 1983). Short muscles crossing few joints would aid resolution of absolute angular positions.

For movements involving joints and muscle groups proximal to the hand the ability to detect passive movements is similar if the parameters of movement are expressed as relative changes in muscle length (Hall & McCloskey 1983). It is unclear whether this finding can be extended to all muscle groups, particularly the intrinsic muscles of the hand. However, observations based on pulling the surgically severed tendon of extensor hallucis longus suggest that it behaves with lower "sensitivity" to changes in muscle length (McCloskey et al. 1983a); other proprioceptive inputs are required to explain full acuity at the metatarsophalangeal joint. Simultaneous measurement of human spindle responsiveness and thresholds for movement detection would allow the neural mechanisms involved in the psychophysical judgments to be examined more precisely. The potential "extrapolation" of signals from receptors in the intrinsic and extrinsic muscles of the hand is exemplified by a study entitled "Tactile discrimi-

Table 2. *Estimates of the threshold for a static position sensation at various human joints*

	Angular displacement degrees required for detection (velocity ≤2°/min)	Reference
Distal interphalangeal joint (muscle engaged)	5.5 ± 0.5	a
Distal interphalangeal joint (muscle disengaged)*	8.1 ± 4.2 (5.0 ± 0.6)	a
Proximal interphalangeal joint	6.8 ± 0.5 (5.6 ± 2.4)	a**
Metacarpophalangeal joint	4.4 ± 1.5 (4.3 ± 1.0)	a
	~2.5	b
Ankle	~3.5	b
Knee	~5.0	c

Note: Criteria for detection differed slightly between the studies. Values in parentheses for joints in the upper limb were obtained when subjects actively co-contracted flexor and extensor muscles after the displacement had been applied.
*Muscles acting on the distal interphalangeal joint were disengaged by positioning the hand.
**cf. Clark et al. (1989).
Source: References used in compiling the Table are set out below:
a   Taylor & McCloskey (1990a): mean ± S.D.
b   Clark et al. (1985): ~80–90% detection levels
c   Clark et al. (1979): ~80%–90% detection levels

nation of thickness," in which the authors were forced to conclude that discrimination of the thickness of objects placed between the thumb and index finger is largely derived from muscle spindle endings rather than cutaneous tactile receptors (John et al. 1989; cf. Edin & Abbs 1991). The discrimination threshold was only 75 μm, corresponding to angular deviations of 0.1°.

1.2.4. Signals underlying static position sense.
To study the ability to perceive small changes in static position of joints, Clark and colleagues moved joints at extremely low angular velocities (~2°/min), below those at which subjects report any sensation of movement (Clark et al. 1985; 1986). An approximate threshold for detection of changes in position is now available for a number of joints using these extremely slow displacements. Although there is disagreement about the extent to which this capacity is available for the proximal interphalangeal joint (Clark et al. 1986; cf. Taylor & McCloskey 1990a), the threshold differs little for a variety of joints (Table 2). This is unlike the detection of faster movements: Distal muscles of the limb are less accurate than more proximal ones if the movement parameters are expressed in terms of absolute angular motion (Hall & McCloskey 1983). This observation alone suggests that different neural mechanisms subserve detection of "position." Given results with anesthesia of the skin and joint, and of mixed nerves, the ability to detect position changes is likely to be dependent in part upon input from muscle spindle endings, at least for the knee and ankle joints. As for sensation of more rapid movements, the responsible mechanisms for static position sense may differ for various joints, particularly since muscle "disengagement" at the distal interphalangeal joint produces little impairment of the thresholds for detection. Furthermore, deliberate muscular contraction (at the final position) fails to enhance the position signals for finger joints (Table 2; Taylor & Mc-

Closkey 1990a). This suggests that the afferent activity evoked by a weak isometric contraction does not convey unique information about position and that any corollary motor signals do not unmask a precise map of joint position.

1.2.5. Interaction between muscle and nonmuscle kinesthetic signals.
Proprioceptive acuity for applied movements in some joints in the lower limb (knee and ankle) is not impaired by local cutaneous (and joint) anesthesia (Clark et al. 1979, 1985; see also Barrack et al. 1983a), but it *is* affected by anesthesia of the finger (e.g., Brown et al. 1954; Ferrell & Smith 1989; Gandevia & McCloskey 1976). This impairment occurs even if the anesthesia is remote from the joint moved; anesthesia of the finger pulp affects acuity for more proximal joints of the finger (Clark et al. 1986) and anesthesia of fingers adjacent to the one being moved impairs its performance (Gandevia & McCloskey 1976; cf. Ferrell & Smith 1988). Proprioceptive acuity for distal joints is facilitated tonically by a digital nerve input, probably derived from SA II and joint receptors with a background discharge (Table 1).

Some specific proprioceptive input can be extracted from the discharge of cutaneous afferents. This was established for the distal interphalangeal joint of the finger when the muscle and joint afferent contributions to movement sensation had been removed (Ferrell et al. 1987). Cutaneous afferents within the radial nerve respond to finger movements and can presumably contribute to detection of applied movements (Edin & Abbs 1991). Microstimulation of occasional cutaneous afferents innervating receptors in the distal phalanx can evoke sensations of movement consistent with the natural behavior of the afferent (Macefield et al. 1990). In the hand, the additional facilitatory role of tonic input from skin and joint receptors in movement detection is likely to be especially important, at least across the usual angular range. The role of cutaneous afferents may vary according to tactile stimuli present around the joint. This is exemplified by the finding that the variance of estimates of knee position may be reduced by cutaneous anesthesia, as if this removed "noise" from detection (Clark et al. 1985). One implication of this work is the need for convergence within the CNS of the different sources of proprioceptive input.

During the reinstatement of muscle spindle afferents as key contributors to kinesthesia, the potential role of joint receptors was often ignored. Evidence set out below, however, acts as a prudent reminder not to expect intramuscular receptors to provide all the relevant proprioceptive cues under every natural condition. Furthermore, the functionally important contribution to proprioceptive performance from intramuscular, joint, and cutaneous signals (specific and nonspecific) may differ for distal, proximal, and truncal joints.

1.2.6. Contributions of joint afferents to movement sensation at the human distal interphalangeal joint. If the hand is postured appropriately, the long finger extensors and flexors and the intrinsic extensor mechanism can be disengaged so that proprioceptive sensation at the distal interphalangeal joint is due only to cutaneous and joint afferents (Gandevia & McCloskey 1976). With this muscle disengagement, the ability to detect the direction of applied flexion and extension movements within the

midrange of joint positions deteriorates markedly, but the residual acuity is in part dependent upon joint afferents. A dependence on joint receptors was suggested by the improvement in acuity with joint capsule expansion, and the further impairment (but not abolition) with intracapsular anesthesia (Clark et al. 1989; Ferrell et al. 1987). The magnitude of the estimated contribution of joint receptors differed slightly in the studies of Ferrell et al. (1987) and Clark et al. (1989), but the data have been pooled in Figure 5A. When muscles were disengaged so that only joint and cutaneous afferents could contribute, performance was unaffected by cutaneous anesthesia over the dorsum of the joint. That the sensitivity and number of cutaneous and joint afferent fibers responding to the imposed movements is small (sect. 1.1) need not be important if the discharge of individual afferents can be perceived. The results of Taylor and McCloskey (1990a) using imperceptibly slow movements to assess "position" sense at the distal interphalangeal joint confirm that performance need not depend upon intramuscular receptors.

Joint removal offers a seemingly direct way to assess whether proprioceptive acuity requires an input from joint receptors. In human subjects, replacement of diseased joints is associated with minimal impairment of position sense and functional performance (e.g., Cross & McCloskey 1973; Grigg et al. 1973; cf. Karanjia & Ferguson 1983). However, it is likely that the impairments

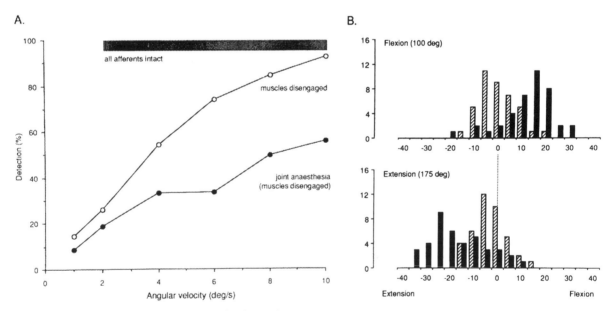

Figure 5. Data emphasize the potential role of joint afferents in sensations of joint movement (panel A) and joint position (panel B).

5A. Detection of 5° movements applied at different angular velocities to the distal interphalangeal joint of the middle finger. Movements were applied from a midposition. Subjects had to nominate correctly the direction of the movement. Mean data from Ferrell et al. (1987) and Clark et al. (1989) have been pooled. Data obtained with the hand postured so as to disengage the flexor and extensor muscles (Gandevia & McCloskey 1976) are shown as open circles. Filled circles represent performance when the distal interphalangeal joint was injected with local anesthetic. Performance is best when muscle, joint, and cutaneous afferents are intact (hatched area); it deteriorates when muscles cannot contribute and deteriorates further when the joint space is injected with local anesthetic.

5B. Distribution of angular errors in the unanesthetized finger (hatched histograms) and during digital nerve block of the matching finger (filled histograms). Subjects (n = 8) were required to align the proximal interphalangeal joint of target finger with that of the matching finger. Data are shown for an angle near full flexion (above) and for one near full extension (below). With the target finger close to full extension (or flexion) subjects hyperextend (or hyperflex) the anesthetized matching finger in an attempt to align them. These errors are less obvious at intermediate joint angles (not shown). Data are redrawn from Ferrell & Smith (1989).

underestimate the potential joint afferent contribution, because tests have usually assessed performance in the midrange of movement, not toward the angular extremes. Barrack et al. (1983b) documented an impairment of movement sensation with aging and this was more marked in patients with degenerative arthritis at the knee (irrespective of joint replacement). These findings suggest a proprioceptive role for joint receptors, but such patients commonly have wasted muscles, presumably due to "disuse," and the possibility remains that intramuscular receptors were also involved. A deterioration in walking was reported for the cat after acute anesthesia of the knee joint (Ferrell et al. 1985).

1.2.7. Signaling the extremes of angular range. The labile "map" set up by muscle spindle endings is unsuited to signal the physiological limits of joint rotation. Artificial activation of muscle receptors is sufficient for subjects to report that joints have moved into anatomically impossible positions (e.g., Craske 1977; Gandevia 1985), although a signal closer to the true position can be extracted if required (Sittig et al. 1985). Against this, the discharge of digital joint receptors is well suited to signal the movement extreme, although a single afferent may not specify uniquely which extreme. One study that examined position sense across a broad angular range for the proximal interphalangeal joint of the index finger documented misalignment between the affected and control sides when the contributions from joint and cutaneous afferents were abolished (Ferrell & Smith 1989). When the joint of the target finger was positioned near the extreme of flexion (or extension), the joint of the anesthetized finger had to be hyperflexed (or hyperextended) to

align the perceived locations of the two fingers (Fig. 5B). This distortion of the extremes of the flexion-extension range remained after anesthesia of the joint alone.

Some unmyelinated and small-diameter myelinated joint afferents from the cat knee respond to joint movement in the midrange, but particularly to twisting movements (Grigg et al. 1986; Schaible & Schmidt 1983). With experimentally induced joint inflammation or effusion, the population of joint afferents is sensitized and the proportion of the full complement of joint afferents responding across the angular range and at the extremes increases (e.g., Ferrell 1987; Grigg et al. 1986). Thus, under these conditions, kinesthetic awareness of the joint would be heightened and protective reflexes that depend on joint afferents would occur for smaller mechanical disturbances.

1.2.8. Microstimulation of joint and muscle afferents. Stimulation of individual large-diameter cutaneous afferents evokes sensations localized to their receptive fields with particular tactile qualities depending on the afferent class (Table 3; Macefield et al. 1990; Ochoa & Torebjörk 1983; Schady & Torebjörk 1983; Vallbo et al. 1984). The probability of obtaining these sensations is virtually 100% for afferents innervating the distal part of the digits, but the probability diminishes for more proximal sites in the hand. An exception is that stimulation of single SA II afferents rarely evokes sensations with a tactile quality. Early indirect arguments that the methodology was faulty (Wall & McMahon 1985) have been refuted (Torebjörk et al. 1987; see also Calancie & Stein 1988).

Microstimulation of joint afferents innervating the in-

Table 3. *Perceptual responses attributed to selective microstimulation of large-diameter cutaneous, joint, or muscle afferents innervating the hand*

Receptor class	Sample size	Occurrence of sensation (%)	Single stimulus perception	Character of sensation
Cutaneous RA (FA I)	(60)	~90	+	tap, flutter, or vibration depending on stimulus frequency and duration
Cutaneous PC (FA II)	(14)	~85	?+	vibration (diffuse)
Cutaneous SA I	(50)	~80	+	local pressure or indentation
Cutaneous SA II	18	~10*	−	joint movement when receptive field over distal interphalangeal joint
Joint	11	~70	−	focal deep pressure, joint movement or stress
Muscle spindle ending	13	<10	−**	sensation of movement at below motor threshold
Tendon organ afferent	3^	?	?	^not known

*The usual lack of perceptual responses to stimulation of SA II afferents is confirmed by Schady & Torebjörk (1983).
**A single stimulus to a group of presumed primary spindle endings evokes a sensation consistent with muscle lengthening (Gandevia 1985).
^One of three tendon organ afferents gave a sensation of muscle lengthening; however, the sample size is inadequate.
Source: Data for sample size are from Macefield et al. (1990) or combined values (in parentheses) with data from Schady & Torebjörk (1983). Perceptual responses for the cutaneous RA, PC, and SA I units taken from the same two studies as well as from Vallbo et al. (1984).

terphalangeal joints of the hand at intensities identical to those eliciting cutaneous sensations can evoke specific sensations referred to the relevant joint (Macefield et al. 1990). These sensations included one of deep pressure projected to the receptive field or one of joint motion or stress in one or more planes. A reasonable correspondence existed between the discharge properties of the afferent and the sensations evoked. For some joint afferents, there must have been perceptual reinterpretation of the afferent signal based on the background discharge of other proprioceptive afferents. Thus, microstimulation of an afferent that discharged at both ends of an angular range, but with a higher frequency at the flexion extreme, elicited an illusion of flexion within the usual physiological range. The extent of the illusory movements was small, increasing with stimulus frequency. Distortions of joint position could also be evoked (Fig. 1B). The capacity of individual joint afferents to evoke proprioceptive sensations at the extremes of joint motion, to some extent, compensates for the relatively poor capacity of these receptors to encode details of natural movements in the mid-range and may obviate the need for the interphalangeal joints to have large numbers of such receptors. The sensation evoked by microstimulation of single cutaneous afferents is weakened or even abolished by movement of the relevant finger (Schady & Torebjörk 1983), but this phenomenon has not been tested for joint afferent sensations.

In contrast to the behavior of joint afferents, microstimulation of muscle spindle endings innervating intrinsic muscles of the hand failed to evoke a perception of movement that would have stretched the receptor-bearing muscle (Macefield et al. 1990). These studies examined perception with microstimulation at the intensities required to evoke sensation from joint and cutaneous afferents, using a range of stimulus frequencies from 1–100 Hz. An increase in the stimulus intensity activated one or more motor axons producing a fasciculation that, if sufficiently large, was detected and could be described accurately as a small twitch or tetanic contraction. This result, although unexpected given the manipulative skills of the hand, is less surprising if the discharge of an individual muscle spindle ending bears a complex relation to external changes in whole muscle length (and thus joint angle) depending upon its fusimotor input, exact intramuscular location, and history of activation. Given this fluidity in spindle coding properties, it is easier to accept that the input from a solitary spindle ending has no perceptual relevance and may even have a negligible reflex effect during a steady muscle contraction (Gandevia et al. 1986a). However, the discharge of a *population* of afferents from a muscle can evoke illusory movements and, in some subjects, even a single stimulus evokes a perceived flicklike movement that would lengthen the muscle (Fig. 3; Gandevia 1985).

Using microstimulation with trains of stimuli we confirmed that most SA II afferents innervating the volar surface of the digit and palm evoked no specific sensation (Macefield et al. 1990). However, stimulation of two SA II afferents with receptive fields over the dorsum of the finger near the nailbed evoked proprioceptive sensations (~10% of the sample, see Table 3); they discharged during flexion of the distal interphalangeal joint of the finger and elicited a comparable sensation on micro-

stimulation. Such afferents could contribute to proprioceptive sensation when stimulated in isolation, and presumably they could also have subserved the proprioceptive acuity remaining after joint and muscle mechanisms had been eliminated (Fig. 5A; Clark et al. 1989; Ferrell et al. 1987).

Clearly, joint, muscle, and even cutaneous afferents can provide meaningful signals of joint position during natural movement. The studies described above highlight the greater spatial summation required for perceptions evoked by muscle spindle compared with other afferents from the hand. The difference probably disappears for more proximal sites in the limb where cutaneous sensibility is less acute.

1.3. Evidence that afferents from skin, joint, and muscle exert reflex effects during natural movements through spinal pathways

That proprioceptive afferents have a critical role in motor control can be supported by demonstration of their reflex effects during voluntary movement. This section reviews this evidence and seeks to emphasize the variety of reflex effects that can be documented in studies on human subjects.

The relative importance of spinal and long-loop (supraspinal) reflex circuits in compensating for phasic disturbances to a posture or movement has been hotly debated (for reviews of human and animal data, see Lee et al. 1983; MacKay & Murphy 1979; Marsden et al. 1983; Matthews 1991; Prochazka 1989). At the spinal level, complex interneuronal machinery governs transmission of sensory feedback and descending commands to the motoneuron pool (Jankowska & Lundberg 1981).

In addition to the group Ia projection from muscle spindles in soleus to homonymous motoneurons, the soleus Ia projections to antagonists (see below), between synergists (such as the different heads of triceps surae), and across joints (such as to thigh muscles) have also been studied extensively (e.g., Bayoumi & Ashby 1988; Mao et al. 1984; Pierrot-Deseilligny et al. 1981b). The pattern of group Ia projections in man has similarities to and differences from that in the cat and baboon (Bayoumi & Ashby 1988; Hongo et al. 1984; Meunier et al. 1990), the differences reflecting the altered requirements for bipedal stance and locomotion.

During voluntary contractions, the activity of spinal reflex pathways can be altered through a number of mechanisms other than direct effects on the α-motoneuron pool.

Voluntary muscle contraction is accompanied by activation of fusimotor neurons sufficient to increase the discharge of spindle afferents (e.g., Vallbo 1971; 1974b; for review see Burke 1981; Vallbo et al. 1979). The data in Figure 6 suggest that this increase in fusimotor drive involves static fusimotor neurons (Vallbo 1973), but there is no convincing evidence on the status of dynamic fusimotor neurons, because appropriate tests cannot be applied in the intact human subject. The fusimotor drive appears to involve γ-motoneurons (Burke et al. 1979) and possibly also β-motoneurons (Aniss et al. 1988; Rothwell et al. 1990; cf. Gandevia et al. 1986a). When voluntary contractions are isometric or produce relatively slow muscle shortening, the fusimotor drive accompanying the

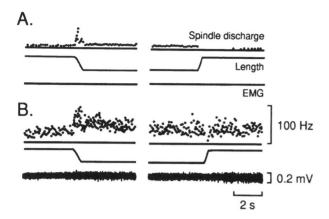

Figure 6. Response of a muscle spindle primary ending to joint movements that would lengthen the receptor-bearing muscle when it was relaxed (A) and contracting (B). The spindle was located in the flexor digitorum muscle acting on the ring finger. Traces are, from above, the instantaneous discharge frequency, angle at the metacarpophalangeal joint, and EMG. Muscle lengthening is shown as a downward deflection with the angle at the metacarpophalangeal joint changing between 140° (top) and 160° (bottom). The interphalangeal joints were kept in full extension. B: Subject exerted constant pressure with his ring finger against the device to which the fingers were fixed. Reproduced from Vallbo (1973).

voluntary contraction can produce a significant increase in spindle discharge (Burke et al. 1978a; Vallbo 1973b), and this is more prominent when the contracting muscle moves a load (Al-Falahe et al. 1990; Burke et al. 1978a; 1978b). The increase in spindle discharge occurs after the onset of EMG activity in the contracting muscle, even when subjects respond in a reaction-time task (e.g., Burke et al. 1980; Vallbo 1971). Clearly, the sensitivity of muscle spindle endings to perturbations occurring during a movement will change as a result of fusimotor drive during the contraction, but there are little data on whether the spindle response to a perturbation would be increased, decreased, or unchanged (Morgan et al. 1984). Moreover, the fusimotor system does not appear to operate as an effective gain-control for the stretch reflex, regardless of whether the disturbance is phasic or tonic (Burke 1981).

The existence of a γ-motor system implies that it ought to be possible to alter spindle discharge without changing skeletomotor activity. Despite the use of many strategies, evidence for this belief has been extremely difficult to obtain in human subjects (Al-Falahe & Vallbo 1988; Aniss et al. 1990a; Burke 1981; Gandevia & Burke 1985; Vallbo & Al-Falahe 1990; Vallbo et al. 1979; see also Ribot et al. 1986). Flexibility in α- or γ-control has been more forthcoming in experiments on cats (Hulliger et al. 1989; Prochazka 1989; Prochazka et al. 1985), perhaps because an even wider range of motor behavior can be studied. There are two main ways for the studies of human fusimotor control to progress: first, to continue the search for α–γ dissociation using novel paradigms and, second, to document the subtle effects of fusimotor reflexes on human spindle discharges under different tasks. Such fusimotor reflexes have been observed during unsupported standing (Aniss et al. 1990a), but their strength has

not usually been sufficient to produce spindle discharge without EMG. One of the rare examples of presumed complete α–γ dissociation was obtained during human stance and is shown in Figure 7.

The strength of spinal reflex pathways can be altered by altering transmission of the afferent volleys to α-motoneurons. In general, the changes in the strength of spinal reflex pathways enhance the excitability of the active motoneuron pool and suppress that of the antagonist muscle, particularly at the onset of the contraction. Examples are given below.

1.3.1. Presynaptic inhibition of Ia afferent terminals on motoneurons of the contracting muscle at the onset of a voluntary contraction is decreased, but presynaptic inhibition of Ia afferent projections to the motoneurons of uninvolved muscles is increased (Hultborn et al. 1987). Stronger contractions of soleus produce greater suppression of presynaptic inhibition of Ia afferent fibers exciting soleus (Iles & Roberts 1987; see also Meunier & Pierrot-Deseilligny 1989). Suppression of presynaptic inhibition is maximal as the contraction ramps up but then returns to control levels (Meunier & Pierrot-Deseilligny 1989). Rapid relaxation of a contraction is associated with temporarily enhanced presynaptic inhibition, which subsequently subsides to the resting level (Schieppati & Crenna 1984; Schieppati et al. 1986). Prior to a rapid voluntary contraction of soleus, the H-reflex is potentiated, presumably by suppression of background presynaptic inhibition (e.g., see Schieppati et al. 1986). During voluntary dorsiflexion of the foot, presynaptic inhibition of Ia fibers projecting to soleus motoneurons is enhanced, and this would decrease the possibility of an unwanted stretch reflex in soleus (Meunier & Morin 1989).

When human subjects stand upright, presynaptic inhibition of Ia fibers to soleus motoneurons increases, while presynaptic inhibition of Ia afferents to quadriceps motoneurons decreases, with little change in presynaptic inhibition of Ia fibers to tibialis anterior motoneurons (Katz et al. 1988). Studies of bipedal locomotion have revealed that the modulation of the soleus H-reflex during the locomotion cycle cannot be attributed merely to a changing level of excitation of soleus motoneurons (Capaday & Stein 1986; 1987; Crenna & Frigo 1987); presumably, this modulation is due to phase-related changes in presynaptic inhibition of the Ia pathway (Morin et al. 1982).

1.3.2. The reciprocal Ia inhibitory pathway has been studied in both the upper and lower limbs of human subjects (e.g., Shindo et al. 1984; Tanaka 1983). Muscle contraction activates Ia inhibitory interneurons during movement so that the Ia inhibitory pathway to its antagonist is enhanced (Cavallari et al. 1984; Day et al. 1984). The disynaptic inhibition of soleus motoneurons from Ia afferents in the peroneal nerve is increased over the resting level during the dynamic phase of a contraction of tibialis anterior, but this is not sustained during tonic dorsiflexion (Crone & Nielsen 1989; Crone et al. 1987; Iles 1986). During contraction of soleus itself, the disynaptic inhibitory pathway appears to be suppressed (Iles 1986). In the upper limb, movements on the contralateral side may enhance reciprocal inhibition, again possibly through

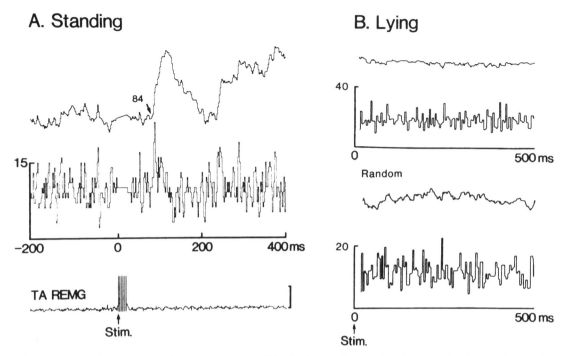

Figure 7A. Peristimulus time histogram (PSTH) of the discharge of a muscle spindle ending in tibialis anterior recorded while the subject was standing. At the arrow, a brief train of stimuli (5 pulses at 300 Hz) was delivered to the sural nerve at the ankle at nonpainful levels. These stimuli produced an increase in discharge of the spindle afferent at ~84 msec. This ending was initially silent and was given a background discharge (9–10 Hz) by direct pressure on the tendon. Upper trace is the cumulative sum (CUSUM), middle trace is the PSTH, and lower trace is the rectified averaged EMG from tibialis anterior. Note the absence of a change in EMG. Vertical calibration is 10 μV. Reproduced from Aniss et al. (1990a).

7B. Data (comparable to those shown in 7A) for a typical muscle spindle ending in tibialis anterior while the subject was lying down, but exerting a dorsiflexor torque to increase the spindle discharge. The upper panels (PSTH and CUSUM) show the response to nonpainful sural stimuli (as in 7A), and in the lower panels, to a randomly generated trigger. Count ratio for the CUSUM is x 2. Reproduced from Aniss et al. (1988b).

changes in the Ia inhibitory interneuron (Delwaide et al. 1988).

1.3.3. An oligosynaptic propriospinal-like pathway has been demonstrated in the human upper and lower limbs, analogous to the C3-C4 propriospinal system which transmits descending commands to motoneurons for forelimb target-reaching movements in the cat (Alstermark et al. 1981). At the onset of and during a weak contraction, transmission in this pathway is more effective (Baldissera & Pierrot-Deseilligny 1989; Burke et al. 1992; Hultborn et al. 1986). Using spatial facilitation, Malmgren and Pierrot-Deseilligny (1988) and Burke et al. (1992) have demonstrated in poststimulus time histograms and reflex studies, respectively, a wide convergence between cutaneous and muscle afferents onto the interneurons responsible for integrating the descending command with the peripheral feedback. The pattern of inhibitory cutaneous control of these "propriospinal-like" neurons is appropriate for the tasks that would lead to stimulation of the appropriate skin region (Nielsen & Pierrot-Deseilligny 1991).

1.3.4. Transmission in excitatory and inhibitory Ib reflex pathways can be modified by voluntary contraction, with facilitation of heteronymous excitatory pathways to a contracting muscle and with suppression of the autogenetic disynaptic inhibitory pathway (Fournier et al. 1983). Suppression via the disynaptic inhibitory pathway in-

creases with contraction strength. Cutaneous influences on the transmission in Ib reflex pathways have been demonstrated in the cat (Lundberg et al. 1977), and in the human lower and upper limbs (Cavallari et al. 1985; Pierrot-Deseilligny et al. 1981a; see also Pierrot-Deseilligny et al. 1982). Cutaneous facilitation of transmission in Ib reflex pathways might be of functional significance in curtailing a movement. For example, in the upper limb, input from skin on the dorsum of the hand facilitates transmission in Ib inhibitory pathways to wrist extensor muscles and in Ib excitatory pathways to wrist flexors, so that hand contact could help to terminate an extension movement of the hand. Evidence for convergence between human knee joint afferents and Ib afferents projecting to thigh muscles has been difficult to obtain during a voluntary contraction of the target muscles (Brooke & McIlroy 1989), possibly because the knee positions studied did not include full flexion or full extension, positions in which joint afferent discharge is likely to be maximal (see sect. 1.1; also Baxendale & Ferrell 1981). However, joint afferents excited by joint effusion inhibit quadriceps motoneurons (e.g., Iles et al. 1984; Wood et al. 1988).

1.3.5. The projection of group II muscle spindle afferents has not been explored in human subjects using the same techniques. On indirect grounds, namely comparing the responses to muscle stretch and shortening with those elicited by the onset and offset of vibration, Matthews

(1984) argued that spindle group II afferents have an excitatory pathway to homonymous motoneurons in the human spinal cord much as in the decerebrate cat (Matthews 1972). Subsequent experiments have caused a revision of his previous views: "No component of the reflex responses, whether excitatory or inhibitory, has so far proved attributable to them" (Matthews 1989).

1.3.6. Stimulation of low-threshold cutaneous afferents in

both the human upper and lower limbs has complex reflex effects on the motoneuron pools of muscles throughout the limb, involving both spinal and supraspinal pathways (Jenner & Stephens 1982). Some of these reflex actions may well be mediated through interneurons that transmit group I volleys such as the Ib inhibitory interneuron and propriospinal-like neurons (see above). When stimulated in isolation, low-threshold afferents from cutaneous mechanoreceptors are insufficient to produce a reflex discharge unless the motoneuron pools are active. Expression of these cutaneomuscular reflexes is task-dependent so that different components of the evoked EMG pattern may change depending upon the movement task (e.g., Burke et al. 1991; Evans et al. 1989; Johansson & Westling 1987; 1988b). For example, in the human upper limb, a long-latency, possibly transcortical response is prominent in tasks requiring independent use of the digits rather than cooperative action in a grip (Evans et al. 1989). Powerful "grasp" compensation oc-

curs when objects slip between the digits (Fig. 4; Johansson & Westling 1984; 1987). In standing subjects, stimulation of the sural nerve evokes responses throughout the limb, the precise pattern depending on whether stance is unipedal or bipedal, or whether subjects stand on a stable or unstable support (Burke et al. 1991). During stepping, such volleys evoke reflex responses in thigh muscles and these vary with the swing or stance phase (Kanda & Sato 1983). There may be relatively more plasticity in the reflex effects of afferents from cutaneous mechanoreceptors on human γ-motoneurons (Aniss et al. 1990b). In subjects standing quietly without support, volleys in low-threshold cutaneous afferents from the foot may increase the discharge of spindle endings in ankle dorsiflexor muscles, even when these muscles are not contracting (Fig. 7A; for cat data see Davey & Ellaway 1989). In previous human studies, such volleys had no effect on spindle endings in relaxed (Gandevia et al. 1986b) or contracting ankle dorsiflexors (Fig. 7B; Aniss et al. 1988a). The appearance of this reflex during standing is consistent with task-dependent gating of reflex pathways acting on γ-motoneurons.

The extensive literature on the reflex compensation for externally perturbed movements is not reviewed here, but there is good evidence that the reflex response to perturbations during isometric conditions and during movement involves both spinal and supraspinal pathways, the latter now conceded for distal muscles acting on

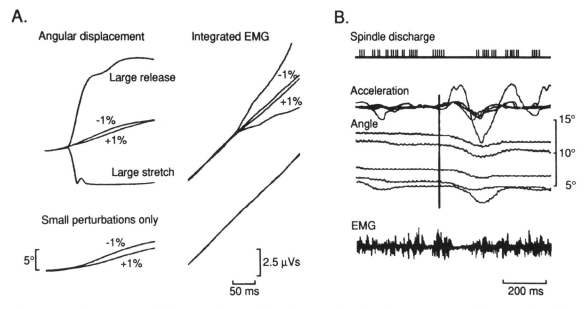

Figure 8A. Compensation following randomized disturbances to thumb tracking movements. The subject flexed the distal joint of the thumb against a constant torque of 0.10 Nm, and in some trials, this torque was changed by ± 1% 50 msec after the start of the recording sweep. This torque change was too small to be detected by the subject (lower record). In the upper records, when larger disturbances were included unpredictably within the sequence, complete positional correction was evident following the ±1% torque changes by the end of the sweep (left traces). There were corresponding changes in the integrated EMG records (right) from surface electrodes over flexor pollicis longus. When small disturbances were given alone (bottom traces), there was no positional correction and no change in EMG. Effective correction was dependent upon the context of the disturbance and the instruction to the subject. Reproduced from Marsden et al. (1983).

8B. Unloading responses resulting from natural irregularities in a slow voluntary shortening movement. Response of a dynamic muscle spindle afferent in tibialis anterior, five traces superimposed. Traces are, from above, standardized spindle potentials; ankle joint angle; tibialis anterior EMG. The event initiating the irregularity appears to be a grouping of EMG potentials. This burst is accompanied by a burst of spindle activity (i.e., apparent α–γ co-activation). The vertical line indicates the onset of the small irregularities in movement accompanying the EMG and spindle afferent burst. Following the shortening there is a reflex reduction in EMG. Reproduced from Burke et al. (1978a).

the hand by the most trenchant of critics (Matthews 1989). As shown in Figure 8A, the intensity of the reflex response is, in general, only adequate to compensate fully for small disturbances (Fig. 8A; Marsden et al. 1983). In similar studies, Cooke (1980) demonstrated that stretch reflexes during active movement of the upper limb may return the limb toward its unperturbed trajectory and, in a trial-by-trial analysis of fast movements, found that the return to the control trajectory was achieved more effectively in those trials in which the perturbation evoked a reflex response. So-called long-latency "reflex" effects operate not only on the stretched muscle but also on functionally-related muscles (Cole & Abbs 1987) as well as on remote muscles to ensure a more appropriate trajectory of the limb (e.g., Gielen et al. 1988; Soechting & Lacquaniti 1988). Reflex feedback also operates during unperturbed movements, to help smooth unintended irregularities in the course of the movement (Burke et al. 1978a; Vallbo 1973b). These irregularities are effectively sensed by spindles in the contracting muscle (Vallbo 1973b) and lead to a rapid alteration in EMG activity (Burke et al. 1978a, 1978b; Vallbo & Al-Falahe 1990). An example is shown in Figure 8B. In accurate position- or force-matching tasks, the gain of the reflex opposition to a very slow disturbance varies appropriately for the task given to the subject (Colebatch & McCloskey 1987); the responsible mechanism is probably a central change in reflex transmission, rather than a change in spindle input due to a fusimotor mechanism (Al-Falahe & Vallbo 1988; Burke et al. 1978b).

The above brief and admittedly selective review of some human reflex studies strongly suggests that during movement afferent feedback plays an important role in the on-going movement. Moreover, for all adequately studied reflex systems, the reflex effects are not immutable, changing in an appropriate direction when the task changes. This plasticity would only be meaningful if the nervous system required feedback control of performance.

2. Motor commands and movement without peripheral feedback

2.1. Sensations of force, timing, and centrally generated motor commands

If we state that centrally generated commands for movement are important in controlling it we may admit no more than that motor programs exist within the CNS and can produce movement (cf. Alexander et al.'s article, this issue). However, because such signals must be relied upon when afferent feedback is removed, brief discussion is warranted of several of their properties, particularly those which influence perception. There seems to be a reluctance to attribute a sensation related to limb movements to endogenous signals generated within the CNS, whereas in other areas of neural control (notably for eye movements), the central action of command signals is well established at neuronal and perceptual levels (e.g., Guthrie et al. 1983). In a recent review Matthews (1988) politely reminds us of the likely conservation in evolution of successful control designs based on motor commands such that "it is naive to suppose that they are not involved and to treat a search for them with the microelectrode as

being rather like trying to find the soul" (see also Robinson's and Fetz's articles, this issue). Their various roles in control of limb muscles have been reviewed (e.g., Gandevia 1987; McCloskey 1981). Although their roles may be difficult to grasp conceptually and to document experimentally, such commands are probably pivotal for understanding motor control. The three experimental approaches described below go some way toward opening up for study what Sherrington (1900) referred to as the "inward action of the neuromuscular machinery."

Centrally generated commands must exist prior to voluntary movement, but it is difficult to define the point at which they give rise to perceptual effects. Based on the relative timing of the "will to move" and an external stimulus, Libet and colleagues have shown that only after several hundred milliseconds does premovement cortical activity become associated with the desire to move (Libet 1985; Libet et al. 1983). Also a temporal delay occurs following cortical arrival of a peripheral sensory input before the time at which perception occurs (Libet et al. 1979), but this does not prevent the CNS from making rapid voluntary movements in response to liminal, even unperceived, stimuli (Taylor & McCloskey 1990b). A similar conclusion on the timing of motor action was reached by McCloskey et al. (1983b), who showed that subjects could time when a muscle contracts using either of two signals: one central, generated *prior* to neural activity in motoneurons, and the other peripheral, due to movement-evoked afferent activity. Subjects often showed a preference for one, but could attend to either signal without training. If making a rapid movement or sequence of movements, the ability to perceive when the motor apparatus is "commanded" is advantageous, rather than waiting for afferent feedback.

An indication of the resolution afforded by centrally generated signals was provided by Gandevia and Rothwell (1987), who showed that subjects have, or can rapidly acquire, the ability to deliver a motor command to specific motoneuron pools at a level subthreshold for activation of motoneurons. Thus, when requested to focus a sustained effort on one of a pair of intrinsic muscles, motor "excitability" was selectively raised as judged by the response to weak, unpredictable stimuli delivered to the motor cortex. This ability was acquired in the absence of useful feedback from the limb, and it could be documented for intrinsic hand muscles, but not for forearm muscles acting on the same fingers. An alternative approach to highlight the potential contribution of preprogrammed control commands was used by Gordon and Ghez (1987a; 1987b; see also Milner 1986). Based on the size of the initial peak "acceleration" accompanying rapid *and* accurate isometric contractions, about 80% of the variance in final force could be predicted. Much of the remaining variance reflected compensatory adjustments for errors in the initial preprogrammed "pulse," presumably determined by monitoring the outgoing motor commands.

2.1.1. Force sensation. Just as there are many mechanisms that contribute to sensation of position and movement, so sensations of muscle force or the perceived heaviness of a lifted object derive from many sources. Golgi tendon organs would seem well placed to provide a stable signal of muscle force (for reservations, see sect.

1.1). In many psychophysical tests involving measurement of weight, force, or stiffness, however, there is little evidence that subjects normally attend to cues about force, though they can do so in appropriate situations. When subjects are not told whether to attend to a signal of absolute force or one related to their central motor "effort," they almost invariably choose the latter. Thus, perceived forces are overestimated under a variety of conditions that effectively "weaken" the muscle, no matter how the weakness is produced, by an effect on the muscle, its motoneurons, or its descending motor pathways. This argument forms the foundation of the contention that signals related to the central motor command or effort bias judgments of force (McCloskey et al. 1974; for review see Cafarelli 1988; Gandevia 1987; Jones 1988; McCloskey 1978; 1981). The exact neural mechanisms are obscure, although the possibility of motor cortical involvement has been considered (e.g., Gandevia 1982; 1987; Roland 1978). An obvious feature of a central signal is that, without an afferent input (e.g., a proprioceptive or visual cue that an object has been moved), the central signal cannot be calibrated for events in the external world (Gandevia & McCloskey, 1978; Ross & Reschke 1982). One study highlighted the tight correspondence between changes in the "excitability" of the motoneuron pool and perceived heaviness (Aniss et al. 1988b). Short-latency reflex excitation (or inhibition) reaching the motoneuron pool of the first dorsal interosseous muscle from afferents in the index finger produced a graded reduction (or increase) in the perceived heaviness of weights lifted by the muscle. The perceived signal of heaviness compensated for a variety of reflex effects over a range of voluntary forces. Some further aspects of the sensation of force raised by recent studies are briefly set out below.

First, subjects can match force accurately despite a documented change in the required central command (produced, for example, by tonic vibration reflexes or partial curarization, see Gandevia & McCloskey 1977a; McCloskey et al. 1974; cf. Roland & Ladegaard-Pedersen 1977). When the sensitivity of tendon organs would have been reduced by a preceding maximal voluntary contraction, subjects generated a greater force to match an initial reference force (Thompson et al. 1990). This was consistent with the subject attending to a signal derived from tendon organ afferents. However, an alternative explanation that needs to be excluded for this and related studies (Hutton et al. 1987) is that subjects attended to a signal of central motor command in the presence of contraction-induced potentiation of the muscle twitch. Under these circumstances, the same central command would generate a greater voluntary force.

The ability to attend to an absolute peripheral force signal does not apply to all experimental situations: Muscle fatigue is an example in which virtually all subjects prefer to indicate that perceived force or heaviness is increasing even when instructed to concentrate on the absolute force or tension (e.g., Jones & Hunter 1983). Second, there is electrophysiological evidence for a cortical projection of group Ib afferents (McIntyre et al. 1984), with segregation of the input from tendon organs and muscle spindles in nucleus Z (McIntyre et al. 1985; 1989). Third, position sense is distorted by progressively loading the contracting muscles (Ferrell & Smith 1989; Roland

1978; Rymer & D'Almeida 1980; Watson et al. 1984). These distortions usually occur in the direction of muscle shortening and have been variously interpreted; tendon organs could well be involved, but corollary motor signals and other peripheral receptors could also provide sufficient explanations (Ferrell & Smith 1989; Matthews 1988). Fourth, one property of Golgi tendon organs that has been given little consideration is their ability to signal muscle length during a contraction (e.g., Fig. 6 of Crago et al. 1982; see also Roll et al. 1989; cf. Horcholle-Bossavit et al. 1988). While the receptors respond to the active force (e.g., Fukami & Wilkinson 1977), both the length and the force of the muscle change simultaneously during active lengthening and shortening. Thus, just as muscle spindle discharge is interpretable centrally in more than one way, an increase in tendon organ discharge could indicate a force increment due to increased neural activation or muscle lengthening. It is interesting that an illusion of muscle lengthening has been observed on microstimulation of one tendon organ afferent; a remarkable observation (requiring confirmation), given that single spindle afferents do not appear to evoke a percept (Macefield et al. 1990). Other ambiguities in perceptual decoding of force signals have been pointed out elsewhere (Gandevia & Mahutte 1982; Ross & Reschke 1982).

Finally, the accuracy of judgments about heaviness has been measured at several joints in the upper limb ranging from first dorsal interosseous to a combination of shoulder, elbow, and wrist muscles (Gandevia & Kilbreath 1990). When the lifted objects had weights that were comparable fractions of maximal voluntary force, the accuracy (expressed as the coefficient of variation) was not greater for distal muscles. Paradoxically, the accuracy of judgments was poorer for small weights compared with large weights (~3% and ~15% maximal force respectively; see also Newell et al. 1984).

2.2. Performance after deafferentation

At a simplistic level, the perceptual consequences of centrally generated motor commands arise in at least three ways: First, they occur directly prior to any movement and without reference to peripheral inputs (e.g., sensation of the "will to move" and the perceived timing of the actual command to move); second, they arise centrally, but require a peripheral signal for calibration or scaling (sensation of effort or force); and third, they may arise indirectly, for example, by measurement of changes in transmission through sensory pathways produced by motor commands. The first two mechanisms would be available following deafferentation, although the second may be impaired, and the third could be lost altogether.

The study of motor performance following deafferentation has long seemed an attractive experimental approach, because "it offers an objective mode of studying the activities of the afferent apparatus of muscular sense by watching the objective effects of it on the character of muscular movements" (Sherrington 1900). More than a century of observations has not led to a consensus about the deficits produced by deafferentation (for review of earlier literature see Goodwin et al. 1972; Nathan & Sears 1960; Phillips 1985; Sanes et al. 1985). This may be traceable to the following factors. First, deafferentation produced by naturally occurring lesions is variable and rarely complete. Peripheral lesions, short of nerve transection, will fail to affect all afferent fibers equally, whereas central lesions will affect more than just sensory pathways. Any residual peripheral sensory cues, even those arising in small-diameter fibers, are likely to be used by patients with loss of large-fiber input. Second, the mode of onset of deafferentation is variable, as is the time between deafferentation and clinical testing. Subsequent performance may differ depending upon whether use of the affected limb has been imperative (Knapp et al. 1963; Wolf et al. 1989). Third, impaired performance after deafferentation may be due to loss of a phasic task-related peripheral signal, loss of a tonic input providing a background excitation to spinal (and other) circuitry, or both. Fourth, performance following deafferentation may depend on the task selected for testing. Thus, deficits of distal movement, not present in one posture, may be revealed when a novel spatial relation between the proximal and distal parts of the limb is required (Polit & Bizzi 1979; for further discussion see Bizzi et al.'s article, this issue). It is difficult to translate deficits with deafferentation into hypotheses about motor control; a deficit in performance could mistakenly be used to highlight the role of sensory feedback, although under normal conditions the tested performance may run largely "open loop," only intermittently relying on peripheral guidance. Conversely, adequate performance may be advanced as evidence for purely central control, whereas an afferent contribution may have been present, but undetected, or made redundant through training.

Performance after deafferentation has been recently reassessed in patients with large-fiber sensory neuropathies and central deafferentation syndromes, and in primates with surgical deafferentation. In the description below, emphasis is given to the consistent deficits observed in a group of six patients with large-fiber sensory neuropathy by Sanes et al. (1985). Similar findings are evident in reports of smaller numbers of patients with this syndrome (e.g., Cole 1986; Cole et al. 1986; Forget & Lamarre 1987; Ghez et al. 1990; Rothwell et al. 1982b) or central lesions producing "deafferentation" (e.g., Jeannerod et al. 1984; Nagaoka & Tanaka 1981; Volpe et al. 1979).

After deafferentation, patients have a wide-based gait, difficulty in making repeated finger movements without vision, and difficulty in maintaining a posture, especially when vision is excluded. Thus, when attempting to maintain angular position against one of a series of flexor or extensor torques with muscles acting across the wrist, the mean angular error during 20 sec of "postural maintenance" was about 6° for a group of six patients whereas it was usually less than 1° in control subjects (Sanes et al. 1985). Errors did not grow with the size of the load (Fig. 9A). Comparable errors developed on removal of visual feedback when movements of specified amplitudes were required against a spring; the initial movement was similar to that of normal subjects when visual feedback of position was permitted but the position accurately attained under visual control could not be maintained (Fig. 9B). Introduction of an unexpected viscous load at the

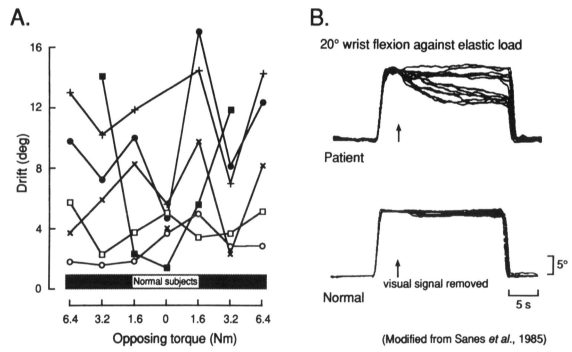

A.

Drift (deg) vs Opposing torque (Nm)

Normal subjects

6.4 3.2 1.6 0 1.6 3.2 6.4

Opposing torque (Nm)

B.

20° wrist flexion against elastic load

Patient

Normal

visual signal removed

]5°

5 s

(Modified from Sanes *et al.*, 1985)

Figure 9A.   Maintenance of a particular flexor or extensor torque at the wrist in six patients with a sensory neuropathy. Data points from each subject are joined by lines. Each point represents the absolute value of the averaged position during a period of 20 sec after visual guidance was removed. Estimated performance from normal subjects is shown as a hatched area. Data recalculated from Sanes et al. (1985).

9B. Voluntary movement of 20° produced by wrist flexors against a (simulated) elastic load in a patient with sensory neuropathy (above) and a normal subject (below). Note that in many trials, once visual guidance was removed, the angle drifted toward the starting position. When expressed in proportion to the amplitude of the movement, relative error increased for smaller (5°) movements compared to large (10°–20°) movements. Reproduced from Sanes et al. (1985).

onset of movement resulted in a slower movement of reduced extent in the patient group, a finding which corroborates data from normal subjects (Day & Marsden 1982). These data are inconsistent with the hypothesis that movement endpoints can be determined solely by the properties of a simple mass-spring system (see Bizzi et al., this issue). In the deafferented patients, unloaded movements, particularly of small amplitude (3° at wrist in the study of Sanes et al., 1985), were associated with large errors in both the dynamic and static phases, and occasionally small movements were made in the wrong direction. These and other data (e.g., Brink & Mackel 1987; Sanes & Evarts 1983a; 1983b) highlight the need for afferent feedback for precise "vernier" contractions. A class of motor cortical cells has input-output properties well suited for such tasks (Fromm & Evarts 1981).

In deafferented patients, abnormally large degrees of co-contraction often developed, but for rapid movements the classical bi- or triphasic EMG pattern occurred in an agonist-antagonist pair. However, although this centrally generated pattern remained after deafferentation (Forget & Lamarre 1987; see also Bizzi et al. 1984), the antagonist burst (leading to limb deceleration) no longer changed in proportion of the initial agonist burst. Furthermore, the timing of the antagonist burst was more variable. In deafferented patients, the ability to match torque (or weight) was extant, but it was not as good as in normal subjects. The deficit appeared especially for low forces, and it showed hysteresis: Forces presented in ascending order were better matched than those in descending order (Sanes 1990). Even in normal subjects, there is poor accuracy for judgment of relatively low forces (Gandevia & Kilbreath 1990; Newell et al. 1984). This impaired sense of motor command or effort may reflect a defective capacity to calibrate this signal by reference to a proprioceptive signal (see Gandevia & McCloskey 1978). Deafferented patients also exhibit a poorer ability to make accurate multijoint pointing movements with the upper limb (Ghez et al. 1990).

Other aspects of motor performance should be measured in patients with deafferentation, including improvement in performance over time (Cole 1986) and the ability to perform sequences of movements in which afferent signals are required to trigger the next component of the sequence (Cordo 1990; see also Benecke et al. 1986; Johansson & Westling 1988a; 1988b). The latter studies highlight the ability of central motor commands to be rapidly recalibrated even within a single movement based on afferent inputs from the limb when the relationship between force required and movement produced is altered unexpectedly. Thus, an unexpectedly light suitcase will be lifted inappropriately, but replaced accurately on the ground. Formal measurements of muscle strength have not been reported yet in these patients although weakness per se has not been a limiting factor for their performance. These measurements, if also obtained prior to deafferentation, could indicate the degree of facilitation mediated through the fusimotor-spindle loop. Unfortunately, the results in individuals would be difficult to interpret because of the wide range of normal strength. Finally, deficits in timing of muscle bursts in highly stereotyped tasks such as respiration and locomotion will probably also be revealed; these have been

documented for the paw-shake response after hindlimb deafferentation in cats (e.g., Koshland & Smith 1989).

An alternative way to dissect the capacity of centrally generated commands and of peripheral inputs to drive motoneurons involves recording the discharge of single motor axons destined for particular intrinsic muscles of the hand. Recordings have been made proximal to a complete anesthetic block of the ulnar nerve so that the capacity of subjects to discharge motoneurons supplying intrinsic muscles of the hand by voluntary effort could be assessed when there was no tonic or phasic feedback from homonymous and heteronymous muscles in the hand (due to the ulnar block) and no phasic feedback from cutaneous and joint afferents innervating the hand (Gandevia et al. 1990). The skin and joints of most of the hand remained sentient, because skin and joint afferents transversed the unanesthetized median nerve. Although this acute muscle paralysis and deafferentation is artificial, any ability to drive intrinsic hand motoneurons represents the minimal capacity of centrally generated commands, whereas any deficit (compared with normally innervated muscle) represents the maximal contribution

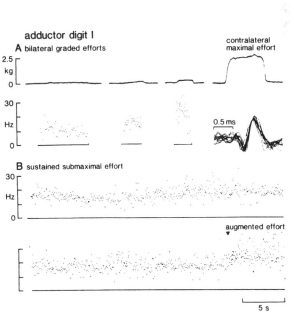

Figure 10A. Graded discharge frequency of a motor axon supplying adductor pollicis during bilateral efforts. Discharge of the motoneuron was recorded from the ulnar nerve at a site 10 cm proximal to a complete anesthetic block of the same nerve. Recruitment of the motoneuron with attempted adduction of the thumb was facilitated by cutaneous stimulation applied to thumb and index finger. This was applied throughout the sequences shown here. Effort on the contralateral side was recorded during matching thumb adduction against a strain gauge. The force generated during contralateral matching effort provides an indication of the low recruitment threshold of the motoneuron. Superimposed spikes recorded during a submaximal effort are also shown.

10B. The same motoneuron as in 10A during an attempt to sustain a constant level of discharge. The upper and lower panels are continuous. Toward the end of the attempt, the subject was instructed to increase his effort. Note that, in the brief and sustained efforts, the peak discharge frequency was ~30 Hz in the absence of homonymous muscle afferent feedback. Reproduced from Gandevia et al. (1990).

attributable to afferent inputs from the hand. The discharge of many axons was facilitated when cutaneous afferents in the hand were stimulated; for example, the ability to discharge a motoneuron supplying adductor pollicis was enhanced if the index finger and thumb were touching (Gandevia & McCloskey 1977b; Kanda & Desmedt 1983). In the absence of feedback from the hand, a motoneuron supplying an intrinsic hand muscle could be recruited in attempted contractions that usually involved that muscle. This ability improved with provision of visual or auditory feedback about the discharge of the motor axon. The discharge frequency of the motoneuron could be graded according to the level of effort (Fig. 10A). The discharge frequencies could be sustained for long periods especially if audiovisual feedback was available (Fig. 10B). Maximal discharge frequencies were below the levels achieved by normally innervated motor units, presumably due to the withdrawal of reflex support to the contraction from muscle spindle endings. Finally, there was no decline in motoneuron firing rates with fatigue, as occurs in normal muscle contractions. This confirms the view that the decline is mediated reflexly by the afferent input from the muscle, via inhibition developing during the contraction (Bigland-Ritchie et al. 1986) or reduced facilitation from muscle spindle endings (Hagbarth et al. 1986).

3. Conclusion

Movements are controlled in "real time," ranging from the production of a single discharge in a low-threshold motor unit of a muscle in the hand to movement of the whole body. The capacity to perform movements is attained by an individual through years of learning, while on an evolutionary scale, it reflects the development of highly "successful" skills. The CNS has access to peripheral kinesthetic signals from intramuscular, joint, and cutaneous receptors; it also has endogenous, centrally generated kinesthetic signals of the timing, grading, and destination of motor output. Although they are not usually the focus of conscious attention, these kinesthetic and endogenous signals can all be attended to at will and the peripheral ones can reflexly influence motoneuron output in a task-dependent way. For many motor skills there may eventually develop a degree of redundancy provided by peripheral feedback, but the full movement repertoire cannot be achieved without the assistance of afferent inputs in adjusting an ongoing movement and in updating the motor controller so that the current and subsequent movements are completed successfully.

NOTE

1. Please send correspondence to: Dr. S. C. Gandevia, Department of Clinical Neurophysiology, Prince of Wales Medical Research Institute, The Prince Henry Hospital, P.O. Box 233, Matraville, N.S.W. 2036, Sydney, Australia.

3

Can sense be made of spinal interneuron circuits?

D. A. McCrea

*Department of Physiology, Faculty of Medicine, University of Manitoba,
Winnipeg, Manitoba, Canada R3E 0W3*
Electronic mail: *dave@scrc.umanitoba.ca*

Abstract: It is increasingly clear that spinal reflex systems cannot be described in terms of simple and constant reflex actions. The extensive convergence of segmental and descending systems onto spinal interneurons suggests that spinal interneurons are not relay systems but rather form a crucial component in determining which muscles are activated during voluntary and reflex movements. The notion that descending systems simply modulate the gain of spinal interneuronal pathways has been tempered by the observation that spinal interneurons gate and distribute descending control to specific motoneurons. Spinal reflex systems are complex but current approaches will continue to provide insight into motor systems. During movement, several neural mechanisms act to reduce the functional complexity of motor systems by inhibiting some of the parallel reflex pathways available to segmental afferents and descending systems. The flexion reflex system is discussed as an example of the flexibility of spinal interneuron systems and as a useful conceptual construct. Examples are provided of the kinds of experiments that can be developed using current approaches to spinal interneuronal systems.

Keywords: flexion reflex; interneuron; motor control; muscle synergy; presynaptic inhibition; reflexes; spinal cord

1. Introduction

This target article addresses some current issues concerning interneurons interposed in spinal motor pathways.[1] Despite the wealth of detailed information about the anatomy, behavior, and synaptic action of spinal interneurons, some feel that a clear understanding of the function of spinal pathways is no closer now than several years ago. Indeed, many simpler models of spinal cord interneuron function have proven inadequate while more comprehensive models have not been forthcoming. The growing realization that spinal interneurons receive convergence from multiple pathways and in turn send axon collaterals to multiple target neurons prompts the question "Can sense be made of spinal interneurons?" In this discussion, I suggest that this massive convergence and divergence is the worst case scenario; it has been revealed in anesthetized or decerebrate preparations but not during movement in the conscious animal. The theme will be developed that during behaviour several neural mechanisms reduce the operational complexity of these systems so that particular interneuron pathways are selected over others. During behaviour, certain connections to interneurons will be dominant and others will be less important; this dominance may change as the movement progresses. It has become increasingly apparent that individual interneurons often receive only a portion of the synaptic input that impinges on the entire population of interneurons involved in a specific motor pathway. Subpopulations of interneurons make it possible for descending and afferent systems to generate fine movements as well as the grosser movements that are often thought

characteristic of reflexes. Sense is being made of spinal interneurons; in addition, techniques used in the past will continue to provide insight into the organization of motor systems (see Fetz's and Robinson's articles, this issue).

The list of spinal interneurons studied in motor systems now includes Ia inhibitory interneurons, Renshaw cells, interneurons mediating reflexes from Ib tendon organ afferents (Harrison & Jankowska 1985a; 1985b), lamina VIII crossed interneurons (Harrison et al. 1986), group II interneurons located in caudal lumbar segments (Lundberg et al. 1987a), group II interneurons in mid-lumbar segments (Edgley & Jankowska 1987) and their activity during fictive locomotion (Shefchyk et al. 1990), interneurons active in the DOPA preparation (e.g., Jankowska et al. 1967a), interneurons involved in presynaptic inhibition (Rudomin et al. 1987), interneurons in lumbar propriospinal pathways (Vasilenko, Kostyuk and co-workers; reviewed in Edgley & Jankowska 1987), interneurons involved in fictive scratching (e.g., Deliagina et al. 1981), and cervical propriospinal cells (Lundberg 1979b). Recent demonstrations of disynaptic cutaneous reflex pathways in cat (Fleshman et al. 1988) and rat (Edgley & Wallace 1989) will no doubt be followed soon by identification of the interposed interneurons. (For more details about the interneurons mentioned, see reviews by Baldissera et al. 1981; Jankowska 1992; Lundberg 1979b; McCrea 1986).

Older views of spinal reflex function suggest that particular sensory afferents have specific reflex actions mediated by discrete populations of interneurons and that the activation of these pathways results in reflex movements. Included is the idea that reflex movements are a class of movements in themselves, and that it is possible to differentiate spinal circuits that serve reflex function from those that serve voluntary movements. As will be discussed, many spinal interneurons and all spinal interneuronal pathways receive convergence from both segmental sensory and descending input. This suggests that the concept of a reflex as an auxiliary or independent component of the motor system needs to be revised. Only those movements elicited by unusual and discrete stimulation of afferents, such as tendon tap, can easily be considered simple reflexes. Natural movements are the result of interactions between many systems located throughout the neuraxis, and all movements occur by the interaction of segmental pathways with afferent and descending systems. Descending and voluntary control does not operate on a backdrop of hard-wired spinal "reflex" systems. Instead, the particular spinal circuitry operating during a movement is dynamically selected and optimized. Dynamic control of movement is also apparent during complex "reflex" movements such as flexion induced by noxious stimulation (the flexion reflex) or cutaneous stimulation during locomotion. Since most reflexes depend upon the state of many groups of spinal interneurons, a description of either afferent activity or the excitability of one group of interneurons will not predict the subsequent movement.

Older descriptions of spinal motor systems often take the view that sensory input pathways and the spinal interneurons they contact monitor or decode single parameters of movement such as velocity, joint angle, stiffness, and so on. This leads to the expectation that spinal interneurons can be categorized as belonging to specific and almost isolated pathways responding to movements readily described in engineering terms. Further experimentation, however, has invariably revealed a more complex organization and interaction between spinal pathways than originally envisioned. All spinal pathways involving interneurons and ending on motoneurons receive input from more than one type of sensory fibre input at one or more points in the interneuronal chain. Consequently, simple models of interneuronal function (involving a single sensory modality) that attempt to explain how sensory feedback can shape ongoing movements are likely to fail. At this point it is useful to review what is meant by the term *interneuron* and briefly consider how interneurons are categorized.

2. Changing views about interneurons: The Ia and Ib

Once it was possible to define spinal interneurons as cells with short axons. The advent of intracellular staining revealed that many spinal neurons thought to be "interneurons" actually project many millimeters. For convenience, these cells now are usually included in the term *interneuron,* and the term *propriospinal* is reserved for those cells projecting several segments. At one time, ascending tract cells with somas in the lumbar cord and axons projecting to higher centres were also considered apart from spinal interneurons. More recent investigations reveal that some ascending tract cells (e.g., ventral spinocerebellar, Bras et al. 1988; spinocervical, Jankowska et al. 1967a) also have axon collaterals and synaptic terminals ending in the lumbar spinal cord. Thus short axoned cells, some ascending tract cells, and long and short axon propriospinal cells all function as segmental interneurons in the sense that they can regulate the activity of other local neurons.

Spinal interneurons are often classified within the experimental context in which they are first investigated. As an example, the interneurons that mediate disynaptic inhibition of antagonist motoneurons from group Ia muscle spindle afferents (i.e., reciprocal inhibition) have been located and extensively characterized (references in Baldissera et al. 1981; Jankowska 1992). Their monosynaptic excitation from Ia muscle spindle afferents and inhibitory projections to motoneurons supplying antagonist muscles make their classification as Ia inhibitory interneurons (IaIns) reasonable. Subsequent investigations, however, have shown that these cells also mediate disynaptic inhibition of motoneurons from the corticospinal tract (Jankowska et al. 1976b) and that they are activated by segmental afferents other than Ia muscle spindle afferents (see Baldissera et al. 1981; Jankowska 1992). If the initial emphasis had been on corticospinal inhibition instead of reciprocal inhibition, these interneurons might have been termed something other than IaIns. In either case, the terminology used for classification would not do justice to an understanding of the function of these cells. When a segmental reflex is evoked, the IaIn distributes inhibition to motoneurons supplying muscles that are antagonists to those activated in the reflex. In the case of the stretch reflex, inhibition of antagonists increases the ability of the stretched muscle to restore limb position. Voluntary movements also may require inhibition of antagonists at a particular time during the movement. Cortical effects on spinal systems probably always involve both excitation and, via interneurons, inhibition of motoneurons. Cortical inhibition exerted through segmental IaIns is an efficient way to inhibit groups of motoneurons, since advantage can be taken of the connections of IaIns to motoneurons. Another situation in which the nervous system uses inhibitory connections from IaIns to motoneurons is fictive locomotion. In the mesencephalic fictive locomotion preparation, rhythmic afferent input is absent and the cortex is removed. IaIns are active in these preparations and show bursts of activity at particular phases of the fictive step cycle. IaIns projecting to flexor motoneurons are active during extension and contribute to inhibition of flexors while IaIns projecting to extensor motoneurons are active during flexion (e.g., McCrea et al. 1980; Severin et al. 1968). These observations show that both the spinal locomotor circuitry and segmental afferents excite IaIns, thus serving as a source of inhibition for multiple systems.

Another example of an interneuronal system that is being considered in a broader context than first envisioned is that between Golgi tendon organ afferents and motoneurons. Activity in Ib Golgi tendon organ afferents results in widespread inhibition of homonymous and synergist motoneurons and excitation of other motoneurons in the limb (references in Jankowska et al. 1981a; 1981b; 1981c). The inhibition of homonymous motoneurons has received the most attention, and the original hypotheses about pathways activated by Golgi tendon organs concerned protective reflexes to prevent damage to the activated muscles (references in Cleland & Rymer

1990). The finding that Ib afferents are exquisitely sensitive and active at even small force levels prompted a reconsideration of their function. Houk (1979) summarizes arguments for the idea that spinal pathways involving Ib interneurons act to oppose motoneuron excitation produced by activity in the Ia reflex arc (but also see Rymer 1984). Further investigation has revealed that interneurons with IB input are also activated by several descending and other segmental systems (Harrison & Jankowska 1985a; Lundberg et al. 1977) including Ia muscle spindle afferents (Jankowska et al. 1981a; Jankowska & McCrea 1983). This convergence of input from Ia and Ib receptors to common interneurons greatly weakens the suggestion of simple opposing actions of Ia and Ib afferents.

A more recent hypothesis concerning Ib pathways (Lundberg & Malmgren 1988; Malmgren 1988) suggests that Ia input to Ib inhibitory interneurons confers a dynamic sensitivity to the Ib system. The background excitability of the interneurons is influenced by increases in Ia input resulting from muscle stretch or gamma drive to the spindles. In this scheme, Ib interneurons are central elements for regulating muscle force and this regulation results from interaction between Ia excitation and Ib inhibition of synergistic motoneurons. Adding a dynamic sensitivity to Ib interneurons would allow precise force regulation throughout a wide range of movements. According to this hypothesis, the Ia excitation of the Ib interneuron population is weak and unlikely to activate all Ib interneurons. This is consistent with the finding that, although some Ib interneurons are readily recruited by Ia input, only about 40% of Ib interneurons have Ia excitation (Jankowska et al. 1981a). Although this hypothesis can account for the role of Ib inhibition of synergists in man and cat (Malmgren 1988), a hypothesis that explains the wide distribution of Ib inhibition and excitation throughout the limb has yet to emerge.

Since individual interneurons are likely to be involved in several aspects of motor control, any interneuron classification that suggests a single function is probably misleading. Terminology for spinal interneurons that does not relate to synaptic input or single functions may be preferable. For example, the name of the Renshaw cell (Eccles et al. 1954) gives tribute and honour to the investigator who predicted its existence in the inhibitory pathway from motor-axon collaterals to other motoneurons. Although succinct taxonomy may be difficult, reasonable interneuronal descriptions are possible. For example, IaIns can be distinguished from other cells with Ia input by their location and the presence of inhibition evoked from stimulation of the motor axon collateral system (references in Baldissera et al. 1981). For the Ib interneuron with its multiple inputs, Ib afferent input is the only input common to all Ib interneurons. One remaining problem, however, concerns differences in the pattern of input between cells originally thought to comprise a homogenous group. For example, only a portion of the interneurons with Ib input receives excitation from Ia muscle spindle afferents. Although it is uncertain whether this observation justifies a reclassification of these interneurons into distinct groups, it is clear that interneuron pools are fractionated according to both the input that individual cells receive and the target cells to which they project. The extent to which interneuron

pools are functionally fractionated is an area of active investigation (e.g., Harrison & Jankowska 1985a; 1985b). Obviously, complete knowledge about the input to individual interneurons is an essential step toward the understanding of spinal motor systems.

Spinal interneurons are subject to a variety of central and segmental influences that work in concert to shape movements. Therefore, a study of the activity of individual afferents during movement will, by itself, offer only limited insight into the resultant reflex action. The divergence of axon collaterals of single afferents to several types of neurons (e.g., group I muscle afferents; Hongo et al. 1987) and the modulation of transmitter release at synapses by afferents (e.g., presynaptic inhibition, see next section) are both essential considerations. Similarly, a simple catalogue of connections to interneurons in anesthetized animals does not predict the effect of activation of a particular afferent or descending system on motor output. The existence of multiple pathways from sensory afferents to motoneurons increases the difficulty of understanding reflex function when studied in isolation in anesthetized preparations. In this regard, the use of a reduced but active preparation such as fictive locomotion has many advantages.

3. Inhibition and excitation of interneurons

Activation of Ia muscle spindle afferents evokes disynaptic inhibition of antagonist motoneurons through IaIns (reciprocal inhibition). The magnitude of this inhibition, however, can be adjusted by activation of other spinal circuitry. One source of inhibition of the IaIn is inhibition from interneurons involved in other segmental reflexes. Renshaw cells, for example, inhibit IaIns as well as motoneurons (references in Baldissera et al. 1981). Since the IaIn mediates disynaptic inhibition from group Ia afferents to antagonist motoneurons, reduced activity of IaIns can result in increased excitability of the antagonist motoneuron pool through disinhibition. Activation of Renshaw cells thus inhibits one set of motoneurons while exciting others. Convergence to IaIns from descending and segmental systems includes cutaneous afferents and those classified as flexion reflex afferents (FRA), the cortico-, rubro- and vestibulo-spinal tracts and propriospinal systems (references in Baldissera et al. 1981; Jankowska 1992). Systems that control IaIn excitability will act to regulate motoneuron inhibition evoked from both peripheral and descending sites.

It is common for interneurons to have collateral actions that inhibit other interneurons in reflex pathways to the same or other motoneurons (e.g., the inhibition of IaIns by Renshaw cells). One such interneuronal organization is mutual inhibition between interneurons with similar functions. Mutual inhibition has been shown for IaIns (references in Baldissera et al. 1981), interneurons mediating inhibition of motoneurons from Golgi tendon organ afferents (Brink et al. 1983), and interneurons in pathways from the FRA (Jankowska et al. 1967a). One consequence of inhibition of interneurons by other interneurons is a reduction in the number of parallel pathways available to descending and segmental afferent commands. Inhibition of interneurons removes them from further participation in that movement. As discussed below in the sections on the FRA and muscle synergies, inhibitory mechanisms between interneurons also may produce a selection of one among several parallel interneuronal pathways.

If an interneuron is already active (i.e., firing action potentials) then additional excitatory convergence may produce little or no added effect. Similarly, sorting out the extensive convergence to interneurons is easier if some segmental afferents or descending systems are not active during that movement. These points sound trivial, but knowledge of how "operational connectivity" to individual interneurons changes throughout the movement is essential to an understanding of spinal motor systems. For interneurons to function as a stable and reliable control system there must be powerful mechanisms for selecting those interneurons required for the movement. Unlike motoneurons, single inputs to interneurons can have very powerful effects and may completely dominate neuronal excitability. For example, EPSPs (excitatory postsynaptic potentials) produced by single group II afferents are very large in interneurons located both in rostral (Edgley & Jankowska 1987) and caudal (Lundberg et al. 1987a) lumbar segments. For rostrally located interneurons with both Ia and group II input, unitary group II EPSPs are much larger than those produced by Ia afferents (Edgley & Jankowska 1987). Individual afferents may even recruit interneurons and thus allow for interneuron activation during movements with minimal afferent feedback. Inhibition of interneurons can also be very effective; for example, Renshaw cell inhibition of IaIns can prevent activation of IaIns by Ia afferents (Hultborn et al. 1971a). Thus, strong inputs to interneurons may momentarily dominate the excitability of the cell.

4. Presynaptic inhibition (PSI)

Activity in many sensory afferents results in decreased monosynaptic postsynaptic actions (presynaptic inhibition; PSI) of either the same or other afferents. While primary afferent depolarization and the subsequent reduction in transmitter release is the best understood mechanism of PSI (reviewed by Davidoff & Hackman 1984; Rudomin 1990a; 1990b), awareness of the postsynaptic modulation of transmitter actions is increasing (e.g., Marshall & Xiong 1991). Whatever the mechanism, a substantial portion of the phenomenon of PSI is due to the actions of spinal interneurons. For the monosynaptic reflex, presynaptic inhibition of the primary afferents and changes in the reflex activation of motoneurons are well studied phenomena. For example, the amount of PSI of group Ia afferents may change as movement commences. In man, PSI of Ia afferents of muscles not involved in a particular movement is increased as the movement begins (Hultborn et al. 1987). Furthermore, at rest there is a tonic level of PSI of group Ia afferents (Hultborn et al. 1987). The level of PSI of Ia afferents changes under different reflex conditions and throughout the locomotor step cycle (e.g., Dietz et al. 1990). Consequently, the gain of the monosynaptic reflex arc can be increased by a reduction in tonic PSI or it can be decreased by increased PSI. Thus, the motor consequences of activity in even the simplest reflex system, Ia afferents and the monosynaptic reflex, can only be under-

stood by knowing about the activity of interneurons in PSI pathways. While the modulation of the gain of the monosynaptic reflex is well accepted, this modulation is often not incorporated into models of other reflex circuitry.

As well as changing the gain of segmental reflexes, selective PSI of a portion of the afferents impinging on interneurons could switch the activation of interneurons from one type of afferent to another. In this way, the particular interneurons recruited by sensory afferents could be controlled and selected without changes in the activity of sensory afferents during movement. Presynaptic inhibition would alter the activity both of interneurons contacted by sensory afferents and of those with which they are in contact. The interneurons which produce PSI are themselves subject to the extensive segmental and descending convergence typical of other spinal interneurons (see Rudomin 1990a). Therefore, during behaviour, there will be a continuous regulation of interneurons producing PSI and of their ability to influence synaptic transmission from segmental afferents onto other neurons.

PSI does not exert its normal regulatory influence in the anesthetized or otherwise reduced preparation in which the convergence to interneurons has usually been examined. Thus the relative strength of individual synaptic inputs to interneurons (i.e., the size of the postsynaptic potentials) has often been assessed in an unphysiological situation. During movement in the intact animal, PSI could change the relative effectiveness of peripheral input to interneurons in reflex pathways, perhaps even to the extent of functionally eliminating certain inputs during certain portions of a movement. During fictive locomotion with no rhythmic afferent feedback, there are phasic modulations of the excitability of muscle afferents (Duenas & Rudomin 1988) and depolarizations of cutaneous (Gossard et al. 1989) and muscle afferents (Gossard et al. 1991). This is evidence for a modulation of postsynaptic effects of these afferents by the actions of intrinsic spinal cord circuitry on interneurons responsible for PSI. Furthermore, the modulation of interneurons in PSI pathways during locomotion may be different from those producing PSI evoked from peripheral stimulation (Dietz et al. 1990; Gossard et al. 1990). This suggests that PSI pathways might be controlled differentially in different movements (Dietz et al. 1990; Gossard et al. 1990).

With changes in the afferent activation of interneurons in reflex pathways comes the possibility of switching between parallel reflex pathways. For example, imagine interneurons (Figure 1, neuron A) with a relatively weak input from both type X and Y afferents that inhibit both motoneurons and other interneurons (Figure 1, interneuron B) with input from Y afferents. Increased PSI of type X afferents would reduce activity of the interneurons with both inputs (e.g., neuron A) and release the target interneurons (Figure 1, neuron B) and motoneurons from inhibition. Activity in Y afferents would now result in a net excitation of motoneurons. In this way one of the parallel pathways available to the Y afferents would have been selected. The utility of inhibition of both motoneurons and other interneurons by interneuron A becomes clear. As mentioned above, there are several examples of interneuron pathways that when activated result in collateral inhibition of other interneurons. Rudomin (1990a; 1990b) describes how selective PSI of

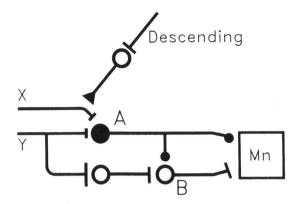

Figure 1. Presynaptic inhibition may allow switching between spinal reflex pathways. Depolarization of an afferent terminal (presynaptic inhibition) is indicated by the filled triangle ending on afferent X. Motoneurons are indicated by a square, interneurons by large circles, inhibitory synaptic effects by small filled circles, and excitatory effects by short lines.

different types of afferents by descending systems would bias reflexes to be more responsive to a particular type of perturbation. For example, interneurons with input from both Ia and Ib afferents could be "prepared" to be more responsive to input from tendon organs during parts of a movement. An appreciation of the organization and activity of PSI pathways is, therefore, essential to an appreciation of the effectiveness of particular afferents in activating interneurons in motor pathways.

The role of PSI in movement is more complex than a simple regulation of afferent feedback to the spinal cord. Selection of specific interneurons from among multiple parallel pathways may be an important role of PSI. Recent evidence (discussed below) indicates that descending systems produce selective presynaptic inhibition of specific types of sensory afferents ending on spinal interneurons. This could be accomplished either directly by actions of descending systems on afferent terminals, or indirectly by activation of spinal interneurons that in turn produce PSI of subsets of afferents.

5. Rostrally located interneurons with group II input

There are groups of interneurons located in rostral lumbar segments with monosynaptic input from both group I and group II muscle spindle afferents with axons that project caudally (Edgley & Jankowska 1987). To date, those located in the L4 segment have received the most attention. These interneurons have monosynaptic actions on motoneurons and serve as disynaptic pathways from group II and group I muscle spindle afferents to motoneurons. Both excitatory and inhibitory interneurons have been identified (Cavallari et al. 1987). The effectiveness of group II afferents in monosynaptically activating L4 interneurons can be reduced by iontophoretically applying monoamines (Bras et al. 1989b) and by electrically stimulating brainstem areas containing monoaminergic neurons (Noga et al. 1992). These interventions have little effect on group I afferent activation of interneurons in the same area. It is not yet known if these actions are due to the direct effects of monoamines on

presynaptic terminals. The resulting reduction of afferent input by descending systems could, however, alter the dominance of the static group II stretch receptors in monosynaptically activating these cells. The inference is that during certain movements, brainstem structures can help to select which type of segmental input will influence segmental reflexes.

These studies of group II interneurons are a good example of how rapidly research on spinal interneurons can progress. Since their first description in 1987 (Edgley & Jankowska 1987), a detailed picture has emerged of the anatomy of these interneurons (Bras et al. 1989a), their effects upon motoneurons (Cavallari et al. 1987), segmental and descending input, and control by monoamine systems (Bras et al. 1989b; Noga et al. 1992). We now know that brainstem areas important for the initiation of locomotion also activate some of these L4 interneurons (Edgley et al. 1988). During fictive locomotion, some of these interneurons are active during flexion while others are inhibited throughout the step cycle (Shefchyk et al. 1990). They also must receive input from the intrinsic locomotion circuits because there is no rhythmic afferent input during fictive locomotion.

Further investigation of the activation of these rostral group II interneurons by afferents throughout the step cycle may offer insights into the extent and manner in which descending and segmental systems modulate motor systems as movements commence and continue. Further investigation of the synaptic pharmacology of these interneurons will allow the testing of specific hypotheses concerning their role in locomotion, other movements, and movement disorders (e.g., spasticity) by providing ways to manipulate their activation and effectiveness. Central to the success of these studies has been a series of testable hypotheses that were developed using standard neurophysiological approaches.

Another system that has been approached with a series of testable hypotheses is the "FRA" system developed by Anders Lundberg and his colleagues. This system has provided substantial insight into motor organization, but because of its complexity it may have also produced some confusion about the organization of spinal interneurons.

6. The FRA system

The flexion reflex involves widespread ipsilateral activation of flexor motoneurons and inhibition of extensor motoneurons. The term "flexion reflex afferents" (FRA) was given to those afferents that may evoke the flexion reflex and this classification includes higher threshold muscle, cutaneous, and joint afferents (Eccles & Lundberg 1959). As a general organizational principle, the FRA concept (e.g., Lundberg 1979a) implies that (1) ipsilateral activation of flexors and concomitant inhibition of extensors should occur with a variety of segmental inputs, (2) there are interneurons with this convergent polymodal input, and (3) there are inhibitory interactions between groups of interneurons interposed in pathways from the FRA, so that (4) under certain conditions the reflex output may change and even reverse to produce ipsilateral inhibition of flexors and excitation of extensors.

Reflex reversals during particular behaviours are examples of the gating of entire reflexes via interconnections between descending pathways and interneurons interposed in segmental reflex pathways. For example, stimulation of the skin can produce either reflex inhibition or excitation of the same motoneurons depending upon the phase of the step cycle (e.g., Rossignol & Gauthier 1980). Reflex reversals during locomotion are consistent with the concept that there exists more than one interneuronal pathway from afferents to motoneurons. The original observations upon which the FRA concept was built revealed that variations in the preparation could dramatically alter reflexes evoked by a given stimulus. Thus, in deteriorating preparations, reflexes evoked by stimulation of the FRA changed from a widespread activation of ipsilateral flexors to an activation of extensors (Eccles & Lundberg 1959). Subsequent studies showed how reflexes evoked from the FRA changed following various lesions of the neuraxis (Holmqvist & Lundberg 1961).

The demonstration of alternative reflex pathways from the FRA was a crucial step in recognizing how spinal circuits function. Parallel reflex pathways involving interneurons are available to segmental afferents; which pathway dominates depends upon the relative excitability of the interposed interneurons. The interactions among several spinal reflex pathways allow the occurrence of reflex modulations, reversals, and other specific reflex actions restricted to certain motoneurons. The particular reflex evoked by sensory input to the spinal cord depends not only on the characteristics of the sensory input but also on the state of spinal interneuronal circuitry. The selection of these spinal pathways by the brain allows segmental afferent information and spinal interneurons to coordinate voluntary movements. Thus, the movements elicited by descending commands depend on the excitability of spinal reflex pathways.

The FRA system consists of interneurons in a polysynaptic pathway that project to motoneurons. Interneurons in these pathways receive multimodal afferent input. While FRA systems have actions on most of the motoneurons in the spinal cord, the projection of individual interneurons in FRA pathways to motoneurons may be much more limited. Limited projection of individual interneurons may enable interneurons in FRA reflex pathways to control fine movements. Little is known about the projections of individual interneurons in FRA pathways to either motoneurons or other spinal interneurons. The polysynaptic nature of these pathways offers multiple opportunities for convergence of afferents, descending commands and other segmental interneurons. Multiple control sites in the FRA system allow for flexibility; the gain and sign of the reflex in motoneurons can be fine-tuned throughout the execution of movements. Whereas Sherrington (1910) mainly described the flexion reflex as a mechanism for withdrawing a limb from noxious stimuli, Lundberg's FRA concept is that these systems are used in normal movements (Lundberg et al. 1987b). At least some afferents in the FRA system will be activated during most movements (e.g., group II muscle spindle afferents, some cutaneous afferents). The extensive convergence of afferent and descending systems to interneurons in FRA pathways almost insures their continuous involvement in all movements (see Gandevia & Burke, this issue).

The FRA framework, with its alternative pathways, has also provided useful hypotheses about the nature and organization of interneurons comprising the spinal stepping generator (Jankowska et al. 1967a; Lundberg 1979a). Intravenous administration of DOPA dramatically alters reflexes evoked by stimulation of the FRA. Short latency (2–3 msec) synaptic potentials in motoneurons are inhibited and replaced by longer latency (30–50 msec) synaptic potentials (Andén et al. 1967; Jankowska et al. 1967b). Rhythmic alternating discharges in flexor and extensor nerves often appear after DOPA administration in these spinal preparations. This suggested to Lundberg and colleagues that DOPA had activated a spinal interneuronal network producing locomotion. These observations resulted in the discovery of reciprocally organized interneurons in reflex pathways capable of producing alternating flexion and extension during locomotion (Jankowska et al. 1967a). These experiments shed considerable light on Graham Brown's hypothesis (1914) that spinal interneurons are organized in "half-centres." The organization of spinal FRA reflex circuitry remains a central framework for current studies of locomotion (e.g., Shefchyk & Jordan 1985; Jordan, McCrea, Noga & Shefchyk, experiments in progress).

Some interneurons interposed in the FRA system have already been located (e.g., Engberg et al. 1968; Lundberg et al. 1987a), reflex output originating in FRA pathways continues to be mapped, and studies on the pharmacological and behavioral control of the interneurons are under way. For many studies of the spinal cord, the FRA concept continues to yield important observations and hypotheses. Observations on reflexes evoked from the FRA and changes following DOPA administration, for example, may also apply to chronic spinal man. Spinal man displays both short and long latency reflexes after high-strength cutaneous stimulation; crossed extension may be present; and volleys in the FRA produce PSI of group Ia afferent transmission (Roby-Brami & Bussel 1990). Despite the obvious successes of the FRA hypothesis in predicting the existence of certain interneurons and describing organizational principles of movement control, some confusion remains. Before further expanding on the contributions of the FRA concept to interneuronal studies, we will address the reasons for this confusion.

7. Problems with the FRA concept

The term FRA has historical origins and is itself problematic. Although strictly speaking FRA denotes the afferents that may evoke the flexion reflex, current usage addresses the entire organization of interneurons producing these widespread spinal reflexes. As will be discussed, the FRA system is used for a wide variety of movements, not just flexion reflexes. It has jokingly been suggested that a semantic solution for the "flexion" aspect of FRA terminology can be achieved by simply moving up one letter of the alphabet to the GRA or General Reflex Afferent system (name offered by E. Kandel; see Lundberg et al. 1987b). The advantage of the term GRA is that it properly removes the emphasis from the "flexion" aspect of movement produced by these systems and places it on general mechanisms for the modification and control of all movements. The disadvantage of the term GRA is that it severs the connection to an 80-year-old history of spinal cord research.

Other concerns about the FRA system and other studies on interneuronal connectivity involve generalizations about reflex actions to "flexors" and "extensors." For many physiologists the classification of muscles as flexors or extensors remains an operational, not an anatomical, definition. Flexors were originally defined as those muscles excited during the flexion reflex (Sherrington 1910). The generalization about output during the flexion reflex (e.g., all ankle and hip extensors are similarly inhibited) stressed the widespread nature of these reflexes. During locomotion, some motoneurons, particularly those innervating muscles spanning more than one joint, are active during both extension and flexion (e.g., Perret & Cabelguen 1976; 1980). Even though such motoneurons cannot be classified simply as flexor or extensor motoneurons, there is no conflict between their activity in locomotion and FRA concepts.

The flexion reflex is but one of the movements evoked by activation of these pathways. The FRA concept suggests an organization of populations of interneurons. Fractionation of these populations into subpopulations projecting to small groups of motoneurons allows for their activation or inhibition. Individual motoneurons receive connections from multiple populations of spinal interneurons in parallel reflex pathways.

Many observations on the FRA have been made using electrical stimulation of afferents. This technique lumps together several classes of receptors and does not permit an appreciation of the reflex subtleties produced by individual afferent systems. Clearly there is the need to consider selective activation of afferents and to assess the extent of common and specialized synaptic actions predicted within the FRA concept (see Lundberg 1982). The original grouping of several inputs when discussing FRA systems stressed the similarity of reflex actions produced by stimuli from different receptive fields and modalities. Painful stimuli evoke flexion, but painful stimuli are not necessary for segmental activation of FRA pathways. Group II muscle spindle afferents are activated throughout many movements and the connection of group II afferents to interneurons in FRA systems indicates that spinal FRA pathways are involved in controlling most movements (Lundberg et al. 1987b). Presumably, nociceptive input contacts spinal circuits involved both in flexion reflexes and perception of pain. There is evidence for pain pathways that are distinct from spinal motor pathways producing flexion (Schouenborg & Sjölund 1983). Again, withdrawal reflexes are only one aspect of the FRA system.

An important FRA concept is that parallel interneuronal pathways are available to the afferents (e.g., Schomberg & Steffens 1986). The existence of parallel pathways might suggest that everything is connected to everything. On the surface, there is apparent disagreement between the generalized reflex actions to many motoneurons evoked by FRA systems and studies which show various degrees of specialization of the reflex pathways (discussed below).

I believe that much of the present misunderstanding of the FRA story centres around two problems. The first is that the FRA organization must explain known movements. The second concerns the classification of reflexes as part of the "normal" FRA, "alternative" FRA, or "private" reflex pathways (discussed in sect. 8). Sherrington

(1910) described flexion withdrawal reflexes, whereas Eccles and Lundberg (1959) used intracellular recording to examine postsynaptic potentials in motoneurons during stimulation that can evoke flexion. Other than the title of Lundberg's first paper on the subject ("Synaptic action in motoneurones by afferents which may evoke the flexion reflex," Eccles & Lundberg 1959), current discussion of the FRA concept rarely addresses withdrawal reflexes. Lundberg's FRA hypothesis was developed to address the organization of spinal interneurons, not to explain a specific behaviour. The organization of the FRA system includes multiple reflex pathways and input from both descending and segmental systems. This organization suggests that these systems function during many types of movements.

The strength of the FRA hypotheses is that they predict an organization of interneurons with testable hypotheses about connectivity. For example, the FRA concept predicts the existence of interneurons excited by several modalities and from large receptive fields. Such cells have been located in the dorsal and intermediate areas of the spinal grey matter (e.g., Engberg et al. 1968). The presence or absence of wide convergent input to interneurons has also been used as an aid in the classification of interneurons as first order or later order cells in FRA pathways to motoneurons (Engberg et al. 1968; Lundberg et al. 1987a; 1987b).

Upon stimulation of the afferents in the FRA system, both EPSPs and IPSPs are seen in many interneurons. This indicates the presence of parallel pathways with multiple control sites. Interneurons can be excited or inhibited at multiple sites with several opportunities for shaping movements in subtle ways. A simple understanding of connectivity in FRA pathways does not predict motor output. As will be discussed, it is only through an understanding of how interneurons in FRA pathways are controlled that muscle activation can be predicted during voluntary movements as well as the flexion reflex.

Lundberg et al. (1987b) hypothesize how descending commands might fit into the spinal FRA organization and use spinal interneurons to modify and evoke volitional movement. In their discussion about caudally located group II interneurons (Lundberg et al. 1987b), they organize spinal interneurons into subpopulations, each with limited convergence from segmental afferents (i.e., fractionation of interneuron populations). Activation of segmental afferents by movement would provide feedback which sets the excitability of these interneurons. The wide variety of afferents included in the FRA provides a multisensory feedback signal that increases the excitability of interneurons in FRA pathways throughout movement. This arrangement would allow descending commands to recruit interneurons appropriate for a particular movement based, in part, on the activity in muscle and cutaneous afferents. Variations in the gamma drive to muscle spindles and mechanical perturbations of the limb would alter group II afferent recruitment. The large unitary EPSPs in interneurons evoked by group II afferents would then alter the populations of interneurons that are activated by descending systems. Thus, higher centres evoke movements through activation of subsets of both excitatory and inhibitory spinal interneurons contacting motoneurons.

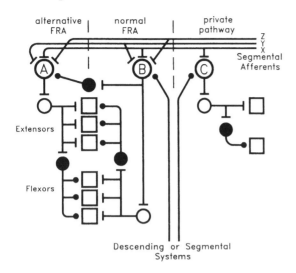

alternative FRA | normal FRA | private pathway

Z
Y
X
Segmental Afferents

Extensors

Flexors

Descending or Segmental Systems

Figure 2. Schematic representation of the differences between the normal and alternative flexion reflex pathways and private reflex pathways. Three segmental afferent fibre systems are labelled X, Y, and Z. Other symbols as in Figure 1. A, B, and C are populations of interneurons.

8. Alternative FRA pathways

A source of confusion about the FRA is the extent to which particular reflex pathways form part of the "normal" FRA pathways, the "alternative" FRA pathways, or "private" reflex pathways. The normal FRA pathways are defined as those that function in the low spinal preparation when widespread peripheral nerve stimulation results in a general excitation of ipsilateral flexors and inhibition of extensors. Postsynaptic potentials in motoneurons produced by stimulation of the very divergent afferent sources comprising the FRA are strikingly similar (e.g., Eccles & Lundberg 1969; see also McCrea 1986). Higher strength cutaneous stimulation and stimulation of group II and III muscle receptors from several areas of the hindlimb produce postsynaptic potential that are qualitatively indistinguishable. The operation of FRA pathways is characterized by the similarity of synaptic actions in many motoneurons; these actions can vary from excitation to inhibition depending on the preparation and excitability of the entire neuraxis.

Operational definitions of the terms alternative FRA, normal (or usual) FRA, and private reflex pathways are illustrated in Figure 2. The FRA system is characterized by widespread convergence from several afferent systems onto common sets of interneurons, labelled A and B. There is also a widespread divergence of excitation and inhibition from these interneurons to motoneurons (squares in Figure 2). During operation of the normal FRA pathways (Figure 2, middle) activity in any of the segmental afferent systems (X, Y, and Z) results in excitation of flexors and inhibition of extensors via a pathway originating with interneuron B. Note that activity in interneuron B prevents interneuron A from mediating reflexes to motoneurons by a collateral inhibitory pathway. When widespread stimuli result in a general activation of extensors and inhibition of flexors, the alternative FRA pathways are in operation. In other words, alternative FRA pathways are similar to normal ones but opposite in sign. Alternative pathways could be selected by

inhibiting interneuron B with descending or segmental systems; interneuron A would be released from inhibition and produce widespread excitation of extensors and inhibition of flexors. The private (or "specialized") reflex pathways are characterized by a more limited convergence of afferents or more limited divergence to smaller numbers of motoneurons. In Figure 2, interneuron C receives more limited afferent input, and its activity results in excitation and inhibition of fewer motoneurons.

As mentioned, the projection patterns of individual interneurons to motoneurons and other interneurons are poorly understood. The interneurons in Figure 2 should be considered as populations of interneurons, not as individual interneurons. It should be noted that sensory afferents have several reflex pathways available to them (A, B, C) including, but not limited to, those of the FRA (pathways A and B). The existence of several spinal pathways has been shown for group II muscle spindle afferents (e.g., Edgley & Jankowska 1987; Harrison & Jankowska 1985a; Lundberg et al. 1987a), low to moderate threshold cutaneous afferents (Schomburg & Steffens 1986; see discussion in LaBella & McCrea 1990), and high threshold cutaneous afferents (e.g., Schouenborg & Sjölund 1983). FRA pathways are characterized by similar synaptic actions in many species of motoneurons evoked from a variety of sensory afferents. In most cases, there also will be several other (i.e., private) reflex pathways available to afferents active during movement.

As shown in Figure 2, alternative FRA and private reflex pathways are different. The terms "private or specialized reflex" should be reserved for those reflex pathways in which the convergence or divergence is much more limited. Engberg (1964) identified reflexes in foot muscles as an example of private pathways. Stimulation of nerves in the foot produced an excitation of extensor motoneurons supplying foot muscles while concomitantly inhibiting motoneurons supplying other extensor muscles in the hindlimb. Recent work by Hongo et al. (1990) has extended these observations by showing that stimulation of the skin over the toes produces different reflexes in the foot depending upon the toe stimulated. These private reflexes seem functionally appropriate for controlling individual digits and maintaining posture (Hongo et al. 1990). Hagbarth (1952) showed that stimulation of the cutaneous area overlying an extensor muscle produced inhibition of other extensor muscles but excitation of the extensor muscle lying under the skin area that was stimulated ("local sign"). This local extensor excitation could also be considered a private reflex mediated by non-FRA circuits.

Other studies also reveal more limited reflex effects evoked by afferents that also produce activity in FRA pathways. For example, stimulation of the sural cutaneous nerve has different effects in the motoneurons supplying the synergist muscles in triceps surae; excitation is dominant in medial gastrocnemius motoneurons and inhibition in lateral gastrocnemius motoneurons (LaBella et al. 1989). Further evidence for the classification of these sural pathways as private comes from the finding that interneurons in the excitatory pathway to medial gastrocnemius motoneurons receive only limited convergence from other afferents (LaBella & McCrea 1990). There are many other examples of reflex actions independent of the

FRA produced by afferents that also may invoke the flexion reflex (e.g., see the diagrams of convergence to interneurons in Baldissera et al. 1981; Harrison & Jankowska 1985a). These observations do not conflict with the FRA concept. In the low spinal preparation, most extensors are inhibited when strong cutaneous stimulus is applied. Of great interest is the fact that these private pathways may be controlled by descending and segmental pathways different from those which control operation of the FRA. Figure 2 shows different descending or segmental pathways controlling FRA and private reflexes.

The generalized reflex systems (GRA or FRA, see above) work in concert with private pathways to produce coordinated movements. During movement, both systems are subjected to continuous and, perhaps, differential control. A combination of reflexes mediated by both FRA and private reflex pathways could result in the appearance of reflexes in individual motoneurons quite unlike those evoked in flexors or extensors in general. In other words, the motoneurons on the right of Figure 2 that receive private reflexes might simultaneously receive input from interneurons A or B (not illustrated).

An important point about multiple inputs to interneurons is that most spinal interneuron systems receive input from both supraspinal and propriospinal motor systems. This is true for all reflex pathways including those of the FRA. Since monosynaptic actions of supraspinal systems on lumbar motoneurons are exceptional, spinal interneurons form an integral part of descending motor control pathways. Some spinal interneurons are "last order interneurons"; that is, they connect descending systems to motoneurons. Therefore, an axiom of spinal cord physiology is that a study of segmental reflex systems is also a study of descending (including voluntary) motor control. The complex organization of spinal interneuronal systems, including the FRA system, has the advantage that descending systems can make use of prewired and flexible spinal circuits to evoke movement. An unchanging descending command could evoke differing movements based upon differences in the excitability of spinal interneurons. The excitability of individual interneurons could be controlled either by other descending pathways or segmental afferent pathways. Complex circuitry in the spinal cord allows for a distributed processing system without the need for multiple descending pathways. Fewer descending pathways could control the actions of entire reflex pathways by appropriate connection to key interneurons.

If these ideas about motor system organization were generally true, there would be little need for a discrete cortical organization to control individual hindlimb muscles (Schieber 1990). Even individual corticospinal neurons have actions on several different muscles (Shinoda et al. 1981; 1986). This supports a role for spinal interneuronal pathways in determining the precise activation of muscles during voluntary movement (this point is discussed in sect. 9). Fine voluntary movements can be produced by the focused action of descending systems on subsets of spinal interneurons.

Sherrington and his contemporaries stressed the idea that the particular reflex response evoked by a stimulus was not invariant, but depended upon several factors. The reflex response resulted from a complex summation of segmental and descending inputs to spinal interneuro-

nal networks connected both in series and parallel. The FRA concept was developed to address certain aspects of this complexity and it remains a viable basis for generating testable hypotheses about spinal interneurons. Interneurons responsible for the long latency effects from the FRA in L-DOPA preparations have been partially identified; the involvement of FRA pathways in spinal stepping is clear, and sense is being made of how descending and segmental information interact in these pathways during voluntary movements.

9. How can a study of spinal interneurons help in understanding general principles of movement control?

Consider the idea of "muscle synergies" as an example of the way in which knowledge of spinal interneuronal circuitry could contribute to an understanding of higher order movement control. According to this idea, originally discussed by Bernstein (1967) and recently developed by Macpherson (1988), descending motor commands are sent to groups of muscles that are then coactivated. This grouping of muscles by the command signals would reduce the variety of ways in which muscles can be activated and thus reduce the variety of motor control signals needed to produce movement. In Macpherson's (1988) experiments, the cat activated limb muscles to stabilize balance on a platform that had been moved in the horizontal plane. One outcome of these experiments was the realization that there was more than one "muscle synergy." In other words, different combinations of muscles could be activated to restore balance. A modified synergy concept was proposed (Macpherson 1988, p. 227): "It is suggested that muscles were recruited in synergy on the basis of their particular action at the joint(s) they spanned and that the specific combination of muscles used to achieve the desired force vector against the ground was dependent on limb position at the end of translation."

Figure 3 is a hypothetical diagram of how the position of a limb might determine the muscle synergy used to restore balance through spinal interneuronal pathways. In Figure 3A and B, three muscle groups are shown along with a common command signal from some portion of the brain. Note that this command signal ends on the interneurons, not directly on motoneurons. In Figure 3A, activity in interneuron B results in an inhibition of interneuron C through interneuron D. Thus the descending command will normally result in excitation of motoneurons I and II but not III since interneuron C is inhibited. Although there might be a brief excitation of motoneuron III by the descending system, the summation of trains of synaptic potentials from the descending system and interneuron D would result in no activity in interneuron C and thus little activation of motoneuron III. In Figure 3B, if there is activity in segmental afferents produced by changes in limb position, interneuron B becomes inhibited via interneuron E. In this case, the muscle synergy produced by a common descending signal changes to an activation of motoneurons I and III instead of I and II. Although this diagram is hypothetical, the general principles of descending connections on interneurons and inhibition of spinal interneurons by other

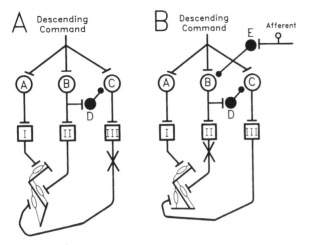

Figure 3. The "motor synergy" resulting from activity in a descending motor command signal can be modified by segmental reflex pathways. Interneurons are the large lettered circles and motoneurons are the squares numbered by Roman numerals. Excitation is indicated by short lines and inhibition by small filled circles. The large X across the axon of a motoneuron represents no activation of that motoneuron by the descending command. The structures in the bottom left of panels A and B represent three muscles in a limb.

segmental interneurons occur in every interneuronal system examined to date.

Figure 3 also serves as an example of how one might develop a testable set of hypotheses about spinal interneuron connectivity. For example, some properties of interneuron C would include: (1) monosynaptic EPSPs and trisynaptic IPSPs from stimulation of the descending command area, (2) monosynaptic excitatory actions of these interneurons on motoneuron III, and (3) a lack of excitation of these interneurons by the segmental afferents that change the muscle synergy. Further experiments in behaving animals could also be designed to see what other afferents can inhibit interneuron B and hence change the muscle synergies. Given such specific questions and probable connectivity, these cells probably could be identified in short order and a clearer understanding of the descending "command signals" would evolve. As mentioned above, the convergence of descending and segmental input onto common spinal interneurons demands a consideration of spinal reflex circuitry before higher-order motor control mechanisms can be fully understood. Although the example in Figure 3 concerned postural adjustment, there is every reason to believe that similar considerations should be used when discussing voluntary movement control.

10. What studies are needed after interneurons are identified?

The identification of the interneuron is usually the initial goal of interneuron studies. This includes identifying the synaptic input to the cell, the axonal projections from the cell, and the synaptic effect on the target cell. Following identification, several other aspects must be investigated before sense can be made of the role of particular interneurons in spinal motor control. The activity of identified interneurons during movements is, of course, at the top of

the list of priorities. For polysynaptic descending and reflex pathways, knowledge of the activity of all of the interposed cells allows inferences about the possibilities for control of transmission through these pathways. Identifying the neurotransmitters released during motor control and affecting the interneurons under study is also of paramount importance. Neuromodulation by interactions between transmitter systems as well as presynaptic inhibition of primary afferents or the terminals of descending systems and spinal interneurons must be considered. Membrane properties that determine cell firing must be understood as well.

Membrane currents that produce sustained repetitive firing of motoneurons have been investigated recently. These "bistable firing" properties or "plateau potentials" allow for continued activation of motoneurons without continued synaptic input (Crone et al. 1988; Hounsgaard et al. 1986). In spinal preparations, serotonin by itself has weak actions on motoneurons. When serotonin administration is paired with a peripheral or central stimulus that produces excitation of the motoneuron, sustained depolarization and repetitive firing can result. In intact preparations, descending systems can probably activate similar mechanisms. The implications of these studies for motoneuron activation during locomotion and other movements are profound: Repetitive firing of interneurons is not necessary for continued motoneuron activity. A demonstration of similar mechanisms for maintained firing in interneurons would require a reevaluation of concepts about interneuron input and output. In terms of synaptic connectivity, further excitatory synaptic input to interneurons made tonically active by bistable properties might be functionally irrelevant. On the other hand, small amounts of synaptic inhibition could have the powerful effect of terminating repetitive firing of neurons during movement.

A further point to consider is that nervous system function and connectivity is often inferred from correlations between the firing of two neurons. Bistable firing behaviour would severely limit the application of correlation techniques. If a single synaptic input results in continuous firing, there would be a low correlation between firing of the presynaptic and postsynaptic cell even though the connectivity initially produced the firing. A strong and short-latency correlation peak could also be masked by multiple firing and variable firing rates occurring after the initial input.

Because of multiple inputs to particular interneurons and multiple sets of interneurons activated by particular primary afferents, records of the activity of primary afferents during movement are not very useful by themselves. Even with a detailed description of the activity of hindlimb afferents during a movement, the synaptic effects on interneurons and motoneurons cannot be known. Presynaptic inhibition of particular afferents is one factor; the other is that multiple parallel pathways with mutual inhibitory interactions could change the reflex effect from inhibition to excitation. It is crucial to estimate the dominating and effective input(s) to particular cells during movement. Although records of single fibre activity cannot be relied upon for this purpose, they help to define the possible modes of movement regulation by segmental reflex pathways. Once the activity of afferents is known during a movement, the synaptic effectiveness of their input to interneurons must be tested. These experiments might involve altering activation of the afferents by direct stimulation or perturbing movements.

11. Summary

Spinal interneurons are an elaborate system of motor control. The simplistic idea that the spinal cord functions as a relay system for descending and segmental afferent commands has been replaced by a dynamic system in which the interaction between spinal interneurons regulates and produces purposeful movements. Voluntary movements depend upon the selection of subsets of spinal interneurons whose excitability is regulated in part by afferent feedback throughout the movement. Standard neurophysiological approaches continue to produce insight into these systems, and the electrophysiological identification of spinal interneurons is becoming more routine. Specific hypotheses concerning spinal interneuron function and interactions during movements can now be proposed; many of these are testable with current technology. The application of information on cat spinal cord interneuron circuitry to the human laboratory is a significant and exciting development. Although many others have contributed to this field, the collaboration between the human lab of E. Pierrot-Deseilligny in Paris and the animal lab of H. Hultborn in Copenhagen has yielded deep insight into the organization of human spinal pathways.

The most significant limitation to progress is not the complexity of spinal interneurons but the few laboratories engaged in these studies. The resurgence of interest in spinal neuropharmacology, the development of specific immunological techniques, the revolution in imaging, and the interest in spinal regeneration will all contribute to rapid progress in this field. The recognition that spinal motor systems can serve as useful models for the general organization of neural systems will bring together those in other areas of neurobiology. As the field of neural networks and modelling matures, interest in spinal motor systems will increase. Spinal motor systems have a well-defined input, the interneuronal elements are becoming better characterized, and unlike sensory systems, they have an easily measurable output in the motoneurons and muscles during behaviour.

There is considerable optimism about arriving at an understanding of spinal motor systems. This optimism is justified by advances in the conceptual framework in which spinal interneurons are now viewed. Not long ago another view was presented: "Those whose experiments have forced us to confront the 'embarrassment of riches' in the workings of the spinal cord must ask whether it is useful to continue to collect more inexplicable data" (Loeb 1987, p. 111). One hopes that the present target article will remove some of the mystery concerning this data and illustrate that sense is being made of spinal interneuron circuits.

ACKNOWLEDGMENTS
The author wishes to thank Elzbieta Jankowska, Allen Jones, Larry Jordan, Anders Lundberg, and Susan Shefchyk for many useful discussions. The efforts of Paul Cordo to organize this meeting and his contributions to this manuscript are greatly appreciated. Sharon McCartney expertly prepared the figures and assisted with the manuscript. Supported by the Medical Research Council of Canada.

NOTE
1. Because of the wide scope of this field and a desire to emphasise general features of these systems, citations are not exhaustive but are given preferentially to recent reviews or those studies that illustrate the principles discussed. For the sake of brevity, discussion is limited to motor systems in the mammalian lumbar spinal cord.

4

Implications of neural networks for how we think about brain function

David A. Robinson

Departments of Ophthalmology, Biomedical Engineering, and Neuroscience, Wilmer Institute, The Johns Hopkins University School of Medicine, Baltimore, MD 21287

Abstract: Engineers use neural networks to control systems too complex for conventional engineering solutions. To examine the behavior of individual hidden units would defeat the purpose of this approach because it would be largely uninterpretable. Yet neurophysiologists spend their careers doing just that! Hidden units contain bits and scraps of signals that yield only arcane hints about network function and no information about how its individual units process signals. Most literature on single-unit recordings attests to this grim fact. On the other hand, knowing a system's function and describing it with elegant mathematics tell one very little about what to expect of interneuronal behavior. Examples of simple networks based on neurophysiology are taken from the oculomotor literature to suggest how single-unit interpretability might decrease with increasing task complexity. It is argued that trying to explain how any real neural network works on a cell-by-cell, reductionist basis is futile and we may have to be content with trying to understand the brain at higher levels of organization.

Keywords: connectionism; coordinate transformations; hidden units; neural networks; oculomotor system; pursuit eye movements; saccadic eye movements; signal processing; vestibulo-ocular reflex

1. Introduction

The function of the brain is to process signals. Whether a group of cells is analyzing sensory signals, storing maps in memory, shaping motor commands, or doing other things, its essential action is to receive a signal, usually distributed over an ensemble of fibers, alter the spatial and temporal patterns of the signal, and pass this new signal, also distributed over its axon ensemble, to downstream networks. How this happens is the main goal of basic neuroscience. Although the recent discoveries of molecular biology reveal fascinating genetic events in assembling molecules and should lead to methods of treating or preventing diseases of the brain, they are, as far as one can see, of little use in illuminating the basic problem of how the brain processes signals or, if you will, how it thinks. This problem is compounded by the suspicion that various subdivisions of the brain may solve similar problems in different ways. This aspect of brain function, probably an *ad hoc* one, suggests that generalized algorithms for neural signal processing are unlikely to emerge. Nevertheless, the die-hard reductionists among us have taken it as a matter of faith that we will someday understand how the brain works, or at least bits and pieces of it, in the same sense that we now understand how an electronic circuit works.

This idea has long been thought naive, but the developments of the last decade in neural networks, which can do clever brain-like things with neuron-like unit behaviors even though they are just electronic circuits (or simulations thereof), offer a reason for taking another look at the problem. This target article makes no attempt to review the history or the scope of artificial neural networks. I am primarily interested in neural network models that simulate real neural systems where much of the neurophysiology is known. There are very few such systems, but the brainstem oculomotor system is one of them. The following is therefore an account of my impressions in trying to simulate parts of the oculomotor system with neural networks.

Unfortunately, the term "neural network" can be applied to many schemes involving artificial, neuron-like elements developed as far back as just after World War II. Within the last decade, however, there has been a surge of interest in networks that, in their most common form, consist of three layers and use a learning algorithm such as back-propagation. It is such networks that have stirred up so much recent interest in the possibility of shedding light on brain function. The following discussion concerns this kind of neural network – the kind that contains hidden units and learns by adjusting synaptic weights in response to some error-driven learning algorithm.

Such artificial neural networks contain neuron-like units connected by things resembling plastic synapses. They are capable of learning extraordinarily complex tasks, but after they have done so we cannot understand the basis of their success at the single-unit level. The same can be said of the brain or any of its parts. It contains neurons connected by synapses, most if not all of which are modifiable, and it also learns extraordinary tasks. These similarities are what give artificial networks their uncanny resemblance to real neural networks. Beyond this, however, there is a sharp departure in attitude. The engineer who uses a neural network to do a job, such as running a chemical processing plant (a young but growing field of neural network applications), chooses the network because the operation is so complex that he despairs of ever creating a designed, hardwired controller. The virtue of the neural network is that it can learn to perform tasks beyond the scope of the design engineer – that is why he chooses it. He has no intention of examining the hidden units. Why? If he could have predicted their behavior, he could have designed the system himself. He knows that their behavior will be largely inexplicable; to examine and worry about them would defeat the whole purpose of using a neural network in engineering. Consequently, recording from single units would be regarded by the applied engineer as amusing but not very constructive.

The main message of neural networks for the neurophysiologist is that the study of single neurons or neuron ensembles is unlikely to reveal the task in which they are participating or the contribution they are making to it. Conversely, even if one knows the function of a neural system, recording from single units is not likely to disclose how that function is being fulfilled by the signal processing of the neurons. A corollary is that being able to describe that function mathematically tells little about what to expect when recording from single neurons.

Examples of these problems abound in the neurophysiological literature concerning single-unit recording in behaving animals. A typical report tells us that in area X, in the alert monkey, 27% of the cells were phasic, 18% were tonic, 38% were phasic-tonic, and 17% did nothing. The conclusion drawn from the study is that in area X, 27% of the cells are phasic, 18% tonic, 38% phasic-tonic, and 17% do not respond. This is not meant as a criticism; I have written similar papers myself. The point is that the problem raised with such painful clarity by neural networks are attested to by a great deal of the relevant literature in neurophysiology.

2. A brief introduction to neural networks

For those who are unfamiliar with neural networks and their capabilities, the following is a simplified description of their essential features. Figure 1 shows a typical scheme (e.g., Anastasio & Robinson 1989; 1990a; Sejnowski & Rosenberg 1987). Each unit is a simple model of a neuron. The output, a_j, of the jth cell, is to be interpreted as a real neuron's discharge rate in spikes/sec. The membrane depolarization of the jth cell is the sum of the activities, a_{ij}, of all cells, $i = 1, 2, \ldots, N$, projecting to it in proportion to their synaptic weights, w_{ij}. This sum is passed through a nonlinearity, NL, which recognizes that cells cut off at zero rate and eventually saturate at some high rate. The exact shape of NL is usually not important.

The cells are customarily arranged with a minimum of three layers, as shown in Figure 1b. Usually, all cells in one layer project to all cells in the next; projections in this type of network are always forward, never sideways or backward, because this would create reverberating feedback loops. Initially, the slate is wiped clean of experience by randomizing all the synaptic weights. A value is applied to each unit of the input layer, forming a spatial input pattern, and the resulting output is compared to a desired output pattern determined by an external

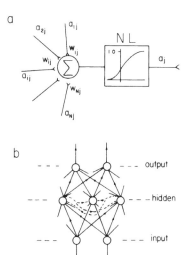

Figure 1.   a: The behavior of each unit in a neural network. Its model membrane potential is the sum of all of its presynaptic inputs, a_{ij}, weighted by the synaptic strength, w_{ij}. This is passed through a bounding nonlinearity, NL, and the output, a_j, is thought of as a discharge rate to be forwarded to downstream neurons. b: Most networks are three layered. Usually, all units in one layer project to all units in the next for spatial transformation networks. When networks are asked to process temporal signals, they are allowed feedsideways (essentially feedback) pathways shown by dashed lines. The number of units in each layer is determined by each application.

"teacher." That is (and this is an important difficulty in proposing these networks as models of brain function), some element outside the network (the teacher) evaluates the output and "knows" whether it is "correct." The difference is the error, which is used to reward good weights and punish bad weights by calculating which weights and units were more or less active when the error was large or small. Usually, one finds some method of estimating the partial derivative of the error with respect to each weight, w_{ij}. The weight is then changed in proportion so that the error is reduced most rapidly in what is called the "method of steepest descent." This process is repeated until the error is driven below some tolerance level and the network has learned to produce the desired output. Usually, there are several input patterns, with each applied in turn until the network has simultaneously learned the correct response to each. It should be appreciated that a steepest descent method is entirely for the benefit of the investigator who would like the network, in computer simulation, to find a solution quickly. There is no indication that the nervous system uses such a method. In some lower animals, escape behavior, which may be required almost immediately after birth, is probably genetically determined; human children, on the other hand, have months to years available to learn many types of motor behavior.

In a similar vein, the issue is often raised whether a particular learning algorithm is physiological. The currently popular, back-propagation method (Rumelhart et al. 1986b) is fairly efficient in converging rapidly but is criticized as being unphysiological. There exists a variety of learning algorithms, however, that are more or less efficient and less or more physiological; all of these end up at the same goal: a network that has learned. In most cases, the properties of the hidden units – our main concern – appear to be similar, regardless of the learning method. This observation cannot be proven rigorously and has not been tested in many specific situations, so it remains only a desired probability. If this is true in general, one could use a rapidly converging algorithm for efficiency, while noting that a more physiological method could have been used without changing the final outcome.

The end result is a network that has learned to give correct responses for a variety of spatial patterns presented as input. Our interest is in the behavior of the hidden units: If we record from 1 to 10 or 100 hidden units, can we tell what the network is trying to do and how each hidden unit is helping to do it?

A nice biological touch is that these networks do not find a unique solution. Each time one randomizes the weights and retrains, a new network is formed, one that still does the same job. Of course, no two cats (or people) are likely to be wired up identically, synapse for synapse, and neural networks reflect this feature, too. The reason is easy to see. Most networks have a much larger number of hidden units than output units. For example, with 40 hidden units and 2 output units, the convergence would be described by 2 equations with 40 unknowns. There is no unique solution. This situation is called the "overcomplete problem" because our university training has always emphasized questions with only one right answer. For neural networks, however, overcompleteness is not a "problem," it is a way of life. The networks are asked only to reduce the error to zero, not to find some hypothetical unique solution. They will use any solution that works. This sounds very biological.

An example of the power of neural networks is the reading machine called NETtalk (Sejnowski & Rosenberg 1987). Engineers have tried for decades to make a reading machine for the blind. They tried to hardwire into it all the complicated rules of English spelling and grammar as well as the many exceptions to those rules that together constitute, as we all know, a morass of contradictions. Engineers declared that a comfortably understandable reading machine could not be built. The NETtalk device receives as input a string of seven letters (and spaces) of English text and is trained to produce as output the phoneme corresponding to the letter in the center of the string. The rest of the string provides context. In Sejnowski and Rosenberg's work, NETtalk told when it did well and when it did badly and, like a child, it learned to read. After training on one text, it could extrapolate to others and eventually cope with an entire English dictionary.

What did the hidden units do? Some preferred consonants, some vowels, but there were just too many tasks and too many synapses (18,600) to attempt any reasonable explanation of how the network achieved its results on a synapse-by-synapse basis. The network learned the task, but observing its hidden units offered only the most arcane clues as to how.

This example illustrates that neural networks can solve exceedingly complicated problems – more complicated than those that can be solved by conventional design techniques. To put it another way: Tell the network *what* you want it to do, but do not even think about telling it *how* to do it.

Neural networks have a variety of applications. They can be used in Artificial Intelligence studies where there is no concern about whether or not they mimic brain function. Neural networks have also been seized upon vigorously by engineers as a new working tool for analyzing and controlling very complicated operations, such as chemical processing plants, where there are so many complex interactions between so many variables that a reductionist analysis is next to impossible. In addition, there have been many attempts to emulate brain function at abstract perceptual levels with no pretense that the network's units have any resemblance to neurons. There are also network models in which one or two of the layers have units that behave like real neurons but the other layers are simply conceptual and correspond to no known neurophysiology.

There are, nevertheless, a few network models in which microelectrode recordings have provided data on all the neurons in a real neural circuit, and it is these models that can begin to give us some credible indications about how the behavior of hidden units becomes more and more divorced from intuitive behavior as their tasks become more and more complicated.

2.1. Spatial and temporal networks. The networks described so far learn to identify spatial patterns presented at the input layer and respond with a spatial pattern of activity at the output units. These particular networks do not accept or generate time-varying signals and so are not very useful in modeling the neural control of movement. This limitation may be remedied by adding a membrane time constant to each neuron. This amounts to inserting the Laplace operator $1/(s\tau + 1)$ just before NL in Figure 1A, where τ is the membrane time constant (about 5 msec) and, if there are N neurons, solving N simultaneous differential equations. There is then no need for the units to project only forward; they may project sideways and backward to form multiple feedback loops that allow the network to generate interesting output waveforms. For a given input waveform, the output waveform is compared to the desired output and the root-mean-square difference over a performance interval is used as the error. Otherwise, learning occurs as before.

To distinguish these two types of neural networks, they will be called *spatial transformation networks* and *temporal transformation networks*. Obviously, some networks can do both types of transformations.

3. Examples of hidden messages in hidden layers

The difficulty in interpreting hidden units is illustrated at one extreme by artificial networks such as NETtalk, and at the other by almost any real network. Neither illustration offers much insight into how obfuscation progresses in hidden layers. Fortunately, the oculomotor system has a number of simplifying features, for example a single "joint," straight muscles, no stretch reflex, and linear behavior in premotor areas. These simplifications allow neural network models of the oculomotor system to have hidden units that, while offering an interesting amount of disarray, are also reasonably comprehensible. Moreover, numerous microelectrode studies have provided reasonably thorough descriptions of the behavior of the units in all three (or more) layers of the real network, allowing us to determine whether the models reflect reality. These models only serve to point out ways in which hidden units will become harder and harder to understand as the complexity of the tasks increases.

3.1. The oculomotor neural integrator.

The vestibulo-ocular reflex (VOR) starts in the semicircular canals, which provide a signal, coded in discharge rate modulation, proportional to instantaneous head velocity. As this signal leaves the vestibular nucleus, it constitutes an eye-velocity command to create an equal, but opposite, compensatory eye velocity so that the images of the visual world remain relatively stationary on the retina in spite of head movements. The eye muscles are mainly position actuators, however, and need to be given an innervation signal proportional to the desired eye position. It has been demonstrated that there is a neural network in the caudal pons (shared by the vestibular and the prepositus hypoglossi nuclei) that integrates (in the sense of Newtonian calculus) input signals with respect to time (see Robinson 1989, for a review), thereby changing the vestibular velocity signal into an oculomotor position signal.

This network is interesting in that it is relatively circumscribed anatomically and has a function that can be specified with precision. Lesion studies suggest that integration is a property of the network as a whole, rather than its individual neurons, and positive feedback seems an appropriate way to model it. In contrast, our relative ignorance about most spinal cord networks in awake animals makes it hard to guess the extent to which similar integrators are needed to change descending phasic commands into tonic innervation for muscles in the control of limb movements.

Since the real integrator network calibrates itself in the first few months of life (Weissman et al. 1989) and requires constant monitoring to maintain accurate function, it retains the ability to learn throughout life and has been modeled by a temporal transformation neural network (Arnold & Robinson 1991). The input signal, illustrated in Figure 2, arrives in a push-pull arrangement from a pair of semicircular canals reflecting a brief head movement. The model is freely connected; the inputs project to all interneurons, which in turn project to all other interneurons and to the output motoneurons. The output is a more or less compensatory eye movement that can be used to predict the error signal, that is, the rate at which visual images would move on the retina. This signal is transduced by direction-selective cells in the retina and distributed to the brainstem by the accessory optic system. The learning algorithm for the network in Figure 2 changed the synaptic weights one at a time and observed how sensitive the error was to this change. This partial derivative was then used to adjust each weight with the steepest descent method to drive the error toward zero. To do this, the network had to learn not only to integrate but to frequency-compensate the plant (eye muscles and passive tissues) by driving it with a combination of an eye-position and an eye-velocity signal. The model network did all this successfully.

One interesting emergent property of our simulation is that every cell in the network carries a combination of the position and velocity signals and no other signals, just as is seen experimentally with microelectrodes. It has been proposed that integration might be done step-by-step, each cell partially integrating the signal and passing it on for improvement. This does not happen in the model or in the real network. There are no partially integrated sig-

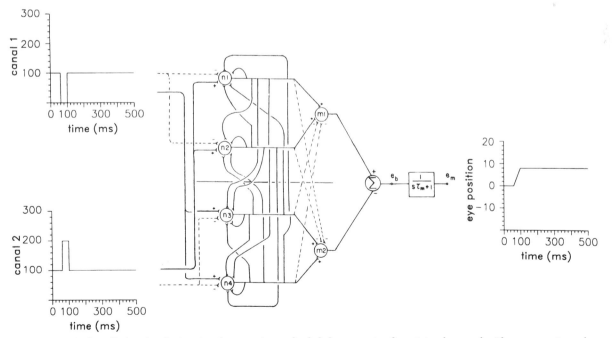

Figure 2. Push-pull, head-velocity signals come in on the left from a pair of semicircular canals. They are rectangular pulses of 100 spikes/sec on a background rate of 100 spikes/sec. These signals project to all interneurons, 4 in this example ($n1$–$n4$). All interneurons project to each other and to the two motoneurons, $m1$ and $m2$. All synapses are modifiable. The difference in their rates is passed through the oculomotor plant with a time constant, τ_m, of about 0.20 sec. The output eye position, right, is trained to become the time integral of the head-velocity input. In some versions of the general model, some projections from the canals and to the motoneurons were restricted to being excitatory (solid lines) or inhibitory (dashed lines). (Reprinted with permission from Arnold & Robinson 1991).

nals. If the eye-position signal appears, it appears fully integrated. This observation is particularly clear in the time domain (e.g., using step inputs), though it would be obscured in the frequency domain, such as when using sinusoidal inputs. For example, a phase lag of 45 deg could give the impression that a particular neuron was half integrating a signal by producing half of the needed 90 deg phase lag. A step input, rich in many frequencies, corrects this misapprehension. The integrated signal arises *de novo* from the circuit properties.

The interesting result, from our standpoint, is the variability in gain with which real and model cells carry these signals. Just as in experiments the sensitivity or gain of each model cell is measured for changes in eye position (k) and velocity (r) in (spikes/sec)/deg and (spikes/sec)/(deg/sec), respectively. Each cell has its own combination of k and r. All the model cells are different, as one would expect in a biological system. There is no correlation between k and r values; the network seems to treat using two signals as separate problems. To put it another way, the position and velocity signals are thoroughly distributed over the cells of the network. Distribution of signals is a natural property of neural networks and derives from the rich connections and the initial randomization of the synaptic weights followed by learning. Since this property imparts a biological flavor to simulations, it is hard not to believe that similar features also shape the real networks.

Occasionally, the model produces a cell that has its velocity and position signals going in opposite directions, or a cell that excites, or inhibits both of a pair of antagonist motoneurons. I call these "rogue" cells: They could be said to hinder rather than help, and one would discount them as model misbehavior if they were not actually observed from time to time with microelectrodes. That rogue cells exist tells us that the system works in spite of such cells; if 90% do right, they can easily overpower 10% doing wrong. The model suggests that such cells do not have a special function. The hidden units offer a menu of behavior patterns and the output layer picks and chooses whatever leads to a solution. When learning stops, because the error has become zero, some hidden units are left in a behavior pattern that seems contrary to the engineering mentality. The network, of course, only seeks a solution and is not concerned with good behavior. The point is that rogue cells are predicted by neural networks and exist in real ones.

3.2. A model with multiple motor commands.

We wanted to use a neural network model to show how signals that create saccades and smooth pursuit movements, as well as vestibulo-ocular movements, could become distributed over the cells in the caudal pons (Anastasio & Robinson 1989). For this purpose, it is unnecessary to include a neural integrator in the model. It is sufficient to use a spatial transformation model in which the input signals represent eye-velocity commands in deg/sec, that is, discharge rates in a pair of push-pull encoding neurons; the output is eye velocity, representing the difference in discharge rate of two output, push-pull motoneurons. Thus, there were six input neurons: a push-pull pair for each of vestibular, pursuit, and saccadic velocity commands and two push-pull output neurons. All but saccadic input units had resting discharge rates of 0.5 (50% of

maximum rate, NL, Fig. 1a) indicating a resting state with tonic cocontraction.

If the vestibular input pair changed from {0.5, 0.5} to {0.6, 0.4} indicating that the head was moving, say, to the left, the output neurons were asked to produce the opposite response of {0.4, 0.6} to produce a compensatory eye velocity to the right. If the pursuit input pair changed to {0.6, 0.4}, the output should respond with {0.6, 0.4} to produce an equal eye velocity in the same direction as the request. Saccades were exceptions, because their velocity commands are a burst of activity for ipsilateral saccades with no firing at any other time. Thus, the background rates of the input neurons were {0.0, 0.0}, saccadic inputs to one side were {1.0, 0.0} and to the other {0.0, 1.0}. The output neurons were required to change from a background rate {0.5, 0.5} to {1.0, 0.0} or to {0.0, 1.0}, respectively.

The network quickly learned all three of these simple tasks. We then looked at the 40 hidden units and found the 3 signal types to be distributed over them in a seemingly random way. As usual, the gains of the hidden units were calculated as the change in their activity divided by the change in one of the push-pull inputs. Just as is found experimentally (Tomlinson & Robinson 1984), each cell could then be described by:

$$R = (kE) + r_p\dot{E}_p + r_v\dot{E}_v + r_s\dot{E}_s \qquad (1)$$

where R is its firing rate, $\dot{E}_p$, $\dot{E}_v$, and $\dot{E}_s$ are eye velocities during pursuit, vestibulo-ocular, and saccadic movements, respectively, and the r's are the corresponding gains. The term (kE) represents the eye position term from the neural integrator (which is being ignored here for simplicity). Each cell (model and real) had a different sensitivity to each type of eye movement and all combinations of r_p, r_v, and r_s could be found. In summary, initial randomization and error-driven learning produced a group of simulated neurons with behaviors closely resembling the premotor oculomotor neurons in the caudal pons.

The network did not treat all the signal types completely independently. If a hidden unit helped move the eye in one direction for pursuit, it usually did the same for a vestibular signal, although with a different gain. There were also rogue cells, however; these might discharge, for example, when the eye went left in pursuit but also when it went right in a vestibulo-ocular movement. Again, these cells seemed to serve no special purpose; they were the result of the initial randomization, and the network found a solution in spite of them. There were similar rogue hidden units for saccades. Most hidden units behaved sensibly, bursting for saccades in one direction and pausing in the other. But, like real interneurons, some burst (or paused) for saccades in one direction and did nothing for saccades in the other, and a few even burst (or paused) for saccades in both directions. Again, the model suggests that it would be a mistake to assign any special function to these rogue cells.

This study has the expected result that a premotor neuron participates in many or all of the motor acts theoretically permitted by its anatomical connectivity. In the oculomotor system there are only four major conjugate systems (we have neglected the optokinetic system); each of these clearly has a different, identifiable function and can be independently activated. The simplicity of this arrangement allows one to appreciate just how commands are mixed on premotor neurons. In the spinal cord, the situation might be much more complex, in part because our wide repertoire of limb movements might not be constructed from the sum of a small number of distinct, functional subsystems.

The equation describing the overall input-output behavior of our vestibulo-oculomotor network provides little or no insight into how these movements are organized. The seemingly randomly distributed signals of the hidden units are pieced back together again carefully at the output by synaptic learning so that the gains from input to output are all exactly 1.0. That is, the output should equal the input (a gain of 1.0) for pursuit and saccadic movements, but should equal minus the input (a gain of −1.0) for a vestibular input. The equation describing this behavior would not suggest that the three signals involved were distributed over the entire hidden layer and then reassembled on the motoneurons. This distribution and reassembling mechanism probably developed on an evolutionary scale, creating redundancy and robustness to lessen the impact of lesions, but the input/output equation makes no such predictions.

3.3. Hidden units in coordinate transformations. So far we have looked at the temporal construction and distribution of vestibulo-ocular hidden signals. Next, we wanted to look at signals with spatial orientations (Anastasio & Robinson 1990a). We chose the vertical vestibulo-ocular reflex (VOR), a two-dimensional coordinate transformation, for simplicity. It creates compensatory eye movements using the cyclotorsional eye muscles for vertical head movements in any combination of pitch and roll sensed by the four vertical semicircular canals. Thus, there were four input units, one for each vertical canal, and four output units, a motoneuron for each of the superior and inferior recti and superior and inferior oblique eye muscles of, say, the left eye. As in the previous section, a spatial transformation network is adequate to model these movements in which the activity levels of units in the input layer represent head velocities as reported by the canals, and the activity levels of motoneurons in the output layer represent eye velocity in the pulling directions of the muscles.

The system was trained by rotating the model about many axes in the horizontal plane to create many combinations of roll and pitch. The canal excitations for each rotational axis are related simply to the geometry of the canals, and for the output motoneuron activity, the rotational axis of the eye is related simply to the muscle geometry (Robinson 1982). Thus, the model concentrates only on the neural portion of the process – what happens between canal inputs and motoneuron outputs. The error signal is represented by a failure of the motoneurons to generate signals that would completely compensate for head movements. Once again, the error signal is a measure of the motion of visual images on the retina.

The network required only 500 iterations to learn to generate an accurate vertical VOR; eye velocity compensated for head velocity in all combinations of roll and pitch. We used 40 hidden units. To inspect the behavior of each hidden unit, we found its sensitivity axis, that is, the axis of rotation that creates the maximum modulation of activity. For the canals, this is the axis perpendicular to

the canal plane. For a motoneuron, this is close to the axis around which its muscle rotates the eye. The sensitivity axes of the hidden units pointed over a wide distribution of directions (Fig. 3, left), many of which clustered loosely near or between the principal axes of the canals and muscles, but many were scattered in other directions. The experimental results of Fukushima et al. (1990) for real interneurons are shown in Figure 3, right. The scatter in the real neurons is not as pronounced as in the hidden units of our model, but clearly there is a distribution of sensitivity axes over a pool of interneurons.

Evolution has designed the semicircular canals to resolve head velocity vectors with maximum accuracy (Robinson 1982); the canals are oriented almost exactly at right angles to each other. It therefore seems odd that their signals would be intermixed among hidden units, which would decrease the signal-to-noise ratio. On the other hand, because the VOR is an adaptive system, it is constantly being optimized, and errors are driven to zero. This could compensate for the signal degradation. But then, why did evolution go to all the trouble to orthogonalize the canals so well? Whatever the reason, information from the canals appears to be distributed and intermixed over a set of interneurons and then reassembled vectorially at the motoneurons. Again, the behavior of the hidden units does not, by itself, suggest that a coordinate transformation is occurring, or how it might occur.

The mathematical description of this transformation is both interesting and amusing. The head rotation vector, $\dot{\mathbf{H}}$, produces a neural output from the canals that is also a vector $\mathbf{C}$ (a triplet of push-pull activities from three pairs of canals). The conversion of $\dot{\mathbf{H}}$ to $\mathbf{C}$ can be described by a 3×3 matrix $[C]$, which describes the canal geometry. (We can revert to all three canals in all three dimensions for this discussion.) Similarly, motoneuron activity of three pairs of push-pull motor nuclei constitutes a neural vector, $\mathbf{M}$, that produces an eye-rotation vector, $\dot{\mathbf{E}}$. This conversion can be represented by a 3×3 motor matrix $[M]$, which describes the geometry of the extraocular muscles. The brainstem, the neural component of the VOR, takes the canal signals, $\mathbf{C}$, and connects them to the motoneurons, $\mathbf{M}$, to effect an appropriate eye movement $\dot{\mathbf{E}}$. This process can be described by a 3×3 brain stem connectivity matrix $[B]$. Thus, the whole process can be described by

$$\dot{\mathbf{E}} = [M]\,[B]\,[C]\,\dot{\mathbf{H}}. \qquad (2)$$

When the VOR is working correctly, $\dot{\mathbf{E}}$ equals $-\dot{\mathbf{H}}$.

The brainstem matrix $[B]$ can be found from Equation 2 (Robinson 1982). Its coefficients represent quantitatively how much each canal pair should excite or inhibit each motoneuron pair. But this seemingly elegant description of brainstem function contains no hint that the vector components of $\mathbf{C}$ become disassembled and scattered over many interneurons with sensitivity axes apparently unrelated to those of the canals or motoneurons. This is an even clearer example of a mathematical description telling us *what* must be done without giving any idea of *how* it is done. This description is even misleading in its beguiling simplicity.

Such mathematical descriptions can be pursued to even further lengths. The vector $\mathbf{C}$ is a covariant vector, because each canal output represents the *projection* of $\dot{\mathbf{H}}$

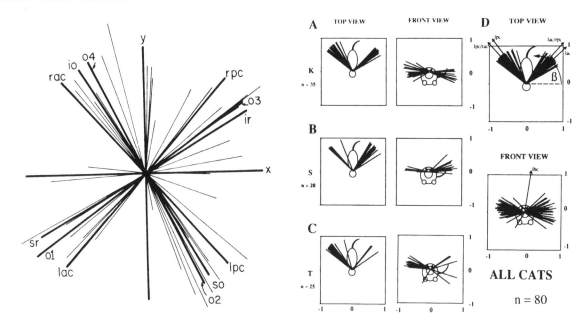

Figure 3. Simulated, left, and actual, right, sensitivity axes of interneurons of the VOR. *Left:* Looking down on the horizontal plane (as in top view at right), y, the roll axis, is anterior; x, the pitch axis, is medial. The axes of the right and left anterior and posterior canals are labeled rac, rpc, lac, and lpc. Those of the left superior and inferior recti and obliques are sr, ir, so, and io. The other vectors show the sensitivity axes, with magnitudes, of the 40 hidden units in the simulation of Anastasio and Robinson (1990). Their axes cluster somewhat around those of the canals and muscles, but there is appreciable scatter and variability. *Right:* Data from Fukushima et al. (1990) of the sensitivity axes of second-order vestibular neurons that project to ocular motoneurons in three cats, A, B, and C. Top view: looking down on the cat (the same view as on the left). Front view: the scatter in the tilt of these vectors out of the horizontal plane. D: the axes of all three cats pooled. The top view should be compared to the model behavior shown on the left. Because recordings were referred to the left side, these axes should be compared to the clusters shown on the left around the lac and lpc quadrants. Note that the cat is upside down compared to the figure on the left. We think the scatter of axes in the model are not dissimilar to those seen experimentally. (Figures reprinted with permission from Fukushima et al. 1990.)

onto the sensitivity axes of the canals. But $\dot{\mathbf{H}}$ is a contravariant vector in which the velocity components along the x-y-z Cartesian coordinates add by the parallelogram rule. To convert the contravariant vector $\dot{\mathbf{H}}$ to the covariant vector **C**, one must multiply the former by the metric tensor of the canal space. Consequently, [C] does two things: It transforms coordinates from Cartesian to canal coordinates and then multiples the coordinate transformation by the metric tensor to produce a covariant vector (see Robinson 1982, for simple, specific examples).

Now **M** is a contravariant vector because muscle forces add via the parallelogram rule, so [B] not only performs a coordinate transformation between canal and muscle coordinates, but multiplies that by the inverse metric tensor. Pellionisz and Llinás (1980) were the first to introduce these concepts in motor control. They further proposed that, since covariant and contravariant vectors appear to be common to all sensorimotor systems, the metric tensor aspects might be handled in a common region of the brain, with the cerebellum being a candidate.

We can now see two extremes in thinking about how the VOR works. In one, basic incontrovertible laws of physics are brought forth, involved in the suggestion that their mathematical expressions are specifically recognized by the brain and dealt with in a compartmentalized manner. In the other, all these considerations are swept aside and a simple network of sensory, motor, and interneurons is asked to wire itself up to eliminate an error signal. The simplicity of the latter scheme, its goal-oriented single-mindedness, its disregard for mathematical elegance, and the naturalness of the sloppy wiring that results draw one to the neural network representation. The wiring is not really sloppy, it is only distributed, but the seeming randomness of hidden unit behavior gives a biological flavor of casualness. Nevertheless, the final solution of the network contains a coordinate transformation and an inverse metric tensor that are very well concealed in the hidden units. A central question is whether anything is gained by recognizing these mathematical transformations. Could these transformations be distinguished neurophysiologically and assigned to different anatomical locations? Given the broad way that signals are distributed over the hidden layer, it would be very difficult, if not impossible, to reconstruct the elements of the [B] matrix by recording from interneurons. Similarly, trying to identify a metric tensor representation in the cerebellum with microelectrodes is likely to be truly impossible. Nevertheless, we can conclude from this example that mathematical descriptions of what a system is trying to do are of little help to the neurophysiologist trying to understand how real neurons do it. It should be added that the laws of physics and mathematical descriptions are perfectly appropriate at the peripheries, in this case the muscles and the canals. This also applies to the recognition of peripheral coordinate systems; canal primary afferents and motoneurons naturally reflect the anatomy of their end organs. The questions raised here, however, pertain to central neurons.

If we were to combine these three models – the neural integrator, multiple command types, and three dimensions – we would produce a Babel of oculomotor spatial and temporal signals in the caudal pons. All interneurons would carry all three types of eye-velocity signals (saccadic, pursuit, and vestibular) as well as the eye-position signal, each with its own sensitivity axis. Based on these models, it is doubtful that the sensitivity axis for pursuit in a given neuron would be exactly aligned with that for vestibular movements or saccades, although some sort of correlation might emerge. The sensitivity axis for the eye-position signal might even differ from that for any of the velocity signals in the same cell. In the extreme, a rouge cell might be activated quite differently for different movements: looking up, pursuit to the left, vestibulo-ocular to the right, and pause for saccades in all directions. One would expect such cells to be rare, but our simulations suggest that they could occur. This picture would be bewildering indeed if we were not lucky enough to know the functions of these three subsystems in all three dimensions and, guided by insights from neural networks, able to allow for the distributed nature of the hidden units.

3.4. Using efference copy to reconstruct the outside world.

As a final example, we can turn to a recent study by Zipser and Andersen (1988). A long-standing problem in oculomotor physiology is whether we track targets using only retinal error (the difference between target image and fovea) or re-create the position of the target in space (with respect to the head, or body image, or inertial space) and use this re-creation to formulate motor commands. The latter idea seems realistic because limb pointing must be done in spatial coordinates. We can see a target, close our eyes, turn our bodies away from or toward it, and then point to it with reasonable accuracy. Thus, the position of the target in space is somehow calculated in the brain and can be used to direct our limbs (see also Fetz's, Bloedel's, and Stein's articles, this issue). We also know that if a visual target jumps from point A to B to C, before the eye can move, we can make saccades in total darkness, from A to B to C, if asked to do so (Mays & Sparks 1980). Note that the retinal error of the target at point C seen from the eye position at B was never available, although the eye went correctly from B to C without it. We can conclude that eye tracking is not directed by retinal error alone.

In another model for generating saccades (see Fig. 4), we used a local (internal) feedback pathway to compare current eye position, E, obtained from an efference copy, E', to desired eye position, E_D (defined with respect to the head, if the head is stationary, or with respect to space itself; Robinson 1975). This model reduced eye position error to zero quickly, thereby producing a goal-directed saccade. There is much evidence both for and against this local feedback model. Stimulation of many parts of the brain (e.g., the superior colliculus, cerebellum, and frontal eye fields) evokes retinotopic saccades that displace the eye in a fixed direction by a fixed amount from any initial position. In contrast, other regions have been discovered (e.g., the supplementary eye fields, Schlag & Schlag-Rey 1987) where stimulation can bring the eye to a fixed orbital position from any initial position (i.e., goal-directed eye movements).

There are two basic problems with his local feedback

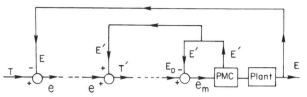

Figure 4. A model of saccade generation that reconstructs the location of a target in space. Retinal error, e, is the difference between physical target position, T, and eye position, E. It is proposed that the CNS adds an internal copy of eye position, E', to e to reconstruct an internal copy of target position, T'. If the CNS decides to make a saccade to T, it sets T' equal to desired eye position, E_D. The difference between E_D and E' is motor error, e_m that enters premotor circuits (PMC) that generate the saccade through the plant (orbital tissues and muscles) and provide the efference copy, E'.

model: It offers no role for the many long-lead burst neurons found in the reticular formation, and it postulates the signal E_D that no one has ever seen. This would require a cell that encoded a visual target position in space, independent of eye position. These problems have caused others to modify the model while essentially retaining the benefits of local feedback (e.g., Scudder 1988). The desired eye-position signal, E_D, in the model of Figure 4, would have to come from another hypothetical calculation: If e is the retinal error (the angle between the target image and the fovea) and E' is an efference copy of eye position, then their sum, T', is an internal re-creation of target position in space. If the brain selects T' as the goal of a saccade, it sets E_D equal to T' and initiates the saccade. As already mentioned, T' probably exists in some form in the brain because we can point to previously seen targets with eyes closed. There remain the questions of how T' is coded and whether it really is used to direct saccades.

Recording from neurons in the monkey parietal lobe, Andersen et al. (1987) found cells that responded to visual stimuli in the usual retinotopic fashion but were also influenced by eye position. These cells responded better to stimulation of a particular point on the retina if the eye was in certain specific positions rather than others. Thus, the cells carried the retinal error signal, e, in the location of its receptive field on the retina. The cell also carried the signal, E, in the form of a multiplicative modulation of the visual response that depended on eye position.

With both signals present, Zipser and Andersen (1988), reasoned that given the power of neural networks, a network could use these signals independently of their format to combine them to yield T', the target position in space. Of course, given the map-like nature of the main input, e, the output would probably be represented in the form of a map with its output signal coded spatially (by location on the map) rather than temporally. Thus, the input layer consisted of a grid representing the retina, with the visual input represented by the activity of units centered around a specific locus on the grid (Fig. 5). The eye position signals were a linear function of E for the four directions: left, right, up, and down. The output was a grid representing external space. Stimulation of the input grid by a target with a constant location in space for many different eye positions had to produce activity at the same point on the output grid. The learning was done by back-

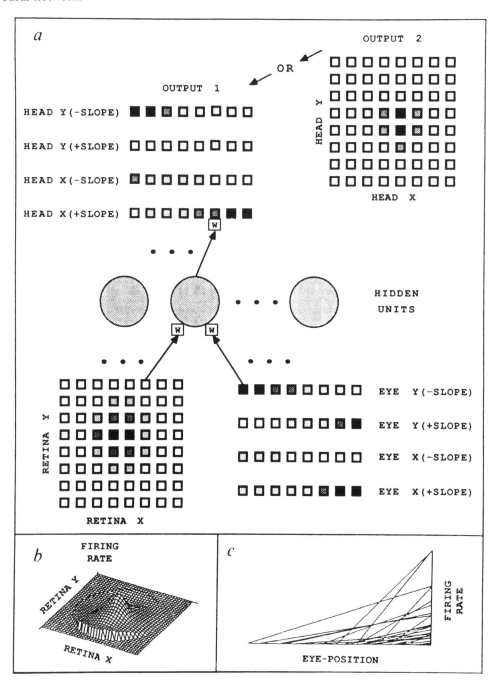

Figure 5. A neural-network scheme (proposed by Zipser & Andersen 1988) uses the behavior of cells in the parietal cortex to calculate the position of a target in space. *a:* At lower left is an 8 × 8 grid of retinal neurons constituting one input layer. The presence of a visual target is indicated by cell activity coded in the intensity of the shading. The second input layer is a set of four rows of eight cells (bottom right) coding eye position for up, down, left, and right. All units in each input set project to all hidden units, which project to all output units. Output units are displayed in two formats. The first, upper left, is an array, similar to the eye position array, lower right, but in head coordinates. A more intuitive display is an 8 × 8 grid representing external space (upper right) showing physical target location. When the network has learned its task, it will combine the target's image location on the retina with the eye position signal and obtain the location of the target in space. *b:* The receptive field properties of a parietal neuron where eye position is held constant and retinal locus is varied. *c:* The dependence of parietal neurons' firing rate on eye position as retinal locus is held constant. (Reprinted with permission from Zipser & Andersen 1988.)

propagation, but as mentioned earlier, the exact learning rule may not be important. The teacher was external – the network was told by the experimenter when it made errors and by how much.

The network learned to make accurate localizations and the hidden units, not surprisingly, resembled neurons in the parietal cortex (Fig. 5). Of course, this network is only a demonstration of how things might be. Just like the cells purported to carry the signal E_D, the cells in this model's output layer have never been observed. One could, of course, press on with the model and design a network to take T' (or E_D) from a spatial code to a temporal code (a burst of spikes) for generating saccades, setting aside for the moment the problem of the apparent nonoccurrence of the signal T'. The immediate question is whether one could build a neural network capable of doing both – finding T' and generating a saccadic pulse – in which T' or E_D never explicitly appeared. This seems likely.

The main message is that hidden units contain signals that are not just hidden from an external observer, but are even hidden from an invasive observer with a micro-electrode. One sees only scraps and pieces of information jumbled together, which can be interpreted only if one knows ahead of time what one is seeking. Again, the mathematical description of what one thinks is being done $(T' = e + E')$ gives no hints of how interneurons might actually do this. If the combined model proves feasible, it will create a problem for those who argue about whether saccades are organized in a retinotopic or craniotopic "coordinate system." Zipser and Andersen's model shows only that craniotopic signals are potentially available. The extended model could make saccades without explicitly realizing such signals. Then what does it mean to ask if the system is craniotopic? The model's hidden units would not be simply represented in retinotopic coordinates either, making the question of coordinate systems an ill-posed one (a frequent result of neural network analysis). The network gets the job done, and for it, coordinate systems are irrelevant; they are a problem of our own making. Obviously, the sensory and motor periphery have their own geometries, which we often call, incorrectly, coordinate systems. (Coordinate systems are human inventions used to measure spatial relationships.) However, we are not concerned with these geometries so much as with whether, in a central neuron, it is even useful to ask in what coordinate system it works. I suggest that it is not.

4. Conclusions

The examples considered here have all been relatively simple because, so far, only a few model networks exist with units that at all resemble real neurons. These "realistic" networks have had to be simple because few people try to model real neural networks with uninterpretable functions. The input signals have consisted of a limited number of types, with only one or two dimensions, and the outputs have been equally simple. The number of simultaneous functions have been few and, more important, they can even be described quantitatively. When one realizes that the network disassembles the input signals, scatters them over the hidden units, and then reassembles them, the behavior of the hidden units becomes understandable. It is not hard to imagine that increasing the number of types of signals, their dimensionality, or the number of functions will lead, long before one gets to memory, language, and consciousness, to a Babel of signal scraps on neurons.

The examples considered above are anecdotal and offer no systematic approach to how this complexity of unit behavior will grow with the complexity of the overall system. The main characteristic to emerge from these simple examples has primarily been the distributed na-

ture of signals. That signals are distributed in systems of neurons is so well known that it is usually brushed aside as an inconvenience. One function of these simple networks is therefore to provide us with concrete examples of signal distribution that will make it more difficult to treat this essential feature of brain organization so lightly. We have also pointed out the phenomenon of rogue units and suggested that to search for coordinate systems in the brain may be a waste of time. One suspects that our attempts to impose our preconceived notions from mathematics, physics, and communication theory (tensors, coordinate systems, sine waves, quaternions, and so on) onto the behavior of central single units is not useful, and indeed, may be holding us back. Modeling increasingly complex systems will probably not reveal anything systematic. Each real network may solve its own problems with its own particular tricks. A related issue is the continuing use of the simplistic lumped, linear membrane summation scheme of Figure 1. Membrane nonlinearities are known to exist that could enrich network behavior, sometimes to our confusion, with such things as oscillations and multiple stable states.

So far, we have concentrated on modeling single-unit behavior, yet much brain modeling has taken place at the black-box level. Has this been useful? There is no simple answer. My complaints about input/output mathematical descriptions that put an entire system into one black box are a case in point, but only an innocuous one. They merely describe the behavior and package it in equations. This is a useful exercise, so long as one realizes that it may have limited predictive ability with respect to single-unit behavior. Modelers rarely stop there, however; they want to guess what goes on inside the black box. So they assemble inner black boxes marked, for example, lead, lag, delay, gain, sample and hold, Fourier transform, and so on, putting them in feedforward and feedback arrangements until the simulation behaves, usually over a limited set of inputs, the way the real system does.

What is gained by this approach? Too often – nothing. How many other arrangements of boxes do the same thing? This question cannot be answered. One can almost never say that there are no equivalent circuits or even no equivalent simpler ones if one likes Occam's razor (a device, it has been proposed, used by biologists to cut their own throats). Usually, the interbox variables have never been observed, and if the forecast of neural networks is correct, they never will be. An example from oculomotor physiology is our ready use of efference copy in our models, signals that reflect desired eye position or velocity. The concept of efference copy is not in doubt, it is just that very few cells in the brainstem, or anywhere else, carry such signals. Usually, the position and velocity signals are intermixed. Neural networks suggest that this might be okay; the signals will probably not be separated until the output layer (motoneurons), so their failure to exist in pure form may be irrelevant. Our jaunty hand waving may be justified after all, but with no thanks to most black-box modelers.

Black box models seldom make testable predictions. The most desirable prediction would concern how single neurons should behave. Even if a black box model were conceptually correct, it seems unlikely that single neuron behavior could ever be predicted, given the considerations of this target article. This poses a real challenge to black-box modeling. If they cannot predict unitary behavior, their usefulness is severely limited.

Nevertheless, some boxes are useful in expressing a concept. There are a few simple examples in the oculomotor system. Direction-selective units in the retina use a small network of cells to transduce the direction and speed of images across the retina. Several plausible models of this network have been proposed, although none have been unambiguously verified. A convenient feature of this network is its anatomical isolation. Its only output must be on retinal ganglion cells. When we set forth a black box with the Laplace operator, s, in it (differentiation) and put image position into it and show neurally encoded image velocity coming out, we feel a useful representation has been made. This is a box not to be thrown out. The same is true of the neural integrator described here: It can be anatomically isolated in the sense that for horizontal head movements its input is a one-dimensional velocity signal from the canals, its output, a one-dimensional signal, one component of which is a position signal funnelled through the motoneurons. All findings so far indicate that the integration is done by a network located in specific regions of the caudal pons. Again, when we set out a box with $1/s$ in it (Laplacian for integration) we feel it is a useful representation.

To throw out these two boxes as part of a grand purge of black boxes would be to throw out the baby with the bath water. However, when one moves centrally and throws in a few more boxes to model the saccadic pulse generator, the optokinetic system, or velocity storage in the VOR (see Robinson 1981, for further descriptions) one's model moves further from reality and becomes less and less testable. Even though this type of model is fairly simple (say 2 or 3 more boxes and 1 positive or negative feedback loop) and based on a fairly well understood system (the oculomotor system, by now subjected to thousands of microelectrode penetrations by dozens of research teams over 15 years), *not a single one has been confirmed or refuted by single-unit recordings*. Over the years, black-box modeling has largely proven to be a blind alley.

Let us return to the applied engineer who uses neural networks but has no desire to look at their hidden units. The engineer must be amused to discover that there is a small army of people in neurophysiology doing just that. What are we learning and what are we not learning by recording from single units? One unrealistic feature of current artificial networks is that they allow all units in one layer to project to all units in the next, thus thoroughly intermixing all inputs. Localization of function in the brain has been one of our major allies, allowing us to concentrate on one modality, such as vision, at a time. Much gross localization had already been done by lesions before microelectrodes came on the scene, but the latter have helped to augment such findings, expand them to their borders, and fill in details in ways that would have been impossible or, at least, tedious to do with lesions. This much localization could have been done by recording field potentials or multiunit potentials, however, since isolating single units addresses a different question – how do neurons process signals – to which there are very few answers. Microelectrode recordings have also pushed localization into other, multimodal areas where lesions would have been uninterpretable. For example, there are 20-odd visual areas fed by primary and second-ary visual cortices, some specializing in color vision, others in motion detection. The supplementary eye field in the supplementary motor cortex and the presence of eye-movement-related activity in the parietal lobe are other examples of multimodal localization.

That is the bright side. The dark side is what single-unit recordings have not told us. In the most general sense, they have not told us how neurons, or groups of neurons, process signals. At a very basic level, single-unit studies have shown that direction-selective neurons exist in visual systems and that they extract image motion. This is a very valuable piece of information, but such recordings apparently cannot tell us how the processing is done. Eye-velocity signals converge in the caudal pons, and suddenly one sees eye-position signals in neural activity there, but unit recording does not tell us how this was done. Perhaps these are neural circuit problems that can be solved in another 20 years.

A slightly more complicated situation is the stretch reflex where, to our embarrassment, we do not know after 50 years the purpose of an animal's proprioceptive signals (see also Gandevia & Burke's article, this issue). Ironically, this is because of a lack of unit recording: Most spinal cord physiologists refuse to bite the bullet and record from proprioceptive neurons in behaving animals. The result is a growing body of electro-anatomy, but no signals. This is a problem that could probably be solved by unit recording, but only in behaving animals.

As we move centrally, the situation deteriorates rapidly. The oculomotor system is no exception. Saccades are created by bursts of activity in premotor cells, but rostral to this, in the superior colliculi and frontal eye fields, attempts to relate unit behavior to saccade metrics (i.e., position, velocity, and direction) have opened a controversy, one that has been fueled rather than quelled by single-unit data. This controversy has led to a plethora of black-box models that have gone nowhere. As mentioned earlier, the great majority of single-unit studies reported in the literature describe signal scraps without interpreting their meaning. Unfortunately, the doubts of the engineer (who uses neural networks as a tool) about whether anything useful will come from examining single-unit behavior, have been painfully corroborated by decades of experience in sensory and motor neurophysiology.

Two future developments might help. One is that applications of neural networks are relatively new in engineering, and although they are used to solve "insoluble" problems in the field, it is unlikely that engineering theorists will be content to let it go at that. This area will no doubt become a new field of study and perhaps in 10 years relationships will emerge that can build a bridge between system function and hidden-unit behavior and tell us how to relate one to the other. The second development is a recent one in neural development research. How does synaptic modification occur? This is a rapidly evolving field that is at the moment noted more for its diversity than for providing any simple, universal answer. We all know that real neural networks learn, but we do not yet know how. In artificial networks, a learning rule is stipulated, the network learns, and one does not worry about details. Perhaps we will be forced to adopt the same attitude: Unable to understand real neural networks at a synaptic or cellular level, we may be forced to wave our hands and say, "Well, we do not know how it works in detail, but we do know that such and such a learning mechanism is at work here to allow synaptic weight changes that are compatible with the observed learning, verified with models, and that's it." This might be as close as we are ever going to get to explaining how a network does its thing. As a result, if we know the learning rules, we may have to accept the inexplicable nature of mature networks.

ACKNOWLEDGMENTS

The author's laboratory is supported by Grant EY00598 from the National Eye Institute, the National Institutes of Health, Bethesda, MD. I thank A. McCracken for preparation of the manuscript and C. Bridges for the illustrations.

5

Do cortical and basal ganglionic motor areas use "motor programs" to control movement?

Garrett E. Alexander,[1] Mahlon R. DeLong and Michael D. Crutcher

Department of Neurology, Emory University School of Medicine, Atlanta, GA 30322

Electronic mail: *gea@sunipc.neuro.emory.edu*

Abstract: Prevailing engineering-inspired theories of motor control based on sequential/algorithmic or motor-programming models are difficult to reconcile with what is known about the anatomy and physiology of the motor areas. This is partly because of certain problems with the theories themselves and partly because of features of the cortical and basal ganglionic motor circuits that seem ill-suited for most engineering analyses of motor control. Recent developments in computational neuroscience offer more realistic, that is, connectionist, models of motor processing. The distributed, highly parallel, and nonalgorithmic processes in these models are inherently self-organizing and hence more plausible biologically than their more traditional algorithmic or motor-programming counterparts. The newer models also have the potential to explain some of the unique features of natural, brain-based motor behavior and to avoid some of the computational dilemmas associated with engineering approaches.

Keywords: basal ganglia; connectionism; cortex; motor control; motor program; motor system; neural network; parallel processing

1. Introduction

There has been a long-standing tendency among neurobiologists and motor control theorists to assume that the process whereby the central nervous system controls voluntary movements is basically serial or hierarchical (Allen & Tsukahara 1974; Bernstein 1976; Brooks 1986; Keele 1968; Rosenbaum 1985). The preoccupation with serial models of motor processing is understandable, for when approached *analytically* each of the various control problems faced by the motor system would seem to require an orderly sequence of computations, that is, a well-defined algorithm, for its solution (Hildreth & Hollerbach 1987; Saltzman 1979). Moreover, computational algorithms of this sort have proven to be a reasonably effective way of using conventional computers (which are serial processors) to control motor outputs in robots (An et al. 1988; Paul 1981). Thus, in the absence of compelling, alternative theories there has been a natural tendency to assume that the sequential/algorithmic approaches that have proven useful in analyzing problems of motor control mathematically and in controlling movement in artificial systems are likely to reflect the brain's approach as well.

Similar considerations have given rise to the now prevalent view that the brain controls movement by executing "motor programs" that are much like software in a conventional computer (Keele 1981; Pew 1989; Poulton 1981; Rosenbaum & Saltzman 1984). The basic assumption is that the motor system effects sequential/algorithmic control over each motor act by executing a specific sequence of computations that converts information concerning the initial conditions (e.g., hand position) and desired endpoint (e.g., a target in space) into a set of muscle activations that will generate an appropriate movement trajectory. The precise sequence of computations is determined by an algorithm that is more or less specific to each particular motor task. To minimize the problem that would be associated with having to store a separate algorithm (or program) for every possible movement a subject might be capable of making, motor theorists frequently assume that only generalized programs (or component subroutines) are stored (Bernstein 1967; Schmidt 1975). Then, when a specific movement is about to be made, the subroutines and variables that are appropriate for that particular movement can be supplied during the "assembly" of the motor program immediately prior to its execution.

The general concept that the motor system uses computational algorithms to control movement, as well as the more specific notion that motor programs are involved, began to become popular at a time when relatively little was known about the structural and functional organization of the motor system. In the beginning, therefore, it was perhaps reasonable to hope that future research would eventually reveal the neural mechanisms through which this type of motor processing might be carried out. However, with the progressive refinement of our knowledge of the motor system's anatomy and physiology, the motor programming concept has become increasingly divorced from any apparent relevance to the neural substrates of motor control.

The central question we raise in this target article is whether the specific concept of the motor program or the more general concept of algorithmic motor processing is an appropriate foundation for the development of *biologically plausible* models of how the brain controls movement. In light of current understanding of how the motor system is organized, it seems to us that both types of models should now be recognized as essentially "blackbox" approaches to the problem of motor control, providing little insight into the underlying neural mechanisms. This is not to minimize, however, the enormously beneficial effects that these approaches have had upon the field of motor psychophysics. There is no disputing the fact that the motor programming and algorithmic processing concepts have been valuable stimulants to motor behavioral research.

After a brief, critical examination of the prevailing algorithmic/motor-programming theories of motor processing, our discussion will turn to the neural circuitry that comprises the cortical and basal ganglionic motor areas. Based on some of the high-level motor deficits that are seen in patients with diseases affecting the basal ganglia, a number of investigators have suggested that the normal role of these structures may be to participate in the selection or execution of motor programs (Brooks 1986; Cools et al. 1984; Marsden 1982). Similar reasoning has led to suggestions that certain cortical motor fields with which the basal ganglia are connected (especially the supplementary motor area, because of its supposed placement at the top of a presumed hierarchy of motor fields; Eccles 1982) may also be involved in "motor programming" (Goldberg 1985; Riehle & Requin 1989; Roland 1984). We will review current knowledge of the anatomy and physiology of the cortical and basal ganglionic motor areas with the express aim of considering whether or not these structures provide the type of neural substrate that might be appropriate for algorithmic approaches to motor processing. (It should be emphasized that in focusing on the cortical and basal ganglionic motor areas we do not mean to imply either that other parts of the motor system are unimportant or that the issues to be addressed are relevant only to cortical/basal ganglia circuitry.)

Recent developments in theoretical neurobiology have provided new insights into the type of information processing that would seem to be required in order to support the motor system's enormous computational burden. These conceptual developments, combined with current understanding of the anatomy and physiology of the cortical and basal ganglionic motor areas, suggest that progress in understanding how these particular structures participate in movement control may require a new perspective on motor processing, one that differs substantially from the traditional notion of the motor system as a sort of Turing machine that uses motor programs to control movement. Instead, we suggest that the motor system, including (but not limited to) the extended network comprising the cortical and basal ganglionic motor areas, is more likely to process information through a mechanism that is essentially "nonalgorithmic," one that depends upon a characteristically biological form of neural architecture that is at once self-organizing, highly parallel, and massively interconnected.

2. Current theories of motor processing

Converging insights from robotics, biomechanics, and motor psychophysics have made it clear that the apparent ease with which humans and other primates are able to make coordinated, goal-directed, limb movements belies the complexity of the underlying computational problems (Atkeson 1989; Bernstein 1967; Hasan 1991; Hildreth & Hollerbach 1987; Hogan et al. 1987; see also Bizzi et al.'s article, this issue). When approached as a conventional engineering problem, the process of transforming the goal of a multijoint limb movement (say, reaching for an object in space) into an appropriate set of muscle activations poses a number of serious difficulties. From an analytical perspective, the problem of controlling goal-directed limb movements is generally recognized as being both complex and ill-posed (Bernstein 1967; Hildreth & Hollerbach 1987; Saltzman 1979).

The problem is complex because of the large number of variables (e.g., visuo-spatial, kinematic, kinetic) that must be taken into account in computing an appropriate set of muscle activations and the fact that there may be significant interactions among some of these variables as the movement unfolds (Hasan 1991; Hogan 1985a; Hollerbach & Flash 1982). Computations of the necessary muscle activations must also include the reconciliation – either explicitly or implicitly – of multiple spatial frames of reference (i.e., retinocentric, head-centered, and body-centered coordinate systems for locating objects in external space, and the separate, multidimensional coordinate systems of joints and muscles; Gielen & VanZuylen 1986; Pellionisz 1988; Soechting & Flanders 1989; see also Stein's, Bloedel's, and Robinson's target articles, this issue). To compute movement control solutions with this approach the motor system would need to be provided, in advance, with quantitatively precise information about: (a) the limb's geometry, including segment lengths and joint ranges as well as the current position and orientation of each limb segment (Mussa-Ivaldi et al. 1985); (b) the entire inertial system comprising the articulated limb as well as any loads carried or obstacles encountered, taking into account both forward and reactive dynamics, and including centripetal and Coriolis forces (Hollerbach & Flash 1982); and (c) the multidimensional muscle plant, with its variable and nonlinear length/tension properties (Gielen & VanZuylen 1986; Houk & Rymer 1981).

The problem of controlling goal-directed limb movements is considered ill-posed because there are no unique solutions to most motor tasks owing to the excess degrees of freedom associated with each of the computational stages – namely, *desired trajectory* → *joint kinematics* → *joint torques* → *muscle forces* → *muscle activations* – that are required to transform the goal of the movement into an appropriate set of muscle activations (Hildreth & Hollerbach 1987; Saltzman 1979). Each of these computations would allow a potentially unlimited number of solutions, so a strategy must be devised to constrain the excess degrees of freedom in order to settle upon one specific solution and execute it in a timely fashion.

The enormity of the problem is further indicated by the rather limited success of engineers in developing robots that can interact smoothly and efficiently with a complex and changing environment. Measured against the relatively rigid, clumsy, and inefficient motor performance of even the most sophisticated of man-made robots, the highly adaptive motor capacities of humans and other mammals appear truly remarkable (Loeb 1983). The problem is of sufficient magnitude that even within the highly constrained environments in which robots are expected to operate, roboticists generally arrange for these computations to be carried out during a planning phase that precedes the initiation of any new goal-directed movement, rather than occurring "on-line" during the course of movement execution (An et al. 1988; Paul 1981). Often, however, the luxury of a substantive planning phase is not available to animals making goal-directed movements, nor are they usually afforded the stable conditions of a highly controlled environment. Thus, it is hard to overestimate the computational difficulties associated with making rapid, targeted limb movements even in the simplest of natural environments.

Such difficulties aside, it has been widely assumed that the brain's approach to the problem of motor control is based ultimately upon analytic, that is, algorithmic solutions (Allen & Tsukahara 1974; Bernstein 1967; Keele 1968; Rosenbaum 1985). In other words, the motor system is thought to control movement by means of ordered sequences of well-defined, rule-based operations that somehow manage to solve the problems of motor control in a manner analogous to that which an engineer or a computer programmer would use, namely, through the systematic manipulation of symbolic information. However, even though this is often the approach used in robotics and despite its providing the most exacting and comprehensive account we have of what the problem of motor control actually entails, the sequential, analytic approach to motor processing seems to us to be a relatively implausible account of how the brain might control movement.

A sequential, symbolic account of motor control is occasionally discounted simply because it requires a lengthy series of complex computations that must be carried out anew for every movement, and because neurons are known to be relatively slow and noise-plagued processing elements (Chandler et al. 1982). It has already been suggested, however, that such computations could be parcellated to some extent for parallel processing (Atkeson 1989; Hinton 1984); such a strategy would increase the speed of computation roughly in proportion to the number of parallel processing elements (Denning & Tichy 1990). Considering the brain's well-known structural parallelism, it would not seem unreasonable, therefore, to expect computations of this sort to be well within the scope of the motor system's processing capacity. Our own reasons for suspecting that the motor system does not use the sequential/analytic approach to control movement are not based on these considerations, however, but on others that will be discussed below in sections 3 and 4.

To many theorists, the concept of the motor program has been attractive because it seems to minimize at least some of the complexities of the sequential, analytic strategy by using prepackaged sequences of a limited set of generic motor commands (or subroutines) to control a large repertoire of goal-directed movements (Keele 1981; Pew 1989; Rosenbaum 1985; Schmidt et al. 1979). The motor-programming concept has also been motivated in part by the fact that at least some movements can be carried out in the absence of sensory feedback, suggesting that the motor system must contain some types of internal representation that are used to control movements of this sort (Keele 1968). Over the past two decades the idea that movements are controlled by motor programs has become so popular that the view is now inherent, in one form or another, in the vast majority of contemporary theories of motor control.

In its most general form, the concept of the motor program is relatively unassailable. It simply refers to an internal representation that makes it possible for a subject to execute a class of related movements in an efficient and reproducible manner (Keele 1981; Poulton 1981; Rosenbaum & Saltzman 1984). When used in this fashion, without implied insight into the underlying neural mechanisms, the concept provides a useful summation of one of the central tenets of motor psychophysics, namely, that most movements represent coherent patterns of behavior whose reproducibility seems to indicate that the information required for their control must be stored and later retrieved by some mechanism residing within the motor system.

The concept of a motor program is appealing because of its apparent simplicity and it has certainly motivated many useful and informative motor behavioral studies. It seems to us, however, that the motor programming metaphor has engendered very little insight into the neural basis of motor control. At best, "motor program" is a convenient but misleading label that serves mainly to obscure our ignorance of the brain's actual approach to motor processing. We suspect that widespread, uncritical usage of this poorly defined term may in fact have impeded progress in understanding the neural substrates of motor control, with ubiquitous references to the brain's "motor programming" functions seeming to imply that the nature of motor processing was already understood far more completely than it really was.

The traditional concept of a motor program, based on a standard computer metaphor, implies a firm distinction between the motor system's hardware (presumably, sets of interconnected neurons) and software (the motor programs themselves, each consisting of a specific set of motor "commands"). But such a model seems to face insuperable difficulties when one tries to imagine how motor programs might actually be represented in the brain's neural architecture. What, for example, would constitute the software in such a model? The conventional answer is likely to be "patterns of neuronal discharge" among the neurons that make up the hardware of the motor system. But this fails to address such fundamental questions as where and how such "software" would be stored while it was not being executed, how the appropriate motor program (or subroutines) would be selected (or "assembled") prior to execution, and, most important, how *new* motor programs would be created. (Innate or reflexive motor behavior can always be readily explained, of course, in terms of hardwired, genetically determined, pattern generators or reflex arcs.) To our knowledge, no one has made a serious attempt to describe the neural mechanisms that might underlie such processes or to provide a sufficiently detailed description of motor programs to allow the model to be tested neurophysiologically.

3. Cortical and basal ganglionic motor circuitry

Although the motor-programming metaphor appears to be vague and untestable from the standpoint of modern neurobiology, the more general hypothesis that the motor system uses algorithmic approaches to control movement is not so easily dismissed. There are, however, a number of anatomical and physiological features of the cortical and basal ganglionic motor areas that seem to argue against both of these models.

In monkeys, several distinct but interconnected cortical motor fields (Dum & Strick 1991), including primary motor cortex (PMC), the supplementary motor area (SMA), and the arcuate premotor area (APA), send excitatory (probably glutamatergic) (Albin et al. 1989; Kitai 1981) projections to contiguous and partially overlapping territories in the putamen (Jones et al. 1977; Kunzle 1975; 1978; Selemon & Goldman-Rakic 1985), which is one of the main "input nuclei" of the basal ganglia (the others being the caudate nucleus and the ventral striatum). The putamen in turn sends inhibitory, GABAergic projections to the internal segment of the globus pallidus (GPi) and to the substantia nigra pars reticulata (SNr) (DeVito et al. 1980; Parent et al. 1984; Precht & Yoshida 1971; Smith & Parent 1986), which together represent the basal ganglia "output nuclei." The output nuclei send their own inhibitory, GABAergic projections to specific subnuclei within the ventrolateral thalamus (Carpenter et al. 1976; DeVito & Anderson 1982; Kim et al. 1976; Parent & De Bellefeuille 1982; Ueki et al. 1977), which then returns excitatory (transmitter unknown) projections back to select portions of PMC and the SMA (Kievit & Kuypers 1977; Matelli et al. 1989; Schell & Strick 1984; Strick 1976; Wiesendanger & Wiesendanger 1985), thereby partially closing what we shall henceforth refer to as the "cortical/basal ganglionic motor circuit" (or, more simply, the "motor circuit"). Given the high spontaneous discharge rates of GPi/SNr neurons (DeLong et al. 1985), it is evident from the gross outlines of the circuit that activation of the corticostriatal inputs to the putamen will tend to disinhibit the thalamic targets of basal ganglia output (Deniau & Chevalier 1985), resulting in further activation of those portions of PMC and SMA that receive thalamo-cortical input from the thalamic subnuclei to which GPi/SNr project (Albin et al. 1989; Alexander & Crutcher 1990a). Overall, then, the circuit might be considered to provide positive feedback to the cortical motor fields. Moreover, since the cortical motor fields are all interconnected by excitatory corticocortical projections, the cortical/basal ganglionic motor circuit has the potential to reinforce the activation of any or all of the cortical motor areas.

On the face of it, this simple, sequential scheme of connections might seem readily suited for the sort of engineering-inspired, sequential/analytic solutions described in the previous section. (For present purposes, we have ignored a substantial number of other important connections that provide impressive amounts of feedforward and feedback control at virtually all levels of the motor circuit.) The sequential aspects of the circuit's organization begin to fade in significance, however, when we consider the anatomy and physiology in more detail.

The architecture of each of the cortical and basal ganglionic motor areas is, of course, "massively parallel" (Alexander and Crutcher 1990a; Dum & Strick 1991) – an overworked phrase, to be sure, but seldom an exaggeration when it comes to brain circuitry. It is not the parallel architecture per se, however, that argues against the motor system's implementing a sequential, analytic approach to motor control. After all, as noted above, parallel implementations of analytic solutions have already been envisioned as potentially feasible strategies for motor control (Atkeson 1989; Hinton 1984). But the sequential/analytic model predicts that even if the various transformations (e.g., from trajectory to joint kinematics, or from joint torques to muscle activation patterns) are computed as a parallel operation (e.g., for all the joints or muscles at once), there is still the logical necessity of performing these separate, parallel operations in a precise sequence. According to this scheme, for example, it is not possible to calculate appropriate muscle activation values until the inverse dynamics has been computed, and that process depends in turn upon prior computation of the inverse kinematics from the desired trajectory (Hildreth & Hollerbach 1987; Saltzman 1979). Thus, if the cortical or basal ganglionic motor areas use a sequential/analytic approach to control movement, there ought to be signs of specialization for such transformations among the different areas, at least at some level of circuit analysis. Thus far, the weight of neurobiological evidence seems to indicate otherwise.

There are, to be sure, clear signs of functional specialization among the different cortical motor areas, but such specializations do not correspond to any of the levels of motor processing that have been defined analytically. In terms of the functional correlates of neuronal activity in the different areas, the most robust differences observed thus far involve the relative proportions of neurons coding for movement preparation versus movement execution. Compared to PMC, the SMA and various premotor areas (lateral area 6) contain much higher proportions of neurons that show sustained preparatory or "set-related" activity that begins long before the initiation of movement; and although all the cortical motor fields contain substantial proportions of neurons with movement-related activity, by far the highest concentrations of such cells are in PMC (Alexander & Crutcher 1990b; Riehle & Requin 1989; Tanji & Evarts 1976; Tanji & Kurata 1985; Thach 1978; Weinrich & Wise 1982; Wise & Mauritz 1985). The SMA may be relatively unique in containing significant numbers of neurons that code selectively for bimanual movements, whereas appendicular representations in PMC are almost strictly contralateral (Tanji et al. 1988). The SMA has also been shown to contain neurons that code selectively for certain sequences of movements, whereas the movement-related activity in PMC shows no such selectivity (Mushiake et al. 1990).

On the other hand, studies that have dissociated some of the analytically defined levels of motor processing have found relatively uniform distributions of neural representations of the different processing levels, not only across cortical motor fields, but in comparisons with basal ganglionic structures as well. These experiments have involved strategies such as the application of opposing and assisting loads to dissociate limb kinematics from limb dynamics. It has been found, for example, that the SMA and PMC (Crutcher & Alexander 1990; Kalaska et al. 1989; Thach 1978), as well as the putamen (Crutcher &

Alexander 1990; Crutcher & DeLong 1984b; Liles 1983) and pallidum (Mitchell et al. 1987), each contain substantial populations of neurons that code for the direction of limb movement in a manner that is either dependent on the external loading conditions (dynamics- and/or muscle-level processing) or independent of such conditions (trajectory- and/or kinematics-level processing). And within the SMA, PMC, and putamen it has been shown (Alexander & Crutcher 1990c) that there are still other populations of neurons that seem to represent the "highest" analytically defined level of processing, that of the target or goal of the movement (in the coordinate system of external space), irrespective of the movement's (body-centered) kinematics or dynamics. Moreover, roughly the same proportion of target-dependent preparatory cells was seen in each of these three motor areas (SMA 36%; PMC 40%; putamen 38%); and the same was true for the limb-dependent movement-related cells (SMA 65%; PMC 71%; putamen 63%) (Alexander & Crutcher 1990c). Such findings seem to argue against the view that motor processing involves a strict parcellation between motor areas of the various sensorimotor transformations that have been defined analytically (of course, not all motor areas have been examined as yet with these approaches).

Within each of the motor areas that have been examined thus far, the separate neuronal populations that seem to represent these different "levels" of motor processing have been shown to be active more or less simultaneously (Alexander & Crutcher 1990c). This suggests that multiple levels of processing may proceed concurrently within each area (Alexander & Crutcher 1990c). Moreover, although the neural onset times for corresponding representations tend, on average, to be somewhat earlier in cortical than in subcortical motor areas, there is enormous temporal overlap *between* areas as well, so that much of the motor processing appears to proceed concurrently even at different stations along the motor circuit (Chen et al. 1991; Crutcher & Alexander 1990). Also noteworthy is the fact that many of the target-dependent cells within the PMC, and a few within the SMA and putamen, have been found to have sensorimotor features that are not generally associated with motor processing levels as "high" as that of the target or goal of the movement. Thus, for example, some of the target-dependent cells were found to have sensorimotor fields restricted to the elbow or shoulder, and some had "muscle-like" responses to external loads (Alexander & Crutcher 1990c). Taken together, such findings are very difficult to explain from a strictly sequential or hierarchical perspective on motor processing. Underscoring this conclusion is the fact that all of the cortical motor fields, including the newly identified cingulate motor areas as well as the SMA, APA, and PMC, send direct projections to spinal levels. Thus, not only do the cortical motor areas seem to lack the sorts of functional differentiations predicted by the sequential/analytic model, they each appear to have relatively direct, parallel access to the motoneuron pools of the spinal cord.

An additional feature of cortical/basal ganglionic motor circuitry that we find difficult to reconcile with engineering-inspired models is the extraordinary degree of convergence and divergence of connections between successive layers of neuronal elements. To a first approx-

imation, perhaps the most obvious convergence occurs along the putamen-GPi pathway. The output elements of the putamen (and the rest of the striatum) are medium spiny neurons, which in monkeys constitute approximately 80% of the total population of striatal neurons (Chang et al. 1981; 1982; Graveland & Difiglia 1985; Wilson & Groves 1980). Approximately 50% of the medium spiny cells are thought to project to the basal ganglia output nuclei (GPi/SNr) (Beckstead & Cruz 1986). Thus, considering the total estimate of about 50 million putamen neurons per hemisphere in humans (Schroder et al. 1975), we might expect that 20 million or so would project to the output nuclei. The total population of GPi neurons has been estimated to be on the order of 0.15 million in humans (Thorner et al. 1975). Considering that in primates about two-thirds of the pallidum receives input from the putamen (the other third from the caudate nucleus; Parent et al. 1984), we can accordingly estimate the convergence ratio along the putamen-GPi pathway to be about 200:1 (ignoring the much smaller projection from putamen to SNr). This is clearly a lower-limit estimate of the degree of convergence along this pathway, for it assumes that each medium spiny neuron projects only to a single pallidal neuron, whereas anatomical studies have shown that this is not the case (Chang et al. 1981). Divergent connectivity within the motor circuit is well exemplified by the projections from cortical motor areas to the putamen. Here, incoming corticostriatal fibers, after reaching the appropriate somatotopic zone (e.g., that of the arm), run longitudinally within that zone for considerable distances (often up to several millimeters), in the process making what appear to be almost haphazard connections with the large number of striatal neurons that happen to lie along each fiber's trajectory (Fox et al. 1971; Jones et al. 1977). This type of architecture seems to us to be ill-suited to the rigidly structured, logically ordered computations envisioned in the sequential/algorithmic model.

It should be noted, however, that the motor circuit does manage to maintain a striking degree of functional specificity and topographic orderliness throughout. For example, in addition to its topographic inputs from PMC, SMA, and APA, the putamen also receives topographic projections from primary somatosensory cortex and from the somatosensory association area (area 5) within the superior parietal lobule (Jones et al. 1977; Kunzle 1977). The coordinated topography of these diverse corticostriatal inputs results in a strict somatotopic organization within the putamen (Alexander & DeLong 1985; Crutcher & DeLong 1984a; Kunzle 1975; Liles & Updyke 1985). Somatotopic organization is maintained throughout the motor circuit, not only at cortical and striatal stages but at pallidal and thalamic stages as well (DeLong et al. 1985; Hamada et al. 1990; Strick 1976). Thus, the separate leg, arm, and face representations within each stage of the circuit can be construed as forming separate, parallel "channels" of information related to these different body parts. Such functional topographic mappings have long been recognized in virtually all of the brain's sensory and motor systems in which there is topographic coding at the periphery, and thus the parallelism inherent in such organizational schemes is now scarcely noticed.

Functional specificity also seems to be maintained throughout the cortical/basal ganglionic motor circuit at the single neuron level. Individual neurons almost invariably show sensorimotor response fields restricted either to leg, arm, or face (usually of the contralateral side of the body). In addition, in various studies it has been shown that many neurons at both cortical (Alexander & Crutcher 1990b; Chen et al. 1991; Georgopoulos et al. 1989; Riehle & Requin 1989; Schwartz et al. 1988; Tanji & Kurata 1985; Weinrich & Wise 1982) and striatal (Alexander 1987; Alexander & Crutcher 1990b; Crutcher & DeLong 1984b; Kimura 1986; Liles 1985) stages of the motor circuit show directionally selective changes in discharge rate either only during the preparation for movement or only during movement execution. This would seem to indicate that the processes subserving the preparation for movement can be carried out by a distributed neural system that is largely separate from that concerned with movement execution, even though some neurons appear to play a role in both processes. The specificity of these responses appears to be maintained not only at cortical and striatal stages, but at the pallidal level as well. A recent study has shown that neurons in the globus pallidus that show preparatory activity receive polysynaptic input mainly from the SMA rather than from PMC (Nambu et al. 1990). This is consistent with the fact that the SMA contains a much higher proportion of neurons that show preparatory activity than does the PMC, and it reinforces the hypothesis that the preparatory and movement-related responses seen at different points in the motor circuit actually represent parallel, functionally segregated channels of information processing.

It seems unlikely that the functional specificity evident throughout the motor circuit results from simple, one-to-one linkages between neurons at successive anatomical stations. As suggested above, there is considerable convergence and divergence of connections at virtually every stage of the circuit (Chang et al. 1981; Francois et al. 1984; Jones et al. 1977; Rafols & Fox 1976). But despite the apparent randomness of these convergent and divergent connections, functional specificity at the level of individual neurons is somehow maintained (Alexander & Crutcher 1990a). In the corticospinal system, intra-axonal injections of anterograde tracers (Shinoda et al. 1981) and the technique of spike-triggered averaging (Fetz & Cheney 1980; Fetz et al. 1989; Lemon et al. 1986) have revealed considerable divergence of projections from a single corticospinal neuron to multiple spinal motor nuclei; retrograde labeling techniques have revealed convergence of inputs from multiple separate cortical motor fields upon a single level of the spinal cord (Hutchins et al. 1988; Martino & Strick 1987). Even so, the divergent projections from a single corticospinal neuron have generally been found to influence muscles with a common functionality (e.g., groups of muscles that are common agonists for a specific movement; Buys et al. 1986; Kasser & Cheney 1985), and the convergent corticospinal projections from different cortical motor areas have been found to be somatotopically specific (Dum & Strick 1991).

Precisely how somatotopy and functional specificity are maintained throughout the cortical/basal ganglionic motor circuitry in the face of substantive convergence and divergence of neuronal connections has yet to be fully clarified. Evidence from other systems, however, would suggest that much of the gross topography/somatotopy within the motor system may be genetically determined,

whereas the fine-tuning (e.g., in terms of the response properties of individual neurons) may depend significantly upon experience and associated activity-dependent changes in local synaptic strengths (Kaas et al. 1990; Kleinschmidt et al. 1987; Roe et al. 1990). There is already evidence that local, activity-dependent changes in neuronal responsivity can occur in the sensorimotor cortex even in adult mammals (Baranyi & Szente 1987; Bindman et al. 1988). It has been shown recently that long-term potentiation of inputs from the somatosensory cortex and the ventrolateral thalamus can be induced in cat motor-cortical neurons (Iriki et al. 1989; 1991). There is also evidence in the motor cortex that rapid reorganization can occur in response to lesioning or stimulation, effects which seem likely to be mediated by treatment-induced modifications of the effectiveness of existing synapses. Thus, either repetitive microstimulation (Nudo et al. 1990) or bicuculline injections (Jacobs & Donoghue 1991) in the forelimb representation of the rat motor cortex will lead to an increase in the size of that representation. Likewise, microstimulation of the vibrissal region of rat motor cortex results in forelimb movements within hours after sectioning branches of the facial nerve that innervate the vibrissae (Donoghue et al. 1990). In addition to this evidence for rapid reorganization, there are also data that indicate changes occurring over a longer time frame may be mediated by axonal sprouting (Keller et al. 1990). It is also noteworthy that the glutamate-responsive NMDA receptor, which is believed to play a major role in activity-dependent synaptic plasticity, has been implicated in synaptic transmission not only at cortical levels but within the neostriatum as well (Cherubini et al. 1988).

The evidence recounted above indicates that cortical/basal ganglionic motor circuitry is characterized by both serial and parallel connectivity, by substantial layer-to-layer convergence and divergence at the level of individual neurons, by topographic organization at a gross level combined with a high degree of functional tuning at the neuronal level, by concurrent, distributed processing of the various analytically defined sensorimotor transformations, and by widespread plasticity with respect to connection strengths and the response properties of individual neurons. These features do not seem to us to be compatible with the engineering-inspired, sequential/algorithmic model of motor processing (and as mentioned earlier, from a neurobiological standpoint we consider the various "motor programming" models to be virtually untestable). Yet such features are precisely those that might be predicted by so-called connectionist approaches to motor processing [See Hanson & Burr: "What Connectionist Models Learn" BBS 13(3) 1990.].

4. Connectionist models of motor processing

To understand the neural mechanisms underlying the motor system's operations would seem eventually to require an explanation of how interconnected sets of neurons manage easily and flexibly to solve motion control problems that appear daunting from an engineering standpoint. Given the apparent limitations of motor-programming theories, the development of alternative, brain-based theories of motor processing would seem to be essential. Fortunately, neurobiologically inspired models that may prove helpful in this respect are now beginning to emerge (Fetz et al. 1990; Houk et al. 1991; Kawato 1990). For reasons to be discussed below, we suspect that these "connectionist" approaches to modeling the motor system will eventually prove to be indispensable to neurophysiologists trying to determine how goal-directed movements are orchestrated at the neuronal level.

The essence of most connectionist models is that they are layered, self-organizing networks of highly interconnected processing units with properties in some ways analogous to those of biological neurons. In connectionist networks, information is stored not in discrete locations (as in conventional computers), but rather in the overall pattern of variable-strength connections among neurons (Feldman & Ballard 1982; Hinton et al. 1986; see also Fetz's and Robinson's articles, this issue). Thus, there is no fundamental distinction between hardware and software in networks of this type.

There are two main categories of connectionist models: those in which learning is *unsupervised,* with changes in synaptic weights (connection strengths) being based entirely on local synaptic conditions; and those in which learning is *supervised* by an intrinsic trainer who makes specific, connection-by-connection adjustments to the network's multidimensional weight space on the basis of a global, error-correcting strategy (Rumelhart et al. 1986d). Unsupervised learning requires so-called Hebbian synapses (of which there are many variants), in which each connection strength is modified in accordance with the degree to which there is correlated activity between the pre- and postsynaptic elements (Brown et al. 1990; Linsker 1990). Our main focus here is on unsupervised learning, which seems to represent the most plausible approximation of learning in biological neural networks. When discussing the general features of connectionist models or the application of such models to the question of how the brain controls movement we refer principally to models in which learning is unsupervised; there are certain situations, however, in which even supervised networks can be useful in addressing some of these issues.

In analyzing the operations of connectionist models, it is useful to distinguish between a learning phase and an execution phase. In the case of motor control networks, solutions to immediate or prospective motor control problems are achieved in the learning phase, during which the strengths of the network's individual connections are adjusted through a repetitive process of trial-and-error (Bruwer & Cruse 1990; Kawato 1990; Kuperstein 1988; Kwan et al. 1990; Massone & Bizzi 1989; Mel 1991; Miyamoto et al. 1988; Ritter et al. 1989). The execution phase for a particular movement consists of the trained motor network's *preconditioned* response to a specific pattern of inputs. The motor network's response during the execution phase is determined in advance by the current pattern of synaptic weights (which reflects, in turn, the cumulative effects of all prior learning); thus, the network's response does not require the computation of a new solution for the specific motor task at hand. This approach effectively circumvents the potentially daunting temporal demands of computing on-line solutions for each new task. Of course, it is not necessary that learning cease during the motor network's execution of previously learned movements; in fact, from a connectionist viewpoint, it is reasonable to hypothesize that the phenomenon of skill enhancement through practice is mediated simply by the reinforcement of successful synaptic weight configurations through repetition of the desired behavior (with learning at the synaptic level being assumed to occur whenever the network is activated).

In connectionist models, where there is no software as such, there is no mystery regarding either the source of the network's response characteristics (these arise through the gradual modification of synaptic weights) or the manner in which this information is stored while the network is not actively engaged in motor processing (it resides in the distributed pattern of connection strengths). It might be tempting to argue that motor programs would be meaningful constructs within a connectionist framework if they were conceived as being represented by the motor system's stored patterns of synaptic weights. This would essentially represent an updated version of the "motor engram" concept (Bernstein 1967). In connectionist models, however, each of the various conditioned motor responses the network had been trained to produce (in response to an appropriate set of inputs) would be represented by the same pattern of synaptic weights, because in each case the network's response would be determined by the overall pattern of connection strengths throughout the entire network. (In connectionist models, the overall pattern of synaptic weights is commonly represented by a matrix that transforms each array [or vector] of inputs into a corresponding output array [Rumelhart et al. 1986a]. The point here is that a single matrix of connection strengths gives rise to many different outputs, depending on the state of each input vector.) Thus, from the connectionist perspective, motor programs could not be considered to exist as discrete, separately identifiable entities (i.e., engrams) within a trained but inactive network, even though, when executed, each movement would indeed be associated with a unique pattern of activation of the different elements within the network. This is implicit, of course, in the principle that information is not stored in discrete locations within a connectionist network, but is distributed instead throughout the network in the pattern of connection strengths.

Connectionist models have an important advantage over algorithmic models in addressing the problem of the motor system's surplus degrees of freedom. As noted above, a major difficulty faced by algorithmic approaches is that the surplus degrees of mechanical freedom provided by multijointed limbs mean that some process is needed for selecting among the infinite number of ways in which each motor task could be solved. For example, to position and orient the hand in space so that an object can be grasped requires a limb trajectory with six degrees of freedom, whereas in combination the joints of the upper extremity (exclusive of those in the hand) provide at least seven degrees of freedom (Saltzman 1979). Although the extra degree of freedom is helpful in insuring that the workspace is fully accessible, it also means that in principle the number of potential solutions to each specific hand positioning task is unlimited.

Yet the motor system must somehow rapidly settle on a single solution in order to execute the movement with dispatch. In addition to biomechanical constraints (e.g., natural limits on joint excursions, muscle forces, etc.), a number of other factors have been suggested as possibly helping to constrain the excess degrees of freedom the motor system must manage. These include the development of fixed muscle synergies (Bernstein 1967) and the adoption of certain optimizing principles such as the minimization of various kinematic or kinetic parameters (e.g., jerk, or work; Flash & Hogan 1985; Nelson 1983). Even in some connectionist models, a number of optimization strategies have been used to assist the network in converging upon a successful structural solution to various problems, including gradient descent methods for minimizing output error; in most cases, however, these strategies have been applied to supervised networks, and their biological plausibility has therefore appeared rather limited.

For the neurobiologist interested in how the brain might control movement, unsupervised connectionist models offer a much more plausible mechanism for dealing with the degree of freedom problem. In the acquisition of motor skills through trial-and-error learning, extra degrees of mechanical freedom are of course quite useful (in fact, they are indispensable in complex, natural environments) in the sense that they enlarge and diversify the repertoire of potentially successful solutions for each specific problem of motor control. At the same time, in connectionist networks the potential solutions become constrained automatically during the course of training through the repetition of those movements that the subject (or trainer) deems successful (Kuperstein 1988; Miyamoto et al. 1988; Ritter et al. 1989). Moreover, as a result of the network's entire history of prior experience, the current pattern of connection strengths already contains an "optimal" solution to whatever new motor task might be presented, even before training on the new task has begun (although through further training that preexisting solution can be modified as necessary). Thus, from a connectionist perspective, there is no difficulty in explaining how subjects may be able to perform novel movements successfully.

There is a crucial distinction between the nature of information processing in connectionist networks as opposed to the type of processing envisioned in traditional algorithmic or motor-programming models. To the extent that connectionist networks are truly self-organizing (i.e., to the extent that their learning is unsupervised) they learn to solve problems in a manner that is essentially nonalgorithmic. Although it is customary within the connectionist literature to refer to the "algorithm" implemented by this or that network, thereby designating the transfer function that is effectively applied by the network to any set of inputs with which it might be presented, there is an important sense in which certain types of connectionist networks (particularly those in which learning is unsupervised) use a nonalgorithmic approach to solve the problems with which they are presented.

This is because the process through which unsupervised connectionist networks arrive at solutions during the learning phase is simply that of trial-and-error. This is analogous to the sorts of nonteleological processes that are thought to underlie biological evolution, namely chance variation and natural selection. While a self-organizing network is being trained, the pattern of connection strengths is constantly being adjusted by an essentially blind process of distributed, trial-and-error learning that tends to select for "successful" solutions, that is, optimal connection weight configurations. Convergence on successful solutions occurs simply because these are the configurations that happen to be associated with satisfying behavioral effects and are therefore most likely to be reinforced, hence selected, through the subject's (or the network's) practiced or habitual repetition of similar behavior. The process of learning is "blind," because for each local connection the training-induced adjustments of its synaptic weight are based solely upon local information concerning levels of pre- and/or postsynaptic activity, and not upon more global information concerning the current or desired configuration of the network or layer as a whole. What connectionist models, genetic structure and, presumably, the distribution of synaptic weights within the motor system have in common is that they all preserve nonalgorithmic (i.e., trial-and-error) solutions that turn out, by chance, to be useful.

Thus, although a logical or mathematical algorithm may be used, after the fact, to describe the problem-solving transfer function that was finally instantiated within a fully trained, self-organizing network, and though even the internal structure and operations of the network may be described in such terms, the blind, trial-and-error process through which the network arrived at its solution can hardly be considered analogous to the type of formal, rule-based, symbolic analysis that an engineer or a mathematician would use in generating an algorithmic solution. Underscoring this principle is the fact that the form of the network's solution (the distribution of connection strengths) typically bears no discernible relation to the formal algorithm that the engineer would use to solve the same problem (see also Robinson's target article, this issue). Moreover, a different set of training inputs, or presenting the training set in a different order, will almost invariably cause the network to arrive at a different structural solution; in other words, a connectionist network's response to new inputs depends on its prior experience as well as on the inputs themselves. This does not mean that the operations of connectionist networks are in any sense indeterminate, that is, that the computations they perform cannot be analyzed mathematically and expressed in the form of algorithms. In fact, it is one of the strengths of connectionist models that their organization and operations can indeed be analyzed and understood in minute mathematical detail.

There is, then, an important sense in which a connectionist network can be viewed appropriately and usefully as implementing an algorithm: during the execution phase. For example, the transfer function implemented by a network's configuration of connection strengths can be, and often is, represented in the form of an algorithm that describes a set of matrix operations, with the structure of the network being formally represented by matrix symbolism (Rumelhart et al. 1986d). It is important to stress, however, that this type of algorithm is rooted in the *structure of the network*, not in that of the physical problem (be it image analysis or motor control) with which the network is engaged. This is why any resulting correspondence between the structure of the problem as defined from an engineering perspective and the structured solution achieved by a self-organizing network through trial-and-error (i.e., through nonalgorithmic processing) can be expected to be largely a matter of chance.

Another distinction inherent in the notion of non-algorithmic processing is the well-known dichotomy between so-called bottom-up and top-down approaches to problem-solving. It has often been noted that self-organizing networks use a predominantly bottom-up approach to solve the problems with which they are presented, in contrast to the formalized, top-down approaches of mathematicians and engineers (Rumelhart & McClelland 1986). There are indications that neural networks can be made more efficient in their problem-solving capacities by incorporating certain types of top-down processing that help to focus and reduce the "blindness" of the search for a successful configuration of connection strengths. One method is to form hybrid networks that include hardwired, symbol-processing modules as well as self-organizing layers of connectionist processing units (Miyamoto et al. 1988). Another is to use supervised learning (e.g., back-propagation, Rumelhart et al. 1986c), in which networks of typical connectionist architecture are trained/shaped by means of learning rules that incorporate error correction and gradient descent strategies to facilitate convergence upon successful solutions (Bruwer & Cruse 1990; Kuperstein 1988). From the neurobiologist's perspective, both the hybrid networks and those in which learning is supervised are unsatisfactory to the extent that they lack plausibility as models of actual brain circuitry. It might also be mentioned that in models of this sort (especially the hybrid variety) the processing can come increasingly to resemble the rule-based, symbolic (algorithmic) approaches that engineers and mathematicians would use to solve the same problem.

5. Future directions: Connectionist models as aids to neurophysiological research

For neurophysiologists interested in the question of how the brain controls movement, the major area in which biologically plausible models of the motor system are likely to prove valuable is in the design and interpretation of experiments involving single-cell recording in awake, behaving animals. Because of its extremely high temporal and spatial resolution, this neurophysiological technique is arguably one of the most powerful methods available for analyzing the functional organization of the motor system. Although essentially descriptive (despite the need for both meticulous internal controls and experimental designs that rigorously dissociate the behavioral variables under study), this approach makes it possible to determine what types and proportions of neural representations (of specific behavioral variables) are found in different motor areas, and how those representations are distributed. On the other hand, the circuitry in the motor system is far too complicated to permit simultaneous monitoring of the activity of a meaningfully large and representative sample of identified neurons (i.e., neurons whose interconnections are known) during any particular motor behavior. Yet only by a strategy such as this could we hope to gain direct insight into the way the network as a whole gives rise to the distributed control of movement.

So instead, constrained by the practical limitations of the methods that are currently available, our only hope in approaching the problem of how circuitry of this complex-

ity actually operates is to gain insight in an *indirect* manner, by testing the predictive accuracy of hypothetical models of such circuits. If it were possible to construct reasonably detailed connectionist models that were consistent with known anatomical and physiological constraints, the validity of such models could in turn be tested by comparing their predictions with biological data obtained in single-cell recording experiments. Thus, when presented with a new motor task, a sufficiently detailed connectionist model of the motor and premotor fields should give rise to explicit predictions concerning the relative proportions and distributions of different neuronal response types that would be expected in the event that an experimental animal were presented with the same task. In this way, the model could be tested, and, one hopes, improved upon, if used interactively to guide neurophysiological experimentation. The models now available are generally too primitive for these purposes, but it seems reasonable to expect that such an enterprise could eventually be realized. Some progress along these lines has already been made in the interactive modeling of motor cortex (Fetz et al. 1990) and posterior parietal cortex (Zipser & Andersen 1988).

Many issues will need to be resolved, however, before we can expect to see realistic models of the motor system or of any of its subdivisions. We have tried to make it clear in the preceding discussion that it may be very misleading to think of the motor system simply as a parallel processor. Its parallel features are undeniable and prodigious, but the motor system's *adaptive* capabilities are also important, as is the fact that these capabilities seem to imply a capacity for *self*-organization independent of any external programmer or trainer. Thus far, the only models we have of this type of process are biological evolution and the unsupervised variety of connectionist networks. There is evidence in other systems, particularly the visual system, that the mechanisms underlying the development, organization, and refinement of connections between layers of neurons may include each of the following processes: (a) purely genetic, nonactivity-dependent; (b) activity-dependent during a critical period of development; and (c) activity-dependent throughout life (Hahm et al. 1991; Rakic 1986; Rauschecker 1991; Roe et al. 1990; Udin & Scherer 1990). We need eventually to learn the degree to which the connections within the different cortical and nuclear fields of the motor system fall into each of these categories, so that our functional models can reflect them appropriately. [See also Ebbesson: "Evolution and Ontogeny of Neural Circuits" *BBS* 7(3) 1984.]

It is clear from psychophysical studies that the motor system uses knowledge of results to guide practice and as an important element in motor learning. The incorporation of neural representations of knowledge of results into connectionist models of the motor system, using appropriate sensory feedback circuitry, could be done in a manner that maintains biological plausibility. In this form, knowledge of results might be comparable to the global reinforcement signals already used in various connectionist networks (Brown et al. 1990). On the other hand, it would be important to avoid, as much as possible, true supervision of synaptic weight adjustments on a connection-by-connection basis if the object is to maintain biological plausibility.

We need to learn more about the transfer functions and learning rules operating in different parts of the motor system. Such information is only beginning to be acquired with the requisite precision for incorporation into functioning connectionist models.

The appropriate inputs and outputs must be considered. Fortunately, connectionist models of the visual (Linsker 1990), proprioceptive (Mel 1991), and vestibular (Anastasio & Robinson 1990b) systems have already begun to be developed; progress in this area should facilitate the development of appropriate sensory feedback networks for models of the motor system. Some investigators have also begun to develop generalized connectionist motor networks whose outputs are interfaced with simulated limbs that incorporate realistic biomechanical constraints (Mussa-Ivaldi et al. 1988). This is a crucial aspect of the overall problem of developing realistic models of the motor system.

The neural substrates underlying the sequencing of motor behaviors should be addressable with connectionist models. Several investigators have begun to incorporate processing units that represent behavioral state into their connectionist networks. These units, which may be analogous to the "preparatory" and "set-related" neurons observed in neurophysiological studies (Alexander & Crutcher 1990b; Kurata & Tanji 1985; Kurata & Wise 1988; Tanji & Kurata 1985; Wise & Mauritz 1985), code for various behavioral contingencies and thus make it possible for the network to generate conditioned sequences of movement without the need for an external sequencer (Arbib 1990; Jordan 1990). Of course, to a limited extent some of this can be accomplished with kinesthetic input. Future research may reveal whether sufficiently complex networks are able to develop such state units through self-organizing processes, as is presumably the case in biological motor systems.

As simulated motor control systems, realistic connectionist models of the motor system ought to be able to mimic at least some aspects of natural motor behavior, such as the ability to adapt to new task conditions and the capacity to generalize from a limited set of training movements to a larger repertoire. Models that meet this precondition, and whose structural features bear at least some resemblance to the neural circuits after which they are designed, can then be tested for their predictive value. For example, do individual processing elements within the network acquire the same types and distributions of response properties as their biological (neuronal) counterparts? Are the transfer functions within the network (i.e., between hidden layers) comparable to those observed in neurophysiological experiments? Other issues that could be addressed include the degree to which the model networks respond to injury, or to simulated electrical stimulation or pharmacologic manipulations. For example, do localized disruptions of processing or discrete lesions within the network lead to the same functional consequences as are seen in their biological counterparts? The process of using such models interactively to guide, and in turn be guided by, neurophysiological investigations of the motor system should prove beneficial to modelers and neurophysiologists alike.

ACKNOWLEDGMENT
The preparation of this review was made possible in part by research grants NS-17678, NS-15417, and NS-23160 from the National Institute of Neurological Disorders and Stroke.

NOTE
1. Please address correspondence to Garrett E. Alexander, Department of Neurology, Emory University School of Medicine, Woodruff Memorial Building, Ste. 6000, 1639 Pierce Drive, Atlanta, GA 30322.

6

Functional heterogeneity with structural homogeneity: How does the cerebellum operate?

James R. Bloedel

Division of Neurobiology, Barrow Neurological Institute, Phoenix, AZ 85013

Abstract: The premise explored in this target article is that the function of the cerebellum is best understood in terms of the operation it performs across its structurally homogeneous subdivisions. The functional heterogeneity sometimes ascribed to these different regions reflects the multiplicity of functions subserved by the central targets receiving the outputs of different cerebellar regions. Recent studies from our own laboratory and others suggest that the functional unit of the cerebellum is the sagittal zone. It is hypothesized that the climbing fiber system produces a short-lasting modification in the gain of Purkinje cell responses to its other principal afferent input, the mossy fiber-granule cell-parallel fiber system. Because the climbing fiber inputs to sagittally aligned Purkinje cells can be activated under functionally specific conditions, these afferents could select populations of Purkinje neurons that would be most highly modulated by mossy fiber inputs responding to the same conditions. These operations may be critical for the on-line integration of inputs characterizing external target space with features of the intended movement, proprioceptive and kinesthetic cues, and the body image.

Keywords: body image; cerebellum; climbing fibers; movement; mossy fibers; posture; proprioception; Purkinje cells

1. Introduction

Motor systems research has placed a strong emphasis on defining the function of central structures either by characterizing deficits following localized lesions or by correlating the discharge characteristics of localized neurons and the behavior performed by the animal. This approach has revealed that the cerebellum participates in several aspects of motor behavior. The classical findings of Chambers and Sprague (1951; 1955) together with early neurological studies of cerebellar patients (Holmes 1917) were among the most important in establishing that the action of the cerebellum in different classes of motor functions may be mediated by different regions of the cerebellum. This inference was based mostly on the fact that lesions in different cerebellar regions produced consistently different motor abnormalities. Both clinical and animal data indicated that midline lesions of the cerebellum affected primarily postural stability and gait, whereas more lateral lesions involving the cerebellar hemispheres produced abnormalities in the coordination of volitional movements. In addition, lesions of the flocculo-nodular lobe produced profound vestibular deficits as well as dramatic modifications in eye movements (for review, see Brooks & Thach 1981; Gilman et al. 1981).

Subsequent studies in Brooks's laboratory (Brooks 1984; Brooks et al. 1973; Uno et al. 1973) supported this general interpretation by showing differences between the deficits produced by cooling the dentate and interposed nuclei. Single unit recording data were also consistent with the differences in the motor deficits produced

by restricted cerebellar lesions. In general, activity in the dentate nucleus was related to motor set or to planning of the movement, whereas modulation of neurons in the interposed nuclei was related to characteristics of the ongoing movement and the EMG activity of the muscles responsible for generating it (Burton & Onoda 1977; Strick 1983; Thach 1970). Together, these historical observations support the view that different functions are ascribable to different cerebellar regions (see Allen & Tsukahara, 1974, for a review of additional evidence supporting this suggestion).

This inference is also consistent with the heterogeneity of the cerebellar efferent projections from the different cerebellar nuclei. Several anatomical studies have clearly demonstrated that there are substantial differences in the predominant output connections between each primary zone and extracerebellar structures. For example, the majority of dentate efferent fibers project to components of the thalamocortical system, and conversely, this nucleus receives its predominant input from cortico-pontocerebellar projections. In contrast, the fastigial nucleus projects to vestibular and reticular nuclei involved in regulating spinal interactions, and it receives its predominant inputs from vestibular and spinal sources (for review see Asanuma et al. 1983a; 1983b; 1983c; Gilman et al. 1981).

In the context of this broad spectrum of functions and output pathways related to different cerebellar regions, one of the most striking features of the cerebellum is its structural homogeneity. This homogeneity is apparent in the similarities of the circuits comprising one of the principal organizational characteristics of the cerebellum, the sagittal zones. The elegant anatomical experiments of Jansen and Brodal (1940) and the subsequent investigations of Voogd and his colleagues (Voogd & Bigare 1980) clearly demonstrated that the cerebellum can be divided into approximately six or seven anatomically distinct sagittal zones, depending on the species (Groenewegen & Voogd 1977; Haines et al. 1982; Voogd & Bigare 1980). Critical to the thesis of this target article is the fact that the basic organizational features of the circuitry in each of these sagittal zones, including the corticonuclear, nucleocortical, and olivocerebellar projections, are remarkably similar (Eccles et al. 1967). Although some subtle differences do exist in the characteristics of the interneurons, the distribution of the infra- and supraganglionic plexes, and the intranuclear circuitry in different cerebellar regions (Beitz & Chan-Palay 1979; Chan-Palay 1973; 1977; Mugnaini 1972), the similarities are appreciably more striking than the differences.

The morphological homogeneity in the context of this functional heterogeneity serves as the basis for the question posed by the title of this target article. The very fact that the cerebellum is involved in several aspects of motor behavior has made it possible to demonstrate a relationship between specific cerebellar lesions or recordings in different cerebellar areas and specific features of motor behavior. These observations have been crucial for "defining" several different aspects of cerebellar function, many of which are reviewed below.

Our thesis is that the past emphasis on elucidating the multiple functions in which the cerebellum participates has resulted in a diminished focus on the *actual* function of this system, which can best be defined as an *operation* that is consistently performed in all cerebellar regions, independent of the modality or source of the afferent inputs being integrated and the organization of the relevant cerebellar efferent projections to extracerebellar sites. The review of our studies and those of other laboratories will attempt to redirect attention from the functions in which the cerebellum *participates* to operations it actually *performs*. (See Edelman & Mountcastle, 1978, for a related discussion pertaining to the cerebral cortex.)

It will be argued that the same basic operation is performed in each cerebellar region, even though the type of information integrated and processed in each cerebellar region may be different and the distribution of activity to other central structures may vary. The sensorimotor processing derived from this basic operation allows the optimal coordination of complex movements through the cerebellum's interaction with many other central systems.

2. Studies supporting functional heterogeneity across cerebellar regions

Chambers and Sprague's suggestion that different regions of the cerebellum participate in different motor functions has been supported by several lines of investigation. Following the historic observations of Holmes (1917), Dichgans and Diener (1985) performed an extensive assessment of neurological deficits in cerebellar patients with pathology localized to different, relatively discrete cerebellar regions. After reviewing the characteristics of

eye movements, limb movements, stability on a balance platform, and long-latency reflexes, they concluded that the functional abnormalities associated with lesions of the major cerebellar subdivisions are reasonably distinct. For example, patients with hemispheric lesions maintained stability on a balance platform at near-normal levels, whereas patients with lesions in the vestibulocerebellum and anterior lobe had very characteristic abnormalities of postural regulation (see also Diener et al. 1984).

More recently, Ivry et al. (1988) compared the motor deficits of patients with lateral and medial cerebellar lesions using clinical findings as well as computerized tomography (CT) scans and surgical reports. The motor task used by these investigators to quantify the patients' deficits consisted of a succession of finger movements at a specified interval (Ivry & Keele 1989). Patients were classified as having a deficit in a "central timekeeper component" or an abnormal "implementation component," based on the analysis of Wing and Kristofferson (1973). Patients with lateral lesions had difficulty in determining the precise time at which the movement was to be generated, whereas patients with more medial lesions had a deficit in executing the finger movement at the specified time. Ivry et al. concluded that the lateral cerebellum participates in regulating the timing of sequential movements, a view consistent with the recent findings of Inhoff et al. (1989), who argued that this structure plays a role in the "translation of a programmed sequence of responses into action."

Other studies have provided evidence for a role of the lateral region of the cerebellum in specific aspects of motor set. Characteristic features of reflex responses to anticipated perturbations are modified substantially by cooling the lateral cerebellar nuclei (Hore & Vilis 1984; 1985). In other studies, patients with cerebellar pathology were unable to adjust the responsiveness or set of their stretch reflexes in order to respond appropriately to successive tilts of the platform on which they were standing (Nashner & Grimm 1978). In another evaluation of motor planning in patients with relatively severe cerebellar pathology, subjects did not display the normally observed increases in the time required to initiate a progressively longer sequence of key presses with the fingers (Inhoff et al. 1989).

The idea that there is a parcellation of function in different regions of the cerebellum has been strengthened by studies designed to link characteristics of single unit data to characteristics of specific behaviors. Several of these experiments compared neuronal activity in the interposed and dentate nuclei during the execution of similar movements. Many of these studies have already been reviewed extensively (Brooks & Thach 1981; Gilman et al. 1981; MacKay & Murphy 1979) and accordingly will only be highlighted here. Thach was among the first to make an extensive comparison between the activity of neurons in these two nuclei during behavior (Thach 1970a). His studies revealed that the activity in the interposed nuclei was more closely related to kinematic parameters of the movement (see also Burton & Onoda 1977; Thach 1978), whereas the activity in the dentate nucleus was more closely related to higher-order features of the motor task such as the anticipated direction of movement. Observations consistent with these findings were subsequently presented by Strick (1983). A para-

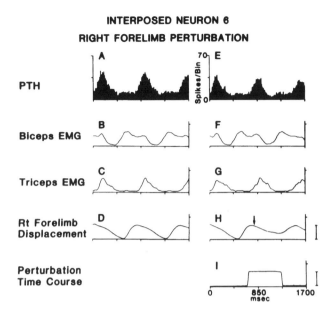

INTERPOSED NEURON 6

RIGHT FORELIMB PERTURBATION

Figure 1. Relationship of the activity of an interposed neuron and the triceps EMG in an unperturbed (A–D) and perturbed (E–I) step cycle. The perturbation was produced by contacting the forelimb during stance phase with a rod that was interjected during swing phase with the time course shown in I. In this and the next figure the forelimb displacement is measured at the animal's wrist for movements in the anterior-posterior direction. The calibration bar = 10.7 cm in H and 7.6 cm in I (from Schwartz et al. 1987).

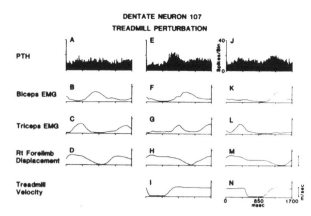

DENTATE NEURON 107

TREADMILL PERTURBATION

Figure 2. Responses of a dentate neuron during unperturbed locomotion (A–D) and during perturbations applied at two different times during the step cycle (E–I and J–N). The perturbation consisted of stopping and restarting the treadmill The responses are averages of 100 sweeps. The calibration bar = 9 cm in M and 0.3 m/s in N (from Schwartz et al. 1987).

digm was used in which the usual direction of an arm movement depended on a combination of the intended direction of the voluntary movement and the direction of an imposed perturbation of the manipulandum. The activity of many dentate neurons was found to be related to the intended direction of the movement independent of the direction in which the limb actually moved. In contrast, the activity of many interposed neurons was related primarily to the actual movement of the limb independent of the intended direction.

Similar findings have also been reported in an experiment in which the locomotor cycle was perturbed in decerebrate cats (Schwartz et al. 1987). These experiments demonstrated that responses recorded from interposed neurons were appreciably more correlated to the electromyographic (EMG) activity of flexors and extensors of the perturbed forelimb than were the responses of dentate neurons. As shown in Figure 1, the discharge of this interposed cell was closely related to the triceps EMG both during normal locomotion and under conditions in which the timing of the triceps activity was altered by a perturbation that impeded the movement of the ipsilateral forelimb. Dentate neurons, however, were responsive to more general features of the animal's response to the perturbation such as the stopping and starting of the locomotor cycle. For example, the responses of the cell in Figure 2 were appreciably more related to the resumption of locomotion than to the EMG activity of either the triceps or the biceps.

In other experiments, Marple-Horvat and Stein (1987) studied the role of the paramedian lobule in goal-directed arm movements. They described populations of Purkinje

cells in this cerebellar region whose activity was highly correlated with limb velocity (Fig. 3) but far less correlated with either position or acceleration. The discharge of these neurons was also tuned somewhat to the direction of the arm movement required to contact one of the four targets. On the basis of these observations the investigators suggested that this region of the cerebellum was specifically involved in regulating dynamic features of limb movement and perhaps also in the computation of velocity vectors responsible for directing movements of the extremity to the target.

These hypothesized functions of the cerebellum in controlling posture and limb movements differ substan-

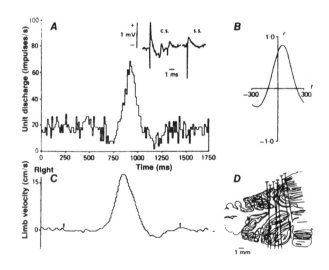

Figure 3. Relationship of a Purkinje cell response recorded in the paramedian lobule to limb velocity during a reaching movement. A: Histogram (bin width = 12 msec) of 20 trials during the reaching movement. B: Cross-correlogram between the unit discharge and limb velocity. C: Average velocity. The portion of the trace between the two arrows was the portion used for the cross-correlation determination shown in B. D: Location of the cell (dot) from which this recording was made. This unit showed a very high correlation ($r = .83$) with the limb velocity at a lag time of 48 msec (from Marple-Horvat & Stein 1987).

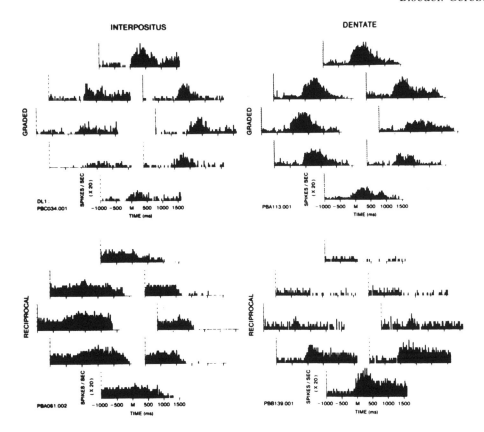

Figure 4.   Similarity of the responses recorded from two interposed neurons (left panel) and two dentate neurons (right panel) during reaching movements made in different radial directions from a center start position to target positions corresponding to the locations of the histograms in each of the four panels (from Fortier et al. 1989).

tially from those proposed for the vestibulocerebellum in regulating eye movements. For example, Lisberger and Fuchs (1978a; 1978b) suggested that this component of the cerebellum is an integral part of a circuit that regulates pursuit eye movements. More recently, Lisberger (1988b) emphasized that this cerebellar region, together with its brainstem target nuclei, may also play a critical role in the neural interactions required for the adaptation of the vestibulo-ocular reflex (VOR).

Other studies indicate that different cerebellar regions regulate various aspects of saccadic eye movements. Lesions in the posterior vermis (Chelazzi et al. 1990; Ritchie 1976) and medial cerebellar nuclei (Optican & Robinson 1980; Vilis & Hore 1981) result in dysmetric saccades characterized by eye movements that either undershoot or overshoot the target. In contrast, saccades made following lesions of the floccular-parafloccular region characteristically have an enhanced postsaccadic drift that produces a substantial mismatch between the initial rapid eye movement and the final position maintained by the eye (Chelazzi et al. 1990; Zee et al. 1981).

3. Similarities in functional properties of different cerebellar regions

Although functional heterogeneity clearly exists among different cerebellar regions, several studies have found strikingly similar behavior among neuronal populations in

the dentate and interposed nuclei during specific types of motor tasks. One of the clearest demonstrations of this similarity was provided by Schieber and Thach (1985), who noted very similar bidirectional discharge patterns for cerebellar neurons in both the dentate and interposed nuclei during the performance of a slow-ramp tracking task. In contrast to several studies reviewed above, they found a rather limited correlation between the activity of neurons in either of these nuclei and the parameters characterizing the movement.

In another study the responses in the fastigial, interposed, and dentate nuclei were compared during direction-specific arm-reaching movements to one of six push buttons on a panel set in front of the monkey (MacKay 1988b). Again, responses in all nuclei were not only poorly correlated with kinematic parameters of the movement but they also had rather similar characteristics. Only differences in response latency were reported. A greater degree of directional tuning among cerebellar nuclear neurons was observed by Fortier et al. (1989) when a monkey was trained to perform a task requiring a succession of reaches in different directions from the start position. In this study similar responses were found in both the dentate and interposed nuclei (Fig. 4). The experiment also provided one of the clearest demonstrations of a relationship between the response properties of a population of cerebellar neurons and cells located in related regions of the motor system. The directional tuning properties of the cerebellar nuclear neurons were remarkably similar to those of motor cortical cells, in which the directional tuning also has been well characterized (Georgopoulos et al. 1986; Schwartz et al. 1988).

There are also similarities in the motor deficits observed in patients following lesions in different cerebellar regions. One of the most dramatic examples is based on a comparison of eye movements in patients with vestibulocerebellar disorders and arm movements in patients with hemispheric deficits. The hypermetric nature of arm movements after ablation of the dentate nucleus has been well characterized (Gilman et al. 1981). An intriguingly similar hypermetria is observed in the eye movements of patients with vestibulocerebellar abnormalities (Selhorst et al. 1976). A substantial overshoot dysmetria was found to be associated with eye movements to specific targets in visual space. This is a remarkable similarity, given the substantially different organization of the motor systems responsible for generating eye movements and limb movements and the differences in the physical characteristics of the eye and arm.

4. The hypothesis: The cerebellum provides a similar neuronal operation across all cerebellar zones

As pointed out in sections 1 and 2, a substantial functional heterogeneity among the various regions of the cerebellum exists despite the remarkable morphological homogeneity of this structure. There is a striking consistency in the organization of the neuronal circuitry across all the cerebellar zones. Each of the primary six or seven cerebellar sagittal zones consists of: (1) a sagittally distributed band of Purkinje cells projecting to a specific region of the cerebellar nuclei that essentially does not overlap with the projections of Purkinje cells from other sagittal zones, (2) the terminations of climbing fiber inputs originating from a specific site in the contralateral inferior olive, (3) the terminations of the nucleocortical projection, the mossy fiber projection from the nuclear neurons to the granular layer of the cerebellar cortex (Dietrichs & Walberg 1979; 1980; Tolbert et al. 1977), and (4) a consistent pattern of interconnections in the microcircuitry of each zone. These functional compartments are remarkably similar from the vermis through the cerebellar hemispheres. Some differences among the zones have been observed recently regarding the distribution of Purkinje cell-specific antigen sites (for examples see Dore et al. 1990; Hawkes & Gravel 1991; Leclerc et al. 1990), as well as certain enzymes (see Boegman et al. 1988). These differences are not believed to be related to differences in the circuitry of the zones, however.

Intracortically, the compartments are interrelated principally through the organization of two components of the mossy fiber system – the parallel fibers and the mossy fibers themselves. (See Eccles et al. 1967, for an overview of the initial contributions of Eccles, Ito, Llinas, Sasaki, and Szentagothai to the organization and synaptic action of elements comprising the cerebellar cortical circuitry.) First, the parallel fibers – the bifurcated axons of the granule cells – can course as far as 6 mm (Brand et al. 1976) along the folium, contacting Purkinje cells in several zones. In fact, the longest fibers can contact cells in both the vermis and the hemispheres. Second, mossy fibers projecting from extracerebellar sites bifurcate many times, contacting granule cells in several zones and producing a somatotopic pattern across the cerebellar cortex referred to as a fractured mosaic (Shambes et al. 1978).

In our view the remarkable homogeneity among the neural elements comprising the sagittal zones suggests that each consists of a basic functional unit that imparts to the cerebellum the capacity to perform a specific type of operation critical to the regulation of a diversity of motor behaviors.

Recent work in our laboratory has focused on characterizing the types of operations occurring within a given sagittal strip. In these studies, multiunit recording techniques were used to record the activity of five to ten Purkinje cells simultaneously in identified cerebellar zones during perturbed locomotion in cats and ferrets. The relationship between the Purkinje cells' simple spike activity and the activation of their climbing fiber inputs has been studied following perturbations of the ipsilateral forelimb during swing phase of locomotion using a new data processing technique, the real time postsynaptic response (RTPR; Lou & Bloedel 1986). It was first shown (see Fig. 5) that Purkinje cells aligned in the appropriate sagittal strip responded with a synchronous activation of their climbing fiber input following perturbation of the step cycle and demonstrated a substantial enhancement of their simple spike modulation (Bloedel & Kelly 1992; Lou & Bloedel 1986). These data indicate that climbing fiber inputs to sagittally aligned Purkinje cells are activated not only during spontaneous activity (Llinas & Sasaki 1989; Sasaki et al. 1989) but also in response to a specific functional condition: perturbation of the forepaw. Furthermore, the findings suggest that the increased simple spike response amplitudes of this small neuronal population may be related to the action of these synchronously activated climbing fiber inputs.

The enhancement of simple spike modulation observed in our multiunit experiments (Lou & Bloedel 1986) was attributed to the same mechanism responsible for the increase in simple spike responsiveness previously demonstrated in our laboratory using a variety of passive paradigms (Bloedel et al. 1983; Ebner & Bloedel 1984; Ebner et al. 1983). With these passive paradigms, various types of data sorting techniques were used to examine the changes in the responsiveness of a Purkinje cell to its mossy-fiber/granule-cell/parallel-fiber input shortly after the occurrence of the same cell's climbing fiber input. Based on these data, a new hypothesis about the heterosynaptic action of climbing fibers was proposed: the gain change hypothesis (Bloedel & Ebner 1985). According to this view, climbing fiber inputs to Purkinje cells are responsible not only for evoking the well-known excitatory response of these neurons but also for increasing the cell's responsiveness to its parallel fiber inputs.

Together, these observations suggest a possible operation for the sagittal zone: the selective enhancement of the responses of the zone's sagittally aligned Purkinje cells to mossy fiber inputs through the action of the climbing fiber inputs to the same neurons. Recently, we proposed the dynamic selection hypothesis (Bloedel & Kelly 1992) to describe how this type of mechanism could be implemented in a task-dependent manner by the cerebellar circuitry during appropriate functional condi-

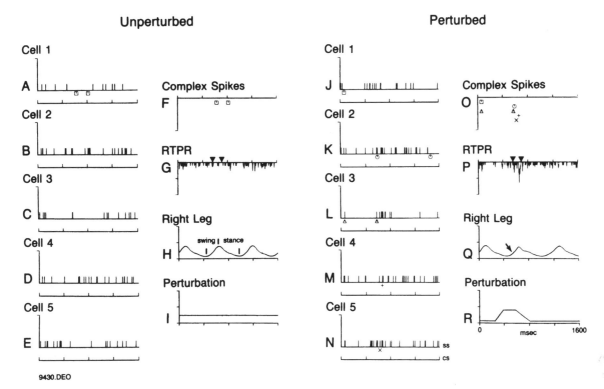

Figure 5. Simple and complex spike responses of five simultaneously recorded Purkinje cells during a single unperturbed (*A–I*) and perturbed (*J–R*) step cycle. Digitized simple spike activity for each of the trials is shown in *A–E* and *J–N*. Different symbols beneath each trace are used to designate the time of occurrence of each cell's complex spikes. They are shown for all five cells in the composite diagrams in *F* and in *O*. *G* and *P*: RTPR (real time postsynaptic response) for the unperturbed and perturbed trials. *H* and *Q*: protraction and retraction of the ipsilateral forelimb. *I* and *R*: movement of the perturbation bar. SS = simple spikes. CS = complex spikes (from Bloedel & Kelly 1992).

tions. According to this view, the distribution of activated climbing fiber inputs would determine the populations of sagittally aligned Purkinje cells that would most dramatically modify the activity of cerebellar nuclear neurons.

Figure 6 conveys the interaction on which the dynamic selection hypothesis is based. Assume that the tap of the forepaw evokes the same patchy distribution of mossy fiber inputs under all three of the conditions indicated in the figure. Based on the additional assumption that the specific population of olivary neurons activated by the tap is task dependent, the spatial distribution of the responding climbing fiber inputs is likely to be different in each of these conditions. Taps encountered in response to a passive stimulus, tripping during locomotion, or contacting an object during a volitional arm movement would activate somewhat different populations of olivary cells, resulting in different regions of overlap between the mossy fiber patches and the activated sagittal zones of climbing fibers. As a result of the mechanism proposed in the gain change hypothesis, those Purkinje cells located in the regions of overlap would be expected to undergo an enhancement of their modulation to mossy fiber inputs. Through this interaction the distribution of climbing fiber inputs would determine the populations of sagittally aligned Purkinje cells to be most highly modulated by the mossy fiber inputs responding to the specific functional condition, in this case the tap of the forepaw. Because of

the organization of the corticonuclear system, this mechanism could provide a way for cerebellar cortical interactions to affect selectively the activity of spatially discrete regions in the cerebellar nuclei despite the wide distribution of mossy fiber inputs activated from discrete sites on the body surface.

Although this hypothesis is speculative and somewhat qualitative, some of its features seem very appealing. First, it provides a unifying functional view of cerebellar integration that implements the circuitry and organizational features of the sagittal zones (Groenewegen & Voogd 1977; Oscarsson 1979; Voogd & Bigare 1980). To date there is virtually no other functional concept that integrates these features into a general view of cerebellar cortical processing. Second, this hypothesis provides a mechanism for task-dependent integration in the cerebellar cortex. Third, it takes into account an enigmatic feature of the organization of mossy fiber systems – the rather wide distribution of mossy fiber inputs originating from a specific site with a somatotopic distribution that is not organized on the basis of the sagittal zones.

It is well known that mossy fiber inputs activated from a specific body region project to several noncontiguous cerebellar regions in a "patchy mosaic" pattern characterized by a multiple representation of different body regions in noncontiguous patches over wide areas of the cerebellar cortex (Shambes et al. 1978). From the point of

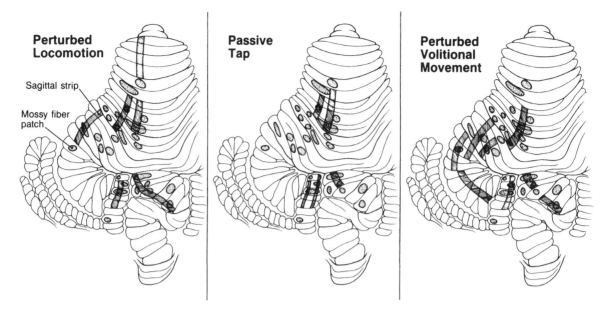

Figure 6. Illustration comparing interactions predicted by the dynamic selection hypothesis under three functionally different conditions (explanation in text; adapted from Bloedel & Kelly 1992).

view of the somatotopic organization of cerebellar efferent projections (Asanuma et al. 1983a; 1983b; 1983c; Gibson et al. 1987), the functional basis for the comparatively wide distribution of mossy fiber inputs is not obvious. It is particularly unclear how a precisely organized cerebellar efferent system can extract the specific information regarding limb position, relevant exteroceptive and teleceptive stimuli, and activity in central pathways from an afferent system that displays so much divergence. The dynamic selection hypothesis proposes a mechanism by which the topographic and somatotopic organization of the climbing fiber system can act to implement the organization of the cerebellar efferent projections while utilizing the multimodal integration provided by the divergent organizational pattern of the mossy fiber system.

The organizational framework for the interactions proposed in both the gain change hypothesis and the dynamic selection hypothesis is shown in Figure 7. Several points should be emphasized before closing this discussion. First, these interactions are clearly not *required* for simple spike modulation to occur. The literature is replete with examples of graded changes in the discharge rate of Purkinje cells in the absence of climbing fiber inputs. It is argued here that the modulation mediated by the parallel fiber system is *enhanced* or *accentuated* when specific functional contexts result in the temporally related activation of mossy and climbing fiber inputs to a given Purkinje cell. The experiments reviewed above illustrate unequivocally that this type of interaction occurs. The hypotheses summarized in Figure 7 may help us understand the functional significance of these findings. Second, these hypotheses do not exclude an important role for the climbing-fiber-evoked excitatory response of Purkinje cells and its effect on nuclear neurons (Llinas & Muhlethaler 1988). Rather, they offer a mechanism by which this excitatory response can be coupled to an associated action of climbing fibers on the simple spike

modulation of Purkinje cells. Third, Figure 7 emphasizes the important role the parallel fibers could play in coupling the interactions in adjacent sagittal zones in a functionally significant way.

Experiments to test the dynamic selection hypothesis have been initiated only recently. Kelly et al. (1990a) compared the responses of up to five Purkinje cells in two different identified sagittal strips during perturbed locomotion in acutely decerebrated cats. The recording sites were selected so that only one of the sagittal strips of Purkinje cells received climbing fiber inputs activated by the perturbation, while both zones received mossy fiber inputs from the perturbed forelimb. According to the hypothesis, those Purkinje cells in the zone receiving the activated climbing fibers should be more highly modulated by mossy fiber inputs responding to the same perturbation. To date, the data support this view. The Purkinje cells whose climbing fiber inputs are activated by the perturbation had a much more dramatically modulated simple spike activity than was observed for the sagittally aligned Purkinje cells in the other zone.

One implication of the dynamic selection hypothesis is that the relationship of neural activity in a region of the cerebellum to a specific behavior may be task- or context-dependent. Taken together with previous studies from their laboratory (Thach 1978; 1970a), the studies of Schieber and Thach (1985) provide an excellent example of the importance of context to information processing in the cerebellum. The characteristics of nuclear neuron modulation were found to be dependent on the nature and speed of the tracking task. When a very slow tracking task was used (Schieber & Thach 1985), the modulation of neurons in the interposed nuclei was associated with higher-order features of the movement. In contrast, during faster tracking the cells' responses were highly correlated with kinematic or dynamic features of the movement (Thach 1978; 1970a).

Another example of context-dependent interactions

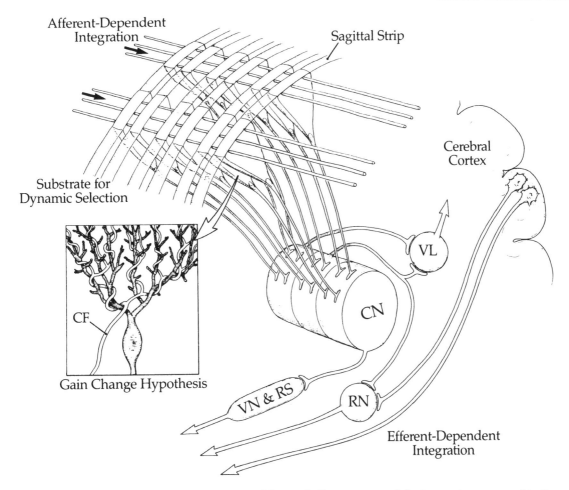

Figure 7. Relationship between components of the cerebellar circuitry and the interactions proposed in the gain change and dynamic selection hypotheses. The wedges in the main diagram represent Purkinje cells projecting to specific nuclear regions. The Purkinje cells in a given sagittal strip are related by the shading. These strips also depict the sagittal distribution of the climbing fiber terminals associated with a given zone. VL: ventrolateral thalamic nucleus. RN: red nucleus. VN: vestibular nuclei. RS: reticulospinal system. CN: cerebellar nuclei. Only a few of the primary projections of the cerebellar nuclei are shown in the diagram.

affecting cerebellar nuclear neuron activity is provided by the recent studies of Chapman et al. (1986). In these experiments monkeys were trained to perform elbow movements in response to different sensory cues. It was interesting to note that the dentate neurons responded to the sensory cues only when the animal actually performed the conditioned movement. In the absence of the movement, no response to the sensory stimulus was observed. This type of "association" exhibited at the level of the cerebellar nuclei may depend on the activation of a specific distribution of climbing fibers because of the convergence of olivary inputs evoked by sensory stimuli and descending motor pathways.

This possibility is supported by evidence that the responses of the climbing fiber system also can be highly task dependent. Some climbing fibers are activated consistently at the onset of a voluntary movement (Mano et al. 1986; Wang et al. 1987). A population of these afferents also respond when a reaching movement is redirected toward another target, even though they are not activated

when the limb traverses a similar location in the absence of any respecified target (Wang et al. 1987). Figure 8 illustrates one cell displaying this response property. The studies of Gellman et al. (1985) demonstrate that the responsiveness of a given olivary neuron to peripheral stimuli depends on the context in which the stimulus is applied. Cells responsive to a passive cutaneous stimulus were unresponsive to a similar stimulus applied during active movements.

These experiments emphasize that the responses of cerebellar cortical and nuclear neurons are context- and task-dependent, including the responses evoked by the climbing fiber afferent system. Consequently, the intra-cerebellar mechanisms proposed in the dynamic selection hypothesis, although highly speculative, are feasible and could produce the set of experimental findings reviewed above. At the very least, this hypothesis illustrates how a homogeneous set of interactions across all of the sagittal zones could contribute meaningfully to the modulation of cerebellar efferent projections.

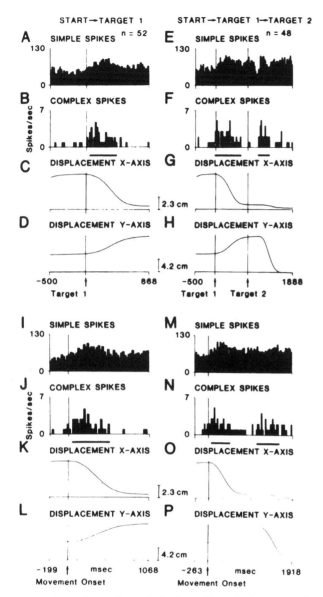

Figure 8. Relationship of climbing fiber discharge to the characteristics of a reaching movement in a monkey. In this experiment, the responses of the simple and complex spike of the Purkinje were compared in two different tasks: a movement from a start position to target 1 (A–L) and a movement initiated toward target 1 that is redirected toward target 2 after the monkey began the movement to target 1 (E–H). A and E: simple spike histogram. C and G: displacement along the x-axis. D and H: displacement along the y-axis. These data are aligned on the presentation of target 1. Using the same format as in A–H, the same data are aligned on movement onset in I–P to show the relationship between the complex spike responses and the time at which the movement was initiated. The bars in B, F, J, and N indicate the bins containing the statistically significant increase in complex spike discharge ($p < .05$). The numbers of responses from which the histograms are constructed are shown in A and in E. Spikes were sampled at 1 kHz and combined in 20 msec bins (from Wang et al. 1987).

5. A unifying view of cerebellar function

Deriving a comprehensive theory of cerebellar function has proven to be a very challenging and elusive objective. One of the major impediments to progress in this area has

been the difficulty in finding a quantitative algorithm for representing specific aspects of cerebellar processing. Studies of the cerebellum's role in regulating vestibulo-ocular and vestibulo-colic reflexes have contributed to progress in this regard (see Lisberger 1988b; Pellionisz 1985; 1989; Pellionisz & Peterson 1988). There have also been several interesting contributions focusing on more general aspects of cerebellar cortical processing (Fujita 1982; Pellionisz & Llinas 1979; 1980; 1982; 1985). Other work has mainly addressed the role of cerebellar efferent projections rather than the properties of the processing actually occurring in the cerebellar cortex (see Flament & Hore 1986; Hore & Flament 1986; 1988; Hore & Vilis 1984; 1985). Notwithstanding these efforts, a generally accepted algorithm describing intracerebellar information processing is not yet available.

Despite the absence of a quantitative understanding of cerebellar processing, the recent literature does provide valuable insights into the overall role of the cerebellum in motor control and the nature of the integration performed by this structure. Critical features of this on-line integration will be reviewed and synthesized into a general qualitative view of cerebellar function following a brief overview of the arguments favoring the cerebellum in a different functional context: as a storage site for the plastic changes underlying motor learning.

6. The issue of memory

During the past decade, one of the most popular theories of cerebellar function has been that this structure plays a primary role in the storage of memory traces for various types of motor learning (Ito 1982; McCormick & Thompson 1984; Thompson 1986; Yeo et al. 1985a; 1985b). This postulate is based on an action of climbing and parallel fibers that differs considerably from the gain change and dynamic selection hypotheses summarized in Figure 7. Consequently, theories favoring a primary role of the cerebellum in on-line processing of sensorimotor information must address the data on which the memory hypothesis is based.

The hypothesis that neuronal interactions in the cerebellar cortex participate in establishing motor engrams in the cerebellum received major support from the initial studies of Ito and colleagues on the adaptation of the VOR (see Ito 1982 and 1984 for review). Supporting this view, studies by Gilbert and Thach (1977) revealed changes in the relation of simple and complex spike activity during the learning of hand movements. In the 1980s, Thompson, Yeo, and their colleagues, working in different laboratories, generated considerable enthusiasm for the learning hypothesis. In support of the idea that the cerebellum is a necessary and sufficient storage site for certain types of motor learning, these investigators demonstrated that specific lesions in the cerebellum selectively eliminate classically conditioned reflex behavior while preserving the related unconditioned reflex (McCormick & Thompson 1984; Yeo et al. 1985a; 1985b). Additional evidence from human studies has also been presented in support of this view (Bracke-Tolkmitt et al. 1989; Leiner et al. 1986; Sanes et al. 1990). Because the major focus of this target article is the operational aspects of cerebellar function, only the most critical observations pertaining to both sides of this controversy will be reviewed. (For a more detailed discussion, the reader is referred to recent reviews by the author: Bloedel et al. 1991; Bloedel & Kelly 1992.)

It should be emphasized that the major controversy concerning the role of the cerebellum in motor learning has focused, not on whether this structure is *involved* in motor learning, but on whether the plastic changes required for processes such as classical conditioning and VOR adaptation actually take place within the cerebellum itself. Substantial evidence challenging this view has now accumulated. Pertinent studies in our laboratory focused initially on determining how essential the cerebellum really is to the acquisition of the classically conditioned nictitating membrane reflex in the rabbit. Experiments in acutely decerebrate rabbits examined whether cerebellectomy exerted selective effects on a conditioned reflex once it has been learned and whether this procedure renders the animal incapable of acquiring the behavior (Kelly et al. 1990b). These studies demonstrated that acutely decerebrate, cerebellectomized rabbits could acquire as well as execute the conditioned nictitating membrane reflex. These findings strongly imply that the cerebellum is not necessary for the plastic changes underlying the acquisition of this reflexive behavior. The data do not rule out the possibility of distributed sites at which plastic changes underlying procedural learning may take place. However, these observations clearly indicate that the data on the effects of lesions restricted to the cerebellum must be interpreted with caution, since it is quite improbable that this structure serves as a necessary storage site for this type of motor learning.

Comparable results have also been reported using intact rabbits. Welsh and Harvey (1989a) demonstrated that cerebellar nuclear lesions in the region presumably required for the plastic changes (see McCormick & Thompson 1984) produce a performance deficit rather than a memory deficit. Furthermore, their most recent studies demonstrated that rabbits are capable of acquiring the conditioned nictitating membrane reflex when the region of the interposed nuclei assumed to be involved in the learning process is anesthetized temporarily during the conditioning period (Welsh & Harvey 1989b). Ongoing experiments in both our laboratory (see Bracha et al. 1991) and Harvey's continue to support this conclusion. These additional findings include data illustrating that the cerebellum is not essential for the acquisition of other types of conditioned behavior. For example, conditioned modifications in limb trajectory during locomotion could be acquired by decerebrate ambulating ferrets that had undergone cerebellectomy 6 to 12 months previously (Lou & Bloedel 1988).

Recent clinical data have suggested that the cerebellum plays a role in cognitive function (Leiner et al. 1986). The most recent studies compared the performance of cerebellar patients and normal subjects either in aspects of cognitive function (Bracke-Tolkmitt et al. 1989) or in tasks requiring motor learning (Sanes et al. 1990). Although the authors interpreted their data as supporting the cerebellar learning hypothesis, the findings do not unequivocally indicate that this structure is a memory storage site. First, in the study reporting learning deficits in patients with cerebellar pathology, there was a sub-

stantial difference in IQ scores between the cerebellar and control subject groups (Bracke-Tolkmitt et al. 1989). As pointed out by the authors themselves, this observation could reflect general motivational abnormalities associated with the decreased motor capabilities of the patients rather than an actual loss of mental skills resulting from a loss of a specific cerebellar function. If so, the results may not be relevant to the issue of the cerebellum's role in learning processes. Second, in the paper of Sanes et al. (1990) arguing that "the cerebellum and its associated input pathways are involved in motor skill learning," their own figures provide substantial evidence favoring the acquisition of skilled movements even in the face of performance deficits resulting from cerebellar pathology. For example, in their Figure 5, clear improvement in the movement performed during mirror-reversed vision is apparent for the patient with the hemispheric deficit. Furthermore, in their Figure 6, average errors were somewhat comparable among the three groups plotted. There were *inter*group variations. These could be due, however, to the differences in the performance deficit between groups rather than a decrease in the capacity of the cerebellar patients to acquire or "learn" the task.

Before leaving this controversy, one other point must be reemphasized. An argument against the establishment of engrams or memory traces in the cerebellum does not and should not imply that the cerebellum and the information processing it performs are not *involved* in the process of skill acquisition and other aspects of motor learning. In fact, clinical studies such as those reported by Sanes et al. (1990), even if interpreted as the authors suggest, do not provide evidence that the cerebellum is actually a *storage site* required for these tasks; they suggest only that it may play a role in the process. In experiments supporting the memory hypothesis this distinction is frequently omitted, even when the relationship of their observations to our findings is discussed (Sanes et al. 1990).

The fact that the cerebellum is involved in the acquisition of some types of motor learning is apparent from several studies. For example, the adaptation of the VOR in cats is substantially impaired following cerebellar removal (Robinson 1976). In addition, patients with cerebellar pathology have considerable difficulty in adapting stretch reflexes evoked by the tilt of a platform so as to minimize the imbalance caused by the stimulus (Nashner & Grimm 1978). Based on the data reviewed above and discussed more extensively in Bloedel et al. (1991), these studies imply that the cerebellum may be involved in the processing that is critical to some types of motor learning or that its removal has a substantial effect on the processing in extracerebellar pathways required for this function. This involvement most probably relates to the importance of the cerebellum for the coordinated *performance* of the movement. Clearly, the progressive improvement in a motor skill requires the neuronal apparatus necessary for the progressive refinement of the task's execution. If the role of the cerebellum in motor learning relates to its importance in regulating motor coordination, lesions in critical areas of this structure would undoubtedly impair motor learning without actually ablating the sites at which the plastic changes take place.

7. New insights into the regulation of motor behavior by the cerebellum

Several experiments during the past decade have provided considerable insight into the overall function of the cerebellum in motor behavior. They suggest that views of cerebellar function restricted to the control of specific movement parameters, error correction, or movement initiation are too narrowly focused and therefore cannot account for several sets of observations that explicitly or implicitly address the complex sensorimotor processing in this structure. The studies of Beppu et al. (1984), which analyzed visuomotor tracking movements in patients with cerebellar pathology, revealed unique aspects of the integration that likely occur in this structure. Patients with cerebellar lesions displayed disorganization of EMG activity, inappropriate amplitude selection and deceleration of the movement, poor pursuit tracking, prolonged reaction times, and movement inaccuracy. It is of interest that these abnormalities are even present in cerebellar patients during the performance of isometric tasks, indicating that limb displacement is not necessary for the manifestation of cerebellar deficits (Mai et al. 1988). In fact, motor abnormalities observed during force tracking as well as during the generation of repetitive alterations in force were closely comparable to those reported for arm displacements in cerebellar patients (see Beppu et al. 1987 and Gilman et al. 1981 for review).

Most critical to the hypothesis that will be presented below are studies examining motor abnormalities observed during complex tasks in which adequate performance depends on the integration of several types of sensory inputs or the simultaneous coordination of more than one type of movement. One experiment (Vercher & Gauthier 1988) examined the effects of dentate lesions on the relationship between oculomotor performance and visuomotor tracking in baboons trained to track visual targets with both eyes and one hand. The authors had previously demonstrated that in these combined eye-hand tracking tasks the activity in efferent pathways controlling arm movements together with the resultant kinesthetic inputs improved the overall tracking capabilities of the oculomotor system (Gauthier et al. 1988; Gauthier & Mussa-Ivaldi 1988).

The effects of dentate lesions in these animals were very revealing. Although the lesion had little if any long-term effect on eye-alone tracking following an adequate recovery period, the quality of eye tracking deteriorated substantially when the task required that the target be tracked by both the hand and the eyes. Joint hand/eye tracking actually decreased the quality of eye tracking to a level below that observed in eye-alone tracking. There was also a decrease in the correlation and an increase in the delay between hand movements ipsilateral to the lesion and the movement of the eyes (Fig. 9). These findings indicate that the dentate nucleus is critical for optimizing the quality of oculomotor tracking when tracking movements of the hand are performed simultaneously. This function includes the temporal coordination of these two simultaneously executed tracking tasks.

The critical role of the cerebellum in this type of integration is further illustrated by additional studies of Beppu et al. (1987). These investigators examined the effect of transiently eliminating the visual cue during a

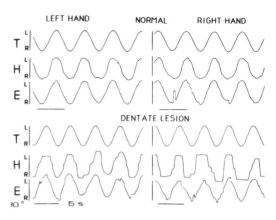

Figure 9. Effect of a dentate lesion in a monkey on the performance of eye movements and hand movements made while tracking the same target. The records on the left were made while tracking with the left hand; those on the right, while tracking with the right hand. In this specific experiment the position of the hand was masked from the subject's vision. The lesion was made in the right dentate nucleus, after which there was an appreciably greater deficit on the side of the lesion (right side) than on the opposite side, although some bilateral effects were observed. Notice the ballistic nature of the movements made by the hand (H) and the irregular eye movements with multiple saccadic components in them. T, H, and E correspond to the target, hand, and eye positions, respectively (from Vercher & Gauthier 1988).

tracking task requiring movements about the elbow joint of normal subjects and patients with cerebellar pathology. Eliminating the visual cue had little effect on the movements performed by normal subjects, but it had a profound effect on patients with cerebellar deficits. This procedure resulted in a smoother, more continuous pursuit movement than when visual information was present throughout the trial (Fig. 10). As in the studies of Vercher and Gauthier (1988), in the absence of the cerebellar processing required for integrating visual cues, the per-

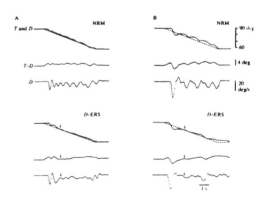

Figure 10. Effects of eliminating the trace showing the displacement of the handle by the subject during a slow ramp-and-hold tracking movement in two cerebellar patients (A and B, respectively). For each patient the top traces show the subject's performance when both the target and the displacement trace can be visualized. The lower two panels show the consequence of eliminating the trace showing the displacement of the handle at the time indicated by the downward arrows in each trace. T = target (broken lines). D = displacement of the handle. D' = the calculated movement velocity (from Beppu et al. 1987).

formance of the overall system was actually worse when the additional information (visual cue) was added. These data indicate that the cerebellum is necessary for the on-line integration of visual cues with other information being generated during movement: Other central structures cannot substitute for this function. Furthermore, this function is so critical that when it is not available to the system the visual cue becomes detrimental to the overall quality of the movement.

More generally, these findings indicate that the cerebellum's primary function may be to integrate information about external target space with information about the body scheme, ongoing activity in central pathways, and the kinesthetic and exteroceptive cues generated by the movement itself. In support of this idea, the specific aspects of motor function either related to the modulation of cerebellar neurons or affected by cerebellar lesions are frequently dependent on this type of integration. These functions include the establishment of motor set, movement initiation, the control of pursuit/tracking movements of the extremities, regulation of pursuit eye movements, and the control of movement accuracy. All of these require the integration of data on specific features of the target or the space in which it is located.

8. A "new" hypothesis of cerebellar function

Based on the inferences drawn from the previous section, it is hypothesized that the cerebellum plays its basic role in motor control by providing the nervous system with the necessary substrate for integrating the properties of external target space with other movement related information in a way that results in an on-line modification of activity in central motor pathways required for the optimal coordination of movement. According to this Vermittler, or "mediator," hypothesis, the cerebellum serves as an active mediator whose output provides the CNS with

an optimized integration of the relevant features of external execution space, internal intention space, body scheme, activity in central pathways, and sensory information resulting from the movement. As a consequence of this integration, the cerebellar output can modify activity in central pathways responsible for motor execution to ensure the specification of the appropriate kinematic and dynamic characteristics of the movement.

Two points are critical to this hypothesis. First, the cerebellum is viewed as uniquely providing the substrate for integrating information required by the nervous system regarding external target space. Second, on the basis of this integration, the cerebellum actively regulates the activity in motor pathways in an on-line manner.

The salient features of this hypothesis are shown in Figure 11, which depicts qualitatively the cerebellar-dependent transformations required when an obstacle is placed in the target space or when the target location is changed slightly after the intent to execute is formulated. The information characterizing external target space is integrated with several types of data, many of which are probably represented in different coordinate systems. The cerebellum uniquely integrates these data with an internal representation of target space and inputs reflecting activity in other components of the motor system. Note that without the cerebellar processing depicted in Figure 11, the motor pathways responsible for movement execution do not receive updated information about the relationship between features of execution space, internal intention space, body scheme (Lestienne & Gurfinkel 1988), and the inputs characterizing the activity in central pathways. It should be emphasized that this process is ongoing throughout the movement, taking advantage of the critical on-line integration performed by the cerebellum *during* movement execution.

This view is a conceptual follow-up to the tensor hypothesis of Pellionisz and Llinas (1979; 1980; 1982; 1985), who used tensor analysis to formulate the transformations

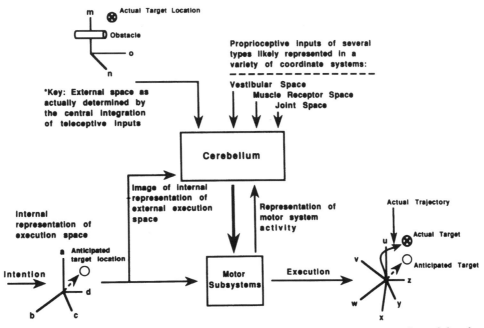

Figure 11. The proposed transformations that depend on the integration performed by the cerebellum (explanation in text).

of an "intention" vector expressed in sensory reference frames into an "execution" vector. The execution vector was expressed in a different reference frame based on the characteristics of the motor system responsible for generating the behavior. The tensor hypothesis played a pivotal role in emphasizing the importance of coordinate transformations, not only to cerebellar function but to concepts of motor function in general.

The Vermittler hypothesis proposed above builds upon the concepts elaborated by Pellionisz (1985; 1989; 1991) and Pellionisz and Llinas (1979; 1980; 1982; 1985) by emphasizing the importance of reference frames to the function performed by the cerebellum. Specifically, the critical classes of information reviewed above, including the internal and external representation of target space, are integrated in the context of reference frames. Clearly the Vermittler hypothesis as outlined in Figure 11 does not propose specific transformations such as the covariant to contravariant transformation elaborated in the tensor hypothesis. Rather, it emphasizes the critical importance of the transformations characterizing internal and external target space to cerebellar integration. Furthermore, this view does not require that the final transformation to a contravariant representation occur at the level of the cerebellar output, although this possibility is certainly not excluded. It is also important to emphasize that the integration the cerebellum performs is probably called upon in a task-dependent manner. This implies that the "decision" to implement the cerebellum's output occurs elsewhere in the central nervous system.

The Vermittler hypothesis is one of several (including the original tensor hypothesis and the "accessory adjustor hypothesis" of MacKay and Murphy 1979) that emphasize that the output of the cerebellum serves as an *intermediate* step in the synthesis of the movement, providing other components of the motor system with information critical for optimizing the coordinated execution of the behavior. Furthermore, the Vermittler hypothesis does not present a new formalism for quantitatively specifying cerebellar operations. Rather, it emphasizes critical features of cerebellar integration that are required by the CNS for formulating a coordinated movement. These features characteristically involve the integration of visuomotor information regarding target location or target movement with critical coordinate transformations necessary for transforming relevant sensory representations into reference frames suitable for the processing performed by other components of the motor system.

The significance of establishing a conceptual link between the cerebellum's basic operation and the interactions occurring in the cerebellar circuitry cannot be overemphasized. In our view, it is very important to approach this problem from the perspective of the processing that takes place in the parallel subsystems comprising the cerebellar sagittal zones. It is likely that the circuitry involving these zones provides the substrate for integrating information related to postural control mechanisms involving proximal musculature with information related to the execution of phasic movements of the extremities. As indicated in Figure 7, this integration occurs at least at three levels. First, input-dependent integration can occur within precerebellar nuclei and the granular layer itself. Second, this integration may involve coupling among the sagittal zones, since parallel fibers are known to be long enough to traverse several sagittal zones (Brand et al. 1976). Third, the convergence among cerebellar output projections may provide output-dependent substrates for integrating the output of the sagittal zones. Patterns of convergence exist at the cortical level of the cerebello-thalamocortical projections and may also exist in some brainstem nuclei contributing to the organization of descending bulbospinal projections. Although most of the above discussion of the substrates for cerebellar information processing has focused on the spatial features of cerebellar integration, the same structural components are undoubtedly critical for both the space- and time-dependent integration required for motor coordination (Pellionisz & Llinas 1982).

In conclusion, this review has emphasized two major points. First, the organization of cerebellar systems and the physiological characteristics of the neuronal interactions in the cerebellar cortex strongly suggest that cerebellar function involves a characteristic operation that is performed across its various components. Second, this operation includes integrating sensory information about execution space, the target, peripheral inputs, and the body scheme with inputs reflecting the properties of the movement being executed. Although the fundamental characteristics of this operation are likely to be similar across the subdivisions of the cerebellum, the organization of cerebellar afferent and efferent systems will result in differences in the behavioral consequences of the processing in various cerebellar regions. This heterogeneity reflects the multiple systems with which the cerebellum interacts rather than a multiplicity of functional operation within the cerebellum itself.

7

Are movement parameters recognizably coded in the activity of single neurons?

Eberhard E. Fetz

Department of Physiology & Biophysics and Regional Primate Research Center, University of Washington, Seattle, WA 98195

Electronic mail: *fetz@locke.hs.washington.edu*

Abstract: To investigate neural mechanisms of movement, physiologists have analyzed the activity of task-related neurons in behaving animals. The relative onset latencies of neural activity have been scrutinized for evidence of a functional hierarchy of sequentially recruited centers, but experiments reveal that activity changes occur largely in parallel. Neurons whose activity covaries with movement parameters have been sought for evidence of explicit coding of parameters such as active force, limb displacement, and behavioral set. Neurons with recognizable relations to the task are typically selected from a larger population, ignoring those cells with complex relations to the task and unmodulated cells. Selective interpretations are also used to support the notion that different motor regions perform different motor functions; again, current evidence suggests that units with similar properties are distributed over widely different regions.

These coding issues are reexamined for premotoneuronal (PreM) cells, whose correlational links with motoneurons are revealed by spike-triggered averages. PreM cells are recruited over long times relative to their target muscles; they show diverse response patterns relative to the muscle force they produce; functionally disparate PreM cells such as afferent fibers and descending corticomotoneuronal and rubromotoneuronal cells can exhibit similar patterns. Neural mechanisms have been further elucidated by neural network simulations of sensorimotor behavior; the pre-output hidden units typically show diverse response patterns in relation to their target units. Thus, studies in which both the activity and the connectivity of the same units are known reveal that units with both simple and complex relations to the task contribute significantly to the output. This suggests that the search for explicit coding may be diverting us from understanding distributed neural mechanisms that operate without literal representations.

Keywords: chronic recording; motor cortex; movement parameters; neural coding; neural computation; neural networks; parallel distributed processing; premotoneuronal cells; representation; spike-triggered averages

1. Introduction

Many systems neurophysiologists record the activity of single units in behaving animals in the hope of understanding the neural mechanisms generating motor behavior. Such "chronic unit recording" experiments are typically designed to test a plausible hypothesis about the function of neurons at some recording site: The animal is trained to perform a behavioral task involving that function and the experimenter searches out relevant task-related cells. Over the last three decades this formula has generated numerous papers illustrating neurons whose activity appears to code (i.e., to covary with) various movement parameters or representations of higher-order sensorimotor functions. Initially, such studies seemed to provide supportive evidence for plausible notions, for example, that motor cortex cells code muscle force and that premotor cortex cells are related to programming movements. With an increasing number of more sophisticated studies it has become clear that the accumulating experimental evidence undermines many of our simplistic notions about neural coding. Moreover, the search for neural correlates of motor parameters may actually distract us from recognizing the operation of radically different neural mechanisms of sensorimotor control.

This article begins with a review of experiments designed to show how various movement parameters may be represented in neural activity. This includes attempts to delineate a functional hierarchy of cells on the basis of their response latencies. We then consider studies of explicit coding of simple movement parameters such as active force and limb displacement and preparation to move. We discuss functional specialization in different cortical regions as well as the possibility that parameters are coded in populations of neurons. Since synaptic connections are an important determinant of the functional consequences of neural activity, we reexamine these coding questions for premotoneuronal cells, which have direct links with motoneurons. Finally, we reconsider these issues in light of results from neural network modeling studies.

2. Representation of movement parameters in neural activity

2.1. Relative timing of cell activity.
To obtain evidence for a causal hierarchy of cells in different motor centers that mediate the programming and execution of movement, it first seemed reasonable to determine the sequential recruitment order of cells in different areas. A particularly useful behavioral paradigm for this purpose is the simple reaction-time response, in which an animal makes a repeatable movement in response to a stimulus such as light. The successive activation of neurons in different regions would then define a causal sequence of neurons mediating the transform between stimulus and response. For a visually triggered key release, for example, the sequence would begin with stimulation of retinal cells followed by propagation of activity to diverse cortical and subcortical centers, which might code the sensory aspects of the stimulus. The conversion of the stimulus-evoked activity into the preparation for movement might occur at intermediate times in cortical association areas. Finally, the neural activity involved in execution would converge in proper combination to activate agonist motoneurons that generate the movement. The peripheral links at the input and output stages of such a sequential scenario have been elucidated, but the central stages have consistently eluded temporal resolution.

The timing of motor cortex cells relative to movement was first studied by Evarts (1968), who showed that pyramidal tract neurons (PTNs) began to change their activity up to 100 msec before the onset of activity in agonist muscles. To determine the relative onset times of other cells that might precede activation of motor cortex neurons, Thach (1978) recorded neural activity in cerebellar nuclei, motor cortex, and muscles during the same responses. The onset times of activity changes of units in cerebellar nuclei were found to largely overlap those of precentral motor cortex cells (Figure 1). The onset times of different neurons in two cerebellar nuclei and motor cortex were distributed over hundreds of milliseconds, with a relatively slight difference in their mean onset times. Comparable overlap in recruitment times has been found in many subsequent experiments. Neurons in the supplementary motor area and primary motor cortex are recruited almost simultaneously in a reaction-time task (Chen et al. 1991) and during a step-tracking task (Alexander & Crutcher 1990c).

The basic problem in attempting to demonstrate serial activation of cells in different motor centers is that each region contains neurons that are recruited over diverse times. The extensive overlap in onset times makes it difficult to assign a sequential order of activation to different regions. Moreover, the duration of most movements as well as the duration of task-related activity greatly exceeds the conduction time between centers, so that recurrent loops could be "traversed" repeatedly during a single response. It is also relevant to note that the focus on the first change in neuronal activity puts undue emphasis on a subtle shift in firing rate that requires statistical determination. Functionally, the initial onset of a change has less to do with the cell's contribution to movement than its maximal activity. In any case, the appealing notion that initiation of movement involves the sequential activation of cells in hierarchically related

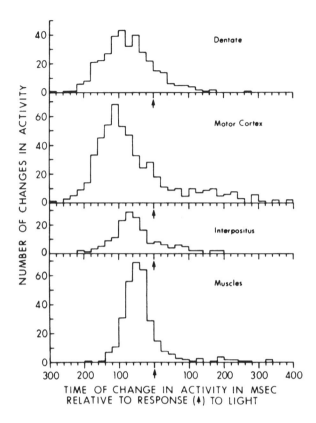

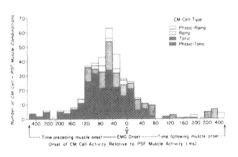

Figure 1. *Top:* Distribution of times of change in neural activity relative to light-triggered wrist movements (from Thach 1978; task is illustrated in Figure 2). *Bottom:* Distribution of onset times of CM cells relative to onset of activity in their target muscles (from Cheney & Fetz 1980).

centers is quite difficult to prove experimentally. In fact, the experimental results suggest that cells in diverse regions are activated more or less in parallel.

2.2. Coding of movement parameters.
Although a cell's onset time provides equivocal evidence for its role in a causal hierarchy, its discharge pattern could provide a more robust indication of its contribution to movement. The hypothesis that parameters of movement are recognizably coded in the activity of motor system cells seems so reasonable that many experiments have been launched on the basis of this assumption. Neural coding, in the sense of covariation, has been amply investigated for a variety of movement parameters (reviewed in Evarts 1981; Fetz 1981; Fuster 1985). Since muscles are ultimately the generators of active force, it seems plausible that central cells controlling muscles could be coding the

force exerted during a movement. On the other hand, since we normally think in terms of moving a limb to particular target positions, it also seems reasonable that cortical cells could code the *displacement* or position of the limb. Evarts's first experiments to determine whether motor cortex neurons code force or displacement provide an excellent example of a behavioral paradigm designed to dissociate these variables. Evarts trained monkeys to make the same movements against different loads and, in some cases, to generate isometric activity without any displacement. In these studies, the activity of selected PTNs was related more to the active force or to the change of force than to displacement (Evarts 1968).

Yet a third variable to which cells could be related was found in monkeys prepared to make a movement: Some cortical cells changed activity long before an intended movement, suggesting that these cells may be involved in the *preparation* to make a movement, as contrasted with its execution, that is, with a behavioral *set*. Experiments designed to reveal set-related activity typically involve behavioral trials beginning with a sensory cue that indicates the correct movement, followed by a delay period and then a go signal to execute the cued motor response. During the delay between the cue and go signal, the monkey is prepared to initiate the movement and neurons in many cortical and subcortical regions exhibit associated changes in discharge.

Numerous other movement parameters have been suggested to be coded in neural activity, such as limb velocity (Gibson et al. 1985), direction of movement (Fortier et al. 1989; Georgopoulos et al. 1984; Schmidt et al. 1975), and target position (Alexander & Crutcher 1990c; Martin & Ghez 1985). The issue of neural coding can be discussed in relation to three parameters: active force, displacement, and behavioral set. Thach (1978) was the first to investigate all three variables systematically for motor cortex and cerebellar cells. He used a task (shown schematically in Figure 2, top) that involved each of these three parameters: The monkey moved the wrist through a sequence of successive hold positions against different loads. The lower trajectories schematize the expected activity patterns of cells primarily related to patterns of muscle force (MPAT), position of the wrist joint (JPOS), and preparation or set for the next direction of movement (DSET). Thach calculated the degree of correlation between these idealized patterns and the recorded activity of cells in motor cortex, cerebellum, and muscles and found that the degree of correlation for different cells was continuously distributed from weak to strong. As expected, many motor cortex cells showed the best correlation with the MPAT sequence. Many other cells in the motor cortex correlated more strongly with joint position and still others correlated with set. Perhaps the most remarkable finding was the relative numbers of cells in each category; as described by Thach, "In summary of the rather astonishing results on neural discharge patterns in motor cortex during holding, all the types of neuron that were looked for were found, in nearly equal numbers" (1978, p. 665). Proponents of coding of movement parameters can only interpret this result as indicating that the motor cortex contains a variety of cells, each coding a different parameter of the movement.

However, such a conclusion would have to be tempered by another remarkable finding in this study: A

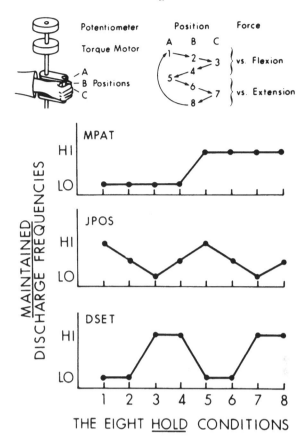

Figure 2. *Top:* Representation of a behavioral task used to test relation of cells to three parameters: pattern of muscle activity (MPAT), joint position (JPOS), and direction of intended next movement (DSET). *Bottom:* Trajectories show discharge frequencies of neurons or muscles having optimal relation to each parameter, plotted as a function of the hold positions illustrated at top (from Thach 1978).

slight change in response conditions could change the parameter that correlated best with a particular neuron. For example, the motor cortex cell illustrated in Figure 17 of Thach's paper had a strong relation to muscle patterns under condition of external load but was better related to joint position when the load was removed. Consistent relations between the activity of motor cortex cells and isometric muscle force have also been dissociated by changing the rewarded response patterns (Fetz & Finocchio 1975). Diehard proponents of coding would have to conclude from such flexible relationships that the same cells can "code" different parameters under different response conditions.

One basic problem with many attempts to relate neural activity to movement parameters is that the data are usually skewed by two types of experimental bias. One is the "task-induced bias" introduced by recording neural activity under particular behavioral conditions. In these experiments animals are performing a specific task designed to test the experimenter's hypothesis, and the activity of the modulated cells is interpreted in relation to that task. The data are further skewed by what could be called a "selection bias." The analysis of neural activity is typically confined to the subclass of neurons that show an interpretable relation to the task. Those cells that best support the hypothesis are illustrated in the figures, and those that are statistically consistent are tabulated. However, in such studies two other groups of cells are invariably encountered: Many neurons are modulated with the task, but in complex ways that seem paradoxical or uninterpretable. In addition, many more neurons are simply unmodulated with the task. Although the latter two classes outnumber the interpretable task-modulated group, they are typically ignored. The paradoxical cells are rarely illustrated in papers, since they would detract from the main hypothesis, provoking the reviewers and confusing the readers. Instead, the uninterpretable and unmodulated cells are usually neglected in the final account of neural coding. Of course, many cells would in fact be marginally relevant to the ongoing task; however, by ignoring all the neurons with complex patterns we risk misunderstanding the real neural mechanisms in favor of dealing only with idealized and simplistic correlates.

This strong selection bias clearly undermines the contention that the functionally interpretable group of task-related cells provides convincing evidence for coding. Observations consistent with a given hypothesis can always be selected from a sufficiently large random data set. A rarely acknowledged fact of life in the neurophysiology laboratory is that neurons in many regions, including the motor cortex, exhibit an enormous variety of responses, a fact that provides an opportunity to find cells related to any given functional hypothesis. Thach (1978) eliminated this bias by objectively correlating the activity of the same population with three different, dissociable parameters. His finding that cells related to all three functions existed in nearly equal numbers suggests that something else may be going on besides preferential coding of particular movement parameters.

2.3. Localization of function. A common notion that is closely related to coding and also turns out to be simplistic in retrospect is the idea that different cortical areas are devoted to computing different motor functions. For example, it is commonly thought that the role of the motor cortex is to execute movement whereas motor association areas, such as the premotor and supplementary motor areas, are supposed to be concerned with motor programming or preparation to move under particular circumstances. Experiments designed to record neural activity in these regions under the appropriate behavioral conditions did indeed discover cells with the appropriate sorts of relationships. However, experiments in which neural activity in different regions was obtained under similar behavioral conditions have revealed that cells of the same type are found widely distributed over many areas. For example, neurons related to activation of muscles are found not only in the motor cortex but in the supplementary motor area (Chen et al. 1991; Crutcher & Alexander 1990c), premotor cortex (Godschalk et al. 1985; Weinrich & Wise 1982), prefrontal cortex (Fuster 1985; Niki & Watanabe 1976), and posterior parietal cortex (Mountcastle et al. 1975). Similarly, experiments involving delayed movements reveal set-related activity in the motor cortex (Tanji & Evarts 1976), premotor areas (Godschalk et al. 1985; Weinrich & Wise 1982), and prefrontal cortex (Fuster 1985; Niki & Watanabe 1976), as well as in the thalamus (Alexander & Fuster 1973) and basal ganglia (Alexander & Crutcher 1990c). Taken to-

gether, these studies suggest a very broadly distributed representation of these motor functions.

An extreme example of cortical specialization, which still remains almost axiomatic, is the presumed dichotomy between the functions of somatosensory and motor cortex. In this view, all precentral cells are thought to be involved in the execution of movement, whereas all postcentral cortex cells are assumed to be involved in somatosensory function. This view was challenged by Woolsey (1958), who noted that the maps of sensory input and motor output are similar and overlapping, in both precentral and postcentral gyri. Chronic unit recordings under active and passive conditions show that cells with similar response types can be found in both areas (e.g., Fetz et al. 1980; Soso & Fetz 1980). If this functional dichotomy is considered to be absolute rather than relative, identical response properties of single units must be interpreted in totally different functional terms. The responses of postcentral cells to passive stimulation are naturally interpreted as subserving somatic sensation, but the equally clear responses of precentral cells to passive joint movement and cutaneous stimulation are thought to subserve unconscious reflex functions. Similarly, the early responses of precentral cells preceding active limb movement are naturally thought to be involved in generating the movement, whereas identical early responses in postcentral cells are interpreted as subserving some sensory "corollary discharge."

The rationale for these diverse interpretations of identical response properties rests on functional presumptions derived in part from the effects of cortical stimulation. Stimulation thresholds for evoking movements are clearly lower in precentral than postcentral cortex (Woolsey 1958). And in conscious humans, cortical stimulation evokes somatic sensations from a larger number of postcentral than precentral sites (Penfield & Boldrey 1937); however, these differences are a matter of degree rather than absolute. In fact, similar effects can be evoked from both gyri, albeit at different thresholds. Nevertheless, the conceptual dichotomy between "sensory" and "motor" cortex is again preserved by applying a double standard to this experimental evidence. The somatic sensations evoked by stimulating precentral "motor" cortex in conscious humans are ascribed to a spread of activity to postcentral sites. The movements evoked by stimulating postcentral cortex are similarly ascribed to mediation via precentral cortex; reports that movements can be evoked from postcentral sites after ablating precentral cortex (Woolsey 1958) are even taken as evidence that the lesions were incomplete.

The assumed functional dichotomy of sensory and motor cortex is further based on their differing output projections. In the macaque the corticospinal axons from postcentral cortex terminate more dorsally in the spinal cord than axons of the precentral PTNs, although there is a good deal of overlap (Coulter & Jones 1977). The postcentral PTNs are undoubtedly more likely than precentral PTNs to affect afferently driven spinal cells, but their target region also contains cells involved in reflex circuitry, as well as dendrites of motoneurons. Perhaps more relevant to the function of individual cortical neurons than the output projections of the descending cells are the strong interconnections between pre- and postcentral cortex. These massive corticocortical connections allow the cells in each region to participate in the functions of the other; indeed, these reciprocal connections would explain the similar response properties of neurons found in these areas.

Thus, the notion that cortical functions are segregated into different cortical areas can be preserved only by imposing different interpretations on similar experimental evidence. Units with the same response properties are imagined to code either sensory or motor parameters, depending on the presumed function of their recording sites. A plausible alternative is to consider the similar response properties of cells in different cortical regions as evidence that they are involved in performing similar functions; the neural substrate for these functions is then distributed correspondingly. This means that a given cortical region would be involved in diverse functions, consistent with the diverse cell types observed. This view provides a basis for distributed interactions between the functionally related sets of cells and helps explain the recovery of function after lesions. Note that this view does not claim equal involvement of all regions in all functions, since cortical areas are undoubtedly specialized. The point is that a region's specialized function need not be its only function, and certainly should not be the only standard for interpreting what each of the cells in this region is coding.

2.4. Population coding. Investigators have recently found that the activities of populations of cells can provide functions that match movement parameters more closely than the firing pattern of any single neuron. The fact that movements are ultimately produced by activity in large ensembles of neurons provides a clear rationale for population coding. Humphrey et al. (1970) first showed that the activities of multiple motor cortex neurons could be added together in the right proportion to match different parameters of wrist movement in an isotonic task. Weighted sums of the cells' firing rates could match the force trajectories and the wrist displacements, as well as their temporal derivatives, if the weighting factors for each cell could be optimally chosen for each parameter. Moreover, the match between the cells' weighted activities and the mechanical parameters improved with the number of cells included. The ability to freely optimize the weighting coefficients, of course, helped to ensure convergence on the movement trajectories; closer matches were obtained with larger populations because each additional nonredundant cell could serve to further reduce the remaining difference.

More recently, Georgopoulos et al. (1984) showed that populations of motor cortex cells could be used to match the direction of limb movement by invoking the "vector hypothesis" to sum the activity of directionally tuned cells. For a given movement direction, each cell was assumed to make a vector contribution pointing in the direction of its maximal activity, and by an amount proportional to the change in its overall mean rate during the given movement. The vector sum of all the unit vectors then approximated the direction of hand displacement. Again, the match improved as more cells with diverse directional preferences were included. This match with movement direction could be taken to suggest that the direction of hand displacement by the arm rather than muscle force is coded in motor cortex populations. The

direct match between the population function and arm displacement is appealing because it conveniently avoids the intervening complexities of synaptic connections and limb mechanics, which present formidable obstacles to a causal explanation. Moreover, the vector hypothesis is virtually guaranteed to work, given a sufficient distribution of cells. For a particular movement the cells whose best direction coincides with the movement will make the largest direct contribution; cells whose vectors point in the opposite direction will make a negative vector contribution, since average rates are subtracted, and therefore also contribute positively to the movement direction. The other cells have off-axis vector components that would tend to cancel with a sufficiently large population. Thus, the vector hypothesis will produce a match with movement direction whether the directionally "tuned" cells have any output effects on muscles or not. The same sorts of matches have been demonstrated for populations of posterior parietal area 5 neurons (Kalaska et al. 1983) as well as cerebellar cortical and nuclear cells (Fortier et al. 1989) and globus pallidus (Turner 1991).

Mussa-Ivaldi (1988) showed that the findings of Georgopoulos et al. (1984) would also result from a population of cells that code muscle shortening, and thus reconciled the apparent coding of limb displacement with the fact that many precentral cells do have effects on muscles. Recent studies by Kalaska et al. (1989) have shown that when the required force is varied independently of movement direction, the population vector of certain motor cortex cells shifts in the direction of active force. This result is consistent with a role in activating the agonist muscles. However, there are other motor cortex neurons whose population vector remains in the direction of movement, independent of force, much like posterior parietal cells (Kalaska et al. 1983). In this case, a key ingredient in making the matches with force or displacement is the ability to select the appropriate cells for each population.

Although one can find good descriptive matches between functions derived from the activity of neuronal populations and particular movement parameters, this correspondence is no proof of neural coding in the causal sense. To demonstrate that the candidate cells actually make a causal contribution requires additional evidence that they have appropriate output effects. Such a direct link is obviously difficult, and often impossible, for many central neurons. Still, a coding theory based merely on a descriptive match with a parameter provides no further basis for dealing with the neural interactions that would mediate the control of that parameter. A useful coding theory should provide some framework for understanding how the observed activity could contribute to the movement. For example, it would be helpful to know how the activity of the population whose "vector" points in the direction of movement is actually transformed into the movement. Descriptive correlates alone do not provide a causal framework for dealing with the underlying neural computation.

2.5. The coding problem. In retrospect, experiments designed to demonstrate coding of movement parameters have provided data that can be interpreted in either of two ways. Proponents of neural coding can point to the slight differences in mean onset latencies in different regions as evidence of a sequential hierarchy of cells; they can point to examples of covariation of neural discharge with movement parameters as evidence of coding and they can ignore the complex and unrelated cells as unlikely to be involved; finally, they can consider different proportions of cell types in different areas as evidence of functional segregation. Alternatively, one could now argue that the accumulating experimental results have largely undermined these simplistic notions. The extensive overlap of activation times in different regions speaks more for parallel than for serial activation. Neural correlates of movement parameters in a particular task can always be selected from what is invariably a much larger variety of response types, but cells with more complex discharge could be just as involved in generating the movement, albeit in more complex ways. The distribution of similar cell types over diverse cortical fields speaks more for distributed representation than for functional segregation.

These issues cannot be resolved by more chronic unit recording data, because observing the activity of single and even multiple units is inherently insufficient to determine the mechanisms that generate movements. These studies usually lack another essential ingredient required to make causal inferences, namely, the connectivity between cells. In addition to the activation patterns generated during task performance, one must also know the output connections of the recorded cells in order to determine the consequences of that activity. The possible output connections are often inferred from independent anatomical evidence on major projections. However, such indirect inference is misleading for many neurons, since the cells encountered at a given recording site typically have diverse projections. If the cells' response properties are correlated with their projections, the functional distinctions described above could have been blurred by lumping them all together.

3. Response coding in premotoneuronal cells

To determine whether the variety of relationships observed in previous studies could be reduced by dealing with cells that directly affect motoneurons, some investigators have focused on those cells that have correlational linkages to motoneurons, as determined by spike-triggered averaging. These premotoneuronal (PreM) cells include the so-called corticomotoneuronal (CM) cells in precentral motor cortex (Buys et al. 1986; Cheney & Fetz 1980; Fetz and Cheney 1980; Lemon et al. 1986), the rubromotoneuronal (RM) cells in red nucleus (Cheney et al. 1988; Mewes 1988), and peripheral afferent fibers recorded in cervical dorsal root ganglia (DRG) (Flament et al. 1992). These PreM cells all produce short-latency postspike facilitation of EMG activity and have been documented in relation to comparable ramp-and-hold wrist movements – a response designed to elucidate the relation of cellular activity to changes in force and sustained force. It is interesting to consider the properties of PreM cells in the context of the four issues discussed above with regard to neural coding.

3.1. Timing. Although previous studies had shown that unidentified motor cortex cells exhibit a broad range of onset times relative to movement, it seemed possible that

CM cells would show a more restricted range of recruitment times relative to onset of activity in their target muscles. This turned out to be only partly true, as shown in Figure 1. The surprising result was that CM cells began to fire up to several hundred milliseconds before the onset of activity in their target muscles. Similar broad distributions of onset latencies have been observed for RM cells (Mewes 1988) and for afferent fibers in DRG (Flament et al. 1992). Since these PreM cells produce postspike facilitation of their target muscles in about 10 msec, the much earlier onset times are presumably related to bringing the motoneurons to threshold; those cells with reciprocal inhibitory linkages to antagonists of their target muscles would also contribute to turning the antagonist muscles off.

The inescapable conclusion is that even connected PreM-motoneuronal pairs are recruited relative to each other over times that straddle hundreds of milliseconds. Thus, connectivity is not a critical factor in restricting relative recruitment times; instead, there may be other relevant variables (if indeed there are any systematic explanations). Even within the same motoneuron pool, motoneurons are recruited in sequential order over extended periods of time. One variable that may be more relevant to recruitment order than the spatial location or the output connections of a neuron is its relative size. An increasing synaptic drive on motoneurons recruits the smallest motoneurons first and then successively larger ones with higher thresholds. Similar size relations may explain the timing of early and late recruited cells in the PreM population, a subject for future investigation. In any case, the distribution of onset times of PreM cells relative to their target muscles is almost two orders of magnitude broader than the latency of their postspike effects.

3.2. Coding of muscle force. The activity of PreM cells clearly has a direct output effect in facilitating their target muscles, which in turn generate active force. In this sense the PreM cells can be said to causally affect force. The relation of CM and RM cell discharge to active force has been confirmed by having monkeys generate different levels of isometric force. The tonic discharge rates of these cells during the static hold period are indeed proportional to active force over a range of torques, as shown in Figure 3 (Cheney & Fetz 1980; Cheney et al. 1985; Mewes 1988). In addition, many of these PreM cells show a phasic discharge at the onset of movement, which is preferentially related to change of force.

These observations pertain to the major subsets of the PreM cells, namely, those that show phasic-tonic or tonic discharge patterns during the ramp-and-hold movement. Figure 4 illustrates the basic response patterns of the three groups of PreM cells and single motor units during the ramp-and-hold movements. Of these patterns, only the tonic pattern is strictly proportional to the ramp-and-hold force trajectory. Other PreM cells show patterns that differ significantly from the active force and from the activation patterns of their target muscles. For example, the phasic-ramp CM cells show a strong burst of discharge at the onset of movement and a gradually increasing discharge during the static hold period. This pattern is totally different from the discharge of its target muscles.

There are also many PreM cells that show more com-

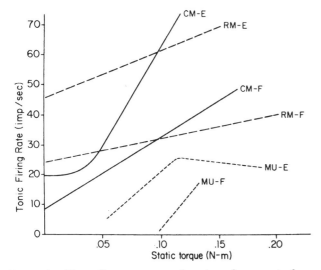

Figure 3. Tonic firing rate as a function of net static force about the wrist for representative CM and RM cells and single motor units. Rates are shown for flexor- and extensor-related cells (F and E, respectively; from Fetz et al. 1989).

plex and even counterintuitive response patterns. Many RM cells and some DRG cells exhibit bidirectional responses, firing during both flexion and extension, even though they facilitate only one set of agonist muscles. In motor cortex, some cells appear to have quite paradoxical relations to muscles: They covary with muscles in which they produce postspike suppression (see Figure 5 in Cheney et al. 1985).

Still another remarkable class of PreM cells discovered in the red nucleus are the unmodulated RM cells (Cheney et al. 1988; Mewes 1988). These increase their activity during wrist movement but are not modulated with alternating flexion and extension movements. Thus, they represent cells that essentially bias their target muscles during movement in both directions. These unmodulated cells, one should note, *are* causally involved in the active movement, as confirmed by their postspike facilitation of agonist muscles.

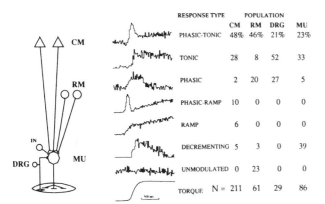

RESPONSE TYPE	POPULATION			
	CM	RM	DRG	MU
PHASIC-TONIC	48%	46%	21%	23%
TONIC	28	8	52	33
PHASIC	2	20	27	5
PHASIC-RAMP	10	0	0	0
RAMP	6	0	0	0
DECREMENTING	5	3	0	39
UNMODULATED	0	23	0	0
TORQUE	N = 211	61	29	86

Figure 4. Response patterns of PreM cells and motor units during generation of ramp-and-hold wrist force (middle records). The proportion of units showing each pattern is tabulated for CM cells (from Cheney & Fetz 1980; Fetz et al. 1989), RM cells (from Mewes 1988), DRG cells producing postspike facilitation (from Flament et al. 1992), and single motor units (from Palmer & Fetz 1985).

Thus, PreM cells show three different relationships to their target muscles (as well as to force): Some PreM cells are simply coactivated with target muscles, others exhibit more complex and counterintuitive patterns, and still others are unmodulated. Despite these various relationships in discharge patterns, the correlational evidence confirms that they are all causally involved in activating their target muscles.

Furthermore, just as central cells can change their discharge patterns in relation to motor parameters under different conditions, single CM cells also can fire differently relative to their target muscles and force for different types of movements. When a monkey performed a finely controlled ramp-and-hold tracking task, the CM cells were strongly modulated with their agonist target muscles. When the monkey made rapidly alternating ballistic movements, however, the same cells were relatively inactive, even though their target muscles were more strongly activated (Cheney & Fetz 1980). Similarly, Muir and Lemon (1983) found that CM cells were preferentially active during a precision grip of a force transducer between thumb and forefinger, but the same CM cells were paradoxically inactive during a power grip, which involved even more intense activity in their target muscles. These results indicate an unexpected variability in the relation between even PreM cells and their target muscles under different movement conditions.

3.3. Localization of function. To determine whether supraspinal cells in motor cortex and red nucleus may have functional specializations that are different from those of afferent cells providing feedback from the periphery, one can compare the CM and RM populations with PreM cells recorded in the DRG. Surprisingly, the response patterns of the DRG units fell into the same categories as the most common supraspinal cells (tonic, phasic-tonic, and phasic; see Figure 4). Moreover, the relative onset times of many afferent cells also preceded the onset of their target muscle activity. This suggests that many PreM cells in radically different locations are recruited in similar ways. A similar result was obtained by Schieber and Thach (1985): During a slow ramp-and-hold tracking task they found similar classes of cells in DRG, motor cortex, and cerebellum.

In addition to PreM cells with similar response properties in all three locations, the supraspinal populations each included some unique types. As indicated in Figure 4, the ramp cells were observed only among cortical cells and the unmodulated neurons were found only in the RM group. This suggests that the three groups of PreM cells are not entirely equivalent but contain subsets of cells whose unique properties suggest some functional distinctions.

3.4. Population coding. Although individual PreM cells exhibited a variety of distinct discharge patterns, the net contribution of all the PreM cells to a target motoneuron would be more relevant to assessing their total effect. The response patterns of the PreM cells can be synthesized into a population average (Fetz et al. 1989). Since the cells were recorded under similar behavioral conditions, the average activity of the population can be obtained by summing the response histograms of individual cells (as well as their target muscles) aligned with the onsets of the movements. This was done in stages, by first compiling subaverages for each response type and then adding these in proportion to the number of cells of each type. The resulting net ensemble averages of the discharge patterns of both the CM and RM population exhibited a phasic-tonic pattern. However, the motor cortex population showed a greater difference in the depth of modulation between opposite directions of active wrist force than the rubral cells, whose population histogram showed tonic activity during both directions of movement.

The net synaptic drive of the PreM cells on their target motoneurons would be proportional to these population histograms. The population histograms could also be used to infer the quantitative effect of the cells on their target muscles; the population activity can be multiplied by the correlational consequences of the postsynaptic potentials evoked from cortex and red nucleus (Fetz et al. 1989). The results provide a causal picture of the population influence on target motoneurons that is based on physiological measures of the synaptic linkages.

It is interesting to note that coding of muscle force in motor units requires a population average. Under normal conditions, single motor units code net muscle force in a highly nonlinear manner, since motor unit firing rates are limited at the lower end by their recruitment threshold and at the upper end by saturation (Figure 3 and Palmer & Fetz 1985). Moreover, the net force generated by the twitch tensions of a motor unit is a nonlinear function of its firing rate. This nonlinear behavior of the individual motor units is resolved by the population sum, which includes the successive recruitment of motor units with larger twitch tensions.

3.5. Implications of PreM cell properties for neural coding.
The properties of PreM cells have significant implications for the coding issue, insofar as their activity is causally related to generating muscle force, but this activity comes in a remarkable variety of discharge patterns. The connectivity of PreM cells to motoneurons is confirmed by cross-correlation methods, yet the response patterns of these PreM cells include all three types of relation observed between central cell activity and movement parameters. Many PreM cells clearly covary with the muscles that they facilitate, as one would intuitively expect. Others, such as the phasic-ramp CM cells and the bidirectionally activated RM cells, show counterintuitive discharge patterns that are distinctly different from the activity of their facilitated target muscles. Moreover, some cortical cells are paradoxically coactivated with arm muscles that they inhibit. In addition, a large group of unmodulated RM cells is tonically active during both phases of movement. This would indicate that the response patterns of neurons alone are not a reliable guide to their causal role in the task and that neural interactions between connected cells involve some highly nonlinear relationships. If the activities of connected PreM neurons and their target motoneurons can show such diverse relations, the chance of finding meaningful correlates of movement parameters would seem even more remote.

The same considerations apply to the relation of PreM cell discharge and the mechanical parameter of force, on which they clearly have a causal effect. Only the activity of the tonic PreM cells is directly proportional to force in this task. Indeed, the entire population of cortical and rubral PreM cells exhibits a net phasic-tonic pattern, suggesting that force is coded in a nonlinear way even in the output cells that generate this force.

4. Computation of movement in neural networks

4.1. Holographic coding mechanisms.
The basic reason that movement parameters need not be explicitly coded in the activity of single neurons is that movement is the consequence of large populations of interacting cells, which can generate an output without requiring any one cell to fire in proportion to the resultant movement parameters. Instead, the activity that is appropriate for a given cell is determined largely by its connections with the rest of the network rather than by any need to code an output parameter explicitly. This point can be illustrated by an apt analogy: the storage of images by holographic mechanisms. Holographic storage is based on a distributed representation of phase relations between wavefronts rather than a literal representation of the stored image. Recall that a holographic plate is constructed by exposing a photographic plate to the interference patterns between two coherent light beams – a reference beam obtained directly from the coherent source and an object beam reflected from the object whose image is to be stored. The spots on a holographic plate record the points of constructive interference, where the two light beams are in phase. These spots are distributed in a pattern that has no recognizable relation to the image. However, when the plate is illuminated with the reference beam, this distribution of spots forms a diffraction grating that reconstructs the object wavefront by the interference patterns in the transmitted beams.

The idea that neural networks may store and process information through holographic mechanisms is based on many salient analogies between the two systems. A small lesion in a holographic plate does not destroy any specific portion of the image but rather degrades the overall image quality; similarly, small lesions in the nervous system typically produce subtle behavioral deficits at most. The association between images of two spatially adjacent objects can be readily demonstrated by creating a hologram from the light reflected from the two objects; illuminating the developed plate with the light reflected from only one of the objects will reproduce a ghost image of the missing one. In this case the light from the remaining object essentially acts as the reference beam for reconstructing the other. This mechanism provides an analogue of associative memory and a model of content-addressable memory (Hinton & Anderson 1981; Pribram et al. 1974). Such a mechanism is likely to be involved in perceptual processes such as figure completion. The ability to execute a skilled movement sequence in a particular context may well involve similar associations between changing sensory inputs and central programs.

The basis for these analogous properties is the distributed representation of the information, using constructive interference between activity propagated in parallel pathways. Activity in a neural network is also propagated by the coincident arrival of sufficient synaptic input to activate the relaying neurons. With regard to coding mechanisms, the relevant point is that the spots on the hologram do not form a literal pictorial representation of the image; instead, the "meaning" of each spot depends on its relation to the rest of the hologram – each point diffracts light in such a way that the net interaction with adjacent points reconstructs the wavefront of the stored image. Similarly, in the nervous system, the activity of a cell need not form a literal representation of a movement parameter; instead, its contribution to movement depends on its diverse connections and interaction with the rest of the network.

Optical holograms clearly represent highly simplified examples of this type of distributed nonliteral coding mechanism, insofar as they store only static images. Nevertheless, the same principles apply to storage and retrieval of dynamic information in neural networks (see papers in Hinton & Anderson 1981). Such analogies between neural and holographic mechanisms have been largely speculative until now. With the advent of neural network modeling, it has become possible to demonstrate these same properties in simulated populations of cells.

4.2. Neural network models. Model networks can be used to simulate the mechanisms operating in populations of cells; they also have unique heuristic value in elucidating the principles of neural computation. The behavior of ensembles of neurons is difficult if not impossible to synthesize by "bottom-up" inferences from single-unit recordings alone, mainly because the relevant connections between the recorded neurons are typically unknown. In contrast, model networks that simulate a particular behavior can be obtained by "top-down" derivations based on examples of the behavior, using training algorithms such as back-propagated error correction (Rumelhart et al. 1986b) or trial-and-error learning (Kuperstein 1988). The resulting dynamic networks can simulate motor activity without explicitly representing movement parameters in the activity of particular units.

For example, to determine what sort of neural network might be able to transform the step change in target position that a monkey sees into the response patterns generated by his agonist motor units, we used back propagation to derive the appropriate dynamic recurrent networks (Fetz & Shupe 1990; Fetz et al. 1990). The input and output layers were connected to intervening excitatory and inhibitory hidden units, as shown by the schematic diagram in Figure 5; an example of a specific weight matrix is shown in Figure 6. Initially, the synaptic weights were assigned randomly; presenting the input produced an initial output that deviated drastically from the desired target output. The difference between the actual output and the desired output was used to change the weights appropriately to reduce the error between actual and target outputs. Successive training iterations produced a network that transformed the temporal input patterns (step and transient inputs for flexion and extension) to the desired output patterns (eight motor unit patterns: tonic, phasic-tonic, decrementing, and phasic, for both flexion and extension). One resulting network is shown in Figure 6; the size of each square represents the strength of connection from the unit identified at the left to the unit at the top. The activation patterns for a flexion-extension cycle are illustrated for each unit. This result represents a complete neural network solution for this simplified sensorimotor transform in that the activity patterns of the

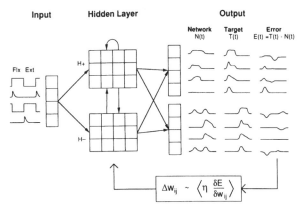

Figure 5. Schematic of dynamic recurrent network that simulates the step tracking task. Input functions represent step and transient signals for flexion and extension; target output patterns represent the discharge patterns of flexor and extensor motor units. The types of recurrent connections implemented are shown schematically. To derive a network that performs the transform, the output activations of an incompletely trained network [N(t)] are subtracted from the target outputs [T(t)] and the weights between units [w_{ij}] are modified to reduce this error.

sorimotor transform in that the activity patterns of the intervening hidden cells, as well as their connectivity, are completely specified.

These network solutions can be analyzed systematically to determine how the output patterns are derived. Relevant to the issue of "coding" we can examine how the response patterns of the output units are represented in the activity of the hidden units. For example, to see how the network in Figure 6 produced activity of the phasic flexor output unit (fp), one can examine its synaptic inputs (represented by the vertical column of weights under fp). The strongest weights indicate that phasic flexion was derived by two different means. As proponents of explicit coding might predict, the phasic output cell had strong excitatory connections from phasically active hidden units (e.g., a1, a8). A second contribution, however, came from excitatory units with tonic activity (a11), in conjunction with a delayed tonic input from inhibitory hidden units (b1). The difference between these two also contributes to the phasic output. Yet a third mechanism has been observed in other network simulations, which were allowed to have tonic biases on the cells. In those cases, the phasic output could also be derived from the sum of excitatory input from a phasic-tonic hidden unit in combination with a negative bias that essentially subtracted the tonic component. Thus, a pertinent lesson from these simulations is that many combinations of hidden unit activity can and do contribute to the same output response pattern.

It is interesting to note that many properties of the hidden units in these networks are analogous to those found in cells in the nervous system. For example, a given hidden unit (e.g., a11) may have divergent excitatory connections to many different types of output cells, just as CM cells facilitate motor units of different response types. Conversely, a given output unit typically receives convergent input from many hidden units, with different activations. Nevertheless, the connections are not equally distributed; this simulation produced preferen-

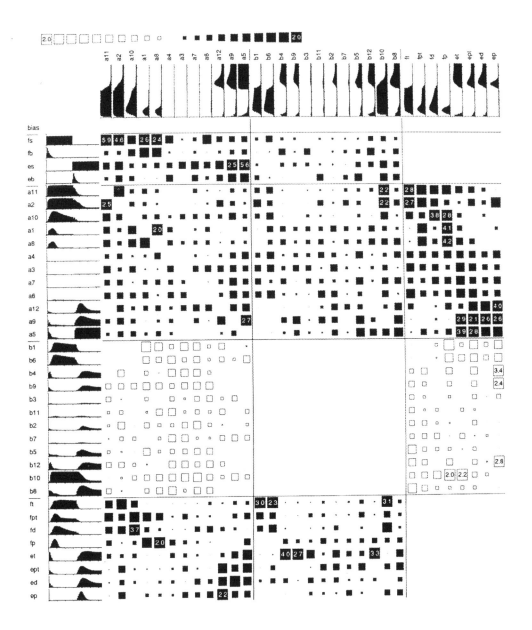

Figure 6.   Neural network transforming step and transient inputs to firing patterns of motor units at output. Unit activations during a flexion-extension cycle are shown at left and along the top. The weight matrix gives the strength of connections from the row to the column units (weight scale at top). The rows represent, from top to bottom, the bias (which was eliminated for this simulation), the inputs (fs to eb), the excitatory hidden units (a's), inhibitory hidden units (b's), and the output flexor and extensor motor units (ft to ep). The target output patterns for both flexion and extension are tonic (ft & et), phasic-tonic (fpt & ept), decrementing (fd & ed), and phasic (fp & ep). To better visualize the relationships between units, the hidden units were sorted in order of the strength of their contribution to the phasic-tonic output units (from Fetz & Shupe 1990).

tially strong connections within the sets of units with sustained and transient activity. For example, the first two flexor hidden units exhibiting tonic activity (a11 & a2) are strongly interconnected; they receive potent input from the flexion step and connect strongly to the tonic output unit. Similarly, the brief flexion input (fb) is most strongly connected to the phasic hidden units (a1 & a8), which are strongly interconnected and which also have strong reciprocal connections with the phasic flexion output unit (fp). Although units with similar response patterns tend to be more strongly interconnected, there are also significant connections between units with quite dissimilar responses; this is even more pronounced in other simulations.

Relevant to the coding issue, one significant result of these simulations is the demonstration that a large number of network solutions can produce the same transform. Starting a given network architecture with different initial weights will usually produce solutions with a different set of final weights and activations. Even within the same network, one can discern a variety of solutions: In Figure 6, the flexion and extension phases involve essentially identical inputs and outputs, but the network utilizes different types of hidden units devoted to each. There is every reason to suspect that biological nervous systems can also utilize a variety of stratagems to perform a given behavior. The variance of experimental data from one animal to the next may be due to the fact that different animals could develop different neural computations to generate the same behavior. In recording data from multiple animals, experimenters typically assume that each animal performs the task using the same network solution; indeed, this assumption is a prerequisite for pooling data recorded from different animals. Network simulations suggest that many different neural network solutions can mediate the same behavior.

In fact, modeling experiments indicate that even in the same network, several different strategies for generating a response pattern can be implemented simultaneously, as discussed above for the phasic output. In this light, the commonly observed variance in a neuron's discharge pattern from trial to trial could well represent a variance in the degree to which different solutions are implemented in each trial. Thus, pooling data from different animals, and perhaps even from different trials in the same animal, would yield some average hybrid of different specific solutions. The common criticism that neural network models do not provide the same solution as biological networks is predicated on the debatable assumption that there is only one "real" biological solution (and, moreover, that it can be found by present experimental methods).

The coding issues that have been discussed above in relation to single-unit recording studies can be reexamined in light of the neural network simulations. In many dynamic network solutions, the *relative timing* of onsets of hidden unit activations can be widely distributed with regard to a given output response (Fetz et al. 1990). This is true even for hidden units that contact the output unit directly. This staggered timing is related to the build-up of recurrent activity in interconnected units; it does not represent sequential, hierarchical stages of processing.

The *response patterns* of hidden units that contribute to the output also show all three types of relation to the output. Some hidden units simply covary with the output unit that they excite, as might be expected intuitively (e.g., a1 and fp). Many other units show activity patterns that differ significantly from their target outputs (e.g., a11 & a10 compared with fp). In some simulations of reciprocal movement, many hidden units have bidirectional responses during both flexion and extension; the inappropriate portion of their activity is simply eliminated by inhibitory units. In addition, the inhibitory hidden units frequently show counterintuitive coactivation with cells they inhibit (e.g., b10), as has been seen in some cortical neurons. Finally, the tonic bias units used by many of these networks are clearly analogous to the unmodulated activity seen in many RM cells. Thus, the activity of output units is not necessarily coded recognizably in the activity of hidden units, even those that provide direct input. Not only do the network simulations reveal all three types of relations between hidden units and output units, but experiments with network lesions (Fetz & Shupe 1990) confirm that each type makes a significant contribution to the output.

The issue of *localization of function* can also be seen in a new light with network simulations. The functional consequence of the activity of any particular hidden unit is determined by its connectivity in the network; its physical location would be entirely arbitrary. Thus, if the hidden units were physically implemented, they could be reorganized in space without affecting the network computation, so long as their connectivity remained intact.

Relevant to functional localization, a common property of representation in cortical fields is the tendency to form topographic maps of the peripheral receptors or muscles. This feature of cortical organization has been simulated in neural network models; using local Hebbian rules to change synaptic strengths will lead to topologically organized feature maps (Kohonen 1982). This type of topographical organization within a cortical field should be distinguished from the segregation of functional computation among different fields. As demonstrated by network models, topographic organization can result from local synaptic interactions; in biological networks this may also have some wiring convenience. In contrast, functional segregation in the form of explicit separation of computational stages does not appear in network simulations.

The issue of *population coding* is also illuminated by these network simulations. The response pattern of any particular output unit is simply derived from the computed sum of all its inputs, weighted by the connection strengths. There is no need for explicit coding of any other sort. One could imagine taking the activity of a population of hidden units and matching some movement parameter by an appropriately weighted sum of their activities. Despite the success of such a mathematical exercise, the weights that are actually significant for the neural calculations are the synaptic links between units, not the mathematical coefficients required to calculate an optimal match. Put another way, the ability to obtain a population function that matches a parameter is quite irrelevant to the neural mechanisms that generate the output.

Clearly, these initial network simulations are still too simplistic in their connectivity and cell properties to be taken as realistic models; nevertheless, they serve to illustrate some of the mechanisms at work in large populations of units interacting in ways analogous to neuronal interaction. Thus, network models provide a useful heuristic tool for investigating network mechanisms and can help to bridge the impasse between single-unit data and behavior. In the future, these network simulations can be improved to provide more realistic models of biological networks by incorporating the activity of more cells recorded in behavioral experiments and by making the connections more appropriate.

5. Concluding comments

We have taken the devil's advocate position on the notion that movement parameters are explicitly "coded" in neural activity. If "coding" is defined simply as covariation with movement parameters, the nervous system will provide ample opportunities to search out cells whose activity correlates with this or that parameter. Given the variety of neural discharge patterns and the ability to select the best examples, one can anticipate further examples of cells that could code some hypothesized variable. Like reading tea leaves, this approach can be used to create an impression, by projecting conceptual schemes onto suggestive patterns. This selective approach ignores two major groups of neurons: those with a complex or paradoxical relation to the task and those that are not modulated. It seems significant that studies in which both the activity and the connectivity of the same neurons are

known – namely, physiological studies using spike-triggered averaging and modeling studies with neural network simulations – reveal that all three classes of units can and do contribute significantly to the output. Thus, the search for explicit coding may actually be misleading, and may divert our understanding of distributed neural mechanisms that operate without literal representations.

If virtually any neuron can potentially contribute to generation of movement, how can we ever hope to understand the underlying mechanisms? Ultimately, systems neurophysiologists can profitably use a combination of single-unit recording techniques and neural modeling to investigate the network mechanisms generating motor behavior. Unit recordings can provide important constraints on the activity of related neurons, but the network models can provide working examples of complete solutions to sensorimotor behavior. To the extent that models can incorporate anatomical and physiological constraints, they can provide plausible explanations of the mechanisms of neural computation.

ACKNOWLEDGMENTS
I thank Paul Cordo for organizing the "Controversies in Neuroscience: I" meeting, which provided a unique opportunity to present provocative perspectives, and the NSF for supporting this meeting. I thank Kate Elias for editorial help and various colleagues for helpful suggestions. This work was supported in part by NIH grants RR00166 and NS12542 and by ONR contract N00018-89-J-1240.

8

The representation of egocentric space in the posterior parietal cortex

J. F. Stein

University Laboratory of Physiology, Oxford University, Oxford OX1 3PT, England
Electronic mail: *stein@vax.oxford.ac.uk*

Abstract: The posterior parietal cortex (PPC) is the most likely site where egocentric spatial relationships are represented in the brain. PPC cells receive visual, auditory, somaesthetic, and vestibular sensory inputs; oculomotor, head, limb, and body motor signals; and strong motivational projections from the limbic system. Their discharge increases not only when an animal moves towards a sensory target, but also when it directs its attention to it. PPC lesions have the opposite effect: sensory inattention and neglect. The PPC does not seem to contain a "map" of the location of objects in space but a distributed neural network for transforming one set of sensory vectors into other sensory reference frames or into various motor coordinate systems. Which set of transformation rules is used probably depends on attention, which selectively enhances the synapses needed for making a particular sensory comparison or aiming a particular movement.

Keywords: attention; connectionism; coordinate transformations; distributed processing; egocentric space; localization; movement; neglect; neural nets; posterior parietal cortex; space perception; topographic maps

We tend to take it for granted that there is a representation of the location of objects around us with respect to ourselves which is common to all our senses, a map of "real" perceptual space, situated somewhere in our brains, since that is what our consciousness presents us with (for further discussion see Morgan 1977). Indeed, there must be some representation of this kind, otherwise we would not be able to transfer so effortlessly between retinal, stereoscopic, auditory, somaesthetic, oculomotor, and limb movement spaces. But, as Arbib (1991) points out, this representation does not have to be an explicit topographical map, like the retinotopic maps found in occipital cortex, as many people seem to assume. In fact, very little evidence has been found for the existence of such a topographic map of perceptual space. [See also Leiblich & Arbib: "Multiple Representations of Space Underlying Behavior" *BBS* 5(4) 1982.]

The hypothesis that real space might be mapped topographically in the brain is suggested not only by introspection, but also because a map like this seems to overcome the difficulties of transferring information between sensory and motor systems. These difficulties fall into three main categories. First, the primary cortical representations or maps of sensory inputs and motor outputs are spatially distorted: The central few degrees of the retina are devoted to two-thirds of striate cortex; the lips and finger pulp are over-represented in the somaesthetic cortex; movements of the fingers receive much more cortical area than those of the spine, and so forth. These distortions make good sense within each modality because they allow a high density of local processing (Barlow 1980; Fox 1988). This arrangement yields

such benefits as high acuity at the fovea or sensitive touch and highly flexible movements of the fingers. But the distortions lead to a nonuniform representation of positional information in primary sensory cortex which makes linear transfer of signals from eye to hand or from finger pulp to arm and finger muscles more difficult, though not impossible.

The second problem is that receptive surfaces are always being moved. So our stable perception of visual space is actually constructed from a series of retinal snapshots by incorporating information about the direction in which the eyes were pointing when each snapshot was taken. Similarly, somaesthetic space is a reconstruction of patterns of cutaneous stimuli integrated according to motor and proprioceptive signals that indicate how the fingers were moving when objects or their contours were encountered. Thus the "reafferent" sensory inputs provided by these receptors must be continuously reinterpreted in the light of how the body part in which these receptors reside is being moved.

The third difficulty facing successful sensory/sensory and sensorimotor transformations is that the different primary sensory and motor maps all use different coordinate systems. The retinal map is centred on the optical axes of each eye; the cutaneous map is a map of the skin; the oculomotor axes pass through the geometric centre of the eye; the coordinate system of each arm is probably centred on each shoulder, and so on. How are signals arriving in one coordinate system transformed into another?

One way these three kinds of difficulty could be resolved would be to transform all the coordinate frames into a common, uniform, axis system to generate a topographical map of egocentric "real" space somewhere in the brain. This would establish correspondence between them all, linearising their different magnifications and compensating for the movements of the sensory surfaces. As we have seen, such a map is an economical way of organising local interactions, such as those necessary for defining lines or edges in the visual system in order to identify objects (Barlow 1980; Fox 1988). What is needed to build up a representation of egocentric space, however, is to locate objects accurately with respect to the observer. For this purpose, comparisons over long distances, rather than local ones, are required. Hence there would probably be no particular advantage to generating a topographical map of real space in the brain, because such a map would only emphasise local relationships. Moreover, it would have the disadvantage that, having laboriously translated all coordinate systems into a common reference frame for motor control or for comparison between the senses, the common coordinate system would then have to be transformed into that of the recipient system all over again. Thus, it might make more sense not to bother with a common "real" space map at all.

The problems described earlier nevertheless do require solutions if sensory signals are to be successfully interconverted or used to control movement. I shall argue that what is required is not an explicit topographical map of real space but a distributed system of rules for information processing that can be used to transform signals from one coordinate system into another.

The brain structure that carries out these transformations should be a multimodal sensorimotor and motivational association area, a region where many different sensory inputs converge and where the over-representation of parts of the sensory surfaces (the fovea, tips of fingers, etc.) in the primary receiving areas can be demagnified. This area should also be informed about the movements of the eyes and the limbs so as to interpret the sense data that the eyes and limbs feed back during movement, thereby building up a stable picture of the world. It should also receive strong limbic connections so as to be informed about the subject's intentions and motives for initiating movements.

A number of regions have been suggested to fill this role. I shall briefly consider the following five, concentrating mainly on the fifth: (1) the superior colliculus, particularly its multimodal deeper layers, (2) the hippocampus, (3) the cingulate cortex, (4) the prefrontal cortex, and (5) the posterior parietal cortex (PPC). I believe that only the PPC meets all the requirements outlined above.

The superior colliculus is certainly important for reflex orienting and directing eye movements (Sparks 1991a), but there is no evidence that it plays any part in the control of voluntary limb movements, nor that it plays a major role in the initiation of voluntary eye or head movements. Its superficial layers contain cells that signal retinal error and its deeper layers contain cells that signal motor error (Mays & Sparks 1980; Wurtz & Mohler 1976) but there seem to be no neurons in the superior colliculus that might signal target location with respect to the observer by combining retinal and eye position signals in a systematic way, whereas there are many such neurons in the PPC.

It has also been suggested that the hippocampus con-

tains a map of real space, but there is no evidence that this is a topographical one. Moreover, it seems probable that the function of the hippocampus is to encode spatial or other relations between one external object and another (Gaffan & Harrison 1989; O'Keefe 1978), rather than between an external object and the observer (or manipulator). Thus, the hippocampus appears to be more concerned with memorising the allocentric relations between objects independently of any observer, rather than with determining the egocentric locations of these objects with respect to the observer. So it may not be a good place to look for a representation of egocentric space. [See also BBS multiple book review of O'Keefe & Nadel's "The Hippocampus as a Cognitive Map" BBS 2(4) 1979.]

The cingulate gyrus receives input from the posterior parietal cortex and projects back to the PPC (Cavada & Goldman-Rakic 1989b; Goldman-Rakic 1988; Mesulam 1981). This gyrus is a limbic structure thought to be concerned with motivation, with major inputs from the hypothalamus via the anterior thalamic nucleus. Another important connection is with the frontal lobe. The cingulate gyrus probably supplies motivational inputs to both parietal and prefrontal lobes, thereby affecting the direction of attention, but it lacks the wide variety of sensory inputs and appropriate neuronal responses (present, by contrast, in PPC) to be considered a likely candidate for the cerebral representation of egocentric space.

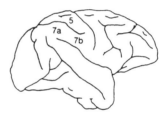

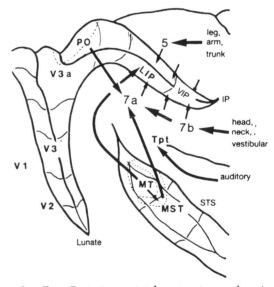

Figure 1. Top: Posterior parietal cortex in monkey (areas numbered). Bottom: Lunate, superior temporal (STS) and intraparietal (IP) sulci spread out. Areas 5 and 7 converge on the lateral (LIP) and ventral intraparietal (VIP) areas. Tpt: temporoparietal area; PO: Von Bonin and Bailey notation middle temporal area (MT); MST: middle superior temporal area.

The prefrontal cortex (PFC) is another region which plays an important role in spatial function, but its neuronal responses are tied to movements or the preparations for movement (Goldberg & Bruce 1990; Niki & Watanabe 1976), and most of these do not seem to occur unless a movement does. Thus, one can consider the prefrontal area more as a motor than a sensory region (even though, as Fetz in his accompanying target article in this issue points out, black and white distinctions between sensory and motor areas are oversimplified). Moreover, the PFC is still to some extent the servant of the PPC; for this reason it is unlikely to play the major role in the perceptual representation of space.

The posterior parietal cortex is the most likely place to find an egocentric map that processes sensory and motor information. Anatomically speaking, the PPC is a multimodal sensorimotor association area supplied with highly appropriate inputs and outputs. Moreover, it has functional properties that make it likely that the representation of egocentric space begins there (Andersen 1987; 1989; Hyvarinen 1982; Paillard 1991; Stein 1989b; 1991a). I shall accordingly now give a brief summary of some neuroanatomical, lesion, and recording results from the PPC to support the view that this region contains a neural network for translating between different coordinate systems by directing attention and that it thereby generates a perceptual representation of space. What follows is not intended to be a comprehensive review of these topics, however (for recent reviews see Andersen 1987; 1989; Goldberg & Colby 1989; Hyvarinen 1982; Stein 1989b; 1991a).

1. Connections of the PPC

MacDonald Critchley referred to the PPC as the "parieto-temporo-occipital crossroads" (Critchley 1953), but even this description does not do justice to the complexity of its connections. More than 100 inputs and outputs of areas 5 and 7 in the monkey have been identified. I will not attempt to describe them all here. I merely wish to emphasise that visual, somaesthetic, proprioceptive, auditory, vestibular, oculomotor, limb motor, and motivational signals can potentially all combine with one another. Hence, the PPC contains the necessary anatomical connections to begin to build up a representation of space that is shared by all the motor and sensory systems.

The evolution of the human brain has been associated with massive enlargement of the PPC, particularly the inferior parietal lobule areas 39 and 40. The PPC in humans is the area lying between the postcentral sulcus anteriorly, the subparietal sulcus on the medial aspect of the hemisphere, the parieto-occipital sulcus posteriorly, and the posterior and lateral part of the superior temporal sulcus laterally. It is divided anatomically into the superior and inferior parietal lobules. In monkeys, the superior lobule consists of Brodman's areas 5a and 5b, but in humans it probably contains both areas 5 and 7 (von Bonin & Bailey 1947).

Area 7 in monkeys can be divided into a number of subregions (Pandya & Seltzer 1982). The intraparietal sulcus separates the superior from the inferior parietal lobule. This sulcus contains several histologically distinguishable areas, of which the lateral intraparietal area

(LIP) has been the most studied; the inferior parietal lobule itself is divided into areas 7a posteromedially, 7b anterolaterally, and 7m situated on the medial surface of the inferior lobule. In man, the interior parietal lobe comprises the supramarginal gyrus (Brodman's area 39) and the angular gyrus (area 40). In addition, the motion-sensitive areas of the superior temporal sulcus (MT and MST) are functionally related to the posterior parietal lobe. The general principle underlying the anatomical relations of the PPC seems to be that it receives from and returns reciprocal projections to all the second-order sensory cortices, as well as motor and limbic structures (Andersen 1987; Hyvarinen 1982).

1.1. Area 5. In the monkey, the main cortical inputs to area 5, the superior parietal lobule, come from somaesthetic areas S1 and S2, that is, Brodman's areas 1, 2, and 3 in front (Andersen et al. 1990a; Cavada & Goldman-Rakic 1989a; 1989b; Jones & Powell 1969; Pandya & Seltzer 1982). The ipsilateral motor and premotor cortical areas 4 and 6 also project to area 5, as does contralateral area 5, via the corpus callosum. The vestibular cortex, lateral area 2, also feeds into area 5 (Cavada & Goldman-Rakic 1989a; Pandya & Seltzer 1982), as well as receiving a strong projection from the cingulate gyrus. The main subcortical input is from the pulvinar and posterior group of thalamic nuclei (Asanuma et al. 1985).

The main efferent projections from area 5 are to the premotor area (lateral area 6), the supplementary motor area (medial area 6), and area 7b posteriorly (Andersen et al. 1990a). Subcortically, area 5 projects to the thalamus, caudate nucleus, and putamen, the tectum, the pontine nuclei, and thence to the cerebellar hemispheres (Brodal 1978; Glickstein 1990; Glickstein et al. 1980).

1.2. Area 7. In the monkey, the main cortical inputs to area 7 come from the somaesthetic, auditory, visual, and limbic systems. Area 7 receives inputs from ipsilateral somaesthetic cortex and from area 5 in front, and from areas 18 and 19 behind (Andersen et al. 1990a; Baizer et al. 1991; Cavada & Goldman-Rakic 1989a; 1989b; Pandya & Kuypers 1969). Laterally, auditory areas 22 and the planum temporale project into area 7 (Divac et al. 1977). There is a large projection from the posterior cingulate gyrus, which is part of the limbic system (Cavada & Goldman-Rakic 1989a). Although the somaesthetic projection may be crudely somatotopic and the visual projection has some residual retinotopic order, there is no evidence of tonotopicity in the auditory projection.

Area 7 has been subdivided into area 7a posteriorly, 7b anteriorly, and the lateral intraparietal area (LIP) on the lateral lip of the intraparietal fissure. Areas 7a and LIP in the monkey are the regions that first receive visual inputs. The dorsal "where" stream of occipitofugal information about the motion and direction of visual targets (Ungerleider & Mishkin 1982) is directed towards these areas, which can therefore be considered the highest centres for visual localisation. Areas 7a and LIP also receive retinal, oculomotor, and neck position information from the superior colliculus by way of the pulvinar (Asanuma et al. 1985).

Area 7b receives somatosensory information about touch, vibration, joint angles, and position from the postcentral gyrus areas 1, 2, and 3 via area 5 (Andersen et al. 1990a). It also receives a few visual inputs from area 7a (Andersen et al. 1990a; Hyvarinen & Poranen 1974). The anatomical principle here seems to be that the visual route through area 7a and LIP and the somatosensory route through area 5 converge on 7b, although the visual projection to 7b seems quite small (Andersen 1987; Andersen et al. 1990a; Cavada & Goldman-Rakic 1989a). The main outputs of LIP and area 7a are to the frontal eye fields (area 8) in the prefrontal cortex and to the cingulate gyrus (Barbas & Mesulam 1985; Cavada & Goldman-Rakic 1989a; 1989b; Pandya & Kuypers 1969). Area 7b projects mainly to supplementary motor cortex and premotor cortex (Cavada & Goldman-Rakic 1989b). Subcortically, all parts of area 7 project to the pulvinar, basal ganglia, and superior colliculus, with a very large projection to the cerebellum, which relays in the pontine nuclei (Glickstein et al. 1980).

In summary, area 5 reciprocally connects somaesthetic with limbic and motor structures, whereas area 7 reciprocally connects somaesthetic, visual, and auditory areas with limbic and motor structures. Thus, the PPC exhibits the requisite connections suggested earlier for combining retinal, somaesthetic, proprioceptive, vestibular, and auditory signals together with information about movements of the eyes, neck, trunk, and limbs. It probably also receives motivational signals from the limbic system. In return, it projects back to the limbic, sensory, and motor areas feeding it. Thus, on anatomical grounds we can consider the PPC a "multimodal sensorimotor association area." It has the appropriate connections for mapping real space, redirecting attention, and guiding movements of the eyes, neck, trunk, and limbs.

2. PPC neuronal responses

Physiological recordings of the responses of single neurons in the PPC help us fill in the details of how these converging connections may help mediate sensorimotor transformations. Such data can show whether these connections interact at the level of single cells, as well as how they do so. Such recordings must be made in awake animals trained to perform tasks that reveal the functions of the PPC. The findings have made one thing very clear: Neuronal responses in the PPC are exceedingly complex! As expected, the activity of PPC neurons is related to many influences: multimodal sensory inputs, a wide variety of different motor outputs, and states of intention and attention. I shall summarise just a few points which are particularly relevant to the representation of egocentric space.

2.1. Area 5. In area 5, the sensory properties of neurons are an order of magnitude more complex than those recorded in S1, the primary somaesthetic area. The discharge of 80% of area 5 neurons is related to joint position, but these neurons are best activated when the animal actively moves its joints in a natural sequence or when appropriate areas of skin are stimulated in the same way as during normal movement (Hyvarinen & Poranen 1974). Bilateral interactions are common; for example, flexion of one hip can inhibit the normal excitatory responses to flexion of the other. Many area 5 neurons are "matching," in the sense that they are maximally stimu-

lated by bringing together two appropriate body parts, for example, rubbing the palms together with the shoulders adducted, or brushing the fur on the left forearm with the right hand.

A most important observation is that even though many PPC neurons can be activated by passive sensory stimulation their discharge is markedly enhanced if the stimulation is "reafferent," that is, caused by the animal's own movement. Thus, PPC neurons must receive signals from both sensory and motor systems. Indeed, some cells in area 5 fail to respond at all to passive cutaneous stimulation but discharge briskly when the animal actively reaches out for an object. Mountcastle called neurons with these characteristics "arm projection units" (Mountcastle et al. 1975).

2.2. Area 7. The great majority of neurons in area 7a and LIP are visually sensitive (Motter & Mountcastle 1981; Mountcastle et al. 1975; Steinmetz et al. 1987). These neurons have very large receptive fields and often respond weakly to stationary stimuli, even to those with no particular significance to the animal. Their receptive fields often cover the entire contralateral hemifield and they expand over the midline well into the ipsilateral field. Often the receptive fields have a region of insensitivity around the fovea itself ("foveal sparing"). If the animal makes an active eye or arm movement towards the stimulus, however, or even if it merely shifts its attention towards the stimulus, the discharge of area 7 neurons is greatly enhanced (Bushnell et al. 1981). Because these neurons have such large receptive fields, there is no obvious retinotopic mapping in their organisation (Andersen et al. 1990a).

An interesting property of area 7 neurons is that, although they do not encode stimulus speed, they are often sensitive to movement in a particular direction, either away from (centrifugal) or, more commonly, towards (centripetal) the midline. This is called "radial opponent vector organisation." The combined output of a population of such neurons could readily encode the true direction of a stimulus (Steinmetz et al. 1987), as well as the intended direction of a limb movement. It has been suggested that these neurons also play an important role in tracking the movement of targets in the visual periphery; as a population, they could also signal the optical flow pattern made by the motion of elements in the whole visual scene, as experienced by an observer moving through the environment.

A most important additional characteristic of about 30% of the light sensitive neurons in areas 7a and LIP is their responsiveness to movements of the eyes and to the position of the eye in the orbit (Andersen et al. 1985b; Andersen et al. 1990b; Andersen & Mountcastle 1983). The discharge rate of these neurons modulates even when the animal moves its eyes around in a darkened room. Andersen and his colleagues (1985b) showed that when these neurons are stimulated at the optimum site ("hot spot") in their receptive fields, discharge rates become maximal when the eyes are in particular position, falling off for all other eye positions, even when the stimulus is kept at the same retinal locus, (i.e., when the visual stimulus is moved by the same amount and in the same direction as the eye movement). In general, the best combination of retinal and eye position sites is that which would place the visual cell's receptive field over the centre of gaze. Such an association of retinal locus with eye position, rather than defining the position of the target with respect to the retina or the observer's head, defines the vector of the saccade that would cause a target in the receptive field to fall on the fovea. In other words, the discharge appears to specify the retinal vector of the saccade that would be required to acquire (foveate) the target. Many neurons in LIP discharge just before a saccade in which this eye movement would foveate the target; this appears to be the case even if, in a two-step paradigm, the stimulus never actually falls on their receptive field, that is, even if it is a memorised or "virtual" target (Barash et al. 1991). LIP neurons seem to operate in a retinotopic, oculomotor frame of reference rather than one that is head-centred or egocentric. There is no evidence that the anatomical distribution of these neurons bears any topographic relation to the outside world.

The eye position signals carried by LIP and 7a neurons are probably derived both from corollary discharges arising in oculomotor centres and from feedback provided by the stretch receptors in the ocular muscles. Many of these neurons respond in relation to saccadic eye movements in the dark, independently of any retinal stimulation. They are selective for the direction of the saccade, but not for its amplitude. Some of these responses (e.g., in area 7a) are postsaccadic, so part of their signal related to eye position might be provided by stretch-receptor feedback from the extraocular muscles. Such cells could not be said to control eye movements, but merely to reflect saccades which have taken place. Others, however, particularly those in LIP, discharge well before saccades begin (Andersen et al. 1990b; Gnadt & Andersen 1988; Lynch 1980; Lynch et al. 1977). Again, these neurons do not need an explicit visual target; they seem to code for the intended direction of an eye movement, as they also discharge during the interval that a monkey has been trained to wait before making an eye movement to a remembered target position. This discharge without movement might correspond to the shifting of the animal's attention to where the target had been, prior to moving its eyes there.

Typically, the responses of these eye movement neurons in PPC to visual stimuli are particularly enhanced when the stimulus is behaviourally relevant (Mountcastle et al. 1975). In most oculomotor regions, such an enhancement occurs only if overt eye movements or limb movements are actually made; but, as mentioned above, in the PPC it is sufficient for the animal merely to direct his attention covertly to the stimulus. Mountcastle and his colleagues (1981) tested the light sensitivity of such neurons under three different attentional conditions. In none did the animal actually move his eyes, yet only in the condition when the monkey's attention was explicitly directed at the target was the light sensitivity of the neuron enhanced.

Bushnell et al. (1981) examined the role of directed attention in these neurons in greater detail by training monkeys to fixate on a central light while attending to the dimming of a peripheral one. The monkeys were trained not to turn their eyes to look at the peripheral target. Bushnell et al. found that, if the peripheral light was in the large visual receptive field of the PPC neuron, its discharge was increased when the monkey was made to attend to the target even though it made no eye movement towards it. Often this enhancement was larger than the discharge preceding an overt saccade, because making a saccade inhibited the cell. Likewise, if the animal attended to the target but then made the wrong response, the discharge still increased. These effects of directing attention contrast with effects on neurons in the superior colliculus or the frontal eye fields, where enhancement only occurs if the animal actually shifts his gaze to the target.

Another expected characteristic of a system for representing real space is demagnification of the over-representation of areas of high receptor density such as the fovea or the tips of the fingers in order to produce a more uniform representation of the dimensions of space. One sign of this demagnification in parietal neurons may be the foveal sparing Mountcastle noted in many neurons with large visual receptive fields. The receptive fields of PPC neurons are very large, so there is little sign of foveal magnification in the representation. Indeed, foveal sparing is the rule. For one of the largest outputs of the PPC – to the cerebellum via the pontine nucleus – there is evidence that the over-representation of the fovea may be "demagnified," as we might expect of a system designed to guide movements uniformly in all parts of extrapersonal space (Glickstein 1990). However, such demagnification is by no means essential. It would be perfectly feasible, though more complicated, for coordinate transformation algorithms to compensate for the expansion of certain areas of sensory space.

In summary, neuronal recording experiments have added greatly to our understanding of the PPC. Area 5 neurons are related to somaesthesia, proprioception, and active limb movements, and these neurons help to mediate "active touch." Area 7 neurons are related to the association of visual, auditory, eye, and limb movements for the visual guidance of movement, and to "active sight." All PPC neurons therefore seem to share two revealing characteristics: They receive combinations of sensory, motivational, and related motor inputs, and their discharge is enhanced when the animal attends to or makes a movement towards a target. Thus, PPC neurons are well suited to transform the requisite information for converting sensory input into signals for directing attention and guiding motor output.

3. Event-related potentials in the PPC

Event-related potential correlates of the activity of neurons in the PPC have been found in humans. A prominent negative wave occurs 100–200 msec after a visual event. The amplitude of this potential is greatest not at the occipital pole but over the PPC, and it is bigger on the contralateral side (Hillyard et al. 1985). Moreover, this potential is greatly enhanced when the subject attends to a particular location within the visual field, rather than when attention is directed elsewhere.

Not only are visual-evoked potentials maximal over the PPC, but the premotor positivity (PMP) which precedes saccadic eye movements by 100–200 msec, is also largest over the parietal cortex (Thickbrook & Mastaglia 1985). This PMP occurs some 100 msec earlier if the target of a saccade is predictable compared to when its location is unpredictable. This result suggests that the PMP in the parietal cortex is related to oculomotor planning.

When subjects direct their attention to visual targets, blood flow in the PPC also increases preferentially (Roland 1982). In particular, there is activation of the right PPC, in agreement with the large number of observations suggesting that both attentional and directional deficits are much more serious following lesions of the right rather than the left hemisphere (Heilman et al. 1985).

4. Parietal lobe lesions in man

Parietal lobe lesions in man give rise to a diversity of symptoms which seem to defy simplification. Nevertheless, these symptoms can all be explained as consequences of impairments in the ability to redirect attention. Lesions of the PPC may disrupt the normal mechanisms by which sensory arrays are transformed into motor coordinate frames. Such transformations are essential for the accurate guidance of attention and movement. This type of lesion often affects not only the patient's ability to attend to and aim at objects in the outside world, but also his ability to move his own body accurately.

4.1. Superior parietal lobule. The most prominent behavioural deficits caused by lesions of the superior parietal lobule are in complex somaesthetic judgements. These are "complex," in the sense that they involve both external stimulation of cutaneous receptors and stimulation caused by the subject's own movements, that is, "active touch."

The best known example of disordered active touch is astereognosis: the inability to recognise the shape of objects by touch alone. It is difficult to separate this symptom from the more generalised deficit of the body image that Denny-Brown and Chambers (1958) termed "amorphosynthesis": the inability to assimilate spatial impressions of the positions of one's own limbs and body to build up an accurate body image (morphosynthesis). Attaining an accurate body image requires one to associate motor signals accurately with reafferent proprioceptive and cutaneous feedback in order to identify that part of the sensory input resulting from one's own movements. This particular set of associations probably takes place in the superior parietal lobule, in both humans (Heilman et al. 1985) and monkeys (Stein 1978), because this cortical region has close anatomical connections with the primary somaesthetic and motor cortical areas.

Asomatognosia, denying the existence of part of the body, is clearly related to amorphosynthesis, although it is a more dramatic manifestation of superior parietal lobule damage. Such a denial may be so complete that the patient does not dress the left side of his body or even denies that his left arm and leg belong to him. On occasion the patient's denial of an affected limb is so extreme that he may attribute pain in the limb, which he presumably feels but cannot localise, to someone else, for example, the patient in the next bed (Critchley 1953).

4.2. Inferior parietal lobule: Visual inattention and neglect. Balint was the first to describe a disorder of visual attention following damage to the posterior parietal lobes (see Husain & Stein 1988). He observed that his patient

had difficulty looking at any object other than the one upon which he was fixating, so much so that he was considered to have paralysis of gaze. More careful investigation revealed that this patient's eye movements were in fact intact, and that his apparent paralysis was due to an inability spontaneously to notice visual objects other than the one on which he was fixating. Thus, once his attention was engaged, the patient had great difficulty in redirecting his attention voluntarily to other objects in the visual surround. Even when prompted, the patient was unable to localise targets accurately, missing them when he reached out for them. Even though Balint's patient had bilateral posterior parietal lesions, he tended to direct his attention first to objects on his right and had great difficulty in redirecting his attention to the left. The particular difficulty of disengaging attention from the right in patients with right parietal lesions has been emphasised by Posner (1986).

Left hemineglect is very commonly found following lesions of the inferior lobule of the right PPC in humans (Heilman et al. 1985) as well as monkeys (Lynch & McClaren 1989; Stein 1978). Patients omit one side of a drawing when asked to draw or copy a picture; when asked to cross out elements in an array, they miss elements on the left; when asked to bisect a horizontal line, they veer towards the right. These patients mislocalise objects on their left, particularly their distance, often bumping into them. They neglect the food on the left side of their plates, and fail to notice people on their left side, even when their visitors are talking. Careful studies have shown that these patients also neglect sounds on their left-hand side (Heilman et al. 1985).

This neglect extends to the sphere of visual imagery (Bisiach & Luzatti 1978). Bisiach's famous Milanese patients with left neglect could only recall the right-hand features of Piazza del Duomo when asked to imagine themselves standing at one end of the square; yet then when they were asked to imagine themselves standing at the other end of the square they could recall buildings on the other side which they had previously been unable to describe. Hence features on both sides of the square must have been stored in their memory, but the patients with left hemineglect could recall only those on their right from any imaginary position in the square. It is as if the patients were unable to inspect the left side of their mind's eye, just as they are unable to inspect the left side of the current visual world surrounding them.

It is clearly important to the theme of this review to discuss the reference point of the coordinate system defining the space neglected by parietal patients. Bisiach and his colleagues (1985) asked patients to explore a tactile array and noted which parts of it were omitted when the patients held their heads at different positions with respect to the body. They tested whether the boundary of neglect moved with the retinal field or whether it was anchored to head or body. Their conclusion was that their patients used at least two coordinate systems, one relating to the body axis and the other relating to the line of sight (oculomotor).

Hence, PPC patients' neglect is seldom purely retinotopic. The space they ignore does not move each time their eyes move; rather, it tends to be centred on a point passing through the centre of the body or the head – the egocentre. The left hemineglect of patients with right

PPC lesions is sometimes partially alleviated by stimulating the vestibular system on the left with caloric methods (Cappa et al. 1987). That this alleviation occurs suggests that a strong signal concerning head movements can help to recalibrate the patients' distorted representation of space.

In humans, left hemineglect is much more common than right. Hemispheric specialisation has led to the concentration of visuospatial functions in the right PPC. Since lesions of the left PPC seldom give rise to neglect, it seems likely that the right PPC duplicates partially the spatial functions of the left PPC for the right hemifield. This inference is supported by the fact that patients with right PPC lesions and left hemineglect often show some degree of inattention to targets even up to 10° in their right (ipsilateral) field. Likewise, visual stimuli in both hemifields cause desynchronisation of the EEG from the right parietal cortex, but only those delivered to the right hemifield desynchronise the left parietal EEG (Heilman et al. 1985). In monkeys, many parietal neurons have receptive fields that extend well into the ipsilateral hemifield. It seems probable, therefore, that more neurons in the right than the left PPC have bilateral receptive fields. This difference might explain how the right PPC has become specialised for the spatial direction of attention.

In contrast, the left hemisphere in humans has probably become specialised for directing attention to temporal order. This role may have been acquired later in evolutionary development (Stein 1989a; Webster 1977). Specifically human attributes such as speaking, logic, and calculation require the skill of being able to sequence events in time accurately. It is well known that these attributes are most impaired by lesions of the left hemisphere, although these functional differences between the hemisphere are by no means absolute, for both hemispheres can be used to do visuospatial and temporal sequencing. For any task, two hemispheres are better than one, and this is particularly true for functions requiring both spatial and linguistic skills such as reading and writing, which can be affected by lesions on either side (Stein 1991b).

So far, this review of the anatomical connections, the neuronal discharge characteristics, and the results of lesions in the PPC has made three main points. First, its anatomical connections and neuronal properties confirm the PPC is a multimodal sensorimotor association area with all the attributes required of a region where many different sensory and motor frames of reference can be correlated with one another.

Second, although it is tempting to suggest that egocentric space may be represented in the brain by a single topographic map in the PPC, there is no real evidence for this idea. No topographic map has ever been found in the PPC. Moreover, there is no reason to suppose that the different sensory and motor reference frames are likely to be converted into a common coordinate system. On the contrary, one of the recurring themes has been that many different frames of reference are utilised, rather than a single map of real space.

Neither anatomical nor recording nor ablation experiments have provided any evidence that a topographical map of physical space is situated in the PPC. Small lesions do not cause "space scotomata" – holes in the patient's mind's eye view of his environment (Heilman et al. 1985).

Instead, PPC patients become inaccurate in localising objects anywhere on their left sides, mildly so after small lesions, grossly so after large lesions, even to the point of neglecting that side altogether. The receptive fields of PPC neurons are extremely large, so the location of an object is unlikely to be encoded by the position of a single neuron in the PPC array. Instead, the ability to localise targets may be an emergent property, generated by the whole ensemble of neurons in the PPC. Thus, the anatomical positions of individual neurons need bear no topographical relationship to the outside world.

Third, the discharge of PPC neurons appears to be enhanced as much by redirecting attention towards an object as by actual movements of the eyes, limbs, or body towards the object; this suggests that the direction of attention may itself be the main function of the PPC. I shall argue that it is this process of redirecting attention which generates correspondences among different frames of reference and allows us to consciously localise objects with respect to ourselves and thus to plan voluntary movements towards them.

5. Coordinate transformation rules and their representations

My next task is to show how the neuroanatomical and discharge characteristics of PPC neurons, together with the effects of PPC lesions, suggest what role this structure may play in the representation of egocentric space as well as how sensorimotor information can be transferred between different coordinate systems. More specifically, how might the PPC cope with the distortions and movements of the primary sensory maps and thereby help to translate between different sensory coordinate systems or to transform sensory information into motor coordinates? By answering this question we may also be able to explain how one gets the subjective impression that real space is represented stably and without distortion in one's brain.

A key concept is that representations need not take the form of explicit topographic maps in which neighbouring points represent neighbouring points in the outside world (Arbib 1991). For example, the cerebral representation of a person's lexicon (meaning of words) is unlikely to be organised in a topographical way or even in an arbitrary alphabetic order like a dictionary. Similarly, there is no absolute requirement for a cerebral representation of points in space to be topographical. Instead, this representation might take the form of an "implicit" map, with no topographical correspondence to the outside world. For example, the representation could consist of a "lookup table" such as the gazetteer found at the back of an atlas. A gazetteer lists place names and their geographical coordinates and stores them in arbitrary alphabetic order, resulting in a random ordering of places with respect to their true topographical position.

6. Neural networks

Such a lookup table could be thought of as providing the input to algorithms that calculate how to get from one place to another. In Barto and Sutton's (1981) neural network for learning landmarks, the strengths of sixteen synapses were shown to be capable of representing an entire set of rules for specifying the vectors defining the direction and distance a robot would need to move to reach a target, starting from anywhere in a field containing four landmarks. Of the landmarks, the field, and the target, none were explicitly stored by means of any kind of map. Instead, a column of four inputs represented each of the landmarks; a row of four outputs represented the required activations of motors to turn the robot north, south, east, or west; and the sixteen synapses represented, by means of the strengths of the connections between columns and rows, the rules governing how the robot should respond to any particular set of landmarks in order to reach the target. The conversion matrix did not take on any of the characteristics of a map; rather, it came to embody a set of rules for converting sensory cues into appropriate actions.

A similar approach has been brought to bear on the PPC. Zipser and Andersen (1988) modelled the PPC as a three-layered neural network. Using the back propagation technique, they trained the network to transform the retinotopic receptive field characteristics of PPC neurons and their eye position inputs to give an output indicating the direction of a target with respect to the observer, independent of gaze direction. Not only did the hidden middle layer units in their simulation develop properties quite similar to those found in PPC recordings, but their simulation, like Barto and Sutton's model, failed to generate an explicit topographical map of retinotopic or real space. Instead, the network represented the set of rules which specified the direction of a target with respect to an observer. This set of rules was distributed over the whole network in a way that was similar to how a holograph transforms a visual scene into a distributed representation of its spatial frequency components (see also Fetz, this issue). Thus, no isomorphic correspondence emerged between the set of hidden units and locations in the outside world.

Parallel distributed processing (PDP) networks such as these are probably what we should be searching for in the PPC. PDP representations have four major advantages (Rumelhart & McClelland 1986). First, the unit of representation becomes not a single neuron but the strength of an individual synapse. Hence this arrangement increases the number of possible combinations of inputs and outputs by many orders of magnitude without requiring an increased number of connections. Second, these systems can be "trained" to recognise spatial, temporal, or logical patterns and their interrelations without having to explicitly analyse the rules relating to them. Successful training of the neural network thus causes it to implement these rules and to "represent" them. Third, the patterns can often be recalled from only partial cues. Fourth, because these algorithms are distributed across a large network rather than being contained in a small area, noise or transmission failures in a limited number of synapses cause only slight impairment in performance. This is the property so delightfully termed "graceful degradation" by the neural network community.

The PPC displays many of the properties one might expect of such a network devoted to coordinate transformations. It is a multimodal sensory motor association area. For the visual system at least, the PPC neuronal network probably reduces the distortion seen in primary

receiving areas to a uniform scaling at all positions. It receives full details of the movements of sensory surfaces so that sensory input may be interpreted in the light of these movements. There is little evidence of topographical mapping in the PPC; but, as discussed earlier, a "super map" probably offers no real advantages and would merely add an unnecessary step in sensorimotor processing. Thus, the PPC neural network meets most of the requirements we have outlined for representing space in a nontopographical way.

Zipser and Andersen's (1988) simulation of the PPC was trained to produce an egocentric representation of the location of objects in space with respect to an observer. But in the same way that a topographical map turns out to be nonessential, so conversion to a head-centred representation may be unnecessary [see also Flanders et al.: "Early Stages in a Sensorimotor Transformation" BBS 15(2) 1992]. To justify this assertion, consider the control of saccadic eye movements.

7. Saccadic eye movements

The saccadic eye movement control system is probably the best understood example of coordinate transformations, and the ways in which retinal information may be used for oculomotor control have been worked out in great detail. For example, an essential feature of Robinson's (1973; see also this issue) model of saccade generation was a stage in which the retinal coordinates of the target image were transformed into an explicit signal coding the position of the target with respect to the head. A motor error signal was then derived by subtracting the current eye position from this measure of target position in head-centred (egocentric) space. This error was then used to drive the saccadic pulse generator, which moved the eyes to the new location. However, no such signal representing the absolute location of visual targets in head-centred coordinates has ever been identified.

An alternative model of how saccades might be generated, one that does not rely upon computing a signal of target position in head-centred space, is a control system which calculates in retinal coordinates the amplitude and direction vector of the saccade which would be required to foveate the target (Goldberg & Bruce 1990). Such a vector is encoded by neurons in PPC areas LIP and 7a. The motor error signal could be calculated by subtracting from the retinotopically defined saccadic amplitude another retinally referenced signal equivalent to how far the current saccade had progressed. Thus the control signal to move the eyes could be derived by subtracting from the retinotopic vector a vector (likewise defined in retinal coordinates) that would represent the amount by which the eye had moved so far. Saccades would accordingly never need to be encoded in real space coordinates, nor would there be any necessity for transformations into and out of this system. Instead, retinal vectors would be directly transformed into oculomotor ones.

It now seems likely that saccades are initially programmed by the PPC in this second way, by direct translation of retinal into oculomotor vectors. Many area 7a neurons fire maximally before those saccades in which the saccadic vector directs the fovea towards the point in space previously occupied by that neuron's receptive field. Moreover, some LIP neurons fire before those saccades in which the saccadic vector links that neuron's receptive field (RF) to the fovea, irrespective of whether a real target is flashed in the RF or the animal merely attends to a "virtual" target situated at that retinal location (M. E. Goldberg, personal communication). Such a virtual target can be produced in double-step experiments by first stepping the target away from the RF and then, before even the first saccade in response to it is made, flashing it where the RF will be positioned after the first saccade has been completed. The target never actually flashes in the RF; it is "virtual" in the sense that it would appear in the RF if the second saccade were delayed until after the first saccade had been completed. Such an LIP neuron discharges if the resultant saccade would cause the fovea to move to the gaze position previously occupied by its receptive field. Thus the programming of saccades by these neurons is probably carried out in retinotopic rather than head-centred coordinates. Multiple eye movements would be achieved by successive additions of such vectors. There may be no need to convert these signals to head-centred coordinates, as envisaged in Robinson's (1973) or Zipser and Andersen's (1988) models.

8. Action spaces

It seems unlikely that a direct conversion from the most appropriate sensory reference frame to a motor output reference frame is confined to the oculomotor system. If we make a targeted movement with the right arm, the transformation is probably from "right arm proprioceptive space" (constructed from motor corollary discharges and feedback from muscle spindles, tendon organs, and joint receptors) to right arm muscle space centred on the shoulder (Soechting & Flanders 1989). If we aim an eye movement with one eye, the reference system is the retina of that eye (Goldberg & Bruce 1990; Ogle 1962), but if we use both eyes, the reference system probably incorporates retinal disparity information and is centred approximately on the bridge of the nose (Barbeito & Ono 1979). If we aim a head movement, the centre of the head probably becomes the reference point for the coordinate system used. Thus, there are a very large number of different action spaces in which we move, and all of these movements use different reference frames (Jeannerod 1988). Thus our conscious impression that all movements use the same egocentric coordinate system is clearly misleading.

The different reference frames used by sensory systems to communicate with each other and to control motor output may also be described in terms of "psychological spaces." "Personal" space is that occupied by our own body; its coordinates are defined mainly by the orientation of the head with respect to gravity signalled by the vestibular system (Berthoz & Grantyn 1986) together with information about the positions of the neck and the limbs. Personal space provides the data point for egocentric localisation and is probably modelled and serviced by the superior parietal lobule; but since all the subregions in the PPC are to some extent interconnected, this designation is not absolute.

"Peripersonal" space is the space immediately surrounding us within which we can reach out and touch objects. Localisation is most accurate in this region,

requiring the association of retinal foveal signals with oculomotor and limb movement information. Peripersonal space might be serviced by area 7b in the monkey and by area 39 in the human.

"Extrapersonal" space is the space beyond, about which we have only teleceptive information giving the location of objects, which is provided by our eyes and ears. Auditory, peripheral, or "ambient" retinal, oculomotor, and whole body signals must be associated for the construction of this space. Extrapersonal space is probably represented mainly by areas 7a and LIP in the monkey and by area 40 in the human.

Thus, there is a great deal of evidence to suggest that we use different sensorisensory and sensorimotor conversions for each different sense and each different movement. This conclusion, however, still fails to explain how we can convert between senses and motor systems so easily.

9. The direction of attention

It is probably the direction of attention that mediates these conversions. Different attributes of an object, such as its noise, colour, shape, texture, and position, whether defined visually, acoustically, or somaesthetically, are analysed by largely separate parallel systems in the sensory regions of the cerebral cortex. It has been suggested that we determine whether these different properties belong to the same objects or different ones by directing our attention towards them. Perhaps we do this by probing the degree of coherence of the signals (Crick 1984; Freeman & Van Dijk 1987) by generating "Gamma" (20–80 Hz) thalamocortical test oscillations. Hence, the act of directing attention towards the signals generated in different sensory areas may "bind" the properties of an object together, so that we can establish which belong to the same one. If so, then the attentional system must be able to select and activate the correct sensorisensory conversions to test for these correspondences. Hence it may not be unreasonable to speculate that, in an analogous way, it is the attentional system of the PPC that selects the required conversions from sensory signals into motor outputs. Thus, the redirection of attention in the PPC may select the sensorimotor association algorithms that are required for different movements, thereby localising targets in the correct sensory coordinate systems for later conversion to motor output if required.

The foregoing description is of course speculative, but there is circumstantial evidence to support it. First, the main function of the posterior parietal cortex in humans seems to be the direction of attention. We should probably make a distinction here between directed attention, which is probably the main function of the PPC, and generalised attention or alertness, which is probably a subcortical process. Although one normally thinks in terms of a single focus of attention, this idea assumes that there exists a topographical map upon which to focus. So the metaphor of "a mental searchlight of attention" (Crick 1984) may be somewhat misleading. The direction of attention is itself neurally distributed, so that it need not be thought of as spatially coherent. So attention can be divided; and it can also be affected by alternative subliminal influences, such as "priming" in the opposite hemifield (Posner 1986). None of these effects undermines the basic arguments presented here, however. Indeed, we might be able to use subliminal effects on attention to elucidate in greater detail how attentional mechanisms switch on particular sensorimotor conversion routines.

The idea that the direction of attention by the PPC helps to select which coordinate transformation rules will be used to help aim movements is also supported by the consequences of PPC lesions. These lesions cause frequent mislocation and misaiming of eye and limb movements, as well as neglect of contralateral space. Likewise, patients often lose their ability to navigate correctly in relation to well known landmarks. The recording experiments described earlier lead to the same conclusion. Neurones in the PPC increase their discharge when attention is directed into their receptive fields. This kind of response is consistent with the idea that these neurons are helping to mediate the direction of attention, and thus helping to perform the first stage of selecting the correct set of synapses to convert from one sensory coordinate system to another in order to locate a target. This process is the necessary antecedent to accurately aiming eye or limb movements towards that target.

10. Summary and conclusions

In this target article I have tried to show that physical space is indeed represented in the brain, but not as an explicit topographical map. The connections, neuronal properties, and effects of lesions of the PPC suggest that it is here that the representation of space commences. A deeper understanding of the signal processing operations which take place in the PPC gives us insight into how the area carries out this function. Neurons in the PPC receive somaesthetic, proprioceptive, vestibular, auditory, and visual sensory inputs together with oculomotor, head, limb, and locomotor movement information. In addition, they are supplied with strong limbic inputs. In this area, sensory stimuli may accordingly be interpreted in the light of one's intentions and current movements to locate objects with respect to the observer, thereby building up a representation of space. In particular, the sensory consequences of moving the eyes or limbs are allowed for, and the signals derived from external stimuli are calibrated in the light of those movements.

Neuronal discharges in the PPC are enhanced just as much when an animal merely attends to a stimulus as when it actually makes an eye or limb movement towards it. I argue that this process of directing attention is probably what brings different frames of reference into register and gives us our perception of their localisation with respect to ourselves. Thus, there is no necessity to postulate a specific area of brain where egocentric space is represented topographically. There is no evidence for such a region; indeed, there is no evidence that different sensorimotor reference frames need ever be converted into a common coordinate system. Instead, it seems that each system operates within its own reference frame. The retina controls eye movements in a retinotopic frame, cutaneous and proprioceptive receptors in the arm direct arm movements in a shoulder-centred coordinate system, and so on.

How are all these different frames of reference brought into correspondence? They must be interconvertible, otherwise we would be unable to direct our arms accurately towards what our eyes are fixating on, or to move our heads to look at the wasps on our arms. The PPC contains, not a map of real space, but a neural network that implements algorithms for converting one set of vectors (e.g., retinal) into another set of vectors (e.g., oculomotor or arm-centred). The set of rules selected for such a conversion is determined by how we direct our attention. This process selectively enhances only the synapses that are going to be needed for directing the next intended movement. Although performing these transformations in one area (the PPC) might, at first sight, seem a less than ideal way of doing things, this is probably more economical in terms of connectivity than the alternative, which would require every point in every sensory map to be connected to every point in a common space map and then to every point in every motor map.

It is probably the sum total of these interconversions between different frames of reference that, in the final analysis, constitutes our mental representation of space, and gives us the impression that our brains contain a map of it.

ACKNOWLEDGMENTS
My thanks to Paul Cordo for organising "Controversies in Neuroscience: I" and for editing this manuscript, and to U.S.N.S.F. for supporting the conference, and to the Wellcome Trust (U.K.) for supporting my research.

Open Peer Commentary
and Authors' Responses

Table 1. *Commentators for special motor issue*

Commentators	Bizzi et al.	Gandevia & Burke	McCrea	Robinson	Alexander et al.	Bloedel	Fetz	Stein
					Target article authors			
Adamovich, S. V.	[EB]							
Agarwal, G. C.	[EB]							
Alexander, G. E.				[DAR]			[EEF]	
Andersen, R. A. & Brotchie, P. R.				[DAR]				[JFS]
Barmack, N. H., Errico, P. & Fagerson, M.						[JRB]		
Berkinblit, M. B., Sidorova, V. Y., Smetanin, B. N. & Tkach, T. V.		[SCG]						
Beuter, A.		[SCG]						
Bischof, H. & Pinz, A. J.				[DAR]				
Borrett, D. S., Yeap, T. H. & Kwan, H. C.					[GEA]			
Bossut, D. F.				[DAR]				
Bower, J. M.						[JRB]		
Braitenberg, V. & Preissl, H.	[EB]					[JRB]		
Bridgeman, B.							[EEF]	
Bullock, D. & Contreras-Vidal, J. L.			[DAM]					
Burgess, P. R.	[EB]	[SCG]						
Burke, D.			[DAM]					
Carey, D. P. & Servos, P.								[JFS]
Cavallari, P.	[EB]		[DAM]					
Cavanagh, P. R., Simoneau, G. G. & Ulbrecht, J. S.		[SCG]						
Clark, F. J.		[SCG]						
Clarke, T. L.				[DAR]				
Colby, C. L., Duhamel, J.-R. & Goldberg, M. E.				[DAR]				[JFS]
Connolly, C. I.					[GEA]			
Cordo, P. J. & Bevan, L.	[EB]	[SCG]				[JRB]		[JFS]
Dawson, M. R. W.								[JFS]
Dean, J.	[EB]			[DAR]				
Dietz, V.		[SCG]						
Duysens, J. & Gielen, C. C. A. M.		[SCG]	[DAM]					[JFS]
Eagleson, R. & Carey, D. P.				[DAR]				
Feldman, A. G.	[EB]	[SCG]	[DAM]					
Flanders, M. & Soechting, J. F.					[GEA]			
Frolov, A. A. & Biryukova, E. V.	[EB]			[DAR]				
Fuchs, A. F., Ling, L., Kaneko, C. R. S. & Robinson, F. R.				[DAR]			[EEF]	
Fuster, J. M.					[GEA]			
Gandevia, S. C.							[EEF]	
Gilbert, P. F. C. & Yeo, C. H.						[JRB]		
Giszter, S.					[GEA]			
Gnadt, J. W.								[JFS]
Goodale, M. A. & Jakobson, L. S.								[JFS]
Gordon, A. M. & Inhoff, A. W.		[SCG]						
Gottlieb, G. L.	[EB]							
Graziano, M. S. & Gross, C. G.								[JFS]
Grobstein, P.				[DAR]	[GEA]		[EEF]	[JFS]
Gutman, S. R. & Gottlieb, G. L.	[EB]							
Hallett, M.	[EB]					[JRB]		
Hamm, T. M. & McCurdy, M. L.			[DAM]					
Hasan, Z.	[EB]							
Heuer, H.				[DAR]	[GEA]			

(continued)

Table 1. (*Continued*)

Commentators	Bizzi et al.	Gandevia & Burke	McCrea	Robinson	Alexander et al.	Bloedel	Fetz	Stein
Horak, F. B., Shupert, C. & Burleigh, A.					[GEA]	[JRB]		
Iansek, R.							[EEF]	
Ingle, D.								[JFS]
Ioffe, M. E.							[EEF]	
Ito, M.						[JRB]		
Jaeger, D.					[GEA]			
Kalaska, J. F. & Crammond, D. J.					[GEA]		[EEF]	[JFS]
Kirkwood, P. A.							[EEF]	
Kuo, A. D. & Zajac, F. E.	[EB]							
Kupfermann, I.				[DAR]				
Lacquaniti, F.	[EB]	[SCG]	[DAM]					
Lan, N. & Crago, P. E.	[EB]							
Latash, M. L.	[EB]	[SCG]	[DAM]					
Lemon, R.							[EEF]	
Levine, D. S.					[GEA]			
Loeb, G. E.	[EB]							
Lundberg, A.							[EEF]	
MacKay, W. A. & Riehle, A.				[DAR]				[JFS]
Masson, G. & Pailhous, J.	[EB]				[GEA]			
McCollum, G.			[DAM]					
Morasso, P. & Sanguineti, V.	[EB]							
Neilson, P. D. & Neilson, M. D.						[JRB]		
Nichols, T. R.	[EB]		[DAM]					
Ostry, D. J. & Flanagan, J. R.	[EB]							
Paillard, J.		[SCG]						
Phillips, J. G., Jones, D. L., Bradshaw, J. L. & Iansek, R.					[GEA]			
Pouget, A. & Sejnowski, T. J.								[JFS]
Pratt, C. A. & Macpherson, J. M.			[DAM]					
Prochazka, A.		[SCG]						
Proctor, R. W. & Franz, E. A.								[JFS]
Quinlan, P.								[JFS]
Rager, J. E.				[DAR]				
Ross, H. E.		[SCG]						
Rudomin, P.			[DAM]					
Schieppati, M.			[DAM]					
Schwarz, G. & Pouget, A.				[DAR]				
Seltzer, B.								[JFS]
Smeets, J. B. J.	[EB]							
Smith, A. M.						[JRB]		
Stein, J. F.						[JRB]		
Stein, R. B.	[EB]							
Summers, J. J.					[GEA]			
Tanji, J.							[EEF]	
Thompson, R. F.						[JRB]		
Tsuda, I.				[DAR]				
Van Gisbergen, J. A. M. & Duysens, J.								[JFS]
Van Ingen Schenau, G. J., Beek, P. J. & Bootsma, R. J.	[EB]							
Winters, J. M. & Mullins, P.	[EB]	[SCG]	[DAM]	[DAR]	[GEA]			

Open Peer Commentary

Commentaries submitted by the qualified professional readership of this journal will be considered for publication in a later issue as Continuing Commentary on these articles. Integrative overviews and syntheses are especially encouraged.

How does the nervous system control the equilibrium trajectory?

S. V. Adamovich

Institute of Information Transmission Problems, Academy of Sciences, Moscow 101447, Russia

[EB] The classical monkey limb perturbation experiments of Bizzi et al. (1982; 1984) greatly influenced motor control studies. Their findings, as well as those of Asatryan and Feldman (1965) on the unloading of the human forearm, were in good agreement with Feldman's hypothesis that the nervous system controls limb movements not by programming EMG bursts or force pulses for the limb acceleration and deceleration, but by defining a new equilibrium of the "limb/external load" system. In their target article, however, **Bizzi et al.** reject the concrete neurophysiological mechanism proposed by Feldman for how the system's equilibrium position is controlled.

1. Alpha-lambda controversy. It is suggested by **Bizzi et al.** that the λ model is a subset of the α model, because the former uses a concrete form of the relationship between joint angle, central control parameters, and the level of α activity. They also reject the necessity of a muscle invariant characteristic for an equilibrium point (EP) model and argue that "the reflex apparatus contributes in a modest way to force generation" (see, however, **Gandevia & Burke,** this issue).

The basic idea of the α model that the muscle is a tunable spring conflicts with the fact that isolated active muscle has a strong static hysteresis, dissipative properties, and responses dependent on the history of activation (see, e.g., Crago et al. 1976; Hill 1938). Moreover, many muscles under a fixed level of activation have negative stiffness (slope of length-tension curve) in a substantial part of the physiological length range (An et al. 1989). The system therefore has a high risk of instability if it is controlled by using "spring" muscle properties.

The main role of reflex activity in the λ model is not to compensate perturbations or to "modify supraspinal commands" but to provide the orderly recruitment of motoneurones as a function of muscle length from the referent point, specified by supraspinal influences. This recruitment, together with intrinsic muscle properties, creates an invariant length-tension characteristic, the "spring" that provides the system's stable equilibrium during interaction with external loads or other muscles. The recruitment causes an increase of muscle activity, stiffness and force as a function of muscle length in the whole physiological range.

To move the invariant characteristic and thus the equilibrium position, the nervous system further shifts the motoneuronal membrane potential. This results in a change (Δλ) of the threshold length for motoneuronal recruitment. The increment in membrane potential and Δλ are position independent. In contrast, the control variable in the α model (the level of α activity) is position dependent and cannot directly define the equilibrium position or virtual trajectory.

2. The timing of equilibrium point shifts. The hypothesis that the nervous system produces a gradual EP shift and controls its velocity to modify movement kinematics was originally proposed by Feldman (see Fig. 1). A similar hypothesis was proposed in the α model. In the model of Hogan (1984), the virtual trajectory of relatively fast movement has to overshoot the final position and then undershoot the actual trajectory in the deceleration phase to stop the limb. The virtual trajectory was obtained without taking into account the dependency of α on kinematic variables. For simple multijoint movements, a straight-line shift with a bell-shaped velocity profile for the EP of the limb endpoint was proposed and successfully used in a computer simulation (Flash 1987).

In the λ model it was assumed and verified by computer simulation that a ramp form of the two main central commands, reciprocal and coactivation (for definitions see Feldman 1979; 1980b), underlie single-joint movements made at fast and moderate speeds (Adamovich & Feldman 1984; Abdusamatov et al. 1987). The shift of the EP at a constant speed along a straight line directed to the target was also used to simulate EMG and kinematics of simple two-joint movements (Feldman et al. 1990).

Thus, good simulation results were obtained with a much simpler control strategy using the λ model versus the α model (see, however, Latash & Gottlieb [1991] for an alternative strategy in the λ model). The reason is that the λ model takes into account the dependence of EMG activity not only on central commands but also on movement kinematics. Alpha activity in the λ model is an increasing function of the difference between the actual muscle length, x, and the dynamic threshold, λ^*:

$$\alpha = f(x - \lambda^*) \text{ if } x > \lambda^*; \; \alpha = 0 \text{ if } x < \lambda^* \quad (1)$$

To a first approximation,

$$\lambda^* = \lambda - \mu dx/dt, \quad (2)$$

where λ is the tonic stretch-reflex threshold and $\mu dx/dt$ reflects the dynamic sensitivity of muscle spindle afferents ($\mu > 0$). Thus,

$$\alpha = f(x + \mu dx/dt - \lambda) \quad (3)$$

At the very beginning of the fast movement, when the speed of muscle shortening is low, the necessary amount of accelerative force is provided by α-activity arising from a rapid, centrally determined decrease in λ. When the speed of shortening increases, the role of the velocity dependent term $\mu dx/dt$ becomes essential and provides a "self-termination" of muscle activity – the agonist muscle becomes silent long before its actual length becomes equal to the equilibrium length.

By using this mechanism, together with the stiffness enhancement caused by co-contraction of antagonist muscles, the nervous system can drive the limb rapidly along a stable trajectory to a new equilibrium position by using equilibrium shifts of constant speed in the case of unconstrained point-to-point movements. In contrast, EP models that do not account for dynamic reflex properties have to introduce a more complex form of virtual trajectory.

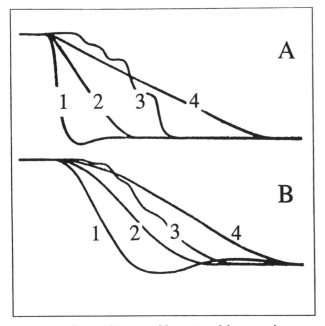

Figure 1 (Adamovich). Possible timing of the control parameter λ for the agonist muscle (A) and resulting changes in joint angle (B). (1) Fast, (2) moderate, (3) step-like, and (4) slow changes in the tonic stretch reflex threshold. (Redrawn with permission from Feldman 1979.)

The constant rate of EP shift readily allows us to scale single-joint movements in amplitude, duration, or both (Adamovich & Feldman 1989; Gottlieb et al. 1989). The rate of the EP shift has been estimated by Adamovich et al. (1984). For fast single-joint elbow movements in humans the rate is in the range of 500–700 deg/sec. Of course, in the case of constraints (accuracy required, obstacles, etc.) the form and timing of the command signal can be modified accordingly.

Because of the simple form of the command signal, the nervous system does not have to define the movement amplitude beforehand. Thus, to correct ongoing movement in response to a target shift without deflection of the movement trajectory, the nervous system can continue the EP shift or arrest it earlier. In this case, the movement kinematics will appear as if the movement to the shifted target were initiated from the very beginning (see Péllison et al. 1986).

There are many reasons to believe that the λ model is physiologically more adequate for intact motor control description than the α model.

NOTE
1. Address for correspondence: S. V. Adamovich, Centre de Recherche, Institut de Réadaptation de Montréal, 6300 Darlington, Montreal, Quebec, Canada H3S 2J4.

Movement control hypotheses: A lesson from history

Gyan C. Agarwal

Department of Electrical Engineering and Computer Science, University of Illinois at Chicago, Chicago, IL 60680
Electronic mail: agarwal@uicbert.eecs.unic.edu

[EB] Whereas the lambda model is defined by Feldman and his colleagues (Asatryan & Feldman 1965; Berkinblit et al. 1986; Feldman 1966a; 1986) is given by a set of invariant characteristics, **Bizzi et al.** have not explicitly provided a similar definition

of their alpha model in terms of some invariant muscle or joint characteristics. A working definition of the alpha model is given in their equation 1, that is, the joint torque T is a function of joint angle θ and a set of α-motoneuron activation:

$$T = \Psi(\theta, \{\alpha\}) \tag{1}$$

As defined in the target article, in the absence of any external torque, the equilibrium joint angle may be expressed as a function of $\{\alpha\}$:

$$\theta_0 = g(\{\alpha\}) \tag{2}$$

Because a subject can hold a joint position with different values of joint stiffness, it is clear that this relationship between motoneuron activations of agonist-antagonist muscle pairs and joint angle is not unique. This is clearly implied in Figure 1 of Bizzi et al. (1982, p. 140). The muscle force generated for a given activation is history dependent (Abbott & Aubert 1952) and known to have a hysteresis in the force-length response (Gottlieb & Agarwal 1978). Because of these inherent nonlinearities (see also Gielen & Houk 1984; Gottlieb & Agarwal 1988; Vaccaro et al. 1988), this relationship may not even be single valued. Bizzi et al. have implied that the lambda model is a subset (or constrained version) of the alpha model. The development and proof in their Note 3 is based on the implicit assumption of uniqueness of various relationships. For example, in their equation 4, $\{\alpha\} = \{\alpha(\theta - \lambda)\}$, for a single-joint movement only one λ value is needed and this must uniquely define a minimum of two muscle activations for agonist-antagonist pairs.

The data in Figure 1 of the target article (which was also published in Bizzi et al. 1982, Figure 3, p. 142 and Bizzi et al. 1984, Figure 3, p. 2740) are interpreted to indicate that the time course of the increase in the amplitude of the initial accelerative transient was "virtually identical" in the intact and the deafferented monkeys. The implication of virtually identical is not clear; however, there is much to be desired in this statement. For the intact animals, this figure shows data for holding time values of 100 to 300 msec and 2 data points at 400 msec. (The abscissa in Figure 1 is not time from onset of movement as shown but time of release or holding time. Unfortunately, the same labeling error was also present in Bizzi et al. 1982 and 1984). This part of the data should be compared to the similar holding time data for the deafferented animals. A second-order fit would clearly indicate a significant difference; it may even be of the order of 10 rad/sec², which is a significant difference, because the acceleration values are only in the range of 10 to 35 rad/sec².

The higher initial acceleration in intact animals is caused by higher initial torque at the time of the release. As noted by Bizzi et al. (1982), the time to peak torque in isometric condition was about 488 ± 92 msec. One possible reason for this difference is a change in the central command, which is contrary to their alpha hypothesis. The other possibility is that if the central commands were nearly equal in normal and deafferented animals, the increased torque in normal animals may be caused by feedback producing increased levels of excitability of the motoneuron pool. This is again contrary to **Bizzi et al.**'s alpha model.

Because the torque is slowly increasing to the peak value, this does not necessarily imply that the CNS had programmed a gradual shift of the equilibrium position instead of a sudden, discontinuous transition to the final position. In linear position servo systems, the time taken to approach the final position (to within some percentage of the final position) to a step input command is defined as the settling time, which is generally taken to be three times the dominant time constant of the system (Kuo 1987). This does not imply that the input command was a gradual transition to the new output state.

The value of virtual positions or equilibria that act as "centers of attractions" must be based on some acceptable values of the final position errors. On the other hand, the target article emphasizes that Bizzi et al. (1984) studied large arm movements with undemanding requirements on the final position accuracy. Does this imply that the chosen value of the virtual position may result in an arbitrarily large final error? Is the applicability of the alpha model limited to such gross movements?

The final comment in section 2 is puzzling at best. **Bizzi et al.** write that "our experiments were not intended to ascertain the relative contribution in the intact animal of feedforward commands versus feedback signals." How could one ascertain that the CNS adopted a final position control strategy in deafferented animals without visual feedback and based on movements that were gross, with undemanding requirements on final position accuracy? One can only conclude that the animal moved in the intended direction.

In considering its strengths in section 7.1, **Bizzi et al.** have noted that the equilibrium-point hypothesis is based only on the static characteristics of the muscles and their reflex connections and requires no knowledge of the dynamic parameters of the limbs (i.e., inertia and viscous coefficients, etc.). This would suggest that the motor control system either cannot or does not exert any control on the velocity of the movement. Only the final position is programmed. This is contrary to the extensive literature on speed-accuracy control in movements.

Bizzi et al. have stated that a major weakness of the equilibrium-point hypothesis is that it is difficult to test. Wilkie's (1954) word of caution for such theories is appropriate: "Misplaced confidence in a theory can effectively prevent us from seeing the facts as they really are" (p. 288). In the same article, Wilkie added: "Even more suspect should be the theory which explains nothing because it can be adapted to explain anything. Such a theory can never be disproved, which gives it an illusory strength" (p. 322).

After nearly 20 years of effort, Merton's (1953) hypothesis was abandoned. There is already ample evidence (some in the target article) to put to rest both the alpha and the lambda hypotheses and concentrate on alternative theories of control of movements. The suggested combined alpha-lambda hypothesis is not different from the already accepted alpha-gamma coactivation. In his summary of the chapter on alpha-gamma coactivation, Granit (1970) wrote: "Exclusive alpha activity in movement has been regarded as a matter of course for well over a hundred years but ignorance, to begin with, of the gamma system and later neglect of it, may have contributed to the preservation of this attitude" (p. 184). **Bizzi et al.**'s target article is contributing to that perception of neglect.

Neurophysiology of motor systems: Coming to grips with connectionism

Garrett E. Alexander

Department of Neurology, Emory University School of Medicine, Atlanta, GA 30322

Electronic mail: *gea@sunipc.neuro.emory.edu*

[EEF, DAR] In their respective target articles, **Fetz** and **Robinson** argue persuasively that insights derived from neural network modeling indicate that we are unlikely to learn how information is processed in biological motor systems through neurophysiological studies alone. They point out that conventional attempts to explain brain-based movement control in terms of the response properties of individual central neurons (considered analogous to the hidden units of artificial neural networks) are predicated on a variety of unwarranted assumptions, including: (a) individual central neurons may make substantial contributions to the generation of motor behavior; (b) in the absence of detailed information about its connections with other units of the network a neuron's ultimate influence upon motor behavior can be inferred from its response properties alone; and (c) only those neurons with simple, readily interpretable response properties need to be taken into account in explaining how a particular brain region contributes to movement control.

Arguing from what might be termed a connectionist heuristic (one that assumes that neural network – i.e., connectionist – models, crude though they may be at this early stage in their development, represent the most promising approaches yet devised for modeling biological nervous systems), both target articles provide compelling and incisive critiques of the naive belief that it is possible to learn how biological networks control movement simply by recording the movement-correlated activity of a sufficiently large sample of each network's neurons. In this respect, both offer useful and sobering reappraisals of the limitations of our conventional neurophysiological methods for studying the neural substrates of motor behavior.

Robinson's critique is moderately pessimistic about the prospects of our ever learning in detail how brains manage to control movement. He could be right. We certainly know little enough now about how signals are processed in central motor networks, even after two and a half decades of recording neuronal activity throughout the brain's motor fields. Nevertheless, a case could be made for guarded optimism about the future of systems research on the neural substrates of motor processing. It is precisely the advent of more realistic, neural-network models of the motor system that should make it possible eventually for neurophysiologists to conduct experiments that will not simply correlate neuronal activity with various motor behaviors, but will serve instead to test detailed connectionist theories of how motor networks process information.

As **Robinson** points out, the complexity of biological networks is such that it may be futile to attempt to explain how any of them processes information on a cell-by-cell basis. Were it possible, however, to devise connectionist models that solved the same problems and were structured in much the same way as their biological counterparts, and if the model networks were found after training to contain hidden units with the same types and distributions of response properties as the neurons in the networks that had been modeled, would it not be reasonable to at least hypothesize that the models and their biological counterparts were processing information in the same fashion? Such a hypothesis could be detailed and mechanistic, and would be *testable* in that any of the specific features of the model (e.g., observed changes in the types and distributions of hidden unit response properties after the network had been trained to perform a different task) could be compared with those of the biological network.

This would obviously be an indirect approach to the problem of discovering how biological networks process information. (The direct approach would be ruled out by the impracticality of characterizing all the connection strengths and activation values of the myriad elements within even the simplest of the brain's networks.) But indirect though it might be, such a strategy has the potential of permitting us to learn in some detail how information is actually processed in the brain's various networks. In fact, this approach is already beginning to yield useful insights, as is well illustrated in both **Fetz**'s and **Robinson**'s target articles. It might be argued, then, that the conclusion we should draw from Robinson's critique is not that information processing in central motor networks is inherently unanalyzable, but rather that we will need better, more realistic models to understand such processing, and that to be taken seriously, those models will need to make detailed, testable predictions about the behavior of neurons in the networks that are modeled.

The development of such models has barely begun, and thus far there have been few attempts to incorporate meaningful biological constraints. This is caused in part by a lack of information concerning some of the relevant neurobiological details (such as the strengths and distributions of local circuit connections, the electrotonic summation functions of complex postsynaptic elements, and the types and distributions of synaptic learning rules). And, as **Robinson** points out, neural network theory is also relatively immature, so that the analysis of network behavior is still a rather tedious and ad hoc affair, as yet resisting the sorts of simplifying principles that make it possible for the behavior of electronic circuits to be analyzed so thoroughly and efficiently. Nevertheless, history suggests that it would be unwise to bet against continued progress in all of these areas. It does not seem unreasonable, therefore, to presume that we may eventually arrive at a fairly detailed understanding of how information is processed in central motor circuits. Such an understanding would be predicated on the progressive, interactive development of connectionist models that would be structured in accordance with known neurobiological constraints and then rigorously tested (for their faithfulness to the real thing) by comparing the properties of their hidden units with the results of neurophysiological samplings of the brain's own "hidden unit" activity.

This general approach appears to be advocated in the target article by **Fetz** who seems reasonably optimistic about the potential for productive interaction between connectionist modeling and neurophysiological studies of the brain's motor networks. At the same time, however, he raises an important issue concerning the interpretability of single neuron recordings from central structures. One of the conclusions he draws from his analysis, were it correct, would seem to me to inspire pessimism, rather than optimism, about the possibility of our eventually learning how information is processed in central motor networks.

Fetz builds a powerful and convincing case for his overall thesis that knowing a neuron's response profile (i.e., the behavioral parameters with which its discharge is correlated) tells us nothing about what the neuron's influence might be on the periphery – unless we also know something about its connections with downstream structures. This is a lesson that neural network modelers have known from the outset, but Fetz is rightly concerned that neurophysiologists have generally failed to take this problem into account when interpreting their data. Unfortunately, in the mammalian motor system it is only near the periphery that substantial progress has been made in gauging the strengths and distributions of connections between sequential layers of neurons. It must be acknowledged, therefore, that Fetz is right in insisting that the response properties of most central neurons generally provide no indication of the contributions those neurons may make to the production of motor behavior.

Nonetheless, it does not follow from this that the response properties of central neurons are meaningless or uninterpretable. **Fetz** goes on to assert that "if the activities of connected PreM [premotor] neurons and their target motoneurons can show such diverse relations, the chance of finding meaningful correlates of movement parameters would seem even more remote." The implication here seems to be that neural correlates of movement parameters (i.e., the response profiles of central neurons) are only meaningful if they covary with the activity of the muscles with which that neuron is ultimately connected. On the contrary, the fact that a particular neuron discharges exclusively in relation to a specific movement parameter (e.g., direction of movement, irrespective of force or amplitude) seems to me to be a meaningful observation, whether or not that same neuron happens to have a net excitatory influence on the agonist muscles that generate the movement. To conclude otherwise would be to conflate a neuron's response properties with its connections to downstream structures. In the analysis of artificial neural networks it has proven essential to distinguish carefully between these two features of the hidden processing units. It would seem to be equally important to keep these two issues (response properties vs. connection strengths) separate if there is to be any hope of using neurophysiological experiments to test connectionist models.

It is a fortunate consequence of the fact that connectionist models are neurally inspired that the response profiles of central neurons can be considered directly analogous to those of the hidden units of an artificial neural network. Moreover, the responsiveness of both types of elements is usually gauged in a similar manner, by measuring the neuron's or the hidden unit's differential responses to a variety of system-level input/output conditions. This is the hallmark of the single-unit recording method, which can be very useful in this regard, provided sufficient care is taken to dissociate the relevant behavioral conditions (e.g., sensory inputs, motor outputs, behavioral sets) that are being used to define each neuron's response profile.

Even without knowing the connection strengths between functional layers of processing elements, it is possible in some multilayer networks – of both the biological and the artificial varieties – to detect distinctive changes in the types of hidden-unit response properties encountered from layer to layer (see, for example: Linsker 1990; Maunsell & Newsome 1987). The strategy of comparing hidden-unit response properties of artificial networks to those of the biological networks being modeled may well suffice therefore as a reasonable test of the modeler's accuracy. Still better, however, would be comparisons that also included some gauge of the strengths and distributions of connections between layers. Perhaps this will some day be possible, if the spike-triggered averaging technique used so effectively by **Fetz** and his colleagues to assess connections between premotoneurons and motoneurons can eventually be adapted to the study of more centrally located connections.

Spatial maps versus distributed representations

R. A. Andersen and P. R. Brotchie
Department of Brain and Cognitive Sciences, Massachusetts Institute of Technology, Cambridge, MA 02139
Electronic mail: *andersen@ai.mit.edu*

[DAR, JFS] Maps imply an orderly topography and conventional wisdom suggests that an egocentric map would be composed of cells with receptive fields that are invariant for a location in egocentric space, arranged in an orderly map of space. The target articles by **Stein** and **Robinson** make the important point that a representation of space in the brain does not have to be in this form. A prime example of a different type of topography can be seen in the posterior parietal cortex, where a distributed representation of ego-centered space may exist.

1. Gain fields. A good summary of the extensive literature on the posterior parietal cortex is provided by **Stein**; the outlines of our results from recordings in area 7a and lateral intraparietal area (LIP) (sect. 1.2, para. 3), however, does not appear to come to the correct conclusion. About two thirds (not 30%) of the visual cells in areas 7a and LIP receive a convergence of retinal and eye position signals and these signals combine in a very specific way (Andersen et al. 1985b; 1990b). For visual stimuli, when eye position is varied and the retinal position of the stimulus is held constant, the activity of the visual response is usually modulated so that the overall activity of the cell varies linearly with eye position. The same result obtains for the saccade response; for saccades of equal amplitude and direction the activity also varies linearly as a function of the initial orbital positions from which the saccades begin. We refer to the variation of the visual and saccadic responses with eye position as the gain field of the cell and we call the gain fields planar because the activity varies linearly for both horizontal and vertical eye positions. Cells in areas 7a and LIP are adding eye and retinal vectors by virtue of the gain fields; if the retinal vector varies, the activity changes, and if the eye position vector varies, the activity changes. The cell's activity does not "specify the retinal vector of the saccade that would be required to acquire (foveate) the target" because the activity would be different for the same retinal vector at different eye positions.

We further found the neural networks trained to map sensory inputs to motor outputs, which require the transformation from retinal (sensory) inputs to head-centered (motor) outputs, also combine eye and retinal position signals in this specific way. In both cases the cells in the parietal cortex and the hidden units in the neural network that perform the coordinate transformation develop planar gain fields. Thus, to state that area "LIP neurons seem to operate in a retinotopic oculomotor frame of reference rather than one that is head-centered or egocentric" misses the point that a population of cells in area LIP can represent the location of a target in head or ego-centered coordinates. In other parts of **Stein**'s target article a distributed representation of ego-centered space is discussed; we wish to make the point that the planar gain fields in areas LIP and 7a can be the mechanism for forming these representations.

2. A receptive field for space is unnecessary. The cells in the parietal cortex do not have to be an intermediate step to the formation of cells somewhere else in the brain that have Gaussian-shaped receptive fields in head or body-centered coordinates. (**Robinson** also makes this point). There is no need to converge the information contained in a population of parietal neurons onto single cells to generate spatially invariant receptive fields. As an alternative, one of our network models contained as the output a simplified set of oculomotor nuclei that coded the desired length of the extraocular muscles in a frequency of firing code (Goodman & Andersen 1989). Thus, all the signals required for coordinate transformations from sensory to motor coordinates exist in the brain, and receptive fields in head- or body-centered space are not necessary to move the eyes or limbs. The network does not require a topographic organization of the gain fields because the information is carried in the weights of their connections and is independent of the ordering of the neurons. This example does not rule out the possibility of maps of space in the cortex, for which there is now evidence from recent microstimulation studies in the ventral intraparietal area (Thier & Andersen 1991) and the dorsomedial frontal cortex (Tehovnik & Lee 1990).

3. The coordinate system for saccades. It is proposed by **Stein** that ego-centered representations of space exist in the posterior parietal cortex in a distributed form. He suggests that this representation is used for reaching movements but that it is not necessary for the programming of saccades. We would like to point out that the presaccadic responses of the posterior parietal cortex exhibit gain fields just like those of the visual responses, thus combining eye and retinal position vectors and, by definition, carrying information in head-centered coordinates. More recent experiments have shown that some cells also carry head position signals, introducing the possibility of a body-centered representation as well (Brotchie & Andersen 1991).

Stein (sect. 7, para. 2) discusses Bruce and Goldberg's (1990) scheme for making coordinate transformations for the frontal eye fields by subtracting retinal vectors. This scheme was proposed to account for the reprogramming of a second saccade when two eye movements are made to remembered visual targets. Such a mechanism would not work in the posterior parietal cortex because it does not take the eye position vectors into account. The activity of an area LIP neuron for the same retinal stimulus will be different for different eye positions. Thus the eye position signals must be subtracted as well.

The generation of a visually evoked saccadic eye movement begins with the target in retinal coordinates and ends with a command that codes the desired location of the eyes in the head in order to foveate the target. This output command is, by definition, in head-centered coordinates. In addition, many shifts in gaze position require both eye and head movements, and for large shifts the eye and head movements are programmed to occur together, requiring a body-centered encoding (Guitton & Volle 1987; Lauritis & Robinson 1986; Tomlinson 1990). In between sensory input and motor output a number of parallel processes are known to occur, and the representation of target position can be in more than one format. We can only discern how the target location is encoded by actually recording from the different sites involved in the generation of saccades. For example, physiological experiments indicate that in the superior colliculus saccadic eye movements are encoded in motor error coordinates. However, recordings from the posterior parietal cortex demonstrate strong effects of both eye and head position, indicating that the area can encode gaze shifts in head- or body-centered coordinates. How this encoding is used by the oculomotor system is not yet fully understood but it is likely to play a role in coordinate transformations by the oculomotor system as is evidenced by the effects of lesions in this area (Karnath et al. 1991).

4. Understanding distributed representations. Although we agree with **Robinson** that the brain's being a distributed system makes it more difficult to analyze, we remain convinced that the most fruitful research strategy is to combine single-cell physiology with realistic models of neural circuits. This approach is made simpler than the target article implies because the brain does in many cases use algorithms to compute information and because the brain demonstrates a high degree of functional localization.

Robinson makes the point that his neural network models for the oculomotor apparatus in the caudal pons are "over-complete." They have many more intermediate units than inputs and outputs and as a result any number of solutions is possible. In his models the signals are mixed in a seemingly random fashion in the hidden units and this haphazard arrangement makes it difficult or impossible to analyze the network by simply observing the activity of the individual units.

Robinson points to our model (Zipser & Andersen 1988) for coordinate transformations as another example of a distributed system in which signals are mixed in a haphazard fashion. Although the coordinate transformation that is performed by this network is distributed, the network performs a computation that we could subsequently analyze and understand (Goodman & Andersen 1990). The interneurons *do not* mix the retinal and eye position signals in a random fashion; rather, they are combined to form the planar gain fields. An analysis of the network shows that the planar gain fields are a clever mechanism for adding two-dimensional vectors in parallel networks (Goodman & Andersen 1990). In our case, there are far fewer units in the middle layer than in the input and output layers. The input signals are provided as high dimensional vectors, but the network learns to represent the eye and retinal signals as two-dimensional vectors in the hidden unit layer and adds them; this process is accomplished through the planar gain fields. It is interesting that we increase the number of hidden units in the middle layer, networks using Gaussian inputs no longer need to compute the coordinate transformation, but rather can use a "lookup" table to memorize the input-output mapping. In this "overcomplete" network the gain fields become nonplanar and random, making the network difficult to analyze, much like Robinson's models.

Why would some brain systems need to "generalize" or compute rather than simply memorize input-output mappings? One reason is that although the brain contains a large number of cells, it does not contain so many that it would be possible to memorize all functions (Poggio 1990).

As **Robinson** points out, the more complicated a neural network, the more difficult it is to understand. Fortunately the brain is not one large, completely interconnected network like the usual neural network models. The high degree of specialization in the brain makes it more amenable to study. This functional localization may have arisen from the fact that neural networks with a modular structure will learn more quickly and generalize better than fully interconnected networks (Rueckl et al. 1989).

Robinson concludes that "each real network may solve its own problems with its own particular tricks." It is these tricks that will be revealed by single-cell recording studies performed in concert with neural modeling and that provide new insights into brain function.

ACKNOWLEDGMENTS
We wish to thank Ning Qian and Pietro Mazzoni for their comments on the articles.

Microzones, topographic maps and cerebellar "operations"

N. H. Barmack, P. Errico and M. Fagerson
R. S. Dow Neurological Sciences Institute, Good Samaritan Hospital and Medical Center, Portland, OR 97209
Electronic mail: *barmackn@ohsu.edu*

[JRB] A central premise implicit in the title of **Bloedel**'s target article, "Functional heterogeneity with structural homogeneity: How does the cerebellum operate?" is that because the cerebellum appears to influence a variety of motor behaviors it is "functionally heterogeneous" and that because the anatomical circuitry is similar throughout the cerebellum it is "anatomically homogeneous." This is more than a half-empty, half-filled glass debate. Grafted onto this notion of morphological homogeneity, Bloedel assigns special importance to the sagittal zones that were first described by Scott (Scott 1964) using a histochemical stain for 5'-nucleotidase. The alternately light and dark sagittal banding pattern has also been observed using histochemical stains for acetylcholinesterase (Boegman et al. 1988; Marani 1981; Marani & Voogd 1977), pseudocholinesterase (Gorenstein et al. 1987), as well as monoclonal antibodies to specific cell surface proteins (Brochu et al. 1990; Hawkes & Leclerc 1987; Wassef et al. 1990). The implication of these anatomical banding studies in Bloedel's target article is that sagittal zones represent functionally discrete regions that have more significance than a mere topographic map. This may be the case, but if so, what is the added feature or abstraction that is contributed by cerebellar zones? Are there clear discontinuities in the representation of body maps across zones? Are there discontinuities in the nuclear projections of the zones that would not be predicted from topography? Are there functional correlates of cellular activity that are clearly confined to particular zones? Do zones change width as a consequence of some functional experimental manipulation?

Presently, sagittal zones appear to be nothing more than a developmental curiosity. It may be that certain zones with distinct chemical gradients are essential for the normal development of the cerebellum, but in the adult cerebellum these zones add nothing to the topographic descriptions of the cerebellum. The same criticism would apply to studies of the cerebellum in which "microzones" are described for movements that are evoked by electrical stimulation or recording (Balaban et al. 1981; Ito 1984; Sato et al. 1983; 1984). Again, one needs to know what additional information is provided by the "microzones" that is not already encompassed by a topographic map.

Why should we accept the deceptive premise that since the basic wiring diagram is similar throughout the cerebellum we should ignore details about the heterogeneous distribution of cell types, transmitters, and receptors? **Bloedel** argues that we might be able to bypass the tedious single unit recording studies that attempt to relate neuronal activity to particular external stimuli and that we should focus on understanding the "operations" of the basic circuitry. What is the advantage of this approach? What are these fundamental operations?

The cerebellum is not the only region in the brain with well-defined circuitry. Do we understand the "operations" of any other circuitry in the brain in this abstract sense: the thalamus, Renshaw cells, hippocampus, or striate cortex? How should we

validate a particular cerebellar operation if not through a refinement of the stimuli used to evoke cerebellar activity or of the measurement of responses correlated with cerebellar activity? Rather, refinements in experimental techniques should enable us to take advantage of the heterogeneity of cerebellar circuitry provided, as revealed by both transmitter and receptor specialization in different cerebellar regions. These particular examples of cerebellar heterogeneity may provide a distinct experimental opportunity to reveal how different sensory and motor commands to the cerebellum engage the basic circuitry. These "details" might also be used to advantage to reveal how the circuitry can be reversibly blocked to reveal its participation in motor performance. Transmitter and receptor heterogeneity might be used experimentally to provide better insight into how the cerebellar circuitry participates in short- and long-term modification of certain motor behaviors.

We would argue that rather than abandoning the attempts to find functional ("real world") correlates of cerebellar activity, investigators should characterize more carefully both the stimuli and motor responses they are trying to correlate with evoked cerebellar activity. If these are inadequately specified and incompletely measured, then it should not be surprising when we are confronted by observations showing the lateral cerebellar nucleus to have a function different from that of the medial cerebellar nucleus. To anyone with a knowledge of the neuroanatomy of the cerebellum as of 1950, this is a trivial conclusion. We need a more functionally precise description about how these structures are different.

To pick an analogous case in the thalamus, one could argue that all sensory relay nuclei are the same; structurally homogeneous. They are composed of principal cells and inhibitory interneurons. Therefore, this thalamic circuitry must perform the same "operations." If one were to use an ill-defined stimulus to evoke thalamic activity, one might be able to discern a functional difference between the medial geniculate nucleus and the lateral geniculate nucleus. However, the reason we have a deeper understanding of these thalamic structures is that experimental inquiries became more detailed and the stimuli used to evoke thalamic activity became better controlled and more specific.

We have known since the 1940s that Renshaw cells participate in the general "operation" of recurrent inhibition (Renshaw 1941). We know the operation, but we still do not understand the functional ramifications of recurrent inhibition for the control of movement. From our perspective, the surface appeal of understanding the neuronal operations of the cerebellum without increased specificity of experimentation has not led, and will not lead, to an increased understanding of how this structure actually contributes to sensorimotor performance.

Afferent influence on central generators and the integration of proprioceptive input with afferent input from other modalities

M. B. Berkinblit, V. Y. Sidorova, B. N. Smetanin and T. V. Tkach
Institute of Information Transmission Problems, Academy of Sciences, Moscow 101447, Russia
Electronic mail: *nskcsbiol@glas.apc.igc.org*

[SCG] To become more and more specialized is a characteristic of contemporary science. The number of facts collected in all specialties is so large that it is becoming difficult to make sense of the information. On the other hand, all investigators wish to know more about other questions, in other fields of research that may be relevant, to find new points of view on the problems in their own fields.

Gandevia & Burke's target article is very interesting and comprehensive, raising a highly specific but important question: the role of skin, muscle, and joint afferentation in the perception and control of posture and movement. The amount of data is so great that the authors could only allude briefly to problems that were "on the border" of different fields of physiology. We would like to focus further attention on two problems of this kind in this commentary:

1. The inhibitory and coordinative roles of afferentation in limb movement control, realized by the influence of central generators. **Gandevia & Burke** raise the question of human movement, but give little attention to describing research on animal movement.

There are not only many similarities between human physiology and animal movement physiology but also some distinctions between them, not because they use different methods but because they have different concepts. The "central pattern generator" is one of the main concepts in the physiology of animal movements. [See also Selverston: "Are Central Pattern Generators Understandable?" *BBS* 3(4) 1980.]

In the discussion of the target article of Berkinblit et al. (1986) in this journal, some commentators working on animal movement physiology recommended avoiding the notion of "reflex," suggesting that it was relevant only in the context of the history of biology. The notion of reflex is still popular in human movement physiology and that of a central pattern generator is not used. This distinction arises more from different scientific traditions than from differences between humans and animals. In any case, research on central pattern generators provides valuable information about the role of afferentation in the control of movements. Here are two concrete examples:

The scratch reflex of cats has two components: a postural component, when the limb is moved to the stimulated area, and a rhythmic component. It has been shown in decerebrate and curarized cats (Deliagina et al. 1975) that stimulation of an ear initiates tonic activity only; there is no rhythmic activity because the experimenter moves the limb to the right position. This shows that the afferent signal from the limb, which takes a "wrong" position, plays an inhibitory role, stopping the rhythmic movement generator.

The other example is the wiping reflex in frogs. This is a composite, multicomponent limb movement. It consists of a series of phases (flexion, placing, etc.). There are two realized motor programs during the movement: The first is "translation," that is, moving the limb to the right position in space; the other is "orientation," a complex of movements of distal joints (turning the ventral part of foot ahead, finger positions, etc.).

From this perspective, wiping reflex movements are similar to human arm movements while grasping an object. The human arm is also controlled by two motor programs: One of these controls translation and the other controls hand orientation as well as the position of the fingers.

There is an interaction between translation and orientation programs during different phases of the wiping reflex. This can be mediated by central interaction or by afferent signals. To answer the question of the role of afferentation in the process of movement coordination we conducted some experiments on the wiping reflex of *Rana temporaria*. The reflex was initiated by chemical stimulation of the skin on the back. We now describe the phase of placing and the transition to the next phase. The placing phase starts from flexion of the metatarsal-tarsal joint and fingers. Then the proximal joints begin their work. The foot is placed over the back at the end of this phase.

We examined whether there is an influence of afferent signals at the end of this phase on this transition to the next phase. Where metatarsal-tarsal joint and fingers are fixed in a straight position the placing phase usually does not start. Absence of a signal about distal joints flexion blocks the realization of the phase.

In the normal realization of the placing phase, after the limb takes its position over the back, it starts to orient its foot to put

its fingers down on the skin. When fingers are fixed in the flexion position by a special "glove," the placing phase occurs in a normal way, but then the movement stops and the foot goes down to the "sitting position." If the fingers have no possibility of straightening, the spinal brain does not receive afferent signals about the normal work of distal joints and it interrupts the movement. It is interesting that this effect is then clearer than in the more caudal stimulus position. The effect was observed on both the intact and spinal frogs.

These results are analogous to the results of scratch reflex studies: Afferent signals about a "wrong" posture block transition to the next phase of the movement. Pearson and Duysens (1976) have shown that the locomotor generator can also be blocked in an analogous way. There are different results from experiments on deafferented cats and frogs. Sherrington has shown that the scratch reflex is preserved after deafferentation.

In contrast, the wiping reflex is disorganized after deafferentation. Sometimes it assumes a position over the back, but the fingers do not straighten, a tremor starts, and the movement is interrupted. This is similar to the movement of a frog with fingers fixed into flexed position by "a glove." We have observed in other experiments that the work of different joints was not coordinated. For example, distal joints make their cycle of changes: first, flexion of both joints, then finger extension, and after that extension of the metatarsal-tarsal joint. Proximal joints, however, remain unchanged during this time. They start their work too late, after the distal joints cycle is finished.

So proprioceptive afferentation is very important for translation and orientation program coordination and for the organization of a complete movement.

2. At any one time the central nervous system reflects the body's status, its posture, the position of the limbs, and the contact between the body and external space. The mechanisms of this remain unclear but they must be very complex because natural movements involve not only fine postural adjustments and the maintenance of specific relative positions of the body segments but also the ability to appreciate body spatial orientation.

Performing natural movements such as pointing, reaching, and other human activities oriented toward a visually detected goal demands a transformation of spatial information related to the target and arm positions into neural commands that eventually move the arm to a given point in external space. The transformation must involve a processing of proprioceptive signals coding both the relative positions of the arm segments and whole body posture. This processing is unlikely to be based on direct interactions between proprioceptive reflexes controlling segmental postures. It is more reasonable to suggest the existence of some central movement organization, operating with highly integrated information from various sensory inputs that could possess much wider functional possibilities than a system based on simple reflexes. We suppose that this function can be performed by the spatial perception system (in its narrower meaning of sensory system as a whole, and central neurophysiological mechanisms responsible for perception of spatial body configuration, body position, and movement) that provides a reference for planning spatially oriented motor actions.

In humans, the fact that proprioceptive signals originating from skeletal muscles contribute to the sensory representation of posture and movement and are processed by the spatial perception system has been clearly demonstrated in experiments with the kinesthetic illusions which can be induced by applying vibration to different muscles (Popov et al. 1986; Roll et al. 1980; 1991; Smetanin et al. 1988; and some others).

Vibration applied to the distal tendons of the biceps brachii and triceps brachii muscles under isometric conditions without viewing the stimulated arm produced different perceptual and motor effects dependent on the frequency of vibration trains and the pattern of the vibratory muscular perturbations. Vibration trains of the same duration and frequency applied alternatively to the biceps and triceps evoked alternating illusory and real flexion-extension movements. Perceptional and motor effects of the arm vibration changed to the opposite ones when the subject was allowed to look at his arm (Roll et al. 1980).

Applying low amplitude mechanical vibration to either the extraocular or neck muscles (or both) of a subject looking monocularly at a small luminous target in darkness resulted in an illusory movement of the target and an appearance of constant errors when performing a pointing task (Roll et al. 1991).

These findings supported the concept that movement perception can be derived from common central processing of proprioceptive and visual afferent signals. The internal spatial representation of the target and the arm (not always an adequate one) formed by such processing can determine the final position of the arm during an aiming movement.

The role of the spatial perception system in performing spatially oriented movements can be illustrated by the effects of vibratory stimulation of muscle on the human vestibulomotor response induced by galvanic stimulation of the labyrinth in a standing subject (Popov et al. 1986; Smetanin et al. 1988). These studies were carried out on normal human subjects using a stabilograph to record the postural changes in response to transcutaneous stimulation of the right labyrinth. The results obtained suggest that illusory perceptions of a distorted position or movement of either the head or the whole body due to vibrative stimulation of muscle proprioceptors resulted in modifications of the vestibulomotor response that were quite similar to those observed under the actual postural changes produced by vibrating different muscles.

Modulation of kinesthetic information can be explored with nonlinear dynamics

Ann Beuter

Neurokinetics Laboratory, Department of Kinanthropology, University of Québec at Montréal, Montréal, Québec, Canada H3C 3P8
Electronic mail: *r11040@ugam.bitnet*

[SCG] **Gandevia & Burke**'s target article is convincing and my goal here is not to discuss the impressive evidence they have presented. Rather, I would like to comment on potential avenues for further research dealing with the interdependence between centrally generated motor commands and available feedback in movement.

Gandevia & Burke consider sensory feedback and motor commands separately. It is important to keep in mind, however, that the CNS is not a passive structure waiting for information, like a simple input/output machine. As Llinas (1990) has pointed out recently, the CNS generates its own activity, which is modulated by input from more peripheral structures. Thus, the internal loop formed by the inferior olive, Purkinje cells, and cerebellar nuclei is modulated but not activated by the outside (Llinas 1990). Gandevia & Burke clearly demonstrate the complex contribution of kinesthetic information and reflex effects during movement, but they do not consider the possible modulation of central commands by kinesthetic signals. According to the authors, the CNS has access to peripherally and centrally generated signals. Peripheral kinesthetic signals arise from intramuscular, joint, and cutaneous receptors and are necessary to adjust ongoing movements and to update the motor controller. Endogenous signals generated within the CNS appear to be involved in the timing, force, and destination of motor output, but their circuitry is not specified except for a possible cortical involvement. Moreover, Gandevia & Burke do not discuss the possible role of central control loops (themselves influenced by peripheral input) in modulating movement output.

Except for simple and well-learned movements that may be controlled by the so-called motor programs, most movement situations (i.e., those involving learning, precision, or perturbations) require the intervention of relatively unknown internal control loops. For example, the basal ganglia are known to modulate the output of the motor cortex through a complex series of segregated circuits (Alexander et al. 1990). One of these basal ganglia-thalamocortical circuits, the "motor circuit," responds to proprioceptive signals evoked by active or passive displacement of individual joints. During a limb movement, movement-related neurons show either a phasic increase or a phasic decrease in their high rates of spontaneous discharges (Alexander et al. 1990). These changes lead to a disinhibition or an inhibition of the ventrolateral thalamus, which in turn gates or facilitates cortically initiated movements. Under normal conditions, a stimulus-triggered movement modifies the activity of the motor circuit at the cortical level with feedback related to target location, limb kinematics, and muscle patterns occurring later through the circuit (Alexander et al. 1990). Under pathological conditions, a disturbance of the thalamic inhibition/disinhibition leads to hypo- or hyperkinetic types of movement disorders. Although current understanding of the functional organization of the basal ganglia is still limited and their physiology is far more complicated than what is described here, the continuum existing between two opposed groups of symptoms or states can be described or modeled by a simple nonlinear mechanism.

As rightly pointed out by **Gandevia & Burke,** the role of centrally generated commands is pivotal for understanding motor control but it is "difficult to grasp conceptually and to document experimentally." In addition, it is "difficult to translate deficits with deafferentation into hypotheses about motor control." In the presence of such complexities and difficulties, nonlinear dynamics may offer an alternative approach to explore the interdependence between centrally generated motor commands, their associated internal loops, and available feedback. Simple nonlinear mechanisms can model relatively successfully physiological phenomena such as the control of respiration, the pupil light reflex, blood cell dynamics, and neural networks with a small number of key parameters (Mackey & Milton 1990). Control parameters are manipulated and the resulting dynamics are compared with experimental results until agreement is reached over a wide range of parameter values between experimental and simulated data.

For example, we conducted studies in which subjects had to maintain a constant finger position using visual feedback for 60 sec. In these simple experiments we systematically manipulated the delay or the noise in the visual feedback loop. An increase in time delay induced oscillations with different morphologies in control subjects and in patients with lesions of the basal ganglia (Beuter et al. 1991). We also showed that when noise was added to the visual feedback loop, control subjects started to perform like some patients with Parkinson's disease. Experimental results were modeled using first-order differential-delay equations (Beuter et al. 1992). We found that experimental results were best modeled when at least two nonlinear, negative feedback loops were included and a stochastic term (noise) was added to the model.

I believe that further research on modulation of kinesthetic signals by the CNS could benefit from the development of simple, noninvasive experimental models amenable to theoretical analysis by nonlinear dynamics. Obviously, the delays used in our equations represent a simplification introduced because a detailed description of the underlying processes is too complicated to be modeled mathematically and also because some of the details are unknown (Cooke & Grossman 1982). By systematically varying control parameters (e.g., delay, gain, noise), however, it should be possible to model the kinesthetic contribution adequately through multiple feedback loops and to characterize in detail the control loops responsible for our rich and complex movement dynamics.

Artificial versus real neural networks

Horst Bischof and Axel J. Pinz

Department for Pattern Recognition and Image Processing, Institute for Automation, Technical University of Vienna, A-1040 Vienna, Austria
Electronic mail: *bis@prip.tuwien.ac.at*

[DAR] **Robinson** argues that "trying to explain how any real neural network works on a cell-by-cell, reductionist basis is futile." In artificial neural networks (ANNs) hidden units are found that are uninterpretable. The conclusion drawn from these findings is that the same thing is also likely to happen in biological neural networks. Though we agree with the conclusion that one cannot explain brain functions reductionistically, we disagree with the line of argumentation.

In the first part of this commentary we analyze in detail several statements made in the target article; the second part discusses the usefulness of single-cell recordings in ANNs.

The behavior of ANNs. In his introduction **Robinson** states that an engineer has no intention of examining the hidden units of neural networks because if he could have predicted their behavior he could have designed the system himself. This statement is wrong for two reasons. First, their learning ability is only one reason for using neural networks; the robustness and the fine-grained parallelism offered by artificial neural networks are other motivations for engineers (e.g., see DARPA 1988). An engineer might choose neural networks because the neural network solution is more robust or better suited for parallel implementation than the conventional algorithm.

Second, although engineers are not able to predict the behavior of hidden units in advance (i.e., they do not know an algorithm for performing a specific task), they might be able to interpret the behavior of hidden units, since it is much easier to tell what an existing system does than to build a system from scratch. In general, engineers at least try to interpret the behavior of their systems.

Learning and complexity. In section 2 **Robinson** states that you should "tell the network *what* you want it to do, but do not even think about telling it *how* to do it." This means that one should solely rely on learning. But because of the complexity of learning (e.g., see Baum 1990; Judd 1988), this way of using neural networks has severe limitations. Hinton (1989) has pointed out that many learning algorithms (e.g., back-propagation) have bad scaling characteristics. For this reason, learning alone is not sufficient. One has to find ways to incorporate a priori knowledge in ANNs. In biological networks such knowledge is supplied by genetic information, acquired over millions of years of evolution. There are many ways to incorporate a priori knowledge in ANNs and to compete against the complexity of learning. One way is to pretrain the network with symbolic rules (e.g., Prem et al. 1992). Another is to use special preprocessing methods, which in turn can also be implemented by neural networks. This leads to modular neural networks (e.g., Bischof 1991; Iwata et al. 1990). The network topology especially can be used to encode information about the task to be learned. Particularly interesting are modular and hierarchical neural networks. This also avoids the unrealistic feature of fully connected neural networks (as pointed out in the conclusion of the target article).

Spatial location of units. Another difference between ANNs and biological neural networks should be pointed out. Units in ANNs have no spatial location, as opposed to biological networks where cells have a position in 3-D space. For this reason **Robinson**'s statement about spatial input patterns (e.g., sect. 2) is misleading. The behavior of an ANN is not changed when the units are interchanged. The only important information is the connectivity (topology) of the network. This property of ANNs is also responsible for symmetries in the solutions found by learning algorithms (Hecht-Nielsen 1989). For these reasons one cannot talk of a nice biological touch (see Robinson, sect. 2),

because these symmetries arise from a difference between ANNs and biological networks.

The behavior of individual units. Let us now turn to the main point of this commentary. **Robinson** argues that single-cell recordings cannot provide enough information about brain functions. Although we agree with this statement we prefer a different line of argumentation.

When one does cellular recordings, one examines the receptive field of a neuron. One cannot examine the projective field of a neuron. Lehky and Sejnowski (1988) have found in their study of ANNs that to interpret the behavior of the unit properly it is also important to examine the projective field of the unit. Pinz and Bischof (1990) have shown the corresponding importance of knowing the surrounding hidden units in the network. In this study we trained neural networks of identical topology with different parameters. After training, several hidden units of one network were replaced by hidden units from another network. The purpose of this process was to improve the predictive accuracy of the network and to eliminate what **Robinson** calls "rogue cells." In some cases the process of "neural network surgery" was successful, so that the new network showed better predictive accuracy than any of its predecessors. Although this result would emphasize a localization of function in the hidden units, there were other cases where a hidden unit that contributed to the classification of the input patterns in one network had opposite effects when moved to another network. From these findings we conclude that the surrounding hidden units in a network should be ascertained because the representation in the hidden units is highly distributed and the hidden units are mutually dependent.

To interpret the behavior of individual hidden units we mainly used visualization techniques (Bischof et al. 1992), especially weight visualization diagrams (similar to Hinton diagrams) (Pinz & Bischof 1990). With these visualization techniques we can examine the receptive and projective field and we can visualize all other units simultaneously in the network, allowing us to interpret most of them.

For biological networks one has not had (at least until now) the possibility of visualizing the cells in a similar manner. The only thing that can be done at the moment is receptive field analysis, and this does not suffice to interpret the behavior of individual cells properly. In this respect **Robinson** is right when he argues against this reductionist approach.

The nonlinear dynamics of connectionist networks: The basis of motor control

Donald S. Borrett,[a] Tet H. Yeap[b] and Hon C. Kwan[c]

[a]Division of Neurology, Department of Medicine, Toronto East General Hospital, Toronto, Ontario, Canada M4C 3E7; [b]Department of Electrical Engineering, University of Ottawa, Ottawa, Ontario, Canada K1N 6N5; [c]Department of Physiology, University of Toronto, Toronto, Ontario, Canada M5S 1A8
Electronic mail: *kwan@utormed.bitnet*

[GEA] **Alexander, DeLong & Crutcher** present an excellent overview of the importance of connectionist modeling in motor control. In this commentary we wish to expand on some of these ideas and to present a connectionist neural network model of movement generation in the central nervous system which fulfills some of their criteria for biological plausibility.

The goal of most neural network modeling is to determine the set of synaptic weights in a specified network that results in a given output upon presentation of a specified input. This type of modeling, by itself, would fail to embed the dynamical nature of movement. With a closed-loop or recurrent network structure in which the output layer feeds back onto the input layer, a neural network computes iteratively until it reaches an equi-

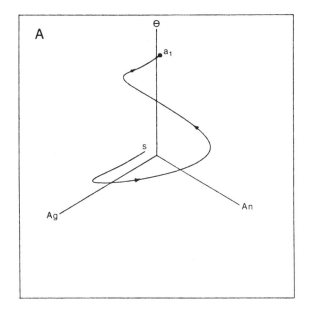

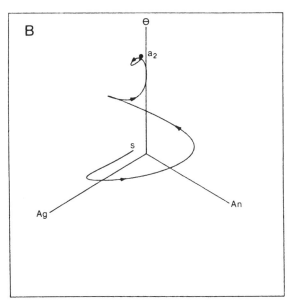

Figure 1 (Borrett et al.).   Nonlinear dynamical conceptualization of movement generation and its disorder. A recurrent neural network (cf. Kwan et al. 1990), modeled after the cortical-basal ganglia loop, was trained to generate the triphasic EMG pattern in normal movements. This pattern of agonist (Ag) and antagonist (An) activations and the associated angular displacement (θ) is represented as an orbit of evolution or relaxation of the network from its initial state (s) toward the attractor a_1 in a 3-dimensional phase-space depicted in 1A. A generalized alteration in excitability to all elements within one layer of the loop, a parametric change that might occur in movement disorders, resulted in an abnormal repetitive agonist-antagonist EMG pattern with the associated bradykinetic movement in Parkinsonism (cf. Hallet & Khoshbin 1980). This abnormality, a near-bifurcation behavior, manifests itself as a multiturn spiralling relaxation orbit toward an attractor a_2 as depicted in 1B. With further reduction in excitability, this near-bifurcation behavior undergoes a bifurcation to become a stable periodic attractor, which may be counterpart of resting tremor in Parkinson's disease.

librium state. Since a recurrent neural network is a dynamical system and is subject to analysis by techniques of nonlinear dynamics, we believe that the theory of nonlinear dynamical systems applied to recurrent neural networks provides an ideal framework for conceptualizing normal movement generation and motor pathophysiology. We have previously suggested (Kwan 1988; Kwan et al. 1990) that a movement can be envisaged in dynamical neural networks as the behavioral correlate in phase space of the relaxation orbit of a recurrent neural network into an equilibrium state or attractor (Figure 1A). As the network relaxes into its equilibrium state, the output of the network drives the limb to its endpoint. **Alexander et al.** indicate that a realistic connectionist model should have structural features that bear at least some resemblance to actual, physical neuronal circuits. We have suggested that the cortical-basal ganglia motor loop may represent a recurrent neural network that iterates repetitively to generate movement (Borrett et al. 1992). In this formulation, descending collateral projections from the loop physically drive the limb to its endpoint as the network relaxes into its attractor. Alexander et al. further indicate that this motor circuit maintains a striking degree of functional specificity and topographic orderliness throughout, and that separate leg, arm, and face representations within each stage of the circuit form separate parallel channels of information. The representational segregation used in iterative recurrent networks for movement generation (Kwan et al. 1990) is again in accord with these experimental observations. Latency studies of electromyographic (EMG), cortical, and basal ganglia units during movement (see Alexander et al.'s target article for review; Mink & Thach 1991a) point to a *coevolution* of activities over a number of anatomical regions rather than a sequential, feedforward passage of information from one structure to another. This observation further argues for the iteration hypothesis of movement generation and control.

Alexander et al. maintain that a realistic connectionist model should not only bear a resemblance to the anatomical circuits on which their design is based; it should also, upon disruption of processing within the network, reproduce the same functional consequences as are seen following lesions in their biological counterparts. The reproduction of a pathophysiological motor state by a biologically plausible alteration in the network would fulfill this criterion. In an attempt to model the bradykinesia of Parkinson's disease, we have recently trained a neural network with four synaptic layers (modeling the cortico-basal ganglia-thalamic-cortical loop) to generate prototypical changes seen in agonist and antagonist activity that produce a displacement about a single joint (Borrett et al. 1992). By introducing a generalized decrease in the excitability of the preoutput layer to mimic the observed increase in inhibitory activity to the thalamus implicated in Parkinson's disease (Bergman et al. 1990), the network required a longer time to complete the movement and accomplished the movement with multiple agonist-antagonist bursts. A phase portrait depicting this abnormality is shown in Figure 1B. This pattern of agonist-antagonist activation produced a bradykinetic displacement and has been observed in Parkinson's disease (Hallet & Khoshbin 1980). When the network was trained to reproduce the agonist and antagonist activities that accompany a repetitive movement, the introduction of a generalized decrease in the excitability of the preoutput layer resulted in an inability to maintain the movement (Borrett et al. 1992), also characteristic of Parkinson's disease.

It is clear in these examples that a parametric change in the network (in this case a decrease in the excitability of one layer) resulted in an alteration of its output behaviors. It is a property of nonlinear dynamical systems that the attractor dynamics remain unchanged over a range of parametric values but that at a critical point a qualitative change of network dynamics (or bifurcation) occurs (see, e.g., Devaney 1986). This was observed in the modeling of repetitive movement in Parkinson's disease noted earlier, where, at a critical value of excitability change, a periodic attractor collapsed into a fixed-point attractor. In addition, the damped EMG oscillations associated with parkinsonian bradykinesia present themselves as near-bifurcation behavior (Figure 1B). This alteration in network dynamics in the neighborhood of a critical parametric value provides an ideal framework to conceptualize movement disorders and to account for such phenomena as tremor, chorea, and myoclonus as dynamical diseases (Mackey & Glass 1977).

It should be emphasized, however, that not all parametric changes such as generalized excitability alterations lead to diseases. Indeed, systematic parametric change within physiological bounds can play a crucial role in motor control. For instance, alteration of the relaxation orbits of the network by parametric change can be utilized to control the final target position, the limb trajectory, and movement speed. Furthermore, parametric control can be used to select a functional structure or "motor program" (i.e., basins of attraction and their boundaries in phase space) within a network appropriate for a specific task (Kwan 1988). Thus, embedded in the parameter space are a range of functional structures punctuated by regions of dysfunctional dynamics we call "diseases."

In sum, we agree with **Alexander et al.** that connectionist approaches in the modeling of motor process have the greatest biological plausibility. However, we must hasten to add that the integration of these approaches with those of nonlinear dynamics provides the greatest chance of accomplishing this task. For connectionism to work, it *must* embed dynamics.

ACKNOWLEDGMENTS
We thank the Toronto East General Hospital Research Foundation for support, and Hoi Nguyen-Huu for assistance in computer programming and graphics.

Implication of neural networks for how we think about brain function

Daniel F. Bossut

Department of Physiology, University of North Carolina, Chapel Hill, NC 27599-7545
Electronic mail: *dfb@med.unc.edu*

[DAR] **Robinson**'s target article provides a short review and a mixed set of remarks and some inconclusive statements about the kind of information the computer simulation of neural networks might contribute. The usefulness of this method of investigation is illustrated by artificial networks that mimic neural activity during oculomotor movements. Robinson argues that current electrophysiological techniques are insufficient to reveal the complexity of brain activity during any given task. He stresses – although with some reservations – that artificial networks provide a new approach, a new concept, and a better tool to explore neural networks. For example, it is possible to envision the global neural task involved in the adjustment of ocular movements. Inputs from the vestibular system, visual targets, or cues for the production of saccades are fed to an artificial network to provide accurate outputs in terms of muscle contraction, speed, and direction. In other words, computer simulation provides a bird's eye view of the neural network for which the precise mechanisms involved at individual synapses are only minimally relevant to the global process and output.

Robinson derives three important findings from this type of model. First, simulated neural networks need to "learn" their way toward the correct response, and during this process they display a variety of responses that sum up to the least error (checked by a "teacher," the investigator himself). While this adjustment takes place, "neurons" are likely to acquire an arsenal of responses, and the response of a given "neuron" is

likely to vary according to the task. This learning process and the variety of responses appear much like random biological learning events striving toward a biologically suitable response to environmental inputs. This appealing concept gives the strongest argument in favor of simulated neural networks. Second, it appears that any given interneuron provides limited information about the global process involved. With regard to this finding, Robinson concludes that little is to be learned by recording from single neurons. Third, the finding hints that the requisite number of neurons may be task-dependent.

The main criticism of computer simulation is that little verification of the theories advanced has been provided. Moreover, **Robinson** reverses his own argument against the narrow limits of electrophysiology (the knowledge of the system acquired from single-neuron recording) to substantiate and support the fact that the output of a given artificial network is what was expected (e.g., observed with single neurons) and therefore correct. Finally, the wiring of artificial networks is a piece of creative art, not necessarily relevant to brain networks.

In essence, electrophysiological studies and artificial networks have similar potential in the study of a given population of neurons, but they realize it by different routes and within their respective limits. The single-neuron recording technique is relevant to the investigation of brain functions only if it is used to provide a population study. The majority of a population of similar neurons (those with comparable afferent paths, efferent paths, and discharge characteristics) is expected to respond very similarly to a given input. One advantage of artificial networks is that new concepts may help make sense of networks' complex communication paths, and of "rogue cells," which are apparent necessary fallouts from any cell population involved in a goal-oriented learning process. But how much influence do these cells and connections have in the natural process? An artificial network can only show that these connections are likely to exist within the limits of the artificial system itself.

The main shortcoming of **Robinson**'s target article is that the author does not appreciate the complementarity of different means of investigation. Without connections to the real world, artificial networks are nothing more than temporary psychological reinforcements for researchers in need. Although, in itself, this may provide new avenues of investigation otherwise unlikely to be sought for, scientific delusions are bound to arise from accepting what is observed from a single viewpoint as the whole picture. The artificial approximation of neural network functions is bound to provide more bias and erroneous concepts than the yield from single-cell recordings. It is therefore imperative to verify the likelihood and usefulness of artificial networks in vivo with critical experiments. The limits are the means of investigation and the genius of the investigators. In the end, the reconciliation in the form of the complementarity of various techniques may prove useful in investigating brain function.

Is the cerebellum a motor control device?

James M. Bower

Computation and Neural Systems Program, California Institute of Technology, Pasadena, CA 91125
Electronic mail: *jbower@smaug.cns.caltech.edu*

[JRB] **Bloedel** is to be commended for his clear identification of what is surely a central and astounding feature of the cerebellar cortex: its remarkably uniform network and cellular architecture in the presence of diverse inputs and outputs. Cerebellar circuitry processes not only proprioceptive input, but also information related to light touch, vision, audition, and even cerebral blood flow (Bloedel 1981; Ito 1984). In fact, the only sensory system that clearly does not provide input to the cerebellum is olfaction. I also agree with Bloedel that this circumstance enhances the importance of the detailed analysis of patterns of afferent and efferent projections in any theory of cerebellar function (cf. Bower & Kassel 1990).

Bloedel presents a general theory to account for the involvement of the cerebellum in several different kinds of motor behavior. His model is thus a variant on the classical view that the cerebellum is involved in the coordination of movement. With similar motivations, and considering the same data, I have recently proposed that this classical description of cerebellar function may be incorrect (Bower & Kassel 1990). While this change in view was primarily motivated by new information concerning the organization of afferent projections to the cerebellar cortex, it also takes into account the fact that experimental animals as well as humans are capable of perfectly adequate motor coordination without a cerebellum (cf. Schade & Ford 1973). The standard surgical procedure for a cerebellar tumor in childhood is simply to remove the cerebellum, with little resulting detectable behavioral effect in adulthood. Given that no other brain structure is organized like the cerebellum, other regions may compensate for its absence but they cannot substitute for its presence. Essentially the same argument is used by **Bloedel** in discussing motor learning and I similarly conclude that motor coordination must be the basic responsibility of brain structures with which the cerebellum is normally involved but for which the cerebellum is not necessary.

Another global theory of cerebellar function. Based on detailed analysis of the regions of the mammalian cerebellum receiving tactile input, I have recently proposed that the cerebellum coordinates the acquisition of sensory information so that the highest quality sensory data are available for a wide range of computations performed by the rest of the nervous system (Bower & Kassel 1990). In this way the cerebellum indirectly affects the efficiency and capacity of other systems without directly participating in their computations. In other words, the cerebellum is not involved in either perception or motor coordination but in the process of assuring good data for both. I believe this is the reason why so many sensory systems project to this structure and why the outputs of the cerebellum are so diverse. The resulting interpretation of the lesion data is that the primary circuits themselves can, under certain circumstances, compensate for poorer input data, but that on closer inspection sensory and/or motor performance should be compromised. This is expected to be especially the case for more complex behaviors based on multisensory data.

Although it was originally proposed in the context of tactile regions of the cerebellum (Bower & Kassel 1990), I believe this idea has general applicability, as the uniform cerebellar architecture requires (Paulin et al. 1989; Rasnow et al. 1989). Its implication for limb movements, for example, is that the cerebellum would monitor and control the acquisition of the proprioceptive and other sensory data on which coordinated motor performance is dependent. The known influence of the cerebellum on the gamma motor system (Ito 1984) is thus interpreted as a mechanism to control the gain of the muscle spindles, and therefore the information they are transducing, rather than to indirectly coordinate movement. Cerebellar lesions should also specifically disrupt complex multijoint movements or movements coordinated by multisensory inputs as these are likely to be less tolerant of poorly coordinated or noisy sensory data. On the other hand, relatively simple movements (rotations around joints) would be expected to be less affected or to recover more rapidly.

With respect to tactile regions of the cerebellum (e.g., the paramedian lobule), cerebellar influence on the fine control of finger movements in monkeys (see **Bloedel**) or on facial movements in rats (Bower & Kassel 1990) is predicted to reflect the use of such movements in tactile sensory acquisition. Thus, it is the function of the movement that is important, not its graceful coordination. This distinction is actually evident in the modern redesign of experimental apparatus intended to study cerebellar involvement in movement control. Where monkeys were once asked to perform large-scale movements of levers (Bauswein et al. 1983; Harvey et al. 1977; Thach 1980), current experiments require finer manipulation of digits over small objects (Dugas et al. 1989; Houk & Gibson 1987).

Finally, this recent hypothesis is most obviously correct with respect to the involvement of floccular regions of the cerebellum in eye movements. These movements are all fundamentally related to the acquisition of sensory information by the visual system. Cerebellar involvement in adjustments to the gain of the vestibulo-ocular reflex (VOR), for example, assure a minimal slip of images on the retina during head movement. As little as $3°/\sec$ of retinal slip significantly degrades visual acuity and thus the ability of the visual system to process information (Westheimer & McKee 1975). Accordingly, the largely accepted role of the cerebellum in the VOR assures that the highest possible quality of visual information is obtained.

Conclusion. Although our interpretation of available data has led to a substantially different conclusion regarding cerebellar function, our theory is consistent with several of the more general points made by **Bloedel**. First, we do not believe that short-term learning is fundamental to function. In addition to the issues raised in the target article, I point out that all interpretations of the effects of cerebellar lesions on learning are absolutely dependent on there not being any sensory-related deficits (e.g., increase in sensory noise). Second, this hypothesis emphasizes the "real-time" nature of cerebellar processing rather than memory storage (Paulin et al. 1989). Third, I also view the cerebellum as a "moderator" that is involved in optimizing brain performance, not as a programmer designing behavioral sequences. Fourth, I have proposed that the algorithm implemented in cerebellar cortical circuitry involves the dynamic generation of a behaviorally relevant context (Bower & Kassel 1990); however, in this case the circuit is responsible for placing information from particular sensory receptors in the context provided by information from other, behaviorally related receptors so as to coordinate the sensory information being obtained. Finally, although we differ on how the cerebellar circuit reflects this context dependence, we also basically agree that it is only through the further exploration of the circuit's details in the context of the patterns of afferent and efferent projection that the general function of the cerebellum will be understood.

Why is the output of the cerebellum inhibitory?

V. Braitenberg and H. Preissl
Max-Planck Institute for Biological Cybernetics, 7400 Tübingen, Germany
Electronic mail: preissl@mpib-tuebingen.mpg.dbp.de

[EB, JRB] We are consumers, not producers of physiology. Like **Bloedel,** who laments the lack of a general theory of the cerebellum, we also feel this problem very keenly and are in the process of putting together something resembling an explanation of the role of the cerebellum in its interaction with the cerebral cortex. Although we are ourselves commentators, we are eager to have the preview we will offer commented upon and possibly criticized by the experimental community.

We are primarily impressed by a set of facts about the cerebellum different from those which seem to have prompted other people's (including Pellionisz's (1985), Marr's (1969), and **Bloedel's**) imaginations. We try to take the anatomy seriously. In particular, if our task is to set off the role of the cerebellum from that of the cerebral cortex, we must provide a reason for three radical differences between the two.

1. Whereas both are flat and about equally large in most species of animals, the cerebellar cortex is not really two-dimensional as the cerebral cortex is, but intrinsically one-

dimensional, in the sense that the fibers mediating specific interactions within the neuropil, whether the excitatory system of parallel fibers or the axons of the inhibitory cells, are strictly oriented in one direction. This suggests an operation somehow related to sequences rather than to two- (or more) dimensional maps.

2. There is no circuitry mediating positive feedback within the grey substance of the cerebellar cortex. In contrast, the cerebral cortex consists almost entirely of positive feedback loops between excitatory pyramidal cells. Whereas the cerebral cortex probably supports internal representations in the form of cooperative activity between neurons widely spread throughout the whole structure, all interactions in the cerebellar cortex must be short-lived and limited in space. The cerebellum does not look like a place in which concepts can find any permanent home.

3. We are impressed by the fact that the entire output of the cerebellar cortex is inhibitory. This is again in contrast to the cerebral cortex, where all output fibers are axons of excitatory pyramidal cells.

We shall not repeat the argument that years ago led to an interpretation of cerebellar morphology in terms of a timing device. It is still true that the shape of the Purkinje cell dendritic tree, the parallelism of the parallel fibers, and their uniform thickness could hardly be explained if not as an optimization of some computation involving space and time. The old timing hypothesis in its original form (Braitenberg & Onesto 1962) had to be abandoned. What is left (Braitenberg 1983; 1987) is the idea that the cerebellar cortex reacts specifically to sequences of input signals when they are presented along a folium or "beam" in the right tempo corresponding to the velocity of parallel fibers. Indeed, if fibers and synapses behave as we think they should, it follows necessarily from the anatomy that sequential activation at the right speed will produce something like a tidal wave of activity, because every additional item in the sequence of inputs will add strength to the volley already travelling in the parallel fibers. Both Purkinje cells and inhibitory interneurons will then be maximally excited, with the consequence that neighboring beams are inhibited while the excited one produces a sequence of output signals in a row of Purkinje cells in synchrony with the tidal wave. What makes this interpretation more likely than our previous one is that it is not limited in time by the maximum delay of about 10 msecs that the individual parallel fiber provides. In fact the tidal wave may spread much farther along the beam, involving ever new parallel fibers in succession.

We are willing to stick our necks out even further, however. There is little doubt in our minds that the cerebellum incorporates knowledge by learning. Our present guess is that learning takes place at different times in two different systems. The first episode of learning is of the "imprinting" kind. Our assumption (totally undemonstrated as yet) is that the mossy fiber input selects its targets (granular cells) at an early stage in such a way that the sequences specified by the mossy fibers along each beam make some sense in a motor (or other) context. This early learning may at one time produce beams that correspond to trajectories through some body map, be it the map of muscles, of masses, or of joints; at another time it may produce some more unpredictable sequence of items or events.

We must assume that the sequences specified by the mossy fibers do not refer only to peripheral input, but equally or perhaps more to central representations of things, mainly in the cerebral cortex. Once such imprinting has taken place, the cerebellar cortex, which is, as is well known, very long in the anteroposterior direction (2 meters in man: Sultan 1992), turns into a long catalogue of sequences relevant in motor contexts, each represented in a latero-lateral strip or beam.

This makes good sense if a second stage of learning is assumed, which we imagine as taking place throughout life. Here we assume that the Purkinje cells along a beam (i.e., the ones

excited by the same input sequence) are individually coupled to or detached from their parallel fibers, presumably by changes in their spine-synapses. Once this has happened, a specific input sequence activates one and only one beam (the inhibitory fibers see to it that the maximally excited beam suppresses neighboring beams that may represent similar sequences) and thus produces sequential signals by activating some selected Purkinje cells determined by the second learning process.

In other words, we envisage an operation in a sequence-in sequence-out mode, ready to accommodate a very large number of input and corresponding learned output sequences, large enough for all the movements that we (including violinists and tennis champions) may learn.

In this scheme, we do have a use for the saggital strips or zones in which the input is organized (see **Bloedel** for references). If the same or similar input is provided for part of the catalogue of sequences, and if the individual input strips are at an angle to each other (in the anterior lobe – they converge frontally), different beams must represent similar sequences presented at different speeds.

What does this have to do with the **Bloedel** and **Bizzi et al.** target articles? Bloedel is not in sympathy with the idea that memory traces relevant for movement are stored in the cerebellar tissue, although he admits that the cerebellum is involved in some way in the process of acquiring these engrams. Our feeling is that if physiological experiments provide no good evidence for learning in the cerebellum, they must have missed the specific nature of the information stored there. Our suggestion (not entirely original) is that what is added in the cerebellum to the motor pattern dictated by the cerebrum is exactly what makes the difference between an elegant and efficient movement and one that is only specified in its essential parameters. Since this implies more calculating power than we are willing to attribute to a nerve net, the solution is most likely in the form of a roster of individual cases acquired through experience and available on line as a look-up table (to use expressions from engineering). What better interpretation for the two meters of cerebellar cortex displaying a vast number of parallel lines in succession, and what better use for the 10^{12} presumably plastic synapses residing on Purkinje cell spines, than to accommodate in this system a catalogue of different movements!

But it is not the movement itself that is stored there: Here we agree with **Bloedel**. Rather, we assume that the cerebellum learns to subtract from the movement all those forces which arise in a parasitic way in an articulated visco-elastic mechanical system: rotatory and linear momentum (yes, linear too – the center of mass of a jointed arm tends to move on a straight line with extension and flexion), elastic forces, viscosity, and friction. We say *subtract*, and not *compensate*, for a good reason. Relaxing a muscle that acts in the direction of an inertial force is obviously a more energy saving strategy than opposing the inertial force through the contraction of an antagonist. Besides, this strategy provides an excellent explanation for the inhibitory output of the cerebellum.

As to the theory proposed by **Bizzi et al.**, we could not be more satisfied. If the cerebellum does physics (in an empirical way), the cerebral cortex is entirely cognitive. It contains a representation of the world, including a representation of the animal's own body, continually updated by sensory input and likely to change at any moment under the influence of external stimuli or internal ruminations. When the idea of one's own body in space changes, this implies a new distribution of activity in cortical neurons devoted to the control of muscle tension. The transition from one body-idea to the next is movement at the cognitive (i.e., cortical) level. The actual movement is nothing but a consequence of the cognitive transition, and we learn from Bizzi et al. that the transition follows a principle of continuity, much like the mental rotations of Shepard's experiments (Shepard & Cooper 1982) or the Abelesian synfire chains (Abeles 1991).

Such cognitive control of movement innocent of the reality of a mechanical system is bound to produce much waste in the form of slinging, overshooting, oscillation. In the economics of information handling in the brain, however, it apparently proved more advantageous to create a separate organ, strongly connected with the cortex (i.e., the cerebellum) for the elimination of these disturbances, rather than to burden the cognitive level with the down to earth problems of mechanics.

We shall be less reticent in a forthcoming publication (Braitenberg & Preissl 1992). In particular, we will have to specify the mechanisms that optimize the compensatory strategies. What measure of optimization does the cerebellum receive while learning? The climbing fiber input is likely to mediate some of this feedback. The connotation of novelty or surprise that has often been associated with it fits the scheme. Also, if the movement is to become physically smooth after the cerebellar learning, we are not surprised to learn that climbing fibers respond to stimuli outside the muscular and skeletal system. Slinging or oscillatory phenomena may well be best detected by receptors residing in the containing wall of the semiliquid mechanical system, that is, in the skin.

Taking distributed coding seriously

Bruce Bridgeman

Program in Experimental Psychology, University of California at Santa Cruz, Santa Cruz, CA 95064
Electronic mail: *bruceb@cats.ucsc.edu*

[EEF] Extending distributed coding to the motor system applies a set of principles that has been accumulating support in sensory systems for some years. Distributed coding is an alternative to the detector cells of the sensory systems, or of the equivalent command cells of motor systems. Distributed codes are revealing a previously unsuspected flexibility of organization. Together with the lateral connectivity and efferent control that are its corollaries, however, distributed coding has even greater consequences than those reviewed by Fetz.

Efferent control magnifies the effects of distributed coding. If a neuron at level n feeds its signals back to a receptor, for example, that receptor combines peripheral information with level n information to create a new level $n + 1$. A higher-level signal moves through the afferent channels of the sensory pathway right from the start. The concept of successive recodings in successive layers breaks down as the logical level of an anatomical structure can no longer be defined.

Along with distributed coding, then, comes the necessity for distributed stages. In the motor system, sensory feedback from the muscles can have analogous effects on control. This new idea contrasts with the classical conception of discrete stages, in which all the neurons at one stage do about the same things on different parts of the input and a transfer function can be written to describe what that stage does to the signals flowing through it.

The idea of distributed stages complements Fetz's finding that even neighboring neurons may show widely different responses in a given situation. In discussing localization of function, he might have added that many more neurons participate in motor activity than are found in the classical precentral motor cortex, for about ⅔ of the pyramidal tract axons originate not in the motor cortex but elsewhere in the cortex. As a psychologist, I am generally happy to be able to explain ⅔ of anything: Thus, to a psychologist's approximation, the pyramidal tract arises from locations other than precentral motor cortex! The neurons and their types of discharge from these other areas are even more varied than those of the precentral motor cortex.

Fetz's observation that neurons in a given region can display nearly every imaginable response pattern is important but too little investigated. It recalls similar evidence that D. N. Spinelli, R. W. Phelps, and I collected in the laboratory of Karl H.

Pribram in 1970. We were interested in the problem of distance constancy, the ability of the visual system to recognize an object as having a certain size regardless of its distance from the observer or the size of its retinal projection. To look at how the striate cortex of the cat contributed to distance constancy we isolated the various distance cues and presented them separately and in combination while recording from single neurons. Our automated receptive field mapping apparatus swept a small spot or bar in a raster pattern across a square area. We could vary the size of the square, the sweep speed, the display distance, and the bar size independently. We could also record monocularly or binocularly, of course.

After factorially varying the parameters of stimulation and recording receptive field properties in a few dozen neurons, we attempted to sort the data into a small number of discrete receptive field types, as was the fashion at the time. We failed miserably – it seemed that every neuron responded in a different way, showing constancy when some parameters were varied and no constancy when others were manipulated. The only consistent finding was that if recording was monocular and if every parameter was varied in a way that kept the retinal image the same, we could change the target distance without changing a neuron's responses. Except for that, no two neurons followed the same rules. The number of receptive field types would have been as large as the number of neurons. We abandoned the study, and we never published the data.

In light of Fetz's target article, it seems we gave up prematurely. Our results, in hindsight, demonstrate a parallel between the organization of the sensory and the motor systems, with each neuron adding a unique signature to the composite population code; groups of neurons with similar receptive fields are found only when incomplete or relatively trivial questions are asked of the neurons.

In subsequent neurophysiological work, aimed at elucidating the processing of a stimulus rather than at the responses of single neurons, a similar pattern has emerged. Working in cat and monkey striate cortex, I attempted to find out how each neuron would respond to a stimulus that had a meaning in a discrimination task. Recording from neurons occurred while the awake animal performed a discrimination for a liquid reward. Rather than trying to find the optimal stimulus for each neuron, and thus driving it far beyond its normal physiological rate, we recorded every neuron with the same stimulus. Single-cell responses were meager compared to those customary in optimal-stimulus studies, and were widely different for each neuron. Only the composite responses, analogous to the population vectors of Georgopoulos et al. (1984), began to look like reasonable codes for the stimulus. The same population of neurons also carried information about the correctness of the response in the behavioral trials (Artim & Bridgeman 1989; Bridgeman 1980; 1982; Bridgeman & Artim 1983).

Finally, a historical note: Although the neural hologram hypothesis originated with Pribram (1971; Pribram et al. 1974) in a sensory context, its recent use has been mostly in models of memory (Eich 1982; Murdoch 1982). Fetz makes an important contribution in expanding the idea to coding in motor regions as well.

Adaptive behavioral phenotypes enabled by spinal interneuron circuits: Making sense the Darwinian way

Daniel Bullock and José L. Contreras-Vidal

Department of Cognitive and Neural Systems, Boston University, Boston, MA 02215
Electronic mail: *danb@cns.bu.edu*

[DAM] From the time of Darwin, the primary path to making sense of biological structures has been through the identification

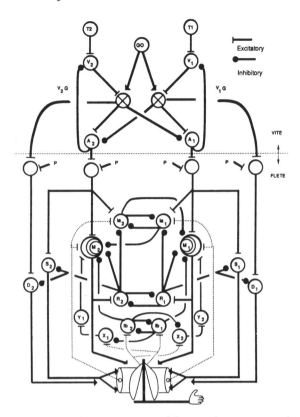

Figure 1 (Bullock & Contreras-Vidal). A schematization of an opponently organized, spino-muscular circuit (lower part) receiving descending signals of two types from a trajectory generator (upper part) and other sites presumed to be located in the higher brain. In the theory, the balance between signals emerging from neuron pools A_1 and A_2 sets desired joint angle, while the signal P sets desired joint stiffness. Signals V_1 and V_2G set desired shortening/lengthening velocities for the opponent muscles during movements. The lower circuit's structure is based on the anatomical and physiological studies summarized in Table 1 wherein indices $i,j \subset \{1,2\}$ refer to the opposing muscle channels. Dotted lines show feedback pathways from sensors embedded in two types of muscles: large, main, or extrafusal muscles and small, parallel, intrafusal muscles. The two lateral feedback pathways (Ia afferent fibers) arise in intrafusal-embedded receptors sensitive to extrafusal muscle stretch and its first derivative. The two medial feedback pathways (Ib afferent fibers) arise in extrafusal-embedded Golgi tendon organs sensitive to main muscle force. The stacked circles surrounding labels M_1 and M_2 symbolize the models' incorporation of the size principle of alpha-motoneuron recruitment. The M_1 and M_2 pools directly activate their respective extrafusal muscles; corresponding S and D pools activate intrafusal muscles. Key to cell types of lower part: Ia = Ia reciprocal inhibitory interneuron pool; M = alpha-motoneuron pool; S = static gamma-motoneuron pool; D = dynamic gamma-motoneuron pool; R = Renshaw recurrent inhibitory interneuron pool; Ib = Ib recurrent inhibitory interneuron pool; X = hypothesized first pool in Ib multisynaptic reciprocal excitatory feedback pathway (sustained cell); Y = hypothesized second pool in Ib multisynaptic reciprocal excitatory feedback pathway (transient cell, sensitive to positive time derivatives of force).

of their adaptive functions. Yet the target article by **McCrea** makes little reference to comprehensive treatments of interneuronal circuit structures in terms of their adaptive functions. Rather, the focus is on a descriptive organizational concept, the flexor reflex afferent (FRA). McCrea is sensitive to the criticism this concept has received, but he seems to believe that it has

been called into question solely because of misunderstanding – perhaps, as he admits, because "flexor reflex afferent" is a misnomer. Our objection runs deeper. Suitably renamed, the FRA may serve as a useful summary of empirical observations that together indicate one scheme of spinal organization. However, until this scheme has been rationalized as part of a critical adaptive competence, it will remain more a descriptive than an explanatory device.

An alternative approach to making sense of spinal circuitry is to pursue an updated Darwinian strategy, which has two phases beyond the primary experimental and descriptive phases to which **McCrea** and other spinal experimentalists have naturally applied themselves. A third phase needs to ask what adaptive behavioral phenotypes the spinal interneuronal circuits have evolved to mediate. However, because of the complexity of the interactions involved – which are often more subtle even than those described in the target article – verbal functional hypotheses and single-pathway assessments are unequal to the task of "making sense." Thus a fourth phase must ask whether the ensemble of cell types and pathways can be shown, in a mathematically rigorous way, to be capable of working together to achieve the proposed adaptive behavioral phenotype.

We are not in a position to illustrate this method as applied to any of the particular circuits that appear in **McCrea**'s figures. However, we can illustrate it by summarizing an evolving theory that encompasses a large subset of the cell types and pathways mentioned in McCrea's target article, in particular alpha motoneurons, Ia reciprocal inhibitory interneurons, Ib interneurons, Renshaw cells, static and dynamic gamma motoneurons, and the known connectivities between these cell types. In a series of papers (Bullock & Contreras-Vidal 1991; Bullock & Grossberg 1988b; 1989; 1990; 1991; 1992; Bullock et al. 1992) we have advanced a set of hypotheses and mathematical arguments about the functional roles of these cells and have reported simulations of increasingly comprehensive circuit models. Figure 1 gives an idea of the complexity of the circuitry now encompassed by the evolving model. Because the target article does not cite primary sources for many of the pathways discussed both in it and here, we also include Table 1, whose citations document anatomical and physiological properties of the modeled system.

Our simulations of the Figure 1 circuit have allowed a rigorous demonstration of the mutual coherence of a set of hypothesized actions of circuit components in service of a high-level behavioral phenotype. Our core hypothesis is that this circuitry has evolved to give the higher brain facile independent control of joint angle (limb configuration) and joint stiffness during posture and movement. If we let θ be the joint angle, the difference between descending commands A_1 and A_2 be the joint angle setting, and descending command P be the stiffness setting, the independent control property requires that

$$\theta(A_1, A_2) = \theta(A_1 + P, A_2 + P) \quad (1)$$

and that the joint stiffness be an increasing function of P. Because joint angle is rigidly linked to muscle length and stiffness to muscle tensions, we have nicknamed the model "FLETE," an acronym for Factorization of LEngth and TEnsion. Because we have been able to make sense of all aspects of the Figure 1 spino-muscular system in terms of this essential competence, it serves as a unifying concept for making sense of an otherwise bewildering array of data. In addition, our model is broadly consistent with many aspects of so-called equilibrium point theories (e.g., **Bizzi et al.**), so *a fortiori* we have shown that the spinal circuitry of Figure 1 can be given a consistent interpretation from the equilibrium point perspective.

The paper by Bullock and Contreras-Vidal (1991) provides a step-by-step reconstruction of the full circuit. In it, we start with the simplest opponent motor-unit structure, and clarify what it can, and cannot, do. Then we add an increment of structure capable of addressing one or another aspect of performance

Table 1 (Bullock & Contreras-Vidal). *Evidence for connectivity and physiology incorporated in the* FLETE

Connection type	Model citations
1. excitatory	Renshaw (1941; 1946)
	Eccles et al. (1954)
α-M N$_i$ $\to$ R$_i$	
2. inhibitory	Renshaw (1946)
	Eccles et al. (1954)
R$_i$ $\to$ α-M N$_i$	
3. inhibitory	Hultborn et al. (1971a)
R$_i$ $\to$ IaI N$_i$	
4. inhibitory	Ellaway (1968)
	Ellaway & Murphy (1980)
R$_i$ $\to$ γ-M N$_i$	
5. inhibitory	Ryall (1970)
	Ryall & Piercey (1971)
R$_i$ $\to$ R$_j$	
6. inhibitory	Eccles & Lundberg (1958)
	Araki et al. (1960)
IaI N$_i$ $\to$ α-M N$_j$	
7. inhibitory	Eccles & Lundberg (1958)
	Hultborn et al. (1971a)
	Hultborn et al. (1976)
IaI N$_i$ $\to$ IaI N$_j$	Baldiserra et al. (1987)
8. excitatory	Hultborn et al. (1971a)
I a$_i$ fiber $\to$ IaI N$_i$	Baldiserra et al. (1987)
9. excitatory	Lloyd (1943)
I a$_i$ fiber $\to$ α-M N$_i$	
10. inhibitory	Laporte & Lloyd (1952)
	Eccles et al. (1957)
IbI N$_i$ $\to$ α-M N$_i$	Kirsch & Rymer (1987)
11. excitatory	Laporte & Lloyd (1952)
	Eccles et al. (1957)
IbI N$_i$ $\to$ α-M N$_j$	
12. inhibitory	Laporte & Lloyd (1952)
	Eccles et al. (1957)
IbI N$_i$ $\to$ IbI N$_j$	Brink et al. (1983)
13. nonspecific excitatory	Humphrey & Reed (1983)
	DeLuca (1985)
P $\to$ spinal motor pools	

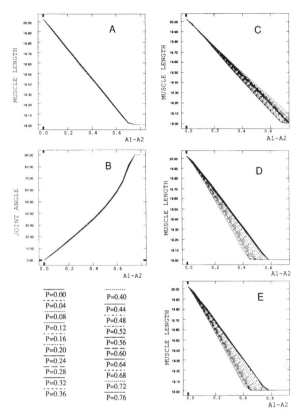

Figure 2 (Bullock & Contreras-Vidal).   Equilibrium values of FLETE model muscle-length and joint-angle variables for the whole circuit (panels A,B) and after removal of various circuit components (panels C,D,E). Twenty curves are plotted in each panel to explore equilibrium positions associated with a full range of combinations of the descending signals ($A_1 - A_2$) and P. Values of the prior signal appear on the ordinate and each curve corresponds to a distinct setting of signal P (see embedded key). Degradation in performance is shown by dispersion among the twenty curves in each panel, all of which would superpose in the ideal case. As shown in panels A and B, the full circuit approaches this ideal. The dispersion is due in (C) to loss of Ib recurrent inhibition, in (D) to loss of Renshaw recurrent inhibition, and in (D) to loss of both. The Renshaw "lesion" is more detrimental than the Ib, as indicated by the greater dispersion at equal ($A_1 - A_2$) values but this would reverse under conditions of muscle fatigue, which was not modeled.

shortfall and reevaluate the improved system's resultant competence. This procedure is iterated until the entire structure is reconstructed and each component's incremental contribution has been clarified. Such a procedure cannot guarantee an accurate reconstruction of the order of evolution, but it can reveal the selection pressures that may have been operative during evolution.

Though not a satisfactory substitute for the step-by-step approach, a useful technique for visualizing different interneurons' hypothetical contributions is a plot that reveals how well independent control of joint angle (or agonist muscle length) and joint stiffness is achieved after various model pathways are selectively "lesioned" (removed from the simulation). That is, we plot system performance as a function of the residual circuit created by removing a component. Figure 2 plots the equilibrium values reached by FLETE circuit variables after the circuit is activated by a full range of combinations of the positioning signal ($A_1 - A_2$) and the stiffening signal P. Figures 2A and

2B illustrate the near-perfect operating characteristic of the whole circuit. They show muscle one shortening and joint angle θ increasing as monotonic functions of $A_1 - A_2$. But they also exhibit the configuration-invariance property of equation (1), because each panel actually plots 20 curves, each corresponding to a distinct setting of the descending co-contractive signal P. The other panels of Figure 2 illustrate the deviation from independent control produced by various simulated lesions. In evaluating the near ideal Figure 2A operating characteristic, it should be noted that the observed linearity and superposition properties exist despite many known neuronal and muscular nonlinearities incorporated into the model equations, for example, the size principle of alpha-motoneuron recruitment. Thus, the network structure discovered by nature appears well adapted to the task of linearizing the system's response to descending signals meant to independently control limb configuration and joint stiffness.

One result of our theoretical studies is that we no longer look upon the interneuronal circuitry in terms of reflexes. Rather, we view it as a sophisticated autocompensatory machine capable of counteracting several types of intrinsic threats to the successful realization of motor intentions. Hence our results are very much in agreement with **McCrea**'s de-emphasis of the historical, reflexive context of discovery of spinal interneurons and his corresponding emphasis on how spinal circuitry provides a sophisticated basis for descending control. The full implications of spinal structure can only be discovered if the experimental and descriptive phases of research, understandably highlighted by McCrea, are complemented by comprehensive circuit modeling conducted under the guidance of strong hypotheses regarding adaptive behavior.

Equilibrium points and sensory templates

P. R. Burgess

Department of Physiology, University of Utah School of Medicine, Salt Lake City, UT 84108

[EB, SCG] *It is unlikely that the human nervous system uses equilibrium-point control to guide voluntary movement.* A reasonable way to evaluate a model of human voluntary movement is to examine its applicability to everyday motor performance. There are several reasons for thinking that the nervous system does not use equilibrium-point control to guide movement.

1. Equilibrium-point models achieve their appealing simplicity by controlling a "virtual" trajectory rather than the actual position of the load. In the commonplace task of lifting an unknown weight, underestimating the load by a factor of two can lead to unacceptable error (elbow flexors; Feldman 1980a). Human subjects lifting unknown weights achieve accurate targeting by making adjustments in muscular torque based on sensory information. This suggests that the nervous system uses sensory-template control (see below) rather than equilibrium-point control in dealing with the pervasive mechanical uncertainties in our environment.

2. Whether human subjects behave like the intact and deafferented monkeys **Bizzi** and his colleagues have studied depends on how they are instructed. If they are simply told to move from position A to position B and the limb is advanced toward the target position by an external force before the subject is instructed to move, the subject does not reverse direction, but simply moves as far as is necessary to reach the target, which may be not at all if the hand is already there. If the subjects are told to duplicate the temporal profile of the unassisted movement, however, they then reverse direction like the monkeys. This indicates that simple end position control is not achieved by controlling a series of equilibrium points but that a sensory template is used to guide the active part of the movement depending on the limb's externally imposed position.

3. Cocontraction (simultaneous activation of agonists and antagonists) is rare during relaxed human voluntary movement. Cocontraction becomes important when unpredictable perturbations may be encountered and is also observed when subjects are apprehensive. The cocontraction that is emphasized in **Bizzi et al.**'s equilibrium-point model for generating the requisite stiffness to move the limb and that is produced by electrical stimulation of the frog spinal cord is unlikely therefore to be relevant to everyday human voluntary movement.

4. When subjects make the same flexion effort at different elbow joint angles, relatively flat torque-angle relationships are generated between 45° and 135°. This is true whether the isoeffort torques are generated isometrically or during isokinetic eccentric and concentric joint rotation at velocities ranging from 5° to 60°/sec. Similar findings are also obtained when subjects are asked to report the effort required to oppose a given extension torque applied at different elbow joint angles; the effort required to oppose a given torque is largely independent of angle over the central portion of the joint's range (Burgess et al., in preparation). Because human subjects can only interpret

an instruction about voluntary movement that has a perceptual counterpart (they can control only what they sense), and since effort is the only efferent perception they have, it follows that isoeffort torque-angle relationships reveal the properties of the involuntary motor mechanisms upon which voluntary commands act. Because the involuntary motor system does not possess much stiffness (isoeffort torque-angle relationships are relatively flat), neither muscle length-tension stiffness (α model) nor stretch reflex stiffness (λ model) can drive voluntary movement. The movements must be organized and controlled in some other way.

The alternative is a model that uses sensory templates to control movement in the face of an uncertain mechanical environment and muscle properties that vary in a complex way over time. A sensory template is a learned representation within the central nervous system of the sensory input that will occur in the course of a movement if the movement is carried out as intended. Because motor tasks are defined exclusively as percepts referred to the body image, the sensory template is the equivalent of the subject's perceptual goals expressed in terms of sensory receptor discharge. This model is developed further below.

Sensory templates and the guidance of voluntary movement. The wealth of cutaneous and proprioceptive sensory information available for the guidance of voluntary movement is documented by **Gandevia & Burke**. This information, together with that from vision, forms the basis for what might be called the "sensory template" model of motor control (MacKay 1982). A sensory template is a central representation of the sensory receptor discharge that would be expected to occur during a movement if the movement is executed according to plan. The need for sensory templates in the guidance of human voluntary movement rests on the even more fundamental assumption that one can control only what one senses (perceives). This assumption can be supported by three lines of argument.

1. Argument from the language of voluntary movement. All the independently controlled parameters of voluntary movement are perceptual attributes referred to the body image, attributes that undergo continuous and largely appropriate modification as movement is carried out. To be intelligible, instructions to subjects have to be expressed in terms of these attributes: Move your hand to this *position* along this *path* at this *average velocity* with this *velocity profile*; reduce the *feeling of strain* in your wrist; *try harder* to raise your shoulder.

How hard one tries is usually referred to as a "sense of effort" and appears to be proportional to the magnitude of the motor command. It is therefore an efferent perception. The other attributes are afferent perceptions because they are well developed when the limb is passively moved. This list is not exhaustive; one of the tasks confronting the sensorimotor physiologist is to complete the list of independently controlled perceptual attributes. For example, there is little doubt that one can sense (control) the rate of change of effort but probably not the contractile state of one's gallbladder, at least not without "biofeedback."

2. Argument from isoeffort torque-angle profiles. As we conduct our everyday motor activities, our upper extremities encounter a variety of frictional and gravitational loads that are in general only imperfectly known. A good example is lifting a covered teapot. The weight of my teapot varies by a factor of three depending on how full it is, yet I can move it with millimeter accuracy from one position to another if I can see it. One solution to the problem of unpredictable loading would be for our limbs to operate like very stiff position servos (the power steering analogy, Merton 1964) so that the hand would move to within a millimeter or two of the desired target regardless of load. This would remove torque generation from direct voluntary control and might well cause injury. In fact, elbow flexor torque does not undergo appreciable involuntary change as the joint is rotated through the central part of its range (45°–135°); that is, the

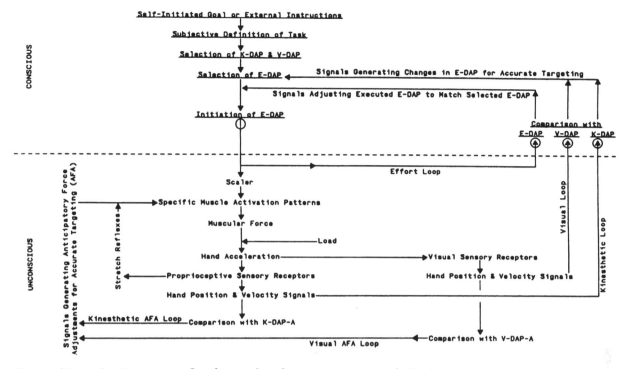

Figure 1 (Burgess). Sensorimotor flow diagram for voluntary movement. Underlined operations are conscious. O indicates points at which information enters or leaves consciousness, perceptually incorporated into the body image. DAP is the desired approach profile. This typically includes end point (target) of movement, movement path, movement time, velocity profile (kinesthetic-DAP and visual-DAP), as well as a preliminary effort amplitude, duration, and profile based on anticipated load (effort-DAP). AFA refers to anticipatory force adjustments needed for accurate targeting. Both K-DAP and V-DAP have unconscious anticipatory components (K-DAP-A and V-DAP-A). The scaler is a circuit that changes the relationship between magnitude of perceived effort and the actual muscular torque generated.

torque does not change much at a given effort level (Burgess et al., in preparation). If voluntary commands act not on a position servo but on the equivalent of an angle independent torque motor, sensory templates become essential for controlling the accelerations and decelerations needed to move unknown loads from place to place.

3. Argument from experimental findings. It can be shown by a simple calculation that if symmetrical force fluctuations appropriate for lifting an initially stationary mass through 10 vertical cm to a stationary target in 1 sec are used to move a mass only 3% less, the load will overshoot the target by many centimeters. In other words, the everyday mechanical problems solved by the nervous system require very precise torque control. How is such torque control achieved when subjects are lifting unknown weights? Let us say the subject has just successfully targeted either a 50 N or a 20 N load and is now asked to lift a 30 N load. Measurements show that the force acting on the load increases initially at varying rates, but shortly after the load begins to move the force peaks and then begins to fall. The 30 N trials are not stereotyped in a situation in which the subject cannot predict the load from trial to trial: The rate of rise of the force varies, which influences the rate at which the load approaches the target position. Similarly, the extent to which the force falls after the initial peak is related to the desired movement time and velocity profile. Because of delays in the positional expression of changes in muscular force (acceleration-to-position delays), the appropriate force adjustments are made in advance; the nervous system has to anticipate where the hand will be at a later time. It is difficult to imagine how such precise yet flexible force control could be achieved without sensory template error correction that produces appropriate adjustments in muscular torque through involuntary anticipatory mechanisms and voluntary changes in effort. That effort is adjusted as a function of load is reported by the subject.

The motor flow diagram in Figure 1 is a hypothetical model of how sensory templates are used to guide voluntary movement. The templates here are called desired approach profiles, or DAPs. DAPs are made up of kinesthetic and visual sensory attributes, hereafter called K-DAPs and V-DAPs. The K- and V-DAPs have two different representations, conscious and unconscious, the latter (K-DAP-A and V-DAP-A) generating the anticipatory force adjustments (AFA) needed for accurate targeting. The temporal features of this anticipatory profile are such that the motor commands are given with the appropriate lead time.

Once the K- and V-DAPs are established, the effort (E-) DAP can be structured with respect to profile, duration, and magnitude. The size of the anticipated load together with the KV-DAP will determine the appropriate E-DAP. The effort loop provides a check on whether the effort that was actually made matches the intended effort with regard to amplitude, duration, and temporal profile. When the selected E-DAP does not match the E-DAP actually sent forward, the perceptual counterpart is that the effort was larger or smaller than intended, improperly profiled, or inaccurate in some other way and the E-DAP can be adjusted accordingly. Such mistakes are most salient when one is trying to make a particular effort under isometric conditions.

The translation of E-DAP into contraction patterns for specific muscles occurs in the involuntary nervous system (e.g., how efferent drive should be partitioned between brachialis and brachioradialis for elbow flexion, etc.). Stretch and other reflexes also operate through the involuntary nervous system.

This model is at best incomplete. Information about contacts with external objects is available from cutaneous receptors and deep somesthetic receptors signal forces generated within the limb. Despite their importance in the control of voluntary movement, these afferent perceptions have not been included in the figure. The model has been put forward not because this commentator has any illusions as to its ultimate accuracy, but in the hope of stimulating progress in sensorimotor physiology.

Movement programs in the spinal cord

David Burke[1]

Department of Clinical Neurophysiology, The Prince Henry and Prince of Wales Hospitals, and Prince of Wales Medical Research Institute, University of New South Wales, Sydney, Australia

[DAM] The thesis presented by **McCrea** is so cogently argued and so eminently reasonable that it is difficult to take exception. In focusing on a few interneuronal circuits McCrea suggests that seemingly paradoxical inputs to interneurons would be quite logical if the motor task that is being undertaken (or is about to be) can determine which circuit is chosen to be active; that is, the reflex circuitry may be task-dependent and as a result the same afferent input may, in a different motor context, evoke a different motor output. This explanation does allow a number of inconvenient difficulties to be rationalized. However, one is left with two questions. First, in the real-life situation, does the interneuronal circuitry of the spinal cord ever change according to task? Second, is it reasonable to generalize from a few well-studied circuits to the rest of the interneuronal machinery?

Task-dependence has become a fashionable concept. The best-documented example (cited by **McCrea**) is that of cutaneous reflex reversals during locomotion. However, the human motor control literature now contains many examples of changes in the gain of different reflex pathways during voluntary movement, with the documented changes generally being appropriate to maximise agonist activity and diminish the activity of antagonists. Some of this literature is discussed by **Gandevia & Burke.** Suffice it to say that McCrea's suggestions are supported by a growing body of literature on both animal and human movement. As an aside, this emphasizes one important role that motor control studies in human subjects have played: to test hypotheses derived from highly reduced animal preparations in an intact cooperative subject.

Does **McCrea** address a fundamental property of all interneuronal circuitry? Alternatively, is it axiomatic that the excitability of and transmission across an interneuronal circuit will differ during different motor acts? Does the sense that he makes of interneuronal circuitry amount to the only way that the system could reasonably operate?

If it is conceded that different motor acts involve different descending commands and generate different afferent feedback, it is inconceivable that the excitability of the relevant circuitry would remain the same for all motor acts. Of necessity there would then be a change in gain of the appropriate pathway. However, a gain change is only one form of task-dependence: Two others are switching between alternative circuits with opposing actions and the paradox of convergent Ia/Ib inputs to common interneurons. To be able to control the relevant interneurons in such pathways would require a high degree of sophistication of supraspinal drives onto the interneuronal apparatus. The required sophistication would not be compatible with the view that movement programs are stored within spinal cord circuitry, so that only a relatively crude supraspinal signal is required to trigger coordinated activity. In addition, the very complexity of spinal circuitry implies that if a supraspinal trigger could generate movement, it would need to be a highly focused drive onto a select population of interneurons, capable of choosing between alternative pathways, if necessary. It is difficult to believe that a crude nonspecific supraspinal input would be sufficient to do this.

Nevertheless, it is well established that the interneuronal circuitry of the isolated spinal cord is capable of generating quite complex movements, such as locomotion and scratching, with appropriately coordinated muscle activity within a limb and between limbs. The movement repertoire of the spinal cord may be limited, and the movements themselves stereotyped and cyclical, but access to such movements would be of great value in patients with spinal cord injury. The restoration of function in paraplegic and quadriplegic patients is more likely to follow restoration of control over spinal cord circuitry than attempts to develop agents capable of promoting nonchaotic regeneration of severed long tracts. One of the major challenges facing clinical neurobiologists is how to exploit the untapped reserves of coordinated movement contained in the spinal cord circuits of patients with a functionally isolated spinal cord or one with impaired supraspinal control. To date the only coordinated or cyclical activity readily obtained from patients with spinal cord injury are flexor spasms and clonus, respectively.

The focus on spinal cord interneurons should not distract the reader from the fact that some descending pathways have direct monosynaptic access to spinal motoneuron pools. It is clear that these connections largely bypass the control systems discussed above. However, all such systems also exert indirect influences on the motor pool through spinal interneurons and through propriospinal pathways. Propriospinal neurons were clearly outside **McCrea**'s agenda, but many of the principles discussed for interneurons apply equally well to them.

One of the most thoroughly studied of propriospinal systems is the one that has its neurons in the C3-C4 segments of the cat, projects monosynaptically to motoneuron pools, and controls a particular form of forelimb movement, "target reaching" (Alstermark et al. 1981). These neurons receive convergent monosynaptic input from cortico-, rubro- and reticulospinal sources, and from muscle and cutaneous afferents in the limb (Illert et al. 1977; 1981). Although peripheral afferents may produce monosynaptic excitation of propriospinal neurons, their dominant action is disynaptic inhibition (Alstermark et al. 1984; 1986). Thus the major role of peripheral feedback seems to be to sculpt the descending motor command. An analogous system has now been identified in human subjects (Gracies et al. 1991; Malmgren & Pierrot-Deseilligny 1988a; Nielsen & Pierrot-Deseilligny 1991) and has been shown to be activated during voluntary contractions (Burke et al. 1992). Although quantitative comparisons have not yet been made for different movements, these propriospinal-like neurons appear to receive descending excitation with movements that cannot be described as "target-reaching," and the possibility remains that this system has evolved to be an important means of integrating descending commands from various higher centres with peripheral afferent feedback about current limb status for a variety of human movements. Whatever the precise role of the propriospinal system in the control of human movement, two lessons come from this work. First, to make sense of intrinsic spinal circuitry and the role it plays in the control of movement requires the detailed mapping of connectivity as undertaken by the Göteborg school, even if the resulting circuit diagram appears extremely complex and confusing. Second, this complexity and confusion may be capable of resolution, or at least understanding, if the hypotheses derived are then tested in a preparation capable of natural movement.

NOTE
1. Address for correspondence: D. Burke, Prince of Wales Medical Research Institute, C/O Department of Clinical Neurophysiology, The Prince Henry Hospital, P.O. Box 233, Matraville, N.S.W. 2036, Australia.

Is "attention" necessary for visuomotor transformations?

David P. Carey[a] and Philip Servos[b]

[a]*Department of Psychology, University of St. Andrews, St. Andrews, Fife KY16 9JU, Scotland and* [b]*Department of Psychology, University of Western Ontario, London, Ontario, Canada N6A 5C2*

Electronic mail: [a]*dpc1@st-andrews.ac.uk;* [b]*servos@cogsci.uwo.ca*

[JFS] **Stein** has done a commendable job in outlining the neuropsychological and neurophysiological evidence for the modular organization of the sensorimotor transformations required for directing eye and arm movements in space. Like Stein and others (e.g., Goldberg et al. 1990) we feel it is unlikely that there exists a unitary representation of space in the posterior parietal lobe. However, Stein's critique of a unitary representation of space in the brain can also be applied to his contention that one region is necessary for transformations across sensory, motor, and motivational domains. A number of recent reviews outline a substantial body of evidence implicating the participation of multiple cortical regions as well as subcortical structures in coordinate transformations (Chapula 1991; Goldman-Rakic 1988; Guitton et al. 1991; Smith 1990; Sparks 1991a). Similarly, attention is now rarely considered to be a unitary phenomenon. There is much evidence that separate attentional processes are mediated by different central nervous system structures (Posner & Rothbart 1992; Rizzolatti et al. 1985). Although the parietal lobe undoubtedly contains components of such attentional and transformational systems, it is unlikely that these networks exist exclusively in the parietal cortex.

A more critical problem with **Stein**'s model is his suggestion that attention mediates these coordinate transformations. He suggests that "it is . . . redirecting attention which generates correspondences among different frames of reference, and allows us to consciously localise objects with respect to ourselves and thus to plan voluntary movements towards them" (sect. 4.2, last para.). This hypothesis is difficult to reconcile with several observations in the visuomotor control literature. For example, Goodale et al. (1986) demonstrated that subjects can make accurate eye and hand movements to targets that are made to jump during primary saccades. Despite accurate compensation for the displacement by oculo- and skeleto-muscular control systems, the subjects are perceptually unaware of the jump and are unable to report it, even in forced/choice conditions. This finding suggests that transformations from retinotopic to oculomotor and limb coordinates do not depend upon Stein's version of "attention." Furthermore, Lee and his colleagues observed compensatory postural adjustments in subjects as a consequence of tilting the walls of a "moveable room." Their subjects were unaware of *both* the visual tilt and their own postural changes (Lee & Thomson 1982). These results contradict Stein's suggestion that movement control is critically dependent upon attending to an effector and the visual stimulus that drives it (Stein 1989b). It could be argued that many of the visuomotor control functions in which parietal subdivisions play a role should be independent of conscious perception (Goodale et al. 1991; Milner & Goodale, in press; Mountcastle & Steinmetz 1990).

The notion that attention is a necessary condition for coordinate transformations is also hard to reconcile with lesion studies. First, visuomotor dysfunction can be seen after parietal damage even when "attentional" mechanisms seem unaffected. For example, although optic ataxia is often attenuated by allowing patients to foveate (Corin & Bender 1972; Ratcliff & Davis-Jones 1972), the syndrome has been seen even when the target is foveated and recognized (Brain 1941; Holmes 1918; Jakobson et al. 1991). Indeed, **Stein**'s own experiments on cooling of area 7 in the monkey revealed that inaccurate reaching movements are made even when it can successfully foveate (and presumably "attend" to) a target (Stein 1978). Second nonparietal lesions can produce disordered visuomotor transformations, again apparently independent of attentional difficulties. For example, neck deafferentation in monkeys results in gross reaching errors even though they have little difficulty foveating the target (Cohen 1961). Taken together, these studies suggest that a concept of attention that is too closely coupled to foveation and/or awareness is unlikely to explain coordinate transformations in the CNS.

Although evidence from lesion and neurophysiological studies suggests that a number of visuomotor processes depend critically on processing in the parietal lobes, it is unlikely that all such visuomotor transformations occur there. A similar argument can be made for the neural substrates of attentional processes. The parietal lobes are likely to contribute to computations that are broadly identifiable as "attentional"; it is less clear that these processes are synonymous with conscious perception or even necessary for coordinate transformations.

From neuron to hypothesis

Paolo Cavallari

Istituto di Fisiologia Umana II, Università degli Studi, Via Mangiagalli 32, 20133 Milan, Italy

Electronic mail: *fisiol@imiucca.csi.unimi.it*

[EB, DAM] *1. The equilibrium point: Hypothesis or necessity?* As far as I understand it, the "equilibrium-point hypothesis" is meant to describe movement in terms of an equilibrium between elastic forces developed by antagonist muscles. Literally speaking, this concept appears to be an obvious application of the laws of classic mechanics (thus, a necessity more than a hypothesis). Or is it an attempt to neglect the influence of the inertial and viscous resistances on movement (and hence, as **Bizzi et al.** acknowledge in section 7.2, unable to describe fast actions)? The label anyhow seems inadequate to denote a special form of motor control. In fact, whatever the mode of motor control (taking into account inertial and viscous forces or not), movement must always respect the balance of forces predicted by the laws of mechanics.

These pedantic remarks may appear as arguments for a mere semantic dispute. This would be the case, if the equilibrium-point hypothesis had not been taken as a starting point for questionable experiments. In my opinion, the effects of spinal microstimulation described in section 5 are rather obvious and predictable on the basis of established notions. First, it is known that the great majority of afferent and descending fiber systems (e.g., Ib, cutaneous, group II and group III afferents, rubrospinal, corticospinal, vestibulospinal tracts) distribute to several different motor nuclei, directly or through interneurons and propriospinal neurons. Only Ia afferences are generally (but not always) funnelled to a single muscle or group of synergists. Second, it is hardly credible that the gray matter in the spinal cord should be a place where electrical stimulation can selectively activate a single motor projection. Third, it is also known that contraction confers elastic properties on the "plastic" resting muscle. Hence stimulation of any given point in the cord will almost necessarily activate several different pathways and excite many groups of motoneurons. The related muscle contractions will in turn generate a distributed pattern of elastic forces that will balance in a single equilibrium point.

Similarly, it is easy to predict that if the stimulating electrode is close to or inside the motor nuclei, the activation will predominantly affect one or a few muscles. In this case (and in all cases where contraction of a group of muscles overwhelms the force expressed by the antagonists) the equilibrium point goes out of the "work space" and, if movement is allowed, the limb is stopped at one extreme position by the osteoarticular constraints.

Another point is worth mentioning. It is by no means clear that the balanced recruitment of motoneurons and the related equilibrium point observed in these experiments are generated through the same interneurons that are used during natural movements to obtain the same limb position. A given pattern of motoneuron activation may be achieved through several different pathways, which may or may not utilize common interneurons (a limb flexion, for example, may be produced by stimulation of cutaneous afferents, FRAs [flexor reflex afferents], the spinal locomotor center, or the corticospinal tract). Moreover, a given equilibrium point may result from different sets of forces, obtained, for example, by substituting for one another the various synergists or single-joint with multijoint muscles.

In fact, if a point of equilibrium is reached even after a blind electrical stimulation of the cord, should one not suspect that this depends only on the biomechanical characteristics of muscles and bones?

2. More on group II pathways. To aid in David **McCrea**'s effort to make sense of spinal circuits, I would like to provide some information and add some comments on the systems that process group II information.

Group II connections have a wide distribution in the spinal cord but interneurons that mediate group II effects can be divided into functional subunits characterized by convergence from a limited number of muscles acting at different joints. L4 interneurons are known to receive group II afferents from leg extensors (sartorius and quadriceps) and from pretibial flexors (tibialis anterior and flexor digitorum longus). Other interneurons, localized in L6-S1 segments, show a convergence from ankle extensors (such as gastrocnemius-soleus and plantaris) and from more proximal muscles (such as quadriceps and gracilis). In both cases, however, group II afferents originate from muscles that are not strictly synergists but that can be simultaneously stretched during different motor acts. For example, during a flexion of the knee, L4 interneurons are excited only by quadriceps afferents. When flexion of the knee is accompanied by an extension of the foot (plantar flexion) there will be a summation of the effects from quadriceps and tibialis anterior.

On the other hand, L4 interneurons have multiple axonal projections which connect motoneurons of muscles that are different from those that contribute to the afferent input. The functional significance of this result is still unclear, but it suggests that muscle groups that are not directly engaged in the movement receive information about the length of the muscles that are stretched simultaneously. It is quite apparent that the complex scheme of convergence on and divergence from these interneurons adds to (and completes) the more stereotyped distribution of group I afferents. This would probably subserve and regulate the contraction of multiple muscles during complex movements. Recent results obtained in collaboration with Lars-Gunnar Pettersson (Cavallari & Pettersson 1991) in spinal cats indicate that excitatory group II effects from quadriceps and sartorius to posterior biceps-semitendinosus and gastrocnemius-soleus can be relayed by interneurons located in L7 (the same segment of the target motoneurons) and/or by midlumbar (L4-L5) interneurons, which in turn project to hindlimb motoneurons.

In different animals, each of the "parallel" pathways may contribute in different proportions to the final effect on motoneurons, indicating that the two systems may undergo independent control. Moreover, in the acute low-spinal state a group of interneurons located in the rostral lumbar segments (L2-L3) is tonically active and exerts an inhibitory action on L4-L5 as well as on L7 group II interneurons (Cavallari & Pettersson 1989).

Spinal tonic control acting on both group II and group III reflex pathways could also be of importance in regulating the access to the "alternative pathways" (see **McCrea**'s section 8) since it has been shown that it does not act with the same strength on all interneurons. In the same preparation the transmission of the effects from a certain muscle can be privileged. The modulation of the tonic inhibition therefore seems to be crucial in regulating and distributing reflex discharges toward some of the possible pathways.

Posture and gait in diabetic distal symmetrical polyneuropathy

Peter R. Cavanagh, Guy G. Simoneau and Jan S. Ulbrecht
The Center for Locomotion Studies, Penn State University, University Park, PA 16802
Electronic mail: *prc@ecl.psu.edu*

[SCG] From a clinical perspective, it is tempting to consider the title of the target article by **Gandevia & Burke** to be somewhat "tongue in cheek." There are very few clinicians who treat patients with disturbances of the peripheral nervous system who would even consider the issue open to doubt. If the question were asked at all, it might be posed as "Under what circumstances are kinesthetic inputs most important in the control of natural movement?" But clinical observation lacks scientific rigor, and despite the minimal role ascribed to proprioceptive feedback by the proponents of "muscle stiffness" (Bizzi et al. 1984), the authors provide us with a comprehensive and compelling argument in favor of a significant role for kinesthetic information in the control of upper limb movements.

In reviewing the studies of movement in deafferented humans, **Gandevia & Burke** report chiefly on the work of Sanes et al. (1985), who studied upper extremity movements in seven patients with subacute onset of idiopathic, predominantly sensory, neuropathy. Three other similar patients were studied by Forget and Lamarre (1987), and case studies of similar single individuals are also available (Cole 1986; Cole et al. 1986; Rothwell et al. 1982a). Whereas diabetes mellitus can cause sensory neuropathy in the lower extremity, the study of lower extremity movement in diabetic patients would seem an obvious supplement to the studies described above. Except for the five patients with diabetes mellitus studied during standing by Ojala et al. (1985), we have been unable to locate other studies on patients with partial deafferentation as a result of diabetes.

In fact, the main topics of interest in our own laboratory are the various consequences of diabetic distal symmetrical polyneuropathy (DDSP) – the typical stocking and glove distribution of sensory loss found in many patients with diabetes mellitus (Dyck & Brown 1987). This research began as a response to clinical needs, since we see many patients with foot problems (such as plantar ulceration), most of which are neuropathic in origin (Cavanagh & Ulbrecht 1991). More recently, in working with such patients, we have been struck by their complaints of lower extremity dysfunction during activities of daily life and we have come to believe that the diabetic neuropathic model has been overlooked as far as its potential for providing insight into motor control is concerned. Approximately 20%–50% of individuals who have had diabetes for at least 10 years experience some degree of peripheral neuropathy (Greene et al. 1990; Pirart 1978). This means that there is a pool of between 1 and 2.5 million individuals available for study in the United States, compared to the rather small and unique group of individuals with idiopathic deafferentation used in the studies mentioned above.

The study of patients with DDSP is not immune to the problems mentioned by **Gandevia & Burke,** however. In particular, the disease is variable in its staging and, as its name suggests, it may affect all divisions of the nervous system – sensory, motor, and autonomic (Dyck & Brown 1987; Dyck et al. 1987a; Greene et al. 1990). Yet it is the sensory manifestations that are predominant in most patients, and recent advances in quantitative sensory testing (Dyck et al. 1987b) allow a reasonable quantitative profile of functional deficits to be obtained

(Dyck et al. 1987b), while motor and autonomic deficits can also be determined (Edwards et al. 1984; Low 1984). Thus, a relatively homogenous sample with well-defined sensory deficits and normal strength can be formed by careful selection and screening.

Because the onset of DDSP is generally gradual and the deafferentation incomplete, this model will necessarily underestimate the role of afferent information in motor control – since patients may have learned to adapt to their disability (cf. Wolf et al. 1989) and since there will generally be some degree of residual sensory information. These two factors should not be seen as essentially limiting, however, especially in light of the availability of potential patients.

We have recently presented data on the performance during standing and walking of a carefully screened group of patients with significant DDSP, an age-matched control group of diabetic individuals without significant DDSP, and a further age-matched nondiabetic control group (Cavanagh et al. 1992; Simoneau 1992; Simoneau et al. 1992). Muscular weakness and a variety of medications were excluded and sensory deficit was quantified by vibration perception threshold, monofilament testing, and ankle movement perception threshold. It should be pointed out that these modalities are all principally served by large fiber afferents and we have no information on small fiber status in these patients.

Our findings highlighted the role of somatosensory input during standing, even in the presence of optimal visual and vestibular inputs. The mean sway (measured from a force platform as total movement of the center of pressure) in the neuropathic group with eyes open looking straight ahead was approximately the same as that measured in the nonneuropathic and control groups while standing in the eyes closed, head back position. Thus, during standing, vision and vestibular inputs were apparently not able to compensate for the loss of somatosensory input.

Gandevia & Burke predict that "deficits in timing of muscle bursts in highly stereotyped tasks such as . . . locomotion will probably also be revealed" (sect. 2.2, para. 6) and although "sensory ataxia" during gait is widely described in the anecdotal literature (Thomas & Brown 1987), we have not been able to locate any other quantitative studies of gait in individuals with sensory deficits. Our findings during gait (based on automated video analysis of treadmill walking) showed minimal (and statistically insignificant) differences between the same groups used above for the posture experiments. This initial study was confined to an examination of variability in sagittal plane kinematic patterns and we intend to extend it to look at frontal planar movements that have been reported anecdotally to be most affected by neuropathy as well as to perturbations during gait, which might be expected to elicit somewhat blunted responses. Nevertheless, the contrast between the marked decrements observed in neuropathic patients during standing and the minimal changes during gait suggest that somatosensory input is critical in the former task and either less important or redundant (perhaps based on learning) in the second. It is tempting to see a dominance of efference over afference in gait in these results. However, the role played by learning during the many thousands of repetitions of the stride pattern that occurred during conditions of declining sensory information (as the natural history of the disease progressed) needs to be further clarified.

There may be some valid debate over whether or not standing posture should be classified as a natural movement – since the movements that occur are small and not so clearly goal directed as in grasping or walking. Nevertheless, the results have indicated that somatosensory input is much more important in postural control than was previously supposed and that there may be good reason to revisit the patient with diabetic distal symmetrical polyneuropathy in the pursuit of further insight into the motor control of human movement.

ACKNOWLEDGMENT
This work was partially supported by a grant from the American Diabetes Association.

How accurately can we perceive the positions of our limbs?

Francis J. Clark
Department of Physiology, University of Nebraska, College of Medicine, Omaha, NE 68198-4575
Electronic mail: *fclark@unmcvm.bitnet*

[SCG] In their target article, **Gandevia & Burke** nicely catalog and describe the many sources of sensory signals that might contribute to kinesthesia and motor control – and there does indeed appear to be ample information available to signal the position and movement of the limbs. With all this kinesthetic information at our disposal, how accurately can we perceive the positions of our limbs? This is the issue I will address in this commentary.

In addressing questions of accuracy, the appropriate measure depends upon the question one is asking. Of the many options available, the commonly used measure of the mean disparity between the perceived position of a limb or joint and the true or target position might seem the best metric. The standard deviation about the mean would, in addition, provide a measure of reliability or precision. In some cases, this mean error might well be the measure of greatest importance. For example, a hunter with shaky hands who shot wildly but whose "average shot" coincided with the target would bag the prize at least once in a while whereas a hunter who shot with great precision but who was consistently off the target might never get the prize. (The latter hunter might do better were he less precise.)

However heuristically useful average measure of error might prove, do real-life motor tasks involve such averages? For the motor control system to program a movement it might do better to know the position of a limb or joint "right now" rather than where it sits "on average." "Where is the joint right now?" and "Where is the joint on the average?" are two different questions and can yield two surprisingly different answers.

Information theory provides a useful way of addressing the question of "where the joint is right now." It provides a measure of the number of bits of information a subject can derive from a unidimensional stimulus array (stimuli that vary along a single dimension such as intensity, frequency, or joint angle). One can also measure a "channel capacity" that can produce a good estimate of the maximum number of stimulus categories, or in our context, the number of different positions over a range that a subject can identify without error (Garner 1962, pp. 74–75). Increasing the number of target positions in the range (stimulus categories) in excess of the maximum indicated by the channel capacity would not increase the number of targets the subject could resolve but would only cause additional errors, with no net increase in information transfer (Hake & Garner 1951).

The measure of information transfer is most closely aligned with variance (Garner & McGill 1956; Miller 1956, p. 81). The noisier the channel, the greater the variability in the responses, the less information that the channel can carry and the fewer the number of stimulus levels that can be resolved without error. Though information theory and the concept of channel capacity existed before the 1950s, they have been used only occasionally in the study of kinesthesia (Durlach et al. 1989; Georgopoulos & Massey 1988; Sakitt et al. 1983; Soechting & Flanders 1989). It is of interest to note that psychophysical studies over the years indicate an upper limit in channel capacity for human sensory systems of seven plus or minus two stimulus categories (for unidimensional stimuli) irrespective of sensory modality, stim-

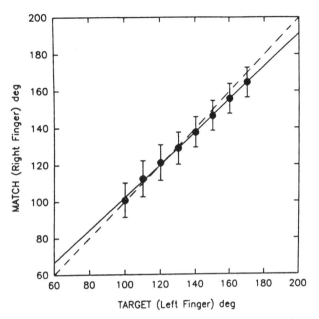

Figure 1 (Clark). Data from a position matching test where subjects move the right index finger PIP joint to match the passively set position of the left index finger PIP point. Eight target positions, spaced 10 deg apart, were used. Each of the 8 targets was used an equal number of times and presented in random order. Shown are mean values +/− 1 SD at each target, a linear regression line (solid line) from the raw data, and the identity line (dashed line). Pooled data from 4 subjects, n = 1280 (160 per target); r = 0.92.

Table 1 (Clark). *Channel capacities expressed in bits and the number of positions within a range that subjects can resolve without error for various joints in a position matching task*

Joint tested	Range (deg)	Bits	Number of positions
PIP	70	1.416	2.67
MCP	70	1.816	3.52
Wrist	70	1.847	3.60
Elbow	70	2.295	4.91
Shoulder	70	3.29	9.83

Note: Subjects matched the angle of the joint on the right to the passively set target angle of the corresponding joint on the left. There were 8 target positions, spaced 10 deg apart in all cases. The PIP, MCP, and wrist moved in a horizontal plane, the elbow and shoulder moved in a vertical plane. Data from individual subjects were not pooled; bits were averaged. (Pooling raw data from individual subjects would add the between-subjects variability factor, which would reduce the measured information transfer. This effect probably accounts for the difference in information transfer for the PIP joint between this Table and Figure 1.)
PIP = Proximal interphalangeal joint (index finger)
MCP = Metacarpophalangeal joint (index finger)

ulus range, and a variety of experimental conditions (Miller 1956). However, work in our laboratory suggests that this apparent universal upper limit may be a consequence of the paradigm used. Most earlier psychophysical studies involved absolute judgments of the stimulus category, where subjects indicated their response verbally with a name or a number. We found that using a different form of output, such as pointing to the target rather than assigning a label to indicate target position, can result in measured upper limits on channel capacities markedly higher than the "magical number seven" categories. Where channel capacities measure below seven levels, however, both the absolute identification (AI) paradigm and alternate methods of output such as pointing are likely to yield similar results. Channel capacities of many human sensory systems may indeed be surprisingly low.

I stated above that assessing the position of a joint or limb "on average" and assessing the position "right now" were different and could give different results. Let me offer an example. Figure 1 derives from a simple position matching task using the proximal interphalangeal (PIP) joints of the right and left index fingers. The data suggest that, on average, subjects matched PIP joint positions with good mean accuracy and reasonable precision. However, information transfer analysis gives another perspective; information transfer over the target span measured 1.24 bits or 2.37 stimulus levels! With a resolution of 2.37 stimulus levels, subjects could do only slightly better than indicating whether their finger was nearer the high end or the low end of the range.

An important difference between our pointing model and the classical AI model is that the pointing model predicts that channel capacity will be dependent on stimulus range, whereas the AI paradigm predicts channel capacity to be nearly independent of stimulus range. This range dependency for pointing would be especially relevant with the shoulder, where the 70 deg range presented in Table 1 is but a portion of the total range of shoulder movement in a plane. The relationship between range and channel capacity is complex, but as a rule of thumb, doubling the range would add between 0.6 and 1.0 bits (an increase of 50% to 100% in the number of resolvable positions).

Which measure should one believe? It depends on the question one is asking. I think that if one asks what kinesthetic information is available for motor control, then the information transfer analysis gives the best assessment. If a person can resolve with a given joint a maximum of three stimulus levels (positions) in a range, then at best the motor system can discriminate whether the joint is in the upper, middle, or lower third of the range.

What about kinesthetic acuity with other joints? Some preliminary findings from our studies of individual joints are listed in Table 1. There is a clear proximal/distal effect, with the proximal joints showing higher positional acuities. However, the overall impression is that joint position sense is rather coarse. Needless to say, there is much more to this story, but our continuing studies appear to confirm that joint position has low resolution.

Mathematics is a useful guide to brain function

Thomas L. Clarke

Institute for Simulation and Training, University of Central Florida, Orlando, FL 32826
Electronic mail: *clarke@acme.ucf.edu*

[DAR] David **Robinson**'s perspective on the "Implications of neural networks for how we think about brain function" is correct as far as it goes, but he does not consider recent developments in mathematics. Although the tensor calculus used by Pellionisz and Llinas (1980) was the basis of fundamental physics early in the century, it has been replaced by more general abstract differential geometry in current physical theory (Singer 1982). The utility of this type of mathematics in brain modeling has only been tentatively explored in the area of vision (Hoffman 1978). Similarly, the theory of dynamical systems has been revolutionized by the advent of chaos theory. Chaos theory also has exciting implications for understanding the brain (Freeman 1990). [See Skarda & Freeman: "How Brains Make Chaos in Order to Make Sense of the World" *BBS* 10(2) 1987.] In both of these examples new mathematics should lead to additional insights, so it is premature to conclude, as Robinson does, that mathematical modeling is limited in its contribution to understanding brain function.

It is difficult to quarrel with **Robinson**'s assertion that the blind probing of neural activity in the brain is not the way to insight. Even the classical successes of neural recording have not greatly illuminated the problems of brain function. Although Hubel and Weisel's (1977) work is a good example of the value of Robinson's exhortation to record from "behaving animals," the retinotopic mapping of neurons in the visual cortex remains a puzzling regularity in brain structure and provides little direct insight into visual behavior. Philosophical considerations suggest an undecipherable pattern of recordings in the brain, so that Hubel and Weisel's discovery of yet another remapping of the visual world was startling. Their discovery moved the locus of the perception problem deeper, but did not solve it; the homunculus was not exorcised, only pushed deeper into the brain.

Hubel and Weisel's observations begin to make sense when viewed from the perspective of the abstract mathematical concepts of differential geometry (Hoffman 1978). The direction-sensitive cells found by Hubel and Weisel can be understood by analogy with the mathematical elements of the Lie group which underlie the symmetries of vision. As Dodwell (1983) points out, the symmetries of vision are continuous and constitute a mathematical group, so Lie group theory must apply. Given the neuron as the basic processing element, the organization of the cortex found by Hubel and Weisel can be seen as a mathematical necessity. Furthermore, the abstract differential geometry provides a path via prolongation and jet-space theory whereby the regress of retinotopic maps may be stopped without resorting to a homunculus. This regress must be stopped if the level of understanding epitomized by technological application is to be reached (Clarke & Ronayne 1991).

As Robinson makes clear, dynamical, as opposed to geometrical, patterns of neural activation cannot be explained by the blind application of simple "black boxes" modeled after engineering practice. Modern developments in theory, however, offer an increased range of possible mathematical models. In much the same way that neural-network connectionist models are an advance over cognitive models based on the von Neuman universal Turing machine, modern nonlinear dynamics offers advances over linear black-box control theory.

Nonlinear dynamic theory allows the analysis of the collective activation of large groups of neurons in terms of higher-order concepts such as the strange attractors of chaos theory. Freeman (1990) discusses how stimulus equivalence in the olfactory bulb may be a manifestation of chaotic neural dynamics. The nonlinear and chaotic neural dynamics of the bulb are characterized by a large number of intricately interleaved strange attractors so that under control of olfactory inputs, the bulb dynamics bifurcate into the attractor state that most closely matches the input. The instability of the nonlinear dynamics provides the necessary sensitivity and the means to learn responses to novel odors. Note that while not identical, the bulb is the same order of complexity as other primitive brain structures such as the lateral geniculate nucleus, which is involved in the oculomotor system studied by **Robinson**.

Just as physical application lags behind mathematical inven-

tion, brain modeling currently lags behind physical application. Thus, whereas **Robinson**'s conclusions concerning the inapplicability of mathematical models to the brain are true of older physical practice, the use of high-order concepts drawn from abstract mathematics and from nonlinear dynamical theory can do much to eliminate Robinson's objections. Using his metaphor of the engineer, it is indeed fruitless for the engineer to analyze recordings of the hidden units in a neural-network controller using linear theory. On the other hand, if the engineer can recognize regularities in the neural activation with the aid of an abstract nonlinear theory of the controlled process, he will be better able to understand the limitations of the neural-network controller. Indeed, just as it can be paradoxically easier to control a chaotic than a nonchaotic system (Ott et al. 1990), he may be able to construct a much simpler and more economical controller by utilizing this abstract insight.

Posterior parietal cortex and retinocentric space

Carol L. Colby,[a] Jean-Rene Duhamel[a,b] and Michael E. Goldberg[a]

[a]*Laboratory of Sensorimotor Research, National Eye Institute, National Institutes of Health, Bethesda, MD 20892 and* [b]*Laboratoire de Physiologie Neurosensorielle, Centre National de la Recherche Scientifique, 75270 Paris, Cedex 06, France*
Electronic mail: *clc@lsr-vax.uunet*

[DAR, JFS] Stein suggests that the brain in general and the posterior parietal cortex in particular do not need an explicit representation of target location in space. This position is supported by our recent evidence that neurons in the lateral intraparietal area (LIP) of the monkey encode spatial locations in terms of distance and direction from the current or intended center of gaze (Duhamel et al. 1992).

Neurons in LIP discharge in a number of disparate circumstances in which their activity can be interpreted as describing a spatial location. They respond to visual stimuli appearing in a retinal receptive field (Andersen et al. 1990b) and they give an enhanced response when the monkey attends to a stimulus in that receptive field (Goldberg et al. 1990). Some LIP neurons also have a presaccadic discharge when the monkey makes purposive saccades to the spatial location of a vanished stimulus that has recently appeared in the receptive field (Barash et al. 1991). They even discharge before saccades of the proper amplitude and direction when no stimulus has recently appeared (Goldberg et al. 1990). Furthermore, LIP neurons can anticipate the appearance of the stimulus and begin to discharge even before the stimulus appears (Colby et al. 1992). Finally, some LIP neurons discharge continuously during the interval between target appearance and the initiation of a saccade to the target, as though the neuron were holding an image of the target location (Colby & Duhamel 1991; Gnadt & Andersen 1988).

Thus LIP neurons respond (1) when the monkey makes a saccade *and* when it is expressly forbidden to do so; (2) when the monkey makes a saccade without a stimulus *and* when a stimulus appears without any probability of the monkey's making a saccade to it. This contrasts with the arcuate frontal eye field, where there are separate classes of neurons that discharge in response to visual stimuli and before non-visually guided saccades (Segraves & Goldberg 1987). In LIP, the majority of neurons have visual, presaccadic, anticipatory, and memory-related activity. All these functions, however, refer to a spatial location: a saccade to that location, a stimulus in that location, anticipation or memory of an event at that location. The intersection of these disparate functions is the description of a location. It is critical to emphasize that location in this case means distance and direction from the center of gaze, not from some body-centered or world-centered reference point.

This retinocentric representation is updated with each saccade. We have recently demonstrated that when a saccade brings the spatial location of a visual target into the receptive field, some LIP neurons begin discharging long before the time that would be predicted by the visual latency, and some discharge even before the saccade begins. The neurons also respond to flashed targets that have disappeared before a saccade that brings the spatial location of the flashed target into the receptive field. Parietal cortex compensates for the saccade by remapping the cortical representation into the coordinate system of the new fixation. The retinal reafferent signal now confirms the shifted signal. The output message of LIP neurons describes a spatial location a certain distance and direction from the current or intended location of the fovea, and these neurons can combine retinal information with information about the saccade to update the cortical representation.

This remapping mechanism is necessary for two reasons: (1) Every eye movement shifts the fovea to a new location and brings each stable visual object into a new set of visual receptive fields. If the system relied only on new excitation of the cortex from the new retinal locations of visual stimuli, the cortical representation would be spatially inaccurate for a visual latency of 70 msec after every saccade. (2) The spatial location of flashed targets could never be accurately represented across saccades, because the new excitation necessary to update the representation would never occur. With remapping, LIP can maintain a spatially accurate retinocentric representation across saccades and never require an explicit representation of target position in space.

Stein asserts that shifts in attention play a major role in coordinate transformation. However, the remapping effect that we see seems to be independent of attention. In the experiment described above the monkey is never asked to use the stimulus that drives the cell. We have demonstrated the effect (Walker & Goldberg, unpublished) in a monkey that was never trained on a task in which it might use the stimulus. Insofar as it is ever plausible to assert that a monkey is not attending to a stimulus, we can make that assertion here. Furthermore, the remapping effect requires a saccade. If the stimulus is in the proper location relative to the saccade target and we ask the monkey to make a saccade to the target, LIP neurons anticipate that the saccade will bring the stimulus into the receptive field and the cells discharge. If we ask the monkey only to attend to the target without making the saccade, the cell does not discharge despite the fact that the stimulus is the proper distance and direction from the attended target. The distinction between the effects of attention alone and the effects of attention plus intention may be important for understanding coordinate transformations in all domains.

The remapping mechanism renders unnecessary the explicit calculation of target position in space postulated by Zipser and Andersen (1988) in their network model and described here by **Robinson.** That model requires that some neurons signal target position in space, a signal that has been difficult to find in precisely that region, the parietal cortex, where it should be easiest to find. Robinson argues that because signals can be distributed across a network, they may not necessarily be seen at the tip of a microelectrode. However, he describes another network, one for the vestibulo-ocular reflex (VOR), that uses input and output signals that have been described in recordings. The hidden layer neurons of this network resemble to a surprising degree the second-order vestibular neurons described by Fukushima et al. (1990). In this example, the network model and the experimental details complement each other nicely. In contrast, the output layer of Zipser and Andersen's model has no established correlate in the brain. **Stein's** analysis suggests that the search for neurons that code target position in space may be futile not because, as **Robinson** suggests, it is a distributed representation but because it is an unnecessary one.

A robotics perspective on motor programs and path planning

Christopher I. Connolly

Laboratory for Perceptual Robotics, Computer Science Department, University of Massachusetts at Amherst, Amherst, MA 01003
Electronic mail: *connolly@cs.umass.edu*

[GEA] The **Alexander, DeLong & Crutcher** target article is relevant, not only to theories of biological motor systems, but to robotics research as well. If we roboticists are to construct versatile robot systems, we need to concentrate on engineering solutions that provide robot systems with the ability to react quickly and appropriately to environmental changes – just as their biological counterparts do. Indeed, biological motor systems are the only existence proofs for this sort of flexibility. I therefore think it is appropriate for roboticists to maintain at least some familiarity with those aspects of neural science that are relevant to their objectives. In light of this opinion, I would like to tie the target article's criticisms to some current robotics literature.

The notion of a "motor program" is prevalent in the robotics path planning literature. This is most apparent in techniques that rely on methods from computational geometry or algebra (see, for example, Canny 1987; Hopcroft et al. 1985; Kapur & Mundy 1989; Lozano-Perez 1981). In these methods, the emphasis is on constructing complete paths from one point to another. Some of these algorithms are of great value to theoretical computer science, but in a practical sense, they succumb to the same criticisms that **Alexander et al.** level at theories of biological motor systems. Most "motor program" algorithms in robotics are not naturally suited to a massively parallel computing architecture. Moreover, with these techniques, any change in the robot's environment requires a "substantive planning phase". With current computer technology, such replanning is a time consuming step and of necessity results in a certain inflexibility in the robot system.

Over the past 10 or so years, a class of "potential field" planning methods has arisen in robotics (see, for example Arkin 1987; Barraquand & Latombe 1991; Khatib 1987; Krogh 1984; and Newman & Hogan 1986). Such methods rely on the gradient descent of a potential function constructed in some state space to drive the robot system away from obstacles and toward goal positions. These algorithms typically use a geometric model of the environment to set up a repulsive vector field around every obstacle and an attractive field around the goal (this is also reminiscent of Arbib's control surfaces, Arbib 1981, and the path-planning model of Arbib and House, Arbib 1987). The robot system then follows the vector field that results from the sum of the repulsive and attractive fields. These techniques are sometimes regarded as being local in nature; it has been assumed that they cannot by themselves be used for generating whole paths from start to goal. The reason for this is that in the case of arbitrary vector fields, spurious local minima can arise. These are points where the individual vector fields cancel, the result being zero. No movement can occur at these points, and once there, the robot is stuck short of the goal. Recently, however, progress has been made in devising techniques that do not exhibit the "local minimum" problem (see Connolly et al. 1990; Koditschek 1987). Moreover, the technique described in Connolly et al. (1990) admits of a very simple and fast hardware implementation (Tarassenko & Blake 1991) that has some connectionist qualities. It consists simply of a programmable resistive network. Voltage levels within the network provide the function to be descended.

Although not strictly connectionist by some definitions, potential field methods can also serve as candidates for partial models of the motor system (Mussa-Ivaldi et al. 1991a). The implementation described by Tarassenko and Blake is indeed "unsupervised, with changes in synaptic weights (connection strengths) being based entirely on local synaptic conditions." In this case, the "weights" correspond to voltages within the grid. Obviously, such a model is a simplification that ignores the plasticity of the motor system. On the other hand, the medium spiny neurons of the neostriatum receive massive cortical afferents yet are the primary output neurons for the neostriatum. The monosynaptic nature of the cortico-striato-pallidal pathway does not suggest the presence of "hidden units" that would support a more traditional connectionist model. The same cells are also relatively quiet (Kandel et al. 1991; Shepherd 1990). Is it possible that these cell properties really reflect some sort of local, passive computation, such as that seen in a resistive grid?

The somototopic divisions of the putamen are consistent with a state-space, especially in light of the results of Crutcher and DeLong (Crutcher & DeLong 1984a; 1984b) and Kimura (1990). The massive striato-pallidal convergence presumably gives the globus pallidus access to information from any point within the neostriatum. Could this somatotopy and convergence reflect some sort of computation of potential functions within different state spaces (corresponding to positions of limbs, face, etc.), which are then sampled by the globus pallidus in order to drive state changes within the system?

Alexander et al. suggest that the transformations required by "a sequential/analytic approach to control movement" (sect. 3, para. 4) would be reflected in specialization within the motor system. I would like to note that in some discrete spatial representations used in robotics, forward and inverse kinematics computations can be performed in a massively parallel manner by simply permuting and combining the inputs using a Boolean "or" function to go from one representation to another. Is it possible for this to be accomplished within the motor system by permutation and arborization of afferent fibers?

In conclusion, I think the target article offers insights and criticisms that are applicable to the field of robotics. Moreover, increased knowledge of biological motor systems should lead to improvements in our own ability to build successful, versatile robot systems. I especially agree with the authors insofar as the only real successes in robotics will come through the use of distributed, massively parallel computing, similar to that found in biological systems.

ACKNOWLEDGMENT
I would like to thank Brian Burns for some crucial discussions during the preparation of this commentary. I would also like to thank Mary Andrianopoulos and Rod Grupen for their insights.

Successive approximation in targeted movement: An alternative hypothesis

Paul J. Cordo and Leslie Bevan

R. S. Dow Neurological Sciences Institute, Good Samaritan Hospital and Medical Center, Portland, OR 97209
Electronic mail: cordop@ohsu.edu

[EB, JRB, SCG, JFS] *Springlike properties do not mandate equilibrium-point control.* As Bizzi et al. remind us, muscles have springlike properties because of their passive mechanical characteristics and local reflex innervation. All tissues are elastic to some degree, and elastic properties are useful in allowing us to move in and interact with our environment. Contact instability in robotics serves as an example of how a *non*elastic system fails to interact appropriately with its environment, and extending the equilibrium-point model to virtual trajectories and contact control tasks is interesting and potentially valuable. Clearly, the central nervous system (CNS) must incorporate the mechanical properties of its actuators to move the limbs accurately. Bizzi et al. demonstrate that the joint has elastic proper-

ties, but this does not mandate that these springlike properties are used explicitly to control targeted limb movements.

Equilibrium-point control is only a model of the initial 100–200 msec of normal movement. Is equilibrium-point control used to plan movement trajectories or are trajectories planned by a different mechanism? What is the significance of this question? Normal limb movements are influenced by both visual and kinesthetic input, yet when it comes to sensory control Bizzi et al. adopt a minimalist position, acknowledging only the potential role of local reflexes. In contrast, Gandevia & Burke have clearly shown that sensory input can play many roles in the control of limb movements, and both Bloedel and Stein describe elegant cortical mechanisms for processing sensory information that play an important role in guiding limb movement. Because the equilibrium-point models do not adequately account for the role of sensory input during normal limb movement, their utility must be restricted to movement initiation.

How is sensory information used during limb trajectories? It is now well established that movements to visually locatable targets are guided during the early stages of a movement by visual information obtained from the target (Cordo 1987; Gordon & Ghez 1987b). This type of sensory control is particularly useful when the target is moving, either predictably or unpredictably. A second stage of sensory control begins 100–200 msec after movement onset when visual and kinesthetic information about errors in movement trajectory result in significant error corrections.

Most normal movements are not made at maximal velocity; they last at least 400–500 msec. Hence the proportion of time that normal movements would be under the *exclusive* control of an equilibrium-point or other trajectory-planning mechanism would be 50% at the extreme. So why all the deliberation over the first 100–200 msec of a movement? One could argue that it is the initial part of a limb trajectory that has the most influence over the direction of movement. It seems equally possible, however, that later phases of movements (controlled by sensory input) could be more influential in shaping the trajectory of targeted arm movements.

Overhead and precision in determination of initial movement trajectories. It is pointed out by Bizzi et al. that the virtue of the equilibrium-point model is computational simplicity – the CNS is not required to determine muscle forces (i.e., to compute inverse dynamics) to produce accurate movements. On the other hand, models in which the CNS computes inverse dynamics (or looks up the answer) are criticized for having too much *overhead*, either in computational load or storage space. Bizzi et al. emphasize that the inverse-dynamics problem must be solved with precision. We would like to suggest that equilibrium-point models also have overhead and that inverse-dynamics determinations do not have to be precise.

Where does the overhead come from in the equilibrium-point model? The force produced by a muscle at a given length depends upon the context of the movement dynamics (i.e., force-velocity relationship) and the activation history of the muscle (i.e., muscles exhibit potentiation, fatigue, and hysteresis). Thus, there is no unique relationship between force and stimulus frequency. Even if we were to assume that muscle force is not activation-history dependent and that hypothetically any movement could be executed by some combination of the length-tension relationship at one of 20 different activation levels, the CNS would have to "know" over 10,000 different length-tension relationships! Thus, for the CNS to maintain an accurate library of length-tension curves for each muscle under all possible conditions of activation history would require significant storage and analysis overhead.

Neither inverse-dynamics computations nor length-tension determinations have to be precisely made for limb movements to be accurate. Hypothetically, computations of joint torques and muscle forces need only be precise when no other information about the movement task is available, a rare condition. When any sensory information is available, even if it is only a visually located target, adjustments can be made to the movement trajectory during the movement.

An alternative: Successive approximations. As an alternative explanation of the events leading to a targeted arm movement, we hypothesize that the target is acquired by the hand through a series of approximations, as first hypothesized by Greene (1972; 1982), who reasoned that "it is often easier for a control system to perform an automatic, even if not quite correct, feedforward, which is then corrected by a simple feedback system, than to try to compute an exact compensation in one stage." In our laboratory we have observed a succession of centrally and sensorily driven approximations in visually guided torque changes (Cordo 1987; Cordo & Flanders 1989).

With successive approximations an equilibrium-point or alternative-trajectory planning mechanism would control the initial phase of a movement. The CNS would rely on previous experience, length-tension relationships, inverse-dynamic computations, and so forth, to generate the first approximation to the movement trajectory. This initial phase only has to get the arm moving in the general direction of the target. In the second phase, visual information about the target location adjusts the trajectory toward the target. In the third phase, visual and kinesthetic input would reduce residual errors in trajectory.

If movement is carried out by successive approximations, the inverse-dynamic computations or length-tension relationship would not have to be very precise, eliminating any major "advantage" of the equilibrium-point models. We still do not know how the CNS controls the initial part of the trajectory, but by removing the need for precise computation the choice between equilibrium-point models and inverse-dynamics models becomes more difficult. If one can accept that initial trajectories do not have to be precisely determined it opens the door to a striking variety of sensory control mechanisms, as described by **Gandevia & Burke, Bloedel,** and **Stein.**

FINSTs, tag-assignment and the parietal gazetteer

Michael R. W. Dawson

Biological Computation Project, Department of Psychology, University of Alberta, Edmonton, Alberta, Canada T6G 2E9
Electronic mail: *mike@psych.ualberta.ca*

[JFS] A major theme of **Stein**'s target article is that space need not be represented in the posterior parietal cortex (PPC) by explicit topographic maps. To illustrate a plausible alternative, Stein suggests that space could be represented in the PPC by an "implicit map" or a nonspatial "lookup table" analogous to an atlas's gazetteer. If one takes a serious look at this proposal, some interesting design issues emerge. In particular, two types of coordinate transformation rules must exist in the PPC: one set of rules to transform coordinates from one frame to another (e.g., from visual to motor coordinates), and another set of rules to perform transformations within a single frame (e.g., within visual coordinates, or within motor coordinates). [See also Flanders et al.: "Early Stages in Sensorimotor Transformation" *BBS* 15(2) 1992.]

Pylyshyn (1989) has proposed a scheme for spatial representation that could serve as an instantiation of **Stein**'s implicit map. In essence, Pylyshyn's model consists of a finite number of attentional tags that can be assigned as labels to individuated visual entities. These tags, which he calls finger instantiations (FINSTs), are analogous to place name entries in a gazetteer: FINSTs serve as references to spatial locations (i.e., places in the visual field), but do not *explicitly* specify spatial coordinates. "A FINST is very different from an encoding of the position of a feature. The FINST itself does not encode any properties of the feature in question, it merely makes it possible to locate the feature in order to examine it further if needed" (p. 70).

To be useful, a gazetteer cannot be only a set of labels that represent entities of interest. In addition, these labels must be associated with reliable spatial references. For instance, in my National Geographic *Atlas of the World*, the entry for Edmonton directs me to page 104, and to coordinates (P, 11). This entry is useful because its spatial reference is fixed – in my atlas, Edmonton is *always* on this page and at these coordinates.

In contrast Pylyshyn's FINST mechanism must solve some difficult coordinate transformation problems to produce reliable spatial references. This is because the locations pointed to by FINSTs are not fixed. The retinal coordinates of an individuated feature cluster can change if the distal object producing the feature cluster moves, if the eye moves, or if the observer moves. In order for his model to work under these volatile conditions, Pylyshyn's attentional tags must be "sticky" – once assigned to an entity, a FINST must continue to point to it even as the entity moves. This suggests that if space in the PPC is represented as a "gazetteer," then a transformation must be computed to map spatial coordinates (e.g., retinotopic position) onto the FINSTs reference frame. This type of transformation does not require an explicit mapping from one coordinate frame to another, and as a result differs from the transformations reviewed by Stein.

Dawson (1991) has argued that mechanisms that solve the so-called correspondence problem in motion perception are perhaps better viewed as the mechanisms that solve what Strong and Whitehead (1989) call the "tag-assignment" problem in their earlier target article *BBS* 12(3) 1989. In short, motion correspondence processing is responsible for making FINSTs "sticky." The importance of this claim to **Stein**'s work is that there is considerable theoretical and experimental evidence to suggest that motion correspondence processing (or tag-assignment) is also carried out by the PPC (see Dawson 1991, part 4). This is to be expected if space is actually represented in the PPC as a "gazetteer," because this representation requires the tag-assignment problem to be solved in order to maintain appropriate references to sensory coordinate systems.

A second theme of **Stein**'s target article is that attentional processing plays an important role in coordinate transformations. He speculates that attention serves to select appropriate algorithms for coordinate transformations; however, this view is somewhat at odds with the fixed properties of models like the feedforward transformation network designed by Zipser and Andersen (1988). Pylyshyn's (1989) FINST model provides an alternate view of attention's role.

Pylyshyn suggests that visuomotor coordination could be accomplished by having an additional set of attentional tags, called ANCHORS. These tags are functionally equivalent to FINSTs, but refer to coordinates in motor space, not visual space. Motor movements are coordinated with visual information by a primitive BIND operation which links a FINST and an ANCHOR together as a unit. This BIND operation would appear to be functionally equivalent to the coordinate transformation operations discussed by **Stein.** Note, however, that in this case attention is used not to select a particular transformation algorithm but rather to designate which FINST and ANCHOR are to be bound. In other words, attentional processing is analogous to selecting specific gazetteer entries as being interesting enough to examine. Furthermore, the coordinate transformation between two bound tags could be fixed – and thus performed by a "hardwired" feedforward network. This is because between-modality transformations would essentially be transparent to sensory coordinates and would involve mappings between two labels that do not explicitly represent such coordinates. The ultimate translation into sensory coordinate frames could be accomplished by the same within-modality mechanisms that mediate tag-assignment [See also Pylyshyn: "Computational Models and Empirical Constraints" *BBS* 1(1) 1978 and "Computation and Cognition" *BBS* 3(1) 1980.]

ACKNOWLEDGMENTS
This work was supported by Natural Sciences and Engineering Research Council of Canada operating grant A2038 and equipment grant 46584.

Is equilibrium-point control all there is to coding movement and do insects do it, too?

J. Dean

Abteilung 4/Fakultät für Biologie, Universität Bielefeld, D-4800 Bielefeld 1, Germany
Electronic mail: *jeff@bio 128.uni-bielefeld.de*

[EB, DAR] Biological selection works on the performance of all animals in similar ways and interspecies interactions such as predator-prey relationships should ensure similar levels of performance. It is therefore interesting to consider to what extent similar control strategies have evolved. **Bizzi et al.** make a strong case for the equilibrium-point model of motor control, but their evidence is based on results from primates and more recent experiments with frogs. As they state, the primary problem is finding a way to calculate muscle activities suitable to move the body and limbs in the desired manner. The allure of equilibrium-point control is that it greatly simplifies the problem because it circumvents the need to solve the inverse dynamics and it does so by relying upon biological characteristics of muscles, characteristics that at first make muscles appear to be poor actuators. On the other hand, as Bizzi et al. admit, equilibrium-point control in its simplest form does not solve all problems. In this comment I would like to discuss this hypothesis in the light of data from insect and humans; I will also briefly address issues raised by **Robinson**.

Evidence from insects. This same control problem arises in an extreme form when control models are required for a walking machine. Because insects are quite skilled walkers and walk in such a way that static stability is maintained throughout the step cycle, they have often been considered as biological models for suitable control schemes. Numerous control models have been developed that generate proper step rhythms and interleg coordination, but all use rather idealized legs and simplified step patterns, ignoring real dynamics (e.g., Cruse & Graham 1985; Dean 1991a). A realistic model of leg dynamics has been developed for the stick insect and used to answer the question of what criteria might be used to constrain redundant degrees of freedom (Pfeiffer et al. 1991). Using this model, the inverse-dynamics problem can be solved. However, the requisite computation would slow any real-time control to an extreme crawl and would surely doom any animal to extinction. Hence current efforts are aimed at developing alternatives to the inverse-dynamic approach.

Behavioral results with the stick insect provide evidence both for and against the simple equilibrium-point hypothesis. One apparent problem concerns the relation between muscle activity and movement. Typical electromyographic (EMG) recordings from leg muscles show a more or less strict alternation of activity in antagonistic muscle groups. When resistance to movement increases or the movement is longer, activity in the agonist is simply increased. This pattern makes it appear that the controller merely strives to move the leg in one direction or the other. Of course, one could argue that the antagonistic spring is just the residual stiffness in the previously active muscles. However, if the antagonist is generally inactive, the simple notion that the equilibrium point is defined by a particular ratio of activity in antagonistic muscle groups appears less suitable and suggests that one should look for alternatives. Whether the various modulators that insects can use to modify muscle relaxation or the elasticity present in the joints themselves adequately replaces the missing antagonist spring is unclear.

In behavioral experiments, external forces have been added to resist or assist leg movement during both return and power strokes (Dean 1984; 1991b). The results are not in accord with movement along a predetermined virtual trajectory. For example, a leg making a return stroke could not be moved ahead of its virtual trajectory by applying a brief external force, as was possible with deafferented monkeys. Assisting forces applied during the power stroke slowed rather than speeded up walking. Similarly, when a leg is stopped during its power stroke, the measured force first increases, as one would expect with an increasing divergence from a virtual trajectory, but then it decreases (Cruse 1985). Together, these results suggest that the insect attempts to move to a specified end point but that it tries to control velocity while moving (Dean & Cruse 1986). This interpretation of the data must be qualified slightly because the action of reflexes and proprioceptive information in the intact animal may conceal the underlying control strategy, as **Bizzi et al.** point out. Deafferentation experiments such as those in monkeys are not feasible in the stick insect. However, behavioral experiments with animals in which a proprioceptive feedback loop was externally closed also support the notion of velocity control during movement (Weiland & Koch 1987). Bizzi et al. mention an extension of the equilibrium-point hypothesis that includes velocity and position feedback, and they suggest that this may be particularly relevant for fast movements. Whether this model is better suited to the insect data remains to be tested.

Other simulation studies, however, do support the notion that leg control does not involve explicit calculation of the kinematics. In the stick insect, the return movement is a true target movement in that the tarsus of the moving leg steps to a position just behind the tarsus of the next adjacent leg. The target position is determined using proprioceptive information on the angles of the leg joints. Thus, this interleg coordination would appear to require a solution of the direct kinematics to determine the tarsus position of the target leg, translation into a coordinate system appropriate to the moving leg, and solution of the inverse kinematics to determine the required joint angles for placing the tarsus at the desired target position. However, a neural network simulation performed to test different coding schemes showed that realistic levels of accuracy can be obtained simply by associating joint angles of the moving leg with joint angles of the target leg (Dean 1990). In accord with the tenor of the equilibrium-point hypothesis, the results show that biological systems may rely on less exact but simpler algorithms in preference to exact calculations.

Moreover, in agreement with the conclusion expounded by **Robinson**, the network simulation also shows that an exact mathematical description of the behavior need not correspond closely to the operations actually carried out by the underlying neural circuits; the former may be implicitly rather than explicitly present in the function of the latter. The simulation was undertaken because intracellular recordings had shown that, contrary to prior expectations, surprisingly few intersegmental interneurons convey a tonic measure of target leg position to the segmental ganglion controlling the moving leg. Moreover, each of these interneurons appears to report the angle at only one of the three leg joints. In this respect, the simulation results help us interpret the physiological results. By showing that the performance level can be realized by simple association of joint angles, they allay the fear that the microelectrode is presenting an unrepresentative picture of the neural mechanisms. Of course, much remains to be learned about the conversion of the target information into motor commands. Recordings of local premotor interneurons in various insects suggest that these mechanisms will be complex and difficult to untangle, as Robinson concludes. The behavior of the hidden units in the simulation was examined only far enough to see that at least two

Commentary/Movement control

qualitatively different patterns of connectivity can be established to perform the task. Obviously, the circuit in the insect, which has many more hidden units, may be entirely different.

Evidence from human pointing movements. Movement control involves both planning a trajectory and generating appropriate muscle activity. The equilibrium-point hypothesis addresses the second problem, but as **Bizzi et al.** indicate, modifications of either trajectory planning or control algorithms may be necessary to realize the full range of natural movements. Here, I would just like to mention several areas in which the equilibrium-point control may need to be augmented.

The hypothesis was originally developed to describe single-joint movements and then extended to multiple-joint movements by replacing the desired joint angle with a virtual trajectory for the hand or end-effector (Flash & Hogan 1985). Most supporting evidence concerns movements using two joints. Movements using three joints, which provide redundant degrees of freedom, are said to be similar, and this is interpreted in support of a virtual trajectory planned in terms of hand position (Abend et al. 1982; Flash & Hogan 1985; Hogan 1988a). In a quantitative comparison of unobstructed movements and movements around hindrances using two and three joints (Brüwer & Dean 1992) we also find that the paths are qualitatively similar, but characteristic path differences do occur and movement times tend to be shorter when three joints are used. For two-joint movements, Flash (1987) showed that dynamic interactions can convert a straight virtual trajectory into curved paths. An analogous explanation for our results cannot be ruled out until the dynamics of the arm and manipulandum are modelled in a similar manner.

Some means of constraining the extra degree of freedom in three-joint movements, however, such as the cost functions proposed by Cruse (1986; Cruse et al. 1990) must still be introduced. Moreover, movements in which three joints are used to avoid hindrances appear more complex than those postulated by a simple equilibrium-point control applied to the end-effector (Dean & Brüwer 1992). In particular, the wrist appears to be used both to move the hand away from the hindrance and to shorten the effective length of the distal arm segments and therefore reduce the joint excursions required at the proximal joints. To describe fully such movements, separate equilibrium-point trajectories may need to be specified, for example, for both the pointer tip and for hand position or wrist angle. Qualitatively, the action of the hand may be somewhat decoupled from that of the two proximal joints; the latter serve to position the hand so it can carry out an action such as grasping. The simple equilibrium-point hypothesis might primarily apply to the proximal joints. None of these points contradict equilibrium-point control; they simply suggest that it may be augmented by other control strategies according to task demands.

Control of natural movements: Interaction of various neuronal mechanisms

V. Dietz

Department of Clinical Neurology and Neurophysiology, University of Freiburg, D-7800 Freiburg, Germany
Electronic mail: trippel@sun1.ruf.uni_freiburg.de

[SCG] **Gandevia & Burke** provide an excellent review of what is currently known about the contribution of afferent input from peripheral afferents to limb movements in the rapidly expanding research on motor control. It incorporates data derived from a series of elegant experiments on human muscle, joint, and cutaneous afferents. The well-balanced and well-founded conclusions drawn from the data make this target article useful for all working in this field.

As **Gandevia & Burke** admit, however, their article is some-what restricted to reflex studies in a great variety of muscles and motor conditions. The discharge behavior of peripheral afferents with their perceptional aspects may therefore differ and can hardly be compared with each other. Differences may exist between distal and proximal muscles (see also Thilmann et al. 1991), between voluntarily guided and automatically performed complex movements such as writing and falling forward on extended arms (Dietz et al. 1987), and between upper and lower limbs (Dietz et al. 1987; 1991a).

The common basis for the various aspects of motor control established in the target article represents one experimental paradigm, that is, recordings of peripheral afferent signals that, for technical reasons, are only applicable for a very limited range of motor activities, for example during slow voluntary muscle contractions. However, biologically useful behavior does not consist of the action of single afferents, neurons, muscles, or limbs. A wide range of sources of afferent input act at several levels in the nervous system to produce a functionally integrated pattern of muscle activity. In this respect the title of the target article is somewhat misleading. It would profit from additional data gained from functional movements, using other methodological approaches such as described below.

1. Limited significance of a specific afferent input to natural movements. One example concerns the modulation of presynaptic inhibition of Ia afferents during stance and gait (cf. sect. 1.3.1 & 1.3.2). It is difficult to assess the functional role of this modulation in the control of natural limb movements. Other methods of investigating neuronal regulation of stance and gait have indicated that the contribution of segmental stretch reflex activity to leg muscle activation is of limited significance (for review see Dietz 1992, pp. 38–40). Irrespective of the method of perturbation used, short-latency stretch reflexes are not of great significance in freely standing humans (Elner et al. 1976; Gurfinkel et al. 1976; Gurfinkel & Latash 1979; Woollacott et al. 1984). During postural tasks the predominantly polysynaptic EMG (electromyographic) responses show direction-specific effects. However, these responses are probably not mediated by group Ia fibers (Dietz et al. 1987; 1989a; 1992).

A further example of the limited value of recordings from single afferents in the assessment of motor control mechanisms is the significance of Ib afferent input from Golgi tendon organs. In motor control the emphasis has for too long been upon length feedback; only recently has the need for force-regulating mechanisms been fully recognized (cf. Taylor & Gottlieb 1985). Little information is available regarding the behavior of this receptor system from recordings of Ib afferents. The potential significance of this afferent input is quite speculative (cf. sect. 1.3.5).

Investigations of functional movements in the cat (Conway et al. 1987; Duysens & Pearson 1980) and man (Dietz et al. 1989b; 1992; see also Hansen et al. 1988) have shown that load receptors in the antigravity muscles (which most probably correspond to Golgi tendon organs) play a crucial role in the regulation of stance and locomotion.

2. Vestibular and visual systems interact with kinesthetic information. There is little doubt that kinesthetic information is a significant factor in the control of natural limb movements. Its relative contribution cannot be assessed, however, without taking into account that during such movements a close interaction exists not only between inputs from skin, joint, and muscle receptors, as established in the target article, but also between the information coming from vestibular and visual systems. This is true for eye-hand coordination during the execution of voluntary movements of upper limbs (e.g., Biguer et al. 1982; Fischer & Rogal 1986; Gielen et al. 1984) and for the contribution of the vestibulo-spinal reflexes to the regulation of upright stance and gait (for review see Dietz 1992, pp. 44–46). The lateral vestibulo-spinal and reticulo-spinal tracts synapse on gamma-motoneurons and interneurons (Wilson & Peterson 1981). This allows functional alpha-gamma-coactivation and interaction with other pathways depending on the bias of interneurons

(Wilson & Melvill Jones 1979; Wilson & Petersen 1981). The labyrinth organ is involved in the modulation of spinal reciprocal inhibition during postural sway via the vestibulo-spinal reflex pathway (Iles & Pisini 1986; Rossi et al. 1988). Conversely, proprioceptive mechanisms may influence labyrinthine function (Lund & Broberg 1983; Nashner & Wolfson 1974).

3. Task-dependent selection of afferent input. For performing a purposeful movement it is reasonable to assume in humans, as in cats (Grillner 1975), that afferent information influences the central pattern (cf. sect. 1.3 *and, conversely,* that the central pattern generator selects the appropriate afferent information, for example, for the generation of manipulative forces during precision grip (see Gordon et al. 1991b).

Hence it is to some extent artificial to separate central from peripheral mechanisms. Muscle stretching during a functional task can result in different EMG patterns: Dorsiflexion of the feet induced by platform rotations during stance result in a small early gastrocnemius activation followed by a functionally essential tibialis anterior activation. Backward platform translations evoke only a strong long-latency gastrocnemius compensatory response (Dietz et al. 1991b). During both tasks the triceps surae becomes stretched. Recent studies have shown that the appropriate EMG pattern is not due to a reflex adaptation, as suggested earlier (Nashner 1976), but to a selection of the appropriate afferent input (Dietz et al. 1991b; 1992; Hansen et al. 1988; Gollhofer et al. 1989).

Consequently, reflex function can only be assessed while taking into account the biomechanical events connected with the actual task, their needs, and restraints. Nevertheless, when applied with caution, recordings from peripheral afferents under restrained motor conditions can serve as a basis for possible neuronal connections and interactions.

4. Interaction between central mechanisms and afferent input. Locomotion also represents a typical experimental paradigm for many automatic functional movements with respect to central programs. For such movements one can hardly agree with **Gandevia & Burke**'s conclusion that "the capacity to perform movements is attained by an individual through years of learning." There are innate programs in addition to those based on experience. For example, steplike coordinated movements are present at birth and can be spontaneously initiated or triggered (but not modified) by peripheral stimuli. A central origin for these movements can be assumed because EMG bursts precede actual mechanical events (Berger et al. 1984; Forssberg 1985). This infant-stepping pattern also occurs in anencephalic children (Forssberg 1985), which suggests that a spinal mechanism coordinates the movement, as it does in cats (Grillner 1975).

Finally, I agree with **Gandevia & Burke**'s conclusion that "for many motor skills there may eventually develop a degree of redundancy provided by peripheral feedback." One should be aware, however, that the effectiveness of active neuronal control of functional movements is difficult to demonstrate unless the test fully exploits the system's capabilities.

Spinal integration: From reflexes to perception

J. Duysens and C.C.A.M. Gielen

Department of Medical Physics and Biophysics, University of Nijmegen, EZ 6525 Nijmegen, The Netherlands
Electronic mail: admin@mbfys.kun.nl

[SCG, DAM, JFS] **The spinal circus of spinal circuits.** Some questions that many reflex physiologists "always wanted to ask but were afraid to" are openly addressed by **McCrea**. A first question concerns the issue of interneuron terminology. Theoretically, there are several ways to classify interneurons physiologically, based either on input, output, or a combination of both. The last possibility was used to coin the term "flexor reflex

afferent" (FRA) system. The problem that activation of FRA afferents can also lead to extensor reflexes has led to the concept of "alternative pathways," "private pathways," and, more recently, to the proposal to change from FRA to GRA (general reflex afferents).

McCrea is critical of the GRA but partly for the wrong reasons. In our view, the GRA is not an improvement, for two reasons. First, the term has the same problem as the FRA, namely, it suggests that there is a set of afferents that subserves predominantly one function, which is to elicit a "general" or flexor reflex, respectively. This is confusing because the same afferents are involved in other reflex pathways as well. Admittedly, this problem is partly solved by a strict use of the term "GRA system," but we know from the past how easily one has shifted from the correct notion of "FRA system" to the problematic idea of FRA. It would therefore have been better to replace the term "FRA system" by "FR system," because the emphasis is on the common output more than on a common type of input. What makes these interneurons special is not that they receive input from groups of afferents belonging to the FRA but that they are in a reflex chain producing generalized flexion.

Defining interneurons in terms of their input is also difficult for another reason. Most interneurons receive extensive convergence from many types of afferents. This is one reason why the "Ib interneuron" terminology is gradually abandoned. A secondary advantage of the FR terminology, proposed here, is that it would allow us to introduce the concept of "ER (extensor reflex) system" to describe the group of interneurons involved in generalized extensor reflexes such as the extensor thrust. This would be better than the "alternative pathways" nomenclature defended by **McCrea**, because it is more precise and simpler. In essence, the proposed FR and ER systems may overlap with what Schomburg (1990, who otherwise defends FRA as a concept) has termed "interneuronal task groups," that is, groups of interneurons mediating a command for movement.

A second reason why GRA is again a misnomer is that "general" is too general. It is certainly wise to make the distinction between local "private" pathways and global "general" systems, but it is again confusing to equate FRA and GRA, because non-FRA pathways such as those originating from Golgi tendon organs are also known to have quite widespread "general" reflex actions. Our proposal for FR and ER systems circumvents this problem; in addition, it would allow for the incorporation of the same type of terminology for the so-called private pathways, for example, in using the term "PR (plantar reflex) system" for the pathway leading to plantar flexion of the toes following stimulation of the lateral part of the foot sole.

In the final sections of his target article, **McCrea** raises a completely different type of question, which relates directly to his title. Should the interneuron researchers, in the words of Loeb (1987), "continue to collect more inexplicable data" (sect. 11, para. 3)? In fact, this type of question is not infrequently (though maybe less outspokenly) voiced by some researchers, who deal with reflexes in behaving animals. It is the frustration of the input-output people, who study reflexes with a black box approach, not to be able to make sense of this black box on the basis of the available interneuronal data, provided by the "hardware" people.

McCrea fails to point out, however, that there are quite a lot of "inexplicable data," produced by us, the input-output people themselves. Most recent studies, including those from Loeb's group and our own, have shown that there is an amazing specificity in the reflexes that can be obtained following electrical stimulation of various cutaneous nerves in both the cat and in humans (Duysens et al. 1991; Pratt et al. 1991). We are still far from understanding the functional significance of this specificity, however. Hence, it seems likely that for some time to come both the "hardware" and the "software" reflexologists will keep producing more data, which will only become explicable when a sufficient "critical mass" is available and when interdisciplinary approaches like the one described by McCrea will have been elaborated.

Cutaneous proprioception: Let's get our hands on the feet. In restoring the balance between the contributions of cutaneous, joint, and muscle afferent input and by rightly pointing out that cutaneous and joint input should not be overlooked with respect to kinesthesia, **Gandevia & Burke** do us a great favor. It is unfortunate, however, that they do not pay tribute to one of the big pioneers of this idea, namely Moberg (1983). In addition, it is striking that a disproportionate amount of attention is devoted to hand proprioception. This is of course mostly because the hand has been a more favored subject of investigation than the foot. One may wonder why. As humans spend a great deal of their time running around, it is obvious that the foot sole is a popular interface between us and the environment. Recordings from the tibial nerve during locomotion in cats reveal that an enormous amount of afferent activity is generated at each touchdown (Duysens & Stein 1978). Foot sole afferents may be equally important in simple standing, because anesthesis of the soles increases body sway (Magnussen et al. 1990). Why do we have so few microneurographic data on skin afferents from the foot then?

Some of the activity from this region is likely to generate not only proprioception but also functionally important reflex responses. **Gandevia & Burke** themselves have studied such responses under static conditions. They were impressed by the task-dependency of the responses but note in their target article that the reflex discharges were present only when the motoneuron pools were active. We can add that the latter is only true when there is no movement. During walking or running, however, these responses are easily elicited, even during periods when the muscle is silent (Duysens et al. 1990; Yang & Stein 1990). In addition, the latter studies illustrate the plasticity of reflex pathways even more clearly than the examples given by Gandevia & Burke because they show that there exist complete reversals in the type of response depending on the phase of application of the stimuli during gait. Moreover, our preliminary data indicate that the gain of these responses is substantially increased during gait as compared to rest.

Hence, it appears that such exteroceptive reflexes may be functionally important during this type of movement. To interpret these data it would be most interesting to know the exact afferent input from the skin of the foot (not only from the sole but also the neighboring areas) during small movements of the foot, which resemble those occurring during stance and during some phases of locomotion (i.e., foot placing). An even more difficult experiment but potentially just as revealing would be to test whether stretching the skin of the foot induces both movement sensation (equivalent to the human tendon stretch experiments described in the review) and reflexes, such as found following nerve stimulation. The appearance of such responses would not be unexpected because Darton et al. (1985) have already provided evidence that some of the responses to muscle stretch may arise from stretch of foot skin.

A second important point about **Gandevia & Burke** concerns the interaction between muscle and nonmuscle kinesthetic signals. As this is a very interesting topic it is unfortunate that the authors leave us with very little speculation about the locus of this interaction. There are at least two extreme possibilities that are not mutually exclusive. The first is that the different modalities underlying kinesthesia (skin, joint, and muscle sense) transmit their information along "labeled lines" up to the somatosensory cortex and that the integration of the resulting kinesthetic representations are integrated at the level of the parietal cortex (see also **Stein,** this issue). It is our impression that Gandevia & Burke adhere to some variation of this view because they propose that "joint and cutaneous inputs could help 'realign' any perceptual maps derived from muscle spindle endings."

There may be an alternative explanation, based on a much more peripheral level of integration. For example, in the rat, convergence of exteroceptive and muscle proprioceptive afferent input has been demonstrated at the level of the lumbar spinal cord in some of the cells giving rise to the spinoreticular tract (Menetrey et al. 1984). If similar types of cells exist in humans, this would explain some of the interaction results described by **Gandevia & Burke,** and it would support the idea that an integrated kinesthetic representation is already present at the lowest level of CNS organization.

ACKNOWLEDGMENT
The authors are supported from ESPRIT (MUCOM 3149) and NATO.

Connectionist networks do not model brain function

Roy Eagleson[a] and David P. Carey[b]
[a]Centre for Cognitive Science, University of Western Ontario, London, Ontario, Canada N6A 5C2 and [b]Department of Psychology, University of St. Andrews, St. Andrews, Fife KY16 9JU, Scotland
Electronic mail: [a]eagleson@uwo.ca; [b]dpc1@st-andrews.ac.uk

[DAR] **Robinson**'s main point is that it is futile to develop models for brain function by examining single-cell behavior. A parallel situation exists in robotic sensorimotor control. It is equally unfeasible to develop models for robotic functionality in terms of the local interactions of transistors in a processor. Sensorimotor control is better described as a hierarchy of control structures, each investigated using different analytical tools, and each with its own explanatory vocabulary. For any given function, one level is sure to provide a more appropriate description than the others. According to computational principles, the information-processing requirements of a particular brain function must first be considered, so that it is possible to ascertain which organizational level (single cell, cell population, cytoarchitectural region, etc.) merits experimental attention.

Robinson argues that it is futile to study single units in order to understand the functional modules they comprise. To be sure, abstract functional descriptions never seem to constrain completely the physical instantiation of a system, but this is a problem of general systems theory – that there are severe limitations on what can be said about a particular level of an abstraction hierarchy from the perspective of another one – especially if they are not adjacent. Robinson has chosen to focus on the related shortcomings of what single-unit recordings can tell us about brain function, especially when the physical architecture is modeled using PDP (parallel distributed processing) techniques.

Other researchers have made recent attempts to compare neurophysiological models to those generated using PDP techniques (Anastasio & Robinson 1990a; Zipser & Anderson 1988). What these comparisons fail to address is that back propagation is an unlikely candidate for a self-organization principle in the CNS (Crick 1989; Moorhead et al. 1989). Similarly, the problem of where the "teacher" of the network would appear in the CNS is left undiscussed. These problems are often dismissed by tacitly suggesting that the majority of feedback projections in the brain support back propagation.

The class of functions that can be trained into artificial neural systems is currently quite limited. The most common are simple transfer functions, which explains their power for modeling coordinate-system transformations, although **Robinson** gives examples of how this class can be extended to include specialized transfer functions such as integrators and time delay functions (see also Williams & Zipser 1989). Even though the ingredients for such systems, and undoubtedly more complex ones, obviously exist in the brain, connectionist theory is deficient in specifying general learning rules for the self-organization of more complicated ones (even for feedback transfer functions, $G/(1 + GH)$). Similarly, the functional organization of the brain is more complicated than can be observed exclusively at the single-cell level. Neuroscientific approaches to system modeling would be seriously limited if circuit-level recordings were the only available technique.

Nevertheless, recent findings in neuroanatomy (Krubitzer & Kaas 1990; Morel & Bullier 1990), neuropsychology (Desimone & Ungerleider 1989) and neuroimaging (Corbetta et al. 1991; Zeki 1990) have confirmed many theories of functional specialization in visual processing that were originally derived from single-unit studies. In the motion perception domain, neurophysiological identification of subregions in STS (superior temporal sulcus) and their functional roles are largely being confirmed using other methods. Lesions in MT/MST (middle temporal/middle superior temporal) result in motion-processing deficits in psychophysical (Newsome & Pare 1988) and visual pursuit tasks (Newsome et al. 1985). Furthermore, Newsome and his colleagues have recorded directional selectivity of particular columns of cortex, and by microstimulation of those regions they have biased the coherence motion judgments of monkeys in the direction preferred by the column (Salzman et al. 1990). Recent findings in motor control are also beguiling; Duhamel et al. (1992) have shown that indexed items in the visual display have their identity maintained using what appears to be efference copy of the motor control signal. In fact, the success of single-cell investigations of brain regions is underscored by their frequent use in constraining the initial construction of neural network models. **Robinson**'s own vestibulo-ocular reflex (VOR) model restricts plasticity to the input stage, based on single-unit findings in the vestibular system (Anastasio & Robinson 1990a). However, it is the functional specification of this system that is most true to the empirical data and could have been proposed solely on the basis of the computational requirements, namely, the coordinate transformations in the VOR.

Suppose an engineer was asked to design and construct a VOR system in a robotics lab and then an outsider was asked to "explain how the system worked." The first engineer would begin with a problem specification; a statement of how the internal gyroscopic velocity measurements should be converted to a servo control output to drive a camera position. The resulting functional specification could be identical to the tensor representation and transformations described by **Robinson**. The technical staff would be given a lot of freedom to choose any particular implementation. The actual implementation would not matter to the design engineer, so long as the specifications were met. A potential misunderstanding arises if another engineer is asked to explain "how the system works." This individual could study the black box with logic probes, establish its connectivity, and basically recreate the circuit and timing diagrams of the hardware. If successful, what sort of description could be given? This is important: At a functional level, it would be exactly the same as that proposed for a biological model with the same physical geometry.

This analogy raises the following important meta-question: What do we mean when we ask how a system works? Are we seeking the overall functional distinction? Or do we really mean to explain the behavior of each subcomponent in the system? The latter can lead to the danger of falling into the infinite regress of reductionism. One must first select the abstract level at which a satisfactory model can be deemed acceptable. Only then can one select the appropriate scientific and theoretical tools.

ACKNOWLEDGMENTS
The authors were supported by the Ontario Information Technology Research Centre, the Canadian IRIS program, and by NSERC operating grants to Z. W. Pylyshyn (A2600) and M. A. Goodale (A6313).

Fundamentals of motor control, kinesthesia and spinal neurons: In search of a theory

A. G. Feldman

Institute of Biomedical Engineering, University of Montreal, H3C 3J7, and Research Centre, Rehabilitation Institute of Montreal, Quebec, Canada H3S 2J4

Electronic mail: *feldman@ere.umontreal.ca*

[EB, SCG, DAM] In this commentary on **Bizzi et al.**, **Gandevia & Burke**, and **McCrea**, I argue that the α model is a misunderstanding of the original equilibrium-point (EP) hypothesis. I then consider position sense in terms of the λ model and give an example of how some properties of motoneurons (MNs) and spinal interneurons (INs) are integrated in the model.

1. Bizzi et al.'s alpha model: A misinterpretation of the EP concept as defined in physics. Having been deprived of the opportunity of going to the West or of regularly publishing articles there, I could only passively observe how an important idea – the EP hypothesis – evolved independently of its originator (Asatryan & Feldman 1965). Many physiologists have considered the hypothesis (λ model) of motor control significant and have taken into consideration the experimental determination of invariant torque/angle characteristics, the independent control variable, λ, and the integration of muscle properties and central and afferent signals into the model (Feldman 1966a). In contrast, **Bizzi**'s group, in creating an alternative model, has basically ignored these experimental and theoretical results as well as other findings that conflict with their interpretations (e.g., Day & Marsden 1982; Hasan & Enoka 1985). In their model, control variable λ was replaced with the level of muscle activation, α. They initially assumed that muscular activation only modulated muscle stiffness (see Bizzi et al. 1978). Obviously referring to intact motor systems, they made calculations of equilibrium trajectories assuming that the level of α activity did not depend on feedback variables (e.g., Hogan 1984). I quite agree with Bizzi and his colleagues that such models fly in the face of physiological common sense and add nothing to our knowledge of motor control (Note 2). Bizzi et al. have also misinterpreted the notion of springlike behavior. According to the λ model and, specifically, the concept of invariant characteristics, such behavior is provided by afferent and central systems rather than by the muscle itself (Feldman 1986). Instead, Bizzi et al. have regenerated the old mechanistic idea popular in the beginning of the century that the muscle is a tunable spring, an idea that was refuted by the classical experiments of Hill (1938).

Recognizing that the reformulation by **Bizzi et al.** was leading to confusion and increasing criticism of the model, I tried to reorient motor control scientists by publishing my 1986 article (Feldman 1986) with the help of Scott Kelso. The most serious criticism was associated not with the above flaws but with the essence of the α model, such that the level of muscle activation was considered a variable that specifies or affects the equilibrium position. For example, according to Hogan (1984, p. 276), "the set of muscle active states may *always* [Hogan's italics] be interpreted as defining a virtual equilibrium position for the limb." One can try to defend this idea by saying that any static position is associated with a specific level of muscle activation and, consequently, defined by it. Unfortunately, however, it is not. In a spring-load system, the equilibrium position is associated with a certain magnitude of spring force. Mathematically, it would be erroneous to say that spring force defines the equilibrium position: If we briefly perturb the load, force will oscillate, but the equilibrium position of the system will, by definition, remain the same. Similarly, the fact that the limb's equilibrium position is associated with a specific level of muscle activation does not mean that position is defined by it. The same level of muscle activation can be observed at any limb position under specified loads (Feldman 1986). It is more important, however, to remember that according to the EP concept accepted in physics, position-dependent variables (in our case, force, stiffness, and the level of muscle activation) do not specify the equilibrium position because *they are functions of position.* Here, cause and effect are physically predetermined.

Note that shifts in the equilibrium state of the system are produced at premotoneuronal (segmental and suprasegmental) levels (Feldman & Orlovsky 1972) and that changes in λ mirror these central processes. Muscle activation ("EMG bursts") is simply a dynamic reaction of the system to the difference between the initial and new equilibrium states. This understanding of the EP concept (absent in the α model) underlies the explanation of kinematic, dynamic, and EMG patterns of different movements (Feldman et al. 1990). The λ model therefore represents a departure from the classical dogma that takes for granted that all neuro-control processes can be measured in terms of the level of MN activation. The fundamental control process – shift in the equilibrium state – accomplished at a premotoneuronal level, is neither affected nor measured by the level of MN activity. In contrast, the α model, consistent with the dogma, has rearranged cause and effect, resulting in a serious misinterpretation of the basic concepts of the EP hypothesis, particularly of the control variables, EPs, and equilibrium trajectories.

In the target article, **Bizzi et al.** strangely reconcile the two fundamentally inconsistent models. First, they reformulate the λ model in a way that makes it primitive. For example, independent control variable, λ, is, in their understanding, a function of α MN activation. They also misrepresent the relation between Merton's (1953) servo hypothesis and the λ model, as well as the role of α-γ linkage in the latter. This relation has been discussed in detail (Feldman 1976; 1979; 1986; Feldman et al. 1990). Briefly, the stretch reflex in the λ model is a powerful mechanism that brings the system not to a unique, load-independent position specified by γ efferents as in Merton's hypothesis, but to any equilibrium position in the range $x > λ$ defined by the parameter λ. This position is load dependent, but control systems can adjust λ to reach the desired final position. Indirect inputs (mediated by γ efferents and muscle spindle afferents) and direct inputs to α MNs can act independently and are additive in terms of changes in their membrane potentials and, as a result, λ. The stretch reflex has sufficient (but not excessive and physiologically unrealistic as in Merton's model) positional gain and effective damping to perform these functions. Second, Bizzi et al. improve some elements of the α model; for example, they now add the constraint that muscle activation is a function of position. Finally, in emphasizing some common features of the two models but ignoring their fundamental differences, Bizzi et al. announce, figuratively speaking, a "marriage" in spite of the reluctance of the groom!

Surprisingly, **Bizzi et al.** suggest a supremacy of the α model in this union. Actually, they try to show that the λ model is a subset of α models (sect. 2.3), but they seem to have forgotten that they do not consider all models but only those relevant to the EP hypothesis or, specifically, α models in which muscle activation underlies shifts in the equilibrium position. Furthermore, they should note, as I did above, that the set of such models is empty, and then, without further effort, conclude that the λ model cannot be a subset of the empty set!

According to the λ model, the level of muscle activation in statics is a function of only one variable (*u*), which is the difference between actual muscle length (*x*) and threshold length (λ). It therefore makes no sense to decompose α into reflex and central components because it can be argued that the same level of *u* can be reached equally by a change in either muscle length, *x* (100% reflex action) or the control variable, λ (100% central action). Nevertheless, the level of α activity has been artificially decomposed into central and reflex components in McIntyre's (1990) model, which **Bizzi et al.** consider to be a combination of the α and λ models. I anticipate an additional reason for a serious criticism of the α model.

The fact that both muscle activation and variables dependent on it, such as force and stiffness, are indivisible into central and reflex components also means that deafferentation experiments are not logically justified as a test of the α model: They just destroy the fine sensorimotor integration existing in intact motor systems without adding to our knowledge of how this integration is actually produced. On the other hand, by creating the situation in which muscle activation is produced solely by central commands, deafferentation experiments make the obviously wrong impression that in intact systems muscle activation is also provided basically by central commands.

One could believe that "the α model of the EP hypothesis" is undoubtedly correct in the case of deafferentation. It has been argued that the effect of descending control input to α MNs in both intact and deafferented conditions can be measured by an independent increment ($δV$) of the MN membrane potential (Figure 1A; Feldman 1986). Alpha MNs are threshold elements with complex dynamics (e.g., Crone et al. 1988). Again, the level of MN activation is thus the dynamic effect of the control input $ΔV$ and consequently cannot, even in case of deafferentation, be considered the cause of EP shifts.

2. Position sense in the λ model: Central and afferent components of kinesthesia. The data on position perception so nicely described by **Gandevia & Burke** teach us several lessons. First, no peripheral receptors convey positional information directly. For example, the activity of muscle spindle afferents is a function of position and, simultaneously, γ influences. In particular, when an increase in γ MN activity is associated with muscle shortening, the firing frequency of muscle spindle afferents can be the same regardless of position (Hulliger et al. 1982). Joint receptors basically signal that movement is near or has exceeded the usual physiological limits. They can also respond to joint pressure and to muscle contraction. Second, practically all peripheral receptors (muscle, cutaneous, and joint) contribute to kinesthesia. Third, internally generated control signals in some way contribute to perception of posture and movement.

The problem of kinesthesia can be considered in terms of the λ model. In Figure 1A, two measures of independent inputs to MNs ($δV$ and $δλ$) are defined based on MN threshold properties and afferent feedback. This analysis shows that: (1) peripheral feedback establishes a correspondence between basic electrical parameters of the MN and space variables so that the current and threshold membrane potentials are associated with the current and threshold muscle lengths, respectively. As a result, MN functioning becomes associated with external space. (2) This correspondence is not one to one and can be modified by independent control signals. Specifically, control signals shift the frame of reference (λ) for positional recruitment of the MN. (3) Control signals are independent of afferent feedback but rely on them for calibration in terms of space coordinates (the dimension of $δλ$ is "length").

In Figure 1B, the equilibrium position is represented as the sum of the central command λ and the deflection from it produced by the load. Perception of the deflection can be based on the total signal level from muscle, joint, and skin afferents. This signal alone is insufficient for adequate positional perception, however, because the same level of afferent signal can be produced when the load is balanced in different positions, as has been observed for muscle spindle afferents (see above). Thus, for adequate perception, the afferent signal should be summed with the efferent signal or its copy. This hypothesis has been used to explain different kinesthetic illusions (Feldman & Latash 1982). The model also predicts essential deficits in motor performance in the case of deafferentation. These deficits are associated not only with the absence of the afferent component of position sense but also with the loss of the natural calibration of the central control signal ($δV$) in terms of the space coordinate, λ.

3. Intermuscular interaction in the λ model. In his excellent review, **McCrea** has shown what remarkable progress has been

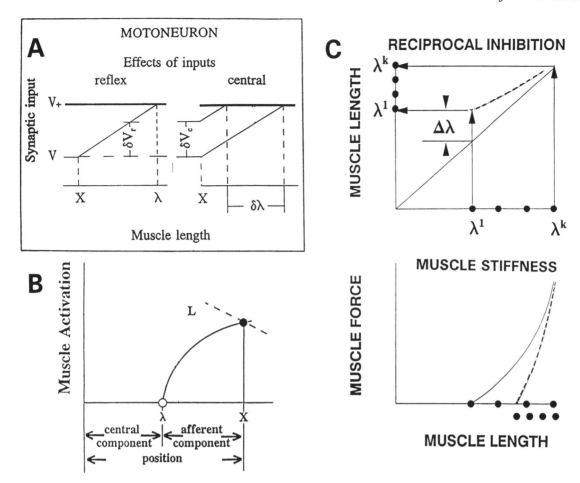

Figure 1 (Feldman). Some principles of the λ model. *A:* Two measures of central control signals. *V* is the initial membrane potential of the MN. When descending control signals are fixed (left panel), a quasistatic stretch of the muscle results in an increasing depolarization (δV_r) of the MN as a function of muscle length *x* because of the proprioceptive feedback from muscle afferents. The threshold membrane potential, V_+, and consequently, the recruitment of the MN, will be reached at a muscle length λ. The effect of a change in the tonic control signal (right panel) can be measured primarily by an independent decrement (δV_c) in the membrane potential at a given muscle length. On the other hand, the same control signal is expressed as a decrement ($\delta\lambda$) of the threshold muscle length at which the MN is recruited. *B:* Two components of position sense. Solid line represents MN recruitment as a function position (*x*) when control parameter λ is fixed. Load (L: dashed line) is characterized by the level of muscle activation that provides its compensation at different positions. Filled circle: EP. *C:* Reciprocal inhibition (RI) in terms of MN recruitment (upper panel) and the shape of force-length characteristics (lower panel; solid line: without RI; dashed line: with RI).

made in our knowledge of spinal IN circuits and their functional role. At the same time, it is clear that we will be helpless in explaining the increasing complexity of spinal structures without meaningful theoretical ideas and physiologically realistic models. The λ model is a step in this direction. As an example, Figure 1C illustrates the role of reciprocal inhibition (RI) of antagonist muscles in the recruitment of MNs and the specification of the control variable λ. Filled circles on the horizontal axis (upper panel) represent the thresholds of MNs in the order of their recruitment in terms of individual MN thresholds λ^k ($k = 1, 2, \ldots$), an alternative formulation of the "size principle." Circles on the vertical axis represent the same thresholds after their transformation under the influence of intermuscular interaction. The diagonal line symbolizes a one-to-one transformation. Inhibition of MNs mediated by the antagonist muscle

spindle afferents and IaINs gives rise to an increase ($\Delta\lambda$) in the thresholds of the agonist muscle (Feldman & Orlovsky 1972), shifting them above the diagonal and resulting in a new transformation function (dashed line). When the agonist muscle lengthens, the antagonist muscle shortens and the effect of the RI decreases. Thus, the transformation function approaches the diagonal. It can be seen that RI gives rise to an increase both in the threshold λ and gradient of MN recruitment. As a result, RI contributes to stiffness in terms of the slope of force/length invariant characteristics (lower panel). It is quite natural that descending control signals to MNs can be mediated by IaINs: The latter thus provide a part of the independent change in the threshold λ associated with voluntary movements. IaIN pool is therefore a multifunctional system (see also Feldman et al. 1990).

Network magic

Martha Flanders and John F. Soechting
Department of Physiology, University of Minnesota, Minneapolis, MN 55455
Electronic mail: *john@neuro.med.umn.edu*

[GEA] The recognition that the extensive pattern of convergence and divergence in the CNS has functional implications and the demonstration that neural networks can be modeled represent welcome progress over the long-entrenched (if unstated) notions of labeled lines and relays. **Alexander, DeLong & Crutcher** provide a thoughtful, critical review of the anatomy and physiology of the motor areas of cortex and basal ganglia, and their own work has contributed significantly to undermining traditional viewpoints concerning the function of these areas. Their criticisms of motor programming and engineering-based models are well taken. But they also seem to believe that neural network modeling (especially based on unsupervised learning) is the magical cure for our lack of understanding of the function of cortical/basal ganglia motor circuitry. We wish to emphasize that neural network models are useful *only* when the inputs and the desired outcome can be defined.

We agree that the concept of motor programs has led investigators down a blind alley, as the target article points out. Like the motor programming approach, some mechanical engineering models may also have served to distract investigators from the biology of motor control.

We also agree that neural network modeling can become a very useful tool. We believe, however, that the distinction **Alexander et al.** make between algorithmic, "black-box" models and connectionist models with unsupervised learning is artificial and, ultimately, detrimental to progress toward an understanding of neural function. The authors claim that connectionist networks "learn to solve problems in a manner that is essentially nonalgorithmic," because they learn by trial and error. Solutions arise by virtue of the repetition of movements that "the subject (or trainer) deems successful." *But the chosen criterion of success determines an algorithm!* If we define success as constituting movements that are accurate, fast, efficient in their use of energy, and so on, we have defined specific criteria for a solution. This solution can be implemented by means of a series of black boxes (as in robotics) or in a distributed fashion by a neural network. Even the process by which the neural network arrives at a solution represents an algorithm (otherwise, one would be unable to simulate it on a digital computer!); therefore, it does not seem appropriate to pose the issue as algorithmic black boxes versus nonalgorithmic neural networks.

A more fruitful approach might be to ask the question: How can a black-box model be implemented by a network of neuron-like elements? This question has not received as much attention as it merits. Nevertheless, there are some instructive examples. Robinson's (1981; 1982; 1989) models of the vestibulo-ocular reflex represent a well-known and highly successful example of the black-box approach [see also Robinson, this issue]. One element of this model is a "neural integrator," that is, a black box that performs the equivalent of a mathematical integration. This black box can be realized by means of positive feedback (Cannon & Robinson 1985), short-term potentiation in neurons (Shen 1989), or distributed, parallel processing (Anastasio & Robinson 1989).

Another familiar example is the work of Andersen and Zipser (1988). They showed that a transformation from a retinocentric to a head-centered representation of target position can be accomplished by a three-layer network with hidden-layer properties resembling the activities of neurons in the posterior parietal cortex. It should be emphasized that this network model was inspired by the hypothesis that a perceptual representation of target position must somehow be stabilized to an earth-fixed frame of reference – a simple black-box transforma-

tion. To give the network model full credit for this proof would be to put the buggy before the horse.

Another example pertains to the jamming avoidance reflex in electric fish. This behavior and the neural processes that govern it have been worked out in great detail by Heiligenberg (1991) and colleagues. At one level, this behavior can also be described by a black-box model. Yet, at the same time it can also be described in terms of the kinds of neural nets espoused by **Alexander et al.** We believe the elegant work of Heiligenberg is instructive. Success was achieved by means of careful behavioral and lesion experiments, anatomical and electrophysiological investigations, and judicious use of more than one type of model.

We argue, therefore, that black-box models are not only "valuable stimulants to motor behavioral research"; they can also provide a guide to determining how the behavior is effected by the CNS. Such models are not as well developed for limb movements as they are for eye movements or for the jamming avoidance reflex. Recently, we proposed a model that attempted to account for the sensorimotor transformations used by the CNS to direct the arm to a visual target (Flanders et al. 1992). Although our model incorporates discrete stages and discrete physical parameters, we postulate that the transformations are represented by the distributed activities of groups of neurons.

In Figure 1 we suggest a way in which our black-box model might be implemented by neural networks. A three-layer network transforms a retinocentric representation of target location into a representation of the distance and direction of the target relative to the shoulder (Figure 1, top). Although there is ample evidence for this type of transformation, it is probable that a

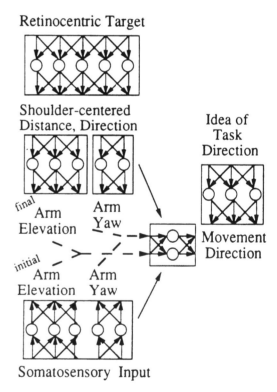

Figure 1 (Flanders & Soechting). A "black-box" model for the sensorimotor transformations subserving targeted arm movement. The retinocentric target input and the output information about movement direction are explicit neural representations. The more intermediate outputs and inputs need not be explicitly represented, but may be useful intermediate goals for training serially arranged three-layer networks. The channels in the transformation to arm angles are suggested by psychophysical data.

discrete representation of shoulder-centered target parameters is *not* formed in the CNS (Alexander 1992; cf. Flanders et al. 1992). Likewise, the subsequent representation of intended, final arm angles is probably *not* a discrete one. Nevertheless, we suggest that a transformation toward a representation of intended arm angles occurs, and furthermore, our work suggests that this transformation occurs in distinct channels (Flanders & Soechting 1990). At the bottom left of Figure 1, we illustrate the hypothesis that somatosensory input is combined with this visually derived representation and that there may also be limited lateral connections in the somatosensory transformation (i.e., channels). Finally, on the right we show two parallel transformations toward the cortical representation of movement direction: from a combination of visual(ly derived) and somatosensory information, and from a cognitive representation of the task or target direction (such as would be given by a visual display).

The retinal representation of target location and the motor cortical representation of movement direction are discrete entities described by the activities of retinal and cortical neurons, respectively (cf. Georgopoulos 1991). There is no evidence for *explicit* intermediate representations of shoulder-centered target parameters or arm angles, but there is evidence for the intermediate transformations (see also Helms Tillery et al. 1991; Soechting et al. 1990). In Figure 1 therefore we have placed the transformations rather than the inputs inside the black boxes. The boxes signify CNS processing.

Future neural network simulations might usefully investigate ways to align serially three-layer networks trained with intermediate goals such as those indicated in Figure 1. Is it feasible to use the output of a hidden layer as the input to a subsequent three-layer network? Realistic models might also explore limitations on lateral connections. Even if this is not the optimal way for the network to solve a problem, such constraints may be imposed by the development of the CNS along the lines of vestibular and ocular coordinates (Soechting & Flanders 1992a). We think that neural network models can indeed perform "magic" if they are combined with anatomy, physiology, *and* black boxes!

ACKNOWLEDGMENT
This work was supported by National Institute of Neurological Disorders and Stroke grants NS-27484 and NS-15018. We thank S. I. Helms Tillery for comments on the manuscript.

Adaptive neural networks organize muscular activity to generate equilibrium trajectories

A. A. Frolov and E. V. Biryukova
Institute of Higher Nervous Activity and Neurophysiology, Russian Academy of Sciences, Moscow 117865, Russia
Electronic mail: *bir@sms.ccas.msk.su*

[EB, DAR] **Bizzi et al.**'s target article gives an in-depth review of the equilibrium-point (EP) hypothesis and its applications. We consider the alpha/lambda discussion (Feldman 1986) to be very fruitful from the viewpoint of the development of an EP-based approach. Although we agree with **Bizzi et al.** that the alpha and lambda models do not contradict each other, we do not agree that the lambda model is a special case of the alpha model. Both models assume that the equilibrium muscle length is the main goal of supraspinal regulation. The question arises, however, as to which central parameter is used by supraspinal levels to set the given equilibrium muscle length. Theoretical and experimental results (Feldman 1979) have made it possible to conclude that this parameter is the stretch-reflex threshold set by the supraspinal inflow of lambda to alpha-motoneurons. As for the alpha model, it simply ignores this question, because

its advocates (Bizzi et al.) "deliberately make no attempt to distinguish between central and reflex effects on the alpha-activity."

Bizzi et al.'s statement can perhaps be explained as arising from their experience with deafferented animals. In the case of intact muscle, however, it is important to distinguish between the two distinct effects mentioned above, although this could be difficult to do experimentally. The lambda model therefore claims to be more advanced concerning the central mechanisms of spinal reflex control. The view of the lambda model as a subset of the alpha model (considered here as the restriction of the mapping from motoneuron activities to equilibrium angle) hence seems to be rather formalistic, because it ignores the feedforward and feedback nature of motoneuron activity.

We would like to make two points with regard to the extension of the Ep-hypothesis to multijoint movements. The first concerns the problem of fast movement control.

According to the EP hypothesis, fast single-joint movement can be implemented in two ways: first, by a fast shift of equilibrium position and, second, by stiffness augmentation. The former, being extended to multijoint movement, results in a substantial deviation of the trajectory from a straight line in external space (Frolov et al. 1992). To avoid this, a gradual shift of the hand equilibrium position can be used (Hogan 1985a), while the necessary movement velocity can be achieved by stiffness augmentation. This augmentation has substantial physiological limitations, however (Flash 1987). Our experience also shows that introducing explicit control of velocity results in a good agreement with experimentally recorded straight-line hand trajectories (Frolov et al. 1992). Our approach is based partially on the Vector Integration to Endpoint (VITE) model (Bullock & Grossberg 1988a) and assumes that joint torques are proportional not only to the difference between the current and the equilibrium angle but also to the difference between the current and desired angular velocity. Theoretical studies show that nonlinear velocity control allows arbitrarily high control accuracy (Piatnitsky 1988).

The second point concerns the overall approach to the study of motor servo-system functions. It is generally accepted that movement plans are represented in the nervous system in the form of functions of the hand's endpoint coordinates in external space (Bernstein 1966; Brooks 1986). A special neural network must hence exist to transform this plan into one for internal space where, for example, muscle lengths could be regarded as coordinates. The outputs of this network would be the signals to the muscles, allowing the desired hand movement to occur (Frolov et al. 1992; Massone & Bizzi 1989). It seems justified to assume that the error used by the neural network to correct these signals in the process of spatial transformation learning is also explicitly specified in terms of coordinates in external space and must be transformed into an error in internal space. In this context the particular supraspinal signal to the muscle does not seem to be the variable of principal importance. Its effect obviously depends on the properties of specific muscles. More important is its dependence on the properties of all other muscles involved in the movement, because those muscles contribute to the total error in the external space the network uses to adjust its parameters in the process of spatial transformation learning. On the other hand, the supraspinal signal can control not only the stretch-reflex threshold, but also movement velocity (Van Gisbergen & Robinson 1977), executive organ delays (Fujita 1982), or, probably, still further factors. The supraspinal signals are modified in the process of neural network learning in such a way as to generate the desired movement in external space; this is the most important fact functionally.

According to this approach, the signals to the motoneurons play the role of hidden units in an adaptive neural network; we share **Robinson**'s point of view (this issue) that "individual hidden units . . . would be largely uninterpretable." It is hence doubtful that even more careful study of individual-unit function will yield enough information to allow us to understand multijoint system behavior.

Network simulations and single-neuron behavior: The case for keeping the bath water

Albert F. Fuchs, Leo Ling, Chris R. S. Kaneko and Farrel R. Robinson
Department of Physiology & Biophysics and Regional Primate Research Center, University of Washington, Seattle, WA 98195
Electronic mail: *fuchs@u.washington.edu*

[EEF, DAR] David **Robinson** has been a guiding force in shaping the direction of oculomotor research for the past two decades. From his control-systems-engineering vantage point, he has formalized vague concepts into testable models that have provided the motivation for many studies of the oculomotor system. Those early models were based on the discharge patterns of actual neurons that were then just being discovered. He gave us such important concepts as the local feedback burst generator for saccadic eye movements and the neuronal integrator for eye position. Many of our colleagues who are interested in the control of somatic movement have noticed the general applicability of his ideas and have incorporated them into their thinking. It is disappointing, therefore, that the pope of eye movements is now worshiping at the altar of a lesser god!

Robinson's target article seems a bit schizoid. On the one hand, he says that unit recording is useless because we cannot see signals in the behavior of single neurons and because the CNS has a variety of solutions to any given problem anyway. On the other hand, he uses published discharge patterns (but only his own) to legitimize his models. For example, the myriad of cell types he and his colleagues found throughout the vestibular nuclei all turn up in the hidden neurons of his network simulation of the vestibulo-ocular reflex (VOR). In particular, he points out that his model predicts the existence of hidden units whose discharge pattern "seems contrary to an engineering mentality." Indeed, such "rogue cells" are occasionally found in the vestibular nuclei.

While decrying unit recording in favor of modeling, **Robinson** fails to point out two important facts. First, in his VOR model the weighting coefficients of the rogue cells are always very small, so those cells contribute little to the resulting network output. Second, only certain neuron types in the vestibular nuclei in fact project to motoneurons and are therefore interneurons of the VOR, that is, the hidden neurons of his network. It is clear that none of those identified interneurons behave like rogue cells. The network simulation therefore suggests a signal processing solution that we already know the CNS does not employ. It suggests this solution because it is not constrained by the anatomy and neurophysiology that is known to underlie the VOR. Therefore, it simply is incorrect that the actual neural "network disassembles the input signals, scatters them over the hidden units, and then reassembles them." Until it includes the appropriate neurophysiological constraints, this type of network model will only be an interesting laboratory exercise to "amuse the applied engineer," who is not concerned with explaining an existing system but only in creating a new one. Indeed, if this were all there was to it, the brain would be easy to build because it would solve all of its problems with a rather simple learning algorithm.

There is a similar myopia in **Robinson**'s arguments concerning the neural integrator. For almost two decades, Robinson has enthralled the oculomotor community with the possible existence of a black box whose function is to integrate the prenuclear

eye velocity signals present in the various oculomotor subsystems and thereby produce the eye-position signal required by extraocular motoneurons. According to Robinson, few neurons in the pons and medulla have the pure eye-position signal expected at the output of such a neuronal integrator. Instead, brainstem neurons with a position signal often carry velocity information as well. The hidden neurons that emerge in Robinson's simulated network integrator also have various combinations of eye-position and velocity information; he therefore concludes that the network performs integration as the brain does. His basic tenet is that biology is messy and that intermediate stages that have position and velocity signals distributed "thoroughly" over all of the interneurons are to be expected. "This property imparts a biological flavor to simulations" and "it is hard not to believe that similar features also shape the real networks." But wait a minute: Were we not told in the Introduction that "the study of single neurons or neuron ensembles is unlikely to reveal the task in which [the units] are participating or the contribution they are making to it"?

As with the VOR, **Robinson**'s network model of the neural integrator ignores some fundamental neurophysiological data. First, lesions of the cerebellum impair gaze-holding ability, suggesting that the integration process is distributed in the brain, not just confined to hidden neurons in the nucleus prepositus hypoglossi (NPH), the site of the putative integrator. Second, some neurons in the simian NPH have pure eye-position signals, whereas probably none have only pure eye-velocity signals. Indeed, the neurons that exhibit a relatively pure eye-velocity sensitivity also have a signal related to head velocity, whereas none of the cells with any eye-position sensitivity do. Consequently, it is again incorrect that there is a "thorough" distribution of signals among *actual* neurons in the NPH.

In sum, unlike **Robinson**, we believe that there is considerable evidence of order in the processing of neuronal signals in the oculomotor system if one considers all of the available neurophysiological data. Furthermore, we believe that black-box models constrained by the rigorous application of available data will continue to be invaluable conceptual tools that will serve to guide oculomotor research. Surely the alternative of network models with no real physiological constraints is merely an intellectual exercise at best. Indeed, whenever the neurophysiology of an oculomotor subsystem is taken into account, the network model is shown to be hopelessly naive. To paraphrase Robinson, perhaps it would be wiser to keep the neurophysiological bath water and just throw out the network babies.

Fetz's approach to network modeling appears to have more promise than that of **Robinson** and his colleagues. First, the connectivity of Fetz's model is constrained by known anatomical connections for, as he says, "the cells encountered at a given recording site typically have diverse projections." It follows that a network such as Robinson's, which finds a use for *all* of the neurons in a structure, represents no great breakthrough and is probably simply wrong. Second, Fetz's model produces temporal discharge patterns of its elements, whereas Robinson's networks can deal only with the coefficients that specify each hidden unit's sensitivities to eye or head movement variables. Fetz's simulation therefore has the potential to allow optimization of the temporal characteristics of discharge patterns.

We do take issue, however, with **Fetz**'s contention that movement parameters are not reliably encoded in the discharge properties of even those neurons connected monosynaptically to motoneurons. This generalization comes from his observations that some cortical cells, identified by spike-triggered-averaging (STA) as connected to certain muscles, exhibit:

> counterintuitive discharge patterns that are distinctly different from the activity of their facilitated target muscles. Moreover, some cortical cells are paradoxically coactivated with arm muscles that [STA says] they inhibit. . . . This would indicate that the response patterns

of neurons alone are not a reliable guide to their causal role in the task.

A similarly "counterintuitive discharge pattern" occurs in the oculomotor system where there is a logical explanation for it. The so-called position-vestibular-pause (PVP) neuron in the vestibular nuclei projects directly to oculomotor neurons and during suppression of the VOR it continues to provide an unwanted signal related to head velocity. (Fortunately, **Robinson** has made us feel comfortable with such rogue cells!) Recent work by Cullen et al. (1992) shows that the motoneuron eventually receives at least two additional inputs, which either cancel or inhibit the PVP drive during VOR suppression depending on the behavior being generated. By Fetz's standards, the PVP discharge during suppression is "paradoxical," but it is easily interpretable when there is a more complete understanding of the system. We suggest that Fetz think of such "paradoxical" responses as enriching the repertoire of his cells' behaviors rather than as presenting irreconcilable problems.

Fetz is also concerned that "neural computation between connected cells involves some highly nonlinear relationships." Nonlinear input/output relations are characteristic of most neurons in most neural systems. In the vestibular nuclei, for example, many neurons have relatively low resting rates, so they are driven to cutoff during part of a sinusoidal cycle of head rotation. Despite this obvious nonlinearity, the resulting eye movements are complete sinusoids because the motoneurons are driven by "push-pull" signals. Therefore, even though a prominent input may have a nonlinear relation to some movement parameter, it may be linearized by some other input to produce a net linear response. The only disadvantage of nonlinear relations is that they tax the mathematical capabilities of the modelers; they are not difficult for the investigators to interpret.

Indeed, many neuron types in the oculomotor system *do* have firing rates that are linear functions of some movement parameter. This is obviously true for most vestibular nerve afferents and for ocular motorneurons operating above their thresholds for steady firing. For most brainstem neurons that discharge a burst of spikes with saccades, the number of spikes increases linearly with saccade size, as does burst duration. Many of these neurons drive motoneurons directly. Another example is the floccular Purkinje cell, which exhibits a monotonic increase of firing rate with eye velocity. Because the discharge characteristics of all of these neuronal populations are relatively homogeneous, one gets an accurate impression of the salient discharge features of the entire population by evaluating the discharge of a single cell. For these examples, it has been possible to use the behavior of archetypical cells in black-box models that produce reasonable replicas of eye movement. As one progresses inward to the dark regions of the brain where the sensorimotor interface and other exotic transformations occur, the relations between firing rate and sensory or motor events may become more obscure. But at least in the oculomotor system, a robust relation seems to exist several synapses from the motoneuron, even as distant as the superior colliculus.

In conclusion, perhaps the complexities resulting from the many muscles and feedback loops controlling the limbs tend to dissociate the relations of "descending" neuronal signals from specific motor parameters. The oculomotor plant, with its constant load and its apparent immunity to signals from muscle receptors, presents a simpler control problem. The eye-movement system is therefore more manageable and the signals carried by its individual neurons at all levels are easier to recognize. We find it just as likely, however, that when the somatic motor system is as well understood as several of the oculomotor subsystems, the paradoxes will disappear and the nonlinearities will become transparent.

ACKNOWLEDGMENT
This work was supported by NIH grants EY00745, RR00166, EY06558, and EY07991.

Brain systems have a way of reconciling "opposite" views of neural processing; the motor system is no exception

Joaquin M. Fuster[1]
Department of Psychiatry and Brain Research Institute, School of Medicine, University of California, Los Angeles, CA 90024
Electronic mail: *joaquin%chango.dnet@loni.ucla.edu*

[GEA] **Alexander et al.**'s point is well made and well taken: Motor control depends on spatial and temporal patterns of neuronal activity in cortical/basal ganglia circuits that are self-organized under experience and that process information largely through parallel channels; connectionist models are neurally plausible and highly promising. The authors argue persuasively for the merits of these concepts. Their argument against "alternative" views, however, is weak. Fortunately, their position is compatible with the positions they consider antithetical to their own. As I see it, the "other side" of each of their theoretical dichotomies is not orthogonal but complementary to theirs.

Nonalgorithmic versus algorithmic processing. The argument against algorithmic processing is somewhat reminiscent of the argument against the molecular approach to the study of brain function. No one can dispute that physiology is theoretically reducible to molecular events. However, the problem with the molecular approach to higher neurophysiology is that it proceeds at the wrong (i.e., impractical) level of discourse and analysis (like trying to understand the written message by studying the chemistry of the ink). Similarly, certain algorithms, albeit probably far different from those of the robotics engineer, surely govern the neural transactions in motor control. The problem here is that neuronal population dynamics is governed by variables that are still unknown and by probabilistic, nondeterministic, rules that defy the formulation of algorithms. Furthermore, any such algorithms would most likely develop at unsuitable levels of discourse and analysis. This is one of the reasons the algorithms of robotics appear simplistic and irrelevant to the motor physiologist. But that alone does not make a "nonalgorithmic" approach more appealing, whatever its definition.

Parallel versus serial processing. The connective anatomy of cortical systems, notably that for vision (Van Essen 1985), has been shown to be so organized as to accommodate both parallel and serial processing. Convergence and divergence of axons can be widely observed, but they occur both between and within channels, leaving ample room for the two processing modes. As in the motor system (Crutcher & Alexander 1990), there is evidence of simultaneous processing in different stages of the visual cortical hierarchy: for example, in striate and inferotemporal cortex (Ashford & Fuster 1985). But there is also evidence, in the form of latency differences, for some serial progression of visual information from one area to another (Ashford & Fuster 1985; Coburn et al. 1990). I see no reason to suspect that things happen in the motor system differently, and certainly no evidence against serial processing in the control of motion. In fact, it is hard to imagine how sequential goal-directed actions could be concatenated without some serial processing. In the organization of a simple, visually instructed, finger movement, Gevins et al. (1989a; 1989b) observed substantial covariation of electrical activity in different cortical fields, indicating that information is processed in parallel, but that there is also some serial progression of covariations from occipital to frontal fields. There is no reason to exclude serial processing "further down" as well, into the cortical/basal ganglia system, both before and during the execution of movement.

Unsupervised versus supervised learning. The unsupervised and self-organized learning of new action that **Alexander et al.** postulate cannot take place on a *tabula rasa* only by trial-and-error. Whereas the self-organization of the connective motor substrate is an adaptive process that owes much to failure and success, this process is inconceivable without a preexisting substrate of inherited and acquired capabilities that can be characterized as motor memory. There is nothing basically wrong with the concepts of "kinemnesis" (Mackay 1954) and "motor engram" (Bernstein 1967); both are useful. Phyletic and individual motor memory guides the acquisition of new action in a manner similar to the way perceptual memory guides the acquisition of novel perceptual experience. Nothing is ever completely new. New action accrues to old action, and the guidance exerted by established motor representations over the new may be rightfully considered "supervision" of the learning process. For this reason alone, it does not make much sense to speak of a *learning phase* as separate from an *execution phase*. In the learning of new skills it is practically impossible methodologically to separate representation from processing, hardware from software. To trace the demarcation between the two, in what is really a blend of the two, is one of the nearly intractable problems of cognitive neuroscience, which encumbers studies of mapping and mechanism alike.

Nonhierarchical versus hierarchical processing. Here again we run into that fuzzy border between memory and process. And here the problem of separating the two looms large, because behind the execution of discrete motor actions lies the control from hierarchically higher representations of action. It is difficult to substantiate the hierarchical motor processing that Jacksonian thinking would lead us to predict, but it is not difficult to conceptualize a representational hierarchy in the motor system and to outline its connective substrate. It is important to recognize, first of all, the hierarchical structure of behavioral action. Just as there is a hierarchy of perceptual categories, there is one of motor categories. Any goal-directed behavior is made of motor subcomponents nested within it and organized in time and space according to goal. There is enough electrophysiological and neuropsychological evidence (summarized in Fuster 1989) to broadly and tentatively outline a hierarchy of motor representations in the cortex of the primate's frontal lobe. The motor cortex would represent discrete skeletal

movements, the lowest level of that hierarchy. The premotor cortex (SMA included) would be an intermediate level of representation for more general movement. At the highest level, the prefrontal cortex would represent, however schematically and imprecisely, global behavior of larger spatial and temporal range and dimensions. In behavioral performance, when representational networks become operational, networks in lower levels would be subject to control from above.

The control of action by supraordinate verbal "schemata" in prefrontal cortex is the basis of what Luria and Homskaya (1964) called the "regulating role of speech" in human behavior. In speech as in other sequential behaviors, the scheme of a higher-order representation of action would govern the chain of acts necessary to reach its goal. Of course, top-down processing would be accompanied by bottom-up feedback and "corollary discharge." (Top down vs. bottom up is thus another reconcilable "dichotomy.") The flow of motor processing would largely cascade down the hierarchy through connective gradients that originate in prefrontal cortex and, through successive subcortical loops, reach M1 (Alexander & Crutcher 1990a; Muakkassa & Strick 1979; Pandya & Vignolo 1971). The operations from the scheme to the "microgenesis" (Brown 1977) of the action would proceed top down, but under the regulating influence of bottom-up feedback constantly adjusting and refining the action.

Top-down and bottom-up processing would thus complete the *perception-action cycle* (Fuster 1989), the cybernetic flow of information, which at its highest level would course from frontal cortex to motor effectors, through the environment, through sensory systems, through posterior association cortex, back to frontal cortex, and so on (Fig. 1). This would be the gross "anatomy" of one of biology's most basic functional principles (Arbib 1981; Neisser 1976; Weizsäcker 1950). If we assume that the "cycle" is closed at various levels of a hierarchy of vertically interactive layers of categorical representation of action, operating by probabilistic rules and subject to environmental constraints, we do not have a problem with potentially infinite

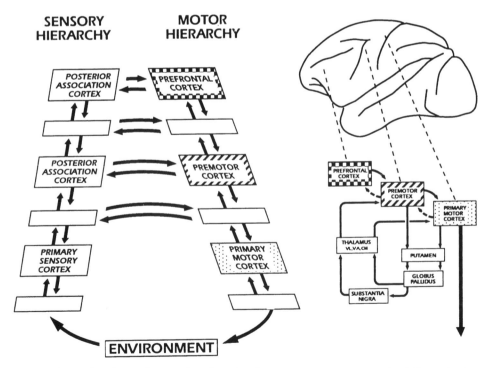

Figure 1 (Fuster). General principles of circuitry between structures processing information in the perception-action cycle. Unlabeled "boxes" in the sensory and motor hierarchies represent intermediate cortical and subcortical stages of processing. At right, some of the known subcortical circuitry supporting motor control in the monkey.

degrees of freedom. Nor do we need "wired-in" programs; and we certainly do not need anything like the rigid and deterministic software of artificial intelligence. Neither perfect schemes nor perfect matches – between action and scheme – are necessary in a circular interplay of action and motor memory like the one we envision, provided it has enough latitude for adjustment to widely changing environments.

In conclusion, whereas "motor programs" may be unnecessary, internal representations of action may well be critical for motor control. Quite conceivably, these representations consist of self-organized associative networks (Kohonen 1984), the product of largely "unsupervised" learning. Some, perhaps most, neural processing is parallel and nonhierarchical. But somewhere, some of the time, that processing is serial and hierarchical. And at some level it is certainly algorithmic, however irrelevant the algorithms at that level may be to understanding the mechanisms of motor control.

NOTE
1. Address correspondence to J. M. Fuster, Neuropsychiatric Institute, UCLA, 760 Westwood Plaza, Los Angeles, CA 90024.

How complex is a simple arm movement?

S. C. Gandevia[1]
Department of Clinical Neurophysiology, Institute of Neurological Sciences, The Prince Henry and Prince of Wales Hospitals, and Prince of Wales Medical Research Institute, University of New South Wales, Sydney, Australia

[EEF] **Fetz** has pointed out neatly some conceptual limitations to recordings from cells in the motor cortex and other putative motor areas of the brain. In particular, there is difficulty in interpreting the discharge of cells that appear unmodulated by the task and the discharge of cells that show a complex and "counterintuitive" relationship to the task in their timing or pattern of discharge. Hence, a model neural network that produces an array of discharge profiles similar to those observed *in vivo* is obviously attractive.

The complexity of the motor tasks must be considered before one reconsiders one's credence in the messages contained in actually observed neural discharges. Although a number of studies have required subhuman primates to produce torque around a particular joint, commonly the wrist (e.g., Fetz & Cheney 1980; Thach 1978), the total "movement" may require much more than simply generating say, flexor torque by the two prime wrist flexor muscles (flexor carpi ulnaris and flexor carpi radialis). In some studies, the hand is also required to grip a manipulandum so that contraction must occur in finger flexor and thenar muscles (for example, see Fig. 2 in **Fetz**'s target article). Should one regard as "rather astonishing" the observation that motor cortical cells correlated for force, position, and next movement direction occurred in equal proportions (Thach 1978)? This would require one to accept either that gripping the manipulandum had no effect on the motor cortical cells or that its effect was constant under the different experimental conditions. Given that corticospinal axons diverge to several muscles and that cutaneous and muscle afferent inputs from the hand can rapidly influence the discharge of motor cortical cells (e.g., Lemon 1981; Lemon et al. 1976; Wannier et al. 1991), the observation may not be so surprising. Small changes in contact between the digits, and between them and the handle, could change the afferent input so much that finding altered correlations between one parameter of the wrist movement and the cell's discharge may be expected rather than "remarkable." Similarly, in studies requiring a precise grip by the digits (e.g., Buys et al. 1986; Muir & Lemon 1983), activity in many intrinsic and extrinsic muscles of the hand must occur. Given the number of intrinsic and extrinsic muscles acting on a digit, the torque

generated by its tip is not specified uniquely by only one pattern of muscle activities. This provides an additional source of variation between the discharge of central cells and mechanical movement. Furthermore, when studies require movements of more proximal parts of the limb, contraction of shoulder and trunk muscles occurs, perhaps with "stabilizing" contractions in leg muscles.

Thus, in what seems to the animal (and possibly the experimenters) to be a simple mechanical task, there is potential activation of a multiplicity of muscles. The variable spatial and temporal activations of many muscles represent suboptimal conditions for examining causal relationships between a single neuron's discharge and force production. The size of this problem can of course be controlled by appropriate electromyographic recordings of seemingly remote and uninvolved muscles (e.g., Tanji et al. 1988), but it may increase when, as is the current trend, the experimental animals make movements of increasing biomechanical complexity, including multijoint movements with the upper limb to a range of positions in three-dimensional space (e.g., Schwartz et al. 1988). One possible experimental strategy would be to examine smaller, more discrete movements such as isolated recruitment of only a few motor units in an intrinsic hand muscle as the initial task. This task could then be grafted onto tasks needing active movement (or just postural stabilization) at the wrist and elbow, while the properties of neuronal discharge were re-examined. Such a strategy may also reveal the importance of cells whose discharge is unmodulated in the initial task, but which are recruited as the "spatial" requirements of the task increase.

Observations reported very briefly for dorsal root ganglion cells are difficult to fit into **Fetz**'s scheme without much more information. Since the cells have an excitatory projection to motoneurons, they are most likely to come from muscle spindle endings, but their parent muscle is not indicated. The discharge of spindle afferents is a complex function of changes in muscle length, location of the spindle within the muscle, fusimotor drives, and intrinsic spindle properties (see **Gandevia & Burke**, this issue). Although it should be easier to understand the role of proprioceptive afferents than that of cells in the motor cortex and other "motor" areas, it remains necessary to know which muscles are contracting and how their lengths are changing.

Finally, it will ultimately be necessary to overlay the knowledge about the discharge of "upper" motoneurons with the properties not just of motoneurons but also of spinal interneurons. For example, some propriospinal neurons activated by descending motor "commands" play a crucial role in forelimb grasping in the cat (Alstermark et al. 1981). Of course, simple linear correlations between cellular discharge and movement parameters may not be sufficiently sensitive for judging how such interneuronal circuits are turned on, switched, and modulated by descending and afferent activity.

NOTE
1. Address for correspondence: S. C. Gandevia, Prince of Wales Medical Research Institute, c/o Department of Clinical Neurophysiology, The Prince Henry Hospital, P.O. Box 233, Matraville, N.S.W. 2036, Australia.

Cerebellar function: On-line control *and* learning

Peter F. C. Gilbert and Christopher H. Yeo
Department of Anatomy and Developmental Biology, University College London, London WC1E 6BT, England

[JRB] The first part of **Bloedel**'s target article reviews evidence for the organization of the cerebellum into sagittal zones and suggests that the functioning of the cerebellum is through sagittally organized units of cells. Bloedel proposes that Purkinje cells in these units can have responses to mossy fibre inputs selectively enhanced through signals from climbing fibres. He suggests that under his "dynamic selection hypothesis" the climbing fibres will determine the populations of sagittally aligned Purkinje cells that would most dramatically modify the activity of cerebellar nuclear neurones. Bloedel supports this speculation with recent, to-be-published data from recordings in an identified cerebellar zone (Bloedel & Kelly 1992). But what does this "dynamic selection" accomplish? A characteristic feature of climbing fibre responses, repeatedly stressed in the target article, is that climbing fibres fire in response to perturbation. Whatever these responses do in terms of modulating Purkinje cell output, they do so after the event. The spectacularly massive cerebellar structure is relegated to assisting with compensatory movements after a motor error. Also, the proposal does not require some of the key features of cerebellar circuitry, in particular, the large numbers of parallel fibres synapsing with each Purkinje cell.

Bloedel claims that his hypothesis provides a unifying view of cerebellar integration that implements the circuitry and organizational features of the sagittal zones. He states that "there is virtually no other functional concept that integrates these features into a general view of cerebellar cortical processing." These claims can be disputed. First, as indicated above, this does not seem to be a particularly convincing explanation of the cerebellar circuitry. Second, it is incorrect to claim that it is the only theoretical proposal to take account of the sagittal organization of the cerebellum: One of us has shown how sagittally organized units of cells could be used in the learning of movements (Gilbert 1974; 1975).

The key test of cerebellar theories and hypotheses is the extent to which they account for empirical data and make testable predictions. On this score, **Bloedel**'s hypotheses are lacking. In particular, the results of Thach and associates (Gilbert & Thach 1977; Thach 1970b) in recording from Purkinje cells of performing animals offer little support for the gain change or dynamic selection proposals.

In contrast, there is a large volume of evidence (Gilbert & Thach 1977; Ito 1984) to support learning in the cerebellum, much of which is not discussed in **Bloedel**'s review (e.g., long-term depression caused by conjunctive stimulation of climbing and parallel fibres; Ito 1989). A major strength of the cerebellar learning hypothesis is that it makes sense of the response properties of the olive. Responses to perturbation, in addition to any immediate corrective function, can be used by the system as an instructor to modify transmission in active synapses for future benefit. Bloedel claims that there is substantial evidence against learning in the cerebellum on the basis of an experiment from his laboratory. He claims that decerebrate, cerebellectomized rabbits can learn eyeblink responses (Kelly et al. 1990b) but fails to point out that there has already been substantial criticism of this experiment. Using Bloedel's protocols, normal rabbits cannot learn an eyeblink response (Nordholm et al. 1991), and using conventional protocols cerebellectomy abolishes conditioned eyeblink responses in decerebrate rabbits (Yeo 1991a).

In any case, the most compelling evidence for cerebellar plasticity has not been from eyeblink conditioning experiments. The very substantial body of experimental evidence for such plasticity is from work on gain and phase modifications of the VOR, primarily by Ito and his colleagues (see Ito 1984; 1989). **Bloedel** demonstrates an extremely limited view of work in this area when he cites only the review by Lisberger (1988a; 1988b) that suggests a brainstem *and* cerebellar engagement in VOR modification.

In his final section, **Bloedel** puts forward his Vermittler hypothesis:

> The cerebellum serves as a active mediator whose output provides the CNS with an optimized integration with the relevant features of external execution space, internal intention space, . . . As a consequence of this integration the cerebellar output can modify activity in

central pathways responsible for motor execution to ensure the specification of the appropriate kinematic and dynamic characteristics of the movement.

This hypothesis is a very general one that is similar to what many others have proposed for the role of the cerebellum. It should be noted that the hypothesis does not have many implications for the detailed functioning of the cerebellar circuitry. In fact, some of the early theories that proposed learning in the cerebellum also hypothesized that the cerebellum would perform in a way similar to that suggested by Bloedel's Vermittler hypothesis (Albus 1971; Marr 1969).

A useful theory of cerebellar function would be one that makes more sense of perturbation response properties of the olive. We suggest that the cerebellum must engage in processes of on-line control based on calibrations and learning that have gone before.

Spinal movement primitives and motor programs: A necessary concept for motor control

Simon Giszter

Department of Brain and Cognitive Science, Massachusetts Institute of Technology, Cambridge, MA 02139
Electronic mail: *giszter@ai.mit.edu*

[GEA] The notion of the motor program as a serial and symbolic algorithm for action is something of a straw man. This is clearly not how the motor program concept has been used by physiologists in the last decades. Indeed, this is admitted by **Alexander et al.** It is true, however, that some artificial intelligence frameworks have initially used the motor program concept in the fashion described. Most users of the concept, specifically in the literature of neuroethology, have viewed the motor program in a very different way. In neuroethology the motor program represents a somewhat distributed entity able to exhibit (and to an extent control) a behavior or a fragment of behavior. This concept can be further sharpened in many instances to a core central pattern generator. [See Selverston: "Are Central Pattern Generators Understandable?" *BBS* 3(4) 1980.] Alexander et al. claim that the concept (in this rather general form) "has engendered very little insight into the neural basis of motor control." This charge seems very extreme and must be addressed.

The general view of the motor program embodies the notion of a limited, but nonetheless real functional capacity or circuitry, which forms a "competence module." It was developed both as a reaction to simple reflex chaining models and as an operationally defined entity (in the case of central pattern generators). Defending the large body of work on central pattern generation is unnecessary, given the breadth of phyla in which these mechanisms have been demonstrated. Perhaps more valid questions are: (1) whether the notion of motor programs is still useful in a framework of neural networks, and (2) whether it is still relevant to higher motor control in vertebrates.

With regard to the first question posed here, it is very clear that the motor program notion is not in any way supplanted by the arrival of the recent neural networks. Indeed, the majority of researchers on central pattern generation have continued to use the motor program concept repeatedly and still find it useful, despite using quite detailed neural modeling techniques for several decades.

Why is the motor program concept useful and even necessary to research on higher vertebrate motor control? In vertebrates many of these movement competencies exist at the spinal level (see **McCrea**, this issue). Are we to suppose that this spinal organization is largely irrelevant to the neural processing used in higher centers? On the contrary, I would suggest that the presence of these spinal competences must affect the ways in which higher centers organize movement (see **Bizzi et al.**, this issue and Bizzi et al. 1991). Work on spinal frogs suggests that the lower vertebrate spinal cord contains a number of pattern generators and what have been termed "movement primitives" able to organize fragments of various behaviors in a coherent way (Fukson et al. 1980; Giszter et al. 1989; 1991a; 1991b; 1992a; 1992b; 1992c; Mussa-Ivaldi et al. 1990; Ostry et al. 1991; Schotland et al. 1989). These primitives are likely to form elements that can either be combined in parallel or chained together serially to generate behavior. They may also form "bootstraps" for higher-level neural network learning processes (see Mussa-Ivaldi et al. 1991a; 1991b).

Is there any reason that evolution would select this particular framework comprised of low-level motor programs for organizing and developing a motor system? Recent research in robotics suggests that there is. Following the abandonment of the classical serial frameworks by what has been termed "New Wave" or "Nouvelle AI," many workers in this area have focused on using the ideas of Minsky (1986) in situated autonomous agents. They have used this framework in the building of action systems based on lower-level autonomous competence modules. This design process is accomplished using an incremental, layered, and somewhat evolutionary methodology (Beer et al. 1991; Brooks 1991; Maes 1991a; 1991b). The methodology has led to robust autonomous systems. The low-level modules used in these schemes often correspond very closely to the motor program concept of the neuroethologists, even to the extent of exhibiting "central pattern generation." [See also Ewert: "The Evolutionary Aspect of Cognitive Functions" *BBS* 10(3) 1987.] Indeed, some of the AI work has drawn inspiration from this area of neurobiology. This burgeoning field, which lends credence to the motor program concept, is completely ignored in **Alexander et al.**'s target article. Many of these frameworks have included neural networks of various types, and often have distributed processing (parallel motor programs). Both of these two independent but converging lines of investigation from engineering and neurophysiology support the importance of elementary competences, motor programs, or movement primitives in sensorimotor action systems.

It therefore seems clear that the motor program is a useful and perhaps fundamental conception. I would further suggest that from this viewpoint the likelihood is that the cortex and basal ganglia will prove to use spinal motor programs to control movement, albeit supplemented in various ways by the other types of mechanisms suggested in the target article.

Area LIP: Three-dimensional space and visual to oculomotor transformation

James W. Gnadt

Department of Neurobiology and Behavior, State University of New York at Stony Brook, Stony Brook, NY 11704
Electronic mail: *jgnadt@sbccmail.bitnet*

[JFS] Inherent in the concept of egocenter is a 3-dimensional spatial representation of surrounding space. There is compelling evidence that the posterior parietal cortex operates in a 3-dimensional parameter space. Lesions in the parietal cortex in humans can produce deficits in spatial perception (Cogan 1953; Godwin-Austen 1965; Holmes 1918; Holmes & Horrax 1919) and eye movements (Manor et al. 1988; Ohtsuka et al. 1988) in depth. In rhesus monkeys, neurons in area 7a and the adjacent superior temporal sulcus have been reported to respond to visual stimulus motion and oculomotor tracking in depth (Sakata et al. 1983; 1985). Neurons responding to stimulus disparity, vergence, accommodation, and binocular eye position in three dimensions have been recorded in area LIP (lateral intraparietal cortex) in the intraparietal sulcus (Gnadt & Mays 1989; 1991). There is also evidence for involvement of suprasylvian cortex in

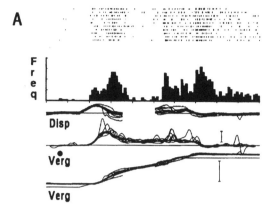

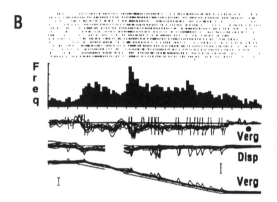

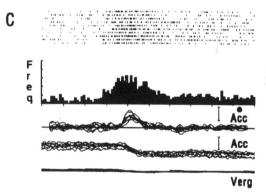

Figure 1 (Gnadt). Examples of visual-motor responses in depth for LIP neurons. A: Response of a cell to crossed disparity during convergence tracking. B: Response of a divergence velocity cell during divergence tracking. C: Response of a cell related to accommodation for a step change in target blur, while vergence was held constant. Each panel shows spike rasters, average frequency histograms and overlayed responses for multiple trials. Frequency = spikes/sec. Time marks (abscissa) = 200 msec. Verg = vergence response and target vergence demand (offset slightly). Targets were blanked during tracking as shown. Verg (dot) = first derivative of vergence position. Disp = disparity inferred from known target and eye positions. Acc = accommodative state of the lens. Acc (dot) = first derivative of accommodative response.

the oculomotor near response in cats (Bando et al. 1984; Toda et al. 1991).

Figure 1 demonstrates three examples of area LIP neurons related to visual-motor behavior in depth (Gnadt & Mays 1989). The cells in Figure 1A and 1B are shown during vergence

A

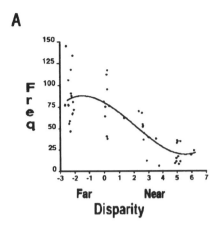

B

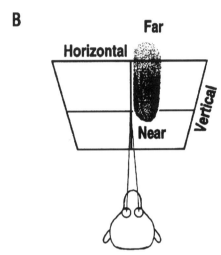

Figure 2 (Gnadt). Three-dimensional tuning of a LIP neuron that would be labeled as "saccade-related" by standard criteria (memory saccades and double saccades). A: The cell's frequency of firing (spikes/sec) prior to a movement into its response field is shown as a function of stimulus disparity (degree). B: The 3-dimensional tuning is modeled schematically as viewed from above by scaling the cell's conjugate response field (modeled as an ellipse) by the tuning curve for depth. The center of the neuron's response field in the plane of fixation was about 10 deg to the right and did not change as a function of depth. At one-half maximal response, the response field was approximately 10 deg of visual angle and did not include the point of fixation.

tracking in depth. The cell in Figure 1A responded to the stimulus property of crossed disparity, whereas the cell in Figure 1B exhibited a response proportional to divergence velocity, independent of the presence of the stimulus. Figure 1C shows a cell related to the velocity of accommodation, independent of ocular vergence. Finally, recent data (Gnadt & Mays 1991) have shown that a majority (86%) of what are thought of as "saccade-related" LIP neurons are tuned for a 3-dimensional volume of space. Figure 2 shows a typical example that exhibited a tuning curve for the distance from the point of fixation (disparity, Figure 2A). A schematic representation of this neuron's 3-dimensional response field is shown in Figure 2B. In addition, I would point out from this figure that despite **Stein**'s assertion to the contrary (sect. 2.2) LIP neurons are tuned for both the direction and amplitude of conjugate movement (see also Andersen & Gnadt 1989, Figure 11).

Another property of posterior parietal neurons makes them appropriate for mediating an egocentric representation of visual-motor space. Since many posterior parietal responses are modulated by eye position (Andersen et al. 1985b; 1990b; Gnadt

& Mays 1989), a solution to the enigma that current eye position is always accounted for in visual perception (Helmholtz 1910) is inherent in their activity as a population. For every combination of current eye position and visual target in 3-dimensional space there is a unique subset of posterior parietal neurons that will be active.

A second theme I would like to emphasize is the role of the intraparietal area as an oculomotor output field of posterior parietal cortex. As reviewed by **Stein**, area LIP has the appropriate input and output connections to process and relay visual information from the "where" visual pathways (areas 7a and middle temporal/middle superior temporal MT/MST) to both cortical (frontal eye fields) and subcortical (superior collicular) oculomotor areas. Stimulation in LIP can produce eye movements (Kurylo & Skavenski 1991; Shibutani et al. 1984), and lesions that include the intraparietal sulcus can produce oculomotor deficits (Lynch & McLaren 1989). Furthermore, evidence indicates that activity in area LIP is encoded in coordinates of "motor error" or "desired change in eye position." A correlation of the neuronal activity with the behavioral state of "motor intention" has been suggested (Gnadt & Andersen 1988). In a contrasting view, a recent paper (Duhamel et al. 1992) interpreted LIP activity as representing a dynamic shift of visual receptive fields based on future motor planning. This study made two controversial assumptions, however: The first was that premovement planning of a second eye movement cannot precede the completion of one in progress. The second was that the absence of an overt eye movement of given metrics can preclude volitional effort (intention) of those same metrics. Since LIP cells do not require a visual stimulus but respond only when the metrics of a potential or realized eye movement match their response field (Bracewell et al. 1991; Gnadt & Andersen 1988), the most parsimonious interpretation is that LIP activity represents a motor planning signal that is distal to the sensory-to-motor transformation (motor error). This interpretation is supported by recent evidence that many of these LIP neurons project directly to the superior colliculus (Gnadt & Mays 1991) where similar motor signals have been demonstrated (Mays & Sparks 1980).

In this respect, it is erroneous to suggest that LIP neurons operate in retinotopic coordinates (**Stein**, sect. 1.2). The frame of reference is oculocentric but not retinotopic. The responses of these neurons are not sensory signals dependent on the location of a stimulus on the retina. By way of further evidence, Bracewell et al. (1991) showed that LIP neurons responded to auditory targets if a saccade was executed into their response field. Only carefully designed behavioral tasks such as the double saccade task or the auditory task can discriminate "retinal error" and "motor error," both of which are oculocentric. This distinction is important, however, because one precedes the sensory-to-motor transformation and the other follows it.

In closing, the study of covert behavior, such as attention and intention, presents some thorny problems for empiricists. By introspection we can consider the condition of suppressing an eye movement to an eccentric visual target while fixating another. The inclination to look toward the eccentric target can be quite compelling, yet one can suppress the behavior of looking toward it. Furthermore, the subjective difference between a suppressed intent to refixate and (nonmotor) attentive vigilance is subtle at best. More important, for investigational science, there is no objective measure of either volitional effort. Only by well-designed constraint of the behavioral significance of the task can one attempt to interpret the lack of overt behavior. The interpretation of unconstrained, spontaneous, or spurious behaviors should be approached with caution. Otherwise, our empirical efforts reduce to little more than high-tech speculation.

ACKNOWLEDGMENTS
Supported by NEI grant EY08217 from the NIH.

Action systems in the posterior parietal cortex

Melvyn A. Goodale and Lorna S. Jakobson

Department of Psychology, University of Western Ontario, London, Ontario, Canada N6A 5C2
Electronic mail: goodale@uwo.ca

[JFS] **Stein** proposes that the brain does not use a single "map of space" for the visual control of actions; instead, he argues, it relies on a distributed set of transformational algorithms for converting sensory vectors into other reference frames or motor coordinate systems. The mechanisms that carry out these transformations are presumed to reside largely within the posterior parietal cortex. So far so good. It has been clear for a long time that the popular characterization of the posterior parietal cortex as a "where" system for localizing stimuli in space was far too vague. In Ungerleider and Mishkin's (1982) original account, there was no recognition of the special role that the posterior parietal cortex plays in the visual control of goal-directed action. Instead, it was assumed that the dorsal stream of projections from striate cortex to the posterior parietal cortex simply computes one visual attribute of a stimulus, its location, while the ventral stream to inferotemporal cortex computes other visual attributes such as size, shape, orientation, and color. Little attention was paid to the reason for this apparent separation in the processing of visual inputs. Stein's account, in contrast, has the virtue of recognizing that the main visual function of the posterior parietal region is not locating stimuli per se but carrying out the required sensorimotor transformations for action.

Nevertheless, **Stein's** account tends to preserve the old distinction between "where" and "what" that was originally proposed by Ungerleider and Mishkin (1982). Thus, in describing the nature of the transformations that occur in the posterior parietal cortex, Stein seems primarily interested in how the brain is able to direct a movement toward a particular location in space. Although he eschews the simplistic notion of a topographic map of egocentric space and invokes instead the idea of a parallel distributed processing network, his main concern is still with "where" the stimulus is located and how the brain computes that location. Most skilled movements of the limb, however, are directed at objects not at locations; that is, the sensorimotor transformations required for an action often reflect the size, shape, and orientation of the goal object as well as its spatial location. In reaching for a coffee cup, for example, we adjust the posture of our hand and fingers to conform to the cup's handle well before we make contact. In other words, in forming our hand we appear to use visual information about the cup – information about its shape, size, and orientation that seems more akin to computations about "what" than about "where." These visual attributes, too, have to be converted into the reference frames and motor coordinates required for the production of a successful grasping movement. Indeed, it has recently been demonstrated that a population of cells in the posterior parietal cortex that fire during grasping movements are also tuned to the goal-object's size and orientation, irrespective of its egocentric location (Taira et al. 1990).

Because information about the intrinsic visual characteristics of objects appears to be coded in both the dorsal and the ventral streams, the question remains as to the function of these two distinct processing systems. In our view, **Stein** is undoubtedly right in suggesting that the dorsal stream is concerned primarily with the sensorimotor transformations required for action. In fact, this emphasis on output requirements is what was missing in the original proposal of Ungerleider and Mishkin (1982). If the posterior parietal system is indeed concerned with action, then it makes sense that there should be transformational rules for converting information about intrinsic object characteristics into reference frames that can be used to orchestrate complex postures of the hand and digits during prehension. Recent neuropsychological evidence supports this conjecture. Some patients with optic ataxia resulting from damage to the posterior parietal region have difficulty not only in generating spatially accurate movements but also in preshaping their hand to accommodate objects of different sizes and orientations (Jakobson et al. 1991; Perenin & Vighetto 1988). It is important to note that when such patients are asked to describe the size, orientation, and spatial location of objects they cannot grasp they often have no difficulty doing so (e.g., Denes et al. 1982; Perenin & Vighetto 1988). The latter observation not only shows that processing of intrinsic object properties is taking place elsewhere in the brain (presumably in inferotemporal cortex), but that profound visuomotor deficits can be clinically dissociated from attentional problems. This would appear to undercut Stein's proposal that directed attention serves as the selection mechanism for the sensorimotor algorithms required for particular actions. Nevertheless, the failure of patients with optic ataxia to control their actions in terms of the intrinsic properties of the goal object as well as its spatial location strongly reinforces the notion that the dorsal stream is primarily involved in the control of action.

If the dorsal stream is mediating action, what is the function of the ventral stream? Some clues can also be found in the neuropsychological literature. Patients with damage in the occipitotemporal region often have great difficulty recognizing common objects on the basis of their visual appearance (Farah 1990). Nonetheless, they have no difficulty grasping objects or navigating through the world, at a local level at least. In short, their perception is disturbed even though their visuomotor behavior remains largely intact. Indeed, some patients with bilateral occipital damage have such profound perceptual difficulties that they are unable to recognize even simple geometric shapes (Benson & Greenberg 1969). We recently had an opportunity to study the visuomotor performance of one such patient (Goodale et al. 1991). Even though the patient was unable to indicate the size or orientation of an object, either verbally or manually, she showed normal preshaping and rotation of her hand when reaching out to grasp it. Again it appears that action systems in the dorsal stream can operate quite independently from object recognition systems, which presumably involve the ventral stream of projections to inferotemporal cortex.

In summary, perhaps a better way of looking at the distinction between visual processing in the posterior parietal and inferotemporal cortices is in terms of the output requirements of the two systems. We agree with Stein's suggestion that the dorsal stream has a special role to play in the sensorimotor transformations required for the spatial control of actions. But we go further and suggest that spatial location is just one aspect of the visual information that is transformed into motor coordinates. In fact, we would suggest that while similar visual information about objects is available to both systems, the transformational algorithms that are applied to these inputs are uniquely tailored to the function of each system (Goodale & Milner 1992; Milner & Goodale, in press). The transformations carried out in the ventral stream enable the formation of perceptual and cognitive representations that embody the enduring characteristics of objects and their spatial relations with each other; those carried out in the dorsal stream, which utilize the instantaneous and egocentric features of the objects, mediate the control of goal-directed actions.

Intermittent use of feedback during movement phase transitions and during the updating of internal models

Andrew M. Gordon[a] and Albrecht Werner Inhoff[b]

[a]*Nobel Institute for Neurophysiology and Department of Pediatrics, Karolinska Institute, S-104 01 Stockholm, Sweden and* [b]*Department of Psychology, State University of New York at Binghamton, Binghamton, NY 13901*
Electronic mail: *inhoff@bingvmb*

[SCG] The nervous system is structured in a manner that makes a wealth of movement-related feedback available, including kinesthetic feedback. This feedback can serve various purposes: It can be used to specify the status of the motor system to control the movement initiation, provide closed-loop control during the ongoing movement, and indicate successful completion of the movement. Although **Gandevia & Burke** acknowledge that kinesthetic information may be used intermittently, as well as to update the motor system, their target article emphasizes the continuous availability and use of kinesthetic feedback.

Several considerations suggest that **Gandevia & Burke**'s view may apply to a limited set of conditions, generally encompassing slow and simple movements. During these movements, such as slow flexion of the elbow or finger, it is likely that kinesthetic information from joints, muscles, and skin is used continuously and is able to provide information on movement position because the simplicity of the movement encompasses sufficient constraints for the interpretation of kinesthetic information. During many natural and complex movements that are performed more quickly, however, kinesthetic information may be insufficient for precise movement specification and it may be used more intermittently to reset open-loop control and to update an internal model for subsequent movements. In the following, we use grasping/lifting movements and typing to illustrate the limited use of continuous monitoring of kinesthetic feedback.

Grasping and lifting. During the dynamic phases of lifting small objects between the thumb and index finger, somatosensory information is used *intermittently* to trigger subsequent motor output and to compensate for slips (cf. Johansson & Westling 1990). For example, tactile afferent information signals the initial contact of the fingers with the object and triggers the onset and termination of force-increase during the start and cessation of vertical movement (cf. Johansson & Westling 1990). Indeed, subjects with sensory impairments (Eliasson et al. 1991; Johansson & Westling 1991) or anesthetized finger tips (Johansson & Westling 1984) have prolonged transitions between phases comprising the grip/lift movement. Likewise, during restraint of active objects exerting unpredictable loads, the normal moment-to-moment sensory control (Johansson et al. 1992b, 1992c) is disrupted following digital anesthesia (Johansson et al. 1992a). It is interesting that frictional and weight information from previous lifts with the contralateral hand can be used to control the force output (Johansson & Westling 1984). Yet, subjects have more difficulty transferring frictional information between digits in the same hand (Edin et al. 1992).

Unequivocal somatosensory information regarding the object's weight is gained only after the object is lifted from its support. Motor programs are therefore used to parametrize the force output in advance (Johansson et al. 1992b; 1992c). Such anticipatory control is based on memories of the object's physical properties acquired even from a single previous lift. These properties include the object's weight and friction (Johansson & Westling 1984; 1988a). Thus, somatosensory information regarding the object's physical properties is used automatically to *update* the internal neural representation, preventing many of the stability problems of continuously monitoring feedback (cf. Rack 1981). Hence, control mechanisms used during grasping are anticipatory in nature, although they make intermittent use of sensory information to update task phase parameters or to shift from one phase of the lift to another. This allows a portion of the sensorimotor apparatus to be disengaged for higher-level exploratory tasks (Johansson 1991).

Sometimes, however, anticipatory control can result in erroneous force programming (cf. Johansson & Westling 1988). For example, the relation between the size and weight of objects is learned in early childhood. As a result, when subjects lift different boxes of equal weight and unequal size, they typically experience a size-weight illusion (Charpentier 1891) in which smaller objects are perceived as heavier. This has long been thought to arise from a mismatch between the expected and the actual weight of the object (Claparéde 1901; Martin & Müller 1899). Indeed, subjects' responses are inconsistent with this perception; the rate of increase of the isometric grip and load forces is greater for the larger boxes, even after many trials (Gordon et al. 1991a; 1991b). This results in grip force overshoots and strong accelerations of the object causing positional overshoots. Yet the force coordination when the object is held in the air is appropriately adjusted to the object's weight. Thus, despite presumably similar somatosensory information during the static phases of previous lifts, the central commands used during the dynamic increase in force result in the illusion, highlighting their importance in weight perception. This supports **Gandevia & Burke**'s contention that the CNS has perceptual access to signals related to central motor commands. It is interesting that the illusion is not experienced by occasional cautious subjects who choose not to program the force output in advance but instead use graduated increases in force (Gordon et al. 1991b). Subjects can therefore flexibly choose strategies that rely more or less on sensory feedback, but movements are considerably *slower* when the strategy involves continuous reliance on kinesthetic feedback.

Typing. In some highly practiced tasks with constant reaching requirements (e.g., skilled typing), kinesthetic information may not be used during movement onset or movement continuation. Instead, it may provide information concerning reaching accuracy after the movement trajectory has been initiated or completed. Lashley (1951) pointed out that latencies between movement elements in skilled typewriting are too short for kinesthetic feedback to guide movement transitions. Recent examinations of reaching movements during typing (Soechting & Flanders 1992b) suggest that skilled typists generally start reaching for a letter key 20 msec after the preceding letter key has been activated. [See also Flanders et al.: "Early Stages in a Sensorimotor Transformation" *BBS* 15(2) 1992.] Kinesthetic information would not be available within this short interval to control the onset or continuation of action. Furthermore, several aspects of recent copy-typing studies demonstrate the representation of movement prior to execution. First, the fixation of to-be-typed letters precedes corresponding key-presses by 2 to 4 characters. Second, manual between-keypress times increase when the visual preview of to-be-typed text to the right of fixation is restricted. Third, word viewing times increase when the fixated word is difficult to type, even though word viewing generally *precedes* word typing (Inhoff & Wang 1992).

Instead, kinesthetic feedback may be used by skilled typists to determine movement success. Typing errors are generally noted even when no visual feedback is available. Errors are also noted quite rapidly as between-keypress times generally increase immediately following the incorrect keystroke. Typists, however, often execute the key-press following the typing error normally (i.e., with normal speed), but the subsequent movements are slower (Salthouse 1984). These delayed-error responses are of diagnostic value as they suggest that the use of kinesthetic feedback concerning movement trajectories *followed* movement specification, movement onset, and, quite often, movement execution.

NOTE
Authorship was determined alphabetically. All correspondence should be addressed to Albrecht Inhoff.

Kinematics is only a (good) start

G. L. Gottlieb

Departments of Neurological Sciences and of Physiology, Rush Medical College, Chicago, IL 60612

[EB] The equilibrium-point hypothesis is based on the premise that in the absence of external forces the following equation is a meaningful way to characterize the control of a muscle's length (x) and thereby of joint angles and from them, limb position.

$$Jx'' + Bx' + \alpha(x - \lambda) = 0 \qquad (1)$$

Bizzi et al. assert: "Centrally planned motor intentions are expressed and transmitted to the periphery using a virtual trajectory" that can be expressed in terms of λ in equation 1. The coefficients α and λ have been sources of controversy. It has been argued that the fundamental premise of the α-model is that *only the parameter* α is controlled by the nervous system (the target article's footnote 2 notwithstanding, Fig. 1 of Bizzi et al. 1982 has led some to an opposite conclusion). On the other side, the λ-model asserts that *only* λ of each muscle is controlled by the nervous system (Feldman 1966a).

To a degree, the target article unifies the two approaches. From the perspective of a joint that always has at least two controlling muscles, we can rewrite equation 1 as:

$$Jx'' + Bx' + C(x - R) = 0 \qquad (2)$$

where C and R are functions of the αs and λs. R and C are both controllable parameters, a viewpoint with which even the most vigorous advocates of the λ side should concur. There is still much room for disagreement over the degree to which α and λ might be viewed as *independently* controlled and especially about the role of feedback in generating those control signals but this is a dispute about the mechanism underlying equation 2, not about its validity or appropriateness.

Bizzi et al. present their arguments clearly and the weaknesses are acknowledged. My point of departure is the suggestion that it is possible to control a movement by defining virtual trajectories with "no knowledge of the dynamic parameters of the limbs." The work of McIntyre (1990) is alluded to as a possible solution to the many problems with that notion. Perhaps so. I will offer an alternative.

Assume that a virtual trajectory is computed that is "a copy of the desired movement" taking no account of speed or dynamics or any of the other problems imposed upon our flesh by physics. Assume too that a knowledgeable central controller exists that understands that fast movements need larger forces in different directions than do slow movements. For such movements, appropriate pulses of excitation to the motoneuron pools are superimposed upon those derived from the virtual trajectory. These excitation pulses "launch" the movements in about the right direction with about the right acceleration. They are not particularly accurate and will have a large degree of variability. Those details are handled by equilibrium point control mechanisms.

How does this smart controller know how to make so many different movements? Because it has been practicing all its life, some it remembers and some it extrapolates. This, however, is only a subset of all the movements we can make. For the rest, it can make slow accurate movements to let equilibrium-point control take over or it can guess and probably make some error. It will do better next time, given another chance. Practice and repetition are universal prerequisites for motor skill.

Is this approach fundamentally different from the one proposed by **Bizzi et al.**? Perhaps not. Let us return to the single-joint case to simplify the discussion and rewrite equation 2 as follows.

$$Jx'' + Bx' + Cx = F_i \qquad (3)$$

The motor controller must generate a specific $F_i(t)$ to produce a desired $x(t)$. The subject denotes different candidates for this function. The original candidate was $F_1 = CR$. For slow movements, R is a copy of the desired movement (R_c). However, in general R will have to be more complex. For fast movements it is N-shaped (Hasan 1986; Hogan 1984b; Latash & Gottlieb 1991c). That the authors find such a virtual trajectory unattractive is not surprising since the computation of a complex virtual trajectory to achieve a specific physical trajectory is not obviously simpler than the problem of inverse dynamics that the equilibrium-point hypothesis was designed to avoid.

Bizzi et al. propose $F_2 = f(R_c', R_c, x', x)$, replacing a complex input function with a complex forward path controller. This appears similar to Feldman et al. (1990). The nervous system can probably compute either kind of function (their comparative computational simplicity seeming an arguable issue).

We suggest $F_3 = f_t(T_i, L_i) + CR_c$, which augments the slowly rising forces generated by the virtual trajectory and achieves a desired acceleration and speed based upon a prior knowledge of the characteristics of the task. T_i and L_i denote a small number of simple, constant parameters of the intended trajectory and expected load such as distance, relative speed, inertia, and so forth (Gottlieb, in preparation). The kinematic transient is mediated by predictive mechanisms embodied in f_t while achievement of stable, accurate posture at the endpoint is mediated by the feedback utilizing second term. This allows fast but inaccurate movements even in the absence of feedback rather than the "sluggish" ones attributed by **Bizzi et al.** to their own model. No doubt generalizing this approach to the multijoint, multidimensional situation of the real world will have its problems but so too have all the alternative hypotheses. Note that $F_1 \equiv F_2 \equiv F_3$. They differ only in how they are computed. Therefore, until we have better information about state variables that go beyond the kinematic ones, we must choose a theory on the basis of parsimony and esthetics.

One of the acknowledged omissions of the α-model that the λ-model has attempted to address is the generation of EMG patterns. For any equilibrium point model to be about the *neural control* of movement rather than just about kinematic descriptions with a nod toward central representation, this issue deserves far more attention than it has received thus far. Feldman's approach accounts for the timing of the antagonist burst incorrectly (Feldman et al. 1990). The approach taken by Latash (Latash & Gottlieb 1991a; 1991b) is slightly better. We have addressed this issue at some length (Gottlieb et al. 1989; 1992) in an approach that is couched in terms of feedforward rather than feedback mechanisms. Where and when feedback mechanisms make important contributions cannot be determined from observations of unperturbed movements.

An outstanding virtue of Merton's (1953) servo hypothesis was its specificity about the nature of the feedback loop. The servo hypothesis was sufficiently specific in its formulation that it eventually succumbed to experimental evidence. On the electromyographic response to perturbations, advocates of equilibrium-point models, which in many regards are reincarnations of Merton's original speculative hypothesis, have not been as clear. Such specifics will be required to allow falsification of one structure or the other. In kinematic terms, the general notion has been successfully defended (e.g., Hogan 1984a) but that test did not address feedback as such.

In summary, **Bizzi et al.** have done an outstanding job in describing the current state of the equilibrium-point hypothesis in terms that should satisfy a mechanical engineer. From the perspective of a neurophysiologist, a great deal of work remains to be done.

Somatotopically organized maps of near visual space exist

Michael S. Graziano and Charles G. Gross

Department of Psychology, Princeton University, Princeton, NJ 08544
Electronic mail: *cggross@phoenix.princeton.edu*

Map me no maps, sir, my head is a map, a map of the whole world.

> Henry Fielding

[JFS] **Stein** claims that there is no evidence for a region where egocentric space is represented topographically. In lieu of a map, he offers us a "neural network," that is, "a distributed system of rules for information processing that can be used to transform signals from one coordinate system into another." Such a computational scheme might indeed work; however, it is quite unnecessary, because a neuronal topographic map of visual space does exist, at least for the region adjacent to the body, that is, immediate extrapersonal space. As there is good evidence for more than one such map in the primate brain, the question would seem to be: What are their different functions? rather than How can we erect a computational network to do without them?

We recently found such a map in the monkey putamen (Graziano & Gross, submitted; Gross & Graziano 1990), and evidence for a similar organization has been found in area 6 and in portions of the parietal lobe. In this commentary, we first summarize our experimental evidence for a somatotopically organized map of visual space. Then we mention the earlier and very similar observations in other regions of the brain.

Neurons in the macaque putamen have tactile receptive fields that are somatotopically organized (Crutcher & DeLong 1984a). We found that cells with tactile receptive fields on the face or arms were often bimodal: They responded both to tactile and to visual stimuli. Furthermore, for each cell the location of the visual receptive field closely matched the location of the tactile receptive field. That is, the cell would respond to an object placed within about 10 cm of the skin and moving toward the tactile receptive field. Thus, these neurons formed a somatotopically arranged visual map of near extrapersonal space. A small proportion of the cells with tactile receptive fields on the arm had movable visual receptive fields: As the arm moved, the visual receptive field moved to remain in register with the arm. This suggests that the map of visual space in the putamen might be a "body-part-centered" one rather than a "head-centered" one.

Bimodal cells with similar properties were previously described in area 6 by Rizzolatti and his colleagues (Fogassi et al., submitted; Gentilucci et al. 1983; Rizzolatti et al. 1981a; 1981b; Rizzolatti et al. 1983). Although the authors did not study the effect of varying limb position, they did study the effect of eye position. When the animal's eye moved, the visual receptive field remained in correspondence with the tactile receptive field. That is, the visual responses were somatotopically and not retinotopically organized. These investigators pointed out that area 6 cells could help program visually guided movements near the body and supported this view by demonstrating that monkeys with lesions of area 6 neglect nearby but not distant visual stimuli (Rizzolatti & Berti 1990).

Similar observations have also been made in area 7b. Cells in 7b were somatotopically organized (Robinson & Burton 1980a; 1980b) and many had visual receptive fields that corresponded to the tactile ones (Hyvarinen 1981; Hyvarinen & Poranen 1974; Leinonen et al. 1979; Leinonen & Nyman 1979). At least in some cases, when the arm was moved, the visual receptive field appeared to move so as to stay in correspondence (Leinonen et al. 1979).

In summary, there are at least three brain regions that contain a somatotopic map of the immediate visual space, namely, the putamen, area 6, and area 7b. These areas are monosynaptically

connected (e.g., Cavada & Goldman-Rakic 1989b; 1991; Kunzle 1978) and appear to form a system for the representation of extrapersonal visual space. All are egocentric maps. Some of their properties suggest that they are body part centered rather than strictly head centered. They presumably receive their visual, somatosensory, and proprioceptive information from the posterior parietal connections summarized by **Stein**. These maps might serve both sensory and motor functions in near extrapersonal space, perhaps each serving primarily one or the other function. They might also provide an input to the allotropic representation of visual space found in the hippocampus.

ACKNOWLEDGMENTS
Supported in part by NIH Grant MH-19420 and NSF Grant BNS-9109743.

Information processing styles and strategies: Directed movement, neural networks, space and individuality

Paul Grobstein

Department of Biology, Bryn Mawr College, Bryn Mawr, PA 19010
Electronic mail: *p.grobstein@cc.brynmawr.edu*

[GEA, EEF, DAR, JFS] *1. Neural networks, brain function and what's in between: Let's indeed have a new baby . . . and not throw out the old ones.* Having myself dared to say in print that the neuron is no longer the centerpiece of neuroscience (Grobstein 1987; see also Grobstein 1986; 1988a; 1988b), I can't argue with **Robinson** (or **Alexander et al.**, **Fetz**, or **Stein**) on that score. "Discomfort, if not always openly acknowledged, has been increasingly expressed by neurobiologists of almost every sort. . . . While diagnoses of the problem vary in detail, a common theme is that characterization of functional populations is not only 'difficult to construct from individual elements' but frequently impossible" (Grobstein 1990). It's been a bit lonely out here, and so I welcome with open arms Robinson's decision (and anyone else's) to come out of the closet and join the community of skeptics.

Mild irony aside, there are real issues to be addressed by those who understand the limitations of not only "bottom-up" approaches to the brain, but "top-down" approaches as well. "Beginning with the computational task is fine if one is dealing with an engineering problem; it can be highly misleading if one is concerned with a preexisting complex information-processing device whose computational tasks, styles, and constraints are in fact part of what needs to be discovered" (Grobstein 1990). **Robinson**'s thoughts are particularly welcome, given his leadership with top-down approaches. I gladly accept his correction to my remark: Not even all engineering problems are necessarily best solved with top-down approaches. Robinson's target article, and those of **Alexander et al.**, **Fetz**, and **Stein**, are important contributions to the requisite discussion of the obvious question: Beyond bottom-up and top-down, what approaches to the analysis of complex information-processing devices are available? I have written elsewhere, in the context of sensorimotor integration (Grobstein 1988b), neuroscience as a whole (Grobstein 1990), and biological systems more broadly (Grobstein 1988c), about an "intermediate level approach" that proceeds from the middle outward. Here I want to test some of those ideas against the insights of my colleagues (and theirs, in turn, against my own) in the specific arena they have defined: Sensorimotor processing and the new insights into it that have been, and will be, gained from the use of artificial neural networks.

My first point is one of whole-hearted agreement with **Robinson** and the others: artificial neural networks have much to offer. Among the things they have to teach is that solutions to information processing problems may have a highly distributed character, which in turn could account both for difficulties in making sense of neuronal behavior that has troubled those using a bottom-up approach and the failure to find particular algorithms in particular neurons or brain regions that has frustrated those using a top-down approach. My second point, however, is that none of this necessarily follows; artificial networks have much more to teach us, including this caveat itself. My third point is that, as always, the brain is cleverer than the clever ways we think up to try and understand it. If we forget that again, we'll go galloping off in the wrong direction. Again.

Many investigators, ourselves included (Carr et al. 1991), have been impressed by the finding that artificial neural networks tend to come up with distributed solutions to input/output problems. This has helped to alter expectations of what one should be looking for in the brain, in both new ways and old (Grobstein 1988a; 1990). What has not been generally appreciated, however, is the reality that artificial neural networks do not *always* come up with distributed solutions (Carr et al. 1991). Even for simple sensorimotor tasks, artifical networks come up with more or less distributed solutions depending on exactly how the task is defined (what investigators think is the important part of the task the brain is doing, and hence how the task is translated into network terms). And even for exactly the same task description, more or less distributed solutions are found depending on details of network structure (the number of hidden units, mentioned by **Robinson**, is one, but only one, example). And even for exactly the same task description and network structure, more or less distributed solutions are found depending on exactly where one starts on the solution landscape (the initial random synaptic weight distribution) and how one searches the landscape from that location (**Fetz** calls attention to solution variability in relation to a different but also important point; see my discussion of **Stein**, below).

Clearly, investigators using artificial neural networks need to be as skeptical about the feeling that everything fits, as do other investigators, and some indication of the ways in which solutions were sought, as well as the frequency with which different kinds of solutions were observed, should become a standard part of any report of artificial network findings. That artificial networks are capable of generating a variety of different kinds of solutions, however, has a more exciting implication (and an aspect of unpredictability and playfulness that needs wider recognition as fundamental both to the nervous system [Grobstein 1992] and to scientific inquiry [Grobstein 1988c]; see **Alexander et al.**, **Fetz**, and discussion of **Stein**, below). The issue of why artificial neural networks come up with particular kinds of solutions raises questions that are both intriguing and experimentally approachable in their own right, precisely by exploration of the variations in solution character with deliberate modifications of the variables mentioned above. We have begun such a research program, aimed at providing new insights into general information-processing rules (Grobstein 1988c) and at testing hypotheses about why the nervous system uses particular forms of organization in particular cases (do artificial networks indeed come up with similar solutions under the hypothetically expected set of circumstances and not under others?). If this is the sense of **Robinson**'s cryptic concluding remark that future work on artificial networks may yield "a bridge between system function and hidden-unit behavior," I couldn't agree more.

An equally important implication of the reality that artificial networks may generate different kinds of solutions (in particular, more distributed and less so) is that available information does not at the moment provide any basis to believe that all of the difficulties in understanding neuronal behavior can be comfortably swept under the blanket of distributed processing (see Grobstein 1988a; 1988c; 1990 for additional reasons for the difficulties). Nor does it give a solid foundation for expectations "as we move centrally," as **Robinson** puts it (and others imply). It's a good thing it doesn't, because whatever the single unit data in particular cases, it is abundantly clear from other methods of investigation, methods more appropriate to the detection of

order at levels of organization above that of the neuron (Grobstein 1990), that one is not in general dealing with a fully distributed system but rather with one possessing discrete "information processing blocks." Lesions in appropriate locations in the frog brain do not yield "graceful degradation" of directed movement; they instead reveal distinct information-processing steps, including an abstract central representation of space itself composed of dissociable, discrete elements (Grobstein 1988b; 1989; 1991; and see below). The same is true of many aspects of brain organization, including neocortical function. There is no argument whatsoever that different neocortical regions do different things. The only question is how to characterize what the distinctive things are that each cortical region does.

In short, let's not, in the enthusiasm over the appearance of distributed solutions in artificial neural networks, give up the idea of "boxes" in the nervous system. Particular boxes, presumed to exist from particular top-down perspectives, may or may not exist, but boxes there certainly are. No, let's *not* throw out the baby with the bath water. **Robinson**'s integrator is real, rigorously demonstrable, and localizable with appropriate techniques. So too are maps, and central pattern generating circuitry, and corollary discharge signals (these two constituting at least part of **Alexander et al.**'s "motor programs"), and an abstract central representation of spatial location. It is the blocks, and not distributed circuits, that are the real intermediary between neurons and global behavior. If artifical neural networks come to be equated with distributed processing, they will hinder rather than help in the primary task for better understanding the brain: the rigorous identification and characterization of information processing blocks. On the other hand, given identification and characterization of such blocks by more "classical" methods (Grobstein 1990), an exploration of the circumstances under which artificial networks do and do not develop solutions that display similar "blockiness" should indeed provide both new insights into how and why the brain does what it does (see below) and new broader principles for complex information processing in general.

2. Egocentric, corticocentric and noncorticocentric views of spatial representation: Respect for diversity, or learning about space and individuality from differences. Stein dismisses the notion of a master topographic representation for a variety of reasons, including my own favorite one: the absence of reports of "space scotomata" following small brain lesions. He also cogently argues that saccadic eye movements do not require any explicit representation of egocentric space: Retinal and oculomotor vectors would suffice, with no intermediate coding in egocentric coordinates. However, Stein's general hypothesis – that signals in different sensorimotor reference frames need never be converted into a common coordinate system, and that the basic representation of space is instead a distributed lookup table in the parietal cortex which directly transforms signals between particular sensory and motor coordinate frames – seems to me off the mark. Stein may be right in his analysis of what the parietal cortex is doing and how, but I think he is wrong in his suggestion that the most fundamental form of spatial representation is a highly distributed, task-oriented one in the parietal cortex. What is at issue is not only how space is represented in the brain and why it is represented that way, but also what the parietal cortex (and neocortex in general) does and why it does it that way. Some points made by **Fetz**, by **Alexander et al.**, and by **Robinson** are relevant.

I work with an animal that precludes getting distracted by the admittedly elegant but clearly overgrown and unnecessary neocortex. The frog simply doesn't have one, and, like many other vertebrates, it performs admirably on a wide variety of behavioral tasks despite its absence. Included are directed movements, which vary with object location in all three spatial dimensions. Subcortical circuitry is clearly adequate for such behavior. The frog might seem a prime candidate for accomplishing directed movements "reflexively," by a direct coupling between sensory and motor circuitry (as Stein has in mind) rather than via an intervening abstract spatial representation (Grobstein et al. 1983). [See also Berkinblit et al. "Adaptability of Innate Motor Patterns and Motor Control Mechanisms" *BBS* 9(4) 1986.] This has proven wrong, however (Grobstein 1988b; 1989; 1991; 1992). Multidimensional input signals representing at least two different sensory modalities converge in the frog's midbrain tegmentum to yield a three-dimensional signal that represents object location in a head- or body-centered coordinate frame experimentally distinguishable from both sensory and motor coordinate frames. It is, for at least some behaviors, a true bottleneck (Grobstein 1991; 1992) of precisely the kind that Stein suggested probably did not exist ("Having laboriously translated all coordinate systems into a common reference frame . . . the common coordinate system would then have to be transformed into that of the recipient system all over again" Introduction, para. 5). This bottleneck also corresponds to what **Alexander et al.** (and others) have characterized as an "ill posed" problem, since it means that a number of different movements can be associated with a given location signal. What appears a problem to particular sorts of top-down theorists seems to me an advantage to organisms, endowing them with the basis of a capacity for "choice" (Grobstein 1992).

As **Stein** expects, the abstract spatial representation in the frog tegmentum is not topographically organized (Grobstein 1991). There are certainly two, and probably three, physically distinct structures on each side of the brain that code independently for stimulus location relative to each of three perpendicular axes (**Alexander et al.**'s and **Fetz**'s distributed processing). Within each structure, the value of the relevant variable is represented not by the location of neural activity but rather by something like the total amount of neural activity. The upshot of this arrangement, and one of the initial and still strongest lines of evidence for its existence, is that appropriately located brain lesions in the frog do not produce "space scotomata," but do produce hemifield disturbances, with at least some similarities to those following unilateral parietal cortex lesions. Unilateral destruction of one component of the representation, coding horizontal eccentricity of stimuli to one side of the midsagittal plane, abolishes a sense of horizontal location on one side of the body, causing stimuli (either visual or tactile) throughout one hemifield to be referred to the midsagittal plane.

Several points follow from this. The first is that if there *is* a primordial and basic abstract spatial representation it is more likely to be in the midbrain tegmentum than in the parietal cortex. Neocortex *is* elegant and attractive, so much so that it must be doing something very special for organisms that have it. In efforts to understand this, it seems to me useful to know what subcortical circuitry is capable of, and hence to avoid becoming distracted by things that are not sufficiently special to warrant adding a neocortex. "Seeing" is not adequately special, whereas knowing that one has seen may be (Weiskrantz 1986). [See also Campion et al.: "Is Blindsight an Effect of Scattered Light, Spared Cortex, and Near-Threshold Vision? *BBS* 6(3) 1983.] Knowing where things are relative to oneself is not sufficiently special, whereas feeling that one knows where they are relative to oneself may be (see Grobstein, 1992, for a likely comparable distinction between choosing and feeling one is making a choice).

My second point is that **Stein**'s arguments for the advantages of a highly distributed processing system to deal with problems of space are apparently wrong. Frogs succeed with a more organized system of identifiable "information processing blocks," and there is increasing evidence that most other vertebrates, including humans (and perhaps even invertebrates) do it that way too (Grobstein 1989; 1991; Flanders et al. 1992, and my accompanying commentary on that article). This raises the obvious question of why space is *not* in general represented in a highly distributed fashion. Although studies of artificial neural networks have

generally impressed people with the capabilities of distributed networks, they can also be used to explore the question of why under particular circumstances less distributed processing is more common (see above). Such explorations are under way (Carr et al. 1991).

These points notwithstanding, it is still possible that **Stein** is right in his interpretation of parietal cortex as a distributed processing system concerned specifically with *egocentric* spatial location, that is, the ability to feel oneself feeling where other things are. If so, then there may be something about this particular kind of information processing that favors distributed networks, and perhaps their creation by something akin to what sometimes goes on in artificial learning networks. Such networks are capable of exploring for solutions to a given problem and of creating a number of different solutions to the same problem (see earlier discussion on **Robinson,** and also the target article by **Fetz;** see also Grobstein, 1988c, on "bounded variance" of which the possible outputs of artificial neural nets are an example). **Alexander et al.**'s distinction between supervised and "unsupervised" learning, with the suggestion that the latter provides even greater exploratory ability, is relevant in this context. True self-organization, however, is independent of any external information. "Unsupervised" seems to me a misleading term for an important intermediate case in which circuits get information about whether they are or are not achieving the task, but none about what direction to go in if they are not. Regardless, if subcortical circuitry is already built to deal adequately with most immediate challenges of life, neocortex may provide the luxury of additional parallel circuitry that can be used to play, to generate and explore alternate solutions to the same problems (see Grobstein, 1988c, for a discussion in a broader biological context). A sense of onself feeling or acting, together with a sense of one's own individuality, may emerge from the interplay of such a parallel cortical system interacting with subcortical circuitry (Grobstein 1992).

ACKNOWLEDGMENTS
Supported by a grant from the Whitehall Foundation and additional support to the Neural and Behavioral Systems Group from the TIDE-POOL project at Bryn Mawr College. The ideas expressed in the first section have emerged from extended conversations with Jeffrey Carr growing out of modeling collaborations with Carr and Despo Louca, for both of which I am deeply grateful.

Virtual trajectory as a solution of the inverse dynamic problem

S. R. Gutman and G. L. Gottlieb
Department of Physiology, Rush Medical Center, Chicago, IL 60612

[EB] **Bizzi et al.** have demonstrated in a very challenging and attractive manner the difficulties associated with the problems they consider. They are the ones who undertook the hard labor of solving these problems. Some of their own results are already classical. An attentive reader, however, may perceive the anxiety arising from the incompleteness of their proposed temporary solutions. Sharing this feeling, we will focus on a model explicitly and noncontradictorily describing some phenomena of movement performance – the λ-model (Feldman 1966b). We will describe an analytical approach to understanding this model that will illustrate connections between muscle-force characteristics, virtual, equilibrium, and actual trajectories, and the inverse dynamic problem in a concise manner. Some of these connections are discussed in the target article but we wish to unfold them in a more direct and explicit form using the λ-model.

Single muscle joint torque-angle characteristics $(T(x))$ have been repeatedly measured in experiments. According to the

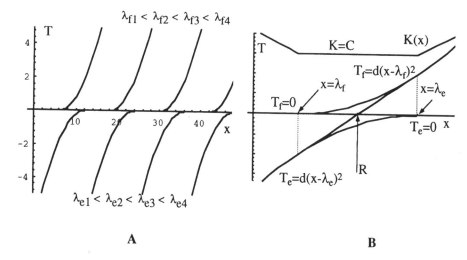

A   **B**

Figure 1 (Gutman & Gottlieb).   A: Semi-parabolic approximation of families of muscle-torque characteristics for the flexor (above the x axis, with the parameter λ_f) and extensor (below the x axis, with the parameter λ_e). λ_e and λ_f are descending commands controlling the muscles.

   B: A torque-angle characteristic of a joint with a pair of muscle-antagonists. See the linear part of the characteristic around the point $x = R$, where R defines the equilibrium position of the unloaded limb. The stiffness $K(x)$ in the vicinity of $x = R$ equals C; according to the λ-model, $R = (\lambda_f + \lambda_f)/2$ and $C = (\lambda_f - \lambda_f)/2$.

λ-model, descending control reduces to the shift of these characteristics by a command, λ. Simplifying the expression $T(x)$, we approximate it with a semi-parabola; for the pair of flexion and extension muscles, the equations of torques are (see Fig. 1A):

$$T_f = d(x - \lambda_f)^2, \quad x - \lambda_f > 0, \quad \text{and} \quad T_f = 0, \quad x - \lambda_f < 0, \quad (1,a)$$
$$T_e = d(x - \lambda_e)^2, \quad x - \lambda_e < 0, \quad \text{and} \quad T_e = 0, \quad x - \lambda_e > 0, \quad (1,b)$$

Note that the derivative of T with respect to x represents the coefficient of stiffness as a function of the angle ($K(x)$). Equation 1 defines two families of characteristics, for the flexor and extensor, with parameters λ_f and λ_e. Torque-angle characteristics of a joint with a pair of muscles-antagonists (Feldman 1986; Latash & Gottlieb 1990) can be found as a sum of T_f and T_e. For the approximation (Equation 1) and denotations $C = (\lambda_f - \lambda_e)/2$ and $R = (\lambda_f + \lambda_e)/2$, it has an expression

$$T_{f/e} = T_f, \quad x > \lambda_e, \quad (2,a)$$
$$T_{f/e} = 4dC(x - R), \quad \lambda_f < x < \lambda_e, \quad (2,b)$$
$$T_{f/e} = T_e, \quad x < \lambda_f. \quad (2,c)$$

Thus, the usually used linear approximation of the two-muscle joint torque-angle characteristics can be exactly derived from the quadratic approximation of muscle force-length characteristics. The coefficient C in Equation 2 reflects a controlled joint stiffness, and R defines the equilibrium position of the unloaded limb; both of these are controlled by the pair of descending commands λ_f and λ_e. In Figure 1B, a graph of this function is represented.

If an external torque T is imposed, the limb moves to a static position where $T = T_{f/e}$. This position is called the "equilibrium point" (x^{eq}). It can be found from Equation 2 for a given C, R, and T. If any three of four parameters T, x^{eq}, C, and R are known, the fourth can be calculated from (2). These simple calculations correspond to not quite obvious physical phenomena illustrated in section 4 of the target article.

Model (1) does not take into account the fact that the torque developed by a muscle depends on the velocity of muscle contraction. Using a linear approximation to this dependence, one must change λ to $\lambda^* = \lambda + \mu x'$, where x' is the velocity and

μ is a coefficient (λ^*-model, Feldman et al. 1990). A semi-parabolic approximation of the torque analogous to (1) leads to the following expression of the torque developed by a pair of muscles:

$$T_{f/e} = 4dC(x - R + \mu x'). \quad (3)$$

The simplified dynamic equation of a linear two-muscle single-joint limb with a moment of inertia I, viscosity B, and external torque T is:

$$Ix'' + Bx' + T_{f/e} = T, \quad (4)$$

where x'' is the acceleration. The simplifications used are: First, we assume that coefficients I and B are constant, and second, we ignore the reflex delays.

Let us now provide three definitions (the target article contains them in a less explicit form):

A. "Virtual trajectory" is the time profile of the command $R(t)$. The virtual trajectory cannot be observed but only estimated.

B. "Actual trajectory" is the sequence of positions of the working point of the limb $x(t)$. The actual trajectory is one that is measured in experiments.

C. "Equilibrium trajectory" $x^{eq}(t)$ is the sequence of equilibrium points, that is, points of the actual trajectory where velocity and acceleration equal zero. During the movement itself, this notion does not make much sense; for a frozen command, the limb eventually achieves a certain equilibrium position

$$x^{eq}(t) = R(t) + T/C. \quad (5)$$

One can say that the equilibrium trajectory is a virtual trajectory distorted by external torque, and that the actual trajectory is an equilibrium trajectory distorted by the offset of inertia and viscosity. Accordingly, both the virtual and the equilibrium trajectories can be estimated from the actual trajectory of the system working according to the principles that have been described. Since in the absence of the external torque $x^{eq}(t)$ and $R(t)$ are identical, we will consider only $R(t)$ for $T = 0$.

Unfolding (4), we obtain a dynamic equation for the λ^*-model:

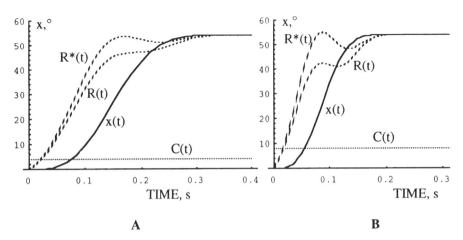

Figure 2 (Gutman & Gottlieb). Analytical reconstruction of virtual trajectories from the actual trajectory x(t) for a constant value of stiffness C. R(t) – virtual trajectory for the λ-model, R*(t) – virtual trajectory for the λ*-model. A: "slow" movement. B: "fast" movement.

$$Ix''(t) + (B + 4dC(t)\mu)x'(t) + 4dC(t)(x(t - R(t)) = 0, \quad (6)$$

which shows that changes of the viscosity during movements have to correlate with changes of the stiffness. To increase the velocity of the movement, we can increase both the R and the C. Note that increasing C we also increase the equivalent viscosity and, with this, the stability of the system.

From (6) we can test different hypotheses of movement generation. As a rough example, let us estimate virtual trajectory R(t) for a point-to-point self-terminated reaching movement assuming C(t) to be known:

$$R(t) = x(t) + (Ix'' + (B + 4dC(t)\mu)s')/4dC \quad (7)$$

(this can be considered a solution to a simplified inverse dynamic problem). The form of this hypothetical virtual trajectory has been studied for different assumptions analytically (Hasan 1986; Hogan 1984) as well as experimentally (Latash & Gottlieb 1991c).

Let x(t) be an analytical expression of this movement (Flash & Hogan 1985; Gutman & Gottlieb 1990, etc.). We will use a formula

$$x(t) = D(1 - \exp(-t^3/\tau)), \quad (8)$$

where D is the distance, τ is the movement time constant defining the movement duration (the utility and verisimilitude of this expression is developed in detail in Gutman et al. [in press]). Substituting (8) into (7), we obtain an explicit expression for a reaching movement virtual trajectory in a system described by an equation (6):

$$R(t) = D\left[1 - \exp(-t^3/\tau)\left(\frac{9I}{4dC\tau^2}t^4\right.\right.$$
$$\left.\left. - \frac{3B + 12dC(t)\mu}{4dC\tau}t^2 - \frac{6I}{4dC\tau}t + 1\right)\right]. \quad (9)$$

Time profiles of virtual trajectories are represented in Figure 2A (slow movement) and Figure 2B (fast movement). R(t) corresponds to the λ-model (i.e., to μ = 0), R*(t) corresponds to the λ*-model. R(t) becomes N-shaped for fast movement (coinciding with results obtained in Hasan 1986; Hogan 1984; Latash & Gottlieb 1990). R*(t) are also slightly N-shaped, whereas in the original works (Feldman 1986; Feldman et al. 1990) monotonic forms of R*(t) were used.

The disparity can be explained at least by two main factors. First, we use a simpler type of nonlinearity of the torque-angle characteristics (this allows us to derive expressions for C and R originally given as definitions, and also to obtain an explicit analytical expression for the differential equation of the movement). Second, we study a simplification of the inverse dynamic problem, whereas the direct dynamic problem was studied originally (i.e., for given commands R(t) and C(t), the actual trajectory x(t) was calculated). However, the corresponding pairs x(t) and R(t) obtained for fast and slow movements in our work and in Feldman (1986; Feldman et al. 1990) are similar.

We hope that this exposition will help clarify the discussion of motor control models.

Operations of the motor system

Mark Hallett[1]
Human Motor Control Section, Medical Neurology Branch, National Institute of Neurological Disorders and Stroke, National Institutes of Health, Bethesda, MD 20892
Electronic mail: *hallett@helix.nih.gov*

[EB, JRB] **1. Testing the equilibrium-point control theory with human arm movement.** One of the most fundamental questions of motor control is: What is the language of the motor system? How does the nervous system talk about movement as it generates the command and sends it to the muscles? There are at least two candidate languages. One is the equilibrium-point control theory, according to which the movement is to be made to a designated place in a coordinate system. To understand this language, it is necessary to know the coordinate system (and whether it might change at different times). According to the second candidate, which we could refer to as the relative control theory, it is necessary to have knowledge of the initial position and target and the movement is generated from place to place. No absolute coordinate system is needed; it is necessary to know only the relative direction and distance from one point to the other. In the first language, the same command can take the limb to the target regardless of its initial position. In the second, a different command will be needed to take the limb to the target, depending on the initial position.

The muscle or nerve, the lowest level of the neuromuscular apparatus, knows only force in a particular direction. This is exemplified by Figure 4D in **Bizzi et al.**'s target article. At the cortical level, cellular activity is strongly related to direction (Georgopoulos et al. 1982; Kalaska et al. 1989). Thus, both the highest and lowest levels of the nervous system use relative control. Equilibrium-point control might still be used in between. It is certainly an extremely attractive idea, and the spinal cord might well be the site where the system is implemented.

This would require that in a simple, single-joint, fixed system, the pattern of electromyographic (EMG) activity in muscles acting at the joint should specify the position of the joint. We have tested this in isolated movements of the elbow (Matheson et al. 1985).

Integrated EMG activity of the biceps and triceps was compared with the angular position of the elbow in several circumstances. One was isometric co-contraction at various amounts of voluntary effort at 90°, 110°, 130°, and 150°. Four types of movements were made to the same angles as the targets: fast flexion, slow flexion, fast extension, and slow extension. EMG was integrated in a 500-msec period, and for the isotonic movements the beginning of the integration was after the limb had come to a steady final position. Biceps EMG, triceps EMG, and the ratio of biceps EMG to triceps EMG were tested for correlation to the four different angles. No correlations were found. The most promising measure, the ratio of biceps EMG to triceps EMG, showed a large variation at each angle with extensive overlap of ratio values at the different angles. For certain movement types, the ratio tended to be larger for more flexed angles, but the pattern of the ratio for the different movements was not the same.

Our results differed from those of Lestienne, Polit, and Bizzi (1981), whose study was similar to ours but did not explore as large a range of situations. The study probably should be repeated, but our results show that the equilibrium-point control theory is not easy to demonstrate. As we pointed out previously, EMG activity is not the only determinant of the length-tension relationship. It will vary with the movement history of the muscle (Ridgway et al. 1983) and with fatigue (Moritani & DeVries 1978). Hence, it would seem difficult for the nervous system to use the length-tension relationship as an invariant. Last, in human studies, perturbed movements made in the absence of feedback will not always achieve the correct final position (Day & Marsden 1982; Sanes & Evarts 1983a).

Given its simple and appealing logic, why might the equilibrium-point control theory fail? One assumption in the mathematics is that the stiffness of the joint is nonzero. The stiffness of human joints is nonzero, but is it sufficiently far from zero to allow the principle to operate?

2. The Vermittler hypothesis needs to be extended to explain all that the cerebellum does. The concepts put forward by **Bloedel** are attractive in that they describe a common function for the common anatomical structure of the cerebellum. The on-line analysis of inputs for aiding movement is an important function, and the evidence is rather persuasive that the cerebellum plays a role in this. Although the cerebellum was for many years thought to be a "motor" structure, it is now clear that it also has a "sensory" function. In a comprehensive study of sensory function, Holmes (1917) had demonstrated the often overlooked fact that patients with cerebellar deficits have difficulty in estimating weights in the unsupported hand. Patients with cerebellar deficits also show abnormalities in estimating the duration of tone bursts (Keele & Ivry 1990). As Bloedel points out, when the cerebellum is malfunctioning, sensory input can be confusing to a patient, and motor performance might even be better without the sensory input. The role of the cerebellum is not merely to regulate the on-line performance of movement, however.

The role for the cerebellum in motor learning is convincing, and it cannot be explained entirely by the cerebellar role in motor performance, as **Bloedel** suggests. In a discussion of motor learning, it is critical to separate its different subtypes, which may well differ in their physiology. The evidence that the cerebellum plays a role in adaptation learning is very strong. The role of the cerebellum in the adaptation of the vestibulo-ocular reflex has been demonstrated by many groups. Adaptation to lateral displacement of vision is impaired with cerebellar lesions, as demonstrated in animals as well as humans (Weiner et al. 1983). Adaptation of limb movement to gain change depends on the cerebellum (Keating & Thach 1991). Classical condition-

ing may have some similarities to adaptation learning in that it deals with relating motor output to environmental changes. A number of investigators have shown that cerebellar lesions cause a deficit in the conditioned eyeblink (Thompson 1986) and we have recently added to this evidence by showing that humans with cerebellar degeneration lose eyeblink conditioning (Topka et al. 1992).

On the other hand, the evidence for a role of the cerebellum in motor skill learning is not as clear. As **Bloedel** points out, the evidence that our group has produced on this issue so far is not strong (Sanes et al. 1990). Bloedel writes that our data show obvious learning for patients with cerebellar hemispheric damage in the task with mirror-reversed vision. Indeed, we stated that this group showed learning; we had suggested that the patients with olivopontocerebellar atrophy were the ones who might be deficient. The data are difficult to interpret, however, because the baseline motor performance of this group was better than that of the normal group. This may well have been an example of a lack of sensorimotor integration in the cerebellum, with these patients less confused by the mirror than the normal subjects were. In subsequent, ongoing work in our laboratory, we are finding that patients with cerebellar deficits may have a problem with skill learning, but only when the movement required is fast (Topka et al. 1991). Because performance is also related to speed (Massaquoi & Hallett 1991), Bloedel's notion that performance aspects are important in skill learning may well be correct.

Although the cerebellum may not play a major role in skill learning, it does appear therefore to be important in adaptation learning and classical conditioning. Whether it stores any of this information, even temporarily, or only mediates the process certainly needs additional research. If there are adaptable synapses, it would seem likely that they are used to store information, at least for short periods of time, well beyond the period of the dynamic selection effects. Moreover, if there is short-term storage, there may also be some long-term storage, given the current understanding of the cellular basis of memory (Schacher et al. 1990). Indeed, such cellular changes have been demonstrated in the cerebellum of animals undergoing classical conditioning (Schreurs et al. 1991).

Bloedel's description of "on-line correction" seems heavily focused on the feedback role of the cerebellum, which may well be the role of its intermediate part. The types of movement illustrated, such as those of Beppu et al. (1984), are the ones in which feedback is critical. The cerebellum appears to play even a more important role in the preparation for the initiation of movement, because in patients with cerebellar deficits the initial part of a ballistic movement is more dramatically impaired than a slow movement (Hallett et al. 1975). There are distinct abnormalities of the beginning of both single joint and multiple joint arm movements (Hallett et al. 1991; Hore et al. 1991; Massaquoi & Hallett 1991). Perhaps these functions can be related to the Vermittler hypothesis by noting that the motor output at the onset of movement requires intimate knowledge of the environment, including the initial positions of the body parts. It does seem, however, that a major problem with the onset of movement is in the timing of the electromyographic components, which may relate to an internal clocking function of the cerebellum that is apparently unrelated to the matching of outputs with inputs.

Last, although the evidence is still fresh, the cerebellum may well play a role in some cognitive functions. Patients with cerebellar deficits may have problems with association learning (Bracke-Tolkmitt et al. 1989), which may even fit with the Vermittler hypothesis if it is extended along the lines suggested by Ito (1990). We have found an abnormality with the Tower of Hanoi task, which suggests a problem with advance planning of sequences of behavior (Litvan et al. 1991).

Bloedel's hypothesis is a step in the right direction, but even with the data now available, it is clearly not comprehensive.

ACKNOWLEDGMENT
With thanks to Dr. Steven P. Wise and the members of the Human Motor Control Section for valuable discussions.

NOTE
1. Address correspondence to: Mark Hallett, Clinical Director, NINDS, NIH, Building 10, Room 5N226, Bethesda, MD 20892.

Making sense of recurrent inhibition: Comparisons of circuit organization with function

Thomas M. Hamm and Martha L. McCurdy

Division of Neurobiology, Barrow Neurological Institute, St. Joseph's Hospital and Medical Center, Phoenix, AZ 85015

[DAM] **McCrea** argues that the complexity of spinal interneuronal circuits, which makes them so difficult to understand, provides them with a flexibility that can be used by descending and propriospinal motor pathways in the initiation and control of movement. He indicates that spinal interneuronal circuits are interpretable to the extent that hypotheses can be made about the function of these circuits as we begin to know something about their organization. Although plausible functions for spinal circuits can be based on what we know about their connections, we are still left with the difficult task of determining how these circuits are actually used in the control of movement. A case in point is the spinal circuit for recurrent inhibition.

The key interneuron for this circuit is the Renshaw cell, which is driven by input from the recurrent collaterals of motoneurons. The axons of Renshaw cells project to alpha motoneurons, gamma motoneurons, and Ia reciprocal inhibitory neurons, and form inhibitory synapses with each. The simplicity of this circuit has prompted several hypotheses on its function. These hypotheses include proposals that this circuit functions to stabilize or limit the rate of motoneuron discharge (Granit & Rutledge 1960), enhance the contrast between the activity in different motor pools (Brooks & Wilson 1959), desynchronize (Gelfand et al. 1963) or synchronize (Renshaw 1941) the discharge of motoneurons, control the extent to which Ia feedback from muscle spindles affects motor output (Hultborn et al. 1971b), and regulate the gain of motoneuron pools (Hultborn et al. 1979). More recently, claims have been made that Renshaw cells generate an efference copy of motor output for use in segmental control systems (Loeb & Levine 1990), that recurrent inhibition filters information to motoneurons from muscle spindles and other muscle receptors (Windhorst 1989), and that recurrent inhibition functions as part of a system that facilitates the independent control of joint position and torque (Bullock & Grossberg 1989). Despite this wealth of hypotheses we do not yet know the extent to which any of these proposed functions is expressed by the spinal circuit. To approach the problem, attention must be given to the organization of the circuit in relation to the function and activity of the motor system with which it is associated.

One prominent feature of recurrent inhibition that has influenced our thinking about its function is its association with projections of Ia afferents from muscle spindles. Is the function of recurrent inhibition the control of Ia input to segmental neurons or is it the control of some other feature associated with the formation of motor patterns? To address this issue, the organization of recurrent inhibition for the motor nuclei of flexor digitorum longus and flexor hallucis longus was considered (Hamm 1990). These muscles have similar mechanical actions and patterns of Ia connections, but they exhibit very different motor activities during locomotion (O'Donovan et al. 1982). The former study (Hamm 1990) showed that recurrent inhibition reflects the pattern of motor activity, not of Ia connections.

Based on this study, recurrent inhibition seems more likely to participate in the control of groups of motoneurons with common patterns of activity than in the regulation of segmental Ia projections.

Further evidence concerning the function of recurrent inhibition has been provided by studies demonstrating that the distribution of recurrent inhibition is not uniform in different limb motor nuclei. Cullheim and Kellerth (1978) have found that recurrent collaterals are absent in motoneurons that innervate intrinsic plantar muscles of the foot. Recent studies from our laboratory and from Illert's have demonstrated that recurrent collaterals and recurrent inhibition are reduced or absent in the motor nuclei that control the digits in the cat hindlimb (Hamm 1990; McCurdy & Hamm 1992) and in those that control the wrist and digits in the cat forelimb (Hahne et al. 1988; Horner et al. 1991). These findings have significance for four of the hypotheses cited above. The hypothesis that recurrent inhibition enhances the contrast in motor signals or that it provides gain regulation presumes that recurrent inhibition assists in fine control (Brooks & Wilson 1959; Hultborn et al. 1979). Yet recurrent inhibition is weak or absent in motor nuclei that control the wrist and digits, just where fine control should be most prominent (see Windhorst 1990). [See also Windhorst et al.: "On the Function of Muscle and Reflex Partitioning" BBS 12(4) 1989.] Similarly, the hypothesis that recurrent inhibition provides efference copy for use in a segmental control system (Loeb & Levine 1990) or that it forms part of a system for the independent control of joint torque and position (Bullock & Grossberg 1989) must accommodate the absence of this signal from motor pools whose muscles act at the digits.

One difficulty in evaluating functional hypotheses of recurrent inhibition is the apparent weakness of this effect. For example, Lindsay and Binder (1991) estimated recurrent inhibitory current arriving at the somas of medial gastrocnemius motoneurons and found values of less than 1 nA during maintained repetitive activation of the lateral gastrocnemius nerve. Current work from our laboratory (McCurdy & Hamm 1991), however, shows that recurrent inhibition is distributed across several motor pools but within a restricted rostrocaudal zone. Since these several motor nuclei would be simultaneously active in most motor activities, the effective strength of recurrent inhibition would be underestimated by confining estimates to input from the homonymous pool, only part of which would contribute to inhibition in individual motoneurons.

We agree with **McCrea** that sense can be made of spinal interneuronal circuits and that more detailed knowledge of the organization and features of these circuits is required. To determine the functional contribution of these circuits, it is essential that we consider them in relation to the function and usage of the motor nuclei, muscles, and motor systems with which they are associated.

ACKNOWLEDGMENT
Supported by USPHS grants R01 NS 22454, F32 NS 08773, T32 NS 07309 and T32 GM 08400.

Is stiffness the mainspring of posture and movement?

Z. Hasan

Department of Physiology, University of Arizona, Tucson, AZ 85724

[EB] At the risk of oversimplification, the following sequence of ideas may be said to underlie equilibrium-point hypotheses. *Muscles have stiffness; thus, joints have stiffness. The same muscles and joints are used for both posture and movement. Stiffness is good for posture. Stiffness must be good for movement.* I accept the second sentence in this sequence, but I have reservations about the rest of the neosyllogism.

Do muscles have stiffness? More than half a century after A. V. Hill's pioneering work (Gasser & Hill 1924), it still bears emphasis that an active muscle cannot be modeled by a parallel arrangement of a spring and a damper. Muscle has two kinds of stiffness: For fast stretch it exhibits high stiffness (attributed to elongation of the series elastic element) and for static stretch it has low stiffness (attributed to change in overlap between the myofilaments). **Bizzi et al.** ignore the former, which is the one that muscle biophysicists call stiffness, and give credence only to the latter; the static stiffness, which outlives the transients, inherits the sole right to be called stiffness.

There is no denying that muscles have some static stiffness, though it may be nearly zero at the peak of the length-tension curve. The force contribution of static stiffness would in any case be washed out by the contribution of the short-term stiffness elicited by any small, fast irregularities in length change. (Inclusion of the stretch-reflex, although it may help linearize the transient responses, would not improve matters. The phasic component of the reflex is greater than the tonic component, which would accentuate the sensitivity to small fast changes.) The short-term stiffness, which may play an important role in stabilization, is ignored in the equilibrium-point hypothesis, whereas the smaller static stiffness, which may not even be observable over the time course associated with a fast movement, is given pride of place.

Do joints have stiffness? Even if a muscle were a spring of some positive stiffness, the angular stiffness of the joint spanned by it may be negative! This is because a change in joint angle that elongates the spring and thus increases its force can at the same time reduce the moment arm so much that the torque decreases. Indeed, in one experimental paradigm, negative values of static angular stiffness have been observed for the elbow joint over a certain range of angles (Hasan & Enoka 1985). Why, then, does the joint not become unstable? The high short-term stiffness of muscle, together with its damping properties, can provide stabilization over a fast time scale, although static stability may not be assured without feedback correction. Contrary to a statement made in section 2.2, the importance of feedback in providing static stability is not vitiated by feedback-loop delays as long as the series elastic element can provide short-term stabilization. (For an example of slowly manifested instability in the absence of feedback, see Sanes et al. 1985. Other examples of how proprioceptive feedback serves to stabilize, not destabilize because of loop delays, are cited in Hasan & Stuart 1988.)

In their discussion of stability, as indeed in the development of their hypothesis, **Bizzi et al.** have chosen to ignore the inconvenience associated with moment-arm variations and the consequent negative stiffness of the joint. The λ-model at least has the virtue that the tonic stretch reflex can save the joint from the horror of negative static angular stiffness.

Is stiffness good for posture? When theories of servo-control of joint angle held sway, it was axiomatic that more accurate regulation was equivalent to higher stiffness of the joint. Thus, the way to stand upright was to stiffen all joints, turning oneself into a rigid mannequin. This strategy prevented the collapse of the joints under one's own weight; when combined with a prayer against strong winds (which had the potential of toppling the mannequin without altering joint angles) the strategy was presumably effective in maintaining posture. It is a welcome change that the equilibrium-point hypotheses have steered our thinking away from the presumed desirability of infinite stiffness and toward the presumed desirability of merely finite stiffness. Perhaps we need to be steered some more, toward less and less stiffness, until we reach the "highest stage of coordinational freedom" (Bernstein 1940/1967) in which we are not afraid of what the motion of one segment can do unto other segments, and we therefore do not need to stiffen the joints.

Setting aside the issues of stability discussed earlier, is stiffness so desirable a property that the nervous system should act to enhance the muscles' intrinsic stiffness? Some discussions of spinal and so-called long-loop reflex mechanisms would lead one to believe so. There is a growing body of evidence, however, that that in multijoint postural contexts the nervous system responds to an external perturbation of the arm in ways that do not follow from the assumption that every muscle should resist stretch (e.g., Gielen et al. 1988; Koshland et al. 1991; Lacquaniti & Soechting 1986). Thus, the observed dynamic responses emitted by the nervous system do not support the notion that arm posture is simply a matter of maintaining adequate stiffness in every joint, supplementing by way of the nervous system the inadequacy of the muscles' intrinsic stiffness.

It may be argued that the measurements of Mussa-Ivaldi et al. (1985) on forces exerted by the hand in extrinsic space in response to perturbations were, after all, static measurements taken several hundred milliseconds after the complicated dynamic responses of the nervous system were emitted, and therefore the stiffness field they reveal has a validity that is not contingent on how the nervous system responds dynamically. I accept this argument but conclude from it that the observed stiffness field need not play a role in neural strategies for maintaining posture. If there is such a role, it needs to be demonstrated, perhaps by a simulation that includes the dynamics of the musculoskeletal system but not the dynamics of the neural response to perturbation. If the latter dynamics are crucial, however, then one would be justified in concluding that the static stiffness fields capture no more than an afterglow of the neural strategies for maintenance of posture.

Is stiffness good for movement? The commonsense answer to the question of whether stiffness is good for movement is no, so it is gratifying that **Bizzi et al.** cite reports showing a drop in stiffness with the start of movement. Theoretically, however, all mass-spring accounts of movement control insist on maintaining stiffness throughout movement and indeed predict higher stiffness for faster movement. (These accounts include, mea culpa, Hasan 1986.) This stems from the physical fact that a mass pulled via a spring moves faster when the spring is stiffer.

Should one abandon all mass-spring models, then, and regard stiffness as an unfortunate byproduct of the contractile mechanism? I would not go so far, because stiffness does seem to increase during some movements (Latash & Gottlieb 1991). The resistance to change that is the hallmark of stiffness may be useful in some circumstances. An analogy may be made with the vestibulo-ocular reflex, which provides a mechanism of resistance to change in gaze. This reflex is suppressed during large gaze shifts to targets in predictable locations but it is present throughout some gaze shifts when the target is unpredictable (for review, see Sparks 1991b). Perhaps in the study of arm movements, also, more attention needs to be paid to possible differences in movement strategy based on task conditions (cf. Gottlieb et al. 1989). Just because certain traffic conditions induce us to drive an automobile with one foot on the brake, we cannot conclude that simultaneous operation of the engine and the brake is the way automobiles are driven. Similarly, stiffness may be used to advantage by the nervous system in some conditions but not in others.

The argument of **Bizzi et al.** that stiffness provides stability of movement trajectory needs careful examination. Specially pertinent are the results of Soechting (1988), who applied force perturbations during arm movement and found that the electromyographic response did not serve to restore the trajectory in its temporal aspects but did serve eventually to restore the path, that is, the spatial aspects alone. As he pointed out, this is not the kind of restoration one would expect from the effect of stiffness as posited in the equilibrium-point hypothesis. Yet the restoration of the path bespeaks a peculiar kind of stability not described by the classical Lyapunov or Poincaré criteria. Perhaps we need to formulate a new criterion of stability. A facile extrapolation from postural stability to movement stability underlies the "unified description of posture and movement" in the equilibrium-point hypothesis. This extrapolation is not a fundamental requirement, as acknowledged in section 7.2; moreover, it is contradicted by experiment.

In conclusion, **Bizzi et al.** have made an inroad into exciting and challenging issues, but a great deal of exploration lies ahead.

Computations, neural networks and the limits of human understanding

Herbert Heuer

Institut für Arbeitsphysiologie an der Universität Dortmund, D-4600 Dortmund 1, Germany
Electronic mail: *heuer@arb-phys.uni-dortmund.de*

[GEA, DAR] **Alexander, DeLong & Crutcher** draw a contrast between motor-program approaches and connectionist models. They favor the latter type of modeling as the one that promises to have a closer relation to the neural basis of motor behavior. In my view the confrontation of the two approaches implies a mixture of two different levels of explanation, and the expected benefits from connectionist models are unlikely ever to accrue.

Computational problems and algorithms. The motor-program concept is not uniquely defined. **Alexander et al.** stress that in its most general form the concept is rather unassailable – largely because of a lack of content. In their preceding characterization of the concept, however, they focus on the computational problems of motor control. The computational analysis seems in fact to be the main ingredient of research in the motor-program tradition. Alexander et al. (sect. 2, para. 5) also state, however, that the motor-program approach implies that the computational problems are solved "in a manner analogous to that which an engineer or a computer programmer would use." There is absolutely no need to accept this apparent implication. In contrast, the motor-program approach is essentially neutral with respect to algorithms and their implementations (cf. Heuer 1991), although certain solutions for the computational problems seem to be preferred currently.

Alexander et al. describe connectionist networks as being nonalgorithmic; connectionist networks, however, belong to the algorithmic level of explanation (drawing, again on Marr's 1982 distinction between the computational, algorithmic, and implementational levels). They not only state a certain computational problem, but they solve it by means of certain representations of input and output variables and certain algorithms for the transformation. Connectionist networks therefore do not stand in opposition to motor-program approaches, but are one way to deal with the algorithmic level of explanation.

Alexander et al. characterize connectionist models as nonalgorithmic, mainly because of the way they can be trained, namely in a trial-and-error mode. Instead of the theorist who puts the wisdom in the model, the model is self-organizing, acquiring its wisdom on its own. The theorist, however, puts the means to acquire the wisdom in the model, the learning algorithms. Self-organization in the sense of adjusting parameters or even in the sense of acquiring rules is not unique to connectionist models. There are several variants of stochastic learning models available in psychology (e.g., Atkinson et al. 1965), rule-based systems that acquire their own rules in engineering (e.g., Kiendl et al. 1991), and various ways of finding a maximum or a minimum of a function of several parameters in mathematics.

Finding maximal benefits (according to whatever criterion) is also what trial-and-error learning does; it is a rather ineffective algorithm, but also a robust one, suited for discrete as well as continuous variables. The relation between a priori and acquired wisdom of a model poses interesting problems. Self-organization in the sense of **Alexander et al.** is not a virtue by itself; rather, the relation between the a priori and acquired wisdom of a model should match that of the human motor

system, which is modifiable to some degree, but not without limits.

Neural networks and the limits of understanding. Connectionist networks can be powerful tools for solving certain computational problems. They have been inspired by the operation of the brain and share with the brain the feature of parallel computations. Nevertheless, the hope that they will exhibit a closer match with brain structures and neuronal behavior is likely to be futile. **Robinson,** in his target article, describes a number of simple networks with units that have a close correspondence to the behavior of real neurons, but he also points out that such correspondence gets lost as the modeled behavior becomes more complex. In my view this state of affairs is not accidental.

There is no generally accepted definition of what it means to "understand" a biological system. One understanding of "understanding" is to be able to build a machine that behaves the same way as the biological system does. A large part of formal modeling is in this spirit. Combining neural networks (and other complex nonlinear systems) with the possibility of implementing them on computers rather than having to analyse them in more traditional ways takes such a meaning of understanding into the realm of absurdity. The thought-experiment to demonstrate this starts with a simple network with units that closely resemble real neurons. This model is expanded step by step to cover more and more behavioral characteristics of some animal. Finally, the model will behave like the real thing. It will in fact be almost identical to the modeled system – and we will not understand either of them. (Of course, the model will be an excellent laboratory animal that can be studied quickly and without ethical limits.)

The thought-experiment is unrealistic, but it points to the fact that understanding requires a certain degree of abstraction from the system that one tries to understand. Nature can be complex, but models have to be simple. Neural networks reach the limits of understanding even with smaller numbers of units than a fraction of the number of neurons in any part of the brain. This is probably a particularly annoying situation: The model reproduces certain input-output relations, but the behavior of the units and their connectivity corresponds neither to real neurons nor to anything that can be formulated in a way that gives us the feeling of understanding (e.g., as some rules that are applied). A pessimist might therefore not expect connectionist models to be of any use for understanding how the brain works (except for small-scale models of simple behaviors); they might rather open a new field of research: the understanding of how connectionist models work.

Implications for human motor control

Fay B. Horak, Charlotte Shupert and Anne Burleigh

R. S. Dow Neurological Sciences Institute, Good Samaritan Hospital and Medical Center, Portland, OR 97209

[GEA, JRB] ***Multitask tests of cerebellar operations.*** We find quite attractive **Bloedel**'s hypothesis that the cerebellum is best understood in terms of a single operation it applies across a wide variety of motor tasks. We agree that some of the apparent functional heterogeneity of cerebellar operation reflects the interaction of different parts of the cerebellum with very different sensorimotor systems rather than different operations in different parts of the cerebellum. This idea needs to be tested, however, both indirectly and directly. Indirectly, we need to determine whether one operation can satisfactorily explain what is already known about cerebellar function and the wide variety of deficits following lesions. Directly, new experiments need to be designed that specifically test the single-operation hypothesis.

The "mediator (Vermittler) hypothesis," in which the cerebellum optimizes movement execution by integrating external execution space with internal intention space and sensory information is supported by our experiments on postural deficits in cerebellar patients (Horak 1990; Horak et al. 1986). Healthy subjects, but not patients with anterior lobe cerebellar degeneration, are able to match their response magnitudes to perturbation size better when postural perturbations are predictable in size, rather than random. For this particular task, the cerebellum uses knowledge of results on previous trials to modify the gain of responses to anticipated perturbations.

Bloedel suggests that the primary role of the cerebellum may be to detect mismatches between the actual sensory feedback generated by a movement and internal models of the anticipated feedback. Failure to detect such mismatches could explain many of the deficits associated with cerebellar lesions. For example, Ivry and Keele observed that some cerebellar patients were unable to match a series of timed auditory signals with finger taps. This could result from an inability to determine whether the sensory information generated during tapping matches the internal model of expected sensory information (Ivry et al. 1988). The same mediator model could be used to explain the important role of the cerebellum in adjusting the gain of the vestibulo-ocular reflex in response to exposure to reversing prisms or magnifying lenses (Lisberger 1988a). It also easily explains deficits in using central set, based on prior experience and expectation, to modify responses to arm or postural displacements observed in cerebellar patients (Hore & Vilis 1984; Nashner & Grimm 1978).

The apparent differences in the deficits following different cerebellar lesions are also at least partly an artifact of disparities among tasks and experimental paradigms. Some tasks, such as walking, tapping, and smooth pursuit, are fairly continuous, allowing for on-line modification of performance by the cerebellum. Other tasks, such as responding to arm or postural perturbations, are more discrete, with enough time between performances for knowledge of results, as well as the sensory information in the trigger and initial conditions, to influence the response. In some tasks, optimization involved modifying movement magnitude and in others it involves modifying movement timing. Task constraints and performer intention influence the optimized variables. Nevertheless, the cerebellum could be performing the same operation or a similar one in all of these tasks.

Normal efficient and effective motor control uses prediction, knowledge of results, and ongoing and prior sensory experiences to modify movements to suit the demands of the particular motor task. The cerebellum could assist in adjusting or tuning movements in a variety of different "time domains." For example, the cerebellum could be involved in ongoing modification of step length and muscle force during continuous locomotion across an uneven surface. It could also optimize the response to a misstep based on previous experience with a similar situation. It could even modify the step cycle to anticipate a visually detected obstacle, perhaps based on cognitive information about the obstacle. The sensory information used by the cerebellum and the movement parameters it attempts to optimize may differ in each case, whereas the basic operation is similar.

To test the hypothesis directly that the cerebellum performs one homogeneous function across a wide variety of tasks, experimenters will need to test the same functional operation in the same lesioned subjects (animal or human) across several different tasks. For example, the ability to use prior experience to modify the gain of triggered responses could be tested in (1) postural tasks, (2) finger/hand grip tasks (Johansson & Westling 1988b), and (3) oculomotor tasks. Experiments need to be designed to measure equivalent characteristics of motor responses (i.e., response magnitude, gain, or timing) across different tasks using similar experimental paradigms (i.e., comparing responses to predictable and unpredictable stimuli). Ideally, subjects with different lesions in different parts of the cerebellum would be asked to perform a variety of tasks to determine whether the cerebellum performs homogeneous or heterogeneous operations. Humans are highly adaptable and do not require extensive training and retraining to perform different experimental tasks. We therefore suggest that some of the most important information concerning the role or roles of the cerebellum may come directly from tests of many different motor tasks in a single well-defined group of human patients with cerebellar disorders.

But what do the basal ganglia do? We would concur that the concept of the basal ganglia as a self-organizing, parallel distributed system is long overdue. However, those of us who try to decipher which motor functions are missing in patients with disorders of the basal ganglia need testable hypotheses about the roles the basal ganglia play in controlling movement. If we reject a role of the basal ganglia as an organizer of "motor programs" in a serial, hierarchical model, we must also suggest new roles that are consistent with a parallel, connectionist model. Although **Alexander, DeLong & Crutcher** tantalize us with generalities such as that the basal ganglia "provides positive feedback to the cortical motor fields" (sect. 3, para. 2) "generates conditioned sequences of movement," and "set-related activity," we are disappointed by the lack of concrete hypotheses regarding what, exactly, the basal ganglia could be doing for movement.

For example, it is possible that the basal ganglia, like the cerebellum, perform a single operation (**Bloedel,** this volume). The cytoarchitecture within the basal ganglia is rather homogeneous, despite discrete inputs and output targets from different areas. Functional heterogeneity observed in patients with disorders of the basal ganglia would then be related to the discrete inputs and output targets to different central systems. This functional heterogeneity would be reflected in diverse, functionally independent deficits in patients, such as (1) reduced muscle force (bradykinesia), (2) set-related inflexibility of motor pattern, (3) freezing and poor initiation of movement, and (4) abnormal postural tone and tremor. Each of these deficits may reflect the specific functions controlled by the four main output targets: (1) primary motor cortex via thalamus, (2) supplementary motor and premotor cortex (SMA) via thalamus, (3) medullary reticular formation/tegmental pedunculopontine nucleus of the brainstem, and (4) motor-related locus coeruleus.

If a single operation is to be hypothesized for the basal ganglia in the control of movement, perhaps it is that they make a critical contribution to the modification of motor networks based on environmental context and central intent. Failure to modify movement patterns based on incoming sensory information specifying environmental context and constraints would result in a variety of deficits. Poor positive feedback to and from motor cortex may then account for the saturation of force output and abnormal EMG activity (Horak & Anderson 1984; Nutt et al. 1992). Reduced output to SMA may account for poor ability to alter movement patterns flexibly depending on initial conditions (Horak et al. 1992). Reduced drive to PPN/MLR may account for poor gait initiation (Smith et al. 1988), and reduced modulation of the locus coeruleus may account for rigidity of postural tone (Fung et al. 1988). It should come as no surprise that each parkinsonian patient shows different amounts of deficits in each of these functions, because dopaminergic loss can differentially affect the various basal ganglia and extrabasal ganglionic circuits.

If the basal ganglia play a critical role in adjusting the overall pattern of variable-strength connections among motor networks, they must do so in close cooperation with the cerebellum. The cerebellum is ideally situated to use "knowledge of results to guide practice" and to provide a "global reinforcement signal" (sect. 5, para. 4). We do not understand **Alexander et al.**'s dismissal of "supervised" network learning on the basis of a global, error-correcting strategy. Without a "supervisor," how does the system recognize "desired behavior" or when "a solution turns out by chance, to be useful"? Could not the cerebellum provide motor networks with an extrinsic trainer who makes specific, connection-by-connection adjustments to the network's multidimensional weight space on the basis of a global error-correcting strategy" (sect. 4, para. 3)? We would not so readily reject the notion that the sensorimotor system uses optimization strategies to arrive at solutions to motor problems just because we do not yet understand how these neural networks are supervised.

Whether or not it acts as a supervisor for motor learning, it seems clear that the cerebellum, like the basal ganglia, plays a role in central set, although these structures play different roles in different types of central set. Our studies of postural control have shown that parkinsonian and cerebellar postural movement patterns are inflexible for different reasons. Parkinsonian subjects do not use sensory information about initial conditions to modify motor output (Horak et al. 1992). In contrast, the cerebellar subjects cannot use prior sensory information and knowledge of results to modify motor output (Horak 1990; Horak et al. 1986). Perhaps the cerebellum provides central set modifications of motor patterns during the "learning phase," whereas the basal ganglia provide central set modification in the "execution phase" of the connectionist model.

According to connectionist models, the precision with which we identify the site of a neural lesion may not be as important as the precision with which we identify the particular questions we ask about neural functions. Because it is difficult to be precise about neural lesions in humans but relatively easy to examine their complex sensorimotor functions, they make ideal candidates to study the adaptability to task conditions and generalization of motor learning proposed by the connectionist models. By asking specific, useful questions about neural function in our patients with lesions of the basal ganglia or the cerebellum we are likely to provide quite valuable information about the adaptive capabilities of these motor systems. Connectionist models must therefore provide us with hypotheses about motor behavior that can be tested in human subjects.

Converging approaches to the problem of single-cell recording

R. Iansek

Neurosciences Department, Monash Medical Centre, Clayton VIC 3168, Australia

[EEF] **Fetz** is to be congratulated for a very informative and critical appraisal of single-cell recording techniques. Any experimenter who has been involved in single-cell recording will understand the dilemmas and methodological problems outlined in this target article. It is a timely criticism of this method, which makes one stop and think about the meaning and interpretation of information obtained from single-cell recording.

A major problem with single-cell recording is that it does not take into account population effects on behaviour. How do we assess the output of a population of cells that have differing discharge patterns yet are directed towards a similar goal or connection? In **Fetz**'s target article the possible role of neural network modelling is detailed in a clear and informative manner. Numerous examples are given of the power of this technique to explain neuronal population function in regulating specific parameters of movement. However, despite all the deficiencies of the single-cell recording technique it is suggested that future research should use neural network modelling, though based on the results of single-cell recorded information. It is hard to reconcile this conclusion given the detailed deficiencies of the single-cell recording technique outlined in

the target article. After all is not the quality of the information provided by the neural network dependent on the quality of information put into it? If that information is wanting, then surely we should question the result.

We welcome new ideas as to how the brain might control movement as long as they can assist in the clarification of the problem at hand. We find that a *combination* of single-cell recording and clinical observations is beneficial (Brotchie et al. 1991a; 1991b). The concept of the motor plan as suggested by Marsden (1987) has provided our best explanation of the possible interaction between the supplemental motor area (SMA) and the basal ganglia to date. The concept of a motor plan was based on astute clinical observations. It is the clinical data that have provided us with the best clue as to the role of structures such as the basal ganglia in movement control, and this should not be overlooked.

We have interpreted our findings from single-cell recordings (Brotchie et al. 1991a; 1991b) in the context of such algorithmic control mechanisms as the motor plan. We have suggested, with experimental evidence, that pallidal phasic activity may signal the end of a movement in a movement sequence. This internal cue might be used by cortical structures such as the SMA to terminate set-related activity and to trigger the execution of previously prepared movements. One can further hypothesise a sequential domino effect that would allow a movement sequence to run once initiated by the SMA.

In addition we have used neural networks to generate possible neural discharge patterns in the basal ganglia and test whether this was indeed its role (Brotchie et al. 1991c). Results suggest that there should be both set sustained premovement-related activity and phasic movement-related activity in the basal ganglia. The patterns predicted by this model fit our findings of neural discharge patterns in the globus pallidus. Our concept of the interaction between the basal ganglia and SMA suggests that a movement sequence is initiated in the cortex. As the first movement in the sequence is executed, a phasic burst is generated in the basal ganglia. This burst is appropriately timed to turn off the set-related activity in the cortex concerning the next movement in the sequence. The next movement then executes, during which time a phasic burst is generated in the basal ganglia and this turns off set-related cortical activity for the third movement. Again, as the third movement is executed, a phasic burst is generated in the basal ganglia to turn off set-related activity for the fourth movement. This hypothesis would allow the sequence to run until it terminated.

Our point here is that a number of *converging approaches* are useful in elucidating the underlying neuronal mechanisms. Researchers do not yet understand how the brain controls movement. In that context, any models and paradigms that help explain the evidence we have so far should be fostered.

Spatial short-term memory: Evolutionary perspectives and discoveries from split-brain studies

David Ingle

Department of Psychology, Boston College, Chestnut Hill, MA 02167

[JFS] *Multiple sensorimotor systems.* First I wish to extend **Stein**'s arguments for the operation of disparate sensorimotor systems that require separate computations of spatial coordinates to include lower vertebrates (such as the amphibians), where evidence exists that separate retinotopic projections have distinct behavioral functions (Ingle 1982; 1983; 1991a). In animals such as frogs and toads which lack a neocortex, we already see visuomotor transformations from retinal to real-world coordinates. One example is size-distance constancy (for selection of bite-size prey, see Ewert & Gebauer 1973; Ingle 1968; Ingle &

Cook 1977). A second example is compensation for changes in head-to-body alignment in aiming a strike at prey. Thus, the parietal cortex need not do all the work in computing action plans from visual, tactile, and kinesthetic data since subcortical mechanisms represent some dimensions of "real world" space. Because size constancy in frogs and toads operates over short distances, however, the ability to perceive "real size" over longer distances may require cortical mechanisms. Indeed, the ability of birds and mammals to prepare skilled motor sequences at relatively large distances is demanded by the greater speeds and distances these animals travel in comparison with those of fishes and amphibians. Although the planning of coordinated movement sequences in large-scale 3-dimensional space may require cortico-cortical linkages between sensory and premotor systems, transformations governing single-action components may sometimes depend on subcortical mechanisms similar to those operating in lower forms.

Recently, we discovered that a robust form of short-term spatial memory in frogs depends upon the integrity of the striatum – the homologue of the mammalian basal ganglia (Ingle & Hoff 1990). Frogs accurately remember the locations of recently seen barriers for at least 60 sec after they have been whisked away, and will not jump in these particular directions during escape from subsequent looming threat. When frogs are passively rotated during this interval between barrier-disappearance and the elicitation of escape, their jumping directions indicate that they remember the barriers according to their real-world coordinates rather than according to retinal locus. Thus, the ability to combine visual and vestibu-lar/kinesthetic information to stabilize memory in real-world coordinates does not require a neocortex.

Our conclusions here are in line with studies of the role of the caudate nucleus of rats in maintaining egocentric spatial orientation (Abraham et al. 1983; Cook & Kesner 1988). In mammals, this striatal role seems to contrast with the well-studied role of the hippocampus in maintaining memory for direction based on "landmarks" in the external environment. Perhaps these two subcortical modules independently encode two types of spatial memory while their integration into a single flexible memory system in higher vertebrates depends upon further synthesis by the parietal cortex.

Cortical mechanisms of short-term spatial memory in man.
In discussing the mechanism of "spatial constancy" in the frog's short-term memory, I recently suggested that the neural representation of barrier location (e.g., the maximum focus of activity within a network) could actually move within a maplike representation as the frog rotated (Ingle 1991b). When we perform an analogous experiment on ourselves (briefly viewing a conspicuous target in a relatively uniform room and then closing our eyes) we maintain for 10–20 sec a distinct mental image of this target, which remains fixed in space as we rotate or walk about. In pursuing my naive hypothesis that neural activity corresponding to the mental image could move within the brain, we conducted the following experiment on subjects lacking a corpus callosum. (The details of this experiment will be submitted to this journal as part of a future target article.)

First, using four subjects with callosal agenesis (Ingle et al. 1990), we presented targets at eye level either 30° or 60° from the fixation point for about one second, asked them to close their eyes, and then to turn slowly either away from the target or past the target. In the first case, when target memory "moved" more caudally within the same hemifield, acallosal subjects pointed accurately at the remembered target locations as did the control subjects. But after turning *past* the target, thus shifting the target-locus between opposite hemifields, the agenesis subjects did poorly, whereas all controls showed good localization. The same results were obtained using subjects with total callosal transections and in two of three subjects with intended cuts throughout the anterior 70% of the callosum. Indeed, by using rotation-plus-walking routes with these new subjects the mem-

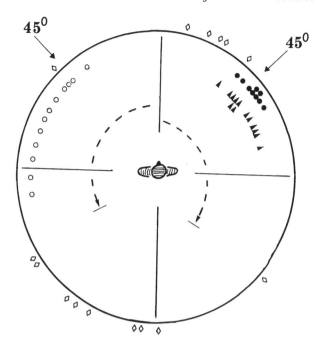

Figure 1 (Ingle).   The subject fixates a midline mark during a 1-2 sec opening of eyes, noting a 4-in-wide red disk set to the left or right side (here depicted by arrows at 45°). After a few seconds of covering eyes with a mask, subject rotates on command slowly to left or right (dashed arrow). Subject then points at remembered target locus. Circles are pointing directions after no-turn trials; triangles are responses after turns in either direction. Dark symbols indicate right-side target; open symbols indicate responses after presentation of left-side targets.

ory deficits were further exaggerated. We conclude from these results that even a primitive form of spatial memory is not accurately maintained when the remembered target moves across the midline of acallosal subjects. We therefore propose that when the callosum is intact some parcel of neural activity representing the remembered target actually jumps across the callosum during self-rotation.

It is of interest that of three patients with intended transection of the anterior 70% of the callosum (by the same surgeon), two showed loss of cross-field transfer of spatial memory and the other was quite normal. Since the caudal boundary of these cuts would (on the basis of anatomical studies in the rhesus monkey) approximate the crossing of the interparietal callosal fibers, it is possible that the extent to which opposite parietal areas are disconnected determines the loss of transfer. In pursuing this hypothesis we were able to test a 16-year-old male patient with total callosotomy who had suffered a right parietal hematoma as a child after a fall on concrete stairs. Although this subject showed good left-side motor skills (and drew pictures well with the left hand) he showed a selective spatial memory loss for left-side targets on our postrotation test, as summarized in Figure 1. The dark circles on the right represent directions of pointing at the remembered target (presented for 1 sec from 20° to 60° to the right of fixation) after 10 sec without rotation. In order to simplify this summary view, we normalized all response directions on this figure *as if* each target had appeared at the same 45° locus. The dark triangles on the right depict pointing directions after self-rotation in either direction (from 90° to 140°); these fell within a normal range for right-side targets. The open circles on the left show less accurate spatial memory after 10 sec without rotation. Open triangles show that rotation in either direction dramatically disrupts spatial memory: Only one out of sixteen trials produced a response within the correct quadrant. Because additional trials with head rotation alone did not disrupt location memory for left-side targets, we conclude that the parietal cortex plays a key role in interfacing short-term spatial memory with knowledge of self-rotation so as to represent target location within extrapersonal space.

The necessity of a complex approach in studying brain mechanisms of movement

Marat E. Ioffe

Laboratory of Motor Learning, Institute of Higher Nervous Activity and Neurophysiology, Russian Academy of Sciences, Moscow 117865, Russia
Electronic mail: *bir@sms.ccas.msk.su*

[EEF] The question posed in **Fetz**'s target article title is very timely. The relativity of movement parameters "coding" in single-neuron activity is very well demonstrated. The question seems to be a much broader one, however. Any attempt to correlate neuronal activity recorded in different brain structures with parameters of movement is a way to analyse the functions of these structures in the organization and performance of movement. In analysing neuronal activity related to movement one can draw conclusions mainly on the level of structures or large populations of neurons. This is also true of Fetz's approach: Three points are discussed in each section, namely, timing, population coding, and localization of function. The general problem can be formulated at a macrostructural level: That is, what brain structures are involved in performance of different kinds of movements and what is the sequence of their activation?

Two main features seem to be important in studying localization of function. First, the specificity of some structures in the control of different kinds of movement, and second, the plasticity of functions that can be changed by learning. These two features appear controversial; however, they are both essential in motor control. Specificity of function can be seen, for example, in lateral and ventromedial descending systems (Lawrence & Kuypers 1968). The lateral system, including the corticospinal and rubrospinal pathways, controls mainly distal muscles, whereas the ventromedial system, including chiefly the reticulospinal and vestibulospinal pathways, is responsible for the control of axial and proximal musculature.

The other example of functional specificity concerns the motor cortex. This structure is well known to control fine discrete movements (see Phillips & Porter 1977). In addition, during the elaboration of new movements through learning, some innate synergies interfering with the new movement have to be suppressed. It has been shown (Ioffe 1973; 1991) that the motor cortex is responsible for generating a command that descends via the pyramidal pathway and inhibits coordinations that interfere with a learned movement. After motor cortical ablation the interfering innate motor patterns become disinhibited and the performance of the learned movement is disturbed. So the inhibition of interfering patterns in motor learning is one of the specific functions of motor cortex.

On the other hand, functional plasticity is also a feature of different structures. It can be seen in the modification of behavioral patterns by learning. For example, it is possible to obtain eating behavior in response to stimulating the ventromedial hypothalamus, which usually produces aversive behavior (Pavlova 1979). It is also possible to rearrange some innate motor patterns by learning. For example, in a dog the flexor reflex to nociceptive stimulation can be suppressed by learning and once conditioned the dog can perform extension to escape stimulation of a limb (Frolov 1983). To take another example, the "diagonal" pattern of postural displacement usually accompanying limb lifting in quadrupeds can be rearranged

by learning into an "ipsilateral" one (Ioffe et al. 1988). [See also Golani: "What are the Building Blocks of the Frog's Wiping Reflex?" *BBS*9(4)1986.]

As mentioned above, the motor cortex is the structure providing such a rearrangement of motor patterns by suppressing interfering innate reflexes or coordinations. These data come mainly from lesioning the motor cortex and other structures. The role of the motor cortex in motor learning and, in particular, the role of different inputs to the motor cortex has also been closely examined using a combination of lesions and microstimulation (Asanuma 1989). Though neuronal plasticity in learning and, in particular, the instrumentalization of neuronal responses were shown in **Fetz**'s classical works (Fetz & Finocchio 1972; 1975), changes in neuronal activity and interneuronal relation during motor learning and rearrangement of the coordinations have unfortunately not been well enough studied. Only a few works on this point are available (Bures et al. 1988; Kotlyar et al. 1983; Mayorov et al. 1977).

The above data were briefly cited to show that the old methods for functionally investigating the brain, namely, stimulation and lesioning, are still available to study the role of different structures in motor control. Methods like lesioning are not without complications. Nevertheless, they have been successfully used by such neuroscientists as Sherrington, Lashley, Sperry, and many others and have allowed us to obtain basic data concerning brain mechanisms of motor behavior.

The new epoch began when Evarts (1981) popularized the method of recording activity of the identified neurons in behaving animals. Much data obtained by many authors allowed us to analyse the timing of different structures in the control of movement, the coding of different movement parameters in neuronal activity, and so on. This was in fact a real breakthrough in neurophysiology. Together with the recording of slow potentials preceding and accompanying movement (Kornhuber & Deecke 1965), this method promoted the appearance of modern ideas concerning organization of movement.

Now, however, the analysis of single-neuron activity yields less and less new basic data. This has stimulated researchers to develop new methods that are more global and effective for explaining the organization of movement. Pribram (1973), for example, has proposed a holographic theory of memory storage which is again cited in the target article. Another modern and promising approach to analysing the organization of movement is neural network modeling. Such models could explain the interrelations among different units in the control of movement as described in the target article.

Other methods are also becoming more popular now, for example, brain imaging, which allows us to represent the general picture of different structures, their timing and interrelations in the preparation, initiation, and performance of movement. The analysis of chemical mechanisms of motor control is also very interesting. For example, the chemical nature of postural asymmetry produced by unilateral lesions of cerebellum, motor cortex and so on is being studied. It has been shown, in particular, that vasopressin is one of the substances responsible for such postural asymmetry (Vartanyan & Klementyev 1991).

In general, one can say that a synthetic approach – including the methods of stimulation and lesioning, the recording of single-neuron and multineuronal activity, neural network modeling, brain imaging, the analysis of chemical mechanisms of motor control as well as other methods – is probably the appropriate way to study brain mechanisms of movement organization.

Function versus synapse: Still a missing link?

Masao Ito

Frontier Research Program, Riken, Wako, Saitama 351-01, Japan

[JRB] The discrepancy between **Bloedel**'s gain-change or dynamic-selection hypothesis and my view of cerebellar learning (Ito 1989) is not on a single point; it resides in a combination of two controversial arguments. First, whereas I take climbing-fiber signals as representing control errors, Bloedel seems to assume they have another meaning. I would like to ask him to define a functional meaning of climbing-fiber signals and to supply supporting data.

Second, whereas I assume that climbing-fiber signals lead to persistent changes in transmission efficacy in parallel fiber-Purkinje cell synapses in the form of long-term depression, **Bloedel** takes the view that climbing-fiber signals enhance coinciding mossy fiber inputs and thereby select a certain set of mossy fiber inputs, which in turn determine ongoing Purkinje cell activity.

Then, if the dynamic-selection hypothesis is combined with the first point of the learning hypothesis (i.e., error representation by climbing-fiber signals), it may be suggested that Purkinje cells provide an urgent rescue against an unexpected control error. This would certainly be the case in an animal's hitting an obstacle on a path. But if climbing-fiber signals represent something else, there must be another suggestion.

The second half of **Bloedel**'s target article seems to be devoted to such suggestions. The author's effort to integrate complicated views and data is indeed admirable. However, I have difficulty in understanding how Bloedel's "new" hypothesis of cerebellar function can be derived from his gain-change or dynamic-selection hypothesis. The cerebellar function he speaks about can be accounted for by a learning hypothesis. Learning-based hypotheses of cerebellar function have a better link with our knowledge at synaptic and neuronal levels. I would like to know whether Bloedel regards the gain-change or dynamic-selection hypothesis as a sufficient basis for constructing his "new" hypothesis, or if there is still an important missing link.

Toward an integration of neurophysiology, performance analysis, connectionism and compartmental modeling

Dieter Jaeger

Division of Biology 216-76, California Institute of Technology, Pasadena, CA 91125
Electronic mail: *dieter@cns.caltech.edu*

[GEA] The central argument of **Alexander, DeLong & Crutcher** pits a traditional algorithmic model of motor programming against more recent connectionist models of parallel processing in distributed systems. To be sure, it is high time we discarded simple-minded notions of equating changes in single-unit discharge in the cerebral cortex or the basal ganglia with specific processes in analytic, engineering-inspired models of motor control. It is counterproductive, however, to describe the field as a dichotomy between models of algorithmic motor-programming on the one hand and models of parallel processing in distributed networks on the other. Instead, understanding the motor system will require a synthesis of neurophysiology, performance studies, analytical approaches, and modeling.

To me, the concept of motor programming is not inherently sequential or engineering-inspired. True, the origin of this concept is in the basic computer metaphor that begot cognitive psychology, postulating a homology between sequential processing by a digital computer and the human psyche. New concepts about motor control are not as much inspired by

engineering, however, as by the results of studies involving reaction-time tasks of high sophistication. These studies provide evidence that human motor control is highly structured and composed of recognizable component processes (Meyer et al. 1985). In this view motor programming can be operationally defined by a "syndrome" of reaction-time effects that result from variations of behavioral tasks. Foremost among these variations is the introduction of a pre-cue (Rosenbaum 1980) that gives advance information about specific aspects of a required movement before the go-cue for the movement is given. For example, pre-cues containing advance information about the direction of movement have demonstrated that it can be "programmed" before the movement begins, that is, there is a specific gain in reaction time when the direction of movement is known beforehand.

Motor programming is by no means the only relevant concept in the analysis of motor control, as its exclusive treatment by **Alexander et al.** suggests. For example, experimental psychology has described with some precision the characteristics of a "foreperiod effect," which consists of a shortening of reaction time following a movement go-cue when this go-cue is preceded by a warning stimulus. This phenomenon requires that the motor system in some way anticipate the time at which movements are to be executed. In a well-trained task, this anticipation is exquisitely tuned to the exact probability distribution of when movement go-cues occur (Niemi & Näätänen 1981).

Many more control mechanisms have been found in studies of human motor performance; these need to be considered in the functional interpretation of single-unit activity in the motor system. It is quite possible that a modularity in information processing will yet emerge in the motor system. After all, taking the visual cortex with its nice division of labor for movement detection (e.g., areas MT and MST) and object recognition (e.g., area V4) as an example, why should the motor cortex, with a similar pattern of multiple anatomical areas (containing multiple somatotopic maps), not also show functional modularity? The modularity in motor areas might be orthogonal to the traditional divisions of preparation versus execution and sensory versus motor aspects of motor control. Cognitive psychology has offered many more possible component processes in recent years. Unfortunately, a distinct function of individual motor areas might not be expressed at the level of single-unit activity but might emerge instead as a product of the connectivity in the network. The activity of the network therefore needs to be investigated with all available techniques (recording of multiple single units, field potential recording, EEG recording, PET imaging).

In the second part of their target article **Alexander et al.** claim that connectionism is a better description of motor control than is the analytical account of cognitive psychology. This judgment is based mainly on the remarkable similarity of activity patterns in the hidden layer of connectionist networks and real neurons in the brain. In addition, both systems are self-organizing through learning, and information is represented in distributed networks. Nevertheless, I surmise that connectionism is hardly more than a replacement of metaphors when we think about brain function. The differences between connectionist models and the brain are staggering: On the one side we have three-layered, feedforward networks with simple elemental units; on the other side we have an extremely rich architecture with multiple interlocked feedback circuits and very complex elemental units. This mismatch in structure and dynamic function is equaled by a mismatch in performance as connectionist models in no way replicate the detailed temporal structure and predictive control of human motor performance. I suspect that Searle's (1980) criticism of rule-based AI to the effect that only a system copying the exact causal structure of the brain can imitate brain function applies to connectionism as well. Alexander et al. acknowledge this criticism to some degree in that they plead for reasonably detailed connectionist models consistent with known anatomical and physiological constraints. The fulfillment of this pledge within the framework of connectionism does not seem a realistic expectation to me. The strength of connectionism lies in enhancing our understanding of basic principles of information processing in distributed networks. Our understanding of such principles is likely to be necessary to explain brain function. It is not likely to be sufficient, however.

In contrast to connectionism, computer simulations of the detailed structure of the brain are carried out in the growing field of compartmental modeling. Inspired by the cable equations for equivalent cylinders of neuronal structures (Rall 1969), single neurons are simulated as many interconnected compartments, each of which can contain a set of ionic conductances (Koch & Segev 1989). Each ionic conductance has a distinct nonlinear activation and inactivation function, often dependent on membrane potential and ionic concentrations. In such simulations, single neurons emerge as powerful computational units. Realistic simulations of single neurons can be combined to construct larger network models. It is my hope that this type of modeling will eventually not only match physiological data but clarify how motor control is organized in the nervous system. To get to that point the integration of knowledge derived from neurophysiology, performance analysis, and connectionism is likely to be necessary.

Neurophysiological mechanisms for the planning of movement and for spatial representations

John F. Kalaska and Donald J. Crammond

Départment de Physiologie, Université de Montréal, Montreal, Quebec, Canada H3C 3J7

[GEA, EEF, JFS] *On the localization of function in distributed systems.* It is argued by **Alexander et al.** that traditional motor-program or serial-algorithm models are of little value for elucidating the physiological mechanisms of motor control. For example, the simple neural hierarchy predicted by those models – a serial cascade of anatomical structures, each containing a homogeneous cell population with distinct response properties – does not appear to exist (see also **Fetz,** this issue). In fact, they conclude that different structures "seem to lack the sorts of functional differentiations" (sect. 3, para. 7) expected of simple serial models. As an alternative, Alexander et al. propose that different analytical levels of movement planning and execution are realized more or less simultaneously in separate parallel "channels" distributed across several neural structures. This hypothesis requires a reevaluation of the concept of localization of function within the motor system. A key finding on which Alexander et al. base this hypothesis is that cells in many different structures show statistically significant correlations with a particular task variable, such as the presence of an external load. As a result, the cells are given a label, such as "muscle-like" cells, implying that they have similar properties and are only implicated in the control of that parameter. Although a parallel-channel model may be closer to reality than simple serial models, we would like to discuss why one must be cautious in building grand schemes on a foundation of relatively simplistic data analysis.

First, it is difficult to see how a movement can be planned efficiently without extensive interaction among representations of movement at different analytical levels. To put it colloquially, how can the "execution" channel generate the proper pattern of muscle activity without knowing what the "preparation" channel is doing? Where is the extensive convergence and divergence of connectivity among levels of representation that is the essential mechanism of information processing in the network models they favor?

Second, cells representing movement at different hypothetical hierarchical levels could still all show a correlation to a particular task variable. For example, many possible control parameters, such as output forces at the hand or handspace "virtual positions," joint torques, and multimuscle or single-muscle activation patterns will all covary with an external load (Kalaska et al. 1989). The tasks typically used cannot distinguish among these or other possible parameters, and so could mask a hierarchical sequence of transformations within a "channel."

Third, the correlation of neural activity with a given parameter may have quite different origins and functions in different areas. A muscle-like representation might be generated in one area, appear in a second area as a corollary discharge of the activity in the first, and be produced in a third region by reafferent peripheral input. One should be very cautious in linking all these neurons together into a single distributed channel with one function. For example, many cells in both the premotor cortex and parietal cortex area 5 show directionally tuned activity changes during the instructed-delay period prior to reaching movements (Crammond & Kalaska 1989; Weinrich & Wise 1982). One could argue that these cells form part of a single functional "channel" for the preparation of movement. Alternatively, the parietal activity could be a corollary discharge about the intended movement that is required, according to several models of kinesthetic sensation, to interpret the complex reafferent input from proprioceptors (Kalaska 1991a). The parallel representations of motor intention in premotor and parietal cortex would thus have different roles.

Fourth, binary classification of cells (correlated vs. noncorrelated) ignores important quantitative differences in their responses. For instance, according to a rigorous analysis-of-variance (ANOVA) test, the activity of many cells in both the motor cortex and area 5 is significantly correlated with external loads (Kalaska et al. 1990). The load-dependent variation of activity in area 5 is several times smaller than that in the motor cortex, however, and its directional tuning bears no consistent relation to the motor task. As a result, single-cell variations cancel out at the population level in area 5 and cannot contribute a meaningful signal for the compensation of external loads, unlike the motor cortex. The ANOVA results on their own give the false impression of a similarity between motor and parietal cortex function.

Conversely, labeling cells according to their best statistical correlation could create a false impression of functional divisions, such as segregated planning "channels," by ignoring the true complexity of cell properties. In fact, most cortical cells show partial correlations with many different attributes of both movement preparation and execution (Hocherman & Wise 1991; Kalaska et al. 1989; Thach 1978; Weinrich & Wise 1982). This is consistent with the complex combinations of response properties predicted by network models (see Fetz and **Robinson,** this issue) but it is not consistent with separate parallel "channels."

Traditional models assume that single cells perform a specific role and that a given function can therefore be attributed to an identifiable population, which could be either a discrete anatomical structure or a distributed but segregated "channel." For the sake of debate, we would like to suggest an alternative hypothesis for the nature of representations within the motor system that is also inspired by network models. The partial correlations of cellular activity with several movement parameters indicate that the activity of each neuron is determined by many different convergent inputs reflecting different aspects of motor planning and execution. Each cell can thus be viewed as a multichannel processor that transmits complex combinations of signals and contributes simultaneously to the representation of several different analytical levels of motor planning (Kalaska 1991b; Kalaska & Crammond 1992). The relative importance or weighting of the contribution to each representation varies from cell to cell. The representation of a particular analytical level of movement planning or of a particular movement attribute is therefore the sum of the corresponding single-cell response components distributed across a population of cells. Furthermore, different representations are distributed across overlapping populations of neurons and do not correspond to separate cell populations. In this model, sequential sensorimotor transformations are produced by the patterns of convergence and divergence of connectivity within the motor system, which result in a gradual change in the relative weighting of different levels of movement representation by single cells across the distributed network. There is no absolute localization of function in the traditional sense, but the representation of different analytical levels of motor planning may still be preferentially concentrated in different cell populations or structures.

On networks and neurophysiology. A superb discussion is provided by **Fetz** of some fundamental problems in interpreting the results of single-unit neurophysiological studies of the cerebral cortical mechanisms of motor control. He argues that they usually do not provide adequate information to demonstrate causality and that attempts to interpret neuronal activity in terms of arbitrary engineering parameters such as force, torque, or velocity are the naive equivalent of pounding square pegs into round holes (see also **Robinson** and **Alexander et al.**). Fetz intended his target article to be provocative. He has succeeded. Although we agree with much of what he says (Kalaska & Crammond 1992), we have been provoked to defend chronic single-unit studies to counterbalance what seems to be an unjustly pessimistic evaluation of their utility.

First, **Fetz** criticizes attempts to identify sequential activation of different neural populations because the latency distributions in different structures show extensive overlap, so that most cells are coactive at any given instant during a movement. Saying that they are active "more or less in parallel" (sect. 2.1, para. 3) does not necessarily mean simultaneous activation, however, and it does not preclude a moment-to-moment serial flow of information among components of the network. Overlapping but sequential recruitment of different neuronal populations within and across neuronal structures is seen repeatedly (Kalaska & Crammond 1992). Because the motor command evolves in real time prior to and during the movement, a sequence of planning transformations can still occur within a distributed network with small time delays between the representations of each planning level and so the neurons contributing to those different planning levels would be coactive during most of the movement.

Second, it is true that many cells show complex or paradoxical relations to movement parameters such as velocity, torque, or muscle motor-unit responses. The usefulness of this observation, however, is only as good as its implicit assumption that these are the actual parameters used by the central nervous system, which is questionable.

A major premise of **Fetz**'s target article is that the ultimate goal of neurophysiological motor-control studies is to have a complete description of the causal mechanisms for the planning and initiation of movement. To attain this "Holy Grail," one must have a description not only of the activity of all the cells in the circuit, but also a complete description of all intercellular connectivity. One could take the even more radical position that we must have intracellular records of all subthreshold membrane events. Attempts at a complete cell-by-cell causal understanding of the circuit, besides being technically impossible, may be a quixotic endeavor (see **Robinson**), however. Fetz himself notes that a broad (if not infinite) range of arbitrary network configurations can be taught to converge on the desired output and that each instantiation of each network configuration results in a unique causal solution. If each biological system likewise represents a unique solution, of what generalizable value is such a profound description? We would argue that this depth of description is unnecessary, and that network models suggest it is pointless. Moreover, this criterion renders futile all neurophysiological studies of sensory systems, because the final

output is a sensation that is experienced introspectively and is not independently measurable, making causal inferences impossible.

Fetz suggests that his network model can provide useful insights into biological motor control and that it transforms a step change in target position into motor unit activity (sect. 4.2, para. 2). It must also be acknowledged, however, that the network is just a curve-fitting matrix that transforms arbitrary input waveforms into a variety of output waveforms that are merely low-pass-filtered and differently weighted combinations of the inputs. We see no evidence, for instance, that it solves any of the sensorimotor transformations required to convert target spatial location into a multidimensional intrinsic reference frame of muscular activity. Given that the hidden units show many different waveforms, some of them will resemble actual cellular response patterns by chance alone. Thus, indiscriminate comparison of the "responses" of elements in a network with the discharge patterns of single cells can be as seductively misleading as correlating that discharge to parameters of Newtonian mechanics. Moreover, the network is interpreted only in terms of the relatively small population of corticospinal and rubrospinal cells that synapse directly on spinal motoneurons. This ignores all the other cells, including the many corticospinal neurons that do not synapse on spinal motoneurons and influence movement indirectly through the intermediary of spinal interneuronal networks. Does Fetz mean to imply that those cells play no significant role in motor control? On the contrary, others have proposed that spinal networks are the critical system that determines muscle activity patterns (Georgopoulos & Grillner 1989).

It is disconcerting that both **Fetz** and **Robinson,** who have made such important contributions to this field, would now appear so critical of the utility of the single-unit method. It is interesting to reflect on the history that has led us to this situation. Traditional motor-control models simplistically predicted that single cells would unambiguously encode a discrete parameter of movement. This assumption justified the single-unit recording method. Paradoxically, the results of those studies did not fit readily into those schemes and so helped invalidate traditional serial models as biologically implausible. Network models have suggested other information-processing mechanisms, in which the resultant single-cell response patterns can be complex combinations of movement-related signals that do not fit readily into any formal physics or mathematics coordinate system. Despite the tone of Fetz's target article, however, this does not invalidate the single-unit recording method. On the contrary, we would like to reemphasize the fact that it was single-unit recording studies that indicated that networks are potentially more valid models for brain processes. Properly designed single-unit recording studies have revealed, and will continue to reveal, important differences in the response properties of neurons in different structures under different task conditions from which one can infer, if not literally then at least symbolically, the nature of the information being processed by neurons in different areas. These data can also be used to develop better computational models. As we have stated elsewhere (Kalaska & Crammond 1992), one of the critical tests of any motor-control model that claims biological validity will be how well it can replicate the response properties of cells observed in single-unit recordings, not the other way around. We have no intention of hanging up our electrodes in the forseeable future.

The role of directed attention in egocentric spatial perception: An artful dodge? The target article of **Stein** addresses the fascinating question of how the central nervous system combines inputs from different sensory systems to generate what we perceive introspectively as a unified supramodal Euclidean representation of our immediate surroundings. Evidence for an explicit map of "real" space is scanty and probably the straw man of this target article. The central issue is not whether there is a single supramodal coordinate system for spatial perception and motor control, but how information encoded in any coordinate system is made available to systems using a different reference frame, be it in perception (sensory - sensory) or action (sensory - motor) systems.

The evidence is indeed compelling that the posterior parietal cortex (PPC) is the site at which information from a sufficient number of sensory and cognitive systems converges in order to allow transformations between different reference frames. For example, some cells in PPC areas 5 and 7b with clearly defined somatic sensory receptive fields may also respond to visual stimuli in a volume of space immediately adjacent to the somatic field, possibly in expectation of stimulation from approaching objects (Duhamel et al. 1991; Leinonen et al. 1979; MacKay & Crammond 1987). The function of such neurons may be to map visual input about objects in peripersonal space onto a body map in personal space, leading to the creation of a multimodal representation of objects in egocentric space (Duhamel et al. 1991). A critical test of **Stein**'s hypothesis is the degree to which directed attention is responsible for this alignment of somatic and visual reference frames. (For instance, repeated visual stimulus presentation can quickly lead to attenuated responses.) This function may not be limited to PPC, because cells with similar properties have also been found in the postarcuate premotor cortex (Gentilucci et al. 1988).

The hypothesis that the deliberate and conscious process of directed attention mediates the alignment of and transformation between different maps is more speculative in the domain of action systems. Moreover, it might be a virtually bullet-proof hypothesis, impossible to disprove. It is difficult to see, for example, how one could completely eliminate some degree of directed attention in a conditioned voluntary motor task or devise experiments to demonstrate the existence of accurate spatial perception in a situation in which we are confident that no attention is involved. For example, the influence of directed attention on parietal activity was demonstrated in a complex behavioral context in which a motivated monkey attended to loci in space at which it was conditioned to expect behaviorally relevant stimuli (Goldberg & Bruce 1985; Mountcastle et al. 1981). During the intertrial interval, when the monkey was idly scanning the task environment and not expecting any stimuli, those effects were absent. Yet it is difficult for us to believe that the animal's spatial perception was in any way degraded during that period. Does this mean that one must begin to invoke different degrees, shades, or flavors of directed attention to account for spatial perception and for sensorimotor transformations in different behavioral contexts? An alternative proposal, which is to some extent merely playing with semantics, is to suggest that the transformations result when signals describing the *intention* to move are relayed to the PPC, where the relevant information in different reference frames is then integrated. This is one possible explanation for the finding that neuronal activity recorded in the PPC during an instructed-delay period prior to arm (Crammond & Kalaska 1989) and eye (Gnadt & Andersen 1988) movements encodes the direction of the intended movement. The problem posed by **Stein**'s hypothesis, however, concerns how one can disprove the possibility that this delay-period activity arises because some form of attention is directed to the target location or even to the body part that will be moved. Furthermore, the direction of attention and of movement are not always colinear, and movement does not always require a high degree of directed attention. Finally, the mere act of directing attention to a locus in space to receive a pertinent instruction leaves open an infinite range of motor responses, each of which will require unique sensorimotor transformations. Hence, we do not see how implicating directed attention clarifies or resolves any of the problems of spatial representations and sensorimotor transformations.

In conclusion, **Stein** is probably right that a cartographic multimodal map of egocentric space does not exist. Neverthe-

less, an egocentric spatial representation may still exist in a form that we do not yet recognize, one that is independent of directed attention. Finally, this hypothesis is still not formulated in a way that is experimentally verifiable.

The identification of corticomotoneuronal connections

Peter A. Kirkwood

Sobell Department of Neurophysiology, Institute of Neurology, London WC1N 3BG, England

[EEF] Within the admirably well-balanced target article by **Fetz**, an old, unresolved issue is lurking: the definition of corticomotoneuronal (CM) cells. In the first paragraph of section 3 these cells are said to "directly affect motoneurons" and in section 3.2 they "can be said to causally affect force." However, as Fetz also states in the first sentence of section 3, these cells are defined by a "correlational linkage," that is, their firing is only statistically related to motoneuron firing. My criticism is that the causal link is not secure: When Fetz (and others) define CM cells, they mostly use postspike facilitations (PSFs), which do not allow an accurate identification of a direct link; hence CM cells may well be misidentified. The issue is important because if Fetz wishes to use connectivities as constraints on his neural models, the connectivities must be accurate.

The problem is that a peak in a correlogram or PSF may result from excitation of the motoneurons not by the cortical neuron recorded but by some other neuron(s) synchronized to it, such as other CM cells, that is, other neurons involved in the common task. To demonstrate that this is not so for a correlogram, criteria for the durations or latencies of the peaks must be met, as described elsewhere (e.g., Davies et al. 1985; cf. Kirkwood & Sears 1991). Although these criteria are not absolute and must be separately argued for each situation, the durations must be short, in the msec or sub-msec range, and latencies are critical within a msec. However, the durations of the PSFs are usually 10 msec or more and the latencies are hard to define by virtue of rather slow initial rates of rise. These durations are inevitable because, in the transformation from somatic motoneuron discharges to the PSF, temporal dispersion occurs in the motoneuron axons and muscle fibres. Latency and duration information is lost and there are therefore *no* criteria which can reliably be used on the PSFs to distinguish a genuine direct connection from one due to synchronization in the cortex.

In his defence, **Fetz** might cite evidence from CM cells that were identified with a PSF and then shown to give a narrow cross-correlogram peak when their spikes were correlated with those of single motor units in the muscles giving the PSF. Such peaks can be good evidence for the existence of a direct connection, but it does not necessarily follow that the PSF represents the same connection. In particular, the narrow peaks seen in illustrated cross-correlograms often sit on top of lower-amplitude but broader peaks (e.g., Fig. 11F of Lemon & Mantel 1989; Fig. 1B of Mantel & Lemon 1987; or Fig. 2 of Smith & Fetz 1989). Because these other components are of lower amplitude and have a quite different time-course, the narrow peaks can be very clearly discriminated from them and the broader ones can be ignored. However, the *area* of the broad peak may well be as large as or larger than the area of the narrow one; so when either effect is measured via the filter represented by the peripheral temporal dispersion, either could give a peak with a time-course similar to that of the PSF, with similar peak amplitudes. Thus, even for those units giving good evidence of a direct connection, one would reasonably suspect that 50% of their PSFs originated in other mechanisms. The rather smaller (and often noisier) PSFs that form the subsidiary projections defining the motoneuronal "fields" of CM cells would seem to be even less secure. Nevertheless, these PSFs still tell us something: At the very least they indicate that the unit concerned and the "muscle field" are closely co-activated during a given movement. I believe, however, that what they have in common may not be a direct connection from the unit to the motoneurons of each muscle but a linkage one or more synapses further back in the network. The issue is equally important in studying changes in the PSF between different states or movements when, because cortical cells may be activated in different combinations, changes in their synchronization would be *expected* to occur.

Fetz and his colleagues have always been aware of the possibility of synchronization effects and, in an important paper, Smith and Fetz (1989) tried to assess them. The conclusion then was that the effects could be considerable. Each spike of a CM cell could be accompanied by, on average, one other spike from the colony of other CM cells converging on the motoneurons of a given muscle. This supports the view expressed above that typically 50% of the strongest PSFs could arise via cortical synchronization. The "sharply rising, later component of the PSF," write Smith and Fetz, "may still be interpreted as evidence for a direct connection." However, in published illustrations, many PSFs do not have such clear sharp components. Smith and Fetz demonstrated that a major problem exists, but Fetz appears to proceed as if it does not.

Could one do better in establishing the connectivities? The general use of single-unit cross-correlations would probably be too difficult. A possible alternative might be to look more assiduously for positive signs of synchronization, that is, for "common input" effects in the PSFs. At present assessments of periodic features in the PSFs (together with consideration of the unit autocorrelograms) are very underused. For these features, applying criteria set out more than 20 years ago (e.g., Moore et al. 1970) might at least allow the recordings most likely to be contaminated by cortical synchronization to be weeded out.

In conclusion, I would not at all wish to dissuade **Fetz** (and others using a similar technique) from pursuing their present course of combining connectivity studies with large-scale modelling. I would only put in a plea for a more critical approach when attempting to establish the connectivities.

What is the nature of the feedforward component in motor control?

A. D. Kuo[a] and F. E. Zajac[b]

[a]*Mechanical Engineering Department, Stanford University, Stanford, CA 94305 and* [b]*Rehabilitation R & D Center, VA Medical Center, Palo Alto, CA 94304*
Electronic mail: *kuo@roses.stanford.edu*

[EB] Although the importance of feedback in human movement is not to be underestimated (Wiener 1961), the importance of feedforward control should not be forgotten either. Even the equilibrium-point hypothesis (of **Bizzi et al.**), in requiring virtual trajectories, uses some sort of feedforward control. The question is: How rich is feedforward control? Is it rather simple, as suggested by the equilibrium-point hypothesis, or is there a full inverse-dynamics representation, or something in between? Experimental evidence in the study of standing posture indicates that the feedforward component may be of considerable complexity. Moreover, we believe that the CNS is capable of learning inverse-dynamics representations.

Studies of arm movement by standing subjects (e.g., Belinkii et al. 1967; Bouisset & Zattara 1987) suggest that the CNS controller may have knowledge of the inertial properties of the body. For example, subjects who perform voluntary arm elevation while standing show anticipatory excitation and inhibition of muscles in both the ipsilateral and contralateral limbs prior to the excitation of muscles raising the arm. These anticipatory

postural adjustments (Bouisset 1990) could therefore be the result of previous learned experiences where, in effect, the CNS has learned a dynamic representation of the muscle forces needed to counter the inertial forces associated with the arm movement. Alternatively, the equilibrium-point interpretation could explain these adjustments in terms of a virtual trajectory. However, such a trajectory would have to include not only a simple straight line describing the desired arm motion but also paths for each of the legs.

Standing subjects who are perturbed by slow movement of the support surface have been shown to respond with motion predominantly about the ankles; for faster disturbances they tend to use motion about both the hips and ankles (Horak & Nashner 1986) probably to avoid large ankle torques, thereby keeping their feet flat on the ground (Kuo & Zajac 1992). This evidence poses two problems for the equilibrium-point hypothesis. First, the two stabilizing strategies cannot be described in terms of the trajectory of a single end point such as the head but rather must be described by trajectories of at least some of the joints (or some function of them). Second, the ability of the body to adjust to control constraints indicates an "awareness" of the body's multijoint inertial properties, which determine the size of ankle torques that cause the heels or toes to lift off the ground. Thus, the feedforward command must adjust to fast disturbances subject to these inertial properties.

Is the CNS capable of such complex behavior? The learning of both inverse and forward dynamics has been demonstrated in artificial neural networks (Kawato 1989). The CNS, with a far greater number and variety of neural elements than any artificial neural network, exhibits the ability to learn coordinate-transformation mappings (e.g. vestibular-ocular reflex; Gonshor & Jones 1976). It seems reasonable to suppose that the brain is also capable of performing inverse-dynamics mappings.

The principal attractiveness of the equilibrium-point hypothesis is that its reliance on feedback simplifies trajectory planning, so that inverse dynamical computations are unnecessary. However, the CNS appears rich enough to be able to perform complex trajectory planning, and such planning appears to account for some degree of "awareness" of the musculoskeletal dynamics.

The challenge, then, is to devise schemes to test for this dynamical awareness. Do motor commands adapt to varying inertial environments, or do they rely purely on error feedback to produce the motion? Are there experiments in which feedback alone can be made unstable, so that stability is only possible given a feedforward adaptation? We must focus on questions concerning the feedforward component in order to understand better the limitations of the equilibrium-point hypothesis.

Neural networks: They do not have to be complex to be complex

Irving Kupfermann

Center for Neurobiology and Behavior, Columbia University, College of Physicians and Surgeons, and The New York State Psychiatric Institute, New York, NY 10032
Electronic mail: *kupferma@nyspi.bitnet*

[DAR] **Robinson** has used examples of networks that control the oculomotor system to illustrate how it becomes increasingly difficult to interpret the activity of single units as the functional task increases in complexity. His conclusions do not necessarily and unambiguously flow from his observations and they leave much room for disagreement and doubt. There is reason to believe, however, that even for very simple parallel circuits it is difficult or impossible to interpret correctly single-unit activity.

We (Kupfermann et al. 1992) have recently been exploring parallel distributed models of the neural circuitry underlying a very simple example of behavioral switching in the invertebrate *Aplysia*. The models consist of sensory, hidden, and output layers of five, five, and three units, respectively. The output layer innervates muscles that produce turns of the head, or a bite response, when food is presented to one or two loci on the lips. As in real animals, the direction and magnitude of a turn, and the magnitude of a bite response, are a function of the loci at which stimuli are presented to the lips or tentacles (Teyke et al. 1990). The hidden units that were generated by various solutions were treated as if they were interneurons whose function we wished to determine via single-unit recording and stimulation methods that are routinely used to understand invertebrate circuits (see Kupfermann & Weiss 1978). Three techniques were applied: (1) determination of the behavior elicited when a unit is fired; (2) determination of the response of the unit to different sensory inputs; and (3) determination of the change of behavior that results when the unit is removed from the circuit and various sensory inputs are presented. The analysis showed that even with this elementary circuitry, it is difficult or impossible to determine the role of a given unit in generating behavior – and that it is easy to be misled by the data.

Robinson offers two hopes for the otherwise pessimistic assessment of our chances of understanding system behavior in terms of component parts. One is that the engineers may come up with something in ten years or so. A second is that we may somehow benefit from discovering how synaptic modifications occur; but he then concludes that even "if we know the learning rules, we may have to accept the inexplicable nature of networks." I suggest that knowledge of the learning rules may provide important constraints on the types of parallel circuits that can be developed, and hence may provide insights far beyond mere knowledge of how the circuits are generated in the first place. Further insights into the constraints that go into developing parallel distributed processing (PDP) networks may come from the study of optimizing methods other than back propagation. Particularly promising are genetic algorithms (Goldberg 1989), which may mimic the evolutionary process by which many circuits evolved in the first place. When appropriately used, genetic algorithms may produce a small subset of the totality of possible solutions, resulting in a relatively manageable set of networks that can be analyzed and related to real circuits.

Reflex control of mechanical interaction in man

Francesco Lacquaniti

Consiglio Nazionale delle Ricerche, Via Mario Bianco 9, 20131 Milan, Italy
Electronic mail: *ifcncnr@imisiam.bitnet*

[EB, SCG, DAM] In this commentary, I would like to address a few issues that are raised in the target articles by **Bizzi et al.**, **Gandevia & Burke**, and **McCrea**. Their reviews present a wide scenario on the contributions of reflex and nonreflex mechanisms to the control of limb movement and posture. The adaptability of the spinal and supraspinal mechanisms is stressed in all three target articles. I will elaborate on some organizational principles concerning the control of mechanical interaction of the limb with the environment (cf. Lacquaniti et al. 1991; 1992). As stressed by Bizzi et al., a preplanned mechanical interaction between limbs and environment poses some special problems for the central nervous system (CNS). It requires the correct specification of both target position and limb compliance. Limb compliance is defined as the dynamic relation between a given change in position and the corresponding change in force. Man has evolved to interact with objects and manipulate tools optimally. Conceivably, he is able to tune limb compliance according to the specific physical properties of different objects. Much thinking has recently been devoted to

the problem of adaptive compliance control (e.g., Bizzi et al. and Gandevia & Burke). However, the neural mechanisms that are actually involved in compliance control are still poorly understood.

Catching a falling ball represents an ideal paradigm for the study of preplanned mechanical interactions with objects. This task requires an accurate control of limb position to intercept the trajectory of the ball. It also requires the control of limb compliance in order to absorb the ball's momentum optimally. Furthermore, subjects cannot self-pace the temporal modulation of these variables (limb position and compliance) but must comply with the timing constraints imposed from outside. By changing the mass of the ball or the height of fall one can study the effect of cognitive set and expectations about the anticipated properties of the forthcoming impact.

The typical findings obtained when catching is performed in the presence of visual stimuli and prior information about both the mass of the ball and the height of fall are the following. The electromyographic (EMG) activity associated with catching is organized in two main segments relative to the time of ball impact: anticipatory (voluntary) activity prior to the impact and reflex activity after the impact. Anticipatory activity is contingent on the presence of vision. Its onset time is locked to the estimated time of impact independent of the height (and duration) of fall. Time-to-contact can be estimated using optical flow information (i.e., the field of the instantaneous positional velocities of the image on the retina). The mean amplitude of the anticipatory activity is scaled in proportion to the magnitude of the expected momentum at impact. The prediction of ball momentum depends on a cognitive operation performed on the basis of heterogeneous information. Instantaneous velocity information is available through vision, but subjects must extrapolate from velocity measured at one instant to the final velocity some time later at impact. The expected mass of the ball can be estimated based on an internal model of the ball's properties. Ball impact on the hand results in a sudden extension of the limb.

According to classical notions about stretch reflexes, one would expect this extension to evoke EMG responses organized reciprocally over antagonist muscles: Flexor muscles that are stretched by the perturbation should be reflexively activated, while extensor muscles that are shortened should relax. Instead, the EMG responses that follow the impact at a short latency violate the law of reciprocal innervation, since both flexor and extensor muscles are coactivated. Although these responses are reflexively triggered by the impact, their amplitude and direction are preset in the CNS prior to impact on the basis of the available information on the forthcoming perturbation. When subjects are provided with accurate information on the impending perturbation, the mean amplitude of the reflex responses generally increases in proportion to the intensity of the perturbation, both in the stretched muscles and in the shortening muscles. Once again this does not accord with the behavior of classical stretch reflexes: The amplitude of muscle activity in shortening muscles should decrease with increasing shortening velocities. In the absence of visual information, however, the amplitude of the mechanical oscillations induced by the impact is much larger in the first trial than in the following trials, indicating fast adaptive calibration of the responses.

This adaptive calibration of the responses has been interpreted as resulting from the operation of a model reference system. According to this hypothesis, the CNS would be endowed with an internal model of the dynamic interaction that is expected to occur at impact. The response of the model to the disturbance is compared with the actual mechanical response of the limb and generates an error signal. This error is then used to calibrate the parameters of the reflex controller and to update the internal model. Initially, the model does not accurately predict the desired performance, possibly because of an erroneous estimate of ball momentum in the absence of vision.

However, kinesthetic and cutaneous information obtained during the first trial is adequate to correct the estimate and achieve the desired performance in subsequent trials.

As for the functional significance of the anticipatory and reflex coactivation, it should be noted that contraction of antagonist muscles results in joint torques with opposite signs, but in additive joint compliances. Thus, the observed scaling of the amplitude of both anticipatory and reflex responses proportional to the amplitude of the perturbation suggests that the compliance of the limb is tuned according to the expected properties of the impact.

Recent experiments have provided conclusive evidence that reflex coactivation does not merely depend on the nature of the peripheral stimulus associated with ball impact but is preset within the CNS. In these experiments torque motor perturbations were randomly applied to the elbow joint at different times before and during catching. It was found that the pattern of the reflexive responses evoked in elbow muscles was not constant but depended on the time of application of the torque pulse. Reciprocal responses, consisting of an increase of activity in the stretched triceps and a decrease of activity in the shortening biceps, were consistently elicited when a flexor pulse was applied at any time in the course of a trial, except during a limited time interval centered on impact. During that interval (from about 60 msec prior to impact up to 60 msec after impact), the pattern of the reflex responses consisted in a coactivation of both stretched and shortening muscles because of a transient reversal of the direction of the short-latency response of biceps.

Since the reflex reversal begins prior to the time of impact, it does not depend on the peripheral stimuli elicited by the impact but is generated within the CNS. Thus it represents a further example of the flexibility of spinal reflexes stressed by **Gandevia & Burke** and by **McCrea**. Reflex coactivation is then part of a prospective control scheme that involves the central gating of stretch reflex responses from the spinal pathways of reciprocal inhibition (via lamina VII Ia interneurons) to the pathways of co-excitation of antagonist α-motoneurons (via lamina V-VI interneurons; see McCrea's target article).

The key to understanding the functional significance of this gating mechanism lies in the experimental finding that the compliance of the hand is transiently minimized during the same time interval in which the biceps reflex reverses direction. Since the torque motor pulses applied at different times during catching were constant, the amount of mechanical oscillation of the hand induced by each such pulse provides a measure of the overall hand compliance: the larger the oscillations, the larger the compliance and vice versa. The variance of the hand oscillations was then computed at different times during catching. In general, the time course of the changes in hand compliance is poorly correlated with that of the changes in overall EMG activity. By contrast, the timing of the minimum in hand compliance is well correlated with the reversal of biceps reflex.

One can therefore hypothesize that the reversal of stretch reflex responses prior to impact corresponds to a transition between two distinct control modes: from a position control, based on reciprocal innervation, to a compliance control, based on coactivation. The switching between these two operating modes could be implemented in a network that changes synaptic weights so as to dynamically optimize the performance criterion corresponding to hand compliance. The minimization of hand compliance suggests that the CNS is able to represent the intended hand compliance internally and to transform it into appropriate patterns of output muscle activities. However, as noted by **Bizzi et al.**, hand compliance is a global variable that depends on both the pattern of muscle activities and the geometrical configuration of the limb. In other words, the same pattern of muscle activity may result in a very different value of hand compliance depending on the values of joint angles. The transformation of intended hand compliance into the appropri-ate muscle activities therefore requires an internal model of limb geometry.

The results obtained on catching can then be interpreted as indicative of the fact that the stretch reflex is gated on the basis of the internal model of limb geometry. This is a striking conclusion, because it leads to the postulation of a linkage between two different domains of neural control that are usually considered to be independent and remote from each other, namely, the domain of the internal models and representations of limb geometry on the one hand, and the domain of muscle reflex control, on the other. According to the prevailing views on the hierarchical organization of motor systems, the level of movement planning and the level of movement control represent two independent stages of processing, dealing with entirely different kinematic and dynamic variables. At the higher level of planning, limb movements are represented in the global terms describing the action of the end effector (e.g., the hand). By contrast, it is often assumed that the level of reflex control deals only with the local variables pertaining to a single muscle (e.g., muscle length, force, or stiffness). The picture that comes out from the series of studies on catching is instead one in which the operation of the reflex control also deals with global variables, many of these being of the same nature as those involved in the process of trajectory formation. Indeed, gating of the stretch reflex is involved in the minimization of one such global variable, the hand compliance, based on an internal model of limb geometry.

Equilibrium-point hypothesis, minimum effort control strategy and the triphasic muscle activation pattern

Ning Lan and Patrick E. Crago

Applied Neural Control Laboratory, Case Western Reserve University, Cleveland, OH 44106

Electronic mail: *nxl5@po.cwru.edu*

[EB] The target article by **Bizzi et al.** tries to outline a unified framework for motor control based on the equilibrium-point hypothesis, although many details of this theory are still the subject of debate. The inherent springlike property of muscles (Hoffer & Andreassen 1981; Nichols & Houk 1976) has provided the neurophysiological basis for the equilibrium-point hypothesis, yet there is still a distance from equilibrium-point control models to the large volume of existing experimental data concerning voluntary arm movements. For example, simulations based on the maximum smoothness of movement control can demonstrate only the kinematic features of arm movements (Flash 1987; Hogan 1984), but they do not provide direct predictions of muscle control signals. The λ-model can generate muscle control signals that are comparable to the EMG activities of antagonistic muscles (Feldman et al. 1990). The λ-model assumes, however, that the equilibrium point shifts at a constant speed toward the target position to generate the movement.

Different optimal control methods have been used to compute muscle control inputs (Hannaford & Stark 1987; Ramos & Stark 1987). These optimization models have shown various degrees of success in predicting the triphasic muscle control signals for fast, single-joint movements. Thus, dynamic optimization methods provide a means of peering into the mechanism underlying CNS control of movements. In relation to the equilibrium-point hypothesis, we should ask: (1) How is the equilibrium point (a central input) specified to guide a movement? and (2) how are muscle stiffnesses modulated (by both central and peripheral inputs) to generate the movement? It is possible that the equilibrium point and muscle stiffnesses are programmed on the basis of a certain objective. Minimizing

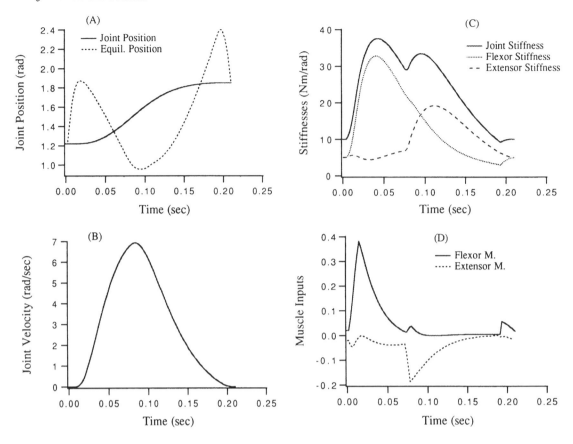

Figure 1 (Lan & Crago).   Diagram shows a fast, point-to-point single joint movement in 200 msec in the horizontal plane. The model is based on the elbow joint with parameters derived from literature. A pair of antagonistic muscles (flexor and extensor) controls the elbow joint movements. The optimal control strategy generates the joint movement trajectory, muscle control inputs, as well as the associated joint stiffness and equilibrium states, from specifications of the initial and final positions and stiffnesses, and the movement duration. In (D), the extensor muscle input is plotted downward for increasing activation to contrast with the flexor muscle input.

the objective function leads to a unique solution for the problem.

With this philosophy, we have considered a dynamic optimization model for the formation of the equilibrium trajectory and for the control of antagonistic muscle stiffnesses in single-joint movements (Lan & Crago 1991). This model minimizes an effort function, defined as the time integral of the multiplication of joint stiffness with the square of the time derivative of equilibrium position, which was proposed by Hasan (1986). The joint stiffness is explicitly controlled by the CNS, according to Feldman (1986), through a feedforward coactivation command. In the spinal cord, the central inputs integrate with peripheral afferents to determine muscle activation. It is assumed that the autogenic feedback regulates the length-tension property of the muscle, so that muscular force (or torque) is an increasing function of length (or angle). The heterogenic feedback (Nichols 1987) coordinates the activation of antagonistic muscles through the difference between the equilibrium position and actual limb position. Joint, muscle contraction, and motoneuron pool activation dynamics are included in the model, which are important factors for the timing of muscle activations.

This optimization model demonstrated qualitatively the major features of fast voluntary arm movements of a single joint. One of the main outcomes of the optimization is the triphasic burst pattern of antagonistic muscle activations. A second main result is a temporal modulation of joint stiffness, which increases with acceleration and deceleration. In Figure 1, we illustrate

these points with the numerical solution for a rapid, point-to-point, single-joint movement. The control strategy produces a normal looking movement, as shown by the solid line in (A). The equilibrium trajectory, dotted line in (A), leads joint movement at the beginning to provide acceleration, but lags behind the joint movement during deceleration. It overshoots the joint position at the termination of movement. Joint velocity in (B) displays a natural bell-shaped profile, with a slight asymmetry. The antagonistic muscle activities are found to have the triphasic burst feature, as shown in (D). This pattern resembles the three bursts in the antagonistic EMG activities observed in voluntary arm and head movements (Ghez & Martin 1982; Hannaford & Stark 1985; Marsden et al. 1983). The first flexor (agonist) burst accelerates the joint toward the final position. The second extensor (antagonist) burst stops the joint at the final position. The final flexor burst raises the joint stiffness to the required value for maintaining the final position. These three phases of muscle activation coincide with the alteration of equilibrium trajectory in (A). Thus, the difference between the equilibrium position and the joint position dictates which muscle to activate, whereas the coactivation command determines the amplitude of activation. As a result of the triphasic activation pattern, the joint stiffness (solid line in C), increases during the movement. The first peak is due to flexor stiffness increase (dotted line); the second peak is added by extensor stiffness (broken line). It is quite striking that all major features of voluntary arm movements are exhibited by the minimum effort control strategy.

Available measurements for joint stiffness during self-initiated arm movements in the horizontal plane also show a dynamic increase of joint stiffness during movements (Latash & Gottlieb 1991). This is compatible with our optimization result. In point-to-point movements, the joint must come to a complete stop at the end of the movement. The joint stiffness required to maintain the posture may be much less than that to accelerate and decelerate the limb. An increase of joint stiffness in the middle of movement is therefore expected.

The optimization model illustrates that the movement planned according to the minimum effort control strategy displays the kinematics and control signals that agree qualitatively with voluntary movements. With this optimization model, the effects of model parameters on the specification of central commands can be studied. Other plausible strategies may also be examined in the context of the equilibrium-point hypothesis.

Are we able to preserve a motor command in the changing environment?

Mark L. Latash

Departments of Physiology and of Physical Medicine and Rehabilitation, Rush Medical College, Chicago, IL 60612

[EB, SCG, DAM] All three target articles (**Bizzi et al., Gandevia & Burke,** and **McCrea**) are brilliant and unique contributions to their fields. They combine extensive reviews with suggestions of solutions for many of the burning problems. Assuming that the authors will be praised in many other commentaries, let us focus on what seems to remain unsolved and controversial.

Are we speaking the same language? No! The first step in discussing or investigating any scientific issue is to agree upon a set of notions that would be appropriate for the desired level of analysis, that is, to choose a language. The field of motor control is apparently too wide to be described with just one language, as is clearly illustrated by the target articles. It can be compared to physics. No one is expecting a researcher in the field of elementary particles to use his particular set of notions for describing the behavior of solid bodies or fluids. Can one describe the functioning of a neural or muscular cell as an ensemble of elementary particles comprising the cell? The answer is probably no. A cell is a unit of a different level of complexity that requires a different language for its analysis. Thus, a neuron is a complex system as compared with elementary particles. The system of motor control of one muscle is complex as compared with individual neural and muscular cells. The system of motor control of multijoint movements is likely to be complex as compared with the control of individual muscles. Nevertheless, there are plenty of attempts to describe the functioning of the system of motor control as an interaction of groups of cells and to describe the control of multijoint movements through signals to individual muscles.

The target article of **McCrea** presents a bulk of data on the relations between spinal interneurons, descending and segmental systems, and peripheral receptors that create a general feeling that in the central nervous system everything is connected to everything, and that all the connections are modulated by all possible inputs. This approach is likely to be productive in the analysis of certain rather simple reflexes and their possible changes during the execution of different motor tasks or in pathological states. This is a separate area of study whose language is very likely to be inappropriate for the motor control studies. One can term it "reflexology." So, my answer to the title question of the target article by McCrea is yes for reflexology and no for motor control. Similarly, the author's conclusion that "the most significant limitation to progress is not the complexity of spinal interneurons but the few laboratories

engaged in these studies" seems to be true for the former but not for the latter area of study. In particular, the equilibrium-point (EP) hypothesis would never have been born if all the laboratories were tracing connections between groups of neurons.

What is controlled in EP control? λ! The target article of **Bizzi et al.** describes an attempt to introduce an adequate language for analyzing the control of single-joint movements. An important source of misunderstanding of different models is in the word "controlled." From everyday experience and experimental studies we know that human beings and animals can be trained to "control" virtually any variable characterizing voluntary movement: joint angle (position), joint torque (force), movement speed, accuracy, electromyographic patterns, and even such exotic variables as the second derivative of torque changes (Ghez & Gordon 1987). All these variables describe voluntary movements at the level of performance. The question is: Can one find a variable at another, higher level of the motor control hierarchy whose time changes would be "controlled" (supplied by the higher levels) for all the variety of motor tasks independently of the external conditions of movement execution?

Adamovich and Feldman (1989) have recently formulated a number of prerequisites for the theories of motor control. In particular, they require that such theories consider movements as shifts in the equilibrium state of the system and indicate variables whose regulation allows the brain to execute the shifts. The λ-model suggests the threshold (λ) of the tonic stretch reflex (TSR) as an independently controlled variable for a muscle. In the target article, **Bizzi et al.** have not explicitly identified the variable that is presumably controlled for a muscle in the framework of the α-model, although the general impression is that total output of the α-motoneuron (α-MN) pool has been implied as such a variable.

The attempt of **Bizzi et al.** to reconcile α and λ versions of the EP hypothesis looks rather artificial. The α-model emerged on the basis of observations of deafferented animals. Then, the solution was generalized to intact animals and humans. The alternative is to observe the behavior of intact humans and suggest a noncontradictory description. This is how the λ-model emerged. As a result, the α-model is more substrate oriented and has more physiological appeal, whereas the λ-model is more cybernetical and does not suggest a definite neuronal substrate. The notion of TSR is introduced in the λ-model as a mechanism giving rise to the observed force-length characteristics of intact muscles. So the suboptimal efficacy of muscle reactions to unexpected perturbations (Allum 1975; Houk 1976; Rothwell et al. 1982c) is no argument against the λ-model since these preprogrammed reactions are assumed to be mediated by a mechanism different from TSR.

In footnote 2, **Bizzi et al.** dismiss the earlier criticisms by Feldman (1986), stating that they hardly share the view that "(1) the muscle activation only modulates the muscle's stiffness but not its rest length and (2) that in an intact preparation, the level of alpha activity does not depend on feedback variables." I agree that both statements fail to make much physiological sense but I still see them as consequences of the α-model as it was described in the earlier publications. Concerning point (1), the idea that muscle activation changes only muscle stiffness was suggested, in particular, by Figure 1 in Bizzi et al. (1982). If both stiffness and zero length can be modulated, one gets a number of families of curves (Figure 1) that can intersect so that each muscle is controlled by at least two variables, and an agonist-antagonist pair is controlled by four variables. This implies, in particular, the possibility of voluntarily inducing intersections of muscle force-length characteristics and very asymmetrical joint stiffness profiles in response to loadings and unloadings for a fixed initial position and level of muscle cocontraction (point A in Figure 1). Neither has been observed in humans or animals with intact reflexes (Feldman 1966a; 1986; Feldman & Orlovsky 1972; Gottlieb & Agarwal 1988; Hoffer & Andreassen 1981; Latash &

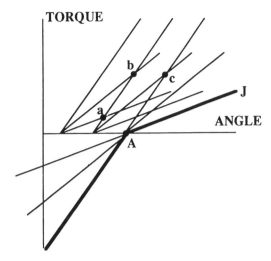

TORQUE

ANGLE

J

Figure 1 (Latash). According to the α-model (see Bizzi et al. 1982), muscle stiffness can be controlled centrally. According to footnote 2 in the target article of **Bizzi et al.**, central commands can also change muscle's zero length. As a result, any equilibrium point can pertain to many muscle characteristics (e.g., points a, b, and c), intersections of force-length muscle characteristics can occur, and joint characteristics (bold line J) can have considerably different stiffnesses in response to loadings and unloadings.

Gottlieb 1990; Matthews 1959; although see Nichols & Steeves 1986). Getting to point (2) of Bizzi et al., if the level of α activity does depend upon the feedback, it apparently cannot be independently controlled. Thus, the question What is controlled? still remains unanswered by the α-model.

It also seems strange that **Bizzi et al.** are so reluctant in accepting the possibility of nonmonotonic equilibrium trajectories for control of fast single-joint movements. In the framework of the λ-model, this possibility has theoretically been predicted by Hasan (1986). Recently, nonmonotonic N-shaped equilibrium trajectories have been experimentally reconstructed in our laboratory (Latash & Gottlieb 1991).

Are we free to choose a motor command? Yes! This question can be reformulated as: Can one fix a value of a control variable at some level of motor control hierarchy independently of the conditions of movement execution? There is no direct experimental way to get an answer. One indirect way is to search for invariants during changes in the external loading conditions when subjects are asked "not to intervene voluntarily." Apparently, hypothetical control variables expressed in terms of actual joint angles, muscle forces, and electromyographic patterns will inevitably change following a change in the external load. This means, in particular, that the trajectory of the working point and the total output of α-MN pool fail to qualify. As far as we know, λ is the only viable candidate. Therefore, if one favors the answer yes, the λ-model does not seem to have an alternative. If one prefers no, human beings become deprived of the inalienable right of free choice, and all the models become equally nondisprovable, since any observation can be explained by subconscious changes in central variables that are no longer independently controlled. Two versions of the EP-hypothesis are not just quantitatively or even qualitatively different, they are philosophically incompatible, the λ-model being an example of yes-attitude to single-joint motor control, whereas the α-model is an illustration of no-attitude. The untestable nature of no and the desire to be able to exert independent control over my own muscles force me to accept yes.

Let us consider a simple hypothetical scheme of single-joint motor control (Figure 2). "Yes" implies that there is a level generating time functions λ(t) for the participating muscles that

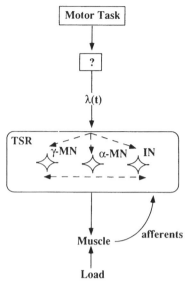

Figure 2 (Latash). A hypothetical scheme of λ control. At some level of the motor hierarchy, a motor command to a muscle is expressed as λ(t). This signal is distributed among many segmental structures including α-MNs, γ-MNs, and interneurons. Its overall effect can be expressed as a shift of the threshold of the tonic stretch reflex (TSR). In deafferented animals, part of the connections becomes useless and the animal needs to relearn how to control its movements with the remaining functional connections (i.e., mainly direct projections to α-MNs).

can be kept independent of the external conditions. Note that the conclusion of **Gandevia & Burke** that central motor commands alone may be sufficient to control simple learned movements implies that the system is able to supply λ(t) in a feedforward manner while the feedback-based mechanism of the TSR generates patterns of α-MN activation based on both control functions and current afferent input. In a deafferented animal, the TSR box loses its function, the hypothetical central controller is forced to use the remaining direct projections to the α-MNs, and λ control degenerates into α control.

Without being champions of any of the EP-models, **Gandevia & Burke** present plenty of evidence that all possible afferents exert reflex effects during natural movements. They also emphasize the problems with interpreting data after deafferentation. The fact that subjects can be trained to deliver a motor command to specific α-MN pools at a level subthreshold for activation of the MNs (Gandevia & Rothwell 1987) also argues against α control. And finally, a statement by **Bizzi et al.** that "sensory feedback was not essential for movement" (sect. 2.1, para. 11) based on their studies of deafferented animals contradicts their later statement that "our experiments were not intended to ascertain the relative contribution in the intact animal of feedforward commands versus feedback signals" (sect. 2.3, para. 6).

Do we know more than Bernstein about control of multijoint movements? No! In 1935, Bernstein suggested that multijoint movements were likely to be planned in terms of the kinematics of the working point in the external 3-dimensional space. He also suggested a concept of synergies as time sequences of control signals to groups of muscles which lead to simple coordinated motor acts. Note that Bernstein stressed the theoretical impossibility to control independently the activity of α-MNs and he has never suggested that synergies are defined as fixed combinations of activation levels of corresponding α-MN pools. These hypothetical control signals are just one factor that can influence actual patterns of muscle activity. Another factor is feedback from the peripheral receptors. The observations by Macpherson (1988) cited by **McCrea** therefore fit well within the original Bernstein (1935) concept of synergies. Recent studies by Nichols (1989) have demonstrated rather unexpected patterns of heterogenic reflexes among cat muscles crossing the ankle joint. One interpretation of these findings, which fits Bernstein's concept of synergies, is that segmental reflexes can be modulated in a way leading to certain preferred patterns of muscle activation. In other words, synergies can be hardwired at a segmental level. For example, if an animal wants to perform a single-joint ankle movement without moving the knee, a certain pattern of heterogenic reflexes between the uni- and bi-articular ankle extensors may considerably simplify control of the task.

The EP approach seems to be a promising way to try to handle the problems of multijoint coordination. However, there is still a big gap. For single-joint motor control, there is a hypothetical variable that is used by the central nervous system to control movements, λ. The existence of the λ-model makes the EP approach experimentally analyzable. For multijoint movements, there are no candidates. The virtual-trajectory hypothesis of **Bizzi et al.** is simply a reformulation of Bernstein's general idea that at some level multijoint movements are controlled in terms of external kinematics.

The meaning for movement of activity in single cortical output neurons

Roger Lemon

Department of Anatomy, Cambridge University, Cambridge CB2 3DY, England

[EEF] **Fetz** has written a really useful and thought-provoking target article. In it he has highlighted a common scientific problem: To what extent are the results obtained simply a consequence of the experimental design? This error is all too easy to make in the field of motor control where the essential operation of the motor system remains a truly hidden unit!

The search for a representation of different movement parameters at the level of single-cell activity has not been as rewarding as some early studies indicated. We may have missed important clues that are present in the organisation of the motor output map, not only in the primary motor cortex but in other frontal motor areas. Hughlings-Jackson hypothesized that the hand muscles in particular would have a large cortical representation because "they serve in more numerous movements." The multiple representation of outputs to individual motoneurons and muscles and the extensive overlap in these representations must mean that the different output units, be they single cells or cell clusters, are likely to make quite different contributions to the measured parameter (Humphrey 1986; Lemon 1990), and this is probably why is it possible to find task-related cells over wide areas of cortex and in different cortical regions (Lemon 1990). In many studies there has been an insistence on the monitoring and measurement of the "encoded" parameter, to the exclusion of more natural movements. The stereotyped nature of most tasks comes up against the versatility of the motor system in finding many different solutions for the same motor task.

The spike-triggered averaging (STA) technique, introduced by **Fetz** and Cheney for the identification of premotoneuronal cells has allowed us to determine the output targets of different units within the motor system. Since most cortico- and rubromotoneuronal (CM and RM) cells facilitate more than one muscle, it is clear that multiple muscle control is the key feature of output organisation for both the motor cortex and the red nucleus. Each cell facilitates a particular combination of muscles, the "muscle field" of Fetz and Cheney (1980).

To understand the output functions of the motor system **Fetz** quite rightly stresses the importance of studying identified output neurons, such as CM and RM cells, because of the causal influence the STA reveals them to have over their target muscles. Here it is very important to distinguish between the central "representation" of a movement and how the activity that is so represented actually controls and communicates with the interneurons and motoneurons involved in the execution of the movement. But in addressing the organisation of these outputs we must carry this point to its logical conclusion: What, if any, is the significance of the "muscle field" that is the signature of a given output neuron? In their original study Cheney and Fetz (1980) stated that they could not find any obvious relationship between the muscle fields of CM cells and their pattern of discharge during a ramp and hold wrist flexion/extension task. Many of the muscles active during these movements, however, are also active for extension or flexion of the digits, and the paradigm used in these studies cannot address the significance of the premotoneuronal unit in question for these other movements.

In a precision grip task, requiring independent control of thumb and index finger, and in which the same group of muscles studied by **Fetz** and his colleagues is active, we have found evidence that the muscle field may be a very good indicator of the way a particular CM unit is recruited (Bennett 1992; Bennett & Lemon 1991). These CM cells tend to be most active when the balance of activity in a number of synergist finger muscles is that corresponding to the relative "weighting" or strength of PSF produced in these same muscles by the CM cell, that is, there is a congruence in the pattern of synaptic connectivity and of recruitment during movement. This relationship, however, only became apparent through examination of the highly varied patterns of muscle activity used by the monkey to shape its hand prior to the onset of the precision grip task. The considerably stronger facilitation that CM cells exert on intrinsic hand muscles than on long forearm muscles (Lemon et al. 1991) also implies that even single cells can exert a significant influence over the level of activity in a given hand muscle.

A problem for many parametric studies is to decide precisely how a neuron might encode a given parameter of movement. In most studies the mean rate of a neuron's discharge has been taken to be a reliable indicator of its activity in relation to that parameter. Much less attention has been paid to the pattern, as opposed to the rate, of cellular discharge. In the case of monkey CM cells, where it is possible, using STA, to assess directly the action of the cell upon its target muscle, there is clear evidence that the pattern of discharge is important. As predicted by earlier work on temporal facilitation at the CM synapse (see Porter 1970), when CM cells fire with very short interspike intervals (<10 msec), these spikes produce strong postspike facilitation (PSF) of EMG activity in their target muscles. Paradoxically, spikes with long interspike intervals (>30 msec) can produce equally strong effects when these are measured as the size of the PSF relative to the ongoing, background EMG level. Cell discharge in midrange intervals (15–30 msec) produced weaker effects than either very short or long intervals (Lemon & Mantel 1989). This shows that both firing pattern and rate have important consequences for muscle activity: It is therefore misleading to equate the output of a given cell simply in terms of mean firing rate alone, yet this has been the approach adopted in most parametric studies.

Toward a genuine theoretical neuroscience of motor control

Daniel S. Levine

Department of Mathematics, University of Texas at Arlington, Arlington, TX 76019-0408
Electronic mail: *b344dsl@utarlg.uta.edu*

[GEA] The importance that **Alexander, DeLong & Crutcher** give to connectionist models heralds a welcome development in neuroscience. Experimental neuroscientists, working in motor control and other areas, are seeing neural network models not as toys or abstract exercises but as an increasingly significant contribution to the interpretation of their data. Yet these authors underestimate the progress that connectionist modeling of motor control has already made (see Levine 1991, Ch. 7).

Broadly speaking, I believe that in motor control, as in vision, conditioning, pattern recognition, or other areas, the most promising approach to models with genuine explanatory power is one that is well developed (see, e.g., Grossberg 1988, for many examples). Typically, the modeler sets out to explain a set of behavioral data using a network that obeys a dynamical system of differential equations. Often, parts of this network incorporate widely used network principles such as associative learning, lateral inhibition, opponent processing, or error correction. Then the network is refined to incorporate known anatomy, physiology, or neurochemistry of brain regions that are good candidates for analogy with some of the network nodes. With this approach, network theory has a "life of its own" as an equal partner with experiment. Experimental data can suggest alterations of models, and models can suggest new experiments to be performed.

I will give two examples of this approach as it has been applied to the role of the cortex and basal ganglia in motor control. Both of the models to be discussed rely on extensive feedback, and thereby synthesize the "bottom-up" and "top-down" approaches (target article, sect. 4). As **Alexander et al.** indicate, this makes such models more plausible as brain models than those that involve supervised learning or hybrids with a hardwired symbol-processing module.

Bullock and Grossberg (e.g., 1989; 1991) have developed a series of models for the performance of planned arm movements, particularly focusing on how positional control is independent of speed and force rescaling. The error-correction in their model requires a vector that computes the difference between a present position and a target position. They have noted that cells have been found in the shoulder-elbow zone of the precentral motor cortex with properties analogous to a difference vector (Georgopoulos et al. 1984; Schwartz et al. 1988). Another element that is required by the Bullock-Grossberg models is a "go" signal that is multiplied by the difference vector to generate a movement command. This enables the movement to be interrupted in the middle, by shutting off the go signal, and later resumed. Also, variable amplitude of the go signal allows the same movement to be performed, with essentially the same trajectory, at varying speeds. The neurophysiological requirements for a go signal are that it should (1) have nonspecific effects on a variety of muscles, (2) affect the rate but not the amplitude of muscle contractions, (3) not affect accuracy, but (4) be necessary for movement performance. Such properties were found in a class of globus pallidus neurons by Horak and Anderson (1984a; 1984b).

Higher-order controls on movement sequences involving the prefrontal cortex were modeled by Bapi and Levine (1990). Data on rhesus monkeys (Brody & Pribram 1978; Pinto-Hamuy & Linck 1965) had shown that monkeys with frontal lobe damage can learn one invariant sequence of movements if it is rewarded, but they cannot learn to perform any one of several variations of a sequence if all are rewarded. Modeling these data requires a sequence-generating circuit located elsewhere than the frontal cortex, but the frontal lobes must exert some sort of classificatory control on this circuit. Such a circuit would incorporate primacy and recency effects: The first of the movement sequence has to be performed first even though the later movement representations have stronger connections to the reward representation. Bapi and Levine have made the tentative suggestion that the sequence circuit is located in the caudate and putamen and that the higher control is exerted by connections from the frontal cortex to the striatum, and indirectly back from the striatum to the frontal cortex via the limbic system, nucleus accumbens, and mediodorsal thalamus (all of which are involved in reward representation).

These two examples are not meant to imply that the brain's motor control circuits are totally understood. Rather, they are meant to give some of the "flavor" of the most promising connectionist models, and of how future models might be generated. **Alexander et al.**'s call for models that are parallel, flexible yet mathematically analyzable, incorporate top-down controls, and can suggest further experiments is already being answered.

Past the equilibrium point

Gerald E. Loeb

Bio-Medical Engineering Unit, Queen's University, Kingston, Ontario, Canada K7L 3N6
Electronic mail: *loeb@biomed.queensu.ca*

[EB] Neurophysiologists are indebted to **Bizzi, Hogan,** and colleagues for having pointed out that muscles are endowed with certain intrinsic mechanical properties that are quite different from those of an "ideal actuator" (i.e., torque motor as used in robotics), and that these properties may actually augment or even simplify the neural control of limb motion under certain conditions. However, their claim that "the hypothesis . . . has been corroborated experimentally" lacks a clear statement of a testable hypothesis and a body of thoroughly analyzed data from well-designed and relevant experiments. I would argue that neither has been presented. Furthermore, Bizzi et al.'s notion has fundamental limitations that have been obvious from the beginning; these have been acknowledged briefly in their Conclusions but not yet addressed satisfactorily.

What is the hypothesis? In the alpha model (sect. 2.1), **Bizzi et al.** present "a central postulate" that the CNS generates signals specifying an equilibrium position. A postulate is a proposition presented without proof, either because it is self-evident or for purposes of exploring its implications. If the CNS outputs in question are merely the levels of recruitment of the muscles, then it is indeed self-evident that the "spring-like properties" of those muscles must define an equilibrium position for the limb as a whole. Perturbations away from this position will lead inexorably to restorative forces (e.g., McKeon et al. 1984). The implicit question of interest to neuroscientists is: Under what conditions and to what extent does the CNS actually rely on this physical fact to simplify the planning and execution of skeletal movements? This question cannot be answered by identifying a few simple situations in which the externally observable behavior happens to be qualitatively consistent with this general notion. Instead, at least two key questions must be addressed:

1. Given the mechanics of a particular task and a particular musculoskeletal system, how many different, equally simple motor strategies would produce adequate performance, and what kinesiological data would be needed to distinguish them?

2. What is the set of tasks and conditions for which the equilibrium-point strategy would be expected to fail and how well does the CNS then cope?

Exploring the implications of this postulate is daunting but necessary if any testable hypotheses are to be generated at all.

What are the behavioral data? Even for the simple tasks selected so far by these researchers, the data are fragmentary and their real implications largely unexplored. In Bizzi et al. (1984), the raw data cannot be examined in detail to separate intrinsic from reflexive responses because the figures are obviously hand-traced (a common necessity for the Visicorder technology the authors used, but not acknowledged in the paper or in the target article), time scales are highly compressed, and EMGs lack calibration bars. The one figure that depicts EMG from an antagonistic pair of muscles (reproduced in the target article as Fig. 2) suggests an unusually high degree of cocontraction in the deafferented limb. Cocontraction is a common default strategy for dealing with novel or difficult tasks. It is not surprising that deafferented monkeys recognize the utility of the "alpha" strategy for coping with their deficit, but this provides little insight into normal sensorimotor behavior.

In dismissing the lambda model (sect. 2.2), **Bizzi et al.** cite the low gain of the stretch reflex in the neck (Bizzi et al. 1978). It is now clear that the segmental feedback from the numerous proprioceptors in the neck is remarkably weak compared to the limbs (Anderson 1977; Brink 1988; Keirstead & Rose 1988; Rapoport 1979; Richmond & Loeb 1992), for reasons that remain obscure. The homonymous feedback in the limbs is much stronger and the total effects of the heteronymous feedback system are just starting to be considered (He et al. 1991; Loeb et al. 1989). Because of the diverse and powerful descending systems for modulating gain in these pathways, any extrapolation between species, limbs, surgical preparations, and behavioral tasks is fraught with uncertainty.

The curved trajectories measured and modeled by Flash (1987) apparently represent the only attempt to simulate the consequences of equilibrium-point planning in a realistic model of a limb. As such, the effort is commendable and the results worth considering in more detail than has been provided in section 3. The agreement that she obtained between the simulated and measured trajectories is often qualitatively impressive, but it is unfortunate and perhaps not accidental that the study was restricted to relatively slow movements where the differences between static and dynamic planning schemes will tend to be small. Any control scheme that undercompensated for intersegmental dynamics might well produce similar results, particularly if the velocity profile within the movement can be selected arbitrarily, as implied by Flanagan et al. (1990). Furthermore, to obtain even these results, it was necessary to increase the joint stiffness matrix beyond that obtained statically by Mussa-Ivaldi et al. (1985). This seemingly innocent "fix" implies an additional cocontraction of antagonist muscles that flies in the face of what is known about the reciprocal control of antagonist muscles during reasonably fast and unconstrained movements. The absence of systematic recording and interpretation of EMG data in the entire body of work on equilibrium-point control frequently leaves its advocates in the tenuous position of trying to infer actuator kinetics from end-point kinematics, a risky business indeed in a system this complex.

Bizzi et al. note that Flash's results (weak as they are) might be explained by CNS programming of the shape and orientation of the stiffness field, presumably by means of reflex gating rather than patterns of cocontraction. They dismiss this possibility by citing Mussa-Ivaldi et al. (1987), in which such changes were not seen in response to various perturbations during a static task. This is a fatally flawed argument in motor psychology. The question is not whether subjects happened to change the shape of the stiffness profile for the chosen task but whether they could change this stiffness if it suited their purposes. In the experiments cited, subjects had no way of knowing whether they were changing the stiffness profile nor did they have any particular incentive to change it.

What are the neurophysiological data? Various parts of the frog spinal cord have been "microstimulated" by **Bizzi et al.** to find neurons that produce the motor patterns associated with the wiping reflex. This seems to be one of the most naive electrophysiological experiments published since the invention of the inductorium. The spinal cord is responsible for many sensorimotor behaviors, none of which is known to be segregated anatomically. What topographic structure there is (and it is particularly weak in the frog; Lichtman et al. 1984) suggests elongated, columnar entities that could not be selectively recruited from a monopolar microelectrode thrust into the gray matter. No data are provided regarding the amount of distribution of EMG activity thus evoked in the various leg muscles, or whether these EMG patterns resemble those developed during naturally evoked wiping reflexes. Given the "spring-like" behavior of muscles, virtually any combination of muscle recruitment that was not strongly polarized toward one group of muscles would produce an "equilibrium point." Apparently, such conditions obtain when stimulating in the intermediate zones (where all the motoneurons and interneurons have densely intermingled dendritic trees) and not in the ventral horn (where there is some clustering of motoneuronal somata into motor nuclei). Bizzi et al.'s Figure 5, showing gradual shifts in this equilibrium point during a stimulation train whose parameters and temporal relationships are not given, is not interpretable.

What is the point? It seems clear that robotics engineers have derived some insights into the shortcomings of torque motors and servocontrollers from this work (although the cryptic allusion to the unpublished work of McIntyre resurrecting position-derivative control does not bode well). When it comes to elucidating biological mechanisms, we are given only the continuing development and promotion of dogma in place of critical thinking and experimentation. The value of simple models is that they focus our thinking on the various fundamental aspects of problems, not that they are likely to constitute universal solutions. The roles of modeling and experimental design are supposed to be complementary in that the latter strives to reveal the limitations of the former. We have here too many examples of experiments and phenomena that seem to be selected to flatter rather than to reveal.

To what extent are brain commands for movements mediated by spinal interneurones?

Anders Lundberg

Department of Physiology, University of Göteborg, S-413 90 Göteborg, Sweden

[EEF] The initial studies of unit recording in behaving animals seemed to be based on the assumption that the motor cortex is interested solely in motoneurones even though it was known at that time that corticospinal volleys influence many spinal mechanisms, including interneurones of almost all spinal reflex pathways to motoneurones, to ascending pathways, and to many of the pathways producing presynaptic inhibition by their action of primary afferent terminals. It was an important advance in this field when **Fetz** and Cheney (1980) introduced the method of recording from cortico-motoneuronal (CM) cells identified by their EMG postspike facilitation. Even so, caution is required and I believe that some interpretations are based on doubtful assumptions. An inherent assumption seems to be that only the *monosynaptic* pathways from the motor cortex and the red nucleus matter. The cat does very well without these connections and I am convinced that the major command also in primates is via spinal interneurones, implying that the monosynaptic pathways, though important, contribute only a fraction of the command, perhaps producing some fractionation of muscle activation and final adjustment of the movement. If so, a causal relationship between cell and muscle activity would not always be expected.

Two examples of command mediating interneuronal systems in cats should be mentioned. The C3-C4 propriospinal system is particularly interesting because behavioural studies have shown that it can mediate the command for target-reaching (Alstermark et al. 1981b; Alstermark & Kümmel 1990). The C3-C4 PNs have powerful connections with forelimb motoneurones and receive monosynaptic convergence from all major descending pathways except the vestibulospinal tract; they are also influenced (mainly inhibitorily) from the forelimb which they govern (Alstermark et al. 1984; Illert et al. 1977; 1978). It is worth noting that the C3-C4 propriospinal system exists in humans (Malmgren & Pierrot-Deseilligny 1988a; 1988b) and that it is activated during movements (Baldissera & Pierrot-Deseilligny 1989) resembling those used by **Fetz** as tests. It is therefore likely that the command for wrist movements is mediated in parallel by CM and RM cells and by the C3-C4 PNs, with the latter contributing the major part. Such parallel processing obviously makes it extremely difficult to determine the contribution of individual systems and a close resemblance of activity in a given premotoneurone and muscle activity may even be fortuitous; consider that the pattern of motoneuronal activation is also influenced by segmental reflexes from proprioceptors activated by the movement.

Another example of a spinal premotoneuronal system is provided by interneurones activated by group II secondary spindle afferents which may servoassist movements commanded from the brain. Some of these interneurones are excited by cortico- and/or rubrospinal volleys (Edgley et al. 1988; Lundberg & Voorhoeve 1961) and it has been suggested that the cortical command via this pathway initially is *phasic*, activating the interneurones and/or the motoneurones. Once the movement has started it might be governed from the brain partly or largely via secondary afferents driven by *tonic* activation of static γ-motoneurones (Lundberg et al. 1987b). The hypothesis is clearly of interest in relation to the finding that the rubromotoneuronal (RM) cells are predominantly phasic. It is possible that some of the tonic rubrospinal neurones, which do not project to motoneurones, instead control static γ-motoneurones and exert tonic action on motoneurones via secondary afferents and group II excitatory interneurones. A sharp division between phasic and tonic neurones would not be expected since the proportion between direct descending α-activation and group II activation may differ; phasic-tonic cells clearly may be incorporated within the framework of the group II hypothesis.

The subdivision by Lawrence and Kuypers (1968) into lateral and medial descending systems may prove fallacious. Cheney et al. (1988) accept this division and assume that the commands for wrist movements are mediated entirely by the lateral system, that is, the cortico- and rubrospinal tracts (CST and RST). It is noteworthy that after complete transection of the CST and RST in cats, reticulospinal pathways can under some conditions mediate the entire command not only for target-reaching (which includes activation of proximal and distal muscles) but also for food-taking, which includes manipulatory toe movements (Alstermark et al. 1987). It can by no means be excluded that reticulospinal pathways contribute to the command even when the CST and RST are available; note that a very large part of the corticofugal output ends in the brain stem.

Playing the role of devil's advocate, **Fetz** remarks about the finding by Muir and Lemon (1983) that some CM neurones are preferentially active during a precision grip between thumb and forefinger but inactive during a power grip. This finding is unexpected only if you believe that each target muscle has its own exclusive cortical neurones through which the brain must operate whatever movement involving this muscle it wishes to command. If you accept that "The brain knows nothing of muscles, it only knows movements" (Hughlings Jackson 1932) then it is not surprising to find different CM cells active during different movement conditions. I also fail to be surprised by the above observation that CM cells strongly modulated in a finely

controlled ramp and hold tracking task were inactive during rapidly alternating ballistic movements that engaged the same target muscles. In cats there is a special pathway from the motor cortex exciting fast twitch but not slow motoneurones (Alstermark & Sasaki 1986; Alstermark et al. 1981a). It would be advantageous to activate fast motoneurones selectively during rapidly alternating movement because lingering tension in slow motor units of antagonists would then not prevent agonist shortening. A ramp and hold task, on the other hand, is better subserved by CM cells contributing to the usual recruitment order.

A recent study of fractionation in another premotoneuronal system – the C3-C4 PNs – has revealed that some PNs project only to motoneurones, others to both motoneurones and reciprocal Ia inhibitory neurones (Alstermark et al. 1990). The latter neurones may be used for commanding reciprocal movement and the former for other movements. It would not be surprising if the same subdivisions exist among CM neurones projecting to a given motor nucleus. Furthermore, a study of the termination of individual C3-C4 PNs has revealed branching to different motoneurones with multiple termination in different combinations; some PNs terminate on both motoneurones to proximal and distal muscles (Alstermark et al. 1990; Tantisira 1990). Such findings indicate that different muscle synergies are laid down in the divergent projection of the PNs. This principle has already been established in a study of CM cell activity related to wrist and finger movements in monkeys (Lemon, personal communication). Premotoneurones to a given motor nucleus may accordingly be differentially used depending on the motor synergy represented in their projection pattern.

I am impressed by **Fetz**'s thoughtful comments on population vectors.

To some extent I find the pessimism in the target article unjustified. The relationships between cell activity and the type of movement described as paradoxical are highly interesting as are the observations regarding convergent activity from different sources and the different response pattern in CM and RM cells. With the new method introduced by **Fetz** the whole field has been transformed and the investigation of unidentified cells now seems almost meaningless.

Fetz closes his essay expressing hope that the combination of two methods will resolve the issue of network mechanisms generating motor behaviour. I hope he is willing to admit that a thorough knowledge of spinal circuitry and its control from the brain, lesion experiments, and the new various activity-dependent histochemical techniques is also required.

The single neuron is not for hiding

W. A. MacKay[a] and A. Riehle[b]

[a]*Department of Physiology, University of Toronto, Toronto, Ontario, Canada M5S 1A8;* [b]*Laboratoire de Neurosciences Cognitives, CNRS, 13402 Marseille Cédex 9, France*
[a]**Electronic mail:** *mackay@utormed.bitnet;* [b]*ariehle@lnf.cnrs-mrs.fr*

[DAR, JFS] Neural networks are a new religion, but each of the faithful beholds a different salvation. In a witty critique of microelectrode glimpses at the inner workings of the brain, **Robinson** presents the network as a realistic medium to model brain mechanisms. **Stein** goes further and boldly hypothesizes that networks embed higher-order mental constructs that emerge under the facilitation of attention. A general problem with neural nets is the lack of concern in most network simulations for self-sufficiency: An external "teacher" (i.e., the investigator) is required to stipulate goals and criteria for success. Robinson explicitly notes this weakness, but does not seem bothered by it. Perhaps the setting of goals is irrelevant to the actual mechanisms of implementation, but we cannot be sure of that. Until a network can supervise itself and establish its own goals, it is difficult to recognize networks as true brain analogs.

Robinson views the hidden unit as an instrument of obfuscation. **Stein** notes, however, that although networks do not provide a topographic map of a data set, they do represent the set of rules for solving a particular problem. The properties of hidden units succinctly capture the underlying structure of the task domain. For example, Rumelhart et al. (1986c) found that hidden units in a family-tree network discriminated nationality, generation, or branch of the family. Far from leading one into a welter of confusion, the feature detection of hidden units reflects the analytic strategy of information processing. If you wish to know *how* a network achieves its results, you must study the hidden units.

Furthermore, **Robinson**'s interpretation of "rogue units" suggests that the preacher is not yet totally converted. The claim that these paradoxical creations are of no value and that the system manages in spite of them sound suspiciously like a traditional linear-systems sermon on control. Can the system do as well without them? That is the pertinent question, and in nonlinear dynamical systems the answer is probably no. Such units are seen in the cerebellum, for example, both in oculomotor and arm-related areas (MacKay 1988a) and can be given a plausible rationale. Ideally, velocity and position signals should be precisely matched to one another in the final motor product. In order to achieve adequate precision using the mathematical idiosyncrasies of neurons it would appear necessary to have *corrective* elements in the system that top up one signal and whittle down another in whatever combination is essential to get the output up to snuff. To call such worthwhile do-gooders *rogues* is the supreme insult. Maybe "fudge units"?

Analysis of single unit activity has led to a much greater understanding of the mechanisms at work in specific brain regions than is acknowledged by **Robinson**. One need only turn to **Stein**'s target article for a sampling of concrete examples in posterior parietal cortex ("PPC neuronal responses," sect. 2). In addition, one can list the purely kinematic nature of area 5, "place cells" in the hippocampus, multiple representation of muscles in motor cortex, "face cells" in the inferior temporal lobe, and a myriad of visual response classes in primary and secondary visual cortices. The list is endless. All of this is critical information, gleaned by single unit recording. Have neural networks yet predicted a neuronal response type prior to its discovery by neurophysiologists? Indeed it is only after the "small army" (really a beleaguered cohort) of neurophysiologists, with their microelectrodes, combs a cortical region that we obtain a detailed functional map. Distinctions between cortical regions can often be much more sharply defined by physiological response properties than by histological criteria (e.g., the boundary between cortical areas 7a and 7b).

Much of the limitation **Robinson** attaches to single-unit recording is not an attribute of the technique but rather of the limited behavioral settings in which neurons have been studied. With more varied and behaviorally realistic experimental paradigms (e.g., freely walking cats and rats, monkeys with freely moving heads), the full potential for microelectrode characterization of neurons can be realized. Although he implicitly recognizes this, Robinson still criticizes the microelectrode's ability to uncover precise mechanisms of neuronal processing. Processing in the CNS has both spatial and temporal aspects. Single unit and local field potential recording give an incomparable measure of the temporal profile of processing. This strength is being gradually exploited to advantage as technologies for long-term unit recording in behaving animals are perfected. Thus the adaptive neuronal changes occurring with behavioral/environmental alterations or learning can be monitored. There is no other methodology to get an accurate temporal profile of brain processing.

Nevertheless, **Robinson** is right that the spatial distribution of CNS processing is the nemesis of single unit recording. Even

here, however, new techniques of multielectrode recording have the potential to reveal strategies of ensemble embedding of information. Already Gray and Singer (1989) have demonstrated the existence of synchronous oscillatory activity in functionally linked ensembles of neurons (again the temporal dimension sneaks to the forefront!). Only a restricted sample of neurons can be simultaneously monitored, but this may be enough to establish general principles of the spatial distribution of neural information processing.

Both **Stein** and **Robinson** discuss the network function of coordinate transformation, and both essentially grope in the dark. Neural networks on their own provide no substantive guidance and we see the resulting leaps of faith. The only reliable guides in this area are physiological, psychological, and neurological data. We believe that the available data favor Robinson's point of view: Global coordinate transformations are indeed made but in a highly distributed and concealed manner so that mathematical descriptions are misleading. First, the conscious perception of the spatial environment is the same regardless of the task to be performed or the body part to be moved, if any. That perception must be subserved by some kind of common neuronal substrate. Second, from an analysis of neurological case studies, Ratcliffe (1991) has noted a progression from egocentric to allocentric transformations. A deficit in the former is always accompanied by a deficit in the latter, but not vice versa. He suggested that the egocentric to allocentric change may correspond to a posterior parietal to hippocampal projection. Since the allocentric conversion seems to require egocentric input, the latter transformation must be put into effect. Furthermore, single cell recording has shown the existence of both egocentric and allocentric referencing of visual responses in the hippocampus (Feigenbaum & Rolls 1991), with allocentric representation predominating. Thus coordinate transformation is manifested at the single unit level. To extend the hippocampal findings, we have conducted a microelectrode study of area 7a neurons (posterior parietal lobe) in a monkey reaching to visual targets displayed on a video monitor; the monkey's head is free to rotate in the horizontal plane so that the gaze angle varies naturally. We found that some neurons responded to the appearance of a visual target in a specific spatial position regardless of the angle of gaze (MacKay & Riehle 1992). Whether the reference is egocentric or allocentric we have not yet determined, but it is not retinal or cranial. Hence we do not see the need at this time to relegate higher-order transformations completely to network patterns of activity. It is more likely that population codes provide *specific* discrimination by ensemble processing of neurons that already possess general discrimination capacity of the same type. Georgopoulos and coworkers (1983) have provided an excellent example in the motor cortex with regard to arm movement direction: A population vector codes direction with relative precision but the individual members of that population necessarily discriminate direction already. Similarly, many hidden units in neural integrator models individually show the integrated signal, as **Robinson** points out.

Stein argues for a direct, local coordinate conversion to action-oriented frames, obviating the need for global transformations. We take the opposing position that there are many stages of processing between sensory stimulus and response. Within area 7a of the parietal lobe we have seen immediate, spatially selective responses to visual targets for arm reach but no evidence of any transformation to an "action space." Target responses were very similar for intended reaches with either hand. Moreover, when a delay period was introduced, visual responses were tightly linked only to the initial target appearance (preparatory stimulus, PS) and not to the "go" signal (response stimulus, RS), even though both stimuli were similar and in the same place (MacKay & Riehle 1992). Only in the frontal lobe does one find responsiveness to the go signal (e.g., Riehle 1991). As shown in Table 1, there is a distinct progression

Table 1 (MacKay & Riehle). *Proportion of visual (signal-locked) responses in sample of task-related neurons*

Signal	Area 7a[a] (n = 185)	Premotor cortex[b] (n = 235)	Motor cortex[b] (n = 221)
PS	32%	4.3%	0.4%
PS & RS	0%	6.8%	0%
RS	0%	15.7%	11.3%

[a]Data from MacKay & Riehle (1992).
[b]Data from Riehle (1991).

from parietal area 7a to frontal areas 6 and 4: A responsiveness to spatial information gives way to a responsiveness to spatial *action*. Of course there are missing links because area 7a does not project directly to area 6, but further rostrally in the frontal lobe, and also to the parahippocampal gyrus (Cavada & Goldman-Rakic 1989a). Our point, however, is that the conversion from stimulus to response progresses through a number of stages in space and in time. The initial stages appear to relate new information to the body or environment and the later ones to precisely formulate appropriate action. At the very least, visual cues intended for an arm response must be initially referenced to the arm. It is not simply a case of going from arm proprioceptive space to arm motor space.

To make neural nets truly useful for elucidating brain mechanisms they must be kept closely related to physiological data. Instead of presenting them with a goal and seeing how they achieve it, the neurophysiologist would be happier to stuff his cell types into them and then determine what functions the network can accomplish. Such a "reverse engineering" approach (e.g., Bower 1990) would require a more evolved network with looser connectivity constraints and more physiological learning rules. It is achievable, however, and worth striving toward. The dream network would provide an extremely valuable interactive framework to study a set of unit data, especially to check for missing pieces. It could be a superb interpretative tool, allowing the investigator to understand the functional potential of his neuronal types and also to *predict* additional neuronal types necessary to fulfill specified aims.

The wise engineer makes acquaintance with his hidden units: They are a gold mine of strategic information. Similarly, the even more fascinating neuron will continue to be studied long after current neural networks have been terminally black-boxed and superseded by the next Perceptron generation.

ACKNOWLEDGMENTS
Supported by MRC of Canada and the Office of Naval Research.

Locomotion, oscillating dynamic systems and stiffness regulation by the basal ganglia

Guillaume Masson and Jean Pailhous
Université d'Aix-Marseille II, Faculté de Médecine, Cognition & Mouvement, URA CNRS 1166, 13388 Marseille, France
Electronic mail: *patre@frmop11.bitnet*

[GEA, EB] **Alexander et al.** point out the obvious divorce that exits between the conventional motor program conception – algorithmic and sequential – and recent anatomical-functional descriptions of the central motor structures that devise and execute the program. In our opinion, however, the authors do not explain clearly how these massively parallel architectures (i.e., the basal ganglia, the various motor cortices, and their

relations) can produce a specific, for example goal-directed, movement. On the other hand, **Bizzi et al.** give a model of such a movement that does not describe how and which central motor structures participate in the movement control. Briefly, these two target articles seem to us relatively hermetic even though several results reported here describe the same motor act. In order to help clarify this question we will concentrate on locomotion for two reasons: (1) It is a typical movement that has given rise to a divergence between neurophysiological and dynamic explanations of motor control; (2) the basal ganglia, one of the structures described in the first article, are involved in locomotion production as shown in particular by locomotor disorders in Parkinson's disease and their subsequent improvement by L-Dopa administration.

The existence of a central locomotor "program" has been demonstrated using a classical model: spinal curarized animal. Well-organized locomotor patterns, also called fictive locomotion, are observed on the spinal efferent roots. This spinal central pattern generator (CPG) sequentially organizes the movement whose rhythm increases as stimulation strength increases. [See Selverston: "Are Central Pattern Generators Understandable?" *BBS* 3(4)1980.] The CPG has been classically described as an example of a hardwired motor program (Grillner 1981). Similarly, based on behavioral data, some locomotor parameters, also called invariants, remain proportionally unchanged even when the walking speed changes (Shapiro et al. 1981). This has been considered a behavioral confirmation of the CPG hypothesis defined by the neurophysiological perspective. On the contrary, dynamic models that attempt to integrate both nervous system and biomechanical properties describe the rhythmic properties of the movement as emerging from a coupling between nonlinear oscillators, that is, oscillating neuronal networks and mass-spring or pendulum actuators (see Taga et al. 1991 for a recent model).

Beyond the discrepancies between the two different points of view it is noteworthy that both of them underline the importance of the link between tonic properties of the effector system (e.g., stiffness) and movement frequency. In particular, the cocontraction of flexor and extensor muscles that determines joint stiffness (but is often wrongly implicated as a cause of movement inefficiency) appears to be an essential parameter to be controlled for the frequency regulation of movement (Pailhous & Bonnard 1992). In dynamic models, the legs are not "forced" to move at a given frequency by a hardwired program. Instead, the leg movement is "sustained" at its characteristic frequency by the neuromotor synergy. According to the nature of the oscillating system (i.e., the way energy dissipation and escape are organized), the locomotor pattern exhibits a well-defined amplitude-frequency relationship (Kay et al. 1987). Thus, stiffness appears to be an adjustable property of the effector, influencing the characteristic frequency of the "sustained" movement. Moreover, for a given stiffness, after a perturbation the effector spontaneously returns to its normal dynamic (angle versus angular speed function) trajectory. In this case, the system behaves as a limit-cycle attractor. In comparison, **Bizzi** and co-workers describe a point-attractor in dynamic system terms. From this point of view, a change in global stiffness of the effector induces a shift in the natural frequency of the rhythmic movement. Without entering into the alpha-delta controversies, we note that from a neurophysiological perspective, reflexive control of stiffness (i.e., stretch reflex) has been shown to be very important (but not sufficient) in controlling postural-locomotor coordination (Grillner 1972); but from the dynamical perspective, stiffness regulation could also offer a useful tool for the central nervous system to control movement frequency.

From this locomotor perspective, we would like to briefly reexamine the involvement of the basal ganglia in motor control. The disturbance of the amplitude-frequency relation in Parkinson's disease suggests that stiffness regulation is disrupted in these patients and that the basal ganglia could be primarily involved in normal motor control via stiffness regulation. In Parkinson's disease, it could be possible to describe both gait initiation and gait execution disorders in terms of a loss of capacity in phasic and tonic stiffness modulation. We have shown in our laboratory that in these patients, L-Dopa intake improves amplitude modulation capacity but not frequency modulation capacity (Blin et al. 1991). Could this selective improvement be more simply explained than by a hypothetical selective action of dopamine on the CPG? We would like to propose here that the basal ganglia may be responsible for tonic or phasic stiffness effector regulation within the motor loops to which they belong. Locomotion again provides an example of how these structures may carry out this coordination. Mori (1987) suggested that the execution of normal locomotion requires the implementation of both locomotor synergy (i.e., a rhythmic organized neuromuscular activity driven by the CPG) and postural synergy (i.e., an adequate level of muscular tonus or limb stiffness). From a neurophysiological standpoint, this requires the parallel activity of three systems: a locomotor synergy generating/releasing system, a postural tonus activating/releasing system, and finally a phase control system ensuring motor coordination. In this perspective, basal ganglia may be highly involved in the regulation of postural-kinetic locomotor coordination, namely, acting upon the level of muscular tonus in the effector limb. In this way, using stiffness regulation, the basal ganglia could assume adaptive locomotor change (i.e., emergent frequency change) and the associated postural-kinetic coordination.

Human pathology has provided us with initial evidence of this. Tonus disorders are frequently encountered in basal ganglia disease. One example is provided by segmental dystonia, resulting from an abnormal level of cocontraction among the agonist and antagonist muscles; this is reproduced in animals following bilateral pallidal lesions (Denny-Brown 1962). Relations to locomotor disorders also encountered in these pathologies are poorly understood because of the difficulty of making observations. However, this could also be explained by the traditional dissociation between posture and movement. In dynamic terms, stiffness deregulation as observed in Parkinson's disease could explain amplitude-frequency link abnormalities. Second, in monkeys, the temporary or permanent chemical inactivation of the pallidum does not alter the reaction time of mono-articular movements but it does trigger an increase in movement time. This argues against the potential implication of the basal ganglia in motor initiation, as **Alexander et al.** point out. Moreover, Mink and Thach (1991b) recently showed that the occurrence of tonic and phasic cocontractions of the flexor and extensor muscles around the joint may be responsible for bradykinesia. In consequence, in healthy subjects these structures may favor the emergence or the execution of a movement by controlling directly the flexor-extensor cocontraction level (i.e., the stiffness of the effector) or by, according to Mink and Thach (1991b), turning off antagonist muscle activity. These two kinds of data briefly presented here suggest that the role of the basal ganglia in motor control should be reexamined by studying their influence on effector stiffness via the regulation of tonic properties of the neuromuscular system.

This interpretation will provide us with the basis for collecting motor behavioral data and attempting to determine not only what tonic parameters must be controlled in order to execute or adapt the movement, but also what structures are likely to interact in this regulation. Indeed, this interpretation refocuses on descending inhibitory relations linking the basal ganglia and the brainstem, which are not considered by **Alexander et al.** In the cat, descending relations between basal ganglia output nucleii and mesencephalic locomotor region have been widely documented (Garcia-Rill 1986) and could support these posturo-locomotor coordinations. Furthermore, this hypothesis alleviates the need to raise the "ill-posed" question of motor programming (how the nervous system performs the sequential or parallel coding of the kinematic and dynamic parameters of movement), while promoting the search for the various means (tonic and phasic) used by the different motor structures to favor the emergence of a well-adapted behavior. Finally, as **Bizzi et al.** state, the dynamic approach of effector stiffness regulation provides us with a better understanding of the posture-movement linkage. The hypothesis presented here (i.e., that the basal ganglia are involved in motor control by means of tonic properties regulation) could explain why these postural-kinetic coordinations are disturbed in basal ganglia disorders. Another example taken from locomotion illustrates this approach. Freezing is a gait disorder often, although not always, found in basal ganglia pathologies. It is characterized by the blocking of the locomotor pattern preceded by an abrupt increase in stride frequency and axial stiffening. Andrews (1973) showed that freezing is accompanied by an exaggerated co-activation of flexor and extensor muscles, that is, a sharp increase in stiffness. Thus, an apparent disorganization of the locomotor-pattern should not be considered a disorganization of the motor program but a dramatic deregulation of the stiffness of the effector limbs.

The speculative hypothesis presented here allows us to describe a nonspecific level (stiffness) as a common denominator of motor disorders described classically as initiation or execution abnormalities of the motor program. In order to test this hypothesis, it will now be necessary to reconsider gait disorders using useful dynamic model tools (phase plane, dynamic analysis of movement, etc.). In this way, it could be possible to redefine central motor programs by describing the means (in specific tonic properties) by which the central nervous system can facilitate the emergence of a well-adapted movement. In our opinion, this approach is in accordance with mass-spring models (such as those illustrated by **Bizzi et al.**) that demonstrate what musculoskeletal means are available to the motor system.

ACKNOWLEDGMENT
Preparation of this commentary was supported by the Centre National de la Recherche Scientifique and by a grant from the French Ministère de la Recherche & de la Technologie.

Global organizations: Movement and spinal

Gin McCollum

R. S. Dow Neurological Sciences Institute, Good Samaritan Hospital and Medical Center, Portland, OR 97209

[DAM] As an aid to making sense of the spinal cord – as **McCrea** claims we can – I would like to suggest formats that allow spinal interneuron organization and interaction to be depicted as movement organizations. If each class of interneurons, by its connectivity, organizes all possible body movements into global patterns, then the patterns could be used in combination to specify particular movements, as hand movements may be shaped by brain structures (Schieber 1990), and as sensory cells can combine to specify a stimulus within an equivalence class (Robertson & McCollum 1991). McCrea describes a large body of experimental results that are already understood according to powerful structural concepts. In order to formalize the structure further mathematically, a global format is required, mapping spinal organization onto a position or movement space such as the restricted hindlimb extension-flexion space shown in Figure 1A. Because the data are complex and there are many possible mathematical formulations, a collaborative interdisciplinary effort may also be required.

If overall flexor reflex afferent (FRA) activity levels correspond to extended and flexed positions of the two hindlimbs, then we should be able to map their formal representations onto each other (Figure 1). I am starting with the FRA because it seems to generate the simplest global pattern of movement

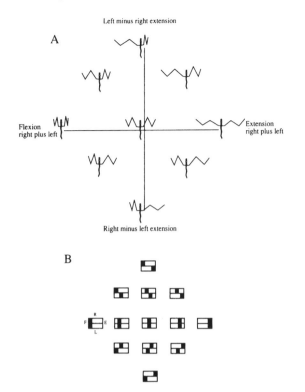

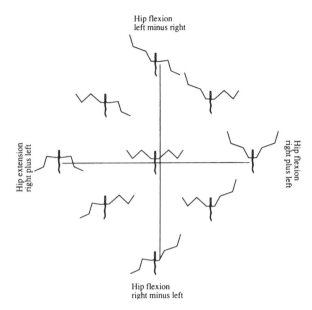

Figure 2 (McCollum). Possible interaction of IaIn with FRA, represented as a transformation of hindlimb extension-flexion space. Instead of each hindlimb changing overall extension-flexion over the space, only the hip extension-flexion changes.

Figure 1 (McCollum). Overall extensions and flexions of both cat hindlimbs. A: Movements shown from above, splayed out. Along the horizontal axis is extension and flexion of both hindlimbs. Along the vertical axis is right-left asymmetry of extension and flexion, with left extension and right flexion at the top and right extension and left flexion at the bottom. B: Hypothetical overall FRA activity levels. Each box has an upper compartment for the right hindlimb (R) and a lower compartment for the left hindlimb (L). A black area shows the level of overall extension or flexion for each hindlimb, with flexion (F) to the left and extension (E) to the right. For example, the box marked with letters (F,E,R,L) has both black areas to the left, indicating both hindlimbs maximally flexed. The black area can be moved continuously in each hindlimb's compartment; only a discrete sample of extensions and flexions are shown. The 2-dimensional array shows the full range of combinations of extension-flexion for the two hindlimbs.

explicable in terms of an interneuron organization. A similar format can be adapted to other organizations and sets of limbs.

The IaIn organization may divide muscles into agonists and antagonists. Synergies can be constructed as combinations of nonantagonist muscles from different segments and these can later be refined (McCollum 1992). Whole-limb extension-flexion may be thought of as one synergy pair, analogous to the hip extension-flexion used by Macpherson's cats in responding to forward and backward perturbations (Macpherson 1988). Perhaps different activity distributions among the IaIns fractionate the FRA, transforming the set of body positions shown in Figure 1A to those in Figure 2, so that any stimulus would be mapped onto a position or vector in the transformed space rather than the original one.

A motor transformation easily depicted as a possible transformation of FRA activity levels is a change in the static right-left slope (Figure 3A). Through either descending or afferent mechanisms the FRA activity levels may shift, favoring extension on the left and flexion on the right (Figure 3B). The motor transformation can be thought of as eliminating positions from the bottom of Figure 1A, steeper slopes eliminating more positions.

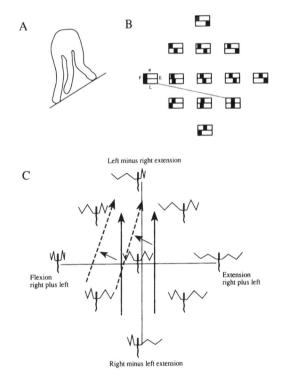

Figure 3 (McCollum). Walking on a slope. A: Cat's posture shown from behind. B: FRA activity levels, each box diagrammed as in Figure 1B. The left hindlimb retains its full range of extension-flexion whereas the right hindlimb only extends half way. In Figure 1B, boxes with equal extension-flexion in the two hindlimbs cross horizontally across the middle of the array; here, the line of equal extension-flexion is lower and sloped, indicated by the dotted line. C: Shifting trajectories. Shifts in relative extension-flexion between the two hindlimbs are represented by long arrows, trajectories in the body position space. Solid, long arrows represent a movement on a horizontal surface. When the surface is sloped as in A, the trajectories shift to those indicated by the dashed, long arrows.

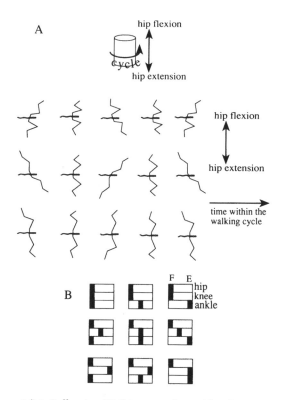

Once global organizations of spinal interneurons are represented clearly as organizations or transformations of motor spaces, more complex interactions of afferents, reflexes, and descending influences can also be represented. Motor spaces avoid single-parameter coding. Both of **McCrea**'s mechanisms can be represented: spinal circuits operating on incoming activity and incoming activity operating on spinal reflexes. Besides the spatial transformations depicted in this commentary, temporal transformations can also be formalized between long and short durations, and between sustained and rhythmic activity. Animals need an explored region of continuously varying parameters, including spatial, temporal, and transformation parameters in which they can modify movements with confidence in their predictions, not just a few exact movements. By mapping spinal interactions onto movement spaces, we may be able to increase our understanding of the way animals provide completeness in their movement modifications.

ACKNOWLEDGMENTS
Thanks to William J. Roberts for reactions and help in writing. The work was supported by NIH grant NS23209.

Equilibrium point and self-organization

Pietro Morasso[a] and Vittorio Sanguineti[b]
Department of Communication, Computer and Systems Sciences, University of Genoa, I-16145 Genoa, Italy
Electronic mail: [a]*morasso@dist.unige.it* and [b]*sangui@dist.unige.it*

[EB] In the concluding section of their target article, **Bizzi et al.** concisely state that "the essence of the equilibrium-point hypothesis is that centrally planned motor intentions are expressed and transmitted to the periphery using the virtual trajectory" but they also warn us that this is a theory which is hard to either prove or disprove because "where should one look for a neural expression of the virtual trajectory?" In fact, the main motivation of the EP (equilibrium point) theory is one of computational simplicity, that is, the theory suggests an organizational principle for capturing the complexity of motor planning and control, but "there remains the formidable problem of how to select an appropriate pattern of neural activity to produce a desired virtual trajectory."

The EP theory self-limits to the "third step" of a posited computational sensorimotor chain (introductory section of **Bizzi et al.**'s target article), where the first two steps are, respectively, the spatial localization of targets and the spatial planning of movement direction, in extrinsic coordinates. The alternative approach, which is outlined in this commentary, is that the stepwise view of the sensorimotor transformations should be abandoned and the three posited steps should be merged into a global, distributed computational architecture (Morasso & Sanguineti 1991; 1992) where EP concepts play a role in a more general sense than the purely muscular one, in the context of a neuronal self-organization strategy.

In particular, we think it is useful to revisit the old neurological concept of body schema (BS) (Head & Holmes 1912) and assume the existence of some kind of *internal model* of the body that represents in a distributed way biomechanical and muscular constraints. Such a model, consistent with the EP theory, would express the well-posed mapping of motor commands into virtual trajectories. Moreover, BS can emerge as a result of a self-organizing process such as the Piagetian *circular reaction* (Piaget 1963) that is based on pseudorandom exploratory movements. Although circular reaction has recently received renewed attention and has inspired a number of neural models (Kuperstein 1991; Ritter et al. 1989) aimed at the so-called direct inverse learning (Jordan & Rumelhart 1991), we think that the correct perspective is to associate circular reaction with the BS concept, yielding a forward (motor-sensory) learning problem instead of an inverse (sensorimotor) one.

Figure 4 (McCollum). Walking transformed by changing the range of hip extension-flexion. A: One cycle of a walk of varying hip extension-flexion range. At the top is a cylinder, showing the cycle repeating around the cylinder, while hip extension-flexion range varies up and down. Below, the cylinder is cut and a sample of positions is diagrammed. The middle row is a normal, forward walk. The top row is a walk with the hip restricted to a flexed range, resulting in backward progression. The bottom row is a walk with hip restricted to an extended range, added hypothetically for the completeness of the diagram. B: One-hindlimb FRA activity levels with hip restricted to flexion. Each box has three compartments, hip on top, knee in the middle, and ankle on the bottom. Flexion (F) is indicated by a black area to the left, and extension (E) is indicated by a black area to the right. For example, the labeled box indicates maximal flexion of both hip and knee and maximal extension of the ankle. Knee and ankle extension-flexion vary over the full extension-flexion range over the array of boxes.

Alternatively, the motor transformation can be thought of as mapping vectors onto a different part of the position space (Figure 3C).

Spinal interactions can be mapped onto motor space to codify the effect of combinations of spinal cord, afferent, and descending effects. Cats transform the same set of hindlimb synergies from forward to backward walking by changing posture, a descending influence that changes afferent feedback and restricts hip extension (Buford & Smith 1990; Buford et al. 1990). The transformation in walking can be represented along a continuous axis of hip extension-flexion range, with the movement along a time axis (Figure 4A). Because the hip movement is greatly reduced, the FRA activity may be effectively fractionated within the limb, with the hip FRA using a smaller part of its range (Figure 4B).

The case of the backward-walking cats apparently involves a transformation that is continuous in the hip extension-flexion parameter, even though the result of the movement switches in direction. Is the switch in direction represented in spinal organization or only in motor outcome?

Our computational BS model is a *neural field* whose formal neurons store composite motor-sensory vectors (with articular, extrinsic, and muscular components) that we call *body-icons*. It self-organizes during exploratory movements by means of unsupervised Hebbian learning, similarly to Kohonen's networks (Kohonen 1982), independently of the degree of redundancy. After training, the nodes in the neural field store a set of motor-sensory *prototypes* that tessellate the motor-sensory spaces in an optimal way. It is worth noting that although we may think of the neural field as a 2D arrangement of computational modules, the lateral patterns of connectivity that may evolve during learning can actually implement a multidimensional feature space.

A salient feature of the model is that *population coding*, reminiscent of recent findings in different cortical areas (Georgopoulos et al. 1982), also emerges from self-organization, so that the neural field as a whole defines a continuous mapping between motor commands and the virtual trajectory space. In this way, smooth variations of the population code correspond to smooth variations of the body-icon. In other words, the BS can operate as a universal planner by transforming smooth arbitrary *navigations* in the neural field into coherent sensorimotor flows. The problem is that task specifications are usually not sufficiently constrained; but a natural solution, in the vein of the equilibrium-point theory, is to let the navigation be a relaxation driven by a potential function defined over the neural field.

A key element of our theory is that the potential function does not have to be limited to the elastic muscular potential but can include any combination of task-related constraints in extrinsic as well as intrinsic coordinates. For example, in planning the transport of a glass of water, an extrinsic task component refers to the distance from the target location and an intrinsic component refers to to the horizontality of the last link. In any case, the potential function can be represented in the neural field by assigning to each neuron an additional activity level that measures, for the corresponding body-icon, its composite *credit*. What is important is that the credit can be assigned at run-time and can be represented with the same population code used for the body-icons.

Summing up, the power and universality of the BS model is based on the possibility of differentiating the satisfaction of two types of constraints (the *biomechanical constraints*, implicitly satisfied by the intrinsic structure of the neural field, and the *task-dependent constraints*, taken into account by the relaxation dynamics) while integrating them in the same neural field. A likely brain localization of such a field is the posterior parietal cortex, particularly regarding association area 5, which has access to intrinsic and extrinsic sensory information and where neurons have been found that are activated in anticipation of intended movements (Crammond & Kalaska 1989) and are insensitive to load variation (Kalaska et al. 1990). [See Kalaska & Crammond's commentary, this issue.]

The BS represented in the neural field has the potential to initiate actual movements by supplying the cerebral motor cortex and the cerebellar cortex with the basic synergies; it can also support force control in constrained movements or mental simulations of planned movements (Decety & Ingvar 1990). In this way, we believe that the EP theory is put in the correct perspective, preserving its main point (the unification of movement and posture) and enriching it with an integrated representation of the motor spaces that blurs the debate over extrinsic versus intrinsic coordinates. Moreover, the theory can suggest a few targets in the investigation of associative areas of the cortex and other regions of the brain involved in motor planning: for example, cues to the existence of body-icons (columns of neurons that code the different coordinates of the same posture), to the connectivity of cortical columns from the point of view of the Kohonenian topology-preserving mappings, to relaxation dynamics in cortical areas, to amplification phenomena or recruitment of new areas contingent on heavy learning of new

tasks, and so on. A combination of single-cell recording with global imaging of brain activity could be used for this purpose.

Adaptive model theory

Peter D. Neilson and Megan D. Neilson
Cerebral Palsy Research Unit, Institute of Neurological Sciences, The Prince Henry Hospital and School of Electrical Engineering, University of New South Wales, Sydney 2036, Australia
Electronic mail: *petern@syscon.sc.ee.unsw.oz.au; megann@syscon.sc.ee.unsw.oz.au*

[JRB] **Bloedel's** target article provides a scholarly review of literature leading to the presentation of a new hypothesis of cerebellar function. It is proposed that the cerebellum integrates information characterizing external target space with several types of data, many of which are probably represented in different coordinate systems. The cerebellum integrates these data with an internal representation of target space and inputs reflecting activity in other parts of the motor system, such as vestibular space, proprioceptive space, and joint space. It provides motor pathways responsible for movement execution with updated information about the relation between features of execution space, internal intention space, body scheme, and inputs characterizing the activity in central pathways. The hypothesis focuses on the critical importance of transformations, such as those described by the tensor transform theory of Pellionisz and Llinas (1979; 1980; 1982; 1985), between internal and external target space, and the overall integration in the cerebellum. It is emphasized that the integration performed by the cerebellum is probably called upon in a task-dependent manner.

At the Cerebral Palsy Research Unit we are excited by the research described by **Bloedel** and colleagues. It relates closely to work at our laboratory. Whereas Bloedel has approached the question of cerebellar function from a physiological point of view, we have approached it from a systems and control theoretic perspective. We have incorporated physiological findings and hypotheses, such as the gain change hypothesis described by Bloedel, into a computational theory about information processing performed by the human central nervous system (CNS) in the control of movement. We refer to this as "adaptive model theory" (AMT) (Neilson et al. 1992a). Using theories from the fields of digital signal processing, adaptive filters, and adaptive systems (see Astrom & Wittenmark 1989; Box & Jenkins 1976; Haykin 1986; Proakis & Manolakis 1989; Widrow & Stearns 1985), we ask the question: What information processing must be performed within the CNS, consistent with known neuroanatomy and neurophysiology, to account for observed motor behavior?

The role of the cerebellum proposed in the AMT is similar, in many respects, to that outlined by **Bloedel**. Whereas Bloedel presents an overall qualitative description of cerebellar function based on physiological data, however, we have presented similar ideas in a quantitative systems theoretic form. For example, Bloedel argues that the same basic operation is performed in each cerebellar region even though the type of information integrated and processed in each region may be different. We argue likewise but go further in postulating that the basic operation consists of nonlinear convolution of neural signals through adaptive neural filter circuits. The advantage of a computational approach is that the theory can be simulated. It can be shown that the physiological characteristics and neuronal interactions described can be characterized in the framework of a computational operation that is physically realizable and functions in the manner hypothesized.

We have incorporated the neural circuitry of the sagittal zones of the cerebellum described by **Bloedel** in the AMT. This includes the terminations of climbing fiber inputs from the contralateral inferior olive, the mossy fiber projections from pontine neurons to the granular layer, the parallel fiber connectivity across multiple sagittal zones, and the corticonuclear projections returning information from the cerebellum to the cerebral cortex via thalamic nuclei. Adaptive neural filter circuits are pictured in the AMT as lines of Purkinje cells in the cerebellar cortex lying along sagittal zones at right angles to the folia. The sensitivities of these Purkinje cells are varied by the low frequency activity of cells in the inferior olive via climbing fiber inputs. The axons of Purkinje cells along each sagittal zone converge onto common target cells in the cerebellar nucleus (CN). Thus, the CN output of each neural filter consists of a weighted sum of the mossy fiber-granule cell-parallel fiber inputs to the Purkinje cells along the sagittal zone. This input varies from region to region in the cerebellum, but in all regions it consists of a convergence of inputs, consistent with the fractured mosaic and mossy fiber patches described by Bloedel. These inputs are weighted by the adaptive gains of the Purkinje cells before being summated by the CN target neurons and communicated to other parts of the CNS. Thus, the sagittal zone circuits of the cerebellum are seen as having a structure similar to that of the linear combiner or transversal filter described in adaptive filter theory (see Haykin 1986; Widrow & Stearns 1985).

Bloedel emphasizes the critical importance of transformations between internal and external reference frames in the function performed by the cerebellum, but it does not elaborate these specifically. In the AMT, we propose that the CNS monitors outgoing motor commands (m), muscle tensions (t), body movements (θ), and sensory consequences (s). It computes, stores in memory, and adaptively maintains the accuracy of sets of parameters (regression coefficients or Wiener kernels) describing the multivariable, nonlinear, dynamic relations between them. A necessary part of this modeling process is the formation of orthogonal coordinate systems (M, T, Θ, S) for the spaces spanned by m, t, θ, and s. These reference frames, and the relations between them, are formed automatically by the self-tuning behavior of the adaptive neural filters described above.

During response execution, the adaptive neural filters are used to transform intended movements preplanned in a sensory reference frame (S^*) into appropriate Θ^*, T^*, and M to drive the muscles and to generate reafference (S) equal to S^*. Appropriate parameters are retrieved from memory and used to modulate (via inferior olive and climbing fibers) the activity of Purkinje cells in the cerebellum. This is equivalent to controlling the weights of the adaptive neural filters, which then function as internal models of the inverse dynamics of muscle control systems, biomechanics, and external systems being controlled.

In testing the explanatory power of AMT, we are currently exploring computer simulations of several key aspects of motor behavior. Using adaptive filter circuits we have simulated the behavior of subjects performing compensatory and pursuit tracking. We have shown that the performance of subjects in adapting their motor behavior to compensate for the dynamic response characteristics of the tracking system (learning the "feel" of the control) and in compensating for reaction time delay by predicting the future position of the target is reproduced by the simulations (Neilson et al. 1988a; 1988c; 1992b). We have also run computer simulations that account for the existence of task-dependent, fast feedforward reflex responses observed in finger, thumb, and forearm muscles during arm movement tasks and in lip, tongue, and jaw muscles during speech (Neilson et al. 1988b). In a recent discussion (Neilson 1991), we address the problem of redundancy in movement control and show that the adaptive neural filters of AMT effectively reduce the number of degrees of freedom of movement through the adaptive tuning of orthogonalizing networks and movement synergy generators.

A systems theoretic approach to the understanding of motor behavior is compelling, but its ultimate value depends on its physical realizability in terms of established biological circuitry. This is currently a vexed issue in the related broad field of neural networks/connection science. In developing AMT we strive to work with biologically plausible circuits. We find the contribution of **Bloedel** and colleagues of extraordinary value in this regard and see our approaches as complementary. Their starting point is the physiology, ours, the computational function (cf. Marr & Poggio 1977). It is gratifying that we seem to be approaching a common middle ground.

Stiffness regulation revisited

T. Richard Nichols

Department of Physiology, Emory University, Atlanta, GA 30322
Electronic mail: *phystrn@emuvm1.cc.emory.edu*

[EB, DAM] The target articles by **McCrea** and by **Bizzi et al.** address the actions of spinal circuitry to coordinate the musculoskeletal system. The perspective of McCrea is based almost entirely on studies of spinal circuitry, with little reference to biomechanics, whereas the arguments of Bizzi et al. are based largely on biomechanical considerations. One might hope that these two perspectives will converge to provide joint insights into the spinal mechanisms of coordination, but the linkage between the two points of view is not apparent. One reason for the difficulty in finding a common ground in the two accounts is that neither target article presents clearcut hypotheses about the role of proprioceptive feedback in coordination at the spinal level.

McCrea argues strongly against hypotheses that did, in fact, link spinal circuitry and biomechanics such as the follow-up-length-servo (Merton 1953) and the stiffness regulation hypotheses (Houk 1979). The thrust of his argument is that these hypotheses depend on the existence of private feedback associated with individual muscles and that the extensive convergence of pathways from different receptors and muscles onto a given motor neuron pool would rule out private feedback.

This argument against private feedback is essentially an anatomical one and is based upon connectivity patterns whose existence has been established by electrophysiological methods. These patterns set constraints on the extent of communication among spinal neurons, but they do not necessarily imply a computational scheme for spinal circuitry beyond postsynaptic, linear summation. It is possible to have identifiable channels of information that return to a given muscle even in the presence of convergence from other muscles. **McCrea** in fact quotes a hypothesis in which input from muscle spindles onto interneurons gates or modulates the transmission of information from Golgi tendon organs (Lundberg & Malmgren 1988). If the tendon organ pathway returned to the muscle of origin then this pathway would control, in some sense, the mechanical properties of the muscle, as long as the pathway was gated by the convergent input. Another example is the monosynaptic pathway from Ia fibers, which has limited divergence and constitutes a private feedback pathway to muscles and close synergists (McCrea 1986).

McCrea does make it very clear that spinal processing goes beyond the control of single muscles, and therefore that Sherrington's notion of composite stretch reflex (Liddell & Sherrington 1924) does not fully reflect the coordinative role of the spinal cord. However, the mere fact of extensive convergence does not carry us much closer to hypotheses about specific coordinative mechanisms in the spinal cord unless the pathways in question can be related to the detailed mechanical actions of the muscles involved. One example of a convergent circuit in which functional arguments can be made is reciprocal inhibition (cf. Lloyd 1946). The convergence of autogenic excitation and inhibition from direct antagonists serves to control joint stiffness in a highly coordinated fashion (Nichols & Koffler-Smulevitz 1991) as well as to support the reciprocal activation of antagonists (Hultborn et al. 1976).

More complex interactions at the interneuronal level such as the convergence of muscle spindle and tendon organ afferents seem to be less amenable to such a straightforward interpretation. In the first place, the conditions under which these interneurons are active in the active spinal cord are unknown. Second, it is not clear which muscles or muscle compartments are most strongly linked by these pathways. Finally, the nature of the computations performed by these interneurons is not known. Clarification of the functions of these pathways will require more information about these biomechanical and computational issues and will not come from electrophysiological mapping studies alone.

Toward the beginning of their target article **Bizzi et al.** argue against a requirement for proprioceptive feedback in establishing equilibrium positions of joints and in favor of the participation of interneuronal circuits in the spinal cord. The understanding of the role of spinal circuitry in mediating the global equilibrium behavior of the motor system depends on the clarification of the degree of participation of neural feedback because of the intimate association between sensory feedback and spinal circuits (McCrea 1986).

The arguments of **Bizzi et al.** concerning the importance of feedback are based on assessments of reflex gain, the importance of muscle properties in determining springlike behavior and observations on deafferented animals. The argument that the stretch reflex is too weak to account for significant control over the mechanical properties of muscle is flawed, because the cited evidence (Vallbo 1973) fails to take into account the complementary nonlinear properties of muscle spindle receptors and motoneurons. The vigorous, transient discharges of primary endings upon initiation of stretch can cause prolonged increases in the firing of motoneurons because of the bistable firing properties of these motoneurons (Hounsgaard et al. 1986). The strength of autogenic reflexes is therefore underestimated in considering the static sensitivity of spindles alone (Vallbo 1973). Furthermore, a number of studies in intact human subjects have shown substantial contributions of autogenic reflexes to the mechanical responses of the limbs (Allum & Mauritz 1984; Carter et al. 1990; Sinkjaer et al. 1988).

Bizzi et al. also cite a seminal paper by Grillner (1972) in which it is argued that the intrinsic stiffness of muscle is sufficient to account for postural stabilization. Grillner's calculations were based on measurements from fully activated muscles under quasistatic conditions, however. These calculations probably overestimate the stiffness presented by partially activated muscles under dynamic conditions (Nichols 1973). In addition, considering normal modes of recruitment and firing rate modulation, stiffness of nonreflexive muscle is a function of operating conditions. It is now believed that stiffness can change independently of operating force and length (Nichols 1987; cf. Capaday & Stein 1986), and this adaptive property requires reflex action.

In a later section **Bizzi et al.** reintroduce proprioceptive feedback as an important element in the motor control of rapid movements when they refer to the studies of McIntyre (1988). These arguments make their prior arguments about the lack of participation of neural feedback in slower movements less compelling, especially in view of the known distribution of proprioceptive feedback onto motor units, which are primarily involved in slower movements (Fleshman et al. 1981). It would be more parsimonious to consider feedback an integral element of motor control for all speeds of movement.

In an attempt to formulate a specific hypothesis that incorporates the control of dynamic properties of muscle as well as the computations necessary for intermuscular coordination, I reevaluated the results obtained by James Houk and myself (Nichols 1973; Nichols & Houk 1976), which showed that autogenic pathways (1) compensate for muscle yield and (2)

govern the dynamic and static force-length relationships for the muscle. These results can be synthesized with more recent information as follows:

Autogenic reflex circuits provide local compensation of muscle yielding. Tonically firing, type S motor units initially present a high stiffness to stretch (Malamud & Nichols 1992; Petit et al. 1990). If the stretch exceeds a critical amount, the active fibers yield, and that protects them from damaging levels of stress. Autogenic reflex action, by recruiting additional motor units, distributes the load over additional active motor units. Hence the muscle as a whole exhibits springlike behavior even though individual motor units do not. Autogenic reflex circuits also set the stiffness of the muscle independently of the operating conditions. More recently it has been found that the stiffness of reflexive muscle can be modulated by supraspinal signals (see Nichols 1987).

The study of Nichols and Houk (1976) did not address the function of intermuscular (heterogenic) feedback. It is now clear, however, that muscular stiffness is further modified by heterogenic reflexes. These pathways form a distributed network in the spinal cord that links muscles acting at the same joint and distant ones (Nichols 1989). The cross-joint reflexes are mainly inhibitory and dependent on contractile force in the muscle of origin (Nichols & Bonasera 1990). Both the force-dependence and the wide distribution of effects suggest that these reflexes promote interjoint coordination and are mediated, at least in part, by pathways originating in tendon organs (Nichols 1989). Additional evidence that heterogenic pathways perform this coordinating role comes from the observation that muscles for which activity patterns are highly synchronized and stereotyped are tightly linked by reflex interactions, whereas muscles with more facultative activity patterns have relatively weak reflex interactions with other muscles of similar mechanical actions (Nichols & Bonasera 1990).

The organization of these heterogenic pathways is consistent with the patterns of convergence which have been found through electrophysiological studies. However, we still have the task of showing how private feedback control of individual muscles can come about in the face of this convergence. As suggested by Houk et al. (1981), the compensatory actions of the stretch reflex can be accounted for mainly by actions of the muscle spindle primary ending. It turns out that of all muscle receptors this receptor may have the most limited divergence (McCrea 1986) and is therefore more likely to constitute a private system. Even in the presence of an additive inhibitory input, the stretch reflex can still effectively compensate for muscle yield (Nichols & Koffler-Smulevitz 1991).

Aspects of the equilibrium-point hypothesis (λ model) for multijoint movements

D. J. Ostry[a] and J. R. Flanagan[b]

[a]*Department of Psychology, McGill University, Montreal, Canada H3A 1B1* and [b]*M.R.C. Applied Psychology Unit and Department of Anatomy, University of Cambridge, Cambridge, England CB2 2EF*

[EB] In this commentary, we consider versions of the λ-model for multijoint arm movement (Feldman et al. 1990; Flanagan et al. 1992) and jaw movement (Flanagan et al. 1990; Ostry et al. 1992) and emphasize differences between the α and λ approach. We begin by pointing out an apparent misunderstanding by **Bizzi** and his coauthors.

According to the λ model, movements arise as a consequence of shifts in the equilibrium state of the system that result from the dynamic interaction of central control signals, segmental mechanisms, muscle properties, and loads. The process is under the control of central commands that establish muscle threshold lengths (λs) for motoneuron recruitment.

Single—Muscle & Load

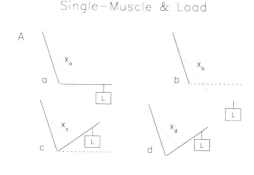

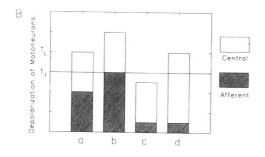

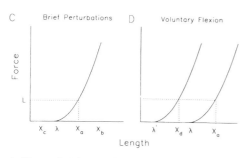

Figure 1 (Ostry & Flanagan). Aspects of the λ-model for a single muscle with load. Limb postures (A), corresponding levels of motoneuron depolarization (B), and muscle force-length curves (C,D) are shown. According to the λ-model, motions arise as a result of λ shifts, which are associated with a direct change to motoneuron depolarization and changes that arise because of segmental reflex mechanisms (from Flanagan et al., 1992, with permission).

Aspects of the model at the level of a single muscle and joint are shown in Figure 1. The top and middle panels show limb configurations and corresponding levels of motoneuron depolarization. The horizontal line designates the level at which the first motoneuron is recruited. The level of depolarization is assumed to be produced by a combination of central and afferent facilitation. In *a*, the load is in equilibrium at muscle length X_a The level of depolarization exceeds threshold and the load is supported by some combination of central and afferent activity. Perturbations in *b* and *c* load or unload the system and lead to increases or decreases in the level of afferent facilitation. The bottom panel shows the effect of perturbations in terms of the muscle's force-length curve. The parameter λ is the muscle length at which the first motoneuron is recruited and active force begins to develop. At muscle length $X_a > λ$ muscle force L balances the load. The force is increased at $X_b > X_a$ and is zero at $X_c < λ$.

According to the model, voluntary movement arises when central facilitation is increased so that the level of motoneuron depolarization increases and more motoneurons are recruited. As the muscle shortens, the length-dependent afferent facilitation decreases until a new equilibrium is established. The lower panel shows this change in central facilitation in terms of muscle force-length curves. The effect of increasing central facilitation is to reduce the recruitment threshold length from λ to λ′. Thus, motions caused by shifts in the centrally controlled parameter λ are associated both with direct changes to motoneuron excitability and changes to excitability based on segmental reflex mechanisms. In the λ-model only part of the control signal can be mediated by factors such as γ-motoneurons and muscle spindle afferents. The parallel proposed between the λ-model and the Merton model thus indicates a misunderstanding of the nature of control in the λ-model.

We have recently reported versions of the λ-model for multijoint arm movements and jaw movement (Feldman et al. 1990; Flanagan et al. 1990; Flanagan et al. 1992; Ostry et al. 1992). As in the one joint λ-model (Feldman 1986), central commands control the equilibrium point (EP) of the multijoint system by controlling the λs of many muscles in concert. Central commands are proposed that control motion in different degrees of freedom as well as the level of coactivation without motion. In the jaw system we posit three basic central commands that control specific motor functions via various combinations of λs. One command is associated with the level of coactivation of all muscles and the other two commands are associated with jaw rotation and jaw translation.

One difference between these multijoint λ-models and the multijoint models described by **Bizzi** and colleagues is that in the λ-model force development is associated with the invariant characteristic and is thus closely tied to the basic physiological mechanisms of force generation. According to the model, force and EMG are developed in proportion to the difference between the actual and equilibrium position and the rate of change in the distance between them. In contrast, in the models described by Bizzi and colleagues forces are generated on the basis of empirically measured hand stiffness fields. There is no clear physiological mechanism for force generation in the Bizzi account.

The α- and λ-models take different approaches to modeling the form of central control signals. Whereas **Bizzi** and colleagues have assumed that the velocity of equilibrium shifts is bell-shaped in form, we have preserved in the multijoint arm and jaw models the notion that constant velocity equilibrium shifts underlie simple voluntary movements (Adamovich & Feldman 1984). In the λ-model, amplitude is not explicitly specified in the central command that initiates movement. In formulations by Bizzi and his colleagues amplitude is fully specified by assuming a bell-shaped virtual trajectory. We have recently suggested that trajectory modifications in reaching to displaced targets are achieved by simply changing the direction of the constant velocity equilibrium shift towards the new target (Flanagan et al. 1992). It is not clear how the bell-shaped equilibrium trajectories proposed by Bizzi et al. would be modified in response to target displacement.

As **Bizzi et al.** point out, one major advantage of the EP hypothesis is that it provides a uniform account of both free or unrestrained motions and compliant or restricted motions. Our jaw model for mastication and speech addresses these problems in the context of the λ-model. The jaw model raises some additional questions concerning the control of motions in multiple muscle systems. The jaw model includes opener, closer, and protruder muscles, each of which contributes to jaw translation and jaw rotation. The basic problem, and one that must also be resolved for models of arm movements, is that central commands must be coordinated to produce motions in separate degrees of freedom. That is, because muscles do not – in general – act in individual degrees of freedom all muscles must be controlled together to produce jaw rotation alone, jaw translation alone, muscle cocontraction without motion, or any combination of these actions. We believe this to be a basic property of all motor function. Examples of solutions to these problems of coordination are found in our papers describing the arm and jaw models (Feldman et al. 1990; Flanagan et al. 1990).

Between perception and reflex: A role for contextual kinaesthetic information

Jacques Paillard

CNRS-LNF 2, 31 Chemin Joseph Aiguier, 13 402 Marseille, Cedex 9, France
Electronic mail: *pailla@frmopll.bitnet*

[SCG] The contemporary study of motor control is characterized by attempts to tackle the perceptual and motor complexity of ecological situations. There are also provisions for a broader and more coherent framework for neurophysiological investigations than the molecular view that has long dominated the reflexologic approach. For this purpose, interfacing "bottom-up" and "top-down" approaches is certainly desirable, although it is still scarcely attempted.

For this reason, I consider **Gandevia & Burke**'s target article a notable step forward. They emphasize the need for "bottom-up scientists" to look at "natural movements." More important, they propose a distinction between a "perceptual" and a "reflex" level for processing kinaesthetic information. In so doing, they might be opening a Pandora's box consisting of a number of intimidating psychological questions that have been, until recently, the exclusive concern of top-down approaches to motor control. I am nevertheless sympathetic with their position and consider the venture worth exploring and highly challenging for future research on kinaesthesia.

My primary concern is that in putting too much emphasis on the role of specific perceptual effects as opposed to reflex effects (involving mainly the spinal circuitry) **Gandevia & Burke** might downplay the important role played by proprioceptive information as a contextual determinant in the modulation of motor commands. The notion of *context* has become crucial to any description of higher-order control systems and of the way in which they are incorporated into neural processing (Paillard 1988). Diverse sources of afferent and reafferent activity intervene at several levels of the nervous system to trigger, tune, guide, and assist *hardwired* local mechanisms. Afferent input may be processed differently depending on the motor command context. It is almost trivial to recall that every "natural" goal-directed act appears to be shaped within a changing context of postural constraints and according to varied environmental requirements. Proprioceptive information is primarily concerned here and intervenes massively in a mode that can hardly be described as resulting from either simple reflex or perceptual effects. Kinaesthetic afferents are widely distributed throughout the whole CNS, pervading large structures such as the cerebellum and cortical associative areas. Proprioceptive information contributes to laying down the engrams of motor programs and motor habits and to elaborating the body frame and the space coordinate systems without which "natural movements" could hardly be generated and controlled. This issue is now being tackled on a sound neurophysiological basis (Paillard 1991).

A specific point I would like to make regards a contextual dimension of "natural movement," once widely used by clinical neurologists: muscle tone. This has been ignored for a long time by neurophysiologists. The work of Ribot et al. (1986) deserves special attention in this connection. Cautious neurographic investigation led Ribot et al. to identify the activity of 12 efferent fibers from the lateral peroneal nerve in man. Their classification as alpha or gamma fibers was based on different criteria such as the presence or absence of associated EMG activity and

spontaneous discharges, as well as their discharge frequency range. The most striking dissociation between the two groups of fibers, however, concerns the exquisite reactivity of the gamma fibers during "reinforcement manoeuvres" (like clenching the fist, twisting the pinna, or performing mental calculation) compared to the rather complete imperviousness of other fibers to these manoeuvres. Confirming earlier results by Hagbarth et al. (1975), Ribot et al. showed that, paradoxically, this potent gamma drive had no effect on the firing rate of all identified Ia fibers recorded during the manoeuvres. They suggested that the gamma fibers concerned with this reinforcement effect were of a dynamic type. These data fit remarkably well with former studies of mine on the human spinal reflex (Paillard 1955; 1959) in which I introduced the concept of "fusorial tone" to designate the tension of intrafusal fibers that was not directly detectable in the EMG and contributed to tuning the sensitivity of the spindle receptor to stretch. The extrafusal muscular tone would then secondarily derive from subsequent reflex activation of alpha motoneurones. The original idea, however, is that a change in the stiffness of intrafusal fibers could result from the gamma dynamic drive without having an immediate effect on the output of spindles unless the muscle was stretched by external forces (see Emonet-Dénand et al. 1985). The enhancement of the mechanically-evoked reflexes (tendon tap) together with a relative stabilization of the H-reflexes that we clearly observed during reinforcing manoeuvres (of the type used by Ribot et al. 1986) can thus no longer be considered as inconsistent with the negative observations on Ia fiber activity (Vallbo 1981). Moreover, we (Hugon & Paillard 1955) have observed a long-lasting depression of the patellar reflex following a reversible passive stretching of the calf muscle, which we related to a long-lasting change in the stiffness of intrafusal fibers. The initial state could be restored following either an active contraction or by merely asking the subject to perform mental calculation. Indeed, both manoeuvres are known to be associated with a potent gamma dynamic drive.

My final comment addresses the problem of "proprioceptive acuity." I refer to a former study on position sense (Paillard & Brouchon 1974) that we recently replicated and extended (unpublished). Using an index approximation task, we were able to show that the ability to discriminate various proprioceptive location cues is far inferior to the capacity to point accurately to different proprioceptive targets. Clearly, the exquisite spatial resolution provided by the CNS to dissociate proprioceptive cues of different origin and nature (movement versus position, dynamic versus static, active versus passive motion cues) contrasts with the crude revolving power of perceptual judgements required to evaluate the misalignment of two fingers. Similar conclusions have been drawn in motor tasks requiring self-paced synchronous hand and foot movements. Under these conditions the CNS automatically adjusts the timing of the central commands to synchronize the reafferent proprioceptive signals issued from both movements. In other words, the command of the finger movement is delayed and starts 30 msec after that of the heel to account for the difference in conduction time between the two pathways. In striking contrast, the resolving power exhibited by subjects in discriminating lack of simultaneity between the two movements on the basis of proprioceptive cues is in the range of 80 to 100 msec (Bard et al. 1992).

Psychophysics uses measurements of perceptual thresholds (through the verbal report of conscious subjects) to evaluate the discriminability of proprioceptive information whereas automatic neural operations process that information at a subconscious level. The experimental observations reported above showed that discriminative capacity is much higher at the later level than at the perceptual level. Hence the question as to the true significance of psychophysical measurements as far as proprioceptive information is concerned.

Levels of explanation and other available clinical models for motor theory

J. G. Phillips,[a] D. L. Jones,[a] J. L. Bradshaw[a] and R. Iansek[b]
Department of Psychology[a] and Department of Neurology,[b] Monash University, Clayton VIC 3168, Australia
Electronic mail: *psy193g@vaxc.cc.monash.edu.au*

[GEA] **Alexander, DeLong & Crutcher** have provided a timely, knowledgeable, and inspiring application of cognitive science and neural networking to the study of motor coordination. Although we generally agree with the broad thrust of their argument that motor programming models must be reconciled with brain anatomy and physiology, we must make the following specific comments:

(a) As a rule, we regard models based upon neural networking with suspicion. The wealth of hidden parameters rapidly renders such models untestable (Bamber & van Santen 1985), their only real value being that of simulations. We believe, however, that the approach of **Alexander et al.** is very promising, because they do demonstrate that it is possible to document the connections and interactions between brain structures so that the number of hidden parameters can be reduced, producing more tractable models of motor coordination.

(b) We disagree with **Alexander et al.**'s pessimism concerning the utility of the motor programming concept. Although at a micro level (single-cell recording) it may prove difficult to identify structures controlling movement parameters, at a macro level (clinical observation and research on patients with movement disorders), it is clear that specific brain structures do play a role in controlling certain parameters. For example, from clinical impressions, the basal ganglia play a role in the scaling or sequencing of movement, the cerebellum helps determine movement endpoints, and the cortex is involved in fixing the angle and orientation of movement. Such observations suggest that parameters are programmed, although whether the mode of programming is serial or parallel has been open to debate for some time (Stelmach & Diggles 1982).

(c) Although parallel and distributed (PDP) models probably match the neural architecture better than the serial models, it remains to be seen whether they will have any real biologically descriptive power. The PDP models themselves will become dated as more information is gathered on pure neurobiology, to be replaced by increasingly biological descriptions. For us, biology must always be the ultimate organizing principle, and models of motor control are to be regarded as just theoretical descriptions until we know better.

(d) Although **Alexander et al.**'s critique is well stated, they offer very few examples of such models reflect biological properties (except those of homology). They offer information on the parallel motor circuits and on various other systems of connections to highlight their distributed nature. The functional implication of these approaches must also be shown, however, so that their true working power can be seen. Put more simply, the applications of such PDP models are not clear.

(e) **Alexander et al.** target the basal ganglia and cortex to dispute the motor program concept while overlooking the cerebellum, a structure that has been linked previously to motor programs. The cerebellum is a highly distributed system that has intrigued theoretical biologists for years. Cerebellar diseases cause disturbances of voluntary movement and balance. The impairments of balance are of particular interest, as they may reflect disturbances of the size and duration of postural (long-latency) reflexes. Because the functionally important long-latency postural reflexes are elicited when *loaded muscles, or voluntary movements are perturbed*, this implies that the cerebellum has a role comparing a *representation* of the intended movement with ongoing movement. It would seem that some useful comparisons could have been made with the cerebellar system, especially because the cerebellum has been reputed to have roles in the timing, learning, and programming of movements (Rothwell 1987).

A distributed common reference frame for egocentric space in the posterior parietal cortex

Alexandre Pouget and Terrence J. Sejnowski
Howard Hughes Medical Institute Research Laboratories, The Salk Institute, La Jolla, CA 92037
Electronic mail: *alex@helmholtz.sdsc.edu*

[JFS] **Stein** has provided a fair and accurate account of what we know about the posterior parietal cortex (PPC) and its possible functions. He concludes that the evidence points toward multiple reference frames rather than one common reference frame and argues further that there is no need for a single one. In this brief commentary we suggest that the existing evidence does not exclude a common representation for spatial transformations in addition to special purpose ones. A general purpose reference frame, however, does not require topographical maps and is likely to be highly distributed.

Stein argues that perhaps the best candidate for a special purpose reference frame is the one for eye movements. Indeed, retinal position vectors can be converted directly into oculomotor vectors without the need for an intermediate head-centered representation. However, this transformation seems to be performed primarily in the frontal eye field (FEF) (Bruce & Goldberg 1985; Goldberg & Bruce 1990), which has connections with, but is not part of, the PPC. Stein suggests that the lateral intraparietal area (LIP) may be using a similar strategy, that is, a direct transformation; but he misses the fact that in LIP, about 70% of the saccadic neurons are modulated by static eye position (Andersen et al. 1990b), compared to only 2% in the FEF (Bruce & Goldberg 1985; Goldberg & Bruce 1990). As shown by Zipser and Andersen (1988), this suggests that LIP contains a distributed representation of head-centered coordinates. This more general spatial representation allows LIP to integrate cues from different sensory modalities such as audition and vision (Bracewell et al. 1991), unlike FEF, which appears to use only retinal coordinates.

Electrical stimulation experiments also suggest that FEF and LIP use different coordinate systems. Focal stimulation of neurons in LIP lead to saccadic eye movements whose directions and amplitudes are a function of initial eye position, unlike the saccades elicited by stimulations in the FEF, whose characteristics are largely independent of initial eye position, often referred to as fixed vector type (Robinson & Fuchs 1969). In other parietal areas, particularly in the floor of the intraparietal sulcus (IPS) and 7a, the eye movements induced by electrical stimulation are of the convergent type (i.e., the saccades tend to terminate in the same zone of space regardless of initial eye position; Kurylo & Skavenski 1991; Thier & Andersen 1991). These results suggest that the representation of saccadic eye movements is quite different in the PPC from that found in the FEF.

This raises the general issue of why multiple representations of the same sensorimotor transformation exist in different parts of the brain. We agree with **Stein** regarding the usefulness of having many special purpose transformations; however, we suspect that the particular purpose of the representations in the PPC is to integrate these into common representations involving synergies between different sensorimotor transformations. Hence, the egocentric representation found in 7a (Brotchie & Andersen 1991; Thier & Andersen 1991) could be useful for coordinating various kinds of movements, such as eye, head, or even arm movements directed toward the same spatial location.

This implies that the same neuronal pool could be used to control many different types of movement. A recent observation by Thier and Andersen (1991) supports this hypothesis, showing that electrical stimulations in 7a and in the floor of IPS can induce simultaneous eye movements as well as shoulder and face muscle movements. A distributed egocentric reference system might not provide the best coordinates for controlling eye movements, but they might be suitable for deciding the extent to which eye movements should be included along with other body movements. Evidence is mounting that there are such representations in the PPC.

A related issue concerns whether the PPC is the first area where a distributed representation of egocentric space can be found. In the visual system, the modulation of responses by static eye position has been observed as early as the lateral geniculate nucleus (Lal & Friedlander 1989), the primary visual cortex area V1 (Trotter et al. 1991; Weyand & Malpeli 1989), and the extrastriate area V3a (Galleti & Battaglini 1989). We have recently shown how these modulatory responses could provide egocentric coordinates for early vision (Pouget et al. 1992). In contrast to the PPC, in the early visual cortex, low-level visual features such as orientation and direction of motion are encoded within retinotopic maps. We have proposed that each location on those retinotopic maps contains a distributed representation of the egocentric positions of these features similar to the one proposed by Zipser and Andersen (1988) for the position of the whole object in the PPC.

The same principle could also apply to other sensory cortices. In the primary auditory cortex, for example, the responses of 62% of the neurons are modulated by the position of the sound source with respect to the head (Ahissar et al. 1992). This is consistent with an early head-centered representation of sound source at early stages of auditory processing.

In summary, we agree with **Stein** that the brain uses specialized sensorimotor transformations. However, these specialized routines are not necessarily implemented in the PPC and may involve other structures such as the FEF. In contrast, the PPC may contain a common reference frame based on distributed representations that would provide a good starting point for mediating and coordinating many sensorimotor routines. Finally, it appears that egocentric transformation might be initiated earlier than previously thought, as early as the primary sensory cortices.

The many disguises of "sense": The need for multitask studies of multiarticular movements

Carol A. Pratt and Jane M. Macpherson
Department of Neurology and R. S. Dow Neurological Sciences Institute, Good Samaritan Hospital and Medical Center, Portland, OR 97210
Electronic mail: *fzgdmac@psudrvm*

[DAM] Once upon a time, small-systems neurobiologists believed that by fully elaborating the synaptic interconnections of the 30 neurons in the lobster's stomatogastric ganglion (STG) they would understand how central pattern generators (CPGs) controlled simple movements of the foregut. This turned out to be yet another fairy tale with a happy ending. Whereas the cellular components, synaptic organization, and intrinsic membrane properties of the two CPGs resident in the STG have been well described (Selverston & Moulins 1985), the synaptic connectivity has proven to be more flexible than originally anticipated (Katz & Harris-Warrick 1990). It was found that neuromodulators could alter the synaptic strengths and intrinsic cellular properties of the CPG components, thereby dynamically reconfiguring a single circuit to produce *different* motor outputs (reviewed in Harris-Warrick 1988). These revela-

tions have prompted a shift in experimental emphasis of small-systems neurobiology toward understanding how extrinsic inputs dynamically modulate or "sculpt" CPG circuits and a new conceptualization of CPGs as "a loose organization of neurons which can interact with each other in many possible ways to generate a family of related behaviors" (Harris-Warrick 1988, p. 286). [See also Selverston: "Are Central Pattern Generators Understandable?" *BBS* 3(4) 1980.]

The modern view of vertebrate spinal interneuronal circuits that is eloquently presented by **McCrea** illuminates many parallels between the functional organization of invertebrate CPG and spinal "reflex" circuits. Thus, it would seem appropriate to use lessons learned from the elegant studies of CPG circuits in "simple" systems to guide future studies of how spinal reflex circuits contribute to the control of movements. What are these lessons? An obvious one is that identification of spinal interneuron inputs and outputs will not reveal how movements are produced or controlled by the spinal cord. Second, emphasis should be placed on understanding the *dynamic* reconfiguring of spinal circuits associated with the production and control of different movements. And, third, as McCrea states, there needs to be a greater emphasis on examining "nonreflex" effects of sensory inputs on motor circuits. Sensory feedback does more than provide fast correction of motor outputs; it is also likely to play a critical role in dynamically sculpting McCrea's interneuronal "operational connectivity" to coordinate motor output and limb mechanics. Some examples of possible roles of sensory feedback that expand on the traditional concept of reflexes are presented below.

McCrea mentions the serotonergic inducement of plateau potentials and bistable firing patterns in cat motoneurons (Hounsgaard et al. 1986) as another example of how descending or segmental inputs can exert long-lasting modulatory effects on motoneuron output. Serotonin also induces plateau potentials in the crab STG (Harris-Warrick 1988). The only source of serotonin to the ganglion is from peripheral *sensory* cells (gastro-pyloric cells) that are activated by increases in muscle tension (Katz et al. 1989). The gastro-pyloric cells appear to colocalize acetylcholine and serotonin and produce both excitatory postsynaptic potentials (EPSPs) and plateau potentials in motoneurons (Katz & Harris-Warrick 1989). Cutaneous afferents can also produce prolonged postsynaptic effects, as evidenced by the recent demonstration that electrical stimulation of the turtle shell evoked increases in the excitability of spinal interneurons lasting several seconds (Currie & Stein 1990). An increased role of sensory feedback in determining some muscle activity patterns during rhythmic motor behaviors has also been suggested recently. In insect walking and flight (Pearson 1985) and cat ear-scratching (Carlson Kuhta & Smith 1990), three behaviors thought to be controlled by CPGs, sensory input appeared to be critical in determining the timing of some muscle activations, leading to the conclusion that "sensory input forms an integral part of the pattern generating system" (Pearson 1985, p. 312; see also Carlson Kuhta & Smith 1990). In a similar vein, in an elegant series of studies (reviewed in Smith & Zernicke 1987), Smith and her colleagues have correlated lability in some muscle activation patterns during paw shaking and locomotion with variable contributions of muscle, gravitational, and inertial forces to net limb dynamics. Smith and Zernicke have proposed that motion-related feedback monitors changes in intersegmental forces and dynamically adapts the output of central programs to compensate for the emergent limb dynamics.

The above examples clearly indicate the powerful global effects sensory feedback can have on spinal circuits. A central theme in **McCrea**'s target article is that "the particular spinal circuitry operating during a movement is dynamically selected and optimized." McCrea argues that the complexity of spinal circuits is not cause for despair but rather an elegant solution for the flexible matching of spinal control to different movement/mechanical requirements. This is a more optimistic but largely heuristic view without a great deal of direct experimental evidence to support it. Of critical importance are studies that compare similar neuromuscular features across a variety of movements or behaviors to determine: (1) the extent of task-related changes in the operational connectivity of spinal interneuronal circuits; (2) which input systems contribute to the observed changes in operational connectivity; and (3) the relationship between moment-to-moment operational connectivity and limb/body mechanics. Examples include studies comparing walking, scratching, and paw shaking in intact cats (Abraham & Loeb 1985; Carlson Kuhta & Smith 1990; Hoy et al. 1985; Hoy & Zernicke 1985; Loeb et al. 1985; Pratt et al. 1991; Pratt & Loeb 1991; Prochazka et al. 1977; 1989; Smith et al. 1977; 1980; 1985; Smith & Zernicke 1987) and embryonic motility, hatching, swimming, and walking in unrestrained chicks (Bekoff 1989; Bekoff et al. 1987; Johnston & Bekoff 1989). Similar studies in preparations exhibiting fictive behavior are needed to compare the connectivity and activity of the same identified interneurons during different behaviors. Examples of this type of experiment are limited but include studies of Ia inhibitory interneuron and Renshaw cell activity during scratching (Deliagina & Feldman 1981; Deliagina & Orlovsky 1980) and locomotion (McCrea et al. 1980; Pratt & Jordan 1987).

Studies of multiarticular movements in intact animals have provided insights into coordinative mechanisms that can be used to formulate specific predictions concerning dynamic sculpting of interneuronal circuits that can be tested in reduced preparations. Almost all natural movements involve more than one segment. The problem of coordinating multiple segments is complex because of the mechanical linkages among even nonadjacent segments – the displacement of any segment is significantly affected by the displacement of all other body segments. This complex physical interaction must be understood in order to appreciate fully the control mechanisms used by the nervous system, since it is quite clear that the nervous system must take into account the physics of its own musculoskeletal system. Thus, a potentially fruitful approach to furthering our understanding of spinal circuits would be to examine interneurons with inputs and outputs from multiple limb segments and, in particular, certain noncontiguous joints. Some examples are presented below.

Postural responses evoked by support surface translation are characterized by muscle synergies that are not fixed but continuously modified as a function of the direction of translation (Macpherson 1988). **McCrea** has shown schematically (in his Figure 3) how this modification of synergy pattern could arise by the selection of new interneuronal pathways through changes of afferent input. Although this fanciful circuit is by no means unique, this kind of analysis can lead to testable hypotheses. One could examine, for example, the set of different muscle activation patterns that are observed in a behavior such as the postural response, and use them to predict interneuronal connection patterns within and across joints. For example, it is remarkable how tightly linked are certain hip and ankle extensors in the postural response (e.g., gluteus medius, adductor femoris, caudofemoralis, lateral and medial gastrocnemius, soleus; Macpherson 1988). The hip and ankle muscles are similarly linked for the paw shake response (Hoy et al. 1985; Smith et al. 1985). Such a tight link across noncontiguous joints leads to the prediction of similar linkages in relevant interneuronal pathways. In addition, both the postural and paw shake studies cited above have shown a certain independence of knee muscle activation compared to hip and ankle. During postural responses, knee muscle activation is greatest for translations in the transverse plane, as opposed to the diagonal plane for hip and ankle. It would be interesting in reduced preparations to compare effects of various species of afferents on interneurons projecting to hip, knee, and ankle motoneurons.

Kinetic analysis of cat hindlimb movement during paw shak-

ing revealed different functional roles for ankle and knee extensor muscles; ankle muscles produced torques that directly controlled paw segment dynamics, whereas knee extensor muscles produced torques that counteracted large inertial forces arising from intersegmental dynamics (Hoy et al. 1985). These results led to the prediction that the timing and amplitude of knee muscle activity during paw shaking would be more dependent on motion-related feedback about limb dynamics than would ankle muscle activity. Indeed, it was shown that manipulating motion-dependent feedback during paw shaking altered the pattern of activity in extensor muscles at the knee but not at the ankle (Sabin & Smith 1984). Although the afferent sources involved in monitoring limb dynamics have yet to be identified, these data predict that they have differential effects on knee and ankle extensor motorneurons. We hope that these well-defined predictions will be tested in reduced preparations in the future.

These examples illustrate the unique power of studies of natural, multiarticular movements to generate predictions of interneuronal connectivity that can account for task-specific, intralimb coordination. Similarly, results from reduced preparations can provide specific predictions that can be tested in behaving preparations. For example, Nichols (1989) reported that the classical short-latency excitation produced in soleus (SOL) by stretch of its synergist medial gastrocnemius (MG) reversed to inhibition when MG was active. The load-dependency of the reversal in reflex sign suggested that convergent input from Ib afferents could be involved. Based on these results, it could be predicted that for some tasks in which MG is loaded, an inhibition of SOL should occur. This has been observed for postural reactions in the standing cat to translations of the support surface (Macpherson, unpublished observation). This finding is consistent with the known convergence of Ia and Ib inputs onto common interneurons (Jankowska et al. 1981a; Jankowska & McCrea 1983) and may represent an automatic, task-related switching between control variables (displacement vs. force) as suggested by Taylor and Gottlieb (1985). Other hypotheses that have important implications regarding spinal interneuronal operational connectivity include the possible differential control of uni- versus biarticular muscles (Nichols 1989; Pratt et al. 1991; Van Ingen Schenau 1989) and possible shifts in dominance of homonymous versus heteronymous reflex effects that appear to be dependent on limb mechanics (Loeb et al. 1989; Nichols 1989).

Thus, although **McCrea**'s question, "How can a study of spinal interneurons help in understanding general principles of movement control?" is appropriate, we would like to emphasize the timeliness of the converse question, "How can the study of multiarticular movements help in understanding principles of spinal interneuron function and organization?" Progress in understanding the role of spinal systems in movement control will be optimized by a close interaction among experimentalists using different approaches. We believe that this kind of interdisciplinary approach will push forward the boundaries of our understanding of spinal cord circuits and justify the optimism we share with McCrea that making sense of these circuits is possible.

A vital clue: Kinesthetic input is greatly enhanced in sensorimotor "vigilance"

A. Prochazka

Division of Neuroscience, University of Alberta, Edmonton, Alberta, Canada T6G 252

Electronic mail: *userapro@ualtamts.bitnet*

[SCG] **Gandevia & Burke** have done an excellent job in bringing together the human and animal data on proprioception and its role in motor control. After some years of apparent contradiction and inconsistency, the human and animal neurographic findings seem to have converged. Thus, in unloaded passive or voluntary movements of moderate speed it is now agreed that muscle spindles signal displacement and velocity. When active forces are appreciable, fusimotor action, which is phasically boosted during alpha motoneuronal activation, significantly raises spindle firing rates, most noticeably during isometric or slow lengthening or shortening contractions. Tendon organ endings signal active contractions, and although individual endings respond in quite nonlinear ways to whole muscle force, the ensemble of tendon organs of a muscle usually signals muscle force well. Indeed, for muscle spindles too, it is the signal conveyed by the whole *ensemble* of a muscle's afferents that is important, because it is what is "seen" by the receiving neurons in the CNS. Variations between individual endings due, for example, to intramuscular location are then rather unimportant. Ensemble firing profiles also remind us of the huge influx of proprioceptive input (up to 200 kilo-impulses/sec: KIPS) that can occur in fast movements (Prochazka et al.).

There is one important disparity between human and animal data, however. It concerns a phenomenon that is very striking in freely moving cats, but that so far has not been seen clearly in human neurographic recordings, as **Gandevia & Burke** point out. This is the very large variation in spindle bias and sensitivity that occurs in different tasks and contexts, and which has been dubbed "fusimotor set." Although a comprehensive description is still lacking, it is clear that fusimotor biasing (presumably mainly via static fusimotor neurons) always increases in the transition from rest to the onset of an activity such as stepping. In complex tasks, or in situations in which the cat is "put on its guard," spindle sensitivity can suddenly increase very dramatically and reach levels consistent with a maximal activation of dynamic fusimotor neurons. Peak spindle Ia firing rates can be quite astonishing (over 600/sec: 30 KIPS from a typical hindlimb muscle). Why does this sensitization happen, and what does it tell us about proprioception and motor control? It presumably happens because the increased spindle gain is "needed" by the CNS, not at the spinal level, where it may actually be destabilizing (Llewellyn et al. 1990), but in the higher centres. If spindle input is "needed" by the brain and cerebellum in crucial motor tasks, this would put proprioceptive input more on a par with efference copy in tasks in which conscious attention is "tuned" to the details of a motor program (i.e., reafference may then assume more importance than Gandevia & Burke suggest). This does not preclude some important segmental effects of proprioception, notably the tendon organ control of flexion onset in stepping (Duysens & Pearson 1980).

Why has task-dependent fusimotor sensitization been so elusive in the human neurographic recordings? Two possibilities have been suggested. The first is that percutaneous neurography in the laboratory environment imposes too many constraints on movement and context. This is still a valid issue, though the most recent experiments have been well designed (e.g., Vallbo & Al-Falahe 1990) and might have been expected to reveal set-related effects. The second is that significant species differences may be involved. In daily life, young domestic cats sleep at lot but when they are active they can exhibit large swings in arousal, alertness, and orientation to their surroundings. Predatory and defensive behavior is often seen, frequently in the absence of prey or of any real external threat. Cats brought into the laboratory for recording sessions are somewhat wary, particularly when being handled, and this may accentuate set-related fusimotor modulation. Perhaps it is the ability of humans to judge better the real need for increased vigilance, and the fact that it is uncalled for in the laboratory situation, that reduces the incidence of observed fusimotor "wind-up." Because the relevant data are not in, one can still posit that fusimotor action in humans, as in cats, is moderate in tasks such as stepping, and greatly elevated in states of vigilance or anxiety and in novel or difficult tasks that really matter. The implication of fusimotor set is that the nervous system depends

on kinesthetic information to a *moderate extent* in stereotyped tasks and to an *increasingly greater extent* as the demands on adaptive motor performance increase. Happily, the other clues discussed by **Gandevia & Burke** have led them to a similar conclusion.

Is the posterior parietal cortex the site for sensorimotor transformation? Cross-validation from studies of stimulus-response compatibility

Robert W. Proctor and Elizabeth A. Franz
Department of Psychological Sciences, Purdue University, West Lafayette, IN 47907
Electronic mail: *proctor@brazil.psych.purdue.edu*

[JFS] **Stein** proposes that the posterior parietal cortex (PPC) is a multimodal associative network for transforming sensory input into the appropriate motor coordinates. In choice reaction tasks, such transformations are attributed to an intermediate stage of information processing called stimulus-response (S-R) translation (Proctor et al. 1990). The characteristics of the translation stage are most apparent in S-R compatibility effects (Proctor & Reeve 1990), which are differences in speed and accuracy of responding as a function of the assignment of a set of stimuli to a set of responses. In our commentary, we evaluate whether **Stein**'s characterization of the PPC can be cross-validated by the research on S-R compatibility.

The prototypical procedure for studying S-R compatibility involves presenting a stimulus in a left or right location to which the subject makes a left or right response. Responses are faster when the assigned response location corresponds with the stimulus location than when it does not. This difference in response latency is attributed to the relative efficiency of the translation processes. It is reasonable to assume that such translation processes fall within the role envisaged for the PPC. For example, Seal et al. (1991) documented specific neural correlates of the S-R translation stage by recording the activity of single cells in the PPC of monkeys performing choice reaction tasks. Similarly, Verfaellie et al. (1990) interpreted S-R compatibility in terms of a model developed initially to explain hemispatial neglect, which is attributed to damage of the PPC. If the PPC is the site of S-R translation, the properties **Stein** ascribes to the PPC should correspond with the properties of the translation stage implicated by compatibility research.

Stein stresses three properties of the proposed sensorimotor transformation network in the PPC that can be evaluated directly from the compatibility research. First, he proposes that there is no single, topographical map of real space. Instead, many reference frames are used, based on such things as the body axis and the line of sight. One of the most prominent conclusions from the compatibility research is that the frames of reference used for coding the stimulus and response sets are flexible and vary as a function of task characteristics (Heister et al. 1990; Umiltà & Nicoletti 1990). Coding can occur with respect to relative location, egocentric location, or anatomical distinctions.

Stein's second assertion is that the PPC uses information processing rules to transform sensory arrays into motor coordinates. Consistent with this assertion, most accounts of S-R compatibility rely on transformation rules. A case in point is Rosenbloom and Newell's (1987) algorithmic model of S-R compatibility and practice that is based solely on production rules. Support for the notion that translation is rule-based comes from studies in which a linear array of multiple stimulus locations is assigned to a linear array of response locations. Responses are almost as fast when the assignment of stimulus locations to response locations is completely reversed as when it is direct (Duncan 1977; Morin & Grant 1955). This implies that translation is relatively efficient when a single "reversal" rule can be applied.

Third, the direction of attention is proposed by **Stein** to mediate S-R transformations. Attention has been shown to be an influential factor in compatibility effects. Nicoletti and Umiltà (1989) displayed six boxes arranged in a row, and instructed subjects to fixate at one end of the row. A small square then appeared between one pair of adjacent boxes to indicate where attention should be focused. One keypress response was to be made if the target stimulus was located to the left of the small square and another response if the stimulus was located to the right. Spatial compatibility was determined to be a function of the location to which attention was directed. Stoffer (1991) proposed that this relation between attentional focus and compatibility occurs because the spatial coding is instituted by a shift in attention.

Stein attributes hemineglect in patients with PPC lesions to an attentional bias toward the unneglected hemispace. Verfaellie et al. (1990) hypothesized that such bias occurs because each hemisphere is responsible for attentional control of the contralateral hemispace. They demonstrated that intact subjects show an enhanced spatial compatibility effect when biased by a precue to prepare a response in one of the two hemispaces (see also Proctor et al., in press). This finding again implicates attentional involvement in S-R compatibility and is consistent with the view that hemineglect and S-R compatibility have a common basis.

To summarize, a remarkable similarity exists between the properties of the PPC that **Stein** proposes based on psychobiological evidence and the characteristics of S-R translation obtained from studies of human performance. Thus, Stein is probably on the right track in his characterization of the PPC as a site of sensorimotor transformation. However, we think it unlikely that the PPC is the only site involved. Seal et al.'s (1991) single-cell recordings suggest that area 5 in the PPC is part of a system for translation that also includes S1, area 6, and area 4, with a continuum of function existing within and between these areas. Also, S-R compatibility effects obtained when the stimulus and/or response sets are defined by nonspatial properties (e.g., Proctor & Reeve 1985) do not seem easily attributable to representations in the PPC. On the whole, though, we are heartened by the general agreement between the psychobiological and performance studies. It seems that progress could be facilitated by better communication between researchers in the two domains.

Real space in the head?

Philip Quinlan

Department of Psychology, University of York, Heslington, York YO1 5DD, England
Electronic mail: ptq1@vaxa.york.ac.uk

[JFS] In a recent textbook, Wade and Swanston (1991) carefully consider the psychological evidence concerning the imposition and use of various frames of reference during normal human visual perception. They conclude with an account of how humans make sense of the ever-changing optic array in which three frames of reference are used to describe visual input. At the level of the *monocular/retinocentric frame* the position of a light source is coded in terms of a coordinate system defined on the retina. Such a retinotopic form of encoding is too constrained, however, to be generally useful. Wade and Swanston therefore discuss the *egocentric frame* (a head-based frame of reference), which codes the position of a light source with respect to the egocenter or cyclopean eye and is established by combining a binocular retinocentric signal and an eye movement signal. As both eyes move by equal amounts during vision, a single eye movement signal can be combined with the two retinocentric signals from the separate eyes to produce an egocentric representation. Such a representation codes angular displacement with respect to the observer independently of eye movement information.

To resolve problems caused by movement of the head during vision, Wade and Swanston (1991) discuss the final *geocentric frame*. This is defined as a representation that takes into account both egocentric distance and movement of the observer. It specifies object motion in 3-dimensional (3-D) coordinates irrespective of the 3-D movements of the observer. The geocentric frame is an environmental one in that object position and motion can be specified relative to the surface of the earth. Wade and Swanston conclude that all normal perception is necessarily geocentric. Psychology therefore tells us that three frames suffice.

Unlike Wade and Swanston, who are interested in a psychological account of vision, **Stein** is more concerned with the underlying neurological apparatus, and primarily with where certain visual processes may be carried out in the head. However, whereas Wade and Swanston conclude with a cogently argued account of vision, Stein fails to provide any clear functional specifications. As a consequence, it is difficult to be sure what his central thesis is.

The title of **Stein**'s target article suggests that the central thesis is that egocentric space is represented in the posterior parietal cortex (PPC). Yet not only have Grossberg and Kuperstein (1989) discussed this possibility, they have also provided functional details of how such a representation of egocentric space is fundamental to the control of saccades. In their account, retinally induced signals and initial eye position signals are joined to compute a target position map defined relative to a system of head-based coordinates. Grossberg and Kuperstein suggest that this takes place in the PPC (although the location of such a target position map may vary from species to species). **Stein** objects strongly to the idea of recoding retinal information into head-based coordinates, because no signal representing the absolute location of visual targets in head-centered coordinates has been found. Indeed he accepts Goldberg and Bruce's (1990) account of saccade control in which retinal vectors are directly transformed into oculomotor vectors. Such an idea fits with the brainstem saccade generator model of Jurgens, Becker, and Kornhuber (1981), in which saccadic control does not rely on a signal for target position in head-centered space. Yet such a model is just one of many, and the question of what role head-centered coordinates play in saccadic control appears controversial. Indeed, one problem with this is that the experimental subjects are typically head restrained during testing. Such conditions are used to discount any possibility of modulation of eye control by the head movement system. Indeed, not surprisingly, when head-free animals are compared with head-restrained animals, head movement plays a critical role in explaining eye movement control (Guitton et al. 1990).

Nevertheless, the whole issue of recoding retinal signals into head-based coordinates has been discussed at great length by Grossberg and Kuperstein (1989). They were concerned with the possibility that saccades can be controlled purely on the basis vector addition and subtraction of stored retinotopic values. However, they were unable to discover any neurally plausible mechanism which carries out direct recoding of retinotopic vectors for eye movement control, arguing instead that "a direct mapping from retinotopic values into difference vectors is more difficult to achieve than an indirect mapping from retinotopic values into (target position in) head-based coordinates" (p. 25). Moreover Grossberg and Kuperstein provide a detailed account of how saccades are controlled by transforming retinal information into a head-based coordinate system.

On the other hand, **Stein**'s central thesis may be that egocentric space is represented in the PPC in a way not previously envisaged. For example, he states that the PPC does not contain a map of real space but a neural network for converting one set of vectors into another; in addition, the PPC has the appropriate connections for mapping real space but the representations need not be explicit topographic maps. The latter point is of interest but has already been discussed in detail by Grossberg and Kuperstein. They have shown how the *implicit* coding of target position in head-based coordinates may take place because combinations of retinotopic and eye position signals can be distributed across a neural network. Indeed, the field of showing how neural nets may account for sensorimotor control is big and growing. Unfortunately, none of this literature is referred to by Stein. Simply stating that a neural network implements algorithms for vector conversion is too vague to discuss further.

Stein also considers the possibility that sensorimotor transformations depend on attention because attention maintains correspondences among different frames of reference. In the absence of any details, it is again impossible to assess this claim. Attentional modulation in the PPC has been discussed in great detail by Grossberg and Kuperstein (1989), but Stein again fails to refer to this work. I am therefore left wondering what can be gained from reading Stein's target article after due consideration of Grossberg and Kuperstein.

There is much information in neural network unit activations

John E. Rager

Department of Mathematics and Computer Science, Amherst College, Amherst, MA 01002
Electronic mail: *jerager@amherst.edu*

[DAR] **Robinson** argues that trying to explain "how any real neural network works on a cell-by-cell, reductionist basis is futile" (Abstract). In a strict sense, his conclusion may be correct. However, it is worth pointing out that there is much information that can be extracted from examining single-cell behavior. The case against so doing is far from clear.

First, it is an admittedly difficult task to analyze the detailed functioning of the hidden units of a neural network given a description of the task being performed. It is, however, neither impossible nor fruitless to do so. **Robinson** mentions NETtalk as an example of a mysterious neural net process. It is. Still, Rosenberg (1987) was able to shed considerable light on NETtalk's internals using principle component analysis and cluster analysis.

In connectionist natural language processing there has been much attention to the learning of language. To verify what has been learned, it is sometimes desirable to examine the hidden unit representations. For example, Elman (1990) analyzes a recurrent network trained on a language prediction task. He discovers that the hidden unit responses for nouns and verbs form distinguishable clusters. In a distributed sense, the network has learned the recognizable concepts of verb and noun, although there are no specific hidden unit patterns that could be designated "noun" and "verb." Such an analysis is important for understanding PDP models and research into techniques of doing analysis.

Second, much less detailed information about units in a neural network can still be a useful clue to the structure and function of the network. For example, if recording, either from real neurons or pseudoneurons, shows that few are simultaneously active, it is reasonable to conjecture that a sparse encoding is being used. Rolls and Treves (1990) discuss an increasing fineness of neural tuning as signals move along the gustatory system. Specifically, they observe increasingly finer tuning along the pathway from the nucleus of the solitary tract (NTS) to the frontal opercular taste cortex to the insula and along the path

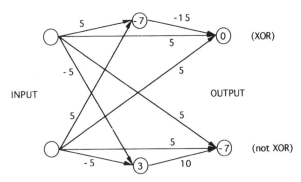

Figure 1 (Rager). A simple network to calculate XOR and not XOR. The numbers inside the units are biases, those on the connections are weights. The input to a unit is the sum of its bias and the weighted sum of the outputs of contributing units. Each unit outputs 1 if its input is > 0, 0 otherwise.

from the operculum to the orbitofrontal cortex. From this they conjecture an increasing sparseness of coding and give reasons why this might be sensible considering the conjectured roles of the regions. This model is based on aggregate properties derived from single cell recording rather than the detailed properties of single cells.

Perhaps it is better to stress a different aspect of **Robinson**'s conclusion. As he argues, neural net implementations of mappings may be different from those envisioned by using a symbolic, programmed computer model of the mind. Mathematical models are to a certain extent judged on their clarity and conciseness. There is no reason to think that neural network representations are crafted by these concerns. They are probably influenced by contrary pressure toward displaying robustness and graceful degradation.

Many researchers in connectionism have remarked that distributed neural representations are really different from symbolic ones, exhibiting aspects of structural composition without explicit symbols (see Touretzky [1991] and Hinton [1990] for discussions). The lesson of this work is that one must be aware that the internal functioning of neural networks is different and requires a different sort of explanation, but one need not give up trying to understand its internal functions.

In light of this understanding, let me lift an example from the target article and build a warning from it. In several places, **Robinson** discusses "rogue cells." In particular, in section 3.1 he mentions a cell "that excites, or inhibits both of a pair of antagonistic motorneurons." He observes that such cells "do not have a special function." In the context of this small system, I am sure he is right. However, it is generally dangerous to assume that cells are unnecessary because they seem to contradict our understanding of how things should work. Figure 1 is a simple example of a neural network which calculates both the XOR and the not XOR of its two inputs. Both the input units excite both of the output units, although only one output unit is active at a time. In complex or recurrent networks it may be impossible or difficult to learn some functions without constructing such "anomalies."

There are other questions that will require a detailed analysis of neural networks if they are ever to be answered. Many linguists, notably Chomsky (1986) have argued that some of the language-processing capability of humans is built into the structure of the brain. [See also Chomsky: "Rules and Representations" 3(1) 1980.] If this is true, it is necessary to discover this innate structure to complete an understanding of human language processing. It is hard to see how this would be possible without detailed analysis of real neural circuitry.

Command signals and the perception of force, weight and mass

Helen E. Ross

Department of Psychology, University of Stirling, Stirling FK9 4LA, Scotland
Electronic mail: *her1@forth.stir.ac.uk*

[SCG] **Gandevia & Burke** (sect. 2.1.1) state that "sensations of muscle force or the perceived heaviness of a lifted object derive from many sources." Many sources are certainly involved in both cases (Jones 1986), but it is not helpful to equate the sensation of muscle force with that of the heaviness of a lifted object. The former always relates to an internal sensation, whereas the latter often pertains to an external object and shows perceptual constancy. The sensation of muscle force usually relates to the weight of an object, but the perception of heaviness is more dependent on its mass. The distinction between weight and mass becomes important in altered force environments, such as under water (Ross et al. 1972), or under high or low gravity (Ross 1981; Ross & Reschke 1982). In such situations, heaviness judgments are usually about midway between the changed effective weight and constant mass; they tend more toward mass when the objects and the method of lifting are visible, and when time is allowed for adaptation to an altered force environment.

Mass constancy may not be perfect, but the fact that it occurs to any extent implicates motor command signals. According to the laws of physics, weight = mass × acceleration. Weight is a force, and it may be judged passively through pressure receptors in the skin, or actively through the muscular effort of holding the object against the force of gravity. Mass is the ratio of two forces, weight and acceleration, and can only be judged by actively imparting an acceleration. The sensory and motor mechanisms corresponding to acceleration are less clear than those for weight: We do not have accelerometers in our hands. Acceleration may be judged partly by the motor command signals to the hand and partly by the reactive force of the object in the hand. In an altered force environment, we can judge mass correctly only if we can distinguish between the acceleration caused by the environment and that caused by our own actions. The monitoring of command signals is critically important for the latter. Mistakes in mass perception may be due to mismonitored command signals, misperceived afferent signals, or a miscalculation of the relation between the two.

There is some evidence that changes in the way in which objects are lifted accompany the size-weight illusion (Davis & Roberts 1976; Mounoud et al. 1979), and that mismonitoring of command signals occurs in conditioned weight illusions (Hershberger & Misceo 1983). Changes in hand movements also occur in altered force environments (Ross 1991). When astronauts shake objects to judge their mass, they apply a greater acceleration under zero gravity than under normal gravity, and a lower acceleration than normal on first returning to normal gravity: This suggests that their command signals take some time to adapt to the changed force environment (Ross et al. 1987). Further analysis of videos of astronauts' hand movements (unpublished) suggests that errors of mass judgments often occur through mismonitoring of command signals: Astronauts believe that a change in the acceleration of the hand is caused by a change in the mass of the object instead of by a change in the force of the shake. There are other sources of error. Misperception of passive pressure occurs under altered gravity (Ross et al. 1987) and subjects may miscalculate the relation among the various forces. In conclusion, the monitoring of command signals is an essential component of mass perception, but one that is very susceptible to error.

One further point: **Gandevia & Burke** (sect. 2.1 and 2.2) find it paradoxical that the accuracy of judgments for relatively small weights or forces should be poorer than for larger forces. It is a well-known fact that Weber's Law breaks down at low inten-

sities for most modalities, including weight (Engen 1971) and mass (Ross & Brodie 1987). A large literature has been devoted to possible explanations of this breakdown (e.g., Laming 1986 [see also *BBS* multiple book review of Laming's *Sensory Analysis, BBS* 11(2) 1988]), and there is no reason to suppose that failure at low levels is specific to the sensorimotor system. Many explanations assume that there is a small level of background noise in the system, which hinders discrimination only when the signal level is low.

Presynaptic inhibition and information transmission in neuronal populations

P. Rudomin

Department of Physiology, Biophysics and Neurosciences, Centro de Investigación y Estudios Avanzados del IPN, México D.F. 07000
Electronic mail: *rudomin@invesmx.bitnet*

[DAM] As time progresses, information on different spinal interneuronal systems keeps accumulating. David **McCrea** is rather optimistic when he argues that current approaches will continue to "provide insight into motor systems." What does he really mean by this? I believe he thinks that with all the available information it is feasible to have models that describe motor systems and that critical experiments can be designed to test them. This may be true, but I am still concerned about the lack of a general hypothesis that goes beyond the flexor reflex afferent system (FRA) concept and that may be useful for experimental design. This is probably because we are mostly dealing with individuals (single neurons) and not with population behavior. An underlying assumption in most available studies is that population responses can be synthesized by adding up the activity of single interneurons. This may not necessarily reflect reality, because of the existing nonlinear interactions between individual actions. I believe that the study of interneurons must also contemplate recordings of more than one element at a time and must consider that the degree of synchronization (correlation) between the activity of individuals (afferent fibers and interneurons) is also a functional variable that is subjected to central control. Changes in the correlation between the activity of single neuronal elements that converge on common elements will determine whether or not there will be further impulse transmission at that particular node, even if there were no changes in the information conveyed by each of these converging inputs (Solodkin et al. 1991).

I believe, as **McCrea** does, that studies of the functional connectivity of a given set of interneurons will make sense only when those cells are performing the particular function we want to investigate. Studies of the behavior of spinal interneurons during the actual performance of a movement are a challenge from both a methodological and a conceptual point of view. Although methodological problems may eventually be solved by finding appropriate models (such as fictive locomotion), there remains the problem of how to characterize the behavior of functionally related sets of spinal interneurons that are intermingled with other sets of interneurons. We clearly need new methods to measure the overall activity of specific sets of interneurons, not only the mean frequency of their activity but also in terms of some variable that expresses the functional relationships between the interneurons in the set. This may reveal changes in the dynamics of populations that have so far escaped from our attention. One interesting contribution in this direction is the analysis of distributed networks performed in the olfactory bulb (Skarda & Freeman 1987).

Another issue pertains to the possible role of presynaptic inhibition (PSI) in motor control. A commonly accepted view is that PSI can work as a gating mechanism. It can do so if there is significant synchronization in the activity of the last-order interneurons synapsing with specific sets of afferent fibers. It has been established that different sets of last-order interneurons mediate PSI of Ia, Ib, and cutaneous fibers and probably also those of group II fibers, but we still do not know all the timings and correlations in the activation of these interneurons during specific motor acts.

PSI may have other roles besides acting as an effective gate control of afferent transmission. One attractive possibility is that the last-order interneurons that mediate PSI also act as a *source of noise* that is introduced into the ensemble of afferent fibers. If a given last-order interneuron has synapses with more than one afferent fiber, this common noise imposed onto many fiber terminals will introduce redundancy in the information transmitted by the ensemble of afferent fibers. But redundancy of what? The signal that is being introduced centrally has probably no direct relation to the information conveyed by sensory inputs because all these interneurons receive strong influences from several descending pathways. The noise introduced by the sets of last-order interneurons synapsing with afferent fibers will change the information conveyed by these fibers. This could be a way to shape the image of the "external world" according to the requirements of the particular movement that will be performed. That is, rather than being a gate control mechanism, PSI may function as an adaptive filter of information.

In this context it is important to point out that the intraspinal terminals of descending fibers do not appear to be subjected to presynaptic modulation of their synaptic effectiveness, which means they have some "priority status" as compared with afferent inputs. This could imply that there will be a better match between the descending activity related to a particular movement and the actual movement (Georgopoulos et al. 1986) than between the sensory projections and motor output.

A final issue is that the last-order interneuronal sets engaged in the control of synaptic effectiveness do not appear to act only on afferent fibers; they also produce postsynaptic (GABAergic) inhibition of motoneurons (and probably also of interneurons; see Rudomin et al. 1987). From the point of view of the control of the efficacy of the afferent fibers, the existence of postsynaptic actions in motoneurons may appear unnecessary and redundant. However, we must consider the last-order interneurons producing presynaptic inhibition of afferent fibers as inhibitory relays for motoneurons in their own right. If so, PSI may function to decrease interference by peripheral inputs during the descending inhibition of motoneurons, which may be important in the restriction of unnecessary movements.

Selection of task-related motor output through spinal interneurones

Marco Schieppati

Istituto di Clinica delle Malattie Nervose e Mentali, University of Genoa, 16132 Genoa, Italy
Electronic mail: *fisiol@imiucca.csi.unimi.it*

[DAM] **McCrea**'s timely target article updates the answer to the old, though often unstated, question of the role(s) of spinal interneurones. His attempt has been to frame the possible function of interneurones within the context of a general theory of movement execution. In this view, one of his remarks seems particularly interesting, namely, that one consequence of mutual interneuronal interaction "is a reduction in the number of parallel pathways available to descending and segmental afferent commands." Freely interpreting his statement, one can imagine that in this way both optimization and economy can be achieved in the motor output, in addition to protection of the intended motor message from unwanted perturbing factors, such as (a) ongoing peripheral input or (b) stochastic activity in nearby neural circuits. At the same time, the selection of prewired pathways is made possible, a mechanism that is

certainly more efficient and time-saving than the instruction of a pattern of cell activity through the richness of spinal neurons. The possibility of selecting prewired pathways, either by the central command or by the input from the evolving movement (or a combination of the two), is a necessary condition that allows meaningful function in the presence of overwhelming convergence at the spinal interneuronal level. In this connection one wonders whether learning can lead to optimization of the pattern of muscle activity by reinforcing transmission through interneuronal pathways.

Probably the most speculative parts of **McCrea**'s review depend upon a (deliberately) loose use of the term "movement." However, it might be of some help, in dealing with the possible role(s) of the spinal interneuronal machinery, to detail as much as possible the parameters of movement for each of which a particular interneuronal function (or a given set of testable hypotheses) might be envisaged. This aim can be better understood in human work, where the required movement is readily obtained and appropriate techniques are available. For example, to what extent are interneurones responsible for (1) the subtle differences between otherwise equal (or so intended) movements, (2) the kinematics and (3) the dynamics of the focal movement, or (4) the associated postural adjustments, be they anticipatory, secondary, or "set-related" – an aspect rarely considered (Schieppati & Nardone 1991)?

The last point can also imply that a given movement is delayed while waiting for the postural adjustment to take place. Therefore, (5) can interneurones also play the role of a delay line? For that matter, the delay could also be an integral part of the command to the focal movement. During head turning to the left (or to the right) in response to a "go" signal, the reaction time of the right (or, respectively, the left) sternomastoid muscle is shorter than the reaction time of either muscle during head flexion, a task in which both muscles are agonistic (Mazzini & Schieppati 1992). It is likely that the delay of muscle activity is accomplished by an inhibitory interneurone called into action by the descending command. In human standing, rising on tiptoes (in a self-paced or in a reaction-time condition) is accomplished by an abrupt increase in soleus activity. Paradoxically, this increase is preceded by silence in the ongoing tonic postural activity of the same muscle (Nardone & Schieppati 1988), which is in turn accompanied by inhibition of the H-reflex attributed to presynaptic inhibition (Nardone et al. 1986).

In addition, one can easily think of a voluntary contraction of a given muscle under different conditions, and ask about the role of interneurones in: (6) allowing the contraction of a particular muscle to start from rest and steadily increase, or to decrease from a given value to rest; or (7) switching the purpose of one or more muscle's activity from a motor to a braking action (as in lifting a weight or slowing its fall); or (8) converting the action of a muscle from a synergistic to an antagonistic one.

There are some hints that the behaviours mentioned in the last three points involve interneuronal action. Schieppati et al. (1985) showed that deliberate relaxation from a tonic muscle contraction is quicker than the onset of contraction, as measured to the termination or, respectively, the beginning of the EMG in a reaction-time condition. This was interpreted as a phenomenon whereby the "unloading reflex" is mimicked by fast descending activation of presynaptic inhibition.

Voluntary lengthening contraction of the triceps surae, as in easing a load (braking action), appears to be performed by activating the fast gastrocnemii (with the soleus almost silent), whereas lifting the same load along the opposite trajectory is performed, as expected, mainly by soleus activation (Nardone & Schieppati 1988). Soleus silence during lengthening is accompanied by a striking decrease in its H-reflex excitability (Romano & Schieppati 1987), which keeps soleus motoneurones from being reflexively activated by the presumably large Ia input (due to stretch of spindles of the active soleus – most likely under gamma drive).

In this vein, one might ask about the role of spinal inter-neurones in the selective recruitment of high-threshold motor units (with silence of low-threshold ones) observed during lengthening contractions of leg (Nardone et al. 1989) and arm muscles (Schieppati et al. 1991b), although not finger muscles (Schieppati & Valenza, unpublished observation). The activation of fast-twitch motor units allows graded decreases in force during fast but controlled lengthening, due to their delivering force for a shorter period of time than the slower-twitch ones (one would not stress here the "active lengthening" per se, but the capability of using the muscle as a damping unit rather than as a stiff spring). It is proposed that different interneuronal types are jointly involved in this task, some of them probably facilitated by inputs connected with the particular lengthening task. As to the possible absence of the selective recruitment in the finger muscles, various observations can be made, among them that the interneurones responsible for this motor behaviour (as well as for others) are not uniformly distributed among all motor pools (see Hörner et al. 1991).

In a limb flexor reflex, afferent input triggered by the evoked movement may secondarily induce the activation of a particular set of interneurones (more or less constant under predictable conditions) that contributed to the full expression of the motor response. In this sense, a given set of afferents may initiate a cascade of interneuronally mediated effects (such as the exaggerated withdrawal response in the paraplegic). One of the interneurones secondarily involved can be the Ia-inhibitory interneurone (not itself belonging to the interneurones associated with the flexor reflex afferents [FRA] family): It has been suggested that Ia afferent fibres from the gastrocnemius can inhibit the soleus motoneurones through Ia inhibitory interneurones (Gritti & Schieppati 1989; Schieppati et al. 1990; 1991a), thereby selecting the leg-flexor action of the gastrocnemius over the ankle extension action of the triceps.

In conclusion, sense can indeed be made of spinal interneuronal circuits, and emphasis is appropriately put on the descending command (see **McCrea**'s figures). Some interneurones can certainly produce complex and meaningful motor patterns (e.g., fictive locomotion), and corticomotoneuronal fibres can be sufficient for a variety of movements. But there are probably some important tasks that can be accomplished *only* by suitable interaction between descending pathways and segmental interneuronal networks.

Signals, brains and explanation

Georg Schwarz[a] and Alexandre Pouget[b]

[a]*Department of Philosophy, University of California, San Diego, La Jolla, CA 92093-0302 and* [b]*Computational Neurobiology Laboratory, The Salk Institute, 10010 North Torry Pines Rd., La Jolla, CA 92037*
Electronic mail: [a]*gschwarz@ucsd.edu;* [b]*alex@helmholtz.sdsc.edu*

[DAR] **Robinson** uses the vestibulo-ocular reflex (VOR) as "an even clearer example of a mathematical description telling us *what* must be done without giving any idea of *how* it is done" (sect. 3, para. 6), while questioning the explanatory significance of single-cell recordings on grounds that they provide "only scraps and pieces of information jumbled together, which can be interpreted only if one knows what one is seeking" (sect. 3.4, para. 7). These and numerous other passages suggest that Robinson agrees with Marr (1982) that the explanation of an information-processing task consists of three components: (1) a computational theory that explains what information is being processed, (2) the representations and algorithms used by the system that explain how the task is solved, and (3) an account of how the computations are implemented in the available physical (or physiological) substrate. Robinson's claim that "single unit recordings have not told us . . . how neurons, or groups of neurons, process signals" and that they "apparently cannot tell

us how the processing is done" (sect. 4, para. 9) further indicates that he assigns neuronal activity to the implementational level.

Robinson's example shows both the need for, and the beauty of, a computational explanation of the VOR. He is much less articulate, however, when it comes to the question what other aspects of brain activity can be similarly explained. He does, at one point, allude to "localization of function in the brain . . . allowing us to concentrate on one modality, such as vision, at a time," but this is about as specific as he gets. We do not, of course, expect Robinson to supply us with a list of successfully identified brain functions – solving the specification problem is likely to be the hardest part of the whole enterprise.[1] But we do think that more should be said about what kinds of systems are potential, or even likely, candidates for computational explanation.

As a matter of fact, more *has* been said about just this issue. Fodor (1983; see *BBS* multiple book review: *The Modularity of Mind, BBS* 8(1) 1985.) isolates a number of properties he thinks are constitutive of what he calls *modular systems*.[2] We agree with Fodor that modules are prime candidates for computational explanation. Whether the brain is modular or not, however, is a different question. In the absence of more concrete alternatives, we will concentrate on modules as systems whose chances of explicability are the greatest. Fodor lists a number of properties characteristic of modular systems. For our present purposes it suffices to emphasize two: Modules are *informationally encapsulated* and their operation is (largely) *automatic*. Associated with the latter feature is the fact that modules are domain-specific and fast; associated with the former, modules produce highly constrained outputs and their immediate representations are in general inaccessible to other processes. Informational encapsulation, however, does not rule out the feedback of information within a module (see Fodor 1983 pp. 76ff); rather, it means that the output of the module is – in general – insensitive to the knowledge a system might have about the domain processed (as illustrated by the persistence of perceptual illusions).

It is easy to see what makes modules attractive candidates for computational explanation. They are tuned to specific domains, they work autonomously, and they deliver their output fast, suggesting that their operation is highly task-specific. What is less clear is the extent to which brain activity is in fact modular. Consider a domain as intensely studied as early vision (Felleman & Van Essen 1991). To put the empirical findings in a nutshell, the situation is a mess: massive lateral connectivity, neurons with multiple selectivity, significant backward projection from later stages in processing whose nature and significance is unknown, and so on. The upshot is that we simply do not know what areas process what functions, if any.[3]

Although it is too early in the game to draw any definite conclusions, we believe there are only a few options available. The visual system might look like a mess, but this is merely because its modules are distributed over several cortical areas. Or the visual system might look like a mess because it *is* a mess. The first option – let us call it the "superimposition theory" of early vision – draws two implications from the highly distributed nature of neural signal processing. First, it concludes that individual areas can participate in the processing of more than one function. Second, functions need not be implemented by individual cortical areas; they can be distributed over an array of such areas. Taken together, these two factors allow for the possibility that several modules are superimposed over a network of cortical areas. If so, it should come as little surprise that things look the way they do. The superimposition theory is perfectly compatible with the empirical data mentioned above.

Compatability is always nice, but it is clear that the superimposition theory can be successful only to the extent that it succeeds in solving what appears to be a particularly nasty version of the specification problem. On the one hand, we have numerous theories about which problems the visual system needs to solve. On the other hand, we have little by way of data that help us confirm our hypotheses. Moreover, if the superimposition theory is right, it may be excruciatingly difficult to ever get adequate empirical evidence. One of the few ways of coping with such difficulties is modeling (e.g., Lehky & Sejnowski 1988; Zipser & Andersen 1988).

The second option mentioned above – let us call it the "messy theory" of early vision – rules out the possibility of a computational explanation of vision altogether. However, we are not quite prepared to accept **Robinson**'s pessimistic conclusion that such a scenario would leave us without *any* hope of understanding brain function. The empirical sciences of the mind continue to accumulate data, and it would appear preposterous to continue to dismiss their findings as irrelevant on grounds that the only adequate explanatory scheme fails to apply to the brain. On the contrary, a "messy theory" of, say, vision would turn things around. Empirical findings ranging from single-cell recordings all the way to experimental results in psychophysics and psychology would gain renewed importance and they would have to be accounted for within a different explanatory framework. It is our contention that ongoing developments in the theory of neural computation (e.g., Smolensky 1988; 1991) could provide such an account of explanation, one that no longer concentrates on functions but attempts to explain the dynamics of brain activity and the emergence of observed effects and regularities instead.

What will ultimately prevail – mess or superimposition – depends on the contributions of everyone involved in studying the mind/brain. But the mere fact that the messy theory is an option for early vision should caution us against being too optimistic about understanding higher-level cognitive functions.

ACKNOWLEDGMENT
We would like to thank Alexander Levine for his corrections and suggestions.

NOTES
1. For a detailed discussion of the specification problem and its consequences for computational explanations see Cummins & Schwarz (1991).

2. For the record, most of Fodor's discussion focuses on input modules, but examples like the VOR indicate how his views can be extended to peripheral modules in general.

3. Even the specification of middle temporal (MT), one of the best-studied visual areas, is not much more detailed than "is involved in the analysis of pattern motion."

An anatomy of parallel distributed processing

Benjamin Seltzer

Department of Psychiatry and Neurology, Tulane University School of Medicine, New Orleans, LA 70112

[JFS] As suggested by **Stein**, posterior parietal cortex (PPC) is a prime candidate for the role of the cerebral region most concerned with the representation of space. But PPC is not a unitary structure, in terms of either architecture or connections (Pandya & Seltzer 1982). The superior parietal lobule (area 5) receives its sensory-related cortical connections exclusively from somatic sensory areas. Cortical visual input, by contrast, is direct to the inferior parietal lobule (IPL) (area 7). The latter also receives somatic sensory afferents but little cortical auditory input. Even within area 7 there are connectional differences: between the exposed cortex of the IPL and adjacent lower bank of the intraparietal sulcus (area LIP) and between the rostral and caudal IPL. Although all these subdivisions of area 7 have complex functional correlates, it is the caudal IPL (area 7a; or caudal area PG and area Opt) that has the most diverse pattern of

cortical afferents and comes closest to fulfilling the connectional requirements for a "multimodal sensorimotor association area." The medial parietal cortex (area 7m or PGm) is not well understood.

In addition to the parasensory and paralimbic cortical input outlined by **Stein**, the PPC has another set of cortical connections of potential importance in understanding its functional role. These are pathways between the PPC and the superior temporal polymodal (STP) cortex in the upper bank of the superior temporal sulcus. Like the PPC, this area has reciprocal connections with parasensory (chiefly auditory and visual), paralimbic, and prefrontal regions of the hemisphere (Jones & Powell 1970; Seltzer & Pandya 1978; 1984; 1989; 1991a; 1991b). We have tentatively subdivided the STP cortex into different rostral-to-caudal zones. The rostral sector interconnects with the caudal IPL; the middle area with area LIP; and the caudal zone with the medial parietal cortex (area 7m). As Stein points out, there is little evidence for a topographical map of the world in PPC and the same appears to be true for the STP cortex (Hikosaka et al. 1988). What is the significance, then, of such a precise and intricate pattern of connections? The answer to this question is obviously unknown, but Stein's hypothesis about the mechanism of PPC function may be relevant. These pathways (among others) may constitute the actual anatomical substrate underlying parallel distributed processing. According to this interpretation, they hold within themselves not a map of the external environment, as may occur at earlier steps in the sequence of cortical sensory connections, but rather a set of rules for integrating the different types of sensory and nonsensory information that reach the parietotemporal junction. These PPC-STP connections are not the only pathways linking heteromodal cortexes. Both PPC and STP also interconnect with the prefrontal cortex and parahippocampal gyrus. Neural networks subserving highly complex functions are part of a hemisphere-wide system (Goldman-Rakic 1988). Nevertheless, as Mesulam (1990) has suggested, there may be certain regional specializations within this large-scale network.

The PPC (particularly the IPL) pole of this neural network has traditionally been associated with attentional mechanisms. This is based in part on clinical observations in humans reviewed by **Stein**. However, as Watson et al. (1985) have pointed out, the human IPL corresponds in architectonic terms to Brodmann's areas 39 and 40, not to area 7 as it does in the monkey. And some authors consider the monkey homologue of human areas 39 and 40 to be STP cortex (Jones & Powell 1970; Mesulam 1985). This does not detract from the significance of the PPC's role in attentional mechanisms but points out the need to consider the interactions between these two polymodal cortical areas (as well as others) in understanding the genesis of attention and other complex behavioral phenomena. As **Stein** writes in his summary, the PPC is where the cerebral representation of space "commences." But to trace it further one must ultimately take into consideration the entire distributed pathway of neural networks.

What do fast goal-directed movements teach us about equilibrium-point control?

Jeroen B. J. Smeets

Vakgroep Fysiologie, Erasmus Universiteit Rotterdam, NL-3000 DR Rotterdam, The Netherlands
Electronic mail: *smeets@fgg.eur.nl*

[EB] Since the work of Feldman (Feldman 1966a; for a recent description, see Feldman et al. 1990), researchers in motor control have been aware that, as the neuromuscular system behaves more or less in a spring-like fashion, the output of this system can be described not only in terms of muscle force and muscle length, but also in terms of equilibrium position and associated stiffness. By this change of variables, an elegant unified description of both voluntary and reflex movement is obtained. The λ-version of the theory is even more elegant, as it also includes a description of the EMG, the final outcome of the neural computations.

The question addressed by **Bizzi et al.**'s target article however, concerns whether equilibrium points are more than a mere description. The claim of Bizzi et al. is that the nervous system *thinks* in equilibrium points. The question "How can we know whether the nervous system uses a certain set of variables for movement control?" is an important issue in motor control, but not a new one (see, for example, Stein 1982). In their target article, Bizzi et al. show how they judge the different possibilities. According to them, the nervous system has limited powers of computation, so a control strategy used by the nervous system should be simple and robust. More specific: The variables used by the nervous system will be simple functions of time. I will follow this line of reasoning and test the equilibrium-point model in a situation that imposes severe time constraints on the computations: fast goal-directed movements.

One of the problems of the equilibrium-point model is that fast goal-directed arm movements cannot be described by a simple shift of the equilibrium point. How can we describe these movements in terms of equilibrium points and stiffness? It was concluded from the EMG patterns during these movements (Gielen & Houk 1986) and from a mechanical analysis (Latash & Gottlieb 1991) that during fast movements the equilibrium point follows a trajectory that is very different from the actual trajectory of the hand. The calculation of this trajectory is of the same complexity as a calculation of the requisite torques. So the argument of simplicity does not hold in situations in which the nervous system would need simplicity the most.

Concerning fast goal-directed movements, the equilibrium-point model makes an explicit prediction: Independent of the load to be moved, the goal will be reached purely because of the stiffness of the neuromuscular system (Schmidt & McGown 1980). Misjudgments of load or changes of load during movements will be compensated without a recalculation of the equilibrium-point trajectory. Any corrections in the EMG patterns associated with a perturbation will thus lag the perturbation. This prediction has recently been tested (Smeets et al. 1990). In this experiment, subjects had to move a mass as fast as possible over a rail to a target position. In some trials however, the mass to be moved was unexpectedly increased or decreased. This resulted (apart from a slower or faster movement execution than planned) in changes in the EMG patterns, as predicted by the λ-theory. By cross-correlating the changes in EMG activity with the changes in velocity and position, it was found that (on average) the EMG response preceded (!) the deviation in velocity and position.

In later experiments (Smeets et al. 1992) it was shown that the response to an unexpected mass consisted of two components. The first component (latency between 25 and 40 msec) is a simple feedback response, as predicted by the λ theory. The second component (latency about 60 msec) preceded errors in movement execution. In equilibrium-point terms this means that not all corrections are due to stiffness; the equilibrium-point trajectory is also changed.

An analysis of stretch-reflex experiments can give some more insight into the strategies used by the nervous system. In experiments in which more than one degree of freedom was studied (Gielen et al. 1988; Smeets & Erkelens 1991) is was found that the direction of the response depended on its latency. If we want to describe these responses with an equilibrium-point model (which is of course possible), we must use an equilibrium point that already changes during the first 100 msec after the perturbation.

From these experiments we can conclude that if the human nervous system uses equilibrium points, it does so in a more complex way than suggested by **Bizzi et al.**. Probably a more

adequate description is that the nervous system tries to find the best feedforward activation (motor program[1]), and uses the elastic properties (including feedback of afferent information) to compensate for misjudgments or perturbations. The afferent information is not only used for direct feedback, however; it is also used to adjust the feedforward part of the motor program.

This "programming hypothesis" does make some predictions that can be tested in neurophysiological experiments like those performed by **Bizzi et al.** If in their experiments on the spinal frog the angles of joints distal to the force transducer are changed, the inertial properties of the limb are also changed. These changes in joint angle can be sensed (for instance by muscle spindles), and the nervous system will use this information to change the feedforward part of the program. The programming hypothesis thus predicts for a given stimulation a change in the measured equilibrium-point trajectory when distal joint angles are changed. According to the equilibrium-point model, however, the nervous system does not use knowledge about inertia, so for a given stimulation, the equilibrium-point trajectory should be independent of the configuration of the distal joints. In this way, it can be tested whether the motor centers in the spinal cord of the frog control equilibrium points.

In conclusion: Measuring the trajectory λ(t) is a way to describe the feedforward part of the control system; measuring the stiffness field is a way to describe the feedback part. I have given experimental evidence that the human motor control system does something more complex than setting equilibrium points, and I have suggested an experiment to test this hypothesis for the spinal frog. The programming hypothesis leaves the most interesting problem open for further research: How is the feedforward part programmed?

NOTE
1. I use the term "motor program" here slightly differently from the way Keele (1968) introduced it, and without the algorithmic interpretation given to it by some authors (e.g., **Alexander et al.**, this issue).

Can the inferior olive both excite and inhibit Purkinje cells?

Allan M. Smith
Centre de Recherche en Sciences Neurologiques, Université de Montréal, Montreal, Quebec, Canada H3C 3J7

[JRB] I enjoyed reading **Bloedel**'s target article, and there are actually only a few aspects upon which I might offer an alternative opinion. From the outset my reading of the neurological history of the cerebellum is somewhat less congenial than Bloedel's, and suggests that cerebellar function has been a bone of contention ever since Gordon Holmes (1927) snubbed the excellent French neurological studies of Babinski (1899) and Thomas (1925). The controversy as to whether the cerebellum coordinates postures and movements through the control of appropriate muscle synergies, or whether these functions are simply the product of cerebellar actions assuring the appropriate timing, strength, and stability of muscular activity is still with us today. Enough history, however; let us now examine Bloedel's very contemporary view of cerebellar operations.

My first and perhaps most fundamental difficulty is with **Bloedel**'s conviction that climbing fibers selectively enhance the parallel fiber inputs to the same Purkinje cells. Although he provides some evidence to support his supposition, he skillfully skirts any neurophysiological explanation of the precise mechanism by which a climbing fiber potentiation of parallel fiber excitation of Purkinje cells might be accomplished. He also appears to ignore some of his own evidence to the contrary: ("Our data show that both the direct excitatory action of climbing fibers on the dendrites of Purkinje cells and the activation of interneurons by climbing fiber collaterals can effect a suppres-

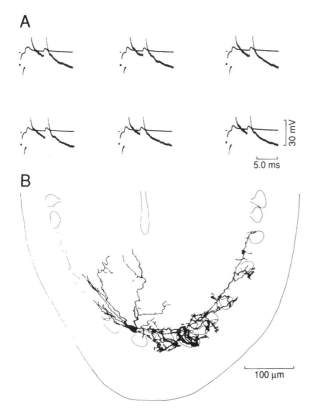

A

B

30 mV

5.0 ms

100 μm

Figure 1 (Smith).   Basket cell receiving olivary excitation. Above are six photographs of the action potential recorded in a cerebellar basket cell after single shock electrical stimulation of the inferior olive. A high and low gain sweep are shown and the calibration of 30 mv applied only to the high gain response in which both vertical peaks of both the stimulus artifact and the action potential are truncated. Below is the reconstruction of this basket cell after intracellular injection of HRP (drawn by Y. Shinoda).

sion of the spike activity in spontaneously active Purkinje cells," Bloedel & Roberts 1971, p. 29). In agreement with this statement, I would contend that either electrical stimulation (Bloedel & Roberts 1971; Murphy & Sabah 1970) or chemical activation of the inferior olive by harmaline (Lamarre et al. 1971) produces a suppression of simple spike activity. It is interesting to note in passing that Figure 1F of Lamarre et al. (1971) illustrates the rhythmic activation of a cerebellar cortical inter-neuron in response to harmaline. In contrast, lesions of the inferior olive appear to increase the tonic simple spike activity many times over (Batini & Billard 1985; Benedetti et al. 1984; Savio & Tempia 1985). Some years ago, we impaled a basket cell that could be activated monosynaptically by inferior olivary stimulation shown here in Figure 1 (Krieger, Shinoda & Smith, unpublished observations; Smith 1990). If this observation is not a total coincidence, then this olivo-basket-Purkinje cell inhibition is a powerful means whereby a single olivary neuron discharging at 2–3 spikes/sec could inhibit approximately 10 tonically active and parasagittally arrayed Purkinje cells (Palay & Chan-Palay 1974). The consequence of this basket cell inhibition would be to reduce climbing fiber Purkinje cell excitation from a parasagittal band to either a "patch" or more likely a medio-lateral strip.

My second point is related to the statement that mossy fibers are "not organized on the basis of the sagittal zones." Although perhaps not the zones **Bloedel** had in mind, the mossy fibers nevertheless terminate in clearly recognizable rostro-caudal

bands (Jasmin & Courville 1987). In addition, the terminal arborizations of individual mossy fibers extend over rostro-caudal distances that exceed the medio-lateral dimensions by a ratio of about 4 to 1 (Krieger et al. 1985). Moreover, the terminal rosettes do not form a continuous band, but appear instead to target specific granule cell regions, skipping or avoiding other intervening regions (see Figure 2 of Krieger et al. 1985). These clustered terminals would no doubt present a "patchy mosaic-like" appearance to a single unit recording electrode.

My final comment is that our knowledge of the olivo-cerebellar story is still far from complete. Our laboratory has been studying the effects of harmaline tremor in three strains of mutant mice with total Purkinje cell degeneration (Milner & Smith, unpublished observations). The harmaline injections caused strong tremor in PCD mice, weak and intermittent tremor in Nervous mice, and no tremor in Lurcher mutants. The degree of tremor could be correlated with the percentage of inferior olivary neurons capable of retrograde transport of WGA-HRP from the cerebellar cortex (Cadoret, Lessard & Smith, unpublished observations). This observation suggests, among other things, that the inferior olive, through the col-laterals of climbing fibers, exerts a powerful direct effect on the cerebellar nuclei that is independent of any cerebellar cortical influence.

In answer to **Bloedel**'s original question concerning what operations the cerebellum might perform, I still feel that Babinski was probably right when he suggested it must be the control of muscle synergy. In essence, this means the coopera-tion between antagonist muscles in either cocontraction or reciprocal inhibition, at single or multiple joints, and in both axial and distal body parts. The control of joint stiffness is a fundamental motor control function that is too complex to be relegated exclusively to peripheral afferents acting through reflexes but too trivial to require the full attention of the cerebral cortex. In the final analysis, as Bloedel suggests, the cerebellum is the perfect intermediary.

The role of the cerebellum in calibrating feedforward control

J. F. Stein

University Laboratory of Physiology, Oxford University, Oxford OX1 3PT, England
Electronic mail: *stein@vax.oxford.ac.uk*

[JRB] Those who attempt to discover the "function" of the cerebellum find it a most frustrating experience. Even though we probably know more about microcircuitry and electrical intercourse between cells in the cerebellum than in any other part of the brain, this knowledge has gained us no great insights into what the cerebellum does. We still know less about this area functionally than almost any other part of the brain. Recent PET and other studies suggesting a cerebellar role in cognition have even undermined our old assurance that it has an exclusively motor function (Leiner et al. 1989).

The most striking feature of the cerebellar cortex is its homogeneous structure. Not even the most experienced histo-logist can state with any certainty where a particular cerebellar section may come from: the anterior or posterior lobe, vermis or hemispheres; this is quite unlike the cerebral cortex. Through-out its 50,000 square centimetres, the cerebellar cortex has an identically simple structure. As **Bloedel** points out, this unifor-mity of course implies very strongly that all parts perform the same basic information processing operation. But it does not tell us what this operation is. A myriad different functions have been ascribed to the cerebellum. However, these are probably a consequence of the different connections of different parts of the cerebellar cortex rather than of different processing functions in those parts. The cerebellum connects with separate regions of the brainstem and cerebral cortex (Allen & Tsukahara 1974) and these regions all have separate functions. This is why the cerebellum has been shown to contribute to physiological control systems as disparate as the baroreceptor, vestibulo-ocular and stretch reflexes, automatic and voluntary movements, feedback and feedforward, as well as parametric and predictive control (Ito 1984). None of this knowledge, however, has told us what the basic information processing function of the cerebellar cortex is.

Bloedel's target article introduces three new concepts to remedy this defect: gain change, dynamic selection, and "Vermittler." Of the three, the details of the gain-change hypothesis are the most controversial. Dynamic selection has many similarities with Eccles's dynamic loop hypothesis (Eccles 1969), and Vermittler translates quite easily into MacKay and Murphy's "metasystem" for the parametric control of reflexes (1979) or Ito's side path model (Ito 1984). But the gain change that Bloedel observes is that synchronous climbing fibre discharge causes a 100–200 msec enhancement of Purkinje cell simple spike activity (Bloedel et al. 1983). This effect is opposite both to the classical view that climbing fibres cause a short-term (10–20 msec) "inactivation" response (Eccles et al. 1966; Granit & Phillips 1956) and to the current belief that paired climbing and parallel fibre stimulation leads to long-term (hours) depression of simple spike activity (Ito 1989). It also conflicts to some extent with the evidence that climbing fibres exert a tonic inhibitory effect on Purkinje cell simple spike activity (Strata 1987). Nevertheless, Bloedel's evidence is convincing and demands an explanation.

Despite **Bloedel**'s opposition to the idea that the cerebellum plays a role in motor learning, I believe that the resolution of these contradictory findings lies in recent discoveries that have been made at the behavioural, biophysical, and system simulation levels about the role of the cerebellum in acquiring motor skills.

The major problem facing motor systems is that negative feedback control, despite its inherent precision, is normally too slow to guide movements successfully. Feedforward control must therefore be used. The big drawback of feedforward control is that it tends to be inaccurate because it cannot cope with changes in the environment. The trick for feedforward control is accordingly to learn to use feedback to adjust the parameters of the feedforward pathway for a movement, to "calibrate" it, so that it precisely meets the requirements of the job. Another way of looking at this calibration process is to consider the transformations of scale and reference points that are required to shift from one coordinate system to another, as in **Bloedel**'s discussion. There is now a wide consensus, with which he would probably agree, that the adjustment of feedforward control parameters is the main function of the cerebellum (Stein & Glickstein 1992). What Bloedel does not admit is that this calibrating operation is precisely the one that is required for learning motor skills. Adjusting the parameters of feedforward processes to adapt to changed circumstances is how we acquire and automaticise new skills.

One of the reasons **Bloedel** resists this idea is that he believes there may be only one site for the long-lasting, structural, "plastic" changes in synaptic strength that are clearly required for adaptation of the vestibulo-ocular reflex or for the conditioning of the nictitating membrane reflex. Without entering into the controversy about the latter (Yeo 1991b), it now seems almost certain in the case of the former that plastic changes occur both in the main disynaptic vestibulo-ocular reflex (VOR) pathway and in the cerebellar side pathway when the VOR is adapted to changed optical conditions (Lisberger 1988b; Miles et al. 1980). We should not be at all surprised if it emerges that the same is true in the case of conditioning the nictitating membrane, or any other reflex.

Bloedel's discovery that synchronous climbing fibre discharge can cause short-term increases in Purkinje cell simple spike activity actually fits quite well with the learning hypothesis he opposes. This is because neural network and other simulations of the requisite properties of a learning machine indicate that separately neither long-term depression alone nor long-term potentiation alone can support efficient learning (Willshaw & Dayan 1990). Both processes must be capable of taking place, if not in the same synapse, at least on the same neurone.

At a biophysical level Ito and his colleagues have shown clearly that the parallel fibre/Purkinje cell synapse is depressed for long periods after combined climbing fibre and parallel fibre stimulation (Ito 1989), even though it has been argued, I think unconvincingly, that his use of GABA blockers invalidates these results. Crepel and his colleagues have both confirmed Ito's results and shown the Hebbian converse that mild depolarisation of Purkinje cells coupled with parallel fibre stimulation can lead to potentiation of simple spike activity (Crepel & Jaillard 1991). The latter result is perhaps analogous to **Bloedel**'s findings. It suggests that single climbing fibre discharges might cause short-term facilitation, whereas multiple discharges may have the opposite effect. Whatever the explanation, the finding that both facilitation and depression can be observed in Purkinje cells lends further weight to the hypothesis that the cerebellum is indeed one site for the kind of synaptic plasticity which underlies motor learning.

In short, the cerebellum is indeed a metasystem, side path, or Vermittler; its function is probably to mediate the plastic changes in feedforward pathway synaptic strengths which are required to adapt to changed external conditions and to acquire new motor skills.

Varying the invariants of movement

Richard B. Stein

Division of Neuroscience, University of Alberta, Edmonton, Alberta, Canada T6G 2S2

[EB] **Bizzi et al.**'s target article is an excellent review of the theoretical controversies and the striking new experimental evidence regarding the equilibrium-point hypothesis. The work cited by Bizzi et al. and reviewed in Georgopoulos (1991) certainly supports the idea that movement trajectories can be specified by the CNS. The authors also provide a useful formulation in which the λ model is seen as a constrained version of the α model and both models are consistent with the notion of servo-assistance put forward many years ago by Matthews (1972) and myself (Stein 1974).

This implies that Feldman's (1986) "invariant characteristic" relating muscle length and tension can be varied as the "gain" of the reflex component of the length-tension characteristic changes. Indeed, much recent work suggests that reflex gain varies widely within a simple task, as in walking, and between one task and another, as in walking and standing (reviewed in Stein & Capaday 1988). Why does this variation take place? Does it not merely complicate the task of computing the target point and the trajectory to reach it? Yes, but such a complication is necessary in a real world where loads and muscle properties are continually changing.

If Bizzi et al. (1984) had applied maintained torques, rather than transient positive (and negative) ones to their deafferented monkeys, the arm would have overshot (or undershot) the target rather than reaching the equilibrium point. Yet we work all the time against a variety of external (and internal) loads and must compensate for the torques generated by these loads if we are to successfully hit our targets. Computation of a "virtual trajectory" to an equilibrium point is not enough.

In principle, there are three solutions to this problem. First, the nervous system might compute the inverse dynamics required, taking into account the load and muscle properties, if these are accurately known. I believe **Bizzi et al.** are right to discard this option as unrealistic.

Second, reflexes can be used to assist in reaching the goal despite differing conditions. Reflex gains in biological systems are inevitably low, however, because of problems caused by delays and potential instability, so load compensation will be incomplete. These potential problems will be more or less acute in different tasks. Thus, by varying gain adaptively, reflexes can be used to compensate for deviations to the maximum extent possible at any time. Patients in whom these reflex adjustments are severely compromised may have extreme difficulty in performing tasks such as walking, although they can still produce approximately correct patterns of muscle activity (Yang et al. 1991).

Finally, by trial and error one can adjust the magnitude and timing of muscle activations and hence the trajectory and equilibrium point, based on knowledge of results from previous trials. This motor learning must take place in skilled movements, such as the ones required to achieve the "right" golf swing. The need for trial and error learning can account for the slow time course and day-to-day variations as muscle properties and external conditions change. I therefore view the equilibrium-point hypothesis as a good but basic starting point from which one can add the task-dependent reflex variations and the variations in central activation patterns needed to achieve accurate, complex movements in a changing environment.

The demise of the motor program

Jeffery J. Summers

Department of Psychology, University of Melbourne, Parkville VIC 3052, Australia
Electronic mail: *jeff=summers%psychology%unimelb@muwaye. unimelb.edu.au*

[GEA] **Alexander et al.**'s central hypothesis is that the view that the motor system contains motor programs and uses algorithmic processing can no longer be sustained in light of current knowledge of the neural mechanisms underlying movement. As such, the authors join a growing number of researchers from both behavioral (e.g., Kelso 1981; Pew 1984; Summers 1989) and physiological orientations (e.g., Seal et al. 1992) to question the concept of a motor program.

The motor program concept was developed within the information processing approach in psychology to explain aspects of skilled motor behavior, such as the planning and central representation of actions, which had no known neural correlates. Since Keele's (1968) initial definition of a motor program as "a set of muscle commands that are structured before a movement begins" (p. 387), the concept has been adopted almost universally by motor control researchers across a number of disciplines (e.g., engineering, human movement science, neuroscience, physiology, psychology). The motor program, however, has been plagued by a lack of distinction between merely metaphoric and literal applications of the concept (Stelmach & Hughes 1984).

Within the psychology domain the status of the motor program as a metaphor or a physical entity has been particularly unclear. Keele and Summers (1976), for example, implicated follow-up servo and alpha-gamma coactivation mechanisms in the translation of motor programs into muscular movement. In recent years, however, the motor program concept has undergone considerable modification and there has been a shift from a literal use of the term to a more metaphoric use emphasizing the abstract nature of programs (for overview, see Summers 1989). Rosenbaum (1991), for example, suggests that a motor program should be viewed "as a functional state that allows particular movements, or classes of movements, to occur" (p. 109). In

contrast, in the neurological literature on movement disorders motor programs continue to enjoy the status of identifiable entities with specific locations in the brain designated for the storage, assembly, and execution of programs (e.g., Delwaide & Gonce 1988; Haaland & Harrington 1990; Marsden 1984). There have also been recent attempts to model neuromotor programming in terms of corticostriatal interactions (Wickens 1991).

The term "motor program," therefore, has been defined and used in a variety different ways by researchers in different disciplines. It may be, as Pew (1984) has suggested, that a unitary integrative concept of a motor program cannot be found because the program may have different representations at different levels in the motor system. Other researchers in the behavior analysis of movement have already abandoned the concept completely (e.g., Kelso 1981; Saltzman & Kelso 1987).

A metaphoric use of the term "motor programming" may still be justified to describe cognitive processes involved in movement planning without specifying the neural correlates. However, in view of the lack of consensus as to what exactly is a motor program and whether it is a metaphoric or literal term, I agree with **Alexander et al.** that continued use of the term may actually impede progress in the field.

Alexander et al. suggest that for an understanding of the neural mechanisms in motor control researchers should look toward "connectionist" models that can be integrated with and tested against neurophysiological data. In particular, the capacity for self-organization independent of any external controller is seen as an important feature of the motor system's operation that needs to be modeled. It is maintained, however, that there are few models of this type of process available. An approach to this issue that the authors neglect is the study of dynamical/emergent properties of motor control exemplified by the work of Kelso, Kugler, and Turvey (e.g., Kelso 1981; Kugler & Turvey 1987). In this approach the emphasis is on self-organizing systems, and motor behavior is modeled using principles derived from the physics of nonlinear limit-cycle oscillator systems. Others have argued for a hybrid view of a multileveled motor system involving a cognitive system able to override essentially autonomous lower-level dynamic systems (see Abernethy & Sparrow 1992; Summers 1992). Some consideration of dynamical theories as an alternative to computational models of movement control would have made a useful addition to the target article.

Cortical area-specific activity not yet found?

Jun Tanji

Department of Physiology, Tohoku University, School of Medicine, Aoba-Ku, Sendai 980, Japan
Electronic mail: *n7aOmid@jpntuvm0.bitnet*

[EEF] Apart from the issue of neuronal coding of movement parameters, **Fetz**'s target article presents a number of views on the organization of multiple motor areas. Although he does not deny the specialization of different areas, the author warns against a simplistic view that different areas subserve different functions. The following points of fundamental importance are clearly made: (1) Studying relative timing of cellular activity does not provide any basis for determining the causal hierarchy among different areas. (2) Neurons related to the activation of muscles as well as to motor set are widely distributed over many areas, suggesting broadly distributed representation of these motor processes. (3) No such thing as preferential coding of particular movement parameters has been found in a single motor area. I agree entirely that a given motor area would be involved in diverse functions, so that cell types of enormous diversity will be observed in each area. It is about time to abandon an excessively simplistic view that motor areas are patchworks of regions involved in totally different functions.

As much as I admire **Fetz**'s article, I must challenge the inference that no area-specific differences in properties of neuronal activity have been reported. Reading this article, one would conclude that only neurons of the same type have been found everywhere in motor areas and in more or less the same quantity. This may be true in relation to muscle activation or "motor set" (for future movements in different directions), as described in detail in the text. One should not understate recent studies, however, which report area-specific preferential relations of neuronal activity to particular aspects of motor behavior (Humphrey & Freund 1991; Rizzolatti & Gentilucci 1988; Tanji & Kurata 1989).

Let us take an example of a recent report from our own laboratory (Mushiake et al. 1991). Neuronal activity was examined while monkeys were performing sequential movements either under visual guidance or on the basis of memorized information. Most neurons in the primary motor cortex were typically related to movements in a similar way, regardless of whether they were guided by visual signals or memory. Exceptions were extremely rare. On the other hand, a majority of neurons in secondary motor areas exhibited preferential relations to either visually guided or memorized movements. Premotor cortical neurons were more active when visually guided, whereas neurons in the supplementary motor area were more active with memorized sequential movements. These differences are not at all trivial and point to the presence of specialization in cortical areas.

There is also a report describing profound differences of neuronal activity in primary and secondary motor areas in relation to the use of one hand or both hands together (Tanji et al. 1988). [See also MacNeilage et al.: "Primate Handedness Reconsidered." *BBS* 10(2) 1987.] I do not believe that secondary motor areas are the *only* essential ones in these specialized aspects of motor behavior. In this sense, I agree with **Fetz** that "a region's specialized function need not be its only function, and certainly should not be the only standard for interpreting what each of the cells in this region is coding." On the other hand, just because one fails to find area-specific differences in neuronal activity in relation to relatively easy motor tasks, it does not follow that one cannot find aspects of regional specialization by observing cellular activity.

The statement in the target article: "Thus, the notion that cortical functions are segregated into different cortical areas can be preserved only by imposing different interpretations on similar experimental evidence" is accordingly not acceptable. I wish to point out the effectiveness of surveying neuronal activity in experimental conditions addressed to well-defined behavioral specificities. The question we pose is: "When the organism needs to perform a particular motor task in a particular behavioral condition, what type of neuronal activity in which area is used?" After such studies (as mentioned in relation to secondary motor areas) have been undertaken, we can proceed to the problem of how the variety of activity found in different areas can give rise to the requisite motor performance. As discussed extensively in **Fetz**'s article, this will require computational modeling.

The cerebellum and memory

Richard F. Thompson

Neurosciences Program, University of Southern California, Los Angeles, CA 90089-2520
Electronic mail: *thompson@neuro.usc.edu*

[JRB] Much of **Bloedel**'s target article is concerned with his "new" hypothesis of cerebellar function:

> The cerebellum serves as an active mediator whose output provides the CNS with an optimized integration of the relevant features of external execution space, internal intention space, body scheme, activity in central pathways, and sensory information resulting from the movement . . . [hence] the cerebellar output can modify activity in central pathways responsible for motor execution. (sect. 8, para. 1)

Although vague, this view of cerebellar function seems largely derivative from the traditional view of the role of the cerebellum in motor coordination that evolved from the classic early studies of Holmes (1917) and Chambers and Sprague (1951; 1955). More contemporary views stress the adaptive nature of cerebellar function in the coordination of movement and in the storage of memories for such movements (Ito 1984; Ivry & Baldo 1992; Thach et al. 1992).

Here, I focus on **Bloedel**'s review of recent work on the essential role of the cerebellum in classical conditioning of discrete, adaptive movements. Bloedel does not cite most of the relevant evidence in the field here or in his other papers (e.g., Berthier & Moore 1990; Clark et al. 1984; Haley et al. 1988; Lavond et al. 1984; 1985; 1990a; Lewis et al. 1987; Lincoln et al. 1982; Mamounas et al. 1987; Mauk & Thompson 1987; Mauk et al. 1986; McCormick et al. 1985; Polenchar et al. 1985; Rosenfield & Moore 1983; Sears & Steinmetz 1991; Steinmetz et al. 1986; 1987; 1989; Thompson 1989; 1990; Yeo et al. 1986) and seems to misconstrue what little evidence he does cite. Thus, he claims that he (Kelly et al. 1990) has "demonstrated that acutely decerebrate, cerebellectomized rabbits could acquire as well as execute the conditioned nictitating membrane (NM) reflex" (sect. 5, para. 5). Kelly et al. (1990) reported results from only a very few of the many animals they tested; they used methods of measurement that would count spontaneous responses as CRs and did not run any of the control groups essential to rule out nonassociative processes such as sensitization or pseudoconditioning. Indeed, the training procedures they used (9 sec ITI) do not result in any learning at all in normal rabbits (Nordholm et al. 1991).

Yeo (1991b) trained acute, decerebrate rabbits with more reasonable procedures and reported that cerebellar lesions completely abolished CRs that were established in the decerebrate animals. It would seem that the responses described by Kelly et al. (1990) were due to sensitization (see discussion in Nordholm et al. 1991). To demonstrate that the memory trace is stored in the brainstem (i.e., not in the cerebellum) in normal animals it is necessary to train them in the normal state and to show (with appropriate control groups) that they retain the CR after decerebration and decerebellation. In fact, we showed just the opposite several years ago: Normal animals were trained and then decerebrated; they retained the eyeblink CR only if the cerebellum and red nucleus (a part of the necessary CR pathway) were not damaged (Mauk & Thompson 1987).

Next, according to **Bloedel**, Welsh and Harvey (1989a) found interpositus lesions that abolished the CR to "produce a performance deficit rather than a memory deficit." This is incorrect. Welsh and Harvey actually reported that effective lesions of the interpositus (that abolished the CR) had *no effect* on the amplitude of the UR at all US intensities they used. They claimed to have shown a very small effect on UR topography at very low intensity USs. In fact, they did not even demonstrate this because they did not report prelesion URs; to demonstrate a deficit in UR performance it is necessary to compare lesion effects on CRs and URs in the same animals pre- and postlesion. We have done so and find no persisting effect of lesions that permanently abolish the CR on any property of the UR over a wide range of US intensities (Ivkovich et al. 1990; Steinmetz et al. 1992). In marked contrast, lesions of the motor nuclei that produce massive and permanent impairment of UR performance have much less effect on CRs (Disterhoft et al. 1985; Ivkovich et al. 1991). Finally, large lesions of cerebellar cortex that markedly impair or abolish the CR result in an *increase* in UR amplitude (Logan 1991; Yeo 1991b). There is thus a double dissociation between lesion effects on the CR and UR; the "performance" argument is decisively negated (see also Ivry & Baldo 1992).

The most critical evidence to date for the cerebellar locus of the essential memory trace for eyeblink conditioning comes from Lavond's cold probe studies (Clark et al. 1992; Lavond et al. 1990b). If naive animals are trained while cerebellar tissue just dorsal to the interpositus nucleus is cooled, they do not learn and subsequently learn (no cooling) with no savings, as though they were naive to the situation. **Bloedel** claims that Welsh and Harvey (1989b), and now Bloedel himself, find that animals do learn if Lidocaine is infused in the cerebellum during training. In fact, just the opposite is true. In current work we find that infusion of Lidocaine in the critical region of the cerebellum during training in naive animals completely prevents learning and the animals subsequently learn with no savings. Cannula location and drug dose (concentration) are critically important. Higher concentrations of Lidocaine may be necessary to prevent CR acquisition rather than to reversibly abolish CR performance, at least with some cannula locations. With appropriate cannula locations, complete prevention of acquisition is a consistent finding.

In sum, evidence to date demonstrates that the cerebellum is absolutely necessary for both learning and memory in classical conditioning of eyeblink and other discrete responses. A wide range of evidence from electrophysiological, microstimulation, lesion, reversible inactivation, and anatomical studies, most of which are not cited by **Bloedel** here or in his other papers, strongly supports this statement and also the further hypotheses that the necessary CS pathway includes mossy fiber projections to the cerebellum, that the necessary US pathway includes the dorsal accessory olive and its climbing fiber projections to the cerebellum, and that the essential CR pathway involves projections from the interpositus nucleus to the magnocellular red nucleus to premotor and motor nuclei (Ivry & Baldo 1992; Steinmetz et al. 1989; Thompson 1990; Thompson & Steinmetz 1992). The fact that reversible inactivation of the critical region of the cerebellum completely prevents learning in naive animals (see above) argues very strongly that essential memory traces are in fact formed and stored in the cerebellum. All of these results are consistent with, and strongly supportive of, modern views of cerebellar function (e.g., Ito 1984; Thach et al. 1992) and with the now classic models of the cerebellum as a "learning machine" (Albus 1971; Eccles 1977; Eccles et al. 1967; Ito 1974; 1984; Marr 1969).

Nonlinear dynamical systems theory and engineering neural network: Can each afford plausible interpretation of "how" and "what"?

Ichiro Tsuda

Department of Artificial Intelligence, Faculty of Computer Sciences, Kyushu Institute of Technology, Iizuka, Fukuoka 820, Japan
Electronic mail: *tsuda@dumbo.ai.kyutech.ac.jp*

[DAR] **Robinson** succeeds in showing that neither "single-unit recordings" nor "black box models" alone are promising in our attempt to understand higher functions of the brain. He especially points out our inability to know "how the (information) processing is done" with single-unit recordings; nor can we describe "how real neurons do it" using mathematical black box models. His theory is persuasive because he describes it by citing his own and other neural network models and theories of the oculomotor system that he views as having succeeded in describing the system's functions. Nevertheless, there seems to be room for commentary. I accordingly will highlight the following two points: the notion of dynamic information processing, which is set aside in the target article, and the question of the methodology for understanding brain function.

In my opinion, before rejecting the possibility of interpreting how interneurons work, we must carefully investigate another process of encoding information by single neurons and by neuron assemblies: the temporal encoding of information. In area 17, neurons with dynamic orientation selectivity were recently found in addition to a conventional type of neuron known as the orientation-selective cell – the simple cell (Dinse et al. 1990). One type of neuron among these dynamic cells develops selectivity following a stage of nonselective excitation and the other changes temporally its orientation selectivity. Because of the abundance of such neurons in area 17, it is difficult to think of such a neuron being "a rogue" cell. Thus, even at the single-neuron level the intermixed information represented by a neuron in multimodal fashion may have another dimension for a code, that is, temporal dimension.

An elaborated analysis of the temporal modulation of neuron activities has been made in both cortical and subcortical areas. In particular, the temporal modulation of the spike trains of neurons not only in the inferior temporal cortex but also in the striate cortex was confirmed by evidence that cells could encode multiple parameters of the input information (Richmond et al. 1987; 1990; Richmond & Optican 1987; 1990). In this case, single-unit recording and principal component analysis were useful as experimental and mathematical tools, respectively. Further analysis of the temporal coding of data obtained in single-unit recordings might give "a very valuable piece of information" concerning how a neuron, even an interneuron, processes information at higher levels. However, even though a single neuron can encode multiple stimulus parameters in the waveform of spike train, the multiplicity of the code will not be so high at the single-neuron level, which will lead to incompleteness of encoding and hence decoding in single-neuron activities.

Thus, as **Robinson** points out, a complete spatiotemporal coding of information is ascribed to neural networks. Besides the oculomotor system described by Robinson, a beautiful example of the coding problem has been investigated by Freeman in measurements of collective activity in the rabbit's olfactory bulb as well as in neural network models (Freeman 1987; Skarda & Freeman 1987; Yao & Freeman 1990). Odor information is encoded in spatial patterns of the bulb's activities. Moreover, through the temporal chaos created in the dynamical processing of sensory information the bulb network acquires variability in the processing and learning of sensory information, depending on the animal's internal state; here, only a collective activity of neurons in the bulb is a meaningful variable, that is, observable.

There is a non-black-box mathematical theory that can extract hidden dynamics, namely, modern nonlinear dynamical systems theory. With this theory we can analyze, for example, how interneurons enhance their computational ability in consequence of the network's behaving in a determinstically chaotic manner (Tsuda 1992). One role of deterministic chaos is then to unravel in the temporal dimension the intermixed information over the spatial dimension, preserving the functional relevance of the overall activity of networks. This dynamic information processing can be generalized with a notion of "chaotic itinerancy" and embodied in a neural network model for dynamic associative memory (Tsuda 1991). Thus, dynamical systems theory may help explain how neural networks process information. It can also be applied to systems that cannot be "anatomically isolated," where most "black box mathematical models" fail, since is does not treat "input-output relations" but the process of information flow.

Nevertheless, dynamical systems theory cannot directly describe what meaning is processed. Contrary to **Robinson**, I would like to emphasize that what seems to be the most crucial but difficult, and almost unsolved, problem concerning the brain is the problem of "what" (see also Marr 1982). In brain theory, we have to deal with a system where the observer is involved. It is thus a theory of an observer who tries to understand the meaning and information structure of the environment, in relation to his internal state. Here, the environment cannot be decoupled from the observer. This process of understanding is inevitably interpretative. Classical hermeneutics may serve as a methodology for the understanding of such a system (Tsuda 1984; Winograd & Flores 1986).

We must interpret what a neuron or a neural network is doing except, perhaps, the pure output unit. It is determined in advance what a computer should do, and we do know the purpose of its whole system and even its parts, whereas we do not know this about the brain. We start with an assumption about "what," which allows a plausible interpretation. In other words, any machine is straightforward in the sense that it functions as a whole in consequence of being assembled of functional parts, whereas in the brain a temporal differentiation of functions seems to progress at the neuronal level in consequence of the system's purpose. This differentiation can be found in neural events with short time scales as well as in phylogenetic and ontogenetic development.

The plausibility of an interpretation cannot be directly substantiated by experiments. An engineering approach that gives an appropriate design for the system can help strengthen the plausibility. Thus, I agree with **Robinson** that "applications of neural networks . . . can build a bridge between system function and hidden-unit behavior and tell us how to relate one to the other."

Coordinate transformations in sensorimotor control: Persisting issues

J. A. M. Van Gisbergen and J. Duysens

Department of Medical Physics & Biophysics, University of Nijmegen, EZ 6525 Nijmegen, The Netherlands

Electronic mail: admin@mbfys.kun.nl

[JFS] Skillfully and successfully **Stein** has sketched a vast panorama of facts and issues that must be dealt with by any comprehensive theory of what may well be the most complex staging and integration area for multiple sensory and motor signals: the parietal cortex. In our opinion, the best we can hope for at the moment is that it will be possible to outline a number of fundamental questions and hot topics and to clarify these views with real data. In this, Stein's target article succeeds quite well. Our own contribution will be to try to bring some of the main issues outlined by Stein better into focus and to argue, based on the discussion of specific examples where we can rely on extensive data, that it may be premature to draw major conclusions at this early stage.

Stein emphasizes that the nonuniformity of sensory neural maps presents a problem for a linear transfer from sensory signals to movements; he argues that there may be a need, in representing real space, to demagnify the overrepresentation of high receptor density areas. The phenomenon of foveal sparing in parietal neurons with large receptive fields is offered as an example. We definitely agree that the nonuniformity of sensory neural maps is an important aspect to be considered in models of sensorimotor transformations. But is it really a problem calling for a radical solution at the level of the parietal cortex? Although we can see demagnification as a potentially useful transformation in some tasks, such as egomotion detection, it may have the consequence of a loss of precision in the control of fine orienting movements. Consider the control of saccadic eye movements, for example. The motor map of the superior colliculus has a nonuniform layout that is very similar to the nonuniform visual map (Ottes et al. 1986). This is the case even though the same neural map is also shared by control signals from the auditory system. The nonuniform collicular motor map is reflected in the greater precision of small saccades (Van Opstal & Van Gisbergen

1989a). In a quantitative model of the motor colliculus, it has been proposed that the transformation from collicular activity to movements is highly nonlinear and is in fact the inverse of the nonlinear mapping from retina to motor map (Van Gisbergen et al. 1987; Van Opstal & Van Gisbergen 1989b). Thus, at least in this case, the demagnification that is required in some form somewhere along the path from stimulus to movement seems to take place in the motor system. The control of fine finger movements may be another example where demagnification is probably deferred until a very late stage. Admittedly, these counterexamples do raise important unresolved questions about signal representations at the level of the parietal cortex.

One of the most interesting topics in **Stein's** target article is coordinate transformations, an area of research where very interesting results have been obtained in the last decade. [See also Flanders et al.: "Early Stages in a Sensorimotor Transformation" *BBS* 15(2) 1992.] Stein highlights the modeling of Zipser and Andersen (1988) as an interesting way of showing how units in the hidden layer of a neural network, trained to construct target location relative to the head from retinal stimulus location and eye position signals, have properties that are similar to those found in the parietal cortex. The parietal neurons Zipser and Andersen were trying to understand had been found in area 7a. These cells have very large receptive fields with a gradient of sensitivity that gives them directional tuning but often scarcely any eccentricity tuning. In many of these receptive fields (shown in detail in Zipser & Andersen 1988) it would be impossible to really identify a "hot spot." When the response of such a cell to the presentation of a given receptive field stimulus (at a fixed retinal location) is tested while the monkey fixates at various spatial locations, the activity of these cells varies as a smooth function of eye position. The gradient of this so-called gain field varies from cell to cell. Another type of cell, not sharply distinguished in Stein's paper and not accounted for by the Zipser and Andersen model, is found in LIP. Many of these cells code the desired saccade vector (motor error), even when the visual target is no longer there. Zipser and Andersen's claim (1988) is that area 7a gain-field neurons, of the type just described, code target location relative to the head. This work raises questions that are still the subject of a lively debate in the oculomotor research community:

1. An explicit representation of target location relative to the head, a coding format imposed as the output of the Zipser and Andersen model, is not found as such in the brain. According to this criticism it is therefore not clear how the hidden unit properties should be interpreted.

2. It is now quite clear that the coding at the level of the superior colliculus has the format of a topographical motor error signal. How could such a signal be derived from a target relative to the head signal?

Recent neural network simulations in our laboratory (Krommenhoek et al. 1992; submitted for publication) have shown that Robinson's model, the Zipser and Andersen model, and the coding format in the superior colliculus can be incorporated in a unified model where eye-position dependent units (similar to those in area 7a) play the role of coding target position relative to the head [see also **Robinson,** this issue]. The subtractive efference copy loop in Robinson's model (which represents instantaneous eye position) is then used to create a topographic map of motor error as we know it from work on the monkey superior colliculus. It should be added that this transformation from head to ocular coordinates (using efference copy) requires an intermediate step that already contains motor error signals but relies on a different (nontopographical) coding format. This might explain the presence of motor error signals at a supracollicular level, such as in LIP, where (as **Stein** notes) a retinotopic oculomotor frame of reference seems to prevail.

It is well established that saccadic control is not just based on the retinal location of the target but may also take into account the occurrence of eye movements. This allows the system to refixate the location of a remembered target, after making a saccade in a wrong direction, even though the experimental paradigm precludes the use of visual feedback. This was first regarded as strong evidence that the goal for saccades is coded in craniotopic coordinates, but several groups have now emphasized, with good reason, that all that is required to explain these responses is the computation of motor error. To compute motor error, it has been argued, does not require the availability of a signal coding target location relative to the head, but could be derived from retinal information by accounting for subsequent eye displacements. Although this is a valid point of view, it leaves us without any idea why gain-field neurons exist.

In conclusion, we agree, based on the elegant study by Zipser and Andersen, that there is no particular reason to expect neat topographic maps coding egocentric space. Thus, topographical coding, which is such a striking feature in sensory and motor systems, may be lost at the core of the system responsible for major signal transformations between these systems. If correct, this would be an important principle. But to assert (see **Stein**'s Summary and conclusions) that different sensorimotor reference frames may not have to be converted into a common coordinate system goes much further and is premature. For example, the existence of a common motor map in the colliculus for command signals with a different sensory origin suggests otherwise. The mapping of vestibular and optokinetic signals into a common coordinate frame which plays a role in the control of the vestibulo-ocular reflex is another example. Similar objects apply to the statement that each system operates within its own reference frame, although this may be correct at the peripheral stages of motor control. The possibility that, at a central level, signals coding target location in body coordinates are available for several motor subsystems remains an intriguing one which deserves further study.

ACKNOWLEDGMENT
The authors are supported by NWO and ESPRIT II (MUCOM 3149).

Is position information alone sufficient for the control of external forces?

G. J. van Ingen Schenau, P. J. Beek and R. J. Bootsma
Faculty of Human Movement Sciences, Vrije Universiteit, 1081 BT Amsterdam, The Netherlands
Electronic mail: *p_beek@sara.nl*

[EB] An essential problem in the study of the organization of motor control concerns the question of how information guides the production of movement (cf. Beek & Bootsma 1991). **Bizzi et al.**'s α-model provides a most valuable contribution to this problem because it demonstrates convincingly that information with respect to actual position and virtual equilibrium position suffices to generate the requisite torques at the joints in reaching tasks without the necessity of any internal representation of these torques. The authors claim that this is also the case for what they define as "contact control tasks" (sect. 4). We see arguments, however, that render this claim problematic.

In most motor tasks the hand or foot that is in contact with the environment has to exert a force on the environment in certain directions. This requires a distinct and functionally specific distribution of torques at the joints. A simple example may serve to illustrate this (Fig. 1).

Imagine a subject sitting at a table and being asked to slide an object across the table in the direction indicated. Let us, for reasons of simplicity, assume that the movement is executed with uniform velocity (inertial effects being ignored), and that the friction between object and table is constant. If the table surface is positioned horizontally with respect to the gravitational vertical then the task requires a constant force F_h in the desired direction of movement (Fig. 1a.). At any position, this external force can be realized only by a specific combination of a flexing torque M_e at the elbow and a flexing torque M_s at the shoulder. Moving through the series of positions depicted, one observes initially an elbow flexion and subsequently an elbow extension. This sequence of flexion and extension shows that in multijoint movements the directions of the torques at the joints that are required to control the direction of the external force need not necessarily be the same as the direction of movement (Van Ingen Schenau et al. 1992).

The required external force and hence the pattern of torques required at the joints can also depend on other forces acting on the object. If, for example, the table is tilted in such a way that

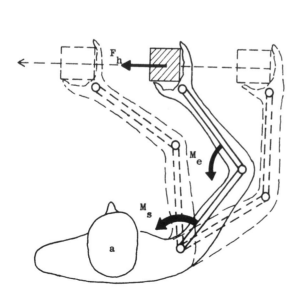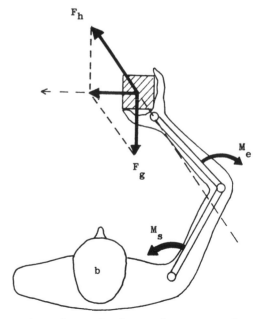

Figure 1 (Van Ingen Schenau et al.).   The control of an external force requires a distinct distribution of torques over the joints that cannot be generated on the basis of position information only.

the object comes under the influence of a component F_g due to gravity (Fig. 1b), the force exerted on the object by the hand has to be changed in both magnitude and direction in order to achieve the resultant force on the object drawn in Fig. 1a. Hence, the requisite torques at the joints depend strongly on the mechanical interaction with the environment, even when the desired displacement remains the same. How these multi-joint contact tasks are accomplished cannot be accounted for on the basis of position information alone as is suggested by Figure 3 of the target article. Additional information (perhaps also of a haptic or tactile nature) seems to be necessary to specify adequately the required configuration of torques at the joints.

As our Figure 1b shows, the required direction of force may deviate from the desired direction of movement. But even in the situation depicted in Figure 1a it seems impossible to generate the requisite torques using position information only. The torque at the shoulder has to remain constant, which would require a virtual equilibrium trajectory that runs in front of the object but cannot at the same time prescribe the requisite torque at the elbow (or at the wrist). At the elbow we need a flexing torque that gradually decreases, while the elbow first flexes and then extends. To assume a springlike behavior of elbow flexors would lead, contrary to what is required, to an increase in flexing torque in the second part of the movement. This problem cannot in general be avoided by activation of the bi-articular part of the biceps (Van Ingen Schenau et al. 1992).

The preceding arguments (which also apply to Feldman's [1986] λ-model and Bullock and Grossberg's [1991] VITE-model) favor explicit specification of torques at the joints, but they do not necessarily contradict the α-model in all aspects. In fact we have found strong evidence that activation of mono-articular leg muscles is indeed mainly based on the required changes in joint position, irrespective of the required torques at the joints associated with the control of the external force. In various leg extension tasks we have also found that bi-articular muscles seem to effect the required distribution of net torques over the joints (Gielen & Van Ingen Schenau 1992; Van Ingen Schenau 1989; Van Ingen Schenau et al. 1992).

Since we agree with the argument of **Bizzi et al.** that muscles should not to be evaluated in their capacity as force generators, the question remains as to how the control of mono- and bi-articular muscles is organized so as to generate appropriate (time-varying, increasing and decreasing) torques at the joints that lead to stable movements and accurately directed external forces. Particularly in the context of multijoint movements in which contact is made with objects in the environment, the control of muscle torques seems to require more than position information alone.

We agree with **Bizzi et al.** that explicit computation of the inverse dynamic problem is neither attractive nor very likely. The experience that the types of complex movements discussed often require some practice might point to another way to avoid these computations, namely, learning through gradual adaptation of neural networks. During this process multimodal sources of information can be linked to movement execution along the lines indicated by, for example, Stein (1986) and Kennedy (1990).

In sum, we are inclined to answer the question posed in the title of this commentary in the negative and we submit that this is particularly true in the context of multijoint contact movements.

Synthesized neural/biochemical models used for realistic 3-D tasks are more likely to provide answers

Jack M. Winters[a] and Patricia Mullins[b]
[a]Biomedical Engineering Program and [b]Department of Psychology, Catholic University of America, Washington, DC 20064
Electronic mail: *winters@pluto.cua.edu*

[GEA, EB, SCG, DAM, DAR] **1. Make the jump toward an inclusive field-based hypothesis for human movement (JMW)**

1.1. Virtual goal and mechanical stability. It is suggested that the strength of the equilibrium-point hypothesis lies beyond the suggested existence of an equilibrium "point" or "trajectory." Rather, it lies in the tie between the concepts of a "virtual goal" and mechanical stability. Arm movements are inherently goal-directed, and most prove to be inherently stable even when in contact with variable external environments (e.g., contact at hand) and with variable proximal conditions (e.g., same arm movement made while sitting or standing). The concept of a centrally commanded "virtual trajectory" that serves to "attract" a dynamic system is intriguing. That the central command takes the form of a position trajectory has been convenient for tasks considered so far, but it does not seem to preclude other forms for fundamentally different classes of movement goals.

1.2. Activation-dependent muscle "springs." Both the active "length-tension" and the series compliance relations show activation-dependent variations in stiffness, with the series element more dramatic for low levels of activation (Winters 1990). There is also a stiction-like contractile element effect near zero velocity and variable springlike effects resulting from the neurocircuitry. In addition, as muscles cocontract, muscle spindle "gains" can increase without loss of stability. In summary, the central command "sees" a very springlike periphery.

1.3. Whole-body "anticipatory postural adjustments." For fast voluntary arm movements while standing, selective muscles within the torso and even the legs are active in "anticipation" of the prime movers movement (e.g., Bouisset & Zattara 1990). This suggests a need for some combination of impedance modulation, inertial dynamics compensation, and/or gravitational compensation.

1.4. Torso muscles strongly influence hand movement. Ongoing sensitivity analysis studies with simple 4- and 5-link torso-shoulder-arm models (mounted at pelvis) show that hand position is much more sensitive to changes in muscle activation or properties of muscles in the torso than to those for the wrist or elbow. Why? Two reasons emerge: (1) because of the longer distance from hand to back, a small rotation at the torso causes a large hand movement; and (2) because of the inherent inverted-pendulum structure, stability challenges are much more formidable. It is intriguing to watch people make everyday "arm" movements in 3-D space: The torso and shoulder can be viewed as driving the arm system in 3-D space, with the elbow relegated to a subservient subsystem and the wrist serving to subtly orientate the hand. The usefulness of planar arm models attached to a shoulder "base" has reached its limits. Any hypothesis about arm movement organization *must* be extendable to the conceptual framework of a dynamic system to *either* side – the hand at one side and a shoulder-torso system (with potentially variable impedance in 3-D) at the other. Better yet, consideration of the torso-shoulder-arm system introduces intriguing and fundamental questions related to posture and movement organization.

1.5. Using orthogonal impedance to guide movement paths. During fast *horizontal* head movements, muscles traditionally considered to be involved in *vertical* movements are fairly active (Winters & Peles 1990). What purpose could this serve? Consider the problem of replicating horizontal movements using a multilink, head-neck model and the conventionally assumed muscles, the splenius and sternocleidomastoid: Vertical deviation is difficult to avoid. By cocontracting vertically oriented muscles, impedance to vertical deviation is increased and a horizontal path emerges.

1.6. Simplicity through "virtual" systems? Many important human movement subsystems are mechanically complex: The shoulder, neck, torso, and hand-wrist come to mind. Indeed, hinge-type joints – the exception – are always situated near a more complex proximal neighbor. Complex proximal systems can set a kinematic orientation and then, through variable impedance modulation, act like "virtual" systems with fewer kinematic degrees of freedom. Perhaps the remarkable feature about horizontal arm movements is that the shoulder and torso keep the path within a plane.

1.7. Recommendations. We need to spend less time thinking about the subtle (task-dependent) details of the movement trajectory of end-effectors (a "signal-centered" approach) and more time on why, in 3-D space, movements rarely get way off course (a "field-centered" approach).

A posture is stable if the potential energy – a field effect – is at a minimum. As one lowers stiffnesses in an inverted pendulum elasto-static system, one part of the field starts to flatten first. This defines a potential direction of movement. For larger-scale systems, this direction can be changed by altering relative muscle stiffnesses. This represents an impedance-based view of movement regulation.

Neural networks should be inherently good at taking advantage of mechanical fields. They should also be good at generating fields – an inherent byproduct of their operation. **Bizzi et al.**'s target article describes "force-field" clusters of interneurons; such mechanically based spinal "compilers" may simplify higher brain function. Higher-brain topographical maps also suggest field concepts, but perhaps closer to levels of planning (setting virtual goals?).

What should these "fields" be? Force? Stiffness? Impedance? Position? Neurons really do not care as long as the goals (defined by a performance criterion or "teacher") are met. Here we come back to the concept of a "virtual goal" and mechanical stability. It is a powerful combination. The target article suggests that field effects start at the interneuronal level, where regulation of multiple muscle groups first become possible. It also suggests that feedback signals compliment feedforward; a position feedback signal (in 3-D space) defines another (e.g., gamma-based) field. Notice that fields "emerge" whenever we think in terms of more realistic large-scale systems in 3-D space.

2. If neural networks provide the answers, what are the questions? (JMW & PM) It is refreshing to see prominent neuroscientists questioning the very foundation of much of the past work in their chosen fields. Our synthesis of the target articles by **Robinson** and **Alexander et al.** is summarized by three statements, with which we agree. First, the process of creating serial "systems" models of neurocircuitry by identifying hypothetical neural signal and subsystem blocks is, by and large, a "blind alley" (Robinson; Alexander et al.). Second, the concept of algorithmic "motor programs," although it certainly has some significance for studies of motor behavior, has often been used as a crutch that has failed to help and has perhaps even hurt our understanding of the neuromotor organization of movement (Robinson; Alexander et al.). Third, single ("hidden") unit studies provide little (often no) insight into neuromotor organization (Robinson). These three areas of study represent the essence of the traditional neuroscientific approach to motor control. If we are to dismiss them, what is left? Approaches based on connectionist neural networks? Current artificial neural network models are structurally simple, typically consisting of one hidden layer representing the entire CNS and using "neurons" that sum algebraically instead of over time. Will the use of such vague networks provide fundamental insight into human movement organization?

It is worth stepping back and considering movement organization from a wider perspective, conceptually dividing it into four levels: (1) a higher self-organizing (unsupervised) level; (2) a middle level (supervised by the upper level) that consists of goal-directed neural networks; (3) a lower neural level with properties intricately connected to the musculoskeletal system; and (4) the musculoskeletal system (which may be dynamically coupled to the environment).

Let us start from the bottom. Unlike the CNS, physiologically based parameters for musculoskeletal models can be identified that provide reasonably accurate predictions of experimental data. This has caused many of us (including JW) to use musculoskeletal models where the "inputs" to the model – "equivalent" motoneurons for each muscle – are the outputs of the CNS. Investigations of human movement from this biomechanically based viewpoint are being carried out by large numbers of scientists, including many former single unit researchers. For example, about 20 chapters of a recent book entitled *Multiple muscle systems: Biomechanics and movement organization* (Winters & Woo 1990) explicitly or implicitly use such models, often quite effectively. These models are helping us understand the effects of inertial dynamics between links, basic principles behind postural control, and the critical importance of the unique properties of muscle, the biological actuator. Yet from the point of view of understanding movement organization, this approach is limited. Why? Because the difficulty of identifying strategies without "help" from "neural networks" is overwhelming.

Just above this level, we suggest that basic spinal neurocircuitry should be identified as an integral part of the motor organizational process. Considerable effort has gone into identifying spinal connections, which are numerous (see the target articles by **McCrea** and by **Gandevia & Burke**). Interpretation seems to require modeling approaches that include both higher and lower levels. There is perhaps more organizational structure here than would be predicted for "hidden units" in the less structured layers used in artificial networks such as those discussed by **Robinson**. McCrea points out that spinal interneurons are more than just relay systems. He also identifies unique needs for presynaptic inhibition. This suggests that both multiplicative and additive types of interaction between cells are of fundamental importance, which has implications in neural networks and in engineering adaptive control. **Bizzi et al.** identify clear relations between neuromotor organization and the musculoskeletal system being controlled; they also identify unique roles for groups of interneurons in postural control. These subsystems have evolved together and need to be considered as a whole.

There has been a tendency on the part of **Robinson** and **Alexander et al.** to minimize neuromechanical interplay. We suggest that neural networks should be controlling such lower-level neuromechanical systems because neural organization strategies are intricately influenced by the properties of the musculoskeletal system being controlled. Robinson suggests – correctly – that a "teacher" (supervisor) can tell the network *what* to do, but not *how* to do it. We believe there need to be middle-level networks with specific goals that are taught.

Consider the analogy of this middle-level process of learning within the framework of the engineering optimization problem: A *goal* is formulated as a performance criterion that is to be minimized (or maximized), a certain set of control parameters are available for *modulation*, and then an optimization *algorithm* finds a solution for the control parameters that minimizes the scalar performance criteria, subject to constraints such as the dynamic system under control. The goal, we suggest, is set by a self-organizing connectionist network, perhaps developed at levels of neurocircuitry similar to those that are the focus of **Alexander et al.** Under this middle level would be the lower stages of organization and execution, developed at the levels of neurocircuitry considered by **McCrea** and by **Gandevia & Burke.** During goal-directed movement tasks, what are the control parameters that are being modified? Neurocontrol parameters. Might not some of these be parameters in a model (e.g., feedback or other spinal gains)? Sure. Might the dynamic model not include lower neurocircuitry, with these parameters being "synaptic" gains? Sure, and furthermore these gains may involve any specified information flow between spinal neurocircuitry. But now we have identified two "types" of modifiable neural synapses, those in our neural network algorithm (i.e., higher up) and those representing control parameters in our neuromusculoskeletal model. The latter may end up close to being "hardwired," or they may not. One set may explicitly involve time (e.g., control parameters); the other may not. In any case, the reasons for modifying synapses may differ. What we have accomplished by carefully defining our system components is a hierarchical neural structure. There is also nothing to prevent further local nesting, in series or in parallel.

This approach suggests common ground between modelers and experimentalists studying peripheral neuromotor interactions during natural movements (see **McCrea** and **Gandevia & Burke**). It gets us closer to some of the flavor of the equilibrium-point approaches and spinal neuron "fields" (see **Bizzi et al.**). We also have not left out the musculoskeletal system, which many of us believe has advantageous properties of fundamental importance (less so for eye movements). It is interesting that we often find faster optimization convergence and more effective strategies when the muscles in the model have highly nonlinear (as opposed to linearized) muscle properties (Seif-Naraghi & Winters 1990); "sloppy" neural networks will take advantage of such properties, whereas traditional control algorithms break down.

At the very top, we have the fascinating cognitive underpinnings of movement, related in some mysterious way to the development of a goal. Most behavior – certainly that necessary for sustaining life – is fundamentally goal directed. Self-organizing systems must be capable of producing such goals. **Alexander et al.** identify many of the salient features that such networks must have, including distributed, parallel, and serial processing with extensive convergence and divergence. Within the framework of theories in cognitive psychology, a synthesis of sensory information and past experience should be combined to formulate goals. We see a need for interdisciplinary scientific research involving cognitive psychologists, neuroscientists, and biomechanical engineers: Psychologists investigate setting goals, biomechanical engineers study goal-directed movement (given goals and a neuromechanical system), and neuroscientists fit somewhere in between.

We have suggested that neural networks have an important role to play, yet not as isolated entities. The field of artificial neural networks is still in its infancy. In contrast to the highly structured white matter of the human brain, the structure of most current neural networks consists of only three stereotyped layers. As noted by **Robinson,** the vast majority of past applications have been static networks involving algebraic summation ("spatial transformation networks"). We suggest that the jump to history-dependent "neurons" with greater interconnectedness, that is, to spatiotemporal transformation networks with purposeful "white matter," represents a much larger leap than is indicated by Robinson. As an analogy, calculus can pose questions and identify solutions that cannot even be developed within the framework of algebra. Robinson closes by suggesting that engineering theorists can build a bridge between system function and hidden-unit behavior. Perhaps psychologists, biomechanical engineers, and neuroscientists can accomplish this if we work together to develop more appropriate network structures that will allow such casual relations – spatial and/or temporal – to evolve more easily than might be expected a priori (for certain systems and tasks).

Authors' Responses

The equilibrium-point framework:
A point of departure

E. Bizzi, N. Hogan,[a] F. A. Mussa-Ivaldi
and S. Giszter

Department of Brain and Cognitive Sciences and [a]Department of
Mechanical Engineering, Massachusetts Institute of Technology,
Cambridge, MA 02139
Electronic mail: emilio@wheaties.ai.mit.edu

R1. Key issues

A number of commentators stress that the equilibrium-point control hypothesis does not solve all problems. We heartily agree.

R1.1. Simplicity. Much of the appeal of the equilibrium-point hypothesis stems from its apparent simplicity, though we agree with several commentators that simplicity is treacherously difficult to quantify. In one extreme form of the hypothesis, the central nervous system (CNS) need only specify the target position of a reaching movement, without knowledge of the limb's initial position. The target article reviews the experimental basis for rejecting this hypothesis and postulating a moving equilibrium point (the virtual trajectory). **Agarwal** argues that the slow increase of the holding torque observed in those experiments might be attributed to the dynamic response of the neuromuscular system. This possibility was considered (Hogan 1984), but could not be reconciled with the available data on neuromuscular dynamic response; the time course of the holding torque had durations significantly more than three times the dominant time constant of excitation-contraction dynamics.

Adamovich and **Ostry** & **Flanagan** argue that the λ-hypothesis yields good results with a simpler control strategy than the α-hypothesis: a shift of the equilibrium position at constant speed along a straight line to the target position. A constant-speed virtual trajectory was considered (Hogan 1984, Fig. 4) and shown to reproduce experimental observations adequately, so it is not obvious which hypothesis is simpler.

The experimentally observed response to perturbations could only be reproduced, however, if the duration of the virtual trajectory was comparable to the duration of the movement (e.g., about 400 msec for a 60-degree movement lasting about 600 msec). **Adamovich** proposes virtual trajectory with a constant speed of about 600 degrees/second, that is, about 100 msec duration for a 60-degree movement. It is not clear that this model would be able to reproduce the observed response to perturbations.

R1.2. Stability. The foundation of the various "equilibrium-point" hypotheses is the observed stability property of the limbs during posture and movement. This stability has a profound influence on the problems of motor control. It is a critical factor in contact tasks and motor learning and gives rise to the appealing computational simplicity of equilibrium-point control. **Masson** & **Pailhous** elaborate a synthetic framework partly based on equilibrium concepts and stiffness regulation in order to account for a variety of experimental data. This demonstrates the utility of the equilibrium-point approach. Several commentators (**Cordo** & **Bevan, Kuo** & **Zajac, Gutman** & **Gottlieb**) propose different versions of feedforward control equivalent to approximate solutions of the inverse dynamics problem. Although we find these proposals plausible, feedforward control strategies are not workable without a significant degree of limb stability (whether reflex generated or due to intrinsic muscle behavior or both).

R1.3. Negative feedback and time delays. The influence of limb stability is so pervasive that it may be taken for granted despite the fact that it is neither an intrinsic property of musculoskeletal mechanics nor guaranteed by negative reflex feedback.

Negative feedback generally has a stabilizing effect, but not always, a fact known to engineers since Watt invented the steam engine. In the presence of complex dynamic behavior or information transmission delays, the stabilizing effect may be severely curtailed. Both phenomena are clearly present in biological systems; how significant are they? A net delay as short as 40 msec means that at a frequency of 2 Hz loop gain must be less than unity, and to ensure robustness to uncertainty, it should be substantially less than unity.

The λ-model is a particular (and physiologically plausible) implementation of the well-known PD (proportional plus derivative) negative feedback controller. A loop gain less than unity at 2 Hz would limit this controller to modest effects at frequencies well within the range of human capability. The magnitude of the feedback gain is clearly a critical test for this hypothesis. Despite **Feldman**'s claim that it "is a powerful mechanism" with "sufficient but not excessive" gain, the aggregate of our commentators' remarks indicates that the evidence is, at best, ambiguous.

R1.4. Negative stiffness. The objection is raised by **Adamovich** that muscles have negative stiffness in a substantial part of the physiological length range. **Hasan** points out that even if muscle stiffness is positive, the angular stiffness of the joint may be negative because of moment-arm variations. We fully recognize that the moment-arm variations due to musculo-skeletal kinematics can induce negative stiffness. Kinematic effects on net stiffness are the basis of a proposed solution to the problem of determining the joint angles of a kinematically redundant mechanism corresponding to a specified end-point position (Mussa-Ivaldi & Hogan 1991; Mussa-Ivaldi et al. 1988). However, the effects of moment-arm variations apply equally to the intrinsic force-length behavior of deafferented muscle and to the static force-length behavior resulting from muscle spindle feedback. Thus it is not clear that the λ-model can "save the joint from the horror of negative static angular stiffness."

Some of that "horror" may be imagined. The stability properties of the limbs are essential to define an equilibrium point or virtual position, but positive stiffness is not. Stiffness refers to the local slope or tangent of a static (steady-state) force-position relation. The equilibrium points of a mechanical system are the zeroes of that force-position relation. If an equilibrium point is to be a stable attractor for the system, it is necessary that the force-position relation be positive-definite throughout a region including the equilibrium point. It is not necessary for the stiffness to be positive at all points in that region, although stiffness must be non-negative *at* a stable equilibrium point. Consequently, observations of a negatively sloped, torque-angle relationship do not necessarily imply instability.

In our view, the profound effects of moment-arm variations highlight the complexity of multijoint posture and movement. The observed stability of the limb, even in the deafferented case, is not an inevitable consequence of musculoskeletal mechanics as **Cavallari** implies. This leads us to hypothesize that activations are restricted to those patterns that yield stable behavior.

R1.5. Limitations of the equilibrium-point hypothesis.
Many commentators view the equilibrium-point hypothesis as a good starting point from which "one can add the task-dependent reflex variations and the variations in central activation patterns needed to achieve accurate, complex movements in a changing environment (see commentary of **R. B. Stein**). **Lacquaniti**, among others, describes an example of task-dependent reflex modulations that are set by the central nervous system (CNS) in parallel with the specification of an equilibrium posture for the arm. In his experimental paradigm, the CNS sets the equilibrium point for the arm and also gates stretch reflexes of the arm in an adaptive way. Lacquaniti demonstrates clearly the presence of reflex responses in both stretched and shortened muscles, which are set by the activity of the CNS prior to the impact with an external disturbance.

Along the same lines, **Dean** suggests that modification of the equilibrium-point control may be necessary when arm movements using more than two joints are executed. As a way to deal with the extra degree of freedom, Dean puts forward the interesting suggestion that the CNS may specify a number of separate equilibrium trajectories for controlling multijoint arm movements. In the case of reaching and grasping movements, Dean suggests that this complex movement results from an equilibrium trajectory for the proximal joints and another supporting the hand motion.

A similar point is raised by **Kuo & Zajac**. The anticipa-

tory postural adjustment of leg muscles that precede the fast elevation of the arm could result from the CNS programming a set of virtual trajectories, one for the arm and another for each of the legs. Although Kuo & Zajac consider multiple virtual trajectories a possibility, they clearly favor the alternative idea that the CNS handles the complex coordination between arms and legs by way of associative neural networks. Specifically, they speculate that the CNS could learn transformation mappings by associating muscle forces in the legs with the anticipated inertial forces generated by arm movements.

Whereas **Dean** and **Kuo & Zajac** have suggested ways to modify equilibrium-point control by directing virtual trajectories to different limb segments, **Smeets** raises the possibility that the CNS modifies the equilibrium-point trajectory as a consequence of sudden changes of the loads encountered by subjects trained to move a mass along a rail.

The complexity of multijoint movements is also the starting point of **Winters**'s commentary. He points out that fast voluntary arm movements performed while standing involve anticipating the activation of torso and leg muscles. Clearly, any control hypothesis for arm motion must be considered in a broader context because in most circumstances the muscles of the torso are going to influence the hypothetical equilibrium-point trajectory of the arm and vice versa. Winters points out that thinking directed at providing a frame for answering these complex movements might benefit from a "force field"-centered approach.

In summary, we feel that the points made by the commentators are plausible. To a certain extent, their suggestions are in agreement with the work of Flash and Henis (1991) on combining virtual trajectories.

Finally, **Gottlieb** suggests a modification of the equilibrium-point hypothesis to account for fast movements. He advances the idea that pulses of excitation to the motoneurons are superimposed upon those specified by the virtual trajectory. These pulses, which send the movement in the right direction, derive from knowledge stored in the CNS as a result of experience. A similar concept is proposed by **Cordo & Bevan**, who postulate that a feedforward system controls the initial phase of the movement and moves the arm in the direction of the target. Additional learned control signals triggered by visual inputs adjust the trajectory.

Kuo & Zajac's and **Cordo & Bevan**'s suggestions represent an attempt to deal with some of the problems inherent to the equilibrium-point hypothesis. With the suggested additions, the original computational simplicity of the equilibrium-point hypothesis vanishes and the notion that springlike properties are used explicitly to control movements is blurred.

R1.6. Alternative hypotheses.
Radically different schemes for controlling movements are described by **Burgess, Lan & Crago**, and **Morasso & Sanguineti**. Some of the commentators, like Morasso & Sanguineti, relate these schemes in some way to the equilibrium-point hypothesis; others, like Burgess, **Hallett**, and Lan & Crago, formulate different models of sensorimotor activity transformations and basically reject the equilibrium-point idea. Morasso & Sanguineti, for example, argue against an algorithmic sequential view of sensorimotor transformations. They favor a distributed computational

architecture characterized by "continuous mapping between motor commands and the virtual trajectory space." Clearly Morasso & Sanguineti's views on motor control are different from those presented in the target article. What transpires from their brief commentary is a novel way of thinking about coordinate transformations. Unfortunately, because their ideas are only briefly sketched, it is difficult for us to react in an informed way. However, we note a similarity between their proposals and the idea of using force fields to produce complex motor behavior (such as navigation without collision in a complex and changing workspace) which has been applied successfully in robotics (Andrews & Hogan 1983; Khatib & Le Maitre 1978; Koditchek 1989).

Like **Morasso & Sanguineti, Burgess** presents a different view on how the CNS guides movements. However, unlike Morasso & Sanguineti, Burgess is quite explicit about one point: The equilibrium-point hypothesis is a failure. This rejection is based on two observations: (1) because the motor system "does not possess much stiffness," neither the α nor the λ model can drive movements and (2) the motor system uses a sensory-template control in dealing with unpredictable loads in the environment. Probably, what Burgess has in mind is that sensory templates are associated with motor outputs via a kind of global neural network. If this is correct, then Burgess's ideas are similar to those expressed by **Kuo & Zajac** and perhaps Morasso & Sanguineti.

The claim that the motor system does not have stiffness is based on experiments in which subjects were asked to make the same flexion effort at different elbow joint angles. The flat torque-angle relationship found by **Burgess** in this context represents the basis for rejecting the equilibrium-point hypothesis. Burgess is basically advocating a torque control model. According to his view, the biological motor apparatus at the elbow behaves like a torque motor. It is certainly of interest that by imposing a "sense of effort" construction, Burgess has managed to obtain results that are consistent with torque rather than equilibrium-point control.

A torque control model is not compatible with a number of experimental results, however. In particular, the findings in both intact and deafferented monkeys reported in the target article, which were obtained by driving the forearm passively from an initial position to a new position, do not support a control scheme based on torque (or effort) control (see Fig. 2, target article). Additional evidence against a torque model derives from experiments in normal human subjects whose hand path was displaced briefly from an intended trajectory. The fact that the displaced hand regained the original path is certainly compatible with an equilibrium-point view.

Burgess's findings lead to an interesting observation related to the issue of coordinate transformations. In the multijoint arm, the effort at the endpoint is related to joint torque via a position-dependent transformation. Thus, a constant pattern of torques across the arm's workspace leads necessarily to a nonconstant pattern of forces at the hand and vice-versa. Burgess's commentary suggests that in a single-joint task the isoeffort task leads to a (nearly) constant joint torque pattern. This observation leads quite naturally to the testable question, "Does the same isoeffort task lead to a constant joint torque pattern in a multijoint situation?"

R2. Criticisms of the equilibrium framework

We will address in this section several substantive criticisms leveled at the equilibrium hypothesis by commentators.

R2.1. Springlike properties. An objection is raised by **Feldman** and **Adamovich** to the idea, essential to our development of the equilibrium-point hypothesis, that isolated muscles can be regarded as "tunable springs." In particular, Feldman suggests that this idea "was refuted by the classical experiments of Hill (1938)." This idea is not rejected by Hill's work for steady state behavior. Hill was mostly concerned with the force/velocity relation of a muscle and proved that this relation cannot be accounted for by a linear viscous term. However, Hill's experiments confirmed that both the active and the passive muscle "contains an undamped elastic element." The idea that muscle exhibits a (nonlinear) springlike behavior "tuned" by neuromuscular activity was supported by later experiments of Rack and Westbury (1967). These experiments involved completely deafferented muscles. Therefore, the tunable springlike behavior is not contingent upon the presence of sensory feedback information.

In our work we have not taken the springlike properties of the neuromuscular system for granted. We were especially concerned with the validity of describing the position-dependent forces in a multijoint system as springlike. For a multijoint spring, the curl of the resulting force field must be zero (Hogan 1985a). In our measurements of the steady-state force-field of the arm (Mussa-Ivaldi et al. 1985), we found the curl to be insignificant or negligibly small.

If there is a valid issue concerning steady-state springlike behavior, this is not related to the velocity-dependent dissipative component but to the time-dependence of the static force that is usually described as hysteresis. As we stress elsewhere, this is a critical issue in any control scheme, including the λ-model, especially if one takes into account that not only the motor output but also the feedback information is likely to be plagued by a time-dependence. Open questions such as this one should serve once again to point out that equilibrium-point models (in all their current flavors) are still far from having established a comprehensive theory of motor control. As **Hasan** points out, "a great deal of exploration lies ahead."

R2.2. Multijoint stiffness. As several commentators observe, the measurements and interpretation of muscle, single-joint, and multijoint stiffness are crucial. Only with these data in hand can we have an understanding of the potential role of equilibrium-point mechanisms in posture and movement. The magnitude of stiffness, the muscle history dependence of stiffness, and the reflex control of limb stiffness are all crucial issues, as pointed out in the commentaries. For simplicity, some models (e.g., Hogan 1985) have been formulated to make no distinction between the influence on limb impedance of intrinsic muscle mechanics and reflex effects. We fully acknowledge their importance, however.

Is multijoint limb stiffness large enough or even of the right sign? **Hasan** discusses our approach to this question at some length. Measurements of stiffness by Mussa-

Ivaldi et al. (1985) were made in the multijoint case. The estimates were made after transients had died out. The measure was thus of static stiffness as Hasan points out. The measure was of endpoint stiffness and thus incorporated all effects of moment arm variations within a short distance of the held posture. However, early experiments (which remain unpublished) showed that the measured stiffness was adequate to fully predict the effect of force pulses applied to the resting arm for the first 150 msec. The return to equilibrium, however, was faster than predicted by the stiffness model; we attribute this to reflex actions in corrections. The initial trajectory and thus the initial resistance to perturbation and postural stability were fully compatible with the measured values. These data suggest the static stiffness estimates are more than "an afterglow of neural strategies." In addition, recent measures conducted with slow ramps over longer distances confirm the initial role of this measured postural stiffness in limb posture stabilization (Shadmehr et al. 1992). A postural force field was identified by Shadmehr of which the static linear stiffness in Mussa-Ivaldi et al. (1985) is a local estimate. The work of Shadmehr also suggests a critical role of the postural force field in slow movements. Thus static stiffness measures seem adequate to predict slow movement trajectories and early postural stability.

What of the history-dependent response of muscle (**Hallett, Cordo & Bevan, Agarwal**)? This is actually a serious potential difficulty for any computational framework for biological motor control, especially since hysteresis may generate the problems of choosing activation commands discussed by Agarwal even in otherwise simple and straightforward movements. Must a complex dynamic history be stored in the CNS to compute the motor commands needed for a particular muscle response? Probably not. A body of evidence suggests that in part the role of reflex mechanisms may be to alleviate this problem, as described in the commentary of **Nichols**. In fact the several means by which reflex effects can help linearize muscle stiffness and joint stiffness go some way toward creating a mechanical plant with response properties that are well suited to equilibrium-point control. It is our view that this kind of "muscle management" is a major function of reflex feedback. The panoply of mechanisms and compensations needed to achieve this in a multijoint redundant limb are of course still under investigation, and may include the mechanisms described in Nichols' commentary, and variants of λ-control.

R2.3. Feedforward control and inverse dynamics. Reflex contributions to motor control are complex and go beyond stiffness regulation. Visual correction must also be included in any general scheme as described by **Cordo & Bevan.** We fully concur with the sentiments expressed in their quotation of Peter Greene (1972). We agree with their suggestion that the "automatic, even if not quite correct, feedforward" in Greene may for multijoint motion be represented by the force-field evolution that is summarized by the equilibrium-point trajectory. As Cordo & Bevan discuss, visual guidance and correction is not unusual; this has not been considered in the simple formulation of equilibrium-point models. However, the work of Flash on target switching is relevant in this context and demonstrates how our model can be ex-

tended to these types of control. Given the history dependence of the musculoskeletal system, such a correction may require some knowledge of limb dynamics and muscle state.

As discussed by several commentators, the inverse dynamics problem, which seemed horrendous a number of years ago, appears to yield easily to parallel network algorithms (Jordan 1990; Kawato et al. 1987). This suggests that other strategies besides simple equilibrium-point control are feasible and such strategies (by avoiding coactivation) are desirable from an energy conservation standpoint. Indeed, as pointed out by several commentators, optimized fast movements require either some knowledge of inverse dynamics or a more complex equilibrium path (see **Lan & Crago**). Is equilibrium-point control only good for slow movements? One scheme explored by McIntyre (1988; 1990) and discussed in the target article uses reflex feedback to tune system performance and is close to the λ-formulation. The command to this system is a simple trajectory. Similar types of schemes are also to be found in Flanagan et al. (1990).

We also believe that the equilibrium-point framework may have a role in motor learning for fast feedforward movement. The force field/equilibrium framework can act as a "bootstrap" for faster movements. This view is suggested by the work of Terence Sanger (in preparation) who has shown that it is possible to use a neural network learning algorithm to learn successively larger modifications to a virtual trajectory as the speed of movement is gradually increased. In this manner, the simple virtual trajectory for control of a slow movement is smoothly deformed into the complex trajectory required for a fast movement. The network's approximate solution at each speed is used to guide learning at a slightly greater speed and in this way learning is "bootstrapped" from a region of easy controllability to one of more difficult controllability. Sanger (in preparation) has shown that a robot arm that uses this technique can learn to copy a signature at nearly full human speed within 15 trials. It is noteworthy that the stability property provided by equilibrium-point control is critical in achieving this result as it ensures rapid convergence to a satisfactory solution.

Thus, it may be possible that the complex dynamic strategies by **Kuo & Zajac** could ultimately derive from bootstrap equilibrium mechanisms. Of course, the learning process has to subvert these mechanisms to adapt to the novel task constraints. A second possibility which we must also acknowledge is that these represent completely different parallel mechanisms.

R2.4. Equilibrium-point organization by the spinal cord. The rich possibilities that can derive from a framework of a few force field types and their organization are discussed by **Braitenberg & Preissl.** We look forward to their detailed publications. Two other commentaries address spinal organization (**Cavallari** and **Loeb**). We acknowledge that our presentation of spinal cord microstimulation in frogs in the target article is sketchy; many of the controls are in a recently submitted articles (Giszter et al. 1992b). For this reason, we will review the critical evidence suggesting that this work is more relevant. In our view, the microstimulation experiments' results could not be simply anticipated for the following reasons: (1) the force fields fall into a few invariant types (Bizzi et al. 1991; Giszter et al. 1992c) and not a random variety, (2) these field types correspond to several field types that have now been recorded in natural behaviors (Giszter 1992; Giszter et al. 1992a; 1992b; 1992c), and (3) simulation of random recruitment of motoneurons, or local recruitment of mononeurons from a known topography cannot predict our results. In addition, the histological identification of activated motoneurons reveals selective recruitment distributions over broad areas of lumbar cord (Bizzi et al. 1991; Giszter et al. 1991a; 1992a). The nonphysiological nature of our activation is, of course, a given, which we do not dispute.

We would now like to discuss the evidence that force fields fall into only a few discrete classes. The fact that there are a few classes is not in agreement with random recruitment of a spinal cord's cells and fibers; if the latter were the case, we would have found a continuous distribution of fields. Measurement and simulation show that random recruitment fields will be biased toward leg extension (see Bizzi et al. 1991). Convergent force fields also cannot be predicted by a model of motoneuron topographic organization assuming local recruitment. Such a model predicts a continuous spatial variation of fields through the cord. Finally, a model in which all pools are activated to some degree predicts only a single field type. Remarkably, the few fields observed in experiments are of types which might be adequate to span the limb workspace.

What do these fields represent? There are two possibilities that we cannot as yet distinguish. The first is that these premotor elements, which we have termed movement primitives, are subunits used in more complex behaviors in isolation or in combination. The second is that these fields represent spinal behaviors recruited via interneuronal (not afferent) pathways. The observation that similar or identical force fields underlie some spinal behaviors shows these fields have roles in natural movement. However, at this juncture, these observations do not allow us to discriminate these interpretations.

Recruitment of motoneurons was examined by looking at sulforhodamine uptake by active cells in deafferented frogs which were microstimulated. These experiments revealed broad activation of motoneurons over as much as a centimeter of lumbar cord. This activation was bilateral. The frog's motoneurons have extensive electrical connectivity. Blockage of gap junctions with heptanol did not alter either microstimulation fields or sulforhodamine uptake patterns.

All of this suggests that a simple hypothesis of random recruitment of several spinal pathways and motoneurons in our microstimulation work is untenable. Our electrical stimulation of a small region (< 200 micro diameter) leads to the emergence of a few well-defined and specific activity patterns and force fields. Simulations show that the muscular system of the frog's limb is biased to extension (not surprisingly). The fields are arranged in flexion, adduction and abduction, as well as extension. Simple limb mechanics is not an adequate account. Finally, such fields occur in frogs executing or attempting to execute spinal behaviors such as wiping (Giszter 1991; Giszter et al. 1992a). Thus, we believe the field organizations are utilized physiologically.

R3. The α/λ debate

Before addressing the issues related to the α/λ debate, we would like to begin with a preliminary note. It is unlikely that in future work we will refer to our conceptual framework as the α-model. This term was introduced in the literature not by us, but by the authors of the λ-model. As we stated in the target article, we do not identify ourselves with the views attributed to this model. However, the α-model has been consistently and repeatedly attributed to our group on the basis of fragmentary information. We have never committed ourselves to some detailed, monolithic, and dogmatic "α"-version of the equilibrium-point hypothesis. Our ideas have changed with time and we hope they will continue to change. In this process, we have collected a good deal of experimental data and we have elaborated a number of concepts that range quite broadly around the notion of equilibrium-point control.

R3.1. Misconceptions. Although the definition of an "α"-model does not belong to us historically, we decided to accept the challenge and to provide our own version of an α-model. This was clearly to some extent a "dangerous" choice, since the α-model had been introduced to describe a clearly flawed theory that reflected, at most, a caricature of our own views. We decided to take this risk because the equilibrium-point hypothesis has been investigated by different groups and in different ways. Thus, we hoped that the discussion engendered could help establishing a useful exchange of opinions.

In this respect, there are a number of points raised by some of the commentaries that do not reflect the content of the target article. For example, **Feldman** and **Latash** would still like to identify us with (or at least to make us "responsible" for) their original formulation of the α-model. In particular, they refer to the exclusive modulation of muscle stiffness and to the independence of α-activity on feedback signals. We are not going to reiterate these points as they are clearly and unambiguously stated in the target article. Furthermore, we do not believe that the λ-model needs comparison with such straw-man hypotheses to survive. In the past, we used strong simplifications solely to illustrate the basic principles of the equilibrium-point hypothesis. One of these simplifications was a sketch (one that was never presented as "real data," however) showing muscle length-tension curves with constant rest-length. Another simplification was introduced in a simulation of the head-neck system as a second-order model with stiffness (but no rest-length) modulation. In doing so, we may have accidentally given the impression that we literally proposed a model in which only muscle stiffness is regulated or, worse, that the length-tension behavior of muscles is linear. These simplifications were not intended to be taken literally as a dogma for motor control, quite the contrary. They were merely instruments to test and communicate some general ideas.

Similarly, **Adamovich, Feldman,** and **Ostry & Flanagan** suggest that we identify the λ-model with Merton's (1953) follow-up servo model. In the target article we stress some points common to the two models, but we also acknowledge that one of the major criticisms leveled against Merton's model is avoided by the λ-model in a manner reminiscent of the servo-assistance model (Eldred et al. 1953; Granit 1955) while the other criticisms of the servo hypotheses still apply. In contrast to Adamovich's perception, we do not "reject" Feldman's model. We merely express some doubts, and a different (but not antagonistic) perspective. We will not elaborate on these issues further because we believe they have been sufficiently clarified by the target article. Instead, we will focus on the critical points that have been raised regarding the target article's formulation of the α-model.

R3.2. The physics behind the equilibrium-point models.
"According to the EP concept accepted in physics, position-dependent variables . . . do not specify the equilibrium position because *they are functions of position*" (**Feldman**'s emphasis). We disagree. About two hundred years ago, Lagrange showed that a conservative physical system has stable behavior with respect to any minimum of its potential energy. About a hundred years later, Lord Kelvin reiterated that position. The existence of a potential energy function with at least one minimum requires a position-dependent force (defined by the potential energy gradient) that vanishes at any minimum. This is the basis of the definition of equilibrium to be found in Goldstein's *Classical mechanics*: A "system is said to be at equilibrium when the generalized forces acting on the system vanish" (1980, p. 243). Thus, a position-dependent force (i.e., a force field) is *required* to define equilibrium. This is the definition we have in mind. We presume that Feldman may have been referring to some other definition, different from that of Kelvin and Lagrange.

Having said that, we agree that the notion of equilibrium point, as we use it in motor control, may sometimes be misleading. If we consider a system that includes a limb coupled with the environment then the equilibrium point of this system depends on the properties of the limb and of the environment. It was with this in mind that the term "virtual position" was introduced (Hogan 1984). A virtual position refers to the equilibrium point of the limb uncoupled from its environment (or, coupled with an environment having zero impedance and exerting no net force on the limb). As a consequence, a given control pattern applied to the limb will achieve different states according to the environment to which the limb is coupled. We should stress, however, that the same "lack of invariance" will affect any model based on the sole specification of a limb's state, including the λ-model.

This issue has been nicely summarized by **Gutman & Gottlieb**, who make a clear terminological distinction between an "equilibrium" point that depends upon the interaction with the environment and a "virtual" position that depends only upon the central command. Gutman & Gottlieb and **Latash** propose to estimate the virtual position using a priori assumptions on the controller's structure. This seems to be a clever and promising way to test a control hypothesis such as the λ-model.

R3.3. Use of deafferented preparations.
Objections to our use of deafferented preparations are raised by **Feldman** and **Latash**. In particular, Latash suggests that "the α-model emerged on the basis of observations of deafferented animals. Then [it] was generalized to intact animals and humans." Use of deafferented preparations in our previous and current work is not justified by the need for establishing a behavioral model for such an abnormal state. On the contrary, the deafferented preparation offers a unique opportunity to test whether or not sensory feedback is a *necessary* condition for any particular behavior observed in the intact preparation. If one states that sensory feedback is necessary to implement an equilibrium posture, one logically expects that deafferentation abolishes equilibrium-point control.

This is the simple and clear rationale behind the use of deafferented preparations. We found that equilibrium point control is by no means suppressed by deafferentation, not only in monkeys but also, more recently, in the spinalized frog. These facts are in our view sufficient to dismiss any claim that the very existence of an equilibrium point is contingent upon the operation of reflexes. Similar considerations should induce some words of caution in the proponents of the λ-model when they propose to account for triphasic EMG bursts on the basis of the interaction of feedforward and feedback signals. The triphasic pattern has indeed been observed in clinically deafferented subjects as well. Obviously, as is implied by the distinction between a necessary and a sufficient condition, these findings do not deny a role for reflexes. **Feldman**'s λ-model represents a sufficient condition for equilibrium-point control. It is a nice framework for integrating feedforward and feedback information based on the formal equivalence of synaptic summation on a threshold unit with a "coordinate transformation" that maps such thresholds into a linear variable. In this context, our finding suggests that such a mechanism does not have to conflict with the biomechanical properties of the muscles. In contrast, it may take advantage of a mechanical machinery that is inherently capable of performing equilibrium-point control.

R3.4. Compatibility of α and λ.
Our discussion about the compatibility of the α- and λ-models has been favorably accepted by **R. B. Stein**, but has encountered strong resistance from the proponents of the λ-model. We cannot find in the commentaries any valid challenge to our formal argument. The main objections seem to be philosophical. We are not sure how to understand **Latash**'s argument that the α-model, contrary to the λ model, would imply a denial of the "inalienable right of free choice." More pragmatically, **Feldman** argues that the two models are incompatible because one (α) is wrong and the other (λ) is right. Perhaps this argument is induced by reluctance to abandon even temporarily the original notion of the α-model and to focus on ours. This attitude is revealed by the statement that in our understanding, the "control variable λ is . . . a function of α MN activation" and that such a (mis)understanding is at the core of our demonstration. We never said this. On the contrary, we merely considered the dependence of α upon muscle length and λ as stated by Feldman (1986). We never implied or assumed that this function can be inverted to express λ as a function of α. Contrary to what is stated by **Agarwal**, our proof of the inclusion of λ in α is not contingent on any "implicit assumption of uniqueness" that is not also implicit in the formulation of the λ-model

(the idea that for any muscle length the variable λ defines a single value of that muscle's α can be found in Feldman 1986, p. 23).

More to the point is the commentary of **Frolov & Biryukova**, which states that our view on the two models "seems to be rather formalistic, because it ignores the feedforward and feedback nature of motoneuron activity." Actually, our argument was a way of establishing a relation between the two models based precisely on a distinction between feedback and feedforward. As we stated in the target article, the λ-model is (legitimately) concerned about this distinction whereas our model (also legitimately) is not. Our proof of inclusion of the λ-model into the α-model is a formal way to express this fact. The distinction between a feedback and a feedforward component in the control signal is equivalent to imposing a restriction upon the set of possible cases (possible equilibrium-point/motoneuron activity pairs) in the α-model. We are surprised that this formulation has been interpreted by **Feldman** as a claim of "supremacy" on our part. On the contrary, a more constrained theory is often called a "stronger" theory because, by virtue of the constraints, it is usually capable of generating more precise predictions. Figuratively speaking, we are not proposing an inappropriate "α/λ" marriage. This would be incestuous, since the α- and λ-models already belong to the same family of ideas.

R3.5. Planning and execution. Some of the commentaries (**Ostry & Flanagan, Latash**) attribute to our approach a rigid constraint on the temporal shape of the equilibrium-point trajectory. In particular, they refer to simulation work by Flash (1987), which showed that unconstrained, two-joint arm trajectories could be accounted for the complex hand kinematics observed in human movements almost all by the mechanical interaction of a simple equilibrium trajectory with the inertial and viscous properties of a limb. On the basis of simple geometrical observation, one can easily see that a bell-shaped velocity profile of the hand is generally inconsistent with a monotonic movement of a single joint. Thus, it appears totally unfounded that, as stated by Latash, we are "reluctant in accepting the possibility of nonmonotonic equilibrium trajectories for control of fast single-joint movements." Hogan (1984) has described a wide repertoire of "N-shaped" single-joint equilibrium trajectories quite similar to those that Latash describes as being typical of the λ-model.

More generally, we would like to stress that in our view the shape of the equilibrium-point trajectory is an issue of motor planning. Bell-shaped trajectories correspond to a smoothness optimization that may be appropriate for certain tasks but not for others. In contrast, the equilibrium-point hypothesis is a model of movement execution that addressed the issue of translating a planned motion (of any spatiotemporal form) into a sequence of postures. This would clearly be a poor execution model if it were only able to generate bell-shaped velocity profiles. **Ostry & Flanagan** state that "it is not clear how the bell-shaped equilibrium trajectories . . . would be modified in response to target displacements." This issue has recently been addressed by a model of Flash and Henis (1991; see also Morasso & Mussa-Ivaldi 1982) that accounts for the formation of complex velocity profiles in response to target-switching by the superposition of simple trajectories with bell-shaped velocity profiles. Other issues of multijoint movement planning related to the coordination of a kinematically redundant system can also be approached in a theoretical context that is consistent with the equilibrium-point hypothesis (Mussa-Ivalda & Hogan 1992; Mussa-Ivaldi et al. 1988).

R3.6. What is controlled? We disagree with **Latash** that the version of the α-model we presented does not address the question "What is controlled?" In our view, what has to be controlled is the mechanical behavior at the interface with the environment. And by mechanical behavior we not only mean limb position, velocity, and direction but also the pattern of forces involved in contact tasks. However, we agree with **Frolov & Biryukova** that the α-model does not address the issue of "what central parameter is used by supraspinal levels" to define the equilibrium position. This issue derives from a classical control-system approach to the problems of motor physiology. This approach has inspired the research in the oculomotor system for a number of years. One should also ask how much sense it makes to search for such a specific parameter represented by some equally specific signal in the context of a biological neural network.

The advocates of the λ-model have proposed a model in which the interaction of feedforward and feedback signals is captured by a modulation of the stretch reflex. Presumably, central and sensory feedback interact in a complex and not yet understood fashion over a wide network of spinal interneurons, not to mention the supraspinal structures. Having to deal with physiological issues, we would rather investigate the organization of the neural circuits subserving the generation of observed mechanical behaviors. In this highly distributed context, the traditional distinction between spatially segregated feedback and feedforward signals is likely to lose some of its original significance.

R4. Contact

It is claimed by **Van Ingen Schenau et al.** that to control the force exerted on a moving object, the torques required at the joints cannot be derived from position information alone. For the particular example cited (see Fig. 1a of their commentary), the force to be applied requires a flexive torque at the shoulder and elbow. The motion requires a continuous flexion of the shoulder, but the elbow must initially flex, then reverse its motion and subsequently extend. This task is by no means incompatible with the equilibrium-point hypothesis because in the presence of contact forces the actual position of the limb will deviate from the virtual position. We see at least two distinct possibilities:

To generate a straight line motion while exerting a force along that line a virtual trajectory may be computed that deviates from the straight line by an amount determined by the compliance of the hand and the magnitude of the force. This deviation need not be oriented in the same direction as either the motion or the force and given the values we have measured for the hand stiffness, we expect that it would differ from both. Appropriately oriented, a position deviation may be determined to satisfy both the path and force requirements of the task.

Alternatively, the task might be accomplished by generating a straight virtual trajectory along the desired path. Because of the force required to move the object, the actual hand path will deviate from the virtual path by an amount determined by the stiffness of the hand and the requisite force. Again, this deviation need not be oriented in the same direction as the virtual trajectory. In the later phases of the example task cited we expect the actual hand path to deviate outwards from the desired path. The resulting inward-directed position error may readily generate the required distribution of torques about the joints.

Other external loads, for example those due to gravity, may be accommodated in a similar manner. Given stable contact, the deviations between virtual and actual position due to contact loads are similar to the deviations due the dynamic (e.g., inertial) loads encountered during unrestrained movement; they imply that a similar compromise may be made by the controller: Either a certain level of error can be tolerated or a computation of greater complexity can be performed.

We agree with **Van Ingen Schenau et al.**, however, that multijoint contact tasks are complex and may reveal limitations of the equilibrium-point hypothesis. Equilibrium-point control basically implies a relation between the execution of motions and the performance of contact tasks. Given the observed values of limb impedance (e.g., stiffness) there may be limitations on the ability to control force in contact tasks. If these limitations are measureable they may provide yet another way to falsify this hypothesis.

R5. Concluding remarks

We would like to reemphasize the point we made in the original target article: Our version of the equilibrium-point framework was proposed as a simplest case hypothesis with which we examine multijoint limb behavior. This simple core model was intended as a basis for experiments from which a more accurate and necessarily more complicated description of motor control could be elaborated. The ultimate goal of our approach has been an incremental incorporation of reflex feedback, motor learning using dynamics, and so on. Our model was intended, as some commentaries (e.g., **R. B. Stein**) have rightly described it, as a point of departure and not as a destination. Of course, this point of departure defines a route of investigation with which many may disagree. What is remarkable to us is not how many types of movements our model cannot explain, but rather the fact that such a deliberately simplified and impoverished framework can explain or approximate so much of motor behavior.

Afferent feedback, central programming and motor commands

S. C. Gandevia and D. Burke
Department of Clinical Neurophysiology, Institute of Neurological Sciences, The Prince Henry and Prince of Wales Hospitals and Prince of Wales Medical Research Institute, University of New South Wales, Sydney 2036, Australia

R1. Introduction. Before we address specific issues raised by commentators, the scope of our target articles requires restatement. It arose from a formal debate about the role of afferent feedback in the control of movement in which we defended the role of feedback signals against those who give them little place. We tried to show that the afferent input does supply contextually important information about an ongoing movement, does alter the reflex contribution to movement, and does provide perceivable signals about the position and movement of the limbs. This alone would set up a logical framework in which the afferent contributions to movement control were assured. Rather than limit ourselves to a consideration of afferent signals, however, we described in the final section some of the evidence that centrally generated signals related to motor commands are used by the nervous system. Our intent was not to consider the relative contributions of the two systems under every circumstance of natural movement (see **Cavanagh et al.**, for example), and we apologize to those who may have expected a broader coverage. Many of the examples we cited came from our particular focus on human motor control; there was no implication that examples from other movement systems in different animals are irrelevant to control in humans. However, some movements are unique to primates, and because the underlying neural strategies may be highly specialized, particular study of them is justifiable. This remark is especially relevant to the commentaries by **Paillard, Prochazka, Burgess, Clark, Cordo & Bevan,** and **Berkinblit et al.**

R2. Specific responses. It is suggested by **Berkinblit et al.** that it is in two areas, one from animal and one from human experiments, that afferent feedback can play a crucial role in formulating the final motor response. Their commentary accordingly supports the position taken in our target article, though it approaches the same problem from a different perspective. An issue raised by these commentators is worthy of further discussion, namely, the extent to which central pattern generators govern the movement of human subjects, the experimental animal to which we restricted our article. For example, there is little evidence that a central pattern generator is of critical importance in human locomotion, and this is a clear difference from other species. In spinalized human subjects it is quite difficult to generate sustained locomotor rhythms, although it is possible that the coordinated contraction that occurs in spasms suffered by spinal patients represents the fragmentary expression of a spinal apparatus normally used to sustain locomotion. There are many possible explanations for the inability to demonstrate a critically important spinal generator in human subjects: These range from a species difference associated with bipedal locomotion and the "encephalization" of movement to the possibility that the subjects were tested under circumstances that were inappropriate for demonstrating the importance of a generator. Nevertheless, the failure to find human spinal generators was not for want of looking, and the emphasis on "reflexes" in human studies rather than on "central pattern generators," as noted by Berkinblit et al., has a simple explanation: The former can be demonstrated readily; the latter cannot. Of course, we do not deny that human subjects have spinal locomotion generators with "half-centers" for the four limbs. The involuntary movement pattern of the arms when walking is evidence of this, although even here some doubts may be raised: Patients with Parkinson's disease do not swing their arms normally when walking (or shuffling); this suggests that the normal spinal coordination may be subject to overriding forebrain control.

A different approach to movement analysis was suggested by **Beuter,** who also criticized the deliberately restricted nature of our target article. As detailed above, we wrote on a specified topic to argue a particular point, restricting our arguments to a brief assigned to us. After reading this commentary, however, we are not overwhelmed by the potential of nonlinear dynamics: Indeed, we are left wondering how this approach advances our knowledge of motor control. Perhaps, the necessary understanding would come from the papers cited by Beuter. The only example given led to the conclusion that "experimental results were best modeled when at least two nonlinear, negative feedback loops were included and a stochastic term (noise) was added to the model," a conclusion from which conventional neurophysiologists will gain little insight. Perhaps this is a further example of how motor control scientists with different approaches fail to speak the same language, the very issue that this journal attempts to address, and one raised by **Latash** in opening his commentary.

Burgess points out the need to control torque accurately and then presents a scheme for controlling it in a figure that would find favor with most in the field. An explicit argument at the beginning of his commentary is that "one can control only what one senses (perceives)" and therefore there must exist "templates" in the central nervous system related to the "expected" movement. One can reject this initial assumption and still accept the remainder of Burgess's arguments. If perception implies the ability to make a verbal description of the perceived element, then it is interesting that neurological disease can produce situations in which there is drastic impairment in the perception of the shape and orientation of objects (presented visually) yet a normal capacity to manipulate them (Goodale et al. 1991). This definition and case study caution against inexorably linking the control of "voluntary" movement to perceived events.

As clinicians, we appreciate the emphasis placed by **Cavanagh et al.** on the experiments wrought by nature: Those who work on human subjects are denied the opportunity of selecting and placing lesions at will, but the wide scope of naturally occurring human pathology presents opportunities to study conditions one could not create for ethical reasons. Given the ability to define a deficit scrupulously, as can now be done with some accuracy for polyneuropathies, the approach of **Cavanagh et al.** can lead to significant insight into the contribution of sensory feedback to motor control. They raise the interesting paradox of greater disturbance to stance than to locomotion in neuropathic patients. This is reminiscent of

the greater suppression of short-latency spinal reflexes and cortical sensory-evoked potentials in walking subjects than in standing subjects documented by **Dietz,** and the phase-dependent modulation of such reflexes during locomotion (e.g., Capaday & Stein 1986; 1987). However, before this analogy is drawn too closely, it would be prudent to ask whether the test conditions were truly comparable: for example, whether walking subjects would be disturbed if subjected to disturbances equivalent to those of normal sway, whether other orienting cues were equally available, and so on.

How should proprioceptive or kinesthetic acuity be catalogued? **Clark** points out that the way this is done can provide qualitatively and quantitatively different answers. When measured using the channel-capacity approach, the number of resolvable positions that can be perceived may be low, but *during* movement (active or passive) movement acuity is greater. Are these different kinesthetic parameters or is the method of measurement critical to the result? The answer is probably yes in both cases. It would certainly be of interest to know how these channel capacities are distributed across the full angular range of joint rotation and whether similar findings occur for conditions in which the muscles are active rather than passive. Detection of an applied movement increases markedly when the muscles contract (e.g., Colebatch & McCloskey 1987; Gandevia & McCloskey 1976), an issue addressed directly elsewhere (Gandevia et al. 1992). Resolution of angular position may be easy near the extremes of the range, partly because of the additional input from joint receptors; and it may differ when assessed not by direct psychophysics but by examining movements triggered by sensory cues of angular position. In studying a movement not unlike throwing a frisbee, Cordo (1990) found that wrist and finger movements could be triggered surprisingly accurately by inputs related to elbow angle.

Cordo & Bevan point out some theoretical limitations to the equilibrium-point hypothesis. Particularly relevant is the need to predict accurately the behavior of the muscle spring system when many nonlinear factors distort the relationship between length and force or between motor command and force. These include the effects of synchronization, length history, potentiation, and exact pattern of motoneuron discharge. Hence it is not surprising that the central nervous system can utilize information from a range of peripheral sources to assess the progress of a movement and that visual and even auditory cues *can* be entirely satisfactory by themselves (Gandevia & McCloskey 1978; Gandevia et al. 1990).

Dietz focuses on the microneurographic data in our target article. We recognize the experimental limitations imposed by this technique (they are also raised by **Prochazka**), but the data base from microneurography is not as restricted as Dietz implies. For example, recordings have been obtained in freely standing subjects in three separate studies and the motor learning tasks developed by Vallbo and colleagues (Al-Falahe et al. 1988; 1990) cannot be dismissed as biologically irrelevant. For someone interested in precision grip, an isometric contraction is biologically relevant, but an abrupt disturbance designed to throw a standing subject off balance (as used by Dietz) may not be. In truth, what is biologically relevant for a particular worker will depend on his particular experimental paradigm; the conclusions may hence fail to apply to some other motor task, a view which Dietz clearly supports in his commentary. Dietz is interested in stance and locomotion: We can provide micro-neurographic data for the former, but not yet the latter.

We take issue with **Dietz**'s suggestion that Golgi tendon organs have been ignored in microneurographic studies, particularly in more recent studies. Early papers may have focused on spindle afferents, but that is to be expected when the fusimotor system was the topic of interest. To equate "load receptors" within antigravity muscles with tendon organs is just based on tentative speculation until the critical recordings have been done from those afferents. We would refer Dietz to the recordings of Aniss et al. (1990b) that show how identified tendon organs respond in freely standing subjects.

Dietz points to documented interactions between different sensory cues, which we do not contest. One of the emphases in our article was that the nervous system will not ignore potentially useful coherent information from any source, external or internal. We focused on inputs from skin, muscle, and joints, but we agree that these data will be integrated with other information sources if and when they are available. Finally, we do not dispute that there are innate programs for a limited repertoire of movement, but as indicated in our response to **Berkinblit et al.**, such programs may be more limited in human subjects. What separates primates, and in particular human subjects, from "lesser" creatures is not really a cyclical activity such as locomotion (though even that differs), but acquired manual skill. Typing, a task discussed by **Gordon & Inhoff**, may be painstakingly learned, but once learned it may be stored as a program that can be performed without feedback control.

Duysens & Gielen emphasize the role of cutaneous receptors as proprioceptors rather than merely as mediators of superficial cutaneous sensation. We acknowledge Moberg as a pioneer of this view, but not for the paper cited by Duysens & Gielen, because we feel that that paper is overshadowed by the back-to-back paper on the same topic by our colleagues, McCloskey et al. (1983). Why has the hand been more favored than the foot in human studies? Were we to have the same motor repertoire with the foot as the hand, perhaps the focus in our article would have been different. However, as implied in our response to Dietz, we demand little of our feet other than support and locomotion, and even then we don shoes and socks to dampen the importance of afferent feedback from the feet. We expect our hands to perform the motor tasks that truly distinguish us from other animals. As noted by Moberg and discussed in detail by **Gordon & Inhoff**, there may be severe disturbance to the motor control of the human hand when it is deprived of cutaneous feedback, much more so than occurs with the human leg (notwithstanding the difficulties discussed by **Cavanagh et al.**).

Duysens & Gielen raise the issue, not of "plasticity" as we use the term, but of the task-dependent expression of cutaneous reflexes; we await with interest their definitive report on the greater gain of cutaneous reflexes in walking subjects than in subjects at rest. Such behavior contrasts with that of short-latency muscle afferent reflexes. What locus do we favor for interactions between muscle and nonmuscle kinesthetic signals? We acknowledge the po-

tential for interaction at all levels of the neuraxis; one need not see these sites as "alternative" possibilities. However, as Duysens & Gielen note perceptively, we favor the higher levels for *integration of sensory cues into a single percept* because this would allow the integration of these sensory messages, not only with themselves but also with internal cues (as discussed by **Gordon & Inhoff**), inputs from other senses (as mentioned by **Dietz**), and internal feedback loops (as noted by **Beuter**).

Several commentaries allude to the tuning of both reflex circuits and central commands to achieve a particular task. **Lacquaniti** specifically emphasizes the ability to preset the stiffness or compliance of the limb appropriately to a task such as catching a ball. That short-latency reflexes occur in wrist flexors as well as extensors in the task may violate the principle of reciprocal inhibition but it is appropriate for the task. Interneuronal machinery probably exists to switch circuits to muscles in this way and to preset them (see our target article and that of **McCrea**). Both wrist flexors and wrist extensors will be activated prior to impact by the ball. Classical notions of reciprocal inhibition must not obscure the observation that during natural movements both so-called agonists and antagonists are co-activated, particularly when limb stiffness may be preset. In addition, whether muscles act as synergists or antagonists will depend on the task; for example, the ulnar and radial wrist flexors will act together as antagonists to the ulnar and radial wrist extensors in flexion/extension movements but they will act against each other in synergy with the appropriate extensor in ulnar or radial deviation. Lacquaniti's challenge concerns how these changes in short-latency reflexes with different tasks occur, because they appear to be adaptively tuned not simply to the mechanics of the arm and hand but also to the changes during movement.

Our findings on the ability of subjects to activate motoneurons, discharge them at graded levels, and sustain their discharge without muscle afferent feedback is relevant to the debate about the equilibrium-point hypothesis (see **Bizzi et al.**; **Feldman**, **Latash**). These findings have recently been confirmed for the lower limb (Macefield et al. 1991). We tried to emphasize that these results may provide some upper bounds on the role of muscle afferent feedback during natural movement. However, the findings cannot be used to distinguish between the λ (lambda) model and the α (alpha) model on which Feldman and Bizzi et al. disagree. Although never designed to address this dispute, our findings nevertheless reveal capacities for controlling the motoneuron pool in circumstances in which reflex feedback to it is lost and in which calibration of motor commands must be produced by signals other than from the deafferented limb, such as visual or auditory cues. Given that the psychophysical evidence strongly indicates that there are potentially redundant sources of information about limb position (and movement), the λ model would probably be favored, since it depends critically upon positional information.

We acknowledge **Paillard**'s preeminent contributions on reflex studies and movement perception, but our views diverge on two issues. The finding by Ribot et al. (1986) that reinforcement maneuvers increase the discharge of putative gamma efferents but do not affect the discharge of spindle endings remains a paradox not adequately explained by designating the effects as "dynamic." We have no problem accepting that there may be gamma drive insufficient to affect the background discharge of spindle endings (and we have presented suggestive evidence of this, see Aniss et al. 1990a). Paillard suggests that the dynamic gamma drive would contribute to tuning the sensitivity of spindle endings to stretch, but this view is not supported by the available data: Hagbarth et al. (1975) have reported that not only was the background spindle discharge unchanged but, in addition, the spindle response to passive stretch was not altered by reinforcement maneuvers (see also Burke 1981). Furthermore, Burke et al. (1981) reported that the sensitivity of muscle afferents to graded tendon percussion was not altered by reinforcement maneuvers, even though those maneuvers decreased tendon jerk threshold, whether the threshold was measured as percussion intensity or as intensity of the afferent volley.

Assuming that the axons identified as gamma efferents by Ribot et al. (1986) were indeed gamma efferents (and there can be no proof of this), what is the function of this gamma drive? Given the data cited above, it is unlikely to be sensitization of spindle endings to stretch (unless the recording techniques were inadequate to detect the sensitization). Could it be that the gamma drive serves not to sensitize the ending to stretch but to sensitize intrafusal muscle to subsequent gamma inputs? Whatever the answer, the action of reinforcement in enhancing tendon jerks and the H reflex is likely to be central, on reflex transmission within the spinal cord, not primarily peripheral, on spindle sensitivity (Burke 1981).

Contrary to the view expressed by **Paillard,** many studies now confirm that the H reflex is potentiated by reinforcement maneuvers and that the degree of potentiation is similar to that of a tendon jerk of similar size (e.g. Bussel et al. 1978; Clare & Landau 1964). Having spent some 15 years trying to gain insight into fusimotor function using the technically demanding technique of microneurography, we would be delighted if such insight could be gained by the relatively simple technique of comparing the behaviors of the H reflex and the tendon jerk! Inferences about fusimotor function from afferent recordings are indirect, but not as indirect as comparisons of two skeletomotor reflexes. To draw conclusions about receptor sensitivity from these reflex comparisons one must assume that the only difference between the reflexes is that the H reflex bypasses receptor mechanisms. This issue has been studied in depth (Burke et al. 1983; 1984). There are many differences other than inclusion or exclusion of the receptor. We believe that fusimotor function cannot be studied this way and that this technique has been responsible for a number of erroneous conclusions (e.g., about the mechanism of action of reinforcement maneuvers and about whether spasticity is due to dynamic gamma overactivity).

It is interesting that spindle endings in the awake intact cat seem to behave in the very way expected from **Paillard**'s commentary. However, as noted by **Prochazka,** human spindle endings appear to behave differently, and perhaps those concepts are valid for cat but not man! Evidence of set-dependent changes in fusimotor biasing of spindle discharge is readily provided in the cat but less readily in human subjects, though such changes have now been described (Aniss et al. 1990a). Prochazka

suggests two reasons for the difference: First, it is possible to study only a limited range of motor tasks in human subjects (an issue also raised by **Dietz**); second, there may be a species difference in fusimotor control. We believe that both factors are important but that colleagues in the motor control field have focused on the first without truly appreciating the importance of the second. We agree with Prochazka that the usual behavior of the cat is very different from that of the human but even if it were not, the much greater size of human subjects and the much slower conduction velocity of their afferents must impose different constraints on their nervous systems and must demand different solutions for optimal control. If it can be generally accepted that reflex connectivity has evolved in different species to be appropriate to the specific motor requirements of each species (and this seems to be the case, see target article), there is no conceptual leap in suggesting that fusimotor function may also differ in some details. Indeed, there are two major justifications for human experimentation: first, that some behaviors can only be studied in human subjects and, second, that some of the design features of the motor control system are unique to human subjects.

Ross provides additional evidence implicating motor command signals in judgments about "heaviness," namely, the tendency toward mass constancy under conditions of altered gravity. In addition, she points out the limitations of the terminology we have adopted. Thus, judgments of the weight of lifted objects may relate more closely to judgments of their mass. The term "sensation of force" requires a careful definition in the context of each experiment; it is certainly used to refer to judgments made of isometric forces – the capacity to make such judgments can depend on peripheral inputs related to force although, as indicated in the target article, central signals related to motor command are used more commonly. Hence, we cannot accept that the sensation of force "always relates to an internal sensation."

In the target article we did not review the mechanisms that may underlie the impairment of accuracy of heaviness judgment when small forces are generated. Clearly, this may reflect the less favorable Weber fraction at low stimulus intensities, a phenomenon mentioned in our original publications and in those of others who report similar findings for isometric forces. What did seem to require mention was that although the size principle of motoneuron recruitment appears optimized for fine control of small forces, other constraints, such as the Weber fraction, may override this for perceptual judgments. However, this effect probably differs for different muscles in a biologically appropriate way (Kilbreath & Gandevia 1992).

R3. Conclusion. Many commentators (e.g., **Winters & Mullins**) have picked up the points in the target article and extrapolated from them to their own experimental paradigms. Several themes emerged repeatedly: the need to recalibrate motor command signals by reference to peripheral information (from the limb, but also from vision), the presence of more than one sensory input that could provide information about active (or passive) limb movement, and the crude capacities that remain when feedback from the limb is removed entirely. We did not try to catalogue all the circumstances in which sensory input guides movement, nor to enter the controversy about the equilibrium-point hypothesis. However, it would be naive to suggest that the central nervous system would choose not to use any property to which it has access – be it the peripheral feedback, the central command signals (sampled at appropriate times), or the built-in mechanical properties of muscles – to produce the widest movement repertoire.

Spinal interneuronal connections: Out of the dark comes a ray of hope

D. A. McCrea

Department of Physiology, Faculty of Medicine, University of Manitoba, Winnipeg, Manitoba, Canada R3E 0W3
Electronic mail: *dave@scrc.umanitoba.ca*

Bullock & Contreras-Vidal present a model of a single joint flexion-extension movement that includes modelled interneuronal connections and proprioceptive feedback. Their commentary stresses the need for a mathematical description in order to arrive at an understanding of the system. I have several problems with this model.

Because much of this work is not yet published, it is impossible to determine how the commentators chose to model each of the synapses. Thus some of these criticisms may be unwarranted. For example, the location of Renshaw cell synapses, the synaptic conductance change in proportion to the resting conductance of the dendritic tree, and intrinsic firing properties differ greatly between motoneurons and Ia inhibitory interneurons. How are the effects of Renshaw cells on the firing probability of these two cell types modelled? Another point is that the Ib system is not organized on a reciprocal basis of agonist-inhibition, antagonist-excitation. Ib effects are widespread throughout the limb and there is no general principle that activation of Ib afferents in a particular muscle results in excitation of the antagonist. Like the group II system (see commentary by **Cavallari**) the Ib reflex system is organized differently from the Ia afferent reflex system. Furthermore, the pattern of Ib effects has been determined primarily in quiescent anesthetized preparations. Hultborn and colleagues have recently shown that during fictive locomotion, Ib afferent recruitment results in excitation of homonymous and synergist motoneurons. Thus the Ib pathway selected may differ in different movements. There is little data on the details of the inhibition between Ib interneurons. The way we did the experiments prevents a quantitative assessment of either the pattern of mutual inhibition of Ib interneurons or the synaptic efficacy of such interactions. There is no attempt to model the actions of group II afferents; these actions are powerful, complex, and probably crucial to the operation of the system.

Although many of my criticisms could be addressed by simple changes in the model, I feel that modellers must take responsibility for asking experimentalists specific questions about the operation of interneuronal systems during particular movements. This is the only possibility for realistic models of spinal circuitry. I know that many who record from these cells are interested in modelling the system and **Bullock & Contreras-Vidal** are to be commended for their efforts. Since I have personally been involved in recordings from motoneurons, Renshaw cells, and Ia inhibitory interneurons during fictive locomotion, I also know that data could be obtained in a format useful to modellers. To my knowledge, this type of interaction has not occurred. This is indeed unfortunate. Mathematical descriptions of a system that seems to work are essential. The attempts at mathematical "lesions" in the commentary are interesting, and with pharmacological tools, they may even be testable. There are, however, an infinite number of models that can be constructed to flex and extend a limb. The notion that any of these mathematical constructs represent evolutionary and biological solutions to motor control is naive.

Burke notes that the spinal cord can produce complex behaviours (e.g., locomotion) and that a relatively crude descending control system can activate spinal interneuronal centres to produce behaviours. Burke asks the germane question of how we can reconcile the dynamic and subtle control of spinal interneuronal pathways by descending systems with the concept of a simple descending switch. My response is that although a discrete descending signal can activate spinal centres, there are complex and parallel actions of other descending systems evoked simultaneously. Just as there is never one class of afferent recruited by a perturbation of the limb and hence an isolated activation of one reflex pathway, there is never activation of a single descending system. Even those stimuli that evoke a spinal-located behaviour such as locomotion may activate multiple spinal interneuronal circuits. A frequently used preparation for evoking fictive and treadmill locomotion involves trains of stimuli in the midbrain locomotor region. As well as evoking stepping, this stimulation produces synaptic events in interneurons and motoneurons that are time-locked to individual midbrain stimuli. The extent to which these stimulus-locked events are related to intrinsic locomotor circuitry or reflect pathways activated in parallel to those involved in locomotion remains an important area of investigation. There are also descending systems that seem to have a primary role in regulating spinal interneuronal circuitry. For example, the dorsal reticulospinal system produces a decrease in transmission in many spinal reflex pathways (see Lundberg 1982). Activation of multiple descending systems in both a tonic and phasic manner will control spinal reflex transmission and produce the desired movement.

Cavallari makes two important points about reflex systems from group II muscle spindle afferents. The first is that there is more than one population of interneurons with particular group II input. Those best characterised are the rostral (L4) group and the caudal (L6-7) group. An important feature is that rostral and caudal group II interneurons are probably under independent control. Future experiments will be required to fully sort out the nature of this control but it seems likely that both descending and segmental systems can control the excitability of these populations independently. This allows for a variety of group II evoked effects in motoneurons under different conditions.

The second point made by **Cavallari** is that in the group II system, interneurons excited by afferents from certain muscles may not influence homonymous motoneurons. Instead, these interneurons project to other motoneurons and even those operating at other joints. Although traditional discussion of segmental reflexes has focused on feedback control of synergists or antagonists acting at the same joint, the group II system transcends this organization. Few models of motor control incorporate these facts (e.g., see the model in the commentary by **Bullock & Contreras-Vidal**). The point I would add is that group II reflexes are only a part of the function of these interneurons. Some rostral group II interneurons are active during the flexion phase of fictive locomotion (see target article). They are thus active in the absence of rhythmic afferent input and, by virtue of their monosynaptic connections to caudal motoneurons, contribute directly to the synaptic drive to motoneurons during fictive locomotion. They are a good example of how the nervous system can utilize connections of interneurons to distribute synaptic drive to groups of motoneurons under different conditions. The complex input to these interneurons (e.g., segmental group II, descending input from the midbrain locomotor region and input from spinal locomotor circuitry) speaks to their use in a variety of motor behaviours.

Duysens & Gielen agree that the term FRA (flexion reflex afferents) is not very good because it places emphasis on the terms "flexion" and "afferents." They suggest using "FR system" for those instances when flexion is evoked and "ER" when extension is evoked. Although I have no objection to this, I feel that changing names will not decrease the confusion concerning the functional significance of these systems. The discussion in the target article was an attempt to help those who misunderstood Lundberg's use of the terms "alternative" and "private" reflex pathways. As mentioned, it is likely that subsets of interneurons in the FR (or FRA or GRA) systems are used to perform particular movements. The strong generalized flexion or extension produced by segmental afferent activation in reduced preparations does not represent the major use of these interneuronal systems. The full operation of the FRA pathways may in fact occur only in acute-spinal, unanesthetized animals. It is the preoccupation with the flexion reflex and the expectation that the involved interneurons fit into one or two subgroups that leads to confusion. In this context, any global term for these systems, including FR and ER, will be misnomers. Lundberg was very aware of these points when he discussed the FRA system and argued for the terminology simply to honour traditional usage (see target article). It was perhaps an error to underestimate the confusion resulting from any terminology that hints at an explanation. As discussed in the target article, interneuron nomenclature suffers from similar problems. Finally, Duysens & Gielen note that the term "Ib interneuron" is gradually being abandoned. I am not so sure this is true. At least Jankowska's, Hultborn's, Lundberg's, and my own laboratories use the term Ib interneuron when discussing those cells with dominant Ib input. We are also aware that there is a wide convergence to these cells and that the function of these interneurons goes beyond the reflex system involving Golgi tendon organ afferents. An understanding of the relevant history reduces the danger in the use of interneuron names and the FRA terminology.

Lacquaniti's commentary presents several interesting observations about motor control during the task of catching a falling ball. The studies on the effect of experience on the movements during subsequent trials are particularly intriguing. There is rapid modification of the strategy for muscle activation that includes changes in the reflex effects evoked by contact with the ball. This almost certainly involves the modification of spinal interneuronal circuitry. Lacquaniti argues against a division of motor control discussions into reflex systems (i.e., systems generating movement subsequent to a sensory stimulation) and systems for internal representation of limb geometry and body space. In other words, he believes there is an intimate association between those aspects of motor control that have often been separated into segmental (reflex) and higher-order voluntary aspects. This view, to which I subscribe, is contrary to that expressed by **Latash** (see below).

The **Lacquaniti** model of reflex modification centres on the idea that hand compliance is the controlled and optimized variable. Although I have no scientific basis for this position, it also seems possible that the minimization of activation of cutaneous receptors in the hand is an optimized variable. Controlling cutaneous receptor activation (i.e., reducing the instantaneous force transmitted to the hand) would result in a system less influenced by hand position. In either case, motor control strategies would involve operational changes of spinal reflex circuitry.

Latash separates "reflexology" (i.e., the types of experiments my colleagues and I do) from "motor control" (i.e., experiments of Burke, Bizzi, Gandevia, Schieppati, Lacquaniti, etc.). Even apart from the negative connotation in the popular use of the term "reflexology," I strongly disagree with this idea. I believe that this type of separation is in fact at the foundation of much misunderstanding between investigators. Those involved in "motor control" studies must be conversant with the language and concepts of spinal motor systems because all motor output from the limbs is determined by spinal circuitry. Clearly, many human neurophysiologists are not only comfortable with "reflexology" but have also designed useful experiments in "motor control." I think that the divisions raised in this commentary concerning language, terminology, and approach to these problems are artificial and outmoded (cf. the discussion in the **Lacquaniti** and **Schieppati** commentaries).

Nichols is right about the poor links between those who map circuitry and those who study biomechanics. He also suggests that there is a role for private feedback of sensory information to particular muscles in motor control. Finally, he notes that the extensive convergence of sensory information to interneurons does not exclude the possibility of such private reflex control. Although I do not disagree with the spirit of these ideas, I have difficulty believing that many natural motor behaviours in the hindlimb can be characterised in these terms. The limited excitatory monosynaptic actions of Ia afferents to agonist motoneurons mentioned in Nichols's commentary is a case in point.

The monosynaptic connections to Ia afferents are a fraction of the actions of Ia afferents on the spinal neurons. Since activity in Ia afferents will recruit inhibitory interneurons projecting to antagonists and help regulate the activity of "Ib interneurons," Ia actions on spinal circuitry can hardly be viewed as private, even though their direct connections to motoneurons are limited. During movement, Ia afferents will never be activated in isolation from other afferents. Presynaptic inhibition of particular types of afferents will change their synaptic actions on both motoneurons and interneurons (see the target article). **Nichols** also uses the convergence of autogenic (monosynaptic group Ia) excitation and inhibition from direct antagonists as an example of a system in which biomechanical sense can be made of the neuronal circuitry. There may be experimental conditions under which these pathways operate to regulate stiffness, but one cannot expect stiffness to be regulated under all behaviors. For example, the direct cortical and other descending input to reciprocal inhibitory interneurons and the convergence of Ia afferents to inhibitory Ib interneurons affecting the agonists both point to the possibilities for the defeat or at least modification of stiffness regulation. There may be situations in which the activity of one afferent type and a subset of interneurons is up-regulated to the point that they dominate the synaptic actions in motoneurons and effectively constitute a private feedback pathway. Such situations probably would be the exception and not the rule.

I would pose another question. There is a large literature on compartmentalization, the division of motoneurons into task groups and reflexes evoked in single muscles. What are the conditions under which the activity of individual hindlimb muscles are regulated in the intact cat? I suspect that these are rare. The identification of such behaviours should be the first task for those wishing to consider the limited actions of sensory feedback on individual muscles or muscle compartments. Some of these points are raised in the commentary by **Pratt & Macpherson**. Finally, I agree with **Nichols** that maps of neuronal connectivity will never describe the operation of reflex and descending pathways, but they are an essential starting point for discussion and the development of testable hypotheses. **Burke**'s commentary presents these issues quite clearly.

Hamm & McCurdy summarize some of the theories about the function of recurrent inhibition. One hypothesis not mentioned comes from Lundberg's lab (Fu et al. 1978). Afferents recruited during the tonic stretch reflex produce both an excitation of motoneurons and, via subsequent Renshaw cell activation, an inhibition of interneurons mediating reciprocal inhibition (Ia inhibitory interneurons, IaIns). In their scheme, the focus is on the recurrent inhibition of IaIns; Renshaw cells can function to limit and regulate the amount of reciprocal inhibition during certain movements. Hultborn and Illert (1991) also summarize some of the possible actions of IaIns and Renshaw cells. There is no reason to presuppose that any of these hypotheses are mutually exclusive. Perhaps recurrent inhibition of motoneurons and interneurons serves several purposes during movement and in some movements may have weak effects. It should be possible to test some of these ideas by blocking the relevant transmitter systems. For example, experiments using anticholinergic agents (to block Renshaw cell activation from motoneuron axon collaterals) have found surprisingly small effects on the patterns and amplitude of motoneuron activation during fictive and treadmill locomotion (Noga et al. 1987). Similar approaches and the identification of synaptic mechanisms in other interneuronal systems will permit the assessment of the functional importance of identified spinal reflex systems during many movements.

Pratt & Macpherson stress the need for studies on the variety of motor behaviours used in particular tasks. We need detailed analysis of limb movements as well as the degree and sequence of motoneuron activation following controlled perturbations or simple voluntary movements. Such experiments will help define the number of functional groups of interneurons that exist during particular classes of movement. In many movements in the cat there is co-activation of gastrocnemius and soleus motoneurons. This implies considerable similarity in the input to excitatory interneurons projecting to soleus and gastrocnemius. During paw shake, however, soleus is silent while gastrocnemius is active. This indicates that there must be at least two populations of excitatory interneurons differentiated by input from either tonic or cyclic afferent activity during paw shake. Observing the varieties of motor behaviours elicited under controlled conditions will contribute greatly to an understanding of the functional organization of spinal interneurons. Similarly, the finding that cutaneous excitation is not distributed homogeneously to soleus, medial and lateral gastrocnemius (see target article) illustrates another subtlety of interneuronal differentiation to the ankle extensors in the cat. An example of interneuronal differentiation to ankle extensors is outlined in **Schieppati**'s commentary during voluntary movement in man.

McCollum presents an interesting way to diagram movements and has chosen the flexion reflex as a simple starting point. Her pictures offer an immediate appreciation of the varieties of limb position and are quite useful for describing the movement. Many motor control strategies, however, are characterized by co-contractions of antagonist pairs of muscles and not by changes in limb position. Thus a description of limb position is only a part of the information required to describe the operation of motor systems. We must eventually include a representation of muscle activity as well as limb position. How this muscle activity matrix can be incorporated into a mathematical system with the three dimensional matrix of limb position is a challenge for which physiologists such as myself eagerly await the successful efforts of McCollum and others.

Rudomin suggests that the variety of convergence and strength of synaptic connections to individual interneurons within a functionally defined population will result in a range of interneuronal activity patterns during movement. He stresses, therefore, the need to record the overall activity of populations of interneurons before making sense of their function during movement. He points out that an analysis of the firing frequency of interneurons during movement gives a very incomplete picture of their activity. These points are well taken, but an analysis of interneurons during fictive locomotion suggests that this may be less of a problem than envisioned. Dr. Jordan's lab in Winnipeg is analyzing the beginning and end of trains of interneuron spikes in relation to the beginning and end of activity in other neurons (e.g., individual motoneurons or populations of motoneurons). Their "start-stop" analyses show strong consistencies within interneuronal populations during fictive locomotion. Drew et al. (1986) displayed the spike activity of a reticulospinal neuron triggered from the onset of EMG activity of limb muscles. This analysis readily permitted a correlation of single-cell activity with activity of a number of different muscles. Clear linkages between the activity of the neuron and one of the muscles were seen. Such analyses, done one interneuron at a time, can provide meaningful information about the activity of the population. I encourage attempts to describe more elegantly the activity of neuron populations; present approaches should provide adequate data to permit modelling of the effect of a population based on the activity of individual cells.

The other point raised by **Rudomin** is more problematic, namely that nonlinear summation of synaptic activity (i.e., spatial facilitation) in the target cells of the interneurons under study may make it difficult to predict the synaptic effect in target cells. It is therefore essential to record from the target cells. Except for accelerated data collection, multiple and simultaneous recordings would add little to the solution of the nonlinear summation problem. I will again mention the possibility that spinal interneurons may display maintained activity in the absence of maintained synaptic input. The best example of this phenomenon is the plateau potential found in motoneurons (citations in target article). If similar postsynaptic mechanisms are found in interneurons they should be detectable with conventional technology. The demonstration of self-sustained interneuron firing will require a complete revision of models of neuronal systems.

Alpha and lambda models of motor control and the equilibrium-point hypothesis are discussed in the target article by **Bizzi et al.** and in the commentary by **Feldman.** Both accept the need for motor systems to use spinal interneurons for multiple purposes. Neither, however, addresses the issue of sustained motoneuron activity in the absence of sustained synaptic input by neuronal or afferent systems. I believe that general recognition of this phenomenon will change several concepts of motor control. Plateau potentials and repetitive firing of motoneurons result from the activation of intrinsic motoneuron ionic currents that are initiated by synaptic input. It is difficult to imagine that those neurons that allow for plateau potential induction "know" directly about the subsequent activity of motoneurons. Thus the exact degree of motoneuron activation that results from centrally programmed motor commands may not always be predictable. During those movements in which plateau potentials contribute substantially to muscle activation, central structures must rely heavily on afferent feedback to determine whether the desired limb position has been attained. This does not imply that stretch receptor activation of short- or long-latency pathways must produce powerful reflex activation of motoneurons. Instead, afferent feedback may modulate activity in central systems that allow plateau potential induction in motoneurons. Afferent feedback can also terminate sustained motoneuron firing by a brief period of motoneuron inhibition. In this sense, experiments measuring reflex responses in motoneurons will not assess the entire contribution of afferent feedback to motor control.

Rudomin's work has shown that descending systems, though evoking presynaptic inhibition in segmental afferents, do not seem to receive presynaptic depolarization of their own terminals. Since this observation may illustrate a general principle of the organization of spinal motor systems I will restate one part of his commentary. Many spinal interneurons receive descending and segmental synaptic input. Presynaptic inhibition evoked by descending systems may act to increase the priority of descending motor commands to interneurons and ultimately to motoneurons. This is a good example of the utilization of complex spinal circuitry to simplify the operational connectivity to interneurons and produce a particular movement while reducing the possibility for movements generated by other sources. Presynaptic inhibition of segmental afferents evoked by descending systems would also reinforce the direct monosynaptic actions of descending systems on motoneurons by reducing extraneous inputs (cf. the commentary by **Burke**).

Both **Schieppati**'s and **Lacquaniti**'s commentaries provide elegant examples of how interneurons are selected during voluntary movement. There seems to be very strong support from the human physiologists for the notion that sense is being made of complex spinal circuitry provided the hypotheses are tested in preparations exhibiting real and natural movements (cf. **Burke**). Schieppati's example of selection of spinal interneuronal pathways acting on particular ankle synergists in humans (see his commentary) may be an example of the type of flexibility outlined in the target article concerning muscle synergies (Fig. 3). Work from many laboratories reveals the striking similarities of spinal reflex circuitry in cat and man. Roby-Brami and Bussel (1990) have shown how reflexes in spinal man are similar to those in the acute-spinal cat treated with L-Dopa. The acute-spinal Dopa cat is the preparation in which many of the FRA concepts were developed by Lundberg and colleagues. There are now many examples of human experiments that relate directly to reduced cat preparations. These developments in the description, interpretation, and understanding of human voluntary movements strongly support the idea that sense is being made of spinal interneurons. Substantial progress is evident.

I would like to reinforce two points raised by **Schieppati**. The first is the exciting possibility that learning can operate at the level of spinal interneurons. There are reports suggesting long-term augmentation of synaptic effects at the level of the monosynaptic reflex (e.g., Wolpaw & Carp 1990), spinal pain transmission (Slosberg 1990) and descending input to the spinal cord (Iriki et al. 1990). The second is Schieppati's belief that "some [I would say all] important motor tasks can be accomplished *only* by suitable interaction between descending pathways and segmental reflexes." I strongly support this view and mention again that the extensive overlap of spinal reflex circuitry and the sites of action of descending systems argue against a division of motor control into the reflexive and the voluntary (cf. **Latash**).

How far into brain function can neural networks take us?

David A. Robinson

Departments of Ophthalmology, Biomedical Engineering and Neuroscience, The Johns Hopkins University School of Medicine, Baltimore, MD 21287-9131

R1. Network analysis of the future

When he cautions us that history suggests it would be unwise to bet against continued progress in any branch of science, **Alexander** makes a good point. The status of space exploration, genetics, or semiconductors 50 years ago should make one think twice about taking a pessimistic view of the future in neurophysiology. Nevertheless, although great strides have been made in molecular, receptor, and membrane neuroscience, I still look at the progress in understanding neural signal processing the last 50 years and see only minor advances on the fringes of a seemingly intractable subject. I am willing to be optimistic, however, and hope the field of biologically plausible connectionism will grow into a discipline that can produce useful results.

Still, **Alexander** has outlined a partial approach, which may make many feel uncomfortable, that of devising a connectionist model that can solve the same problem as one confronting a real neural network (we already have a problem – defining the problem), using as much architecture as the known anatomy will bear, and demonstrating (one hopes) that the model units will look like real, recorded neurons. Many neurophysiologists who like to see their hypotheses confirmed (not just illustrated) will not accept this. They might deride it as the "gee whiz" method: The neurophysiologist, looking over the shoulder of his connectionist collaborator, says "Gee whiz, I saw a neuron just like that last week in area _____" (fill in the blank). Obviously, Alexander means a more extensive approach and suggests a test by recording the types and distributions of hidden units and neurons when the two networks (model and real) are required to learn a different task. The enormity of this task in most parts of the mammalian CNS needs no comment. Nevertheless, neural network theory is the only reasonable tool at hand and is far better than no tool at all, as in the past.

R2. Hidden units: How many and how functionally segregated?

Two excellent points are made by **Andersen & Brotchie**. One concerns localization. Obviously, if each brain function were neatly separated anatomically, we would have less difficulty in understanding the brain. The main problem is in even defining function. I doubt that it is possible to define a function for subsets of cells in the supplementary motor cortex, basal ganglia, or even the red nucleus. It would be nice if I were wrong. I agree that wherever a function can be isolated and identified, half the battle is won. Andersen & Brotchie are perhaps more optimistic than I. I would leave it to the reader to speculate on the functional identification problem in any particular region of interest.

The second point made by **Andersen & Brotchie** concerns the number of hidden units. I have gained the impression, not contradicted by any experience in the oculomotor system, that in a brain that allows half of its neurons to die around the time of birth, the vertebrate CNS has never felt the need to be stingy with neurons. Consequently, I have tended to be generous with the number of hidden units (e.g., 40 with only 4 units in each of the input and output layers) and I have always been puzzled by those who starve this layer. Certainly there are thousands of cells in the caudal pons mediating oculomotor signals and one should note that although there are thousands of input and output fibers, they all carry just four signals in and one signal out. For example, the many fibers in the vestibular nerve all carry the same signal, qualitatively, and so do the fibers of the motoneurons.

This is obviously an important area for future research, both experimental and theoretical, and I am happy that **Andersen & Brotchie** have focused attention on it because it is a bio-connectionist question that, unlike so many others, is addressable with current techniques.

R3. Do engineers really want to know what hidden units do?

I still doubt it. **Bischof & Pinz** argue that engineers do care about hidden units. They point out that a neural network might be more robust. I agree. I still would not care what the hidden units did. I agree that it is easier to tell what an existing system (i.e., one that has learned) does than to design one from scratch, but this is irrelevant. I agree that engineers try to interpret the behavior of their *systems*, but if they waste time looking at hidden units, they are wasting the company's money.

R3.1. When architecture is known. It is pointed out by **Bischof & Pinz** that it is silly to connect everything to everything else ("full interconnectedness") if you know from anatomy that some connections are fixed and others nonexistent. I am sorry I did not mention this specifically in the target article. In our own attempts to model the vestibulo-ocular reflex (VOR), the models were not fully connected; some weights (on motoneurons) were fixed. This is described in our publications, but I skipped over this point in the target paper. Similarly, in our efforts to model the neural integrator, we realize that ignoring genetic wiring in a reflex as ancient as the VOR is an error (D. B. Arnold & D. A. Robinson, unpublished) and have again fixed several of the pathways. I thank the commentators for bringing this point forward.

R3.2. Maps. A neural network can be rearranged conceptually without changing its function whereas real networks have cells with anatomical locations as, **Bischof & Pinz** point out. If one wants to put onto the input layer a pattern such as $(0, 1, 1.5, -3)$, however, it is necessary to number the input units 1, 2, 3, and 4, thereby imposing an order on them if not a spatial location. This is all I meant by a "spatial pattern" in section 2.1. The term was only meant to distinguish such networks from dynamic networks that generate functions of time. My comment about a "nice biological touch" had to do with the overcompleteness of our networks leading to the lack of a unique solution; it had nothing to do with spatial locations.

An interesting point is raised here however. In some brain areas there is a map and a fairly well defined spatial organization of cell features (e.g., in all primary sensory areas and to a lesser extent in motor cortex and cerebellum) but there are many areas where no such map exists. For example, oculomotor neurons in the caudal pons carry an eye position signal and eye velocity signals that are different for pursuit, the VOR, and saccades. There is little evidence that these four signals are segregated anatomically on premotor neurons. Consequently, as far as we know, anatomical location in this network is as irrelevant as "location" in an artificial network. This situation is found in many other areas of the brain.

R3.3. Unit behaviors. I certainly agree that it would be lovely to know the receptive and projective fields of a neuron. That is a major problem when one records from single neurons in alert animals. In motor areas we cannot even record the receptive field. In some instances, spike-triggered averaging can help us determine some projections, but hardly a projective field. I am curious to know why **Bischof & Pinz** think it is important to know what the surrounding hidden units do in light of the previous comment that, in a computer simulation, units have no spatial location. I think most investigators have not found it very helpful in interpreting the behavior of one neuron in the oculomotor caudal pons to know what its neighbors do.

R4. So what is the alternative?

The suspiciousness of **Bossut** about neural networks may reflect the reaction of many who record single units but do not bother to protest in print – a sort of silent majority. This is the same reaction I allude to in my reply to **Alexander.** Bossut reiterates that single-unit recordings are the *sine qua non* of neurophysiology (and no one will dispute that). But he goes on to call neural networks a piece of creative art and a temporary psychologic reinforcement for researchers in need, predicting that scientific delusions will emerge along with more bias and erroneous concepts than one would get just from looking at single neuron recordings. What does it prove if one's neural network has units that look like observed neurons? The rebuttal to this is the target article itself. I will restate two major points against Bossut's charges.

First, the record of those recordings from single units in explaining brain function during the last 20 years is abysmal. This is because (with some admitted oversimplification) these investigators record data but do not use them to explain anything. Fortunately, this sad message is beginning to come through louder and clearer than ever before.

Second, attempts to model neural systems with block diagrams are, in my view, inadequate and downright misleading. Neural networks (artificial) may be primitive at the moment but at least they are an attempt to model real networks and to *interpret* and *explain*. If **Bossut** has a better way, *BBS* is an appropriate vehicle for telling us about it.

R5. Mathematics: promises, promises

Differential geometry, chaos theory, Lie group theory, and nonlinear dynamics, **Clarke** suggests, represent a new breed of sophisticated mathematics that will help us understand brain function. Such enthusiasm is to be commended and no analytical path should remain unexplored. Without more specific examples pertaining to single-unit behavior, however, it is difficult to make any further comment except to say that many neurophysiologists, who tend to be curmudgeons, do not respond at all to such sweeping proclamations. They tend to wait for real predictions to be made in real neural systems, having noted that if they wait long enough, 99% of all the claims go away. I think chaos theory has a role to play in understanding brain function, but I am not sure that at the moment it can do much more than add to our troubles in interpreting hidden unit behavior.

R6. Coordinate systems – our own invention?

Many reasons are given by **Colby et al.** for why the saccadic system is not concerned with where targets really are (in space) but only where they are on the retina. They do not review the evidence that stimulation of some regions of the brain (e.g., supplemental eye fields) produces goal-directed saccades. Going beyond this, of course, is the problem of how one can reach out, in darkness, and touch a once seen location in space regardless of intervening eye and body movements. Endlessly updating its location in a retinal logbook seems, somehow, less than satisfactory.

Colby et al. refer to our model (Anastasio & Robinson 1990, see target article) of the vestibulo-ocular reflex and were pleased with the similarity between model hidden units and real, second-order, vestibular neurons, but I fear they did not appreciate the conclusion I draw from that model. The hidden units in the model operate under an algorithm to eliminate retinal image slip and know nothing about coordinates and coordinate transformations. I would say that the question of what coordinate system such a neuron was operating in is an ill-posed one. Coordinate systems are things we use to make quantitative descriptions. The assumption that the CNS is obliged to recognize them as well is, in my view, unsupported, yet the idea is seldom questioned. I feel that the idea of coordinate systems in the brain should be put on the defensive, and it was neural networks that led me to this opinion.

R7. The view from the right

I am glad that **Fuchs et al.** chose to comment because it will give the reader a chance to hear from a group on the fringes of the conservative right wing in the area of modeling in the nervous system. This group has earned a reputation over the years as being fairly hostile to modeling. They insist on anatomical correctness and will accept no model unless every cell and every synapse has been identified electrophysiologically and anatomically. At this point, a model is no longer needed, suggesting that this group may not understand why people make models in the first place. This attitude is not difficult to understand. If one has spent two years using laborious techniques to show that cell type A does (or does not) project to cell type B, one might well be put out by some modeler riding roughshod over such hard-won results. But this implies that to understand the brain we must claw away at it, synapse by synapse and cell by cell. That this will take more millions of years than the solar system has left daunts advocates not at all.

Fuchs et al. have not addressed the larger issues raised in the target article because those concern what we might learn from modeling, and this group has a blind spot for modeling. Even though they are only of concern within the narrow confines of the oculomotor system, I will try to address some of the issues raised, taking them up in the order in which they arise.

I am accused of publishing only my own discharge patterns for oculomotor neurons in the caudal pons. Because I did not illustrate any discharge patterns, this comment must refer to Equation 1 in the target article. This equation is deliberately made so loose and general that it encompasses all the discharge patterns I know about in the literature, including those from **Fuchs's** group. Next, I am accused of decrying unit recording in favor of modeling. I hope the target article makes it clear that this is just not the case.

Fuchs et al.'s comments on rogue cells are especially illustrative. These commentators evidently agree that such cells exist but they claim that second-order vestibular cells are not among them. I would probably agree. In our model of the neural integrator (Arnold & Robinson 1991) we thought of our hidden units as representing a mixture of second-, third-, and perhaps even fourth-order neurons. The idea was to see, as a first step, whether a fully-connected network could be taught to integrate. The details of specific connections were not important at this stage. But allowing all hidden neurons to receive an input from the vestibular nerve causes all cells to become, technically, second-order cells and some of them also become rogue cells. This situation is trivially easy to correct and in subsequent models (unpublished) we have done so, but the real point is that such details are simply irrelevant for the purposes of the model.

This illustrates better than anything I can say how these critics fail completely to understand what a model is and why we make them. We were delighted to discover that we could teach a network to integrate, and their reaction is just to pick nits, completely missing the point. In line with the title of their commentary, these critics cannot even seem to see the baby, so, they have thrown it out with the bath water. I am amazed that they did not object equally strongly to our violation of Dale's law, allowing a cell to both excite and inhibit its targets (another minor point, likewise easily corrected, as we have done in later simulations).

Suddenly, in mid-paragraph, the subject switches to distributed signals. It is simply *not* true, **Fuchs et al.** insist, that components of oculomotor signals are distributed in the caudal pons. The fact that everyone, including them, sees these signal components, and even measures and reports on them, is beside the point because I have not included every known cell and every known connection in the model. That this is probably unknowable is one of my points in the target article.

Again, these commentators have missed the whole point of why we make models.

Fuchs et al.'s attitude about the neural integrator is equally telling. We see an eye-velocity command leaving the vestibular nucleus and emerging on motoneurons as an eye-position command. We may not know how that happens, but that it does happen is not open to question. Because the operation is that of integration, to call it a neural integrator is to do no more than give it a name. Yet **Fuchs et al.** talk about its "possible existence"; one must wonder about the reason for this apparent denial of a plain fact. Evidently, if one cannot explain how something works on a cell-by-cell, synapse-by-synapse basis, one is free to doubt its existence. Worse than that, despite the fact that we do not know how the integrator works, some of us have proposed a model for it. Perhaps under these conditions the integrator itself, not just the model, must join the list of the only "possible."

Next, **Fuchs et al.** point out that because the real neurons in the possible integrator carry a combination of an eye-position and eye-velocity signal, these are good indications of what the network is trying to do, contrary to my worry that we may not be able to deduce much from hidden-unit behavior. First, these signals give us strong hints only because we already know the input signal, the motoneuron signal, and the plant transfer function. Thus, our hindsight is 20/20. As I pointed out in the target article, one must start with models of simple systems that we do understand in order to look ahead to see the likely problems trying to understand really complicated systems.

Next, **Fuchs et al.** point out, quite properly, that I have not tried to deal with the cerebellum in my models. Quite true. The target article is not intended to be a review of oculomotor physiology – I was aiming at larger issues. Then we are reminded that some neurons in the caudal pons have pure eye-position signals. Fine. Next comes a strange statement that none of the cells with an eye-position sensitivity has a signal related to head velocity. This is so patently incorrect that I can only conclude there must be some problem here based on definitions and terminology.

R8. The problem of understanding

The problem of the meaning of understanding is pointed out by **Heuer** perhaps more clearly than in my own attempt to express it. In his Gedanken experiment, once one gets above the peripheral level (sensory and motor) where functions can be identified and, with luck, anatomically isolated, the model system will become so complex that the behavior of its elements will be incomprehensible. We do not "understand" turbulent flow by describing each vortex. We do not "understand" weather by insisting that we know what is happening in each cubic millimeter of the atmosphere. We do not "understand" the ecology of a barrier island one grain of sand at a time.

Sad to say, it is necessary to face the certain fact that we will not "understand" the brain on the basis of its individual neurons. This will not stop many from trying and succeeding in various peripheral and simple areas such as the vestibulo-ocular reflex, although even there no one seriously proposes to account for every cell and synapse.

Eventually, however, we will have to accept a broader concept of understanding. The problem is that we cannot now envision the next step up to a new level that we can still call understanding, and we are also held back by the microelectrode and the single unit, which tend to tie us down to a reality with which we are at least familiar. Perhaps, as **Heuer** suggests, a new discipline of neuroconnectionist theory can help us.

R9. Sexy channels and sexy synapses

It is refreshing to read a commentary that is as down to earth as **Kupfermann**'s. Fancy that: Five hidden units and one task, yet the hidden-unit behavior is inexplicable. First, let me clarify that when I said we may have to accept the inexplicable nature of networks, I was thinking of hundreds of cells and thousands of synapses making the system inexplicable by sheer numbers.

I would be less inclined to worry about the learning rules than the nature of the cell membranes and synapses of the hidden units. I have gained the impression that creatures without an overabundance of neurons, such as *Aplysia*, even if they can play the violin, tend to use what neurons they have in clever ways so that single cells can perform tricks that require a whole network in us lazy, spoiled vertebrates. Such cells can have highly nonlinear properties and can act like threshold devices, gates, one-shot multivibrators, or oscillators and they can put to shame the dumb units that we use in neural networks. Five such cells is more than enough to create chaos (I use the word in its technical sense). I categorize such cells as having sexy synapses and channels, and I worry that the real *Aplysia* cells might fall into this category. These properties are most likely to be genetic. This possibility is an unmet challenge for neural networkers who are understandably daunted at the thought of making the inexplicable even more inexplicable.

R10. Will hidden units yield their secrets?

The good, if obvious, point made by **MacKay & Riehle** is that a serious weakness at the moment in neural network applications is their dependency on an external teacher. I protest that, contrary to accusation, I am bothered by it. I was able to make an end run around the problem because two major oculomotor subsystems (vestibular and optokinetic) evolved to reduce image motion on the retina, and cells in the retina provide that error signal directly. Obviously, retinal slip per se is neither "good" nor "bad," but animals with a lot of retinal slip dropped out of the gene pool. In our current struggle to understand hidden units, I do not think we need to be distracted by such issues, but in proposing neural network models of other systems, it becomes embarrassing when the investigator has to tell the network what is good and what is bad. Fortunately, some people are studying truly self-organizing systems and may be able to help us out someday.

R10.1. Rogue cells revisited. In response to **MacKay & Riehle,** let me try to clarify my vision of the creation of rogue cells. If, in the model of the neural integrator described in the target article, one located each hidden unit as a point on a plane that plotted the amplitude of its position signal against its velocity signal, one would, after initial randomization of the synaptic weights, probably see a fuzzy cloud of points showing that position and velocity signals are intermixed. After each iteration and adjustment of the synaptic weights, each point on the plot would move. They would tend to move to regions (the first and third quadrants) where the position and velocity signals have the same sign, because that is the most useful combination to send to motoneurons. When the error became sufficiently small, the process would stop. By then, most cells would be in the first and third quadrants (this is a result, not a speculation) but a few would be trapped in the other quadrants and so would become rogue units.

This is probably not a good name for them because they are not "good" or "bad"; they are just there. The overall system does not care if they are there or not. Their existence neither aids nor hinders the system's ability to drive the error to zero. The only reason for singling them out is that they appear very curious to us because their position signal drives the eyes, say, to the left, while their velocity signal drives them to the right. You do not have to be a systems engineer to think that odd. My point is that when one finds such cells, one should hesitate before inventing a whole new hypothesis to explain their existence.

MacKay & Riehle ask whether the system can do as well without rogue cells, suggesting that the answer is no. Well, if you reinitialize and retrain the network, you will get some mature networks that do not have any rogue cells, so I think the answer is that the network does not need them. If you remove a rogue cell from a specific mature network, the error will obviously increase (it would increase if you did anything to the network), but this does not justify rogue cells as a class. Again, justification is an inappropriate word. Within the context of the networks I described, rogue cells just are. The problem with them is in our heads.

MacKay & Riehle make a point that at higher levels, such as the cerebellum, one might expect to see bits of signals needed to trim up or tailor a motor signal to adjust its metrics. Well, if I saw a cell in the cerebellum that carried position and velocity signals with opposite polarities, I would still be intrigued and, because this was internally self-contradictory, I would not just pass it off as a signal used for adjustment. I would not conclude that the CNS had deliberately made such signals because they are necessary. They are not necessary. But, in addition, the rogue cells in our models are thought to project to motoneurons, which make their signal confusion seem that much stranger.

R10.2. Hidden units I have known. Many central neurons have fascinating signals that tell us lots of things, **MacKay & Riehle** feel. Some cells are motor, some sensory, some like motion, others, color. Some are said to like faces and some seem involved with memory for place. It is fortunate indeed that the brain can be divided up into functional areas, else we should be really lost; but I had in mind something slightly more specific. In none of these examples do we have any clear idea of the signal processing that led to the cell's behavior or how that behavior will be used downstream in subsequent decision processes.

Obviously, we are not ready for that sort of thing, which is what I worry about; and a recitation of such neurons is like admiring pretty flowers in a garden. One cannot fail to be impressed, but where does one go from there? Record more of them?

In the target article, I allude to a specific example in the saccadic system. Saccadic amplitude and direction are coded by place in the superior colliculus, but this is somehow transformed to a temporal code in the form of a burst of activity on cells in the reticular formation that goes on, through a fairly well known chain of events, to produce a saccadic eye movement of the appropriate size and direction. A "small army" of people has been recording in this system for almost 25 years and has identified and characterized a variety of cell types (pause cells, long-lead burst cells, quasi-visual cells, medium-lead and short-lead burst cells, etc.) yet we still do not know how this simple event takes place. Knowing that a cell bursts for saccades enables us to put it in the garden and admire it, but it is just not enough for what I have in mind.

I agree with **MacKay & Riehle** that we often greatly limit the information we can get by restricting the animal's behavior (immobilizing the head or the arm except for a single joint, for example). The problems of doing otherwise are fairly obvious. In a freely moving monkey, one might have a hard time deciding what a neuron was responding to. Note also that by increasing the dimensionality one increases the complexity of the signal recorded, making it that much more difficult to decode. In general, MacKay & Riehle are very upbeat and that is good. In my target article, I chose to be the pessimist, emphasizing the negative to counterbalance what I feel to be an often unjustified optimism. I wanted to call attention to the enormity of what we do not know, and felt that neural networks, an imperfect tool at best, had the feature of holding up a mirror, by giving us specific, if crude, models that allow us to begin to really appreciate that enormity.

R11. Interneuron behavior: It's all the information there is

R11.1. Parsing in hidden units.
Despite the deliberately negative views expressed in the target article, one will never be able to explain whatever is explainable about hidden units without recording from them and trying to develop models. I am still doubtful that 20 or 50 years from now we will be successful in penetrating the real interior of this jungle, but there may well be a lot of mopping up operations that can take place on the peripheries. One can only hope that by using analogs like neural networks as guides we can "understand" these areas in a fairly mechanistic and satisfactory way. In this effort, the signals on hidden units will be just about all we have to work with.

Deeper in the jungle (and I would agree that that would include language) I still worry about what we will have to accept as "understanding." Is the "considerable light" that Rosenberg (1987) has shed, and the appearance of units preferring consonants and vowels, acceptable? Is the emergence of nouns and verbs in a network acceptable?

I am a bit skeptical about **Rager's** idea of sparse coding. The impression I have gained in the cat and monkey oculomotor system is that the brain will never do anything with 100 neurons if it can be done just as well with 1,000 (see my reply to **Andersen & Brotchie**). In mammalian motor systems, I am unaware of any area in which "few neurons are simultaneously active." In sensory systems, the concept of selection for features that are more and more globally specific leads, in the visual system, to the well-known concept of the grandmother cell. One wonders whether Rolls and Treves (1990) have found grandmother cells for taste. This concept is generally discredited in the vertebrate visual system.

R11.2. Rogue cells again; maybe some are not roguish.
In the context of the small systems I have described in the target article, **Rager** agrees that the rogue cells do not have any special function (see my reply to **MacKay & Riehle**). But, he warns, let us not just dismiss every weird cell we come across as "just another rogue cell." Subsequent experimental work in our laboratory has focused my attention on what I call cross-axis plasticity. It is most simply described in the context of the vestibulo-ocular reflex (VOR). When the head rotates horizontally, so do the eyes. This is not a default situation in which the vertical muscles are just passively inactive. If any lesion causes a misalignment between eye and head motion, motor learning quickly corrects the situation by using signals from the horizontal semicircular canals to create compensatory vertical eye movements. This correction does not require the creation, *de novo*, of new fibers and synapses but uses preexisting connections that are normally not needed. Their existence is easily demonstrated by moving a large optokinetic screen vertically in front of a cat, synchronously with its horizontal head rotation. Soon (within a half hour) a significant vertical eye component has been added to the horizontal VOR seen during rotation in the dark (Robinson 1982).

This phenomenon must be mediated by neurons that carry a signal from the horizontal canals to vertical eye muscles (not necessarily monosynaptically) in such a way that their signals normally cancel out. Thus, they probably project to the motoneurons of both muscles of a vertical antagonistic pair. In this respect, they resemble the rogue cells I mention in the target article that excite both of a pair of antagonist motoneurons but, in this case, they serve a necessary function. Thus, rogue cells in artificial neural networks suggest that we may not need to worry about every weird cell we see with a microelectrode, but, as **Rager** points out, we should not dismiss a seemingly weird cell just because we are not clever enough to see what its function might be.

R12. Drawing boundaries is difficult

My thoughts on the meaning of explanation are expressed by **Schwarz & Pouget** better than by me: If one cannot account for every cell, one must settle for the three features mentioned – the nature of the information being processed, the algorithms that could perform the task, and how they are implemented. The problem is how specific one needs to be in describing the "nature" of the information for the explanation to be satisfying. "Vision" is not good enough. Is color, or color contrast, or motion of color edges specific enough or too specific?

In any neural system it is more and more difficult to distinguish and isolate tasks, either sensory or motor, as one moves centrally. As **Schwarz & Pouget** point out, Fodor's modular system is a good candidate for modeling with a neural network because its boundaries can be defined so it can be reasonably isolated. Unfortunately, few systems can be so bounded. In sensory systems, one may know the inputs, but the outputs are very difficult to specify. They are usually diffuse and, because they are inputs to subsequent stages, the decision as to where one stage leaves off and another begins is arbitrary even if one is lucky enough to discern a candidate boundary. In motor systems, it is the opposite; we may know the outputs, but the input codes are a mystery.

Schwarz & Pouget's "superimposition theory" is probably correct but perhaps in a more vicious way than they suggest. There is no doubt at all that signals fan out considerably from any given region and converge from many regions to a given locus. This phenomenon increases as one moves centrally, creating a mess, as these commentators point out. I do not think we need the truly messy theory they propose. Convergence and divergence is enough to destroy any hope of modular systems or of drawing a boundary around a system that denotes functional isolation.

I am happy that **Schwarz & Pouget** are upbeat in feeling that the empirical sciences will come to our rescue, but I should point out that compared to solid state electronics, space flight, genetics, and so on, understanding neural information processing has really dragged its feet. This should not be surprising, since the complexity of the brain surpasses the complexity of any other empirical science by several orders of magnitude. These commentators actually cheer on the notion of total messiness because it forces us to look for a "different explanatory framework." Go for it.

R13. Will chaos make things easier?

Attention is called by **Tsuda** to the "temporal dimension" in hidden unit behavior. I did not think I had implied anything to the contrary. A cell's dynamic behavior is the very signal that worries me. Tsuda goes on to suggest that "modern nonlinear dynamical systems theory" leading to "deterministic chaos" will come to our rescue in explaining the signals on hidden units. I wish I could share his enthusiasm. We have made much progress in the oculomotor system in the caudal pons, in large measure because of the linearity imposed on its cells by the

behavior of the semicircular canals. Moreover, one is not struck by the need to invoke seriously nonlinear cell properties (except for the obvious cutoff at zero firing rate) in most studies of mammalian motor control or in sensory pathways. The work of Optican and Richmond in the visual system, cited by Tsuda, depends heavily on the idea of linear vector spaces.

We will probably be forced, sooner or later, to deal with some cells with essential nonlinear properties. This might occur more readily with "simpler" nervous systems, such as invertebrates (see my reply to the commentary of **Kupfermann**), where signal processing seems to depend more on cell properties than on network properties. Nevertheless, it is a little hard to see how nonlinearities and chaos are going to make it easier to unravel the signals on hidden units.

R14. More on what we can learn from single-unit recordings

The important issue of presynaptic inhibition is raised by **Winters & Mullins**. This phenomenon is apparently used extensively in real neural networks, but I am unaware of its being incorporated in any of the recent model networks. It raises a host of fascinating questions: Are such synapses modifiable? If so, how should they be introduced into the learning algorithm? Presumably, networks with such synapses can do things that networks without them cannot do. What kind of things? And so on. This is a fruitful field for investigation.

One pleasure of working in the oculomotor system is that it is easy to identify the goals of its various subsystems, such as the pursuit and saccadic systems. Defining goals is not so easy in other systems. One's right arm, for instance, is a tool shared by a seemingly uncountable set of systems with different goals. Much of its spinal circuitry, pyramidal, and extrapyramidal pathways are probably also shared. Will it be possible to distinguish functionally separable signals converging onto this shared system? If not, how will we ever be able to speak of goals?

Another worry is that the biomechanics of the arm can be overwhelmed by the complexity of the load, be it a pen, a glass of water, the garbage can, or a bag of cement. We can grasp an object and register its mechanical impedance in a fraction of a second and then our motor systems can issue goals specific to that load (of which there are an infinite number). One can only wonder what one might see, recording from a neuron participating in this incredible capability. Would you recognize what it was doing?

Naturalizing motor control theory: Isn't it time for a new paradigm?

Garrett E. Alexander, Mahlon R. DeLong
and Michael D. Crutcher

Department of Neurology, Emory University School of Medicine, Atlanta, GA 30322

Electronic mail: *gea@sunip.neuro.emory.edu*

The commentaries occasioned by our target article were thoughtful and constructively provocative, and for this we thank those who responded. Only a few of the commentators (namely, **Borrett et al., Connolly, Grobstein, Levine**) appeared to be in sympathy with our call for more

neurally inspired, bottom-up models of the motor system. Many (including **Flanders & Soechting, Fuster, Giszter, Horak et al., Jaeger, Masson & Pailhous, Phillips et al., Winters & Mullins**) expressed concern, either directly or indirectly, that modeling efforts of this sort may be attacking the problem at a level of discourse that is too low (in the reductionist sense) to tell us what it is that the motor system is actually doing. This is an important concern, one that we did not really address in the target article; we do so now in section R1. A second point of contention focused on the distinction we drew between algorithmic and nonalgorithmic information processing. Several commentators (**Flanders & Soechting, Fuster,**

Heuer) took issue with our claim that the learning process in unsupervised connectionist networks (and, by extension, in biological networks) is inherently nonalgorithmic. In section R2, we attempt to clarify our reasons for believing this to be a real and important distinction.

Perhaps because we placed some emphasis on the motor system's parallel and distributed organization, a few commentators (**Fuster, Heuer, Kalaska & Crammond**) accused us of ignoring or downplaying the motor system's obvious serial or hierarchical features. We address this criticism in section R3. Two commentaries (those of **Giszter** and **Masson & Pailhous**) attempted to defend the motor program concept with an existence proof, arguing that various central pattern generators (whose existence, we would agree, is supported by a large body of experimental evidence) are considered by many investigators to be examples of motor programs. We deal with this issue in section R4. We concur with those commentators (**Borrett et al., Jaeger, Summers, Winters & Mullins**) who called for models of the motor system that would deal with dynamical aspects of movement control. In section R5, we suggest ways in which connectionist models might be developed to represent some of the temporally contingent operations that are presumed to underlie the motor system's control of complex movements evolving in time. And finally, some commentators (**Horak et al., Masson & Pailhous**) were distressed by our failure to assign specific behavioral functions to the various cortical and basal ganglionic motor areas. We try to address those concerns in section R6.

Several of these issues raised by the commentators seem to us merely to reinforce the central theme of our target article: that the field of motor control is in serious need of a new motivating paradigm (in the Kuhnian sense) to replace the engineering-based perspective that is now dominant. What is needed, in our view, is a more thorough-going, bottom-up approach to the study of biological motor systems, one which considers each system's anatomical and physiological features to be the necessary *basis* for developing plausible theories about its mode of operation, rather than minor details to be incorporated into such theories after the fact. Another key consideration is that whatever strategies the motor system uses to solve its various tasks must be capable of emerging through natural mechanisms, presumably through both natural selection and trial-and-error learning, rather than through extrinsic engineering. Just as natural selection provides an explanation for the weeding out of genetically controlled systems that are unsuccessful in dealing with their environments and the preservation of those that are successful, so does the connectionist notion of trial-and-error learning through activity-dependent modification of synapses provide an explanation for the emergence of successful motor behaviors through experience and practice. Engineering-inspired theories, on the other hand, offer no biologically plausible explanations for the motor system's chief functional role, that of supporting the acquisition of each individual's unique repertoire of learned motor behaviors.

R1. Levels of analysis: How do we keep from talking past one another? A number of commentators (including **Flanders & Soechting, Fuster, Giszter, Horak et al., Jaeger, Masson & Pailhous, Phillips et al.,** and **Winters**

& **Mullins**) expressed varying degrees of concern about the utility of using connectionist models as a means of gaining insight into the operations of the motor system. For example, **Heuer, Flanders & Soechting,** and **Fuster** argue, quite appropriately, that merely understanding the lowest level of analysis, that of hardware implementation, tells us next to nothing about the information processing problem that the motor system has managed to solve, let alone how that solution was achieved. One could analyze ionic conductances, cable properties, and the sensorimotor responses of every neuron in the motor system and still have no idea what the system as a whole was actually doing.

To fully understand how the motor system processes information may require, for example, that we develop interlocking accounts appropriate to each of the three levels of descriptive analysis that Marr (1982) defined in his landmark monograph on the visual system. He considered the top level description of any information processing device (such as the motor system) to be one which was framed in terms of abstract *computational theory*: What is the goal of the computation, that is, the nature of the problem to be solved? At an intermediate level is a description in terms of *representation and algorithm*: In what form are the input and the output represented, and what is the algorithm that transforms one into the other? At the lowest level, that of *hardware implementation*, is a description of the physical mechanisms that underlie the realizations of both the representation and the algorithm.

From an engineering perspective, Marr's analytical approach may be capable of providing a reasonable description of what a biological network is doing in terms of information processing. Thus, one might say that in generating goal-directed limb movements the motor system is working within the constraints of Newtonian mechanics to transform a set of input parameters (e.g., visually derived information about the location of the target and the current position of the hand, as well as proprioceptive information about the current position and orientation of the limb) into an appropriate set of outputs (muscle activations) to drive the limb along a particular trajectory. This could be construed as an analysis of the problem at the level of computational theory, and it might include a series of black boxes that represented the various intermediate transformations (e.g., inverse kinematics, inverse dynamics, etc.) posited in the sequential/analytic model. To each black box could be assigned a specific algorithm that effected the corresponding transformation. And since virtually any algorithm can be implemented by a neural network, one could propose (as do **Flanders & Soechting,** for example) that the motor system's networks implement the algorithms corresponding to the various black boxes. And finally, the networks themselves could be analyzed at the level of hardware implementation, in terms of the biophysical properties of the constituent neurons.

Thus, as **Heuer** points out, connectionist models might be viewed from this engineering-inspired, top-down perspective as representing Marr's algorithmic level of description. And as **Jaeger** urges, the integration of standard connectionist models – in which the constituent "neurons" are typically treated as highly oversimplified, abstract processing units – with neuronal compartment models that include simulations of ionic conductances and axonal/dendritic cable properties – would certainly enhance the biological plausibility of the resulting hybrids. This could effectively result in models that represented both the algorithmic and the hardware levels of descriptive analysis. (The main practical limitation here, of course, is that of the enormous increase in computational demands placed upon any conventional computing system that might be used to carry out the simulation. Recent developments in which some of these compartmental properties have been simulated in hardware (Mahowald & Douglas 1991) may eventually make such approaches more feasible than is currently the case.)

We need to ask ourselves, however, whether high-level, black-box theorizing is likely to bring us any closer to understanding the actual operations through which the motor system accomplishes its various tasks. Marr (1982) himself recognized that his three levels of description were only loosely related one to the other. We do need to understand the nature of the problem that the motor system is solving if we are to make sense of the particular mechanisms that are used in achieving that solution. However, just as insight into the underlying mechanisms (at the hardware level) does not tell us which black boxes or algorithms are being implemented, the reverse is also true: Insight at the level of computational theory does not allow us to deduce the particular algorithm or hardware implementation the motor system uses.

We have no quarrel with the development of descriptive accounts of information processing in the motor system at the level of computational theory. In fact, such an account would be an essential element in any comprehensive theory of how the motor system operates. And one can certainly posit, as do **Flanders & Soechting,** that some form of the traditional engineering strategy (computational theory) is implemented by the motor system via neural networks that operate according to connectionist principles. Hypotheses of this sort have the virtue that they might eventually be made sufficiently explicit, at least in principle, to become testable with existing neurophysiological methodologies. Unless they are made explicit, however, by assigning particular black-box functions to specific brain structures, or by positing the particular algorithm that is implemented by a given network, models of this sort simply cannot be evaluated with respect to the question of how the brain controls movement.

An alternative to the engineering-inspired, top-down approach would be to consider some of the low-level constraints the motor system faces as a biological system. If we recognize that each motor system has to somehow acquire its capacities to control goal-directed movements, and we are not prepared to invoke supernatural factors as the basis for that acquisition, then hardware level considerations begin to impose serious constraints on any hypotheses that might be generated at the levels of algorithm or computational theory. Unless one takes the indefensible position that all adult capacities for the control of goal-directed movements are determined solely by genetics, and that learning and experience play no role in the development of new skills, one's theory of how the motor system operates would seem to require an underlying mechanism that could plausibly account for self-organization through experience with the outside world. Thus far, the only such mechanism that has been

proposed is that of local, activity-dependent synaptic modification, that is, Hebbian learning.

This and other considerations outlined in our target article suggest to us that it may be time for neuroscientists to begin exploring some bottom-up approaches to the analysis of motor systems. Artificial neural networks clearly work in this way, and there is every possibility that biological networks may do the same. Of course, if biological networks (such as the motor system) do operate according to connectionist principles, this would present one serious difficulty for neuroscientists: Few powerful tools or methodologies have been developed as yet to facilitate the analysis of this type of information processing. Nature, however, is not constrained by our capacity to understand natural processes. Merely the fact that certain theories (e.g., those inspired by engineering principles) promise greater ease of analysis and comprehension does not make them inherently more plausible as descriptors of nature. As we have no reason to believe that natural selection has produced motor systems that operate according to principles of human invention, why should we lock ourselves into such assumptions with respect to the neural substrates of motor control?

From a strict evolutionary standpoint, it is debatable whether engineering-inspired computational theories have more than passing relevance to the question of how the motor system accomplishes its various tasks. If natural selection (of genetically determined patterns of brain wiring) and Hebbian learning are the only mechanisms through which the motor system acquires its capacity to control goal directed movements, this poses some significant plausibility constraints at the level of computational theory. After all, both natural selection and Hebbian learning are fundamentally trial-and-error processes, which would seem to indicate that the motor systems strategy for information processing must be capable of being developed by trial-and-error. This in turn raises the distinct possibility that the motor system's various networks, like typical connectionist networks, may solve the problem of controlling goal-directed movements through a series of arbitrary input/output mappings that cut across – and may thus bear no correspondence to – the various black boxes postulated by engineering-inspired theorists.

If this last possibility were taken seriously, would that mean we should despair of ever being able to develop a comprehensive connectionist theory of how the motor system operates? Not necessarily. It simply means that we may need to pursue a predominantly bottom-up approach to the problem, by first characterizing the algorithms that are implemented by the various neuronal networks that comprise the motor system and then devising a suitable computational theory whose black boxes correspond to the algorithms identified experimentally. At the most abstract level, the outlines of this type of computational theory (generated from a connectionist perspective) might be more or less the same as for one based on engineering principles: that is, motor psychophysical studies define for us the motor system's overall input/output transforms, and the motor system's interactions with the periphery are appropriately characterized within the conceptual framework of classical physics. The differences between the top-down (engineering) and bottom-up (connectionist) approaches would be expected to lie rather in the types of algorithms or black boxes that are hypothesized to transform the motor system's inputs into outputs.

For the bottom-up, connectionist strategy to be successful, however, would require developing analytical tools we do not yet have. In order to characterize the algorithms computed by the motor system's various hidden layers, we must be able to represent the inputs as well as the outputs of any given layer, and these representations must capture the spatial as well as the temporal ordering inherent in the multidimensional patterns of neuronal activity that comprise inputs and outputs of this sort. For this to be possible we will also need to develop new experimental strategies, including neurophysiological recording techniques that take into account the topography and connectivity of the networks under study much more thoroughly than is now customary. Simultaneous recordings of behavior-correlated neuronal activity from adjacent hidden layers, if coupled with correlational analyses (e.g., spike-triggered averaging) that could provide an index of the connection patterns linking the sampled neurons, would be an important first step in addressing these questions. Of course, since only small samples of any biological network's total population of neurons can ever be studied directly in this way, it will also be necessary to develop connectionist models that incorporate the limited empirical observations and can then be used to generate testable predications about the operations of the network as a whole.

If carried to its logical conclusion, the potential payoff of the bottom-up approach advocated here would be the development of black-box schemes, at the level of computation theory, that would be *testable* because the various black-box functions would be assigned to specific layers of neurons (e.g., nuclei or cortical areas) within the motor system. Selective lesioning and/or functional imaging studies could then be used to carry out independent tests of the resulting functional predictions.

R2. Algorithmic versus nonalgorithmic processing. Though we tried very hard to be clear on this issue, it would seem that some restatement is in order, as several commentators – including **Flanders & Soechting, Fuster,** and **Heuer** –seem to have missed our main point concerning the distinction between algorithmic and nonalgorithmic information processing. In networks where learning rules are strictly local, that is, where activity-dependent changes in connection strengths depend only upon events occurring in the immediate vicinity of the connection or synapse, we would argue that as the network learns to solve the problem at hand it does so not in accordance with a predetermined algorithm but rather through an essentially blind process of trial-and-error. It is true, of course, that at the synaptic level the local learning rule can be described as an algorithm (as can the cable properties and membrane transfer functions of the individual processing elements); but there is no sense in which knowledge of these local algorithms can impart to an observer any secure insight into the computational "solution" that the network will eventually achieve once the learning process has been completed. It is also true that one can describe, after the fact, the overall transfer function of the fully trained network by means of an algorithm (simply by analyzing the correspondence between inputs and outputs at the network level after

training has been completed). Such a transfer function would be an algorithm, of course, and so in a trivial sense one could say that the network is thereby processing information by means of an algorithm. But the deeper question is: How did that algorithm arise? We would argue that it arose through an essentially blind process of trial-and-error, that is, nonalgorithmically.

Why do we say that such a process is blind, when it is obvious that at the level of the organism (or the external trainer) there is, in fact, a deliberate selection of successful weight configurations – those that are associated with desired outputs – to be reinforced by repetition (i.e., practice) of associated behaviors? As **Flanders & Soechting** point out, this would seem to indicate that the trainer's selection criteria are what determine the network's final solution. And since the selection criteria can be described as an algorithm, surely the whole process can then be construed as algorithmic. But this misses the deeper point about how the learning actually proceeds at the network level. For in strictly local (i.e., unsupervised) networks the selection criteria play no direct role in determining how the individual connection strengths within the hidden layers will be adjusted from trial to trial, nor do they determine, in any direct or predictable way, the precise configuration of connection strengths toward which the network will eventually converge. The selection criteria are applied directly to the *outputs* of the network, not to the connections within the hidden layers. The latter are determined primarily by local factors which, from a network level, appear to represent more or less random processes.

We should also note that we never meant to imply that learning could be easily separated from execution in biological networks, as **Fuster** suggests. We were simply pointing out that the principle of Hebbian learning predicts that during skill acquisition, that is, during motor learning, the repetition of successful movements will tend to strengthen those connections which, during those same movements, show correlated pre- and postsynaptic activity. Once the skill has been acquired, it can then be executed at will without having to relearn it. Obviously, there is no reason to necessarily assume that learning ceases, either at the network level or at the behavioral level, with those subsequent executions of the previously learned movement. It might well be, however, that in the course of learning to make some movements particularly well, the network reaches what approximates a global minimum in its weight space, from which subsequent changes in connection strengths are likely to be quite small whenever those particular movements are repeated. This is not because local learning rules have been turned off, but because the weight adjustments have already been optimized through practice.

R3. Serial/hierarchical versus parallel/distributed processing. In several of the commentaries (especially those of **Fuster, Heuer, Kalaska & Crammond**) we were chastised for seeming to slight the serial and hierarchical features of the motor system in the course of emphasizing some of the recent evidence for parallel and distributed motor processing. This was certainly not our intent. As we indicated in the target article, it has become obvious by now to most neuroscientists that the brain in general, and the motor system in particular, are characterized by both parallel and serial processing, and that these two aspects of the brain's organization are inherent at virtually every level of analysis ranging from the individual neuron to large scale, transganglionic networks. The motor system's serial and hierarchical features have been repeatedly emphasized for decades, so in emphasizing some of its newly appreciated parallel features we were merely attempting to redress this longstanding imbalance.

In our view, however, the chief difference between biological motor systems and engineering-inspired models – which was the main focus of our target article – has relatively little to do with the serial/hierarchical versus parallel/distributed distinction: Rather, the difference centers mainly on the distinction between non-algorithmic and algorithmic processing as a mechanism for the acquisition of new behavioral capacities. We suspect that unless the latter principle is understood, even the most faithful attempts to model the motor system's serial and parallel connectivity would be unlikely to capture the biological basis of adaptive motor behavior.

R4. Motor programs versus central pattern generators. In their respective commentaries, **Giszter** and **Masson & Pailhous** argue that our brief against the motor programming concept is undermined by the fact that motor programs have been shown to exist in the form of central pattern generators (CPGs). It is true that those who study CPGs often refer to them as "motor programs." This notion of a motor program seems to us to be entirely respectable from a neurophysiological point of view, for in this case the nature of the putative neural substrate can be made reasonably explicit. But the type of motor activity associated with CPGs is not what we were addressing in the target article. We were concerned there with the question of how we can account for the motor system's ability to control *purposeful* movements. We argued merely that the motor programming concept does not help us explain how the brain controls purposeful – that is, goal-directed – movements.

As generally conceived, CPGs are not purposeful. They are considered to be the neural substrates of relatively fixed patterns of movement – often repetitive – that are triggered and/or sustained by an effective stimulus. The movement patterns may be modifiable to some extent by afferent feedback or by descending commands. In both respects, then, CPGs are very much like spinal reflex arcs, the chief difference being that the latter are generally associated with movements that are phasic and brief (e.g., the phasic stretch reflex) and the former are associated with sustained, sequential, or repetitive movements (e.g., locomotion). We find the concept of CPGs to be every bit as useful as the concept of a spinal reflex arc.

The main problem faced by motor control theorists, however, is that of accounting for the production of goal-directed movements. It is for this type of motor behavior that the degrees of freedom (or overcompleteness) problem arises, and for which the issue of computational feasibility must be faced. Such is not the case for nonpurposeful motor responses like stretch reflexes and fictive locomotion, generally thought to be mediated by spinal reflex arcs and CPGs, respectively. The latter responses do not represent solutions to new problems, as do purposeful motor behaviors; rather, nonpurposeful motor responses are simply fixed, predictable reactions based on hardwiring that is, for the most part, genetically determined. For these types of responses the issue of motor *control* is quite limited. There is no target or goal to be achieved; there is simply a fixed reaction to some effective input. Thus, it seems to us that the existence of CPGs (whether or not they are called "motor programs") tells us next to nothing about the larger question of whether voluntary, goal-directed movements can be accounted for in any biologically meaningful sense by the traditional motor programming concept. Our concern is not with the concept of neurally instantiated motor programs in the form of CPGs, but with the uncritical positing of disembodied motor programs to somehow "explain" how the motor system controls complex, purposeful movements.

R5. Connectionist models should address dynamical issues of motor control. Several commentators (including **Borrett et al.**, **Jaeger**, **Summers**, and **Winters & Mullins**) either urged that the future development of connectionist models of the motor system should consider dynamical aspects of movement control or criticized current connectionist efforts for failing to do so. This issue should not be confused with that of computing the inverse dynamics transformation, one of the analytically defined levels of motor processing. Rather, the point raised by these commentators is that the manner in which model systems control movements evolving in time will be a key test of their ability to simulate biological motor control systems.

Borrett et al. present some interesting results of connectionist modeling of a simulated motor control system with dynamical properties. The findings illustrate that there is no fundamental difficulty in using connectionist networks to model or control dynamical processes. Processes that evolve in time, such as movements, can be controlled by conventional neural networks (even though they produce stationary, rather than time-varying, outputs) by effectively segmenting the time-varying process in accordance with the update frequency (cycling rate) of the network. In this way, the complete movement is parsed into segments, each of which is controlled, in sequence, by a correspondingly discrete output state of the controlling network.

In order to model the dynamical properties of biological networks, it will be necessary to include in the hidden units such characteristics as conduction times and membrane time constants. Efforts of this sort are already well advanced among those who have been developing compartmental models of biological neurons (Koch & Segev 1991). The challenge for the future will be to combine these efforts with connectionist models of motor circuitry and to apply these hybrid networks in turn to the control of simulated multijoint limb movements.

R6. What do the cortical and basal ganglionic motor areas do? Let's face it: For the present, nobody really knows. We understand the concerns of those, such as **Horak et al.** and **Masson & Pailhous,** who would like to attribute specific functions to different motor areas either as an aid to understanding how the system operates or as a stimulus for neurobehavioral research. At the same time, however, it should not be necessary to remind most readers that it may be hazardous – indeed, it is generally fallacious – to impute to the lesioned structures those behavioral functions that happen to be impaired by specific lesions. Are we to conclude, for example, that the dopaminergic neurons of the substantia nigra control movement speed because patients with Parkinson's disease, in whom these neurons are lost or damaged, are usually bradykinetic? Neuroscientists now know better than this, because they have learned that even damage to a single nucleus or cortical area is likely to have far-reaching effects upon any or all of the brain regions with which the lesioned structure is directly or indirectly connected. It therefore makes more sense to attempt to ascribe behavioral functions to entire circuits, or networks, of interconnected brain structures (nuclei, cortical areas) than to the individual structures themselves.

The problem, of course, is that dealing with the operations of whole circuits is vastly more complicated than dealing with the operations of individual nuclei. To devise suitably sophisticated theories of circuit operations and their associated behavioral correlates will require a new generation of functional models that will, one hopes, be generated in accordance with the bottom-up strategy advocated above. At present, neurally inspired models of this sort are as primitive as they are scarce. Until or unless we are able to develop some biologically plausible models of central motor circuits and their operations, it seems to

us to be relatively futile to attempt to ascribe specific behavioral functions to individual motor structures. The reason for this, as suggested earlier, is that without a detailed understanding of the physiological operations performed by the motor system's various hidden layers of neurons, one cannot hope to be able to identify the various algorithms they implement. And since we have no reason to assume that those algorithms necessarily correspond to the categories of transformations invented by engineers (e.g., inverse kinematics, inverse dynamics, etc.), or to the categories of motor functions invented by motor control theorists (e.g., movement scaling, movement braking, etc.), attempts to address the problem by means of top-down theorizing seem inherently unlikely to succeed.

In spite of these difficulties, we are quite optimistic about the future of motor systems research. This is due primarily to the advent of modern connectionism, which seems to us to hold great promise as a framework for the development of realistic models of the motor system's complex neural circuitry. With the aid of such models, it may eventually become possible to address in detail the sorts of questions raised by **Horak et al.** and **Masson & Pailhous.** Nevertheless, we feel that much has yet to be accomplished before neuroscientists will be able to make meaningful assignments of functions to the different cortical and basal ganglionic motor areas.

Concepts of cerebellar integration: Still more questions than answers

James R. Bloedel

Division of Neurobiology, Barrow Neurological Institute, Phoenix, AZ 85013

The commentaries on my target article were interesting, insightful, and in some cases provocative. The specific comments generally fell into three categories: (1) specific issues regarding either the hypotheses presented or the generalizations made from the literature review, (2) comments directed toward the cerebellum's role in motor learning, and (3) discussions pertaining to broad aspects of cerebellar function, including the presentation of some new views by the commentators.

R1. Specific issues regarding details of cerebellar organization. The important issue of cerebellar homogeneity is raised by **Barmack et al.**, who implicitly challenge the view expressed in the target article that the cerebellum has adequate homogeneity to support the notion that a comparable operation is performed in all regions of this structure. Barmack asks why we should ignore details and then misrepresents the target article in stating that I argue in favor of bypassing single unit recording studies.

First, I have never indicated that we should ignore details. Furthermore, I have never advocated abandoning studies that assess pertinent details, including single unit recordings. The point is not to choose whether to examine details of this structure but rather to determine what details are relevant and how they are related to an understanding of this interesting system. A focus on defining a cerebellar operation does not eliminate the need for detail. In fact, it requires that these details be well delineated and subsequently incorporated into internally consistent models capable of characterizing the interactions among the neuronal elements in this structure.

Barmack et al. also discuss the relevance of the sagittal zone. I take issue with characterizing these structures as a mere "developmental curiosity." Our recent studies have provided strong support for the functional relevance of these zones under at least one behavioral condition, perturbed locomotion. These studies showed that perturbations of the locomotor cycle activate a sagittally distributed band of climbing fiber inputs (Lou & Bloedel 1992) that can influence the simple spike discharge of Purkinje cells on which they impinge (Lou & Bloedel 1986). These findings and observations from Llinas's laboratory (see target article) provide strong evidence that these zones play a critical spatial role in specifying the action of climbing fibers.

J. F. Stein found that the gain change hypothesis is not only controversial but that it is "opposite" the classical view that climbing fibers cause a short-term inactivation of the Purkinje cell. Although I admit to its controversial nature, I do not accept this hypothesis as inconsistent with the inactivation response related to the excitatory action of climbing fibers. The enhancement Ebner and I described (Bloedel & Ebner 1985; Bloedel et al. 1983; Ebner & Bloedel 1984; Ebner et al. 1983) exerts its effect on simple spike activity *after* the brief inactivation response (approximately 15–18 msec). In other words, both the inactivation response and the response enhancement result from the action of the climbing fiber. Although not discussed by Stein, it should also be pointed out for completeness that a reduction of impulse frequency evoked by mossy fiber inputs can also be enhanced. For reasons not yet understood, the response enhancement is just that – an enhancement of the response evoked by the parallel fiber system rather than an evoked excitability change. Decreases as well as increases in impulse activity evoked by mossy fiber inputs can be enhanced.

J. F. Stein also asserts that this interaction conflicts with the observations reported following tonic removal of climbing fiber inputs. In our view, these tonic changes are probably produced by mechanisms other than those responsible for the short-lasting enhancement we reported. The substantial differences in the paradigms by which these two effects are produced as well as their time course support this view. The unique properties of the climbing fiber afferents may include both a tonic and a phasic effect on Purkinje cell activity, each being relevant to different functional conditions.

Smith also expressed a fundamental difficulty with the enhancement effects serving as the basis for the gain change hypothesis. Apparently he finds an inconsistency between this set of interactions and those we reported several years ago supporting the excitation of cerebellar cortical inhibitory interneurons by the collaterals of climbing fibers. Smith supports this projection with his own data in the commentary. In my view, these facts regarding connectivity are not incompatible with the enhancement interaction we demonstrated. However, to interpret the significance of these details it is necessary to determine: (1) the weighting factors characterizing the importance of the climbing fiber projection to the inhibitory interneurons and (2) the way in which this projection contributes to Purkinje cell output during ongoing information processing in the cortex. It should be emphasized that the effect of activating an inhibitory projection under functional conditions may produce an effect on the system's output very different from what can be predicted from studies in which it is activated pharmacologically or electrically. The collateral system to inhibitory interneurons is imbedded in a network of multiple collateral projections from the Purkinje cells themselves. This entire circuit, working together under behavioral conditions, may actually contribute to the enhancement we described.

Smith is right in pointing out that the mechanism for this enhancement is not known. As a supposition, it is possible that climbing fibers could produce a change in membrane conductance resulting in an overall increase in the input impedance of the Purkinje cell due to the activation of second messenger-dependent systems. As addressed above, it is also possible that the cortical collaterals from climbing fibers as well as from Purkinje cell axons could produce an increase in Purkinje cell excitability following the activation of these afferents (Bloedel & Roberts 1969; 1971; Bloedel et al. 1972).

Smith also raised an extremely important point regarding the direct action of climbing fiber collaterals to the cerebellar nuclei. Physiologically, very little is known about this system. In fact, its predominance has also been debated anatomically. In our studies of this projection, we simultaneously recorded from a Purkinje cell and a nuclear neuron we defined as being physiologically related on the basis of electrophysiological criteria (McDevitt et al. 1987a; 1987b). In the context of this discussion the data suggest that a single climbing fiber exerts a rather weak effect on a given nuclear neuron, certainly much weaker than its effect on Purkinje cells. Although considerably more work needs to be done on this system, the action of climbing fiber collaterals to the cerebellar nuclei appears to be highly dependent on the convergence of these inputs onto individual nuclear neurons.

R2. The cerebellum and memory storage. The most pointed criticisms of the target article clearly were triggered by the brief discussion regarding the cerebellum's role in storing the engrams required for motor learning. **Gilbert & Yeo** also assert that the target article demonstrated an "extremely" limited view on the subject of cerebellar plasticity in motor learning. To repeat my statements in the target article: This overview was not intended as a review of the cerebellum's role in motor learning. Consequently, only certain key manuscripts that we consider critical to the development of our view on this issue were cited. We hope that our previous manuscripts, together with the discussion below, will demonstrate that we have more breadth in this area than inferred by these commentators.

Throughout the commentaries two specific issues surfaced often: (1) the role of the climbing fiber system in generating long-lasting changes in Purkinje cell responses, and (2) the interpretation of the effects on learned motor behavior produced by surgically or pharmacologically manipulating cerebellar systems.

Gilbert & Yeo as well as **Ito** emphasized their view that the climbing fiber system induces a long-lasting change in the excitability of Purkinje cells and that this mechanism is responsible for establishing memory traces in the cerebellum. Our views on this issue have been presented several times (Bloedel 1987; Bloedel et al. 1991; Bracha et al. 1991; Kelly et al. 1990b); consequently, it is sufficient to emphasize only a few critical points. Proposals that the climbing fibers are responsible for establishing long-term changes in the responses of Purkinje cells to parallel fibers are based principally on experiments that used the conjunction paradigm, the qualitative behavioral study of Gilbert and Thach (1977), and the studies of the vestibulo-ocular reflex (VOR) plasticity performed by Ito and his colleagues (Ito 1984).

We have previously argued that the conjunction paradigm, although responsible for several interesting findings, is not physiological in terms of either the pattern of evoked climbing fiber and mossy fiber inputs or the sustained frequencies of climbing fiber stimulation required to produce it. In contrast, the short-term enhancement we reported was observed under several conditions in which either natural stimuli or behavioral preparations were used. In addition, Crepel and Jaillard (1991) recently showed that lower frequencies of climbing fiber activation actually result in an enhancement of Purkinje cell responsiveness. Consequently under physiological conditions we remain committed to the interaction originally described by Ebner and Bloedel (see Bloedel & Ebner 1985, for review) as the physiologically relevant heterosynaptic action of climbing fiber inputs on Purkinje cells. As reviewed in the target article, this interaction consists of a short-lasting enhancement of Purkinje cell simple spike responses to mossy fiber inputs.

The effects produced by surgical and pharmacological manipulation of cerebellar systems on motor learning were addressed principally by **Thompson,** who implies in his first paragraph that our view is not a contemporary one. Rather, he argues that it is derived largely from the "traditional view of the role of the cerebellum in motor coordination that evolved from the classic early studies of Holmes." On this standpoint Thompson is right: I do believe that the cerebellum plays a role in motor coordination! In fact, I insist on it! Thompson implies that by not accepting his "contemporary" view that storage required for motor learning takes place predominantly in the cerebellum (at least for some behaviors), the views we express are somewhat outdated. I hope to show in this section that Thompson has represented our work in his discussion inappropriately and that the views we hold, including those regarding the substrates for motor learning, are not only contemporary but offer viable suggestions about the role of the cerebellum in adaptive motor behavior.

Thompson, like **Gilbert & Yeo,** was highly critical of the thoroughness with which the cerebellum's role in motor learning was covered in the target article. I must again emphasize that, as stated in the target article, this subject was not treated extensively in order to focus on three hypotheses related to our views on cerebellar integration. In other words, in this article I wished to describe what I think this system is doing rather than review an area of research that I feel shows what the cerebellum does *not* do. Thompson chose to support one statement in his commentary with 21 references. Using the same approach, I certainly could have reviewed more manuscripts. However, I do not prefer to use this approach. I also wish to state that many of the articles cited by Thompson in his commentary were not published at the time I was asked for this Response.

At least as pertinent to this discussion are several articles **Thompson** fails to cite from laboratories other than Harvey's or our own, which raise interesting questions regarding Thompson's view. Guillaumin et al. (1991), using rats in a passive avoidance conditioning test, recently reported data showing that the cerebellum may be involved in some aspects of the acquisition and consolidation processes related to this type of motor learning but that it is not a required storage site for this type of conditioning. In another study using avoidance conditioning in rats, cerebellectomized rats learned as fast as intact rats (Dahhaoui et al. 1990). Cerebellectomized animals also required fewer trials to reacquire the behavior than animals trained only after the surgical procedure. In fairness to this argument it should also be pointed out that there was some loss of absolute retention following cerebellectomy in comparison with unoperated controls tested for retention after a comparable delay period. Marchetti-Gauthier et al. (1990) found that bilateral lesions of the interposed nuclei affected only the acquisition and not the retention of a conditioned forelimb flexion reflex in mice. These relatively recent studies add to several others in the literature indicating that the cerebellum plays its major role in several types of motor learning during task acquisition even though it is not essential for task retention, that is, removal of the critical regions of the cerebellum after acquisition does not substantially reduce the magnitude of the conditioned behavior. The cerebellum clearly plays an important, specific role in the modification of these motor behaviors. However, in each of these specific conditions this structure is not required for the storage of the memory trace required for the execution of the learned behavior.

In my view, the implications of these studies are also consistent with the current literature on VOR adaptation.

Cerebellar lesions can impair the adaptation of the VOR (Robinson 1976). However, in goldfish, adaptive modification of VOR gain can occur following complete removal of the cerebellum (Weiser et al. 1988). Furthermore, recordings in the flocculus of monkeys in which eye movement inputs had been eliminated failed to demonstrate modifications in Purkinje cell modulation consistent with plastic changes occurring in the cerebellar cortex (Miles et al. 1980). More recently, Lisberger's laboratory has provided convincing evidence based on detailed latency measurements of eye movement velocity profiles that the actual plastic changes underlying VOR adaptation occur in the vestibular nuclei (Lisberger 1988b; Lisberger et al. 1990; Lisberger & Pavelko 1988).

Thompson also states that I have misconstrued the evidence I do cite, including our study (Kelly et al. 1990b) showing that classically conditioned nictitating membrane reflexes can be acquired in rabbits that have been both decerebrated and decerebellated. He argues that his own work (Nordholm et al. 1991) actually refutes our data and that many of our conclusions were based on inappropriate interpretations of our findings. He specifically asserts that the findings we report were due to sensitization rather than conditioning of the reflex.

We strongly disagree with **Thompson**'s conclusions, including those serving as the basis for his arguments in Nordholm et al. (1991). One of the primary arguments in this manuscript is that we scored an inappropriately large number of conditioned responses incorrectly. Nordholm et al. asserted that our criterion for a conditioned response resulted in scoring many spontaneous eyeblinks as conditioned responses.

This assertion is wrong for a very simple reason: There were no spontaneous eyeblinks in our decerebrate preparations! In addition, unlike in Nordholm et al.'s experiments, we used an apparatus for holding the transducer and airpuff tube that was rigidly coupled to the head, which was in turn attached to a headholder. Consequently, not only were there no spontaneous eyeblinks, there were no movement artifacts that could be confused with conditioned responses. The experiments Nordholm et al. performed to refute our data used intact rather than decerebrate animals. As a consequence, they undoubtedly had both spontaneous blinks and movement artifacts, conditions that did not exist in our animals. Clearly, the premise that our assessment of the conditioned responses was flawed because spontaneous eyeblinks were included is incorrect. Very simply, not only did Nordholm et al. choose to challenge our findings using a different preparation, but their preparation had characteristics that they erroneously applied to an analysis of our data.

Thompson also argues that we did not perform the proper controls for sensitization. Although the results from the application of unpaired controls were not presented graphically in Kelly et al. 1991, this assessment was performed and failed to demonstrate sensitization.

Before leaving the issue of sensitization it is important to point out that both **Thompson**'s commentary and Nordholm et al.'s (1991) study failed to mention a critical set of data in our manuscript (Kelly et al. 1990) supporting the view that responses we reported were not the result of sensitization but rather the consequence of classical conditioning. We demonstrated that the onset of the conditioned response was time-locked to the onset of the unconditioned stimulus, an observation that is highly characteristic of the conditioned response (Gormezano et al. 1983) and one that Thompson has acknowledged as characteristic of this type of classically conditioned behavior (Thompson 1989).

Last, **Thompson,** in this commentary, as well as Nordholm et al. argue that we made our critical observations in only a few animals; they further imply that we discarded a large amount of relevant negative findings. This is completely incorrect. As indicated in our manuscript, our criterion for including a decerebrate animal in the study was the capacity of the preparation to acquire the conditioned reflex. The criterion had nothing to do with any effect of cerebellectomy. This type of exclusion is no different from excluding animals from a locomotor study if they cannot locomote after decerebration. Unquestionably the latter observation would not prove that only a small fraction of decerebrate animals can locomote on a treadmill! The same reasoning holds true for studies of the rabbit nictitating membrane reflex. It would generate misleading data to include decerebrate animals with abnormalities below the lesion in any study. In summary, the unreported decerebrate animals were simply unsuitable for testing the effects of cerebellectomy because of the poor condition of the preparation.

The important point is this: *All* animals in which the decerebration was successful and the preparation was not impaired were used in the study. *All* animals in which the effects of cerebellectomy were examined were reported. It should be emphasized that **Thompson** based his conclusions from his own studies regarding the retention of conditioned responses in decerebrate rabbits on a series of only four animals (Mauk & Thompson 1987).

Later in his commentary **Thompson** describes an experiment he considers better for testing the importance of he cerebellum in the conditioned nictitating membrane reflex than the study we performed. He suggests further that his laboratory has completed such a study and obtained results opposite ours (Mauk & Thompson 1987). In our view, Thompson overstates the results of these experiments. The effects of cerebellar ablations or damage to the red nucleus were not assessed systematically in that study. Only lesion data pertinent to the four decerebrate animals showing retention were presented.

In the remainder of his commentary **Thompson** discusses his disagreements with the findings from John Harvey's laboratory. Clearly Harvey and Thompson have very different interpretations of Harvey's observations. I refer the readers to the original articles for their own assessment of this part of the controversy. I find it disconcerting that Thompson apparently chooses to conclude that Harvey's data are basically incorrect rather than accepting them as a set of challenging findings that must be synthesized with his own. This seems to be particularly true for the critical studies of Harvey examining the effects of lidocaine infusion in the interposed nuclei on the acquisition of the conditioned behavior. Welsh and Harvey showed that infusion of lidocaine into the interposed nuclei during training blocks the expression of the conditioned reflex but not its acquisition (Welsh & Harvey 1991). In the cooling studies cited by Thompson from his laboratory (Clark et al. 1992; Lavond et al. 1990) as well as

in work that is apparently in progress, Thompson finds just the opposite: No acquisition was observed. This difference can't be ignored. Rather, it has to be resolved.

It is also important to point out that alternate sites for the plastic changes required for nictitating membrane reflex (NMR) conditioning have been presented. Recent studies in our laboratory (Bracha et al. 1991) have shown that lidocaine blockade of neurons in the brainstem just medial to the pars oralis of the spinal trigeminal nucleus can selectively decrease the conditioned nictitating membrane response without reducing the unconditioned response, suggesting that neuronal populations in this region may play a role in this process.

In summary, we do not accept **Thompson's** conclusion that "the cerebellum is absolutely necessary for both learning and memory in classical conditioning of eyeblink and other discrete responses." We feel that there is now ample evidence, not only from our laboratory, but from several others, indicating that this is not true as a general principle.

I feel strongly that this controversy will not be resolved until the laboratories involved make a commitment to: (1) synthesize their data with other observations in the literature, and (2) apply whatever new directions result from this synthesis to meaningful discussions of this subject and to the design of future experiments. In our recent publications on this subject (Bloedel et al. 1991; Kelly et al. 1990b), we have attempted to do this by contending that the effects of cerebellar lesions on the conditioned nictitating membrane responses were due to changes in the excitability of brainstem nuclei critically involved in mediating these reflexes rather than the removal of a memory trace. Billard et al. demonstrated that tonic changes in the excitability of cerebellar neurons following lesions of the inferior olive were reflected in the discharge of cells in the red nucleus, a brainstem nucleus that probably plays a role in mediating or regulating the conditioned nictitating membrane reflexes (Billard et al. 1988). Because it is well known that decerebration in general tends to increase the excitability of brainstem circuits and that the output of the interposed nuclei to many brainstem structures is excitatory (with the notable exception of the inferior olive), decerebration may counteract the tonic effects of cerebellar lesions in some brainstem nuclei critical for the conditioned response. Consequently, decerebration could result in the expression of a conditioned reflex that had been reduced or abolished following the cerebellar lesion.

This argument is also consistent with Welsh and Harvey's (1991) observations, which **Thompson** also rejects in his commentary, regarding the effects of lidocaine infusion in the interposed nuclei on the acquisition of the conditioned behavior. The expression of the conditioned behavior after lidocaine withdrawal may be analogous to the effects of the decerebration: During lidocaine infusion the conditioned response is not expressed because of the decreased excitability of brainstem nuclei involved in the reflex; the learned response can be expressed after the infusion is discontinued because the excitability of critical brainstem circuits has been restored.

I now challenge those with opposite views regarding these issues to offer interpretations which include the observations from our laboratory as well as the laboratories of Baker, Lisberger, Llinas, and Harvey. Continuing to ignore these findings is not constructive. Growth and progress in this interesting area of research depends on synthesizing relevant findings, not reasserting evidence from only a selected group of investigators.

I also wish to comment on the inference made by **Thompson** as well as **J. F. Stein** that we have argued against a role for the cerebellum in adaptive behavior. This is simply incorrect. We have made two specific arguments in the literature: (1) that the climbing fiber system is not responsible for inducing a long-lasting change in the responses of Purkinje cells and (2) that the cerebellum is not a necessary and sufficient storage site for the plastic changes underlying the conditioned nictitating membrane reflex in the rabbit.

In my opinion, there is clear evidence that the cerebellum is involved in and may in some cases be necessary for the acquisition (learning) of some motor tasks. Contrary to the assertions of **Hallett** in his commentary, we have never contended that the cerebellum does not play a role in motor learning. Through its involvement in the acquisition process, it clearly participates in the adaptation of motor behavior. For example, the adaptive feedforward processes proposed in **J. F. Stein**'s commentary could be mediated by the cerebellum and be critical for establishing the plastic changes required for motor learning in *other* central structures.

Figure 1 illustrates our current working hypothesis regarding the way the cerebellum contributes to task acquisition and modification during motor learning. The diagram suggests that information processing occurring in the cerebellum during task acquisition is necessary for generating the adequate coordination of the task to be learned and for providing an input to other structures critical for establishing the engram at sites outside the cerebellum. Furthermore, contrary to **Stein**'s assertion concerning our view, we support the existence of multiple sites at which plastic changes required for motor learning can occur (see Bloedel et al. 1991).

R3. Conceptual issues. In the final analysis the target article's merit, if not its intent, is related to its capacity to evoke constructive discussion regarding critical features of cerebellar organization and function. I accordingly found the comments of several colleagues concerning

general issues to be very stimulating. In my more optimistic reflections I perceived a convergence of ideas on certain critical issues. This is perhaps best exemplified in the comments of **Bower.** In his conclusions he indicates several important points on which we agree. It is interesting that these points concern some of the most debated topics in this field: memory, real time processing, and the cerebellum's role as a "moderator." I only disagree with Bower's contention, also stated in his conclusion, that his data have led to a conclusion regarding cerebellar function substantially different from the one expressed in the target article. Bower's comments emphasize the importance of the sensory processing performed by the cerebellum. As indicated in the target article, I accept this emphasis. However, I disagree with his implication that the cerebellum is unlikely to be directly involved in coordinating motor behavior. In general, the brain processes sensory information for two reasons, perception and the organization of behavior. I contend that the purpose of the unique multimodal sensory processing performed by the cerebellum is to provide this structure with the inputs required for the information processing that determines the effect of the cerebellar output projections on components of the motor system. This transformation of multisensory data into a motor reference frame is critical for motor coordination. In virtue of its role in this transformation, the cerebellum is clearly involved in regulating motor behavior.

I also disagree with **Bower** that humans perform "perfectly adequate motor coordination without a cerebellum." Neurosurgically, full recovery after substantial cerebellar lesions most often occurs when there is some nuclear sparing. Patients with complete absence of the cerebellum do have motor abnormalities (Dow & Moruzzi 1958). Despite these differences in interpretation, Bower and I agree fully regarding the critical nature of the sensory processing performed by the cerebellum.

The comments of **Horak et al.** were especially insightful. Of particular interest is their reference to cerebellar function in the context of its role in correcting for mismatches between the sensory information generated as a consequence of the movement and the internal model of expected sensory information. It is interesting that this type of process may underlie the organization of rhythmic

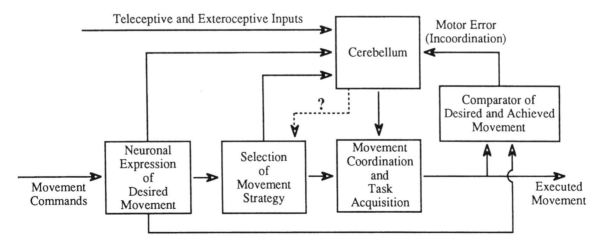

Figure 1.   A proposed mechanism for the cerebellum's role in the acquisition of learned movements without serving as a critical storage site for the plastic changes underlying this process.

limb movements was one of the proposals in a recent review by Baev and Shimansky (1992). The cerebellum could use the feedforward control mechanisms discussed by **J. F. Stein** in mediating the corrections required by this type of mismatch. This type of control may be particularly important in implementing processes related to the control of continuous movements, as required by the successive approximation hypothesis proposed by **Cordo & Bevan.**

Horak et al. also emphasize the importance of "task constraints and performer intent" in determining the type of motor function requiring the processing of the cerebellum. Several of their examples would have been excellent to include in the target article!

Neilson & Neilson presented a stimulating introduction of their simulation efforts in modeling the information processing performed by the cerebellar circuitry. Using an approach they refer to as "adaptive model theory," they have actually derived an algorithm to express a basic operation performed across different cerebellar regions. They define the operation as consisting of "nonlinear convolution of neural signals through adaptive neural filter circuits." Their description clearly indicates that they have incorporated many of the organizational features among the cerebellar neural elements. It is of interest that they propose a heterosynaptic action of the climbing fiber system consisting of a change in Purkinje cell sensitivity, a proposal that is relatable to the gain change hypothesis reviewed in the target article. Neilson & Neilson's model also addressed the critical transformations between internal and external reference frames. I look forward to exploring the interrelationships between specific predictions of this model and experimental observations on this system.

The ideas expressed by **Braitenberg & Preissl** are both provocative and interesting. A more detailed presentation than is possible in a commentary is required to generate a full appreciation of their proposal. For example, it is unclear how the "tidal wave" is mediated, including the involvement of "new" parallel fibers as the wave progresses. The occurrence of beams associated with a lateral suppression of Purkinje cell activity has not been shown under physiological conditions. In fact, we demonstrated several years ago (Bloedel & Roberts 1969) that this type of lateral interaction depends on the connectivities of the collateral systems and the level of spontaneous activity among the cortical neurons. It is also unclear whether the first type of learning proposed by Braitenberg & Preissl is related to developmental processes of whether it is intended to reflect the establishment of fundamental engrams that can be changed even in adults as required by alterations in the task-related objectives of movements.

Finally, several commentators justifiably found the Vermittler hypothesis to be very reminiscent of other views of cerebellar function. As indicated in the target article, this is not surprising. At this point in the evolution of ideas about cerebellar operations there are certain facts about this system that are widely accepted: (1) features of the cerebellar circuitry, (2) its rich multisensory inputs, (3) the organization of cerebellar efferent systems, (3) cerebellar ablation syndromes, and (4) the wide spectrum of motor behaviors to which the cerebellum contributes. However, there are three features of the Vermittler

hypothesis that are not commonly emphasized in other proposals: (1) the uniqueness of the sensory processing in this structure, (2) the likelihood that cerebellar processing contributes to and participates in critical coordinate transformations, and (3) the role of the cerebellum in comparing sensory data acquired in external coordinates during movement execution with the sensory inputs characterizing the internal representation of the movement, body scheme, and the expected properties of the target.

At this point in the ongoing discussion of cerebellar function novel proposals are constantly required to stimulate the experiments and models to assess these new ideas. The field is clearly far from providing a definitive, well accepted viewpoint regarding the "function" of the cerebellum. All of us who are offering testable hypotheses must accept **Ito**'s contention that there is still a missing link: No single current hypothesis is sufficiently supported by direct, experimentally derived observations to justify its acceptance as the definitive, accepted view of how the system works, at least in my opinion. Consequently, the three hypotheses presented in the target article are undoubtedly too simplistic. However, viewpoints like these, based on a synthesis of as much data as possible from all sides of the debates regarding cerebellar function, are beginning to converge. There is at least some agreement regarding the relevance of specific organizational and functional characteristics of cerebellar systems. The next decade should offer a unique opportunity to interface experimental approaches with well-defined models of this system in further delineating the operation performed by the cerebellar circuitry and the functions in which the cerebellum participates.

Saving the baby: Toward a meaningful reincarnation of single-unit data

Eberhard E. Fetz

Department of Physiology & Biophysics and Regional Primate Research Center, University of Washington, Seattle, WA 98195
Electronic mail: *fetz@locke.hs.washington.edu*

Several commentators (**Alexander, Fuchs et al., Kalaska & Crammond, Lundberg, and Tanji**) described my target article as unduly pessimistic about the utility of single-unit recording in behaving animals. Alexander suggests I have concluded that "the response properties of central neurons are meaningless or uninterpretable." Having performed many chronic recording experiments, and having every intention of continuing such studies, I should first clarify that I think chronic unit recording is not only useful but quite essential for understanding neural mechanisms controlling movement. Without these experiments we would not be in the present situation of having a wealth of information for evaluating their significance. My skepticism concerns more the utility of simplistic interpretations of neural response patterns in terms of correlated movement parameters and the assumption that neural networks operate through explicit coding of abstract conceptualized variables. In sensory systems the coding of stimulus features seems more secure, although **Bridgeman** raises corresponding concerns even there; the idea that movement parameters are similarly coded in the motor areas is intuitively seductive but not convincingly supported by the accumulating experimental evidence. This does not mean, however, that chronic unit recording should be discarded along with the dubious interpretations. Although neural mechanisms controlling movement may not operate through explicit coding of movement parameters, the responses of neurons still offer crucial clues to neural computation.

The basic limitation of single-unit data is that they provide a very selective sample of a complex system, leaving a wide gap between highly restricted unit activity and overall system behavior. I would question the utility of bridging this gap by inference or leaps of faith. To better explain how neural networks control movement, we can fill this gap more effectively with neural network simulations. So, rather than being pessimistic about the utility of single-unit recording, I would be optimistic in hoping that single-unit data can finally be put into a causal framework: By incorporating the observed responses of units into neural network simulations we can generate more complete working models that can help explain the functional meaning of neural patterns.

The target article by **Alexander et al.** (this issue) seems to argue much the same point on the systems level that **Robinson**'s target article (this issue) and mine have argued on the cellular level, namely, that the operations of biological neurons do not conform to popular conceptual schemes that are imposed upon the recorded data. In his commentary **Alexander** suggests I would argue that "neural correlates of movement parameters . . . are only meaningful if they covary with the activity of the muscles with which that neuron is ultimately connected." Although I chose premotoneuronal (PreM) cells and their target muscles as useful examples of elements in the motor system with a correlational linkage that can be proven independently of their firing patterns, I do not believe that the relation of central neurons to movement parameters must all be analyzed in terms of these elements. In principle, higher-order cells could code parameters very well, and the literature is full of suggestive evidence for such coding, including, ironically enough, my target article. The tonic firing rate of certain PreM cells codes static muscle force much better than the motor units that produce that force, as shown in Figure 3 of my target article. Whereas individual motor units have a highly nonlinear relation to net force because of their recruitment threshold and saturation, many rubromotoneuronal (RM) and corticomotoneuronal (CM) cells have a linear relation to static force over a considerably wider range.

Alexander's commentary raises the important point that the evidence for the coding of a movement parameter can be strengthened when the relation between unit activity and a candidate parameter is tested under "a variety of system level input/output conditions." This strategy is crucial for determining whether a particular behavioral variable is reliably correlated with the activity of a neuron. For example, the contention that a central neuron codes direction of limb movement is strengthened when it correlates best with that direction independently of the force required to make the movement, or the part of extrapersonal space in which the movement is made. Unfortunately, there do not appear to be many

motor cortical cells that pass such a battery of tests. As more behavioral situations are used to confirm the coding of a particular parameter, fewer neurons appear to correlate consistently with that parameter.

The use of multiple behavioral tasks is related to another excellent point raised by several commentators, namely, that additional techniques are important for providing a full picture of the role of cells in controlling movement. **Lundberg** and **Ioffe** mention the classical strategies of stimulation and lesions, which can be performed both electrically and pharmacologically. **Iansek** points out that the converging approaches should also include evidence from clinical observations. I agree that converging evidence from different techniques strengthens the functional arguments (Fetz 1981), but in the role of devil's advocate I would have to point out the pitfalls involved for interpreting unit data: Stimulation and lesions at best reveal the net effects of the majority of the affected cells and cannot reflect the function of every neuron in the affected region. Relying on the consequences of lesions and stimulation for interpreting the meaning of unit data has led to systematically underestimating the diversity of neural types in a region and the multiplicity of possible functions in which a particular region can participate.

Additional evidence from a number of procedures, including lesions, stimulation, and a repertoire of behaviors, would also be helpful in constructing realistic neural networks that could replicate the range of observations. The inability of biological neurons to consistently code a specific parameter under variable circumstances is bad news for simplistic concepts of neural coding, but these additional behavioral conditions are grist for the neural network mill. Simulating a battery of behavioral situations and responses to lesion and stimulation provides important additional constraints for neural networks. A neural network model becomes more plausible to the degree that it can simulate a range of experimental conditions.

In light of their curious antipathy to neural network modeling **Fuchs et al.** are relatively generous regarding my model and instead take issue with my comments about coding. They state that my target article describes paradoxical unit responses as presenting some "irreconcilable problems" for single-unit recording. In fact, I welcome those paradoxical responses as significant clues that mechanisms other than explicit coding are operating, and I agree entirely that these responses are "easily interpretable when there is a more complete understanding of the system"; this was precisely my point in discussing the insights provided by neural network simulations. We have two ways to deal with such paradoxical units: **Robinson**'s target article and the commentary of Fuchs et al. suggest that neural networks operate successfully despite the existence of so-called rogue cells because these cells are simply outnumbered by those with more appropriate activity. On the basis of simulations I would suggest the possibility that such rogue cells, which have counterintuitive properties individually, may have some rationale in the context of other such cells; that is, their inappropriate components may cancel, resulting in an appropriate contribution.

I would also take issue with the implication of **Fuchs et al.** that I consider nonlinear relations to be a problem.

Their statement that "the only disadvantage of nonlinear relations is that they tax the mathematical capabilities of the modelers" might apply to those theoreticians who try to capture the behavior of the networks in analytical form. In fact, most modelers are simulating networks with nonlinear units and have no problem with nonlinear behavior.

In comparing the oculomotor and somatomotor systems, this commentary indulges in the common conceit of some oculomotor physiologists that their system is somehow superior for being simpler. Thus it is claimed that "at least in the oculomotor system, a robust relation seems to exist several synapses from the motoneuron." Of course, one can find examples of such relations, if one is looking for them and is willing to ignore cells with more complex combinations of signals. However, not all oculomotor physiologists (e.g., **Robinson**) are convinced that the coding is as simple and clean and unequivocal as **Fuchs et al.** like to argue.

Gandevia points out that much of the complexity of neural activity in relation to movement may arise from the complexity of muscle activities involved in performing a motor task. This point is well taken and underlies the rationale for dealing first with much simpler alternating flexion/extension movements. As Gandevia further notes, even such simple movements may involve complicated coactivation patterns in proximal stabilizing muscles and distal finger muscles that can differ from reciprocal agonist/antagonist activity. This again is the rationale for focusing on those PreM cells that have a demonstrable correlational effect on the agonist muscles. Despite such restrictions, even PreM cells exhibit a variety of response patterns relative to their facilitated target muscles; one would think that inferring the relation of central cells to other muscles in more complex tasks would quickly become prohibitively complex.

Iansek raises the concern that if chronic unit recording data are presumed to be deficient, neural network models based on those data would also be flawed. In fact, the deficiency is not so much in the recorded data per se, which must be accepted as an experimental fact, but in the missing information about the rest of the system. Neural network simulations can help fill in the picture in a relatively objective manner, by tying the scattered observations together into a complete working model. So, neural network simulations actually provide a complementary method of analyzing and understanding the significance of the recorded single-unit data, even when the data alone provide a hopelessly selective sample of the system.

Kalaska & Crammond challenge the usefulness of obtaining a "complete description of the causal mechanisms for the planning and initiation of movement." This "Holy Grail" is dismissed as unattainable, which it is, if one considers the term "complete" to mean a comprehensive description of the state and connectivity of every relevant neuron. Obviously, such an exhaustive compilation of all the details in the biological nervous system is not only impossible to obtain but impossible to synthesize, and a description to this depth is unnecessary. Some simplifications must certainly be made, but we can still pursue solutions that are "complete" in the sense intended in my target article, namely, having a sufficient number of elements to implement a working dynamic solution. The network simulations are obviously simplified in many ways, but they can still provide a more complete representation of the neural mechanisms that can generate a behavior than a patchwork of scattered observations of individual neurons. Obtaining a causal model is ultimately a worthier scientific goal than intuitive reading of selected neural patterns.

Kalaska & Crammond appear to dismiss the neural network simulation in the target article as merely a "curve-fitting matrix" that transforms the inputs by low-pass filtering and weighted combinations of the inputs. Such networks, including this one, actually do more than simple linear operations on the inputs. As reviewed elsewhere (Fetz, in press), dynamic, recurrent neural networks can simulate the nonlinear operations of the oculomotor system (cf. **Robinson**'s target article, this issue), reflex responses with local sign (Lockery et al. 1990), autonomous oscillations that resemble central pattern generators (Rowat & Selverston 1991), and short-term memory tasks (Zipser 1991). My particular model was intended as an example of the method, rather than a physiologically realistic model of the sensorimotor system. Therefore, the fact that Kalaska & Crammond "see no evidence that it solves any of the sensory transformations required to convert target spatial location into a multidimensional intrinsic reference frame of muscular activity" is obviously because the network was never designed or trained with this task in mind. However, there is no fundamental limitation precluding the development of such a model.

Kalaska & Crammond make a good point that looking for parallels between activity patterns of hidden units and responses of biological neurons can involve the same sort of selection bias as correlating neural discharge to parameters of movement, and may be just as misleading. This concern can be addressed by determining whether the analogous neurons and hidden units each typify a representative set of elements involved in the task. In some cases the response properties are so distinctive that there is little question of the uncanny similarity of hidden-unit patterns (for example, the discharge patterns of cortical neurons and network units involved in short-term memory tasks; Zipser 1991). There is a very important difference between interpreting the meaning of response properties of single neurons in behaving animals and interpreting the response patterns of hidden units in the network. In the animal data, one is only guessing about their function. In the network models, one can demonstrate explicitly what that function is by tracing their connections. Therefore, to the degree that hidden-unit activity resembles activity of neurons in animals performing the same task, one can make inferences about the possible function of the biological patterns.

Kirkwood has a very particular bone to pick regarding the relation between postspike facilitation (PSF) and anatomical connections. We have taken PSF to indicate a correlational linkage between the PreM neuron and its facilitated target muscles, and have used it to define both operationally. But as Kirkwood points out, it is conceivable that some of these facilitations may be mediated not by a monosynaptic connection of the triggering cell, but by some other cells whose spikes are sufficiently synchronized with the trigger cell. As he indicates, we have analyzed this issue directly by cross-correlating CM and neighboring cortical cells and found their cross-correlation peaks to be too broad to mediate the primary postspike effects (Fetz et al. 1991; Smith & Fetz 1989). Kirkwood suggests that more careful analysis of the PSF would be helpful in discriminating effects likely to be mediated by direct connections versus those mediated indirectly by synchrony, as previously described by Davies et al. (1985). This sort of analysis has been applied in a recent paper on the postspike effects of afferent fibers (Flament et al. 1992). A useful criterion is the latency of *poststimulus* effects, which can help define the minimal latency of postsynaptic effects of the trigger unit. Whether all the PSF in previous reports were mediated by direct monosynaptic connections or only some lesser proportion is not critical to the main point of my discussion, which concerns the response patterns of PreM cells. This issue would be relevant to my argument that PreM cells include a variety of response types if all examples of some particular response type were erroneously identified as PreM cells. This is highly unlikely, because each response type had many units with clear PSF that met the most stringent criteria.

The commentary by **Lundberg** makes several important points. Spinal cord neurons form a very important component of the motor system and we are indebted to Lundberg and his colleagues for their prodigious work in elucidating the segmental circuitry involved in reflex and voluntary movements. Although my target article describes primarily cortical and rubral PreM neurons, we do not assume that "only the monosynaptic pathways from the motor cortex and the red nucleus matter." Our experimental emphasis on the CM and RM cells is based primarily on their greater accessibility in behaving primates and on the fact that they do form a significant component of the supraspinal neurons that affect motoneurons directly. However, we do not believe that they are the only important controllers of motoneurons (Cheney et al. 1991). Regarding the points in the target article, they do provide a significant example of the sort of coding that can appear in cells directly linked to motoneurons. A major remaining experimental challenge is to elucidate the response properties of PreM cells in the spinal cord and we are currently trying to document cervical interneurons that affect motoneurons. The extensive work of Lundberg and colleagues (e.g., Alstermark et al. 1984; Baldissera et al. 1981; Illert et al. 1977; Jankowska & Lundberg 1981) will provide an essential context for identifying cells in such studies. Although the circuitry in the cat will provide important guidance in the monkey studies, significant differences between the cat and the primate, such as the existence of the monosynaptic corticomotoneuronal pathway, should also be remembered. The propriospinal neurons (PN) form an important disynaptic linkage from cortex to motoneurons in the cat (Illert et al. 1977), yet spike-triggered averages of EMG from cortex have so far failed to show that this linkage is sufficiently potent to mediate PSF. Determining whether similar PNs exist to the same extent in the monkey is a primary goal of our current investigations. We agree with Lundberg's suggestion that the CM cells may resemble the C3-C4 PNs in having two subgroups: those cells that project only to synergistic motoneurons and others that project both to motoneurons and Ia inhibitory interneurons (Fetz et al. 1990; Kasser & Cheney 1985).

Another difference between primate and cat may be the degree to which the lateral and medial descending systems play separable roles. **Lundberg** says it is noteworthy that complete transsection of the corticospinal tract and the rubrospinal tract in cats does not abolish commands for target-reaching and manipulatory movements. This result differs from those reported by Lawrence and Kuypers (1968) for the monkey, again suggesting significant interspecies differences. As Lundberg points out, reticulospinal pathways may also play a significant role in controlling primate motoneurons, another challenge for the spike-triggered averaging technique.

The relation of CM cells to movement is further discussed by **Lemon**, whose work has elucidated the function and connections of CM cells related to distal hand and finger muscles in natural precision grip movements. His work provides new insights into the relations between PSF patterns and responses of the monkey hand. Lemon raises the important point that it is helpful to document the response of PreM cells in relation to relatively normal limb movements in addition to a simple flexion-extension task. He says that when activity of CM cells is related to free hand movements "there is a congruency in the pattern of synaptic connectivity [shown by the PSF] and of recruitment during movement." An important caveat here follows from the fact that PSF can be detected only in those muscles coactivated with the cell; thus, during free movements involving variable activation of different muscles, this condition itself will tend to produce a congruence between the cell's facilitated muscles and the coactivated muscles.

Tanji raises an objection that calls for clarification. According to Tanji, my target article suggests "that no area-specific differences in properties of neuronal activity have been reported" and implies that "only neurons of the same type have been found everywhere in motor areas in more or less the same degree." In fact, my paper does state, perhaps not emphatically enough, that different cortical regions clearly show differences in the relative proportions of cells involved in different functions. Indeed, the work of Tanji and colleagues is particularly exemplary in providing persuasive evidence for regional specializations (e.g., Mushiake et al. 1991). Their work has documented the proportions of cells in different areas under behavioral tasks designed to elucidate these functional differences, and has provided ample data supporting specializations. My point is rather that cells of the same response type can be found distributed over many regions and that these like-minded cells probably form functional groups. The degree to which any particular cell type is found experimentally can also be proportional to the degree to which it is sought; therefore, recording bias should also be carefully controlled in experiments involving a search for different types of neurons. I agree entirely with Tanji's point that the experimental data "point to the presence of specialization in cortical areas" and with **Grobstein**'s similar point that we are "not in general dealing with a fully distributed system but rather with one having discrete 'information processing blocks.'"

What may have led to this misunderstanding is my statement that given a sufficient variety of cell types one can find examples to support any hypothesis from a completely random data set. This statement was designed to make a point but does not reflect a belief that the neural data are in fact random. There is good evidence for preferential relations between neural discharges and components of movement, and efforts to resolve what those discharge patterns actually mean are certainly worthwhile. My main point is to caution against inferring their meaning by conceptual projections, as opposed to determining their meaning in a causal framework; the latter can now be approached by neural network simulations that replicate these patterns and provide a mechanistic basis for interpreting their computational significance.

Real spatial maps?

J. F. Stein

University of Physiology, Oxford University, Oxford OX1 3PT, England
Electronic mail: *stein@vax.oxford.ac.uk*

The purpose of my target article was to be controversial and controversial it was, based on the commentaries received. I tried to convey four main ideas: First, there are no topographic maps of "real" space independent of eye, head, or body position to be found in the brain; second, contrary to what is so often assumed, there is probably no need even to convert the separate coordinate systems used by the different sensory and motor systems into a common reference frame (of which an explicit topographical map would be an example); third, if my first two propositions are correct then sensorisensory and sensorimotor interconversions are probably "direct," translating one into another without any intermediate step, and mediated by distributed associative networks situated in the posterior parietal cortex (PPC); and fourth (this was certainly my most speculative point), which interconversion rules are active at any time may be a function of the direction of attention. The intensity of opposition to these ideas increased in the same order, so my suggestion that attention has anything to do with selecting coordinate transformations was definitely the least popular.

R1. A map of real space in the brain? Few disagreed with my assertion that there is no explicit, topographical map of real space to be found in the brain. However, **Andersen & Brotchie** drew attention to their report that in the ventral region of the intraparietal sulcus (VIP) stimulation appeared to drive the eye to a specific egocentric location defined in head-centred coordinates irrespective of its orbital starting position. They found that the locations of these goal zones shifted systematically when the stimulating electrode was advanced through the cortex; they accordingly suggest that this region does indeed map real space topographically. However, there is no reason to suppose that this map can be used for other movements, for example, of the limbs. Hence it does not meet the requirements of a generalised map of real space; rather, it supports the idea that it is a special purpose region for the control of eye and perhaps head movements.

Graziano & Gross were the only commentators to disagree completely with my hypothesis by stating unequivocally that they believe a topographic map of visual space does in fact exist. What they describe are bimodal tactile/visual cells in the putamen, similar to neurones in area 6 and also in the PPC. These are somatotopically

organised and also "matching" in the sense that the location of the tactile and visual receptive fields is similar. But like most other maps of this sort that have been described, they turn out not to be stable maps of real space using a common reference frame. Instead, when the arm moves, so does the visual receptive field of the neurones, so they remain in register with the arm. Thus, they are remapped each time the arm moves. These characteristics are not what you would expect of a coordinate transformation converting all sensory and motor maps into a common reference frame; rather, this mechanism would be useful for relating vision to a specific movement. In other words I think Graziano & Gross have provided further evidence against rather than in favour of a general purpose, real spatial map.

R2. A common reference frame bottleneck? A recipient of much criticism was my suggestion that LIP neurones operate in a basically retinotopic frame of reference (**Pouget & Sejnowski, Gnadt, Andersen & Brotchie**). This is an important issue because it bears on whether all sensory signals are converted into a common reference frame in order to convert between modalities and into motor coordinate systems. I was not attempting to deny the importance of the eye position inputs to these cells; indeed, a large proportion of the target article was devoted to discussing these inputs. Rather, I was pointing to the fact that the basic organisation of a lateral intraparietal (LIP) neurone's receptive field has the standard two-dimensional Gaussian shape in retinotopic coordinates. What eye position inputs seem to do to this basic organisation is to scale the amplitude of the retinotopic visual responses according to orbital position. The quantitative details of this scaling generally produce the outcome that the maximum amount of modulation of the neurone occurs when a target appears on the receptive field hotspot and the eye is at an orbital position such that the hotspot is over the animal's egocentre. As stated in my target article, these are the conditions required to define the metrics of the saccade which would bring the fovea back to a target at the egocentre; and in other papers Andersen and his colleagues (1990b) have shown that the dependence of saccadic activity on orbital position behaves in just the way predicted by this analysis. My calling this a "retinal" vector was clearly misleading, however; my intention was only to call attention to the fact that notwithstanding the clear dependence of LIP responses on orbital position, the main influence on their firing is still the retinal position of a visual target, or, as **Colby et al.**, among others, have shown, the retinal locus of the memory or expectation of such a visual target.

The majority of cells in LIP have visual, presaccadic, anticipatory, and memory related activity in addition to orbital position sensitivity. The convergence of all these signals in LIP serves to allow neurones there to specify the location of a target. Basically, this location is still defined in terms of retinotopic coordinates; but their orbital position sensitivity provides these neurones with an additional signal which indicates how far away the current position of the eye is from orbital straight ahead. **Colby et al.** view the anticipatory activity of LIP neurones as remapping the representation of visual space each time the eyes move; so the retinal locus of the cell comes to represent the same point in head-centred space after eye movements. Clearly, eye movement signals, which would probably derive initially from corollary discharge and later from feedback from orbital proprioceptors, must play an important role in this remapping process.

The output of the whole array of LIP cells could represent the position of a target in head-centred coordinates, as suggested by Zipser and Andersen's (1988) simulations. But there is actually no evidence that such wholesale remapping takes place. For eye movement control it appears quite likely that there is direct conversion of retinotopic vectors, appropriately calibrated by eye position signals, into the required saccade commands. This system does not require an intermediate stage of head-centred representation. The basic retinotopic representation could be regularly updated by eye position signals to take account of each eye movement. Similarly, for limb movement control it is possible that the locations of visual targets are specified primarily in retinotopic coordinates and that these are transferred directly to motor regions such as the frontal lobe and cerebellum.

I am more than happy to have it confirmed by **Gnadt** that LIP neurones represent the third dimension, particularly as it has long been known by neurologists that some of the most common and troublesome symptoms of posterior parietal "neglect" are misjudgments of depth and distance. These symptoms are also common in children with developmental dyslexia, many of whom, as I have been able to show, exhibit mild symptoms of the posterior parietal syndrome (Stein 1991c).

Van Gisbergen & Duysens criticise my suggestion that the spatial nonuniformity of sensory maps may need to be linearised before they can be applied to motor output or to other spatial maps with different regions of magnification. In particular, they suggest that removing foveal magnification might reduce spatial precision of movement. This problem would only arise, however, if demagnification occurred before visual analysis were complete, which is not my suggestion. The point is that the movements to which such visual analysis gives rise must all have the same scale. Those around the midline could not be allowed to be ten times the size of those in the periphery, which is what would happen if a uniform motor programming stage were superimposed on the decidedly nonuniform visual cortical magnification factor.

Ingle describes his very interesting finding that acallosal subjects may mislocate a remembered target when they rotate past it and thus put its remembered location into the opposite hemifield. This again supports my idea that the localisation of objects is a distributed function, under these circumstances involving transfer of a token across the corpus callosum.

Goodale & Jakobson took me to task for preserving what they call the "old" distinction between the "what" and "where" components of the visual processing pathways. They quite rightly draw attention to the fact that actions are directed towards objects not locations, so that one not only has to consider the position of an object with respect to oneself but also its shape. Hence neglect in parietal lobe patients is not always simply egocentric; in some it is object centred. They neglect the left side of objects (Driver & Halligan 1991) as well as things on their left. It is clear therefore that the PPC provides information needed for localising features within objects as well as

for locating them with respect to the observer. These facts, however, do not alter my main point that no form of localisation depends on conversion into real spatial coordinates. Indeed, they support my contention because they show that the location of what is neglected can vary with the shape and size of an object rather than being fixed in space as would be expected if they were plotted on a real space map.

Translated into the requirements for motor control these considerations mean that we must study not only how the arm is aimed towards a target but also how the grasp is modified in the light of the shape of the object. **Goodale & Jakobson** have shown that these two aspects of motor control, aiming and hand shaping, can be differentially affected by lesions. Moreover, disconnections can leave the perception of the shape and orientation of an object intact when appropriate shaping of the hand reaching out to grasp it may be lost. So one should not forget the role of "what" pathway in formulating a movement, because important information for its programming has to be communicated by the pattern and shape recognition system to motor control centres. These points are well taken. But they call for more research into how the "what" and "where" pathways communicate with each other. They do not alter my fundamental point that sensorimotor conversions do not require an intermediate real-spatial map. Indeed, they make the latter less likely. A real space map that was also coding the orientation of objects within it would be even more unwieldy.

Cordo & Bevan point out that the kind of sensory information I conjecture the PPC provides to motor centres is consistent with their findings on successive approximation as a control strategy. Unfortunately, however, the characteristics of the PPC do not help us to choose between the competing claims of inverse dynamics, equilibrium point control, or successive approximation because it is clear that the contribution of the PPC comes at a stage prior to the detailed programming of a movement.

R3. Direct transformations? The basic question, however, still remains: How are these interconversions carried out? Most commentators seemed to assume that an explicit calculation of the position of a target in space (as postulated by **Andersen & Brotchie**) is most likely to be responsible for linking primary sensory inputs with each other and for using them for motor output.

Grobstein points out that the frog lacks a neocortex; so it might be a prime candidate to perform direct transformations, "reflexively," between sensory and motor centres, without transformation via a real spatial map. Yet the frog's tectum does seem to compute a signal which represents object location in a special body-centred coordinate frame which is distinguishable from that of either its sensory inputs or motor outputs. Moreover, this computation is a true bottleneck of the kind I argued would defeat its own purpose. Thus, in the frog's tectum the location of an object seems to be represented by the amount of activity in three structures representing three orthogonal axes; and these have to be accessed by any sensory or motor structures which need to use this information. **Ingle** shows, however, that even the frog exploits its striatum to get over this limitation to some extent.

I am in complete agreement with **Grobstein** that these

findings in the frog make extremely clear the benefits of possessing a neocortex which can do more than just mechanically relay sensory signals onwards to motor output. Funnelling all sensory information about object location to a tripartite bottleneck greatly constrains the repertoire of possible frog perceptions and actions; this perhaps helps to explain why his relatives, the dinosaurs, are unlikely to have been very intelligent! One great advantage of an area such as the posterior parietal cortex is that it can probably facilitate a large number of different interconversions simultaneously, after evaluation and decision, and thus avoid such a bottleneck.

Grobstein's commentary thus provides compelling reasons for believing that explicit rerepresentation of object location in head-centred space does not occur in primates. Instead, the rules for each and every possible interconversion may be laid out in distributed fashion, mainly in the posterior parietal cortex, but assisted by earlier visual and other sensory processing areas, as **Pouget & Sejnowski** point out. This idea is supported by a wealth of clinical observation about the wide variety of symptoms affecting all aspects of perception and motor control which follow lesions of the PPC. **Andersen**'s recent finding (Brotchie & Andersen 1991) that electrical stimulation in VIP induces not only eye but also head movements points to the same conclusion.

At this point, although I disagree with their conclusions to some extent, I should register my indebtedness to Grossberg and Kuperstein's (1989) discussion of the question of whether there is an intermediate stage recoding positional signals into head-centred coordinates or direct transfer from primary sensory maps into appropriate motor coordinates. My shameful failure to cite their work was the main point of **Quinlan**'s commentary. I continue to believe that the evidence favours the view that these interconversions are direct, however.

That there is a common motor representation in the colliculus for command signals originating from different sensory systems does not mean, as argued by **Van Gisbergen & Duysens**, that they are converted into a common coordinate system. Indeed, the way in which auditory receptive fields in the deep layers of the superior colliculus shift when the eyes move suggest just the opposite: not that there is a common spatial map into which all sensory inputs are translated, but rather that the local signs of auditory receptive fields are recalibrated each time the eyes move – that is, there is "direct" translation from one into the other. This is also the point of view supported by **Proctor & Franz** from their studies of stimulus-response compatibility effects.

The hypothesis that coordinate transformations are specific to the sensorimotor system being used and that they do not use a general purpose, real spatial map does not necessarily imply that there is only one processing stage between stimulus and response. This view is attributed to me by **MacKay & Riehle** in their otherwise excellent defense of single-unit recording against the nihilistic tendencies of Robinson's target article. It is clear that the PPC is only one of many areas involved in such sensorimotor transformations, as **Seltzer** also points out. What my use of the word "direct" was meant to suggest was that there is no intermediate representation in a real

spatial coordinate system, different from that used by either stimulus or response.

My suggestion that the PPC may be more like a "gazetteer" than a topographical map provoked some interesting responses, in particular, a very interesting commentary by **Dawson**. He describes Pylyshyn's (1989) concept of FINSTs. These turn out to be analogous to the names of places in my gazetteer; they do not themselves code the coordinates defining their location, but they point to where the coordinates can be found; for motor control by binding them to "anchors." Since these binding operations are between the names and not the locations of objects, they can be hardwired. Thus, they seem to be very similar to the hardwired algorithms that I postulate perform sensorimotor interconversions in the PPC.

In summary, it seems to me that most of my critics do in fact believe that the location of targets depends on a distributed system of coordinate transformation algorithms and that these are neither topographically mapped nor dependent on conversion into and out of a common "real spatial" reference frame. In short, most of us seem to agree that "direct" sensorimotor conversions are the rule rather than some form of transfer via a cerebral map of real space.

R4. The role of attention. My speculation that attention selects which particular transformation or "binding operation" is going to be performed by the PPC provoked howls of protest. Nobody was discourteous enough to mention directly the main weakness of this idea, namely, that it merely passes the buck to a system that we know even less about than the PPC. **Carey & Servos**, among others, pointed out that attention is itself a distributed system. It can no longer be considered a simple "searchlight," not only because what is being illuminated is not itself topographically laid out, but also because many regions other than the PPC have been shown to be involved in attentional processes. These were points I too tried to make, but clearly they did not get across well.

Most of the other criticisms of my speculations were addressed to the fact that many accurate sensorisensory and sensorimotor interconversions can be made without conscious awareness (**Kalaska & Crammond, Carey & Servos, Colby et al.**). If the direction of attention has to be a conscious process then clearly the fact that many sensorimotor tasks can be done automatically contradicts my suggestion. I should have made a clear distinction between learned and novel tasks for this part of my target article. Many sensorimotor tasks may become automatic with practice. While learning them attention must be directed towards selecting the correct routines; but as they become more skillful (probably with the help of the cerebellum) attention becomes less and less necessary. This is perhaps because the conversion routines required for the task become hardwired into the cerebellar cortex. On the other hand, new tasks still require the conscious direction of attention to select the sensorimotor conversion algorithm which is appropriate for the job. Thus, directing attention is necessary for selecting the correct sensorimotor coordinate transformation routines for novel tasks, but this is done automatically for well-learned skills, probably because the necessary algorithms have been transferred to the cerebellum.

References

Letters *a* and *r* appearing before authors' initials refer to target article and response respectively; if those authors also wrote a commentary, *c* precedes their initials.

Abbott, B. C. & Aubert, X. M. (1952) The force exerted by active striated muscle during and after change of length. *Journal of Physiology (London)* 117:77–86. [GCA]

Abdusamatov, R. M., Adamovich, S. V., Berkinblit, M. B., Chernavsky, A. V. & Feldman, A. G. (1988) Rapid one-joint movements: A qualitative model and its experimental verification. In: *Stance and motion: Facts and concepts*, ed. V. S. Gurfinkel, M. E. Ioffe, J. Massion & J. P. Roll. Plenum. [SVA]

Abdusamatov, R. M., Adamovich, S. V. & Feldman, A. G. (1987) A model for one-joint motor control in man. In: *Motor control*, ed. G. N. Gantchev, B. Dimitrov & P. Gatev. Plenum. [SVA]

Abeles, M. (1991) *Corticonics. Neural circuits of the cerebral cortex.* Cambridge University Press. [VB]

Abend, W., Bizzi, E. & Morasso, P. (1982) Human arm trajectory formation. *Brain* 105:331–48. [aSCG, JDe]

Abernethy, B. & Sparrow, W. A. (1992) The rise and fall of dominant paradigms in motor behavior research. In: *Approaches to the study of motor control and learning*, ed. J. J. Summers. North-Holland. [JJS]

Abraham, L. D. & Loeb, G. E. (1985) The distal musculature of the cat: Patterns of normal use. *Experimental Brain Research* 58:580–93. [CAP]

Abraham, L., Potegal, M. & Miller, S. (1983) Evidence for a caudate nucleus involvement in an egocentric spatial task. *Physiological Psychology* 11:11–17. [DI]

Adamovich, S. V., Burlachkova, N. I. & Feldman, A. G. (1984) Wave nature of the central process of formation of the trajectories of change in the joint angle in man. *Biophysics* 29:130–34. [SVA]

Adamovich, S. V. & Feldman, A. G. (1984) Model of the central regulation of the parameters of motor trajectories. *Biophysics* 29:338–42. [SVA, DJO]
(1989) The prerequisites for one-joint motor control theories. *Behavioral and Brain Sciences* 12:210–11. [MLL, SVA]

Ahissar, M., Ahissar, E., Bergman, H. & Vaadia, E. (1992) Encoding of sound-source location and movement: Activity of single neurons and interactions between adjacent neurons in the monkey auditory cortex. *Journal of Neurophysiology* 67:203–15. [APo]

Albin, R. L., Young, A. B. & Penney, J. B. (1989) The functional anatomy of basal ganglia disorders. *Trends in Neuroscience* 12:366–75. [aGEA]

Albus, J. S. (1971) A theory of cerebellar function. *Mathematical Bioscience* 10:25–61. [PFCG, RFT]

Alexander, G. E. (1987) Selective neuronal discharge in monkey putamen reflects intended direction of planned limb movements. *Experimental Brain Research* 67:623–34. [aGEA]
(1992) For effective sensorimotor processing must there be explicit representations and reconciliation of differing frames of reference? *Behavioral and Brain Sciences* 15:323–24. [MF]

Alexander, G. E. & Crutcher, M. D. (1990a) Functional architecture of basal ganglia circuits: Neural substrates of parallel processing. *Trends in Neuroscience* 13:266–71. [aGEA, JMF]
(1990b) Preparation for movement: Neural representations of intended direction in three motor areas of the monkey. *Journal of Neurophysiology* 64:133–50. [aGEA]
(1990c) Neural representations of the target (goal) of visually guided arm movements in three motor areas of the monkey. *Journal of Neurophysiology* 64:164–78. [aGEA, aEEF]

Alexander, G. E., Crutcher, M. D. & DeLong, M. R. (1990) Basal ganglia-thalamocortical circuits: Parallel substrates for motor, oculomotor, "prefrontal" and "limbic" functions. *Progress in Brain Research* 85:119–46. [AB]

Alexander, G. E. & DeLong, M. R. (1985) Microstimulation of the primate neostriatum. II. Somatotopic organization of striatal microexcitable zones and their relation to neuronal response properties. *Journal of Neurophysiology* 53:1417–30. [aGEA]

Alexander, G. E. & Fuster, J. (1973) Effects of coding prefrontal cortex on cell firing in the nucleus medialis dorsalis. *Brain Research* 61:93–105. [aEEF]

Al-Falahe, N. A., Nagaoka, M. & Vallbo, Å B. (1990) Response profiles of human muscle afferents during active finger movements. *Brain* 113:325–46. [aSCG]

Al-Falahe, N. A. & Vallbo, Å B. (1988) Role of the human fusimotor system in a motor adaptation task. *Journal of Physiology (London)* 401:77–95. [arSCG]

Allen, G. I. & Tsukahara, N. (1974) Cerebellar communication systems. *Physiological Reviews* 54:957–1006. [aGEA, aJRB, cJFS]

Allum, J. H. J. (1975) Responses to load disturbances in human shoulder muscles: The hypothesis that one component is a pulse test information signal. *Experimental Brain Research* 22:307–26. [aEB, MLL]

Allum, J. H. J. & Mauritz, K.-H. (1984) Compensation for intrinsic muscle stiffness by short-latency reflexes in human triceps surae muscles. *Journal of Neurophysiology* 52:797–818. [TRN]

Aloisi, A. M., Carli, G. & Rossi, A. (1988) Response of hip joint afferent fibers to pressure and vibration in the cat. *Neuroscience Letters* 90:130–34. [aSCG]

Alstermark, B., Johannisson, T. & Lundberg, A. (1986) The inhibitory feedback pathway from the forelimb to C3-C4 propriospinal neurones investigated with natural stimulation. *Neuroscience Research* 3:451–56. [DBur]

Alstermark, B. & Kümmel, H. (1990) Transneuronal transport of wheat germ agglutinin conjugated horseradish peroxidase in last-order spinal interneurones projecting to acromio- and spinodeltoideus motoneurones in the cat. 2. Differential labelling of interneurones depending on movement type. *Experimental Brain Research* 80:96–103. [AL]

Alstermark, B., Kümmel, H., Pinter, M. & Tantisira, B. (1990) Integration in descending motor pathways controlling the forelimb in the cat. 17. Axonal projection and termination of C3-C4 propriospinal neurones in the C6-Th1 segments. *Experimental Brain Research* 81:447–61. [AL]

Alstermark, B., Lindström, S.,. Lundberg, A. & Sybirska, E. (1981a) Integration in descending motor pathways controlling the forelimb in the cat. 8. Ascending projection to the lateral reticular nucleus from C3-C4 propriospinal neurones also projecting to forelimb motoneurones. *Experimental Brain Research* 42:282–98. [AL]

Alstermark, B., Lundberg, A., Norrsell, U. & Sybirska, E. (1981b) Integration in decending motor pathways controlling the forelimb in the cat. 9. Differential behavioural defects after spinal cord lesions interrupting defined pathways from higher centres to motoneurones. *Experimental Brain Research* 42:299–318. [aSCG, DBur, cSCG, AL]

Alstermark, B., Lundberg, A., Pettersson, L.-G., Tantisira, B. & Walkowska, M. (1987) Motor recovery after serial spinal cord lesions of defined descending pathways in cats. *Neuroscience Research* 5:68–73. [AL]

Alstermark, B., Lundberg, A. & Sasaki, S. (1984) Integration in descending motor pathways controlling the forelimb in the cat. II. Inhibitory pathways from higher motor centres and forelimb afferents to C3-C4 propriospinal neurones. *Experimental Brain Research* 56:293–307. [rEEF, DBur, AL]

Alstermark, B. & Sasaki, S. (1986) Integration in descending motor pathways controlling the forelimb in the cat. 14. Differential control of fast and slow motoneurones from C3-C4 propriospinal neurones. *Experimental Brain Research* 63:530–42. [AL]

An, C. H., Atkeson, C. G. & Hollerbach, J. M. (1988) *Model-based control of a robot manipulator.* MIT Press. [aGEA]

An, K. N., Kaufman, K. R. & Chao, E. Y. S. (1989) Physiological considerations of muscle force through the elbow joint. *Journal of Biomechanics* 22:1249–56. [SVA]

Anastasio, T. J. & Robinson, D. A. (1989) The distributed representation of vestibulo-ocular signals by brain-stem neurons. *Biological Cybernetics* 61:79–88. [aDAR, MF]
(1990a) Distributed parallel processing in the vertical vestibulo-ocular reflex:

Learning networks compared to tensor theory. *Biological Cybernetics* 63:161–67. [arDAR, RE]

(1990b) Distributed parallel processing in the vestibulo-oculomotor system. *Neural Computation* 1:230–41. [aGEA, rDAR]

Andén, N. E., Jukes, M. G. M. & Lundberg, A. (1967) The effect of DOPA on the spinal cord. I. Influence on transmission from primary afferents. *Acta Physiologica Scandinavica* 67:373–86. [aDAM]

Andersen, R. A. (1987) Inferior parietal lobule function in spatial perception and visuomotor integration. In: *Handbook of physiology*, ed. F. Plum & V. B. Mountcastle. American Physiological Society. [aJFS]

(1989) Visual and eye movement functions of posterior parietal cortex. *Annual Review of Neuroscience* 12:377–404. [aJFS]

Andersen, R. A., Asanuma, C. & Cowan, W. M. (1985a) Callosal and prefrontal associational projecting cell populations in area 7a of the macaque monkey: A study using retrogradely transported fluorescent dyes. *Journal of Comparative Neurology* 232:443–55. [aJFS]

Andersen, R. A., Asanuma, C., Essick, G. & Siegel, R. M. (1990a) Corticocortical connections of anatomically and physiologically defined subdivisions of inferior parietal lobule. *Journal of Comparative Neurology* 296:65–113. [aJFS]

Andersen, R. A., Bracewell, R. M., Barash, S., Gnadt, J. W. & Fogassi, L. (1990b) Eye position effects on visual, memory, and saccade-related activity in areas LIP and 7a of the macaque. *Journal of Neuroscience* 10:1176–96. [arJFS, RAA, CLC, JWG, APo]

Andersen, R. A., Essick, G. & Siegel, R. M. (1985b) Encoding of spatial location by posterior parietal neurons. *Science* 230:456–58. [aEB, aJFS, RAA, JWG]

(1987) Neurons of area 7 activated by both visual stimuli and oculomotor behavior. *Experimental Brain Research* 67:316–22. [aDAR]

Andersen, R. A. & Gnadt, J. W. (1989) Posterior parietal cortex. In: *The neurobiology of saccadic eye movements*, ed. R. H. Wurtz & M. E. Goldberg. Elsevier. [JWG]

Andersen, R. A. & Mountcastle, V. B. (1983) The influence of the angle of gaze upon the excitability of light sensitive neurones in the posterior parietal cortex. *Journal of Neuroscience* 3:532–48. [aJFS]

Andersen, R. A. & Zipser, D. (1988) The role of the posterior parietal cortex in coordinate transformations for visual-motor integration. *Canadian Journal of Physiology and Pharmacology* 66:488–501. [MF]

Anderson, M. E. (1977) Segmental reflex inputs to motoneurons innervating dorsal neck musculature in the cat. *Experimental Brain Research* 28:175–87. [GEL]

Andrews, C. J. (1973) Influence of dystonia on the response to long-term L-Dopa therapy in Parkinson's disease. *Journal of Neurology, Neurosurgery & Psychiatry* 36:630–36. [GMa]

Andrews, J. R. & Hogan, N. (1983) Impedance control as a framework for implementing obstacle avoidance in a manipulator. In: *Control of manufacturing and robotic systems*, ed. D. E. Hardt & W. J. Book. ASME. [rEB]

Angel, R. W. & Malenka, R. C. (1982) Velocity-dependent suppression of cutaneous sensitivity during movement. *Experimental Neurology* 77:266–74. [aSCG]

Aniss, A. M., Diener, H.-C., Hore, J., Burke, D. & Gandevia, S. C. (1990a) Reflex activation of muscle spindles in human pretibial muscles during standing. *Journal of Neurophysiology* 64:671–79. [arSCG]

Aniss, A. M., Diener, H.-C., Hore, J., Gandevia, S. C. & Burke, D. (1990b) Behavior of human muscle receptors when reliant on proprioceptive feedback during standing. *Journal of Neurophysiology* 64:661–70. [arSCG]

Aniss, A. M., Gandevia, S. C. & Burke, D. (1988a) Reflex changes in muscle spindle discharge during a voluntary contraction. *Journal of Neurophysiology* 59:908–21. [aSCG]

Aniss, A. M., Gandevia, S. C. & Milne, R. J. (1988b) Changes in perceived heaviness and motor commands produced by cutaneous reflexes in man. *Journal of Physiology (London)* 397:113–26. [aSCG]

Appenteng, K. & Prochazka, A. (1984) Tendon organ firing during active muscle lengthening in awake, normally behaving cats. *Journal of Physiology (London)* 353:81–92. [aSCG]

Araki, T., Eccles, J. C. & Ito, M. (1960) Correlation of the inhibitory postsynaptic potential of motoneurones with the latency and time course of inhibition of monosynaptic reflexes. *Journal of Physiology (London)* 154:354–77. [DBul]

Arbib, M. A. (1981) Perceptual structures and distributed motor control. In: *Handbook of physiology: Nervous system*, vol. 2, ed. V. B. Brooks. American Physiological Society. [CIC, JMF]

(1987) Levels of modeling of mechanisms of visually guided behavior. *Behavioral and Brain Sciences* 10(3):407–65. [CIC]

(1990) Programs, schemas, and neural networks for control of hand movements: Beyond the RS framework. In: *Attention and performance*

XIII. *Motor representation and control*, ed. M. Jeannerod. Erlbaum. [aGEA]

(1991) Interaction of multiple representations of space in the brain. In: *The brain and space*, ed. J. Paillard. Oxford University Press. [aJFS]

Arkin, R. C. (1987) Towards cosmopolitan robots: Intelligent navigation in extended man-made environments. Technical report 87-80, Department of Computer and Information Science, University of Massachusetts. [CIC]

Arnold, D. B. & Robinson, D. A. (1991) A learning network model of the neural integrator of the oculomotor system. *Biological Cybernetics* 64:447–54. [arDAR]

Artim, J. & Bridgeman, B. (1989) The physiology of attention: Participation of cat striate cortex in behavioral choice. *Psychological Research/Psychologische Forschung* 50:223–28. [BB]

Asanuma, H. (1989) *The motor cortex.* Raven Press. [MEI]

Asanuma, C., Andersen, R. A. & Cowan, W. M. (1985) The thalamic relations of the caudal inferior parietal lobule and the lateral prefrontal cortex in monkeys: Divergent cortical projections from cell clusters in the medial pulvinar nucleus. *Journal of Comparative Neurology* 241:357–81. [aJFS]

Asanuma, C., Thach, W. T. & Jones, E. G. (1983a) Brainstem and spinal projections of the deep cerebellar nuclei in the monkey, with observations on the brainstem projections of the dorsal column nuclei. *Brain Research Reviews* 5:299–322. [aJRB]

(1983b) Anatomical evidence for segregated focal groupings of efferent cells and their terminal ramifications in the cerebellothalamic pathway of the monkey. *Brain Research Reviews* 5:267–97. [aJRB]

(1983c) Distribution of cerebellar terminations and their relation to other afferent terminations in the ventral lateral thalamic region of the monkey. *Brain Research Reviews* 5:237–65. [aJRB]

Asatryan, D. B. & Feldman, A. G. (1965) Functional tuning of the nervous system with control of movement or maintenance of a steady posture. I. Mechanographic analysis of the work of the joint or execution of a postural task. *Biofizika* 10:837–46 [English translation 925–35]. [GCA, SVA, AGF]

Ashford, J. A. W. & Fuster, J. M. (1985) Occipital and inferotemporal responses to visual signals in the monkey. *Experimental Neurology* 90:444–66. [JMF]

Astrom, K. J. & Wittenmark, B. (1989) *Adaptive control.* Addison-Wesley. [PDN]

Atkeson, C. G. (1989) Learning arm kinematics and dynamics. *Annual Review of Neuroscience* 12:157–83. [aGEA]

Atkinson, R. C., Bower, G. H. & Crothers, E. J. (1965) *An introduction to mathematical learning theory.* Wiley. [HH]

Babinski, J. (1899) De l'asynergie cérébelleuse. *Revue Neurologique* 7:806–16. [AMS]

Baev, K. V. & Shimansky, Y. P. (1992) Principles of organization of neural systems controlling automatic movements in animals. *Progress in Neurobiology* 39:45–112. [rJRB]

Baizer, J., Ungerleider, J. G. & Desimone, R. (1991) Organisation of visual inputs to inferior temporal and posterior parietal cortex in macaques. *Journal of Neuroscience* 11:168–90. [aJFS]

Balaban, C. D., Ito, M. & Watanabe, E. (1981) Demonstration of zonal projections from the cerebellar flocculus to vestibular nuclei in monkeys (*Macaca fuscata*). *Neuroscience Letters* 27:101–5. [NHB]

Baldissera, F., Cavallari, P., Fournier, E., Pierrot-Deseilligny, E. & Shindo, M. (1987) Evidence for mutual inhibition of opposite Ia interneurones in the human upper limb. *Experimental Brain Research* 66:106–14. [DBul]

Baldissera, F., Hultborn, H. & Illert, M. (1981) Integration in spinal neuronal systems. In: *Handbook of physiology: The nervous system II*, ed. J. M. Brookhart, V. B. Mountcastle, V. B. Brooks & S. R. Geiger. American Physiological Society. [aDAM, rEEF]

Baldissera, F. & Pierrot-Deseilligny, E. (1989) Facilitation of transmission in the pathway of non-monosynaptic Ia excitation to wrist flexor motoneurones at the onset of voluntary movement in man. *Experimental Brain Research* 74:437–39. [aSCG, AL]

Bamber, D. & van Santen, J. P. H. (1985) How many parameters can a model have and still be testable? *Journal of Mathematical Psychology* 29:443–73. [JGP]

Bando, T., Yamamoto, N. & Tsukahara, N. (1984) Cortical neurons related to lens accommodation in posterior lateral suprasylvian area in cats. *Journal of Neurophysiology* 52:879–91. [JWG]

Bapi, R. S. & Levine, D. S. (1990) Networks modeling the involvement of the frontal cortex in performance of flexible motor sequences. *Proceedings of the Twelfth Annual Conference of the Cognitive Science Society.* Erlbaum. [DSL]

Baranyi, A. & Szente, M. B. (1987) Long-lasting potentiation of synaptic transmission requires postsynaptic modifications in the neocortex. *Brain Research* 423:378–84. [aGEA]

Barash, S., Bracewell, R. M., Fogassi, L., Gnadt, J. W. & Andersen, R. A. (1991) Saccade related activity in the lateral intraparietal area. *Journal of Neurophysiology* 66:1095–1124. [aJFS, CLC]

Barbas, H. & Mesulam, H.-M. (1985) Cortical afferent input to principalis region of rhesus monkey. *Neuroscience* 15:617–37. [aJFS]

Barbeito, R. & Ono, M. (1979) Four methods for locating the egocentre. *Behavioral Research Methods Instruments & Computers* 11:31–36. [aJFS]

Bard, C., Paillard, J., Lajoie, Y., Fleury, M., Teasdale, N., Forget, R. & Lamarre, Y. (1992) Role of afferent information in the timing of motor command: A comparative study with a deafferented patient. *Neuropsychologia* 30:201–6. [JP]

Barlow, H. B. (1980) Cortical function: A tentative theory. In: *Neural mechanics of behaviour*, ed. D. McFadden. Springer. [aJFS]

Barrack, R. L., Skinner, H. B., Brunet, M. E. & Haddad, R. J. (1983a) Functional performance of the knee after intra-articular anesthesia. *American Journal of Sports Medicine* 11:258–61. [aSCG]

Barrack, R. L., Skinner, H. B., Cook, S. D. & Haddad, R. J. (1983b) Effects of articular disease and total knee arthroplasty on knee joint-position sense. *Journal of Neurophysiology* 50:684–87. [aSCG]

Barraquand, J. & Latombe, J.-C. (1991) Robot motion planning: A distributed representation approach. *International Journal of Robotics Research* 10(6):628–49. [CIC]

Barto, A. G. & Sutton, R. S. (1981) Landmark learning, an illustration of associative search. *Biological Cybernetics* 42:1–8. [aJFS]

Batini, C. & Billard, J. M. (1985) Release of cerebellar inhibition by climbing fiber deafferentation. *Experimental Brain Research* 57:370–80. [AMS]

Baum, E. B. (1990) On learning a union of half spaces. *Journal of Complexity* 6:67–101. [HB]

Baumann, T. K., Emonet-Dénand, F. & Hulliger, M. (1982) After-effects of fusimotor stimulation on spindle Ia afferents' dynamic sensitivity, revealed during slow movements. *Brain Research* 232:460–65. [aSCG]

Bauswein, E., Kolb, F. P., Leimbeck, B. & Rubia, F. J. (1983) Simple and complex spike activity in cerebellar Purkinje cells during active and passive movements in the awake monkey. *Journal of Physiology (London)* 339:379–894. [JMB]

Baxendale, R. H. & Ferrell, W. R. (1981) The effect of knee joint afferent discharge on transmission in flexion reflex pathways in decerebrate cats. *Journal of Physiology (London)* 315:231–42. [aSCG]

Bayoumi, A. & Ashby, P. (1988) Projections of group Ia afferents to motoneurons of thigh muscles in man. *Experimental Brain Research* 342:1–6. [aSCG]

Beckstead, R. M. & Cruz, C. J. (1986) Striatal axons to the globus pallidus, entopeduncular nucleus and substantia nigra come mainly from separate cell populations in cat. *Neuroscience* 19:147–58. [aGEA]

Beek, P. J. & Bootsma, R. J. (1991) Physical and informational principles in modelling coordinated movements. *Human Movement Science* 10:81–92. [GJvIS]

Beer, R. D., Hillel, J. C. & Sterling, L. S. (1991) A biological perspective on autonomous agent design. In: *Designing autonomous agents*, ed. A. P. Maes. MIT Press. [SG]

Beitz, A. J. & Chan-Palay, V. (1979) A golgi analysis of neuronal organization in the medial cerebellar nucleus of the rat. *Neuroscience* 4:47–63. [aJRB]

Bekoff, A. (1989) Continuity of pattern generating mechanisms in embryonic and posthatching chicks. In: *Neurobiology of early infant behavior*, ed. V. von Euler, H. Forssberg, C. von Hofsten, H. Lagerkranz & R. Zetterstrom. Macmillan. [CAP]

Bekoff, A., Nusbaum, M. P., Sabichi, A. L. & Clifford, M. (1987) Neural control of limb coordination. I. Comparison of hatching and walking motor output patterns in normal and deafferented chicks. *Journal of Neuroscience* 7:2320–30. [CAP]

Belinkii, V., Gurfinkel, V. & Paltsev, Y. (1967) Elements of control of voluntary movements. *Biofizika* 12:135–41. [ADK]

Benecke, R., Rothwell, J. C., Day, B. L., Dick, J. P. R. & Marsden, C. D. (1986) Motor strategies involved in the performance of sequential movements. *Experimental Brain Research* 63:585–95. [aSCG]

Benedetti, F., Montarolo, P. G. & Rabachi, S. (1984) Inferior olive lesion induces long-lasting functional modification in the Purkinje cells. *Experimental Brain Research* 55:368–74. [AMS]

Bennett, D. J. (1990) The control of human arm movement: Models and mechanical constraints. Doctoral Dissertation, Department of Brain and Cognitive Sciences, MIT. [aEB]

Bennett, D. J., Xu, Y., Hollerbach, J. M. & Hunter, I. W. (1989) Identifying the mechanical impedance of the elbow joint during posture and movement. *Society for Neuroscience Abstracts* 15:396. [aEB]

Bennett, K. M. B. (1992) Corticomotoneuronal control of precision grip tasks. Doctoral Dissertation, Cambridge University. [RL]

Bennett, K. M. B. & Lemon, R. N. (1991) The activity of monkey corticomotoneuronal (CM) cells is related to their pattern of post-spike facilitation of intrinsic hand muscles. *Journal of Physiology* 435:53p. [RL]

Benson, D. F. & Greenberg, J. P. (1969) Visual form agnosia. A specific defect in visual discrimination. *Archives of Neurology* 20:82–89. [MAG]

Beppu, H., Nagaoka, M. & Tanaka, R. (1987) Analysis of cerebellar motor disorders by visually-guided elbow tracking movement. 2. Contribution of the visual cues on slow ramp pursuit. *Brain* 110:1–18. [aJRB]

Beppu, H., Suda, M. & Tanaka, R. (1984) Analysis of cerebellar motor disorders by visually guided elbow tracking movement. *Brain* 107:787–809. [aJRB, MH]

Berger, W., Altenmuller, E. & Dietz, V. (1984) Normal and impaired development of children's gait. *Human Neurobiology* 3:163–70. [VD]

Bergman, H., Wichmann, T. & DeLong, M. R. (1990) Reversal of experimental Parkinsonism by lesion of the subthalamic nucleus. *Science* 249;1436–38. [DSB]

Berkinblit, M. B., Feldman, A. G. & Fukson, O. I. (1986) Adaptability in innate motor patterns and motor control mechanisms. *Behavioral and Brain Sciences* 9:585–638. [aEB, GCA, MBB, AP]

Bernstein, N. (1935) The problem of interrelation between coordination and localization [in Russian]. *Archives of Biological Sciences* 38:1–35. [MLL]
　(1940/1967) Biodynamics of locomotion. In: *The coordination and regulation of movements* (originally published in Russian in 1940). Pergamon Press. [ZH]
　(1967) *The coordination and regulation of movements*. Pergamon Press. [aGEA, aEB, aSCG, aDAM, JMF]
　(1966) *Notes on the movement physiology and physiology of activity* [in Russian]. Medicina. [AAF]

Berthoz, A. & Grantyn, A. (1986) Neuronal mechanisms of eye head coordination. In: *Progressive brain research 64*, ed. U. Freund. Elsevier. [aJFS]

Beuter, A., Belair, J. & Labrie, C. (1992) Feedback and delays in neurological diseases: A modeling study using dynamical systems. *Bulletin of Mathematical Biology*, in press. [AB]

Beuter, A., Labrie, C. & Vasilakos, K. (1991) Transient dynamics in motor control of patients with Parkinson's disease. *Chaos* 1(3):279–86. [AB]

Bigland-Ritchie, B., Dawson, N. J., Johansson, R. S. & Lippold, O. C. J. (1986) Reflex origin for the slowing of motoneurone firing rates in fatigue of human voluntary contractions. *Journal of Physiology (London)* 379:451–59. [aSCG]

Biguer, B., Donaldson, I. M. L., Hein, A. & Jeannerod, M. (1988) Neck muscle vibration modifies the representation of visual motion and direction in man. *Brain* 111:1405–24. [aSCG]

Biguer, B., Jeannerod, M. & Prablanc, C. (1982) The coordination of eye, head, and arm movements during reaching at a single visual target. *Experimental Brain Research* 46:301–4. [VD]

Billard, J. M., Batini, C. & Daniel, H. (1988) The red nucleus activity in rats deprived of the inferior olivary complex. *Behavioral Brain Research* 28:127–30. [rJRB]

Binder, M. D. & Stuart, D. G. (1980) Motor-unit muscle receptor interactions: Design features of the neuromuscular control system. In: *Progress in clinical neurophysiology, vol. 8: Spinal and supraspinal mechanisms of voluntary motor control and locomotion*, ed. J. E. Desmedt. Karger. [aSCG]

Bindman, L. J., Murphy, K. P. S. J. & Pockett, S. (1988) Postsynaptic control of the induction of long-term changes in efficacy of transmission at neocortical synapses in slices of rat brain. *Journal of Neurophysiology* 1053–65. [aGEA]

Bischof, H. (1991) Modular, hierarchical and geometrical neural networks. Technical Report (PRIP-TR-9, December), Department for Pattern Recognition and Image Processing, Technical University of Vienna. [HB]

Bischof, H., Pinz, A. & Kropatsch, W. G. (1992) Visualization methods for neural networks. *Proceedings of the 11th International Conference for Pattern Recognition* (ICPR92), The Hague. [HB]

Bisiach, E., Capitani, E. & Porta, E. (1985) Two basic properties of space representation in the brain. *Journal of Neurology, Neurosurgery and Psychiatry* 48:141–44. [aJFS]

Bisiach, E. & Luzatti, C. (1978) Unilateral neglect of representational space. *Cortex* 14:129–33. [aJFS]

Bizzi, E., Accornero, N., Chapple, W. & Hogan, N. (1982) Arm trajectory formation in monkeys. *Experimental Brain Research* 46:139–43. [aEB, GCA, SVA, GLG, MLL]
　(1984) Posture control and trajectory formation during arm movement. *Journal of Neuroscience* 4:2738–44. [aEB, aSCG, GCA, SVA, PRC, GEL, RBS]

Bizzi, E., Dev, P., Morasso, P. & Polit, A. (1978) Effect of load disturbances during centrally initiated movements. *Journal of Neurophysiology* 39:435–44. [aEB, AGF, GEL]

Bizzi, E. & Mussa-Ivaldi, F. A. (1990) Emergent issues in the control of multi-joint movements. In: *From neuron to action*, ed. L. Deecke, J. C. Eccles & V. B. Mountcastle. Springer-Verlag. [aEB]

Bizzi, E., Mussa-Ivaldi, F. A. & Giszter, S. (1982) Arm trajectory formation in monkeys. *Experimental Brain Research* 46:139–43. [aEB]

(1991) Computations underlying the execution of movement: A biological perspective. *Science* 253:287–91. [aEB, SG]

Bizzi, E., Polit, A. & Morasso, P. (1976) Mechanisms underlying achievement of final head position. *Journal of Neurophysiology* 39:435–44. [aEB]

Blin, O., Ferrandez, A. M., Pailhous, J. & Serratrice, G. (1991) Dopa-sensitive and Dopa-resistant gait parameters in Parkinson's disease. *Journal of Neurological Sciences* 103:51–54. [GMa]

Bloedel, J. R. (1987) Cerebellum and memory storage. Technical comments. *Science* 238:1728–29. [rJRB]

Bloedel, J. R., Bracha, V., Kelly, T. M. & Wu, J.-Z. (1991) Substrates for motor learning: Does the cerebellum do it all? In: Activity-driven changes in learning and development. ed. J. R. Wolpaw and J. T. Schmidt. *Annals of the New York Academy of Science* 627:305–18. [arJRB]

Bloedel, J. R. & Courville, J. (1981) Cerebellar afferent systems. In: *Handbook of physiology. Section I: The nervous system*, ed. V. B. Brooks. American Physiological Society. [JMB]

Bloedel, J. R. & Ebner, T. J. (1985) Climbing fiber function: Regulation of Purkinje cell responsiveness. In: *Cerebellar functions*, ed. J. R. Bloedel, J. Dichgans & W. Precht. Springer-Verlag. [arJRB]

Bloedel, J. R., Ebner, T. J. & Yu, Q.-Z. (1983) Increased responsiveness of Purkinje cells associated with climbing fiber inputs to neighboring neurons. *Journal of Neurophysiology* 50:220–39. [arJRB, cJFS]

Bloedel, J. R., Gregory, R. S. & Martin, S. H. (1972) Action of interneurons and axon collaterals in cerebellar cortex of a primate. *Journal of Neurophysiology* 35:847–63. [rJRB]

Bloedel, J. R. & Kelly, T. M. (1992) The dynamic selection hypothesis: A proposed function for cerebellar sagittal zones. In: *The cerebellum revisited*, ed. R. Llinas & C. Sotelo. Springer-Verlag, in press. [aJRB, PFCG]

Bloedel, J. R. & Roberts, W. J. (1969) Functional relationship among neurons of the cerebellar cortex in the absence of anesthesia. *Journal of Neurophysiology* 32:75–84. [rJRB]

(1971) The action of climbing fibers in the feline cerebellar cortex. *Journal of Neurophysiology* 34:17–31. [rJRB, AMS]

Boegman, R., Parent, A. & Hawkes, R. (1988) Zonation in the rat cerebellar cortex: Patches of high acetylcholinesterase activity in the granular layer are congruent with Purkinje cell compartments. *Brain Research* 448:237–51. [aJRB, NHB]

Borrett, D. S., Yeap, T. H. & Kwan, H. C. (1992) Neural networks, movements and Parkinson's disease, submitted. [DSB]

Botterman, B. R., Binder, M. D. & Stuart, D. G. (1978) Functional anatomy of the association between motor units and muscle receptors. *American Journal of Zoology* 18:135–52. [aSCG]

Bouisset, S. & Zattara, M. (1987) Biomechanical study of the programming of anticipatory postural adjustments associated with voluntary movement. *Journal of Biomechanics* 20:735–42. [ADK]

(1990) Segmental movement as a perturbation to balance? Facts and concepts. In: *Multiple muscle systems*, ed. J. M. Winters & S. L.-Y. Woo. Springer-Verlag. [ADK, JMW]

Bower, J. M. (1990) Reverse engineering the nervous system: An anatomical, physiological, and computer-based approach. In: *An introduction to neural and electronic networks*, ed. S. F. Zornetzer, J. L. Davis & C. Lau. Academic. [WAM]

Bower, J. M. & Kassel, J. (1990) Variability in tactile projection patterns to cerebellar folia Crus IIA in the Norway rat. *Journal of Comparative Neurology* 302:768–78. [JMB]

Box, G. E. P. & Jenkins, G. M. (1976) Time series analysis: *Forecasting and control*. Prentice-Hall. [PDN]

Boyd, I. A. & Roberts, T. D. M. (1953) Proprioceptive discharges from stretch-receptors in the knee joint of the cat. *Journal of Physiology (London)* 122:38–58. [aSCG]

Bracewell, R. M., Barash, S., Massoni, P. & Andersen, R. A. (1991) Neurones in the macaque lateral intraparietal cortex (LIP) appear to encode the next intended saccade. *Society for Neuroscience Abstracts* 17:1282. [JWG, APo]

Bracha, V., Wu, J.-Z., Cartwright, M. & Bloedel, J. R. (1991) Selective involvement of the spinal trigeminal nucleus in the conditioned nictitating membrane reflex of the rabbit. *Brain Research* 556:317–20. [arJRB]

Bracke-Tolkmitt, R., Linden, A., Canavan, A. G. M., Rockstroh, B., Scholz, E., Wessel, K. & Diener, H.-C. (1989) The cerebellum contributes to mental skills. *Behavioral Neuroscience* 103:442–46. [aJRB, MH]

Brain, W. R. (1941) Visual disorientation with special reference to lesions to the right cerebral hemisphere. *Brain* 64:244–72. [DPC]

Braitenberg, V. (1983) The cerebellum revisited. *Journal of Theoretical Neurobiology* 2:237–41. [VB]

(1987) The cerebellum and the physics of movement: Some speculations. In: *Cerebellum and neuronal plasticity*, ed. M. Glickstein, C. Yeo & J. Stein. Plenum. [VB]

Braitenberg, V. & Onesto, N. (1962) The cerebellar cortex as a timing organ. Discussion of an hypothesis. In: *Proceedings of the First International Conference on Medical Cybernetics*, Naples, Italy. [VB]

Braitenberg, V. & Preisel, H. (1992) The cerebellum and the physics of movement, in preparation. [VB]

Brand, S., Dahl, A. L. & Mugnaini, F. (1976) The length of parallel fibers in the cat cerebellar cortex. An experimental light and electron microscopic study. *Experimental Brain Research* 26:39–58. [aJRB]

Bras, H., Cavallari, P. & Jankowska, E. (1988) Demonstration of initial axon collaterals of cells of origin of the ventral spinocerebellar tract in the cat. *Journal of Comparative Neurology* 273:584–92. [aDAM]

Bras, H., Cavallari, P., Jankowska, E. & Kubin, L. (1989a) Morphology of midlumbar interneurones relaying information from group II muscle afferents in the cat spinal cord. *Journal of Comparative Neurology* 290:1–15. [aDAM]

Bras, H., Cavallari, P., Jankowska, E. & McCrea, D. (1989b) Comparison of the effects of monoamines on transmission in spinal pathways from group I and II muscle spindle afferents. *Experimental Brain Research* 76:27–37. [aDAM]

Bridgeman, B. (1980) Temporal response characteristics of cells in monkey striate cortex measured with metacontrast masking and brightness discrimination. *Brain Research* 196:347–64. [BB]

(1982) Multiplexing in single cells of the alert monkey's visual cortex during brightness discrimination. *Neuropsychologia* 20:33–42. [BB]

Bridgeman, B. & Artim, J. (1983) Information flow through single cells of cat striate cortex. *Neuroscience Abstracts* 9:619. [BB]

Brink, E. (1988) Segmental organization of the upper cervical cord. In: *Control of head movement*, ed. B. W. Peterson & F. J. R. Richmond. Oxford University Press. [GEL]

Brink, E., Jankowska, E., McCrea, D. A. & Skoog, B. (1983) Inhibitory interactions between interneurones in reflex pathways from group Ia and group Ib afferents in the cat. *Journal of Physiology (London)* 343:361–73. [aDAM, DBul]

Brink, E. & Mackel, R. (1987) Sensorimotor performance of the hand during peripheral nerve regeneration. *Journal of the Neurological Sciences* 77:249–66. [aSCG]

Brochu, G., Maler, L. & Hawkes, R. (1990) ZebrinII: A polypeptide antigen expressed selectively by Purkinje cells reveals compartments in rat and fish cerebellum. *Journal of Comparative Neurology* 291:538–52. [NHB]

Brodal, P. (1978) Principles of organisation of the monkey corticopontine projection. *Brain* 64:244–72. [aJFS]

Brody, B. A. & Pribram, K. H. (1978) The role of frontal and parietal cortex in cognitive processing: Tests of spatial and sequence functions. *Brain* 101:607–33. [DSL]

Brooke, J. D. & McIlroy, W. E. (1989) Effect of knee joint angle on a heteronymous Ib reflex in the human lower limb. *Canadian Journal of Neurological Sciences* 16:58–62. [aSCG]

Brooks, R. A. (1991) Elephants don't play chess. In: *Designing autonomous agents*, ed. A. P. Maes. MIT Press. [SG]

Brooks, V. B. (1984) Cerebellar functions in motor control. *Human Neurobiology* 2:251–60. [aJRB]

(1986) *The neural basis of motor control*. Oxford University Press. [aGEA, AAF]

Brooks, V. B., Kozlovskaya, I. B., Atkin, A., Horvath, F. E. & Uno, M. (1973) Effects of cooling dentate nucleus on tracking-task performance in monkeys. *Journal of Neurophysiology* 36:974–95. [aJRB]

Brooks, V. B. & Thach, W. T. (1981) Cerebellar control of posture and movement. In: *Handbook of physiology*, section I: The nervous system, vol. 2: *Motor control*, ed. J. M. Brookhart & V. B. Mountcastle. American Physiology Society. [aJRB]

Brooks, V. B. & Wilson, V. J. (1959) Recurrent inhibition in the cat's spinal cord. *Journal of Physiology (London)* 146:380–91. [TMH]

Brotchie, P. R. & Andersen, R. A. (1991) Body-centered coordinate system in posterior parietal cortex. *Society for Neuroscience Abstracts* 17:511.1. [rJFS, RAA, APo]

Brotchie, P., Iansek, R. & Horne, M. (1991a) The motor function of the monkey globus pallidus: I. Neuronal discharge and parameters of movement. *Brain* 114:1667–83. [RI]

(1991b) The motor function of the monkey globus pallidus: II. Cognitive aspects of movement and phasic neuronal activity. *Brain* 114:1685–1702. [RI]

(1991c) A neural network model of neural activity in the monkey globus pallidus. *Neuroscience Letters* 131:33–36. [RI]

Brown, J. (1977) *Mind, brain, and consciousness.* Academic Press. [JMF]

Brown, K., Lee, J. & Ring, P. A. (1954) The sensation of passive movement at the metatarso-phalangeal joint of the great toe in man. *Journal of Physiology (London)* 126:448–58. [aJRB]

Brown, T. G. (1914) On the nature of the fundamental activity of the nervous centres; together with an analysis of the conditioning of rhythmic activity in progression, and a theory of the evolution of function in the nervous system. *Journal of Physiology* 48:18–46. [aDAM]

Brown, T. H., Kairiss, E. W. & Keenan, C. L. (1990) Hebbian synapses: Biophysical mechanisms and algorithms. *Annual Review of Neuroscience* 13:475–511. [aGEA]

Bruce, C. J. & Goldberg, M. E. (1985) Primate frontal eye field. I. Single neurons discharging before saccades. *Journal of Neurophysiology* 53:603–35. [APo]

Brüwer, M. & Cruse, H. (1990) A network model for the control of the movement of a redundant manipulator. *Biological Cybernetics* 62:549–55. [aGEA]

Brüwer, M. & Dean, J. (1992) Control of human arm movements in two dimensions: A comparison of movements with 2 and 3 mobile joints. *Proceedings of the 20th Gottingen Neurobiology Conference.* Thieme Verlag, in press. [JDe]

Buford, J. A. & Smith, J. L. (1990) Adaptive control for backward quadrupedal walking. 2. Hindlimb muscle synergies. *Journal of Neurophysiology* 64:756–66. [GMc]

Buford, J. A., Zernicke, R. F. & Smith, J. L. (1990) Adaptive control for backward quadrupedal walking. 1. Posture and kinematics. *Journal of Neurophysiology* 64:745–55. [GMc]

Bullock, D. & Contreras-Vidal, J. L. (1991) How spinal neural networks reduce discrepancies between motor intention and motor realization. CAS/CNS Technical Report 91-023. In: *Variability and motor control,* ed. K. Newell & D. Corcos. Human Kinetics. [DBul]

Bullock, D., Contreras-Vidal, J. L. & Grossberg, S. (1992) Equilibria and dynamics of a neural network model for opponent muscle control. CAS/CNS Technical Report 92-017. In: *Neural networks in robotics,* ed. K. Goldberg & G. Bekey. Kluwer Academic. [DBul]

Bullock, D. & Grossberg, S. (1988a) Neural dynamics of planned arm movements: Emergent variants and speed-accuracy properties during trajectory formation. In: *Neural network and natural intelligence,* ed. S. Grossberg. MIT Press. [AAF]

(1988b) Neuromuscular realization of planned trajectories. *Neural Networks* 1(suppl. 1);329. [DBul]

(1989) VITE & FLETE: Neural modules for trajectory formation and postural control. In: *Volitional action,* ed. W. A. Hershberger. North-Holland/Elsevier. [DBul, TMH, DSL]

(1990) Spinal network computations enable independent control of muscle length and joint compliance. In: *Advanced neural computers,* ed. R. Eckmiller. North-Holland. [DBul]

(1991) Adaptive neural networks for control of movement trajectories invariant under speed and force rescaling. *Human Movement Science* 10:3–53. [DBul, DSL, GJvIS]

(in press) Emergence of tri-phasic muscle activation from the nonlinear interactions of central and spinal neural network circuits. *Human Movement Science* 11. [DBul]

Bures, J., Buresova, O. & Kr˙vanek, J. (1988) *Brain and behavior.* Academia. [MEI]

Burgess, P. R. & Clark, F. J. (1969) Characteristics of knee joint receptors in the cat. *Journal of Physiology (London)* 203:317–35. [aSCG]

Burgess, P. R., Wei, J. Y., Clark, F. J. & Simon, J. (1982) Signaling of kinesthetic information by peripheral sensory receptors. *Annual Review of Neuroscience* 5:171–87. [aSCG]

Burke, D. (1981) The activity of human muscle spindle endings in normal motor behavior. *International Review of Physiology* 25:91–126. [arSCG]

Burke, D., Aniss, A. M. & Gandevia, S. C. (1987) In-parallel and in-series behavior of human muscle spindle endings. *Journal of Neurophysiology* 58:417–26. [aSCG]

Burke, D., Dickson, H. G. & Skuse, N. F. (1991) Task-dependent changes in the responses to low-threshold cutaneous afferent volleys in the human lower limb. *Journal of Physiology (London)* 432:445–58. [aSCG]

Burke, D., Gandevia, S. C. & Macefield, G. (1988) Responses to passive movement of receptors in joint, skin and muscle of the human hand. *Journal of Physiology (London)* 402:347–61. [aSCG]

Burke, D., Gandevia, S. C. & McKeon, B. (1983) The afferent volleys responsible for spinal proprioceptive reflexes in man. *Journal of Physiology* 339:535–52. [rSCG]

(1984) Monosynaptic and oligosynaptic contributions to human ankle jerk and H-reflex. *Journal of Neurophysiology* 52:435–48. [rSCG]

Burke, D., Gandevia, S. C., McKeon, B. & Skuse, N. F. (1982) Interactions between cutaneous and muscle afferent projections to cerebral cortex in man. *Electroencephalography and Clinical Neurophysiology* 53:349–60. [aSCG]

Burke, D., Gracies, J. M., Mazevet, D., Meunier, S. & Pierrot-Deseilligny, E. (1992) Convergence of descending and various peripheral inputs onto common propriospinal-like neurones in man. *Journal of Physiology (London)* 449:655–710. [aSCG, DBur]

Burke, D., Hagbarth, K.-E. & Löfstedt, L. (1978a) Muscle spindle activity in man during shortening and lengthening contractions. *Journal of Physiology (London)* 277:131–42. [aSCG]

(1978b) Muscle spindle responses in man to changes in load during accurate position maintenance. *Journal of Physiology (London)* 276:159–64. [aSCG]

Burke, D., Hagbarth, K.-E., Löfstedt, L. & Wallin, B. G. (1976a) The responses of human muscle spindle endings to vibration of non-contracting muscles. *Journal of Physiology (London)* 261:673–93. [aSCG]

(1976b) The responses of human muscle spindle endings to vibration during isometric contraction. *Journal of Physiology (London)* 261:695–711. [aSCG]

Burke, D., Hagbarth, K.-E. & Skuse, N. F. (1979) Voluntary activation of spindle endings in human muscles temporarily paralysed by nerve pressure. *Journal of Physiology (London)* 287:329–36. [aSCG]

Burke, D., McKeon, B. & Skuse, N. F. (1981) Dependence of the Achilles tendon reflex on the excitability of spinal reflex pathways. *Annals of Neurology* 10:551–56. [rSCG]

Burke, D. McKeon, B., Skuse, N. F. & Westerman, R. A. (1980) Anticipation and fusimotor activity in preparation for a voluntary contraction. *Journal of Physiology (London)* 306:337–48. [aSCG]

Burton, J. E. & Onoda, N. (1977) Interpositus neurons discharge in relation to a voluntary movement. *Brain Research* 121:167–72. [aJRB]

Bushnell, M. C., Goldberg, M. E. & Robinson, D. L. (1981) Behavioral enhancement of visual responses in monkey cerebral cortex. I. Modulation in posterior parietal cortex related to selective visual attention. *Journal of Neurophysiology* 46:755–72. [aJFS]

Bussel, B., Morin, C. & Pierrot-Deseilligny, E. (1978) Mechanism of monosynaptic reflex reinforcement during Jendrassik Manoeuvre in man. *Journal of Neurology, Neurosurgery and Psychiatry* 41:40–44. [rSCG]

Buys, E. R., Lemon, R. N., Mantel, G. W. B. & Muir, R. B. (1986) Selective facilitation of different hand muscles by single corticospinal neurones in the conscious monkey. *Journal of Physiology (London)* 381:529–49. [aGEA, aEEF, cSCG]

Cafarelli, D. (1988) Force sensation in fresh and fatigued human skeletal muscle. *Exercise and Sports Science Review* 16:139–68. [aSCG]

Calancie, B. M. & Stein, R. B. (1988) Microneurography for the recording and selective stimulation of afferents: An assessment. *Muscle and Nerve* 11:638–44. [aSCG]

Cameron, W. E., Binder, M. D., Botterman, B. R., Reinking, R. M. & Stuart, D. G. (1980) Motor unit-muscle spindle interactions in active muscles of decerebrate cats. *Neuroscience Letters* 19:55–60. [aSCG]

Caminiti, R., Johnson, P. B. & Urbano, A. (1990) Making arm movements within different parts of space: Dynamic aspects in the primate motor cortex. *Journal of Neuroscience* 10:2039–58. [aEB]

Cannon, S. C. & Robinson, D. A. (1985) An improved neural network model for the neural integrator of the oculomotor system: More realistic neuron behavior. *Biological Cybernetics* 53:93–108. [MF]

Canny, J. F. (1987) *The complexity of robot motion planning.* Doctoral Dissertation, Massachusetts Institute of Technology. MIT Press. [CIC]

Capaday, C. & Stein, R. B. (1986) Amplitude modulation of the soleus H-reflex in the human during walking and standing. *Journal of Neuroscience* 6:1308–13. [arSCG, TRN]

(1987) Difference in the amplitude of the human soleus H reflex during walking and running. *Journal of Physiology (London)* 392:513–22. [arSCG]

Cappa, S., Sterzi, R., Villar, G. & Bisiach, E. (1987) Remission of hemineglect and anosognosia during vestibular stimulation. *Neuropsychologia* 25:774–80. [aJFS]

Carlson-Kuhta, P. & Smith, J. L. (1990) Scratch responses in normal cats: Hindlimb kinematics and muscle synergies. *Journal of Neurophysiology* 64:1653–67. [CAP]

Carpenter, M. B., Nakano, K. & Kim, R. (1976) Nigrothalamic projections in the monkey demonstrated by autoradiographic technics. *Journal of Comparative Neurology* 165:401–16. [aGEA]

Carr, J. N., Louca, D. & Grobstein, P. (1991) Directed movement in the frog: Explorations using back propagation networks. *Society for Neuroscience Abstracts* 17:1578. [PG]

Carter, R. R., Crago, P. E. & Keith, M. W. (1990) Stiffness regulation by reflex action in the normal human hand. *Journal of Neurophysiology* 64:105–18. [TRN]

Cavada, C. & Goldman-Rakic, P. S. (1989a) Posterior parietal cortex in rhesus monkey. I. Parcellation of areas based on distinctive limbic and sensory corticocortical connections. *Journal of Comparative Neurology* 287:393–421. [aJFS, WAM]

(1989b) Posterior parietal cortex in rhesus monkey. II. Evidence for segregated corticocortical networks linking sensory and limbic areas with frontal lobe. *Journal of Comparative Neurology* 287:422–45. [aJFS, MSG]

(1991) Topographical segregation of corticostriatal projections from posterior parietal subdivisions in the macaque monkey. *Neuroscience* 42:683–96. [MSG]

Cavallari, P., Edgley, S. A. & Jankowska, E. (1987) Post-synaptic actions of midlumbar interneurones on motoneurones of hind-limb muscles in the cat. *Journal of Physiology* 389:675–89. [aDAM]

Cavallari, P., Fournier, E., Katz, R., Malmgren, K., Pierrot-Deseilligny, E. & Shindo, M. (1985) Cutaneous facilitation of transmission in Ib reflex pathways in the human upper limb. *Experimental Brain Research* 60:197–99. [aSCG]

Cavallari, P., Fournier, E., Katz, R., Pierrot-Deseilligny, E. & Shindo, M. (1984) Changes in reciprocal Ia inhibition from wrist extensors to wrist flexors during voluntary movement in man. *Experimental Brain Research* 56:574–76. [aSCG]

Cavanagh, P. R., Simoneau, G. G. & Ulbrecht, J. S. (1992) Ulceration, unsteadiness, and uncertainty: The biomechanical consequences of Diabetes Mellitus. *Journal of Biomechanics*, in press. [PRC]

Cavanagh, P. R. & Ulbrecht, J. S. (1991) Biomechanics of the diabetic foot: A quantitative approach to the assessment of neuropathy, deformity, and plantar pressure. In: *Disorders of the foot & ankle*, 2nd ed., ed. M. H. Jahss & W. B. Saunders. [PRC]

Chambers, W. W. & Sprague, J. M. (1951) Differential effects of cerebellar anterior lobe cortex and fastigial nuclei on postural tonus in the cat. *Science* 114:324–25. [aJRB, RFT]

(1955) Functional localization in the cerebellum. II. Somatotopic organization in cortex and nuclei. *Archives of Neurology and Psychiatry* 74:653–80. [aJRB, RFT]

Chandler, C., Hewit, J. & Miller, S. (1982) Computers, brains and the control of movement. *Trends in Neuroscience* 5:376. [aGEA]

Chang, H. T., Wilson, C. J. & Kitai, S. T. (1981) Single neostriatal efferent axons in the globus pallidus: A light and electron microscopic study (abstract). *Science* 213:915–18. [aGEA]

(1982) A golgi study of rat neostriatal neurons: Light microscopic analysis. *Journal of Comparative Neurology* 208:107–26. [aGEA]

Chan-Palay, V. (1973) A light microscopic study of the cytology and organization of neurons in the simple mammalian nucleus lateralis: Columns and swirls. *Zeitschrift für Anatomie und Entwicklungsgeschichte* 141:125–250. [aJRB]

(1977) *Cerebellar dentate nucleus: Organization, cytology and transmitters.* Springer-Verlag. [aJRB]

Chapman, C. E., Spidalieri, G. & LaMarre, Y. (1986) Activity of dentate neurons during arm movements triggered by visual, auditory, and somesthetic stimuli in the monkey. *Journal of Neurophysiology* 55:203–26. [aJRB]

Chapula, L. M. (1991) Visual function of the pulvinar. In: *The neural basis of visual function*, ed. A. G. Leventhal. CRC Press. [DPC]

Charpentier, A. (1891) Analyse experimentale de quelques elements de la sensation de poids. *Archives de Physiologie Normale et Pathologique* 3:122–35. [AMG]

Chelazzi, L., Ghirardi, M., Rossi, F., Strata, P. & Tempia, F. (1990) Spontaneous saccades and gaze-holding ability in the Pigmented Rat. II. Effects of localized cerebellar lesions. *European Journal of Neuroscience* 2:1085–94. [aJRB]

Chen, D.-F., Hyland, B., Maier, V., Palmeri, A. & Wiesendanger, M. (1991) Comparison of neural activity in the supplementary motor cortex and in the primary motor cortex in monkeys performing a choice-reaction task. *Somatosensory and Motor Research* 8:27–44. [aGEA, aEEF]

Cheney, P. D. & Fetz, E. E. (1980) Functional classes of primate corticomotoneuronal cells and their relation to active force. *Journal of Neurophysiology* 44:773–91. [aEB, aEEF]

Cheney, P. D., Fetz, E. E. & Mewes, K. (1991) Neural mechanisms underlying corticospinal and rubrospinal control of limb movements. *Progress in Brain Research* 87:213–52. [rEEF]

Cheney, P. D., Fetz, E. E. & Palmer, S. S. (1985) Patterns of facilitation and suppression of antagonist forelimb muscles from motor cortex sites in the awake monkey. *Journal of Neurophysiology* 53:805–20. [aEEF]

Cheney, P. D., Mewes, K. & Fetz, E. E. (1988) Encoding of motor parameters by corticomotoneuronal (CM) and rubromotoneuronal (RM)

cells identified by spike-triggered averaging in the awake monkey. *Behavioral Brain Research* 28:181–91. [aEEF, AL]

Cherubini, E., Herrling, P. L., Lanfumey, L. & Stanzione, P. (1988) Excitatory amino acids in synaptic excitation of rat striatal neurones in vitro. *Journal of Physiology (London)* 400:677–90. [aGEA]

Chomsky, N. (1986) *Knowledge of language: Its nature, origin and uses.* Praeger. [JER]

Claparéde, E. (1901) Experiences sur la vitesse du soulevement des poids de volumes differents. *Archives de Psychologie de la Suisse Romande* 1:69–94. [AMG]

Clare, M. H. & Landau, W. M. (1964) Fusimotor function, part V. Reflex reinforcement under fusimotor block in normal subjects. *Archives of Neurology* 10;123–27. [rSCG]

Clark, F. J. (1975) Information signaled by sensory fibers in medial articular nerve. *Journal of Neurophysiology* 38:1464–72. [aSCG]

Clark, F. J. & Burgess, P. R. (1975) Slowly adapting receptors in cat knee joint: Can they signal joint angle? *Journal of Neurophysiology* 38:1448–63. [aSCG]

Clark, F. J., Burgess, R. C. & Chapin, J. W. (1986) Proprioception with the proximal interphalangeal joint of the index finger. Evidence for a movement sense without a static-position sense. *Brain* 109:1195–1208. [aSCG]

Clark, F. J., Burgess, R. C., Chapin, J. W. & Lipscomb, W. T. (1985) Role of intramuscular receptors in the awareness of limb position. *Journal of Neurophysiology* 54:1529–40. [aSCG]

Clark, F. J., Grigg, P. & Chapin, J. W. (1989) The contribution of articular receptors to proprioception with the fingers in humans. *Journal of Neurophysiology* 61:186–93. [aSCG]

Clark, F. J., Horch, K. W., Bach, S. M. & Larson, G. F. (1979) Contributions of cutaneous and joint receptors to static knee-position sense in man. *Journal of Neurophysiology* 42:877–88. [aSCG]

Clark, R. E., Zhang, A. A. & Lavond, D. G. (1992) Reversible lesions of the cerebellar interpositus nucleus during acquisition and retention of a classically conditioned behavior. *Behavioral Neuroscience*, in press. [rJRB, RFT]

Clarke, T. L. & Ronayne, T. (1991) Categorical approach to machine learning. *Proceedings of the 1991 IEEE conference of systems, man and cybernetics.* Institute of Electrical and Electronics Engineers. [TLC]

Cleland, C. & Rymer, W. (1990) Neural mechanisms underlying the clasp knife reflex in the cat. I. Characteristics of the reflex. *Journal of Neurophysiology* 64:1303–18. [aDAM]

Coburn, K. L., Ashford, J. W. & Fuster, J. M. (1990) Visual response latencies in temporal lobe structures as a function of stimulus information load. *Behavioral Neuroscience* 104:62–73. [JMF]

Cogan, D. G. (1953) Ophthalmic manifestations of bilateral non-occipital cerebral lesions. *British Journal of Ophthalmology* 49:281–97. [JWG]

Cohen, L. A. (1961) Role of eye and neck proprioceptive mechanisms in body orientation and motor coordination. *Journal of Neurophysiology* 24:1–11. [DPC]

Colby, C. L. & Duhamel, J.-R. (1991) Heterogeneity of extrastriate visual areas and multiple parietal areas in the macaque monkey. *Neuropsychologia* 29:497–515. [CLC]

Colby, C. L., Duhamel, J.-R. & Goldberg, M. E. (1992) The analysis of visual space by the lateral intraparietal area of the monkey: The role of extraretinal signals. *Progress in Brain Research*, in press. [CLC]

Cole, J. D. (1986) Observations on the sense of effort in a man without large myelinated cutaneous and proprioceptive sensory fibres below the neck. *Journal of Physiology (London)* 382:80P. [aSCG, PRC]

Cole, J. D., Katifi, H. A. & Sedgwick, E. M. (1986) Observations on a man without large myelinated fibre sensory input from below the neck. *Journal of Physiology (London)* 376:47P. [aSCG, PRC]

Cole, K. J. & Abbs, J. H. (1987) Kinematic and electromyographic responses to perturbation of a rapid grasp. *Journal of Neurophysiology* 57:1498–1510. [aSCG]

Colebatch, J. G. & McCloskey, D. I. (1987) Maintenance of constant arm position or force: Reflex and volitional components in man. *Journal of Physiology (London)* 386:247–61. [aSCG]

Colgate, J. E. (1988) The control of dynamically interacting systems. Doctoral Dissertation, Department of Mechanical Engineering. MIT Press. [aEB]

Colgate, J. E. & Hogan, N. (1988) Robust control of dynamically interacting systems. *International Journal of Control* 48:65–88. [aEB]

Connolly, C. I., Burns, J. B. & Weiss, R. (1990) Path planning using Laplace's equation. In: *Proceedings of the 1990 IEEE International Conference on Robotics and Automation* 3. Institute of Electrical and Electronics Engineers Computer Society Press. [CIC]

Conway, B. A., Hultborn, H. & Kiehn, O. (1987) Proprioceptive input resets central locomotor rhythm in the spinal cat. *Experimental Brain Research* 68:643–56. [VD]

Cook, D. & Kesner, R. P. (1988) Caudate nucleus and memory for egocentric localization. *Behavioral Neural Biology* 49:332–43. [DI]

Cooke, J. D. (1980) The role of stretch reflexes during active movements. *Brain Research* 181:493–97. [aSCG]

Cooke, K. L. & Grossman, Z. (1982) Discrete delay, distributed delay and stability switches. *Journal of Mathematical Analysis and Applications* 86:592–627. [AB]

Cools, A. R., Jaspers, R., Schwarz, M., Sontag, K. H., Vrijmoed-deVries, M. & van den Bercken, J. (1984) Basal ganglia and switching motor programs. In: *The basal ganglia. Structure and function*, ed. J. S. McKenzie, R. E. Kemm & L. N. Wilcock. Plenum. [aGEA]

Corbetta, M., Miezin, F. M., Shulman, G. L. & Petersen, S. E. (1991) Selective attention modulates extrastriate visual regions in humans during visual feature discrimination and recognition. In: *Exploring brain functional anatomy with positron tomography*, ed. D. J. Chadwick & J. Whelan. Wiley. [RE]

Cordo, P. J. (1987) Mechanisms controlling accurate changes in elbow torque in humans. *Journal of Neuroscience* 7(2):432–42. [PJC]

(1990) Kinesthetic control of a multijoint movement sequence. *Journal of Neurophysiology* 63:161–72. [aSCG]

Cordo, P. J. & Flanders, M. (1989) The sensory basis of target acquisition. *Trends in Neuroscience* 12:110–17. [PJC]

Corin, M. S. & Bender, M. B. (1972) Mislocalization in visual space. *Archives of Neurology* 27:252–62. [DPC]

Coulter, J. D. & Jones, E. G. (1977) Differential distribution of corticospinal projections from individual cytoarchitectonic fields in the monkey. *Brain Research* 129:335–40. [aEEF]

Courville, J. & Diakiw, N. (1976) Cerebellar cortico-nuclear projection in the cat. The vermis of the anterior and posterior lobes. *Brain Research* 110:1–20. [aJRB]

Crago, P. E., Houk, J. C. & Hasan, Z. (1976) Regulatory actions of the human stretch reflex. *Journal of Neurophysiology* 39:925–35. [SVA]

Crago, P. E., Houk, J. C. & Rymer, W. Z. (1982) Sampling of total muscle force by tendon organs. *Journal of Neurophysiology* 47:1069–83. [aSCG]

Crammond, D. J. & Kalaska, J. F. (1989) Neuronal activity in primate parietal cortex area 5 varies with intended movement direction during an instructed-delay period. *Experimental Brain Research* 76:458–62. [JFK, PM]

Craske, B. (1977) Perception of impossible limb positions induced by tendon vibration. *Science* 196:71–73. [aSCG]

Crenna, P. & Frigo, C. (1987) Excitability of the soleus H-reflex arc during walking and stepping in man. *Experimental Brain Research* 66:49–60. [aSCG]

Crepel, F. & Jaillard, D. (1991) Pairing of pre- and postsynaptic activities in cerebellar Purkinje cells induces long-term changes in synaptic efficacy in vitro. *Journal of Physiology* 432:123–41. [rJRB, cJFS]

Crick, F. (1984) Function of the thalamic reticular complex. The searchlight hypothesis. *Proceedings of the National Academy of Science* USA 81:4586–90. [aJFS]

(1989) The recent excitement about neural networks. *Nature* 337:129–32. [RE]

Critchley, M. (1953) *The parietal lobes.* Hafner. [aJFS]

Crone, C., Hultborn, H., Jespersen, B. & Nielsen, J. (1987) Reciprocal Ia inhibition between ankle flexors and extensors in man. *Journal of Physiology (London)* 389:163–85. [aSCG]

Crone, C., Hultborn, H., Kiehn, O., Mazieres, L. & Wigström, H. (1988) Maintained changes in motoneuronal excitability by short-lasting synaptic inputs in the decerebrate cat. *Journal of Physiology (London)* 405:321–43. [aDAM, AGF]

Crone, C. & Nielsen, J. (1989) Spinal mechanisms in man contributing to reciprocal inhibition during voluntary dorsiflexion of the foot. *Journal of Physiology (London)* 416:255–72. [aSCG]

Cross, M. J. & McCloskey, D. I. (1973) Position sense following surgical removal of joints in man. *Brain Research* 55:443–45. [aSCG]

Cruse, H. (1985) Which parameters control the leg movements of a walking insect? I. Velocity control during the stance phase. *Journal of Experimental Biology* 116:343–55. [JDe]

(1986) Constraints for joint angle control of the human arm. *Biological Cybernetics* 54:125–32. [JDe]

Cruse, H. & Graham, D. (1985) Models for the analysis of walking in arthropods. In: Coordination of motor behaviour. *Society for Experimental Biology Seminar* Series 24, ed. B. M. H. Bush & F. Clarac. Cambridge University Press. [JDe]

Cruse, H., Wischmeyer, E., Bruwer, M., Brockfeld, P. & Dress, A. (1990) On the cost functions for the control of the human arm movement. *Biological Cybernetics* 62:519–28. [JDe]

Crutcher, M. D. & Alexander, G. E. (1990) Movement-related neuronal activity selectively coding either direction or muscle pattern in three

motor areas of the monkey. *Journal of Neurophysiology* 64:151–63. [aGEA, aEEF, JMF]

Crutcher, M. D. & DeLong, M. R. (1984a) Single cell studies of the primate putamen I. Functional organization. *Experimental Brain Research* 53:233–43. [aGEA, CIC, MSG]

Crutcher, M. D. & DeLong, M. R. (1984b) Single cell studies of the primate putamen II. Relations to direction of movement and pattern of muscular activity. *Experimental Brain Research* 53:244–58. [aGEA, CIC]

Cullheim, S. & Kellerth, J.-O. (1978) A morphological study of the axons and recurrent axon collaterals of cat alpha-motoneurones supplying different hindlimb muscles. *Journal of Physiology (London)* 281:285–99. [TMH]

Cummins, R. & Schwarz, G. (1991) Connectionism, computation, and cognition. In: *Connectionism and the philosophy of mind*, ed. T. Horgan & J. Tienson. Kluwer. [GS]

Currie, S. N. & Stein, P. S. G. (1990) Cutaneous stimulation evokes long-lasting excitation of spinal interneurons in the turtle. *Journal of Neurophysiology* 64:1134–48. [CAP]

Dahhaoui, M., Caston, J., Auvray, N. & Reber, A. (1990) Role of the cerebellum in an avoidance conditioning task in the rat. *Physiology & Behavior* 47:1175–80. [rJRB]

DARPA Neural Network Study (1988) AFCEA International Press. [HB]

Darton, K., Lippold, O. C. J., Shahani, M. & Shahani, U. (1985) Long-latency spinal reflexes in humans. *Journal of Neurophysiology* 53:1604–18. [JDu]

Davey, N. J. & Ellaway, P. H. (1989) Facilitation of individual γ-motoneurones by the discharge of single slowly adapting type 1 mechanoreceptors in cats. *Journal of Physiology (London)* 411:97–114. [aSCG]

Davidoff, R. & Hackman, J. (1984) Spinal inhibition. In: *Handbook of the spinal cord*, ed. R. A. Davidoff. Marcel Dekker. [aDAM]

Davies, J. G. M., Kirkwood, P. A. & Sears, T. A. (1985) The detection of monosynaptic connexions from inspiratory bulbospinal neurones to inspiratory motoneurones in the cat. *Journal of Physiology* 368:33–62. [rEEF, PAK]

Davis, C. M. & Roberts, W. (1976) Lifting movements in the size-weight illusion. *Perception and Psychophysics* 20:33–36. [HER]

Dawson, M. R. W. (1991) The how and why of what went where in apparent motion: Modeling solutions to the motion correspondence problem. *Psychological Review* 98:569–603. [MRWD]

Day, B. L. & Marsden, C. D. (1982) Accurate repositioning of the human thumb against unpredictable dynamic loads is dependent upon peripheral feedback. *Journal of Physiology (London)* 327:393–407. [aEB, aSCG, AGF, MH]

Day, B. L., Marsden, C. D., Obeso, J. A., Rothwell, J. C. (1984) Reciprocal inhibition between the muscles of the human forearm. *Journal of Physiology (London)* 349:519–34. [aSCG]

Dean, J. (1984) Control of leg protraction in the stick insect: A targeted movement showing compensation for externally applied forces. *Journal of Comparative Physiology* 155:771–81. [JDe]

(1990) Coding proprioceptive information to control movement to a target: Simulation with a simple neural network. *Biological Cybernetics* 63:115–20. [JDe]

(1991a) A model of leg coordination in the stick insect, Carausius morosus. II. Simulation of normal step patterns. *Biological Cybernetics* 64:403–11. [JDe]

(1991b) Effect of load on leg movement and step coordination of the stick insect Carausius morosus. *Journal of Experimental Biology* 159:449–71. [JDe]

Dean, J. & Brüwer, M. (1992) Control of human arm movements in two dimensions: Paths and joint control in avoiding simple linear obstacles. *Proceedings of the 20th Gottingen Neurobiology Conference*. Thieme Verlag, in press. [JDe]

Dean, J. & Cruse, H. (1986) Evidence for the control of velocity as well as position in leg protraction and retraction by the stick insect. In: *Generation and modulation of action patterns*, ed. H. Heuer & C. Fromm. Experimental Brain Research Series 15. Springer-Verlag. [JDe]

Decety, J. & Ingvar, D. H. (1990) Brain structures participating in mental simulation of motor behavior. A neuropsychological interpretation. *Acta Psychologica* 73:13–34. [PM]

Deliagina, T. G. & Feldman, A. G. (1981) Activity of Renshaw cells during fictive scratch reflex in cat. *Experimental Brain Research* 42:108–115. [CAP]

Deliagina, T. G., Feldman, A. G., Gelfand, I. M. & Orlovsky, G. N. (1975) On the role of central program and afferent inflow in the control of scratching movement in the cat. *Brain Research* 100:297–313. [MBB, AP]

Deliagina, T. G. & Orlovsky, G. N. (1980) Activity of Ia inhibitory interneurons during fictive scratch reflex in the cat. *Brain Research* 193:439–47. [CAP]

Deliagina, T., Orlovsky, G., Pavlova, G. & Popova, L. (1981) Activity of propriospinal neurons of ventral horn of upper lumbar segments during fictive scratch reflex. *Nejrofiziologija* 13:647–48. [aDAM]

DeLong, M. R., Crutcher, M. D. & Georgopoulos, A. P. (1985) Primate globus pallidus and subthalamic nucleus: Functional organization. *Journal of Neurophysiology* 53:530–43. [aGEA]

DeLuca, C. J. (1985) Control properties of motor units. *Journal of Experimental Biology* 115:125–36. [DBul]

Delwaide, P. J. & Gonce, M. (1988) Pathophysiology of Parkinson's signs. In: *Parkinson's disease and movement disorders*, ed. J. Jankovic & E. Tolosa. Urban & Schwarzenberg. [JJS]

Delwaide, P. J., Sabatino, M., Pepin, J. L. & La Grutta, V. (1988) Reinforcement of reciprocal inhibition by contralateral movements in man. *Experimental Neurology* 99:10–16. [aSCG]

Denes, G., Caviezel, F. & Semenza, C. (1982) Difficulty in reaching objects and body parts: A sensorimotor disconnexion syndrome. *Cortex* 18:165–73. [MAG]

Deniau, J. M. & Chevalier, G. (1985) Disinhibition as a basic process in the expression of striatal functions. II. The striato-nigral influence on thalamocortical cells of the ventromedial thalamic nucleus. *Brain Research* 334:227–33. [aGEA]

Denning, P. J. & Tichy, W. F. (1990) Highly parallel computation. *Science* 250:1217–22. [aGEA]

Denny-Brown, D. (1962) *The basal ganglia and their relation to disorders of movement.* Oxford University Press. [GMa]

Denny-Brown, D. & Chambers, R. A. (1958) The parietal lobes and behavior. *Research Publications for the Association of Research in Mental Disease* 36:35–117. [aJFS]

Desimone, R. & Ungerleider, L. G. (1989) Neural mechanisms of visual processing in monkeys. In: *Handbook of neuropsychology*, vol. 2, ed. F. Boller & J. Grafman. Elsevier. [RE]

Devanandan, M. S., Ghosh, S. & John, K. T. (1983) A quantitative study of muscle spindles and tendon organs in some intrinsic muscles of the hand in the bonnet monkey (Macaca radiata). *Anatomical Record* 207:263–66. [aSCG]

Devaney, R. L. (1986) *An introduction to chaotic dynamical systems.* Cummings. [DSB]

DeVito, J. L. & Anderson, M. E. (1982) An autoradiographic study of efferent connections of the globus pallidus. *Experimental Brain Research* 46:107–17. [aGEA]

DeVito, J. L., Anderson, M. E. & Walsh, K. E. (1980) A horseradish peroxidase study of afferent connections of the globus pallidus in *Macaca mulatta*. *Experimental Brain Research* 38:65–73. [aGEA]

Dichgans, J. & Diener, H. C. (1985) Clinical evidence for functional compartmentalization of the cerebellum. In: *Cerebellar functions*, ed. J. R. Bloedel, J. D. Dichgans & W. Precht. Springer. [aJRB]

Diener, H. C., Dichgans, J., Bacher, M. & Gompf, B. (1984) Quantification of postural sway in normals and patients with cerebellar diseases. *Electroencephalography and Clinical Neurophysiology* 57:134–42. [aJRB]

Dietrichs, E. & Walberg, F. (1979) The cerebellar corticonuclear and nucleocortical projections in the cat studied with anterograde and retrograde transport of horseradish peroxidase. I. The paramedian lobule. *Anatomy and Embryology* 158:13–39. [aJRB]

(1980) The cerebellar corticonuclear and nucleocortical projections in the cat as studied with anterograde and retrograde transport of horseradish peroxidase. II. Lobulus simplex, Crus I and Crus II. *Anatomy and Embryology* 161:83–103. [aJRB]

Dietz, V. (1992) Human neuronal control of automatic functional movements: Interaction between central programs and afferent input. *Physiological Reviews* 72:33–69. [VD]

Dietz, V., Berger, W. & Quintern, J. (1987) Task-dependent gating of somatosensory transmission in two different motor tasks in man: Falling and writing. *Neuroscience Letters* 75:288–92. [VD]

Dietz, V., Faist, M. & Pierrot-Deseilligny, E. (1990) Amplitude modulation of the quadriceps H-reflex in the human during the early stance phase of gait. *Experimental Brain Research* 79:221–24. [aDAM]

Dietz, V., Gollhofer, A., Kleiber, M. & Trippel, M. (1992) Regulation of bipedal stance: Dependency on "load" receptors. *Experimental Brain Research* 89:229–31. [VD]

Dietz, V., Horstmann, G. A. & Berger, W. (1989a) Interlimb co-ordination of leg muscle activation during perturbation of stance in humans. *Journal of Neurophysiology* 62:680–93. [VD]

Dietz, V., Horstmann, G. A., Trippel, M. & Gollhofer, A. (1989b) Human postural reflexes and gravity—an underwater simulation. *Neuroscience Letters* 106:350–55. [VD]

Dietz, V., Trippel, M. & Berger, W. (1991a) Reflex activity and muscle tone during elbow movements in patients with spastic paresis. *Annals of Neurology* 30:767–79. [VD]

Dietz, V., Trippel, M., Discher, M. & Horstmann, G. A. (1991b) Compensation of human stance perturbations: Selection of the appropriate electromyographic pattern. *Neuroscience Letters* 126:71–74. [VD]

Dimitrov, B., Hallett, M. & Sanes, J. N. (1989) Differential influence of posture and intentional movement on human somatosensory evoked potentials evoked by different stimuli. *Brain Research* 496:211–18. [aSCG]

Dinse, H. R., Krueger, K. & Best, J. (1990) A temporal structure of cortical information processing. *Concepts in Neuroscience* 1:199–238. [IT]

Disterhoft, J. F., Quinn, K. J., Weiss, C. & Shipley, M. T. (1985) Accessory abducens nucleus and conditioned eye retraction nictitating membrane extension in rabbit. *Journal of Neuroscience* 5:941–50. [RFT]

Divac, I., Lavail, J. H., Rakic, P. & Winston, K. R. (1977) Heterogeneous afferents to the inferior parietal lobule of the rhesus monkey revealed by the retrograde transport method. *Brain Research* 123:197–207. [aJFS]

Dodwell, P. C. (1983) The Lie transformation group model of visual perception. *Perception & Psychophysics* 34:1–16. [TLC]

Donoghue, J. P., Suner, S. & Sanes, J. N. (1990) Dynamic organization of primary motor cortex output to target muscles in adult rats. II. Rapid reorganization following motor nerve lesions. *Experimental Brain Research* 79:492–503. [aGEA]

Dore, L., Jacobson, C. D. & Hawkes, R. (1990) Organization and postnatal development of zebrin II antigenic compartmentation in the cerebellar vermis of the grey opossum, *Monodelphis domestica*. *Journal of Comparative Neurology* 291:431–49. [aJRB]

Dow, R. S. & Moruzzi, G. (1958) *The physiology and pathology of the cerebellum.* University of Minnesota Press. [rJRB]

Drew, T., Dubuc, R. & Rossignol, S. (1986) Discharge patterns of reticulospinal and other reticular neurons in chronic, unrestrained cats walking on a treadmill. *Journal of Neurophysiology* 55:375–401. [rDAM]

Driver, J. & Halligan, P. (1991) Can visual neglect operate in object-centered coordinates? *Cognitive Neuropsychology* 8:475–96. [rJFS]

Duenas, S. & Rudomin, P. (1988) Excitability changes of ankle extensor group Ia and Ib fibers during fictive locomotion in the cat. *Experimental Brain Research* 70:15–25. [aDAM]

Dugas, C., Picard, N. & Smith, A. M. (1989) Changes in simple and complex spike activity in Purkinje cells induced by slip of an object held between the thumb and forefinger. *Society for Neuroscience Abstracts* 15:613. [JMB]

Duhamel, J.-R, Colby, C. L. & Goldberg, M. E. (1991) Congruent representations of visual and somatosensory space in single neurons of monkey ventral intra-parietal cortex (area VIP). In: *Brain and space*, ed. J. Paillard. Oxford University Press. [JFK]

(1992) The updating of the representation of visual space in parietal cortex by intended eye movements. *Science* 255:90–92. [CLC, RE, JWG]

Dum, R. P. & Strick, P. L. (1991) Premotor areas: Nodal points for parallel efferent systems involved in the central control of movement. In: *Motor control: Concepts and issues*, ed. D. R. Humphrey & H.-J. Freund. Wiley. [aGEA]

Duncan, J. (1977) Response selection rules in spatial choice reaction tasks. In: *Attention and performance*, ed. S. Dornic. Erlbaum. [RWP]

Durlach, N. I., Delhorne, L. A., Wong, A., Ko, W. Y., Rabinowitz, W. M. & Hollenbach, J. (1989) Manual discrimination and identification of length by the finger-span method. *Perception & Phychophysics* 46:29–38. [FC]

Duysens, J. & Pearson, K. G. (1980) Inhibition of flexor burst generation by loading extensor muscles in walking cat. *Brain Research* 187:321–32. [VD, APr]

Duysens, J. & Stein, R. B. (1978) Reflexes induced by nerve stimulation in walking cats with implanted cuff electrodes. *Experimental Brain Research* 32:213–24. [JDu]

Duysens, J., Tax, A. A. M., Doelen, B. van der, Trippel, M. & Dietz, V. (1991) Selective activation of human soleus or gastrocnemius in reflex responses during walking and running. *Experimental Brain Research* 87:193–204. [JDu]

Duysens, J., Trippel, M., Horstmann, G. A. & Dietz, V. (1990) Gating and reversal of reflexes in ankle muscles during human walking. *Experimental Brain Research* 82:351–35. [JDu]

Dyck, P. J. & Brown, M. J. (1987) Diabetic polyneuropathy. In: *Diabetic neuropathy.* ed. P. J. Dyck, P. K. Thomas, A. K. Asbury, A. I. Winegrad & D. Porte. Saunders. [PRC]

Dyck, P. J., Karnes, J. & O'Brien, P. C. (1987a) Diagnosis, staging, and classification of diabetic neuropathy and association with other complications. In: *Diabetic neuropathy*, ed. P. J. Dyck, P. K. Thomas, A. K. Asbury, A. I. Winegrad & D. Porte. Saunders. [PRC]

Dyck, P. J., Karnes, J., O'Brien, P. C. & Zimmerman, I. R. (1987b) Detection thresholds of cutaneous sensation in humans. In: *Peripheral neuropathy*, ed. P. J. Dyck, P. K. Thomas, E. H. Lambert & R. Bunge. Saunders. [PRC]

Ebner, T. J. & Bloedel, J. R. (1984) Climbing fiber action on the responsiveness of Purkinje cells to parallel fiber inputs. *Brain Research* 309:182–86. [arJRB]

Ebner, T. J., Yu, Q.-X. & Bloedel, J. R. (1983) Increase in Purkinje cell gain associated with naturally activated climbing fiber inputs. *Journal of Neurophysiology* 50:205–19. [arJRB]

Eccles, J. C. (1969) The dynamic loop hypothesis of movement control. In: *Information processing in the nervous system*, ed. K. N. Leibovic. Springer-Verlag. [cJFS]

(1977) An instruction-selection theory of learning in the cerebellar cortex. *Brain Research* 127:327–52. [RFT]

(1982) The initiation of voluntary movements by the supplementary motor area. *Archiv für Psychiatrie und Nervenkrankheiten* 231:423–41. [aGEA]

Eccles, J. C., Eccles, R. M. & Lundberg, A. (1957) Synaptic actions on motoneurones caused by impulses in Golgi tendon organ afferents. *Journal of Physiology (London)* 138:227–52. [DBul]

Eccles, J. C., Fatt, P. & Koketsu, K. (1954) Cholinergic and inhibitory synapses in a pathway from motor-axon collaterals to motoneurones. *Journal of Physiology (London)* 126:524–62. [aDAM, DBul]

Eccles, J. C., Ito, M. & Szentagothai, J. (1967) *The cerebellum as a neuronal machine*. Springer-Verlag. [aJRB, RFT]

Eccles, J. C., Llinas, R. & Sasaki, K. (1966) The excitatory synaptic action of climbing fibres on the Purkinje cells of the cerebellum. *Journal of Physiology* 182:268–96. [cJFS]

Eccles, R. M. & Lundberg, A. (1958) Integrative pattern of Ia synaptic actions on motoneurons of hip and knee muscles. *Journal of Physiology (London)* 144:271–98. [DBul]

(1959) Synaptic action in motoneurones by afferents which may evoke the flexion reflex. *Archives Italiennes de Biologie* 97:199–221. [aDAM]

Edelman, G. M. & Mountcastle, V. B. (1978) *The mindful brain: Cortical organization and the group-selective theory of higher brain function.* MIT Press. [aJRB]

Edgley, S. & Jankowska, E. (1987) An interneuronal relay for group I and II muscle afferents in the middle lumbar segments of the cat spinal cord. *Journal of Physiology* 389:647–74. [aDAM]

Edgley, S. A., Jankowska, E. & Schefchyk, S. (1988) Evidence that mid-lumbar neurones in reflex pathways from group II afferents are involved in locomotion in the cat. *Journal of Physiology* 403:57–71. [aDAM, AL]

Edgley, S. & Wallace, N. (1989) A short-latency crossed pathway from cutaneous afferents to rat hindlimb motoneurones. *Journal of Physiology* 411:469–80. [aDAM]

Edin, B. B. & Abbs, J. H. (1991) Finger movement responses of cutaneous mechanoreceptors in the dorsal skin of the human hand. *Journal of Neurophysiology* 65:657–70. [aSCG]

Edin, B. B. & Vallbo, Å. B. (1990a) Dynamic response of human muscle spindle afferents to stretch. *Journal of Neurophysiology* 63:1297–1306. [aSCG]

(1990b) Classification of human muscle stretch receptor afferents: A Bayesian approach. *Journal of Neurophysiology* 63:1314–22. [aSCG]

(1990c) Muscle afferent responses to isometric contractions and relaxations in humans. *Journal of Neurophysiology* 63:1307–12. [aSCG]

Edin, B. E., Westling, G. & Johansson, R. S. (1992) Independent control of human finger tip forces at individual digits during precision lifting. *Journal of Physiology*, in press. [AMG]

Edwards, R. T. H., Wiles, C. M. & Mills, K. R. (1984) Quantitation of muscle contraction and strength. In: *Peripheral neuropathy*, ed. P. J. Dyck, P. K. Thomas, E. H. Lambert & R. Bunge. Saunders. [PRC]

Eich, J. (1982) A composite holographic associative recall model. *Psychological Review* 89:627–61. [BB]

Eklund, G. (1972) Position sense and state of contraction: The effects of vibration. *Journal of Neurology, Neurosurgery & Psychiatry* 35:606–11. [aSCG]

Eldred, E., Granit, R. & Merton, P. A. (1953) Supraspinal control of the muscle spindles and its significance. *Journal of Physiology (London)* 122:498–523. [rEB]

Eliasson, A. C., Gordon, A. M. & Forssberg, H. (1991) Basic coordination of manipulative forces in children with cerebral palsy. *Developmental Medicine and Child Neurology* 33:661–70. [AMG]

Ellaway, P. H. (1968) Antidromic inhibition of fusimotor neurons. *Journal of Physiology (London)* 198:39–40. [DBul]

Ellaway, P. H. & Murphy, P. R. (1980) A quantitative comparison of recurrent inhibition of alpha- and gamma-motoneurones in the cat. *Journal of Physiology (London)* 315:43–58. [DBul]

Elman, J. (1990) Finding structure in time. *Cognitive Science* 14:179–211. [JER]

Elner, A. M., Gurfinkel, V. S., Lipshits, M. I., Mamasakhlisov, G. H. & Popov, K. E. (1976) Facilitation of stretch reflex by additional support during quiet stance. *Agressologie* 17:15–20. [VD]

Emonet-Dénand, F., Hunt, C. C. & Laporte, Y. (1985) Effects of stretch on dynamic fusimotor after-effects in cat muscle spindle. *Journal of Physiology (London)* 360:201–21. [JP]

Engberg, I. (1964) Reflexes to foot muscles in the cat. *Acta Physiologica Scandinavica* 62:Suppl. 235. [aDAM]

Engberg, I., Lundberg, A. & Ryall, R. W. (1968) Reticulospinal inhibition of interneurones. *Journal of Physiology* 194:225–36. [aDAM]

Engen, T. (1971) Psychophysics. In: *Experimental psychology*, ed. J. W. Kling & L. A. Riggs. Holt, Rinehart & Winston. [HER]

Evans, A. L., Harrison, L. M. & Stephens, J. A. (1989) Task-dependent changes in cutaneous reflexes recorded from various muscles controlling finger movement in man. *Journal of Physiology (London)* 418:1–12. [aSCG]

Evarts, E. V. (1968) Relation of pyramidal tract activity to force exerted during voluntary movement. *Journal of Neurophysiology* 31:14–27. [aEEF, MEI]

(1981) Role of motor cortex in voluntary movement In: *Handbook of physiology*, vol. 2. American Physiological Society, Waverly Press. [aEEF, MEI]

Evarts, E. V., Fromm, C., Kroller, J. & Jennings, V. A. (1983) Motor cortex control of finely graded forces. *Journal of Neurophysiology* 49:1199–1215. [aEB]

Ewert, J.-P. & Gebauer, L. (1973) Grossenkonstanzphanomene im Beutefangverhalten der Erdkröte. *Bufo bufo* L. *Journal of Comparative Physiology* 85:303–15. [DI]

Farah, M. (1990) *Visual agnosia*. MIT Press. [MAG]

Feigenbaum, J. D. & Rolls, E. T. (1991) Allocentric and egocentric spatial information processing in the hippocampal formation of the behaving primate. *Psychobiology* 19:21–40. [WAM]

Feldman, A. G. (1966a) Functional tuning of the nervous system during control of movement or maintenance of a steady posture II. Controllable parameters of the muscle. *Biofizika* 11:498–508 [English translation 565–78]. [GCA, AGF, GLG, MLL, JBJS]

(1966b) Functional tuning of the nervous system during control of movement or maintenance of a steady posture. III. Mechanographic analysis of the execution by man of the simplest motor task. *Biophysics* 11:766–75. [aEB, SRG]

(1976) Control of postural length and force of a muscle: Advantages of the central co-activation of alpha and gamma static motoneurons. *Biophysics* 21:187–89. [AGF]

(1979) *Central and reflex mechanisms in the control of movement* [in Russian]. Nauka. [SVA, AAF, AGF]

(1980a) Superposition of motor programs. I. Rhythmic forearm movements in man. *Neuroscience* 5:81–90. [PRB]

(1980b) Superposition of motor programs. II. Rapid forearm movements in man. *Neuroscience* 5:81–90. [SVA]

(1986) Once more for the equilibrium point hypothesis (λ model). *Journal of Motor Behavior* 18:17–54. [aEB, GCA, AAF, AGF, SRG, MLL, NL, DJO, GJvIS, RBS]

Feldman, A. G., Adamovich, S. V., Ostry, D. J. & Flanagan, J. R. (1990) The origin of electromyograms – Explanations based on the point hypothesis. In: *Multiple muscle systems: Biomechanics and movement organization*, ed. J. Winters & S. L.-Y. Woo. Springer-Verlag. [SVA, AGF, GLG, SRG, NL, DJO, JBJS]

Feldman, J. A. & Ballard, D. H. (1982) Connectionist models and their properties. *Cognitive Science* 6:205–54. [aGEA]

Feldman, A. G. & Latash, M. L. (1982) Interaction of afferent and efferent signals underlying joint position sense: Empirical and theoretical approaches. *Journal of Motor Behavior* 14:174–93. [AGF]

Feldman, A. G. & Orlovsky, G. N. (1972) The influence of different descending systems on the tonic stretch reflex in the cat. *Experimental Neurology* 37:481–94. [AGF, MLL]

Felleman, D. J. & Van Essen, D. C. (1991) Distributed hierarchical processing in the primate cerebral cortex. *Cerebral Cortex* 1:1–47. [GS]

Ferrell, W. R. (1980) The adequacy of stretch receptors in the cat knee joint for signalling joint angle throughout a full range of movement. *Journal of Physiology (London)* 299:85–99. [aSCG]

(1987) The effect of acute joint distension on mechanoreceptor discharge in the knee of the cat. *Quarterly Journal of Experimental Physiology* 72:493–99. [aSCG]

Ferrell, W. R., Baxendale, R. H, Carnachan, C. & Hart, I. K. (1985) The influence of joint afferent discharge on locomotion, proprioception and activity in conscious cats. *Brain Research* 347:41–48. [aSCG]

References

Ferrell, W. R., Gandevia, S. C. & McCloskey, D. I. (1987) The role of joint receptors in human kinaesthesia when intramuscular receptors cannot contribute. *Journal of Physiology (London)* 386:63–71. [aSCG]

Ferrell, W. R. & Smith, A. (1988) Position sense at the proximal interphalangeal joint of the human index finger. *Journal of Physiology (London)* 399:49–61. [aSCG]

(1989) The effect of loading on position sense at the proximal interphalangeal joint of the human index finger. *Journal of Physiology (London)* 418:145–61. [aSCG]

Ferrington, D. G., Nail, D. S. & Rowe, M. (1977) Human tactile detection thresholds: Modification by inputs from specific tactile receptor classes. *Journal of Physiology* 272:415–33. [aSCG]

Fetz, E. E. (1981) Neuronal activity associated with conditioned limb movements. In: *Handbook of behavioral neurobiology*, vol. 2: Motor coordination, ed. A. L. Towe & E. S. Luschei. Plenum Press. [arEEF]

(1992) Dynamic neural network models of sensorimotor behavior. In: *The neurobiology of neural networks*, ed. D. Gardner. MIT Press, in press. [rEEF]

Fetz, E. E. & Cheney, P. D. (1980) Postspike facilitation of forelimb muscle activity by primate corticomotoneuronal cells. *Journal of Neurophysiology* 44:751–72. [aGEA, aEEF, cSCG, AL]

Fetz, E. E., Cheney, P. D., Mewes, K. & Palmer, S. (1989) Control of forelimb muscle activity by populations of corticomotoneuronal and rubromotoneuronal cells. In: *Peripheral control of posture and locomotion*, ed. J. A. H. Allum & M. Hulliger. Elsevier [aGEA, aEEF]

(1990) Control of forelimb muscle activity by populations of corticomotoneuronal and rubromotoneuronal cells. *Progress in Brain Research* 80:437–49. [rEEF]

Fetz, E. E. & Finocchio, D. V. (1972) Operant conditioning of isolated activity in specific muscles and precentral cells. *Brain Research* 40:19–23. [MEI]

(1975) Correlations between activity of motor cortex cells and arm muscles during operantly conditioned response patterns. *Experimental Brain Research* 23:217–40. [aEEF, MEI]

Fetz, E. E., Finocchio, D. V., Baker, M. A. & Soso, M. J. (1980) Sensory and motor responses of precentral cortex cells during comparable passive and active joint movements. *Journal of Neurophysiology* 43:1070–89. [aEEF]

Fetz, E. E. & Shupe, L. E. (1990) Neural network models of the primate motor system. In: *Advanced neural computers*, ed. R. Eckmiller. Elsevier/North-Holland. [aEEF]

Fetz, E. E., Shupe, L. E. & Murthy, V. N. (1990) Neural networks controlling wrist movements. *Proceedings of the 1990 International Joint Conference on Neural Networks* 2:675–79. [aGEA, aEEF]

Fetz, E. E., Toyama, K. & Smith, W. (1991) Synaptic interactions between cortical neurons. In: *Cerebral cortex*, vol. 9: Altered cortical states, ed. A. Peters & E. G. Jones. Plenum Press. [rEEF]

Fischer, B. & Rogal, L. (1986) Eye-hand-coordination in man: A reaction time study. *Biological Cybernetics* 55:253–61. [VD]

Flament, D., Fortier, P. A. & Fetz, E. E. (1992) Response patterns and post-spike effects of peripheral afferents in dorsal root ganglia of behaving monkeys. *Journal of Neurophysiology* 67:875–89. [arEEF]

Flament, D. & Hore, J. (1986) Movement and electromyographic disorders associated with cerebellar dysmetria. *Journal of Neurophysiology* 55:1221–33. [aJRB]

Flanagan, J. R., Feldman, A. G. & Ostry, D. J. (1992) Equilibrium trajectories underlying rapid target-directed arm movements. In: *Tutorials in motor behavior*. II, ed. G. E. Stelmach & J. Requin. North-Holland. [DJO]

Flanagan, J. R., Ostry, D. J. & Feldman, A. G. (1990) Control of human jaw and multi-joint arm movements. In: *Cerebral control of speech and limb movements*, ed. G. E. Hammond. Elsevier Science. [arEB, DJO]

Flanders, M. & Soechting, J. F. (1990) Parcellation of sensorimotor transformations for arm movements. *Journal of Neuroscience* 10:2420–27. [MF]

Flanders, M., Tillery, S. I. H. & Soechting, J. F. (1992) Early stages in a sensorimotor transformation. *Behavioral and Brain Sciences* 15(2):309–20. [MF, PG]

Flash, T. (1987) The control of hand equilibrium trajectories in multi-joint arm movements. *Biological Cybernetics* 57:257–74. [aEB, SVA, JDe, AAF, GEL, NL]

Flash, T. & Henis, E. (1991) Arm trajectory modifications during reaching towards visual targets. *Journal of Cognitive Neuroscience* 3:220–230. [rEB]

Flash, T. & Hogan, N. (1985) The coordination of arm movements: An experimentally confirmed mathematical model. *Journal of Neuroscience* 5:1688–1703. [aGEA, JDe, SRG]

Fleshman, J. W., Munson, J. B. & Sypert, G. W. (1981) Homonymous projection of individual group Ia fibers to physiologically characterized medial gastrocnemius motoneurons in the cat. *Journal of Neurophysiology* 46:1339–48. [TRN]

Fleshman, J., Rudomin, P. & Burke, R. (1988) Supraspinal control of a short-latency cutaneous pathway to hindlimb motoneurons. *Experimental Brain Research* 69:449–59. [aDAM]

Fodor, J. (1983) *The modularity of mind*. MIT Press. [GS]

Fogassi, L., Gallese, V., de Pellegrino, G., Fadiga, L., Gentilucci, M., Luppino, G., Matelli, M., Pedotti, A. & Rizzolatti, G. (submitted) Space coding by premotor cortex. [MSG]

Forget, R. & Lamarre, Y. (1987) Rapid elbow flexion in the absence of proprioceptive and cutaneous feedback. *Human Neurobiology* 6:27–37. [aSCG, PRC]

Forssberg, H. (1985) Ontogeny of human locomotor control 1. Infant stepping, supported locomotion and transition to independent locomotion. *Experimental Brain Research* 57:480–93. [VD]

Fortier, P. A., Kalaska, J. F. & Smith, A. M. (1989) Cerebellar neuronal activity related to whole-arm reaching movements in the monkey. *Journal of Neurophysiology* 62:198–211. [aJRB, aEEF]

Fournier, E., Katz, R. & Pierrot-Deseilligny, E. (1983) Descending control of reflex pathways in the production of voluntary isolated movements in man. *Brain Research* 288:357–77. [aSCG]

Fox, C. A., Andrade, A. N., Hillman, D. E. & Schwyn, R. C. (1971) The spiny neurons in the primate striatum. A golgi and electron microscopic study. *Journal für Hirnforschung* 13:181–201. [aGEA]

Fox, G. C. (1988) *Solving problems with concurrent processors*. Prentice Hall. [aJFS]

Francois, C., Percheron, G., Yelnick, J. & Heyner, S. (1984) A golgi analysis of the primate globus pallidus. I. Inconstant processes of large neurons, other neuronal types, and afferent axons. *Journal of Comparative Neurology* 227:182–99. [aGEA]

Freeman, W. J. (1987) Simulation of chaotic EEG patterns with a dynamic model of the olfactory system. *Biological Cybernetics* 56:139–50. [IT]

(1990) Nonlinear neural dynamics in olfaction as a model for cognition. In: *Chaos in brain function*, ed. E. Basar. Springer-Verlag. [TLC]

Freeman, W. J. & van Dijk, B. W. (1987) Spatial patterns of visual cortical fast EEG during conditioned reflex in a rhesus monkey. *Brain Research* 422:267–76. [aJFS]

Frolov, A. A., Biryukova, E. V. & Roschin, V. Y. (1992) Neural model of multijoint movement learning and control. In: *Proceedings of 1992 RNNS/IEEE Symposium on Neuroinformatics and Neurocomputers*, Rostov-on-Don, in press. [AAF]

Frolov, A. G. (1983) The effect of instrumentalization of inborn reaction on its transformation into contrary directed escape response in dogs and the problem of reinforcement. *Acta Neurobiologiae Experimentalis* 43:1–14. [MEI]

Fromm, C. & Evarts, E. V. (1981) Relation of size and activity of motor cortex pyramidal tract neurons during skilled movements in the monkey. *Journal of Neuroscience* 1:453–60. [aSCG]

Fu, T., Hultborn, H., Larsson, R. & Lundberg, A. (1978) Reciprocal inhibition during the tonic stretch reflex in the decerebrate cat. *Journal of Physiology* 284:345–369. [rDAM]

Fujita, M. (1982) Adaptive filter model of the cerebellum. *Biological Cybernetics* 45:195–206. [aJRB, AAF]

Fukami, Y. & Wilkinson, R. S. (1977) Responses of isolated Golgi tendon organs of the cat. *Journal of Physiology (London)* 265:673–89. [aSCG]

Fukson, O. I., Berkinblit, M. B. & Feldman, A. G. (1980) The spinal frog takes into account the scheme of its body during the wiping reflex. *Science* 209:1261–63. [aEB, SG]

Fukushima, K., Perlmutter, S. I., Baker, J. F. & Peterson, B. W. (1990) Spatial properties of second-order vestibulo-ocular relay neurons in the alert cat. *Experimental Brain Research* 81:462–78. [aDAR, CLC]

Fung, S. J., Pompeiano, O. & Barnes, C. C. (1988) Coerulospinal influence on recurrent inhibition of spinal motonuclei innervating antagonistic hindleg muscles of the cat. *Pflugers Archiv-European Journal of Physiology* 412(4):346–53. [FBH]

Fuster, J. M. (1985) The prefrontal cortex and temporal integration. In: *The cerebral cortex*, vol. 4, ed. A. Peters & E. G. Jones. [aEEF]

(1989) *The prefrontal cortex*, 2nd ed. Raven Press. [JMF]

Gaffan, D. & Harrison, S. (1989) Place memory and scene memory effects of fornix transection in the monkey. *Experimental Brain Research* 74:202–12. [aJFS]

Galleti, C. & Battaglini, P. P. (1989) Gaze-dependent visual neurons in area {V}3a of monkey prestriate cortex. *Journal of Neuroscience* 9:1112–25. [APo]

Gandevia, S. C. (1982) The perception of motor commands or effort during muscular paralysis. *Brain* 105:151–59. [aSCG]

(1985) Illusory movements produced by electrical stimulation of low-threshold muscle afferents from the hand. *Brain* 108:965–81. [aSCG]

(1987) Roles for perceived voluntary motor commands in motor control. *Trends in Neurosciences* 10:81–85. [aSCG]

Gandevia, S. C. & Burke, D. (1985) Effect of training on voluntary activation of human fusimotor neurons. *Journal of Neurophysiology* 54:1422–29. [aSCG]

(1988) Projection to the cerebral cortex from proximal and distal muscles in the human upper limb. *Brain* 111:389–403. [aSCG]

Gandevia, S. C., Burke, D. & McKeon, B. (1984) The projection of muscle afferents from the hand to cerebral cortex in man. *Brain* 107:1–13. [aSCG]

(1986a) Coupling between human muscle spindle endings and motor units assessed using spike-triggered averaging. *Neuroscience Letters* 71:181–86. [aSCG]

Gandevia, S. C., Hall, L. A., McCloskey, D. I. & Potter, E. K. (1983) Proprioceptive sensation at the terminal joint of the middle finger. *Journal of Physiology (London)* 335:507–17. [aSCG]

Gandevia, S. C. & Kilbreath, S. (1990) Accuracy of weight estimation for weights lifted by proximal and distal muscles of the human upper limb. *Journal of Physiology (London)* 423:299–310. [aSCG]

Gandevia, S. C. & Macefield, G. (1989) Projection of low-threshold afferents from human intercostal muscles to the cerebral cortex. *Respiration Physiology* 77:203–14. [aSCG]

Gandevia, S. C., Macefield, G., Burke, D. & McKenzie, D. K. (1990) Voluntary activation of human motor axons in the absence of muscle afferent feedback: The control of the deafferented hand. *Brain* 113:1563–81. [aSCG]

Gandevia, S. C. & Mahutte, C. K. (1982) Theoretical requirements for the interpretation of signals of intramuscular tension. *Journal of Theoretical Biology* 97:141–53. [aSCG]

Gandevia, S. C. & McCloskey, D. I. (1976) Joint sense, muscle sense, and their combination as position sense measured at the distal interphalangeal joint of the middle finger. *Journal of Physiology (London)* 260:387–407. [aSCG]

(1977a) Sensations of heaviness. *Brain* 100:345–54. [aSCG]

(1977b) Effects of related sensory inputs on motor performances in man studied through changes in perceived heaviness. *Journal of Physiology (London)* 272:653–72. [aSCG]

(1978) Interpretation of perceived motor commands by reference to afferent signals. *Journal of Physiology (London)* 283:493–99. [aSCG]

Gandevia, S. C., McCloskey, D. I. & Burke, D. (1992) Kinaesthetic signals and muscle contraction. *Trends in Neurosciences* 15:62–65. [rSCG]

Gandevia, S. C., Miller, S., Aniss, A. M. and Burke, D. (1986b) Reflex influences on muscle spindle activity in relaxed human leg muscles. *Journal of Neurophysiology* 56:159–70. [aSCG]

Gandevia, S. C. & Rothwell, J. C. (1987) Knowledge of motor commands and the recruitment of human motoneurons. *Brain* 110:1117–30. [aSCG, MLL]

Garcia-Rill, E. (1986) The basal ganglia and the locomotor regions. *Brain Research Reviews* 11:47–63. [GMa]

Garner, W. R. (1962) *Uncertainty and structure as psychological concepts.* Wiley. [FJC]

Garner, W. R. & McGill, W. J. (1956) The relation between information and variance analysis. *Psychometrika* 21:219–28. [FJC]

Gasser, H. S. & Hill, A. V. (1924) The dynamics of muscular contraction. *Proceedings of The Royal Society* B96:398–437. [ZH]

Gauthier, G. M. & Mussa-Ivaldi, F. (1988) Oculo-manual tracking of visual targets in monkeys: Role of the arm afferent information in the control of arm and eye movements. *Experimental Brain Research* 73:138–54. [aJRB]

Gauthier, G. M., Vercher, J.-L., Mussa-Ivaldi, F. & Marchetti, E. (1988) Oculo-manual tracking of visual targets: Control learning, coordination control and coordination model. *Experimental Brain Research* 73:127–37. [aJRB]

Gelfand, I. M., Gurfinkel, V. S., Kots, Y. M., Tsetlin, M. L. & Shik, M. L. (1963) Synchronization of motor units and associated model concepts. *Biofizika* 8:475–86. [TMH]

Gellman, R., Gibson, A. R. & Houk, J. C. (1985) Inferior olivary neurons in the awake cat: Detection of contact and passive body movement. *Journal of Neurophysiology* 54:40–60. [aJRB]

Gentilucci, M., Fogassi, L. Luppino, G., Matelli, M., Camarda, R. & Rizzolatti, G. (1988) Functional organization of inferior area 6 in the macaque monkey. I. Somatotopy and the control of proximal movements. *Experimental Brain Research* 71:475–90. [JFK]

Gentilucci, M., Scandolara, C., Pigarev, I. & Rizzolatti, G. (1983) Visual responses in the postarcuate cortex (area 6) of the monkey that are independent of eye position. *Experimental Brain Research* 50:464–68. [MSG]

Georgopoulos, A. P. (1991) Higher order motor control. *Annual Review of Neuroscience* 14:361–77. [aEB, MF, RBS]

Georgopoulos, A. P., Caminiti, R., Kalaska, J. F. & Massey, J. T. (1983) Spatial coding of movement: A hypothesis concerning the coding of movement direction by motor cortical populations. *Experimental Brain Research* (suppl.) 7:327–36. [aEB, WAM]

Georgopoulos, A. P., Crutcher, M. D. & Schwartz, A. B. (1989) Cognitive spatial motor processes. III. Motor cortical prediction of movement direction during an instructed delay period. *Experimental Brain Research* 75:183–94. [aGEA]

Georgopoulos, A. P. & Grillner, S. (1989) Visuomotor coordination in reaching and locomotion. *Science* 245:1209–10. [JFK]

Georgopoulos, A. P., Kalaska, J. F., Caminiti, R. & Massey, J. T. (1982) On the relations between the direction of two-dimensional arm movements and cell discharge in primate motor cortex. *Journal of Neuroscience* 2:1527–37. [aEB, MH, DSL, PM]

Georgopoulos, A. P., Kalaska, J. F., Crutcher, M. D., Caminiti, R. & Massey, J. T. (1984) The representation of movement direction in the motor cortex: Single cell and population studies. In: *Dynamic aspects of neocortical function,* ed. G. E. Edelman, W. E. Gall & W. M. Cowan. Wiley. [aEEF, BB]

Georgopoulos, A. P., Kettner, R. E. & Schwartz, A. B. (1988) Primate motor cortex and free arm movements to visual targets in three-dimensional space. II. Coding of the direction by a neuronal population. *Journal of Neuroscience* 8:2928–37. [aSCG]

Georgopoulos, A. P. & Massey, J. T. (1988) Cognitive spatial-motor processes. *Experimental Brain Research* 69:315–26. [FJC]

Georgopoulos, A. P., Schwartz, A. B. & Kettner, R. E. (1986) Neuronal population coding of movement direction. *Science* 233:1357–1440. [aJRB, PR]

Gevins, A. S., Bressler, S. L., Morgan, N. H., Cutillo, B. A., White, R. M., Greer, D. S. & Illes, J. (1989a) Event-related covariances during a bimanual visuomotor task. I. Methods and analysis of stimulus-and response-locked data. *Electroencephalography and Clinical Neurophysiology* 74:58–75. [JMF]

Gevins, A. S., Cutillo, B. A., Bressler, S. L., Morgan, N. H., White, R. M., Illes, J. & Greer, D. S. (1989b) Event-related covariances during a bimanual visuomotor task. II. Preparation and feedback. *Electroencephalography and Clinical Neurophysiology* 74:147–60. [JMF]

Ghez, C. & Gordon, J. (1987) Trajectory control in targeted force impulses. I. Role of opposing muscles. *Experimental Brain Research* 67:225–40. [MLL]

Ghez, C., Gordon, J., Ghilardi, M. F., Christakos, C. M. & Cooper, S. E. (1990) Roles of proprioceptive input in the programming of arm trajectories. *Cold Spring Harbor Symposia in Quantitative Biology,* vol. 55, in press. [aSCG]

Ghez, C. & Martin, J. H. (1982) The control of rapid limb movement in the cat 3. agonist-antagonist coupling. *Experimental Brain Research* 45:115–25. [NL]

Gibson, A. R., Houk, J. C. & Kohlerman, N. J. (1985) Relation between red nucleus discharge and movement parameters in trained macaque monkeys. *Journal of Physiology (London)* 358:551–70. [aEEF]

Gibson, A. R., Robinson, F. R., Alam, J. & Houk, J. C. (1987) Somatotopic alignment between climbing fiber input and nuclear output of the cat intermediate cerebellum. *Journal of Comparative Neurology* 260:362–77. [aJRB]

Gielen, C. C. A. M. & Houk, J. C. (1984) Nonlinear viscosity of human wrist. *Journal of Neurophysiology* 52:553–69. [GCA]

(1986) Simple changes in reflex threshold cannot explain all aspects of rapid voluntary movements. *Behavioral and Brain Sciences* 9:605–07. [JBJS]

Gielen, C. C. A. M., Ramaekers, L. & Van Zuylen, E. J. (1988) Long-latency stretch reflexes as coordinated functional responses in man. *Journal of Physiology (London)* 407:275–92. [aSCG, ZH, JBJS]

Gielen, C. C. A. M. & Van Ingen Schenau, G. J. (1992) The constrained control of force and position by multilink manipulators. *IEEE Transactions on Systems, Man, and Cybernetics,* in press. [GJvIS]

Gielen, C. C. A. M. & Van Zuylen, E. J. (1986) Coordination of arm muscles during flexion and supination: Application of the tensor analysis approach. *Neuroscience* 17:527–39. [aGEA]

Gielen, C. C. A. M., van den Heuvel, P. J. M. & van Gisbergen, J. A. M. (1984) Coordination of fast eye and arm movements in a tracking task. *Experimental Brain Research* 56:154–61. [VD]

Gilbert, P. F. C. (1974) A theory of memory that explains the function and structure of the cerebellum. *Brain Research* 70:1–18. [PFCG]

(1975) How the cerebellum could memorize movements. *Nature* 254:688–89. [PFCG]

Gilbert, P. F. C. & Thach, W. T. (1977) Purkinje cell activity during motor learning. *Brain Research* 128:309–28. [arJRB, PFCG]

Gilhodes, J. C., Roll, J. P. & Tardy-Gervet, M. F. (1986) Perceptual and motor effects of agonist-antagonist muscle vibration in man. *Experimental Brain Research* 61:395–402. [aSCG]

Gilman, S., Bloedel, J. R. & Lechtenberg, R. (1981) *Disorders of the cerebellum.* Davis. [aJRB]

Giszter, S. (1992) Force fields and muscle use strategies that underlie reflex behaviors in the spinal frog. *Society for Neuroscience Abstracts,* submitted. [rEB]

Giszter, S. F., Bizzi, E. & Mussa-Ivaldi, F. A. (1991a) Motor organization in the frog spinal cord. In: *Analysis and modelling of neural systems,* ed. F. H. Eeckman & C. D. Deno. Kluwer. [rEB, SG]

(1991b) Movement primitives in the frog spinal cord. In: *Analysis and modelling of neural systems 2,* ed. F. H. Eeckman & C. D. Deno. Kluwer Press, in press. [arEB]

(1992a) Movement primitives in the frog spinal cord. In: *Analysis and modelling of neural systems II,* ed. F. H. Eeckman. Kluwer, in press. [SG]

Giszter, S. F., McIntyre, J. & Bizzi, E. (1989) Kinematic strategies and sensorimotor transformations in the wiping movements of frogs. *Journal of Neurophysiology* 62:750–67. [aEB, SG]

Giszter, S. F., Mussa-Ivaldi, F. A. & Bizzi, E. (1991c) Equilibrium point mechanisms in the spinal frog. In: *Visual structures and integrated functions,* ed. M. A. Arbib & J. P. Ewert. Plenum. [SG]

(1992b) The organization of limb motor space in the spinal cord. *Experimental Brain Research* Series 22, in press. [arEB, SG]

(1992c) Convergent force fields organized in the frog spinal cord. *Journal of Neuroscience,* submitted. [rEB]

Glickstein, M. (1990) Brain pathways in the visual guidance of movement. In: *Brain circuits and functions of the mind,* ed. C. Trevarthen. Cambridge University Press. [aJFS]

Glickstein, M., Cohen, J. L., Dixon, B., Gibson, A., Hollins, M., LaBossiere, E. & Robinson, F. (1980) Corticopontine visual projections in macaque monkeys. *Journal of Comparative Neurology* 90:209–29. [aJFS]

Gnadt, J. W. & Andersen, R. A. (1988) Memory related motor planning activity in posterior parietal cortex of macaque. *Experimental Brain Research* 70:216–20. [aJFS, CLC, JWG, JFK]

Gnadt, J. W. & Mays, L. E. (1989) Posterior parietal cortex, the oculomotor near response and spatial coding in 3-D space. *Society for Neuroscience Abstracts* 15:786. [JWG]

(1991) Depth-tuning in area LIP by disparity and accommodative cues. *Society for Neuroscience Abstracts* 17:1113. [JWG]

Godschalk, M., Lemon, R. N., Kuypers, H. G. J. M. & Van der Steen, J. (1985) The involvement of monkey premotor cortex neurons in preparation of visually cued arm movements. *Behavioral Brain Research* 18:143–57. [aEEF]

Godwin-Austen, R. B. (1965) A case of visual disorientation. *Journal of Neurology Neurosurgery and Psychiatry* 28:453–48. [JWG]

Goldberg, D. E. (1989) *Genetic algorithms.* Addison-Wesley. [IK]

Goldberg, G. (1985) Supplementary motor area structure and function: Review and hypotheses. *Behavioral and Brain Sciences* 8:567–616. [aGEA]

Goldberg, M. E. & Bruce, C. (1985) Cerebral cortical activity associated with the orientation of visual attention in the Rhesus monkey. *Vision Research* 25:471–81. [JFK]

(1990) Primate frontal eye field. III. Maintenance of a spatially accurate saccade signal. *Journal of Neurophysiology* 64:489–508. [aJFS, APo, PQ]

Goldberg, M. E. & Colby, C. L. (1989) The neurophysiology of spatial vision. In: *Handbook of neuropsychology,* vol. 2, ed. F. Boller & J. Grafman. Elsevier. [aJFS]

Goldberg, M. E., Colby, C. L. & Duhamel, J.-R. (1990) Representation of visuomotor space in the parietal lobe of the monkey. *Cold Spring Harbor Symposia on Quantitative Biology* 55:729–39. [CLC, DPC]

Goldman-Rakic, P. (1988) Topography of cognition: Parallel distributed networks in primate association cortex. *Annual Review of Neuroscience* 11:137–56. [aJFS, DPC, BS]

Goldscheider, A. (1889) Cited by Sherrington (1900). [aSCG]

Goldstein, H. (1950) *Classical mechanics.* Addison-Wesley. [rEB]

Gollhofer, A., Horstmann, G. A., Berger, W. & Dietz, V. (1989) Compensation of translational and rotational perturbations in human posture: Stabilization of the centre of gravity. *Neuroscience Letters* 105:73–78. [VD]

Gonshor, A. & Jones, M. (1976) Extreme vestibulo-ocular adaptation induced by prolonged optical reversal of vision. *Journal of Physiology* 256:381–414. [ADK]

Goodale, M. A. & Milner, A. D. (1992) Separate visual pathways for perception and action. *Trends in Neuroscience* 15:20–25. [MAG]

Goodale, M. A., Milner, A. D., Jakobson, L. S. & Carey, D. P. (1991) A neurological dissociation between perceiving objects and grasping them. *Nature* 349:154–56. [rSCG, DPC, MAG]

Goodale, M. A., Pelisson, D. & Prablanc, C. (1986) Large adjustments in visually guided reaching do not depend on vision of the hand or perception of target displacement. *Nature* 320:748–50. [DPC]

Goodman, S. G. & Andersen, R. A. (1989) Microstimulation of a neural-network model for visually guided saccades. *Journal of Cognitive Neuroscience* 1:317–26. [RAA]

(1990) Algorithm programmed by a neural network model for coordinate transformation. *Proceedings of the International Joint Conference on Neural Networks* 2:381–86. [RAA]

Goodwin, G. M., McCloskey, D. I. & Matthews, P. B. C. (1972) The contribution of muscle afferent to kinaesthesia shown by vibration induced illusions of movement and by the effects of paralysing joint afferents. *Brain* 95:705–48. [aSCG]

Gordon, A. M., Forssberg, H., Johansson, R. S. & Westling, G. (1991a) Visual size cues in the programming of manipulative forces during precision grip. *Experimental Brain Research* 83:477–82. [AMG]

(1991b) The integration of haptically acquired size information in the programming of precision grip. *Experimental Brain Research* 83:483–88. [VD, AMG]

Gordon, A. M., Huxley, A. F. & Julian, F. J. (1966) The variation in isometric tension with sarcomere length in vertebrate muscle fibres. *Journal of Physiology (London)* 184:170–92. [aEB]

Gordon, J. & Ghez, C. (1987a) Trajectory control in targeted force impulses. II. Pulse height control. *Experimental Brain Research* 67:241–52. [aSCG]

(1987b) Trajectory control in targeted force impulses. III. Compensatory adjustments for initial errors. *Experimental Brain Research* 67:253–69. [aSCG, PJC]

Gorenstein, C., Bundman, M. C., Bruce, J. L. & Rotter, A. (1987) Neuronal localization of pseudocholinesterase in the rat cerebellum: Sagittal bands of Purkinje cells in the nodulus and uvula. *Brain Research* 418:68–75. [NHB]

Gormezano, I., Kehoe, E. J. & Marshall, B. S. (1983) Twenty years of classical conditioning research with the rabbit. In: Progress in psychobiology and physiological psychology, ed. J. Sprague & A. N. Epstein. *Academic Press.* [rJRB]

Gossard, J., Cabelguen, J. & Rossignol, S. (1989) Intra axonal recordings of cutaneous primary afferents during fictive locomotion in the cat. *Journal of Neurophysiology* 62:1177–88. [aDAM]

(1990) Phase dependent modulation of primary afferent depolarization in single cutaneous primary afferents evoked by peripheral stimulation during fictive locomotion in the cat. *Brain Research* 537:14–23. [aDAM]

(1991) An intracellular study of muscle primary afferents during fictive locomotion in the cat. *Journal of Neurophysiology* 65:914–26. [aDAM]

Gottlieb, G. L. & Agarwal, G. C. (1978) Dependence of human ankle compliance on joint angle. *Journal of Biomechanics* 11:177–81. [GCA]

(1988) Compliance of single joints: Elastic and plastic characteristics. *Journal of Neurophysiology* 59:937–51. [GCA, MLL]

Gottlieb, G. L., Corcos, D. M. & Agarwal, G. C. (1989) Strategies for the control of voluntary movements with one mechanical degree of freedom. *Behavioral and Brain Sciences* 12(2):189–210. [SVA, GLG, ZH]

Gottlieb, G. L., Latash, M. L., Corcos, D. M., Liubinskas, T. J. & Agarwal, G. C. (1992) Organizing principles for single joint movements: V. Agonist-antagonist interactions. *Journal of Neurophysiology* 67(6). [GLG]

Gracies, J. M., Meunier, S., Pierrot-Deseilligny, E. & Simonetta, M. (1991) Pattern of propriospinal-like excitation to different species of human upper limb motoneurones. *Journal of Physiology* 434:151–67. [DBur]

Granit, R. (1955) *Receptors and sensory perception.* Yale University Press [rEB]

(1970) *The basis of motor control.* Academic Press. [aEB, GCA]

Granit, R. & Phillips, C. G. (1956) Excitatory and inhibitory processes acting upon individual Purkinje cells of the cerebellum in cats. *Journal of Physiology* 133:520–47. [cJFS]

Granit, R. & Rutledge, L. T. (1960) Surplus excitation in reflex action of motoneurons as measured by recurrent inhibition. *Journal of Physiology (London)* 154:288–307. [TMH]

Graveland, G. A. & Difiglia, M. (1985) The frequency and distribution of medium-sized neurons with indented nuclei in the primate and rodent neostriatum. *Brain Research* 327:307–11. [aGEA]

Gray, C. M. & Singer, W. (1989) Stimulus-specific neuronal oscillations in orientation columns of cat visual cortex. *Proceedings of the National Academy of Science* 86:1698–1702. [WAM]

Graziano, M. S. A. & Gross, C. G. (submitted) A bimodal map of space: Somatosensory receptive fields in the macaque putamen with corresponding visual receptive fields. [MSG]

Greene, D. A., Sima, A. A. F., Albers, J. W. & Pfeifer, M. A. (1990) In: *Diabetes Mellitus theory and practice*, ed. H. Rifkin & D. Porte, 4th Ed. Elsevier. [PRC]

Greene, P. H. (1972) Problems of organizing motor systems. *Progress in Theoretical Biology* 2:303–38. [rEB, PJC]

(1982) Why is it so easy to control your arms? *Journal of Motor Behavior* 14:260–86. [PJC]

Gregory, J. E., Morgan, D. L. & Proske, U. (1988) Aftereffects in the responses of cat muscle spindles and errors of limb position sense in man. *Journal of Neurophysiology* 59:1220–30. [aSCG]

Griffiths, R. I. (1987) Ultrasound transit time gives direct measurement of muscle fibre length in vivo. *Journal of Neuroscience Methods* 21:159–65. [aSCG]

Grigg, P., Finerman, G. A. & Riley, L. H. (1973) Joint-position sense after total hip replacement. *Journal of Bone and Joint Surgery* 56:1016–25. [aSCG]

Grigg, P. & Greenspan, B. J. (1977) Response of primate joint afferent neurons to mechanical stimulation of knee joint. *Journal of Neurophysiology* 40:1–8. [aSCG]

Grigg, P., Schaible, H.-G. & Schmidt, R. F. (1986) Mechanical sensitivity of Group III and IV afferents from posterior articular nerve in normal and inflamed cat knee. *Journal of Neurophysiology* 55:635–43. [aSCG]

Grillner, S. (1972) The role of muscle stiffness in meeting the changing postural and locomotor requirements for force development of the ankle extensors. *Acta Physiologica Scandinavica* 86:92–108. [aEB, GMa]

(1975) Locomotion in vertebrates: Central mechanisms and reflex interaction. *Physiological Reviews* 55:247–304. [VD]

(1981) Control of locomotion in bipeds, tetrapods and fish. In: *Handbook of physiology – the nervous system*, vol. 2, Motor control, ed. J. M. Brookhart & V. B. Mountcastle. American Physiological Society. Waverly. [GMa]

Gritti, I. & Schieppati, M. (1989) Short-latency inhibition of soleus motoneurones by impulses in Ia afferents from the gastrocnemius muscle in humans. *Journal of Physiology (London)* 416:469–84. [MS]

Grobstein, P. (1986) Review of the brain machine. *Journal of the American Medical Association* 255:2677–78. [PG]

(1987) The nervous system/behavior interface: Levels of organization and levels of approach. Commentary on target article by J.-P. Ewert. *Behavioral and Brain Sciences* 10;380–81. [PG]

(1988a) On beyond neuronal specificity: Problems in going from cells to networks and from networks to behavior. In: *Advances in neural and behavioral development*, vol. 3, ed. P. G. Shinkman. Ablex. [PG]

(1988b) Between the retinotectal projection and directed movement: Topography of a sensorimotor interface. *Brain Behavior and Evolution* 31:34–48. [PG]

(1988c) From the head to the heart: Some thoughts on similarities between brain function and morphogenesis, and on their significance for research methodology and biological theory. *Experientia* 44:961–71. [PG]

(1989) Organization in the sensorimotor interface: A case study with increased resolution. In: *Visuomotor coordination: Amphibians, comparisons, models and robots*, ed. J.-P. Ewert & M. A. Arbib. Plenum. [PG]

(1990) Strategies for analyzing complex organization in the nervous system. I. Lesion experiments, the old rediscovered. In: *Computational neuroscience*, ed. E. Schwartz. MIT Press. [PG]

(1991) Directed movement in the frog: A closer look at a central representation of spatial location. In: *Visual structures and integrated functions, research neural computing*, ed. M. A. Arbib & J.-P. Ewert. Springer-Verlag. [PG]

(1992) Directed movement in the frog: Motor choice, spatial representation, free will? In: *Neurobiology of motor programme selection: New approaches to mechanisms of behavioral choice*, ed. J. Kien, C. McCrohan & B. Winlow. Manchester University Press, in press. [PG]

Grobstein, P., Comer, C. & Kostyk, S. K. (1983) Frog prey capture behavior: Between sensory maps and motor output. In: *Advances in vertebrate neuroethology*, ed. J.-P. Ewert, R. R. Capranica & D. J. Ingle. Plenum Press. [PG]

Groenewegen, H. J. & Voogd, J. (1977) The parasagittal zonation within the olivocerebellar projection. In. Climbing fiber distribution in the vermis of cat cerebellum. *Journal of Comparative Neurology* 174:417–88. [aJRB]

Gross, C. G. & Graziano, M. S. A. (1990) Bimodal visual-tactile responses in the macaque putamen. *Society for Neuroscience Abstracts* 16:110. [MSG]

Grossberg, S. (1988) *Neural networks and natural intelligence*. MIT Press. [DSL]

Grossberg, S. & Kuperstein, M. (1989) *Neural dynamics of adaptive sensory-motor control*. Pergamon. [rJFS, PQ]

Guillaumin, S., Dahhaoui, M. & Caston, J. (1991) Cerebellum and memory: An experimental study in the rat using a passive avoidance conditioning test. *Physiology & Behavior* 49:507–11. [aJRB]

Guitton, D., Munoz, D. P. & Galiana, M. L. (1990) Gaze control in the cat: Studies and modelling of the coupling between orienting eye and head movements in different behavioral tasks. *Journal of Neurophysiology* 64:509–531. [PQ]

Guitton, D., Munoz, D. P. & Pelisson, D. (1991) Spatio-temporal patterns of activity on the motor map of the superior colliculus. In: *Brain and space*, ed. J. Paillard. Oxford University Press. [DPC]

Guitton, D. & Volle, M. (1987) Gaze control in humans: Eye-head coordination during orienting movements to targets within and beyond the oculomotor range. Journal of Neurophysiology 58:427–59. [RAA]

Gurfinkel, V. S. & Latash, M. L. (1979) Segmental postural mechanisms and reversal of muscle reflexes. *Agressologie* 20B:145–46. [VD]

Gurfinkel, V. S., Levik, Y. S., Popov, K. E., Smetanin, B. N. & Shlikov, V. Y. (1988) Body scheme in the control of postural activity. In: *Stance and motion: Facts and concepts*, ed. V. S. Gurfinkel. Plenum. [aSCG]

Gurfinkel, V. S., Lipshits, M. I., Mori, S. & Popov, K. E. (1976) Postural reactions to the controlled sinusoidal displacement of the supporting platform. *Agressologie* 17B:71–76. [VD]

Guthrie, B. L., Porter, J. D. & Sparks, D. L. (1983) Corollary discharge provides accurate eye position information to the oculomotor system. *Science* 221:1193–95. [aSCG]

Gutman, S. R. & Gottlieb, G. L. (1990) Nonlinear "inner time" in reaching movement trajectory formation. *Abstracts of 1st World Congress of Biomechanics*. [SRG]

Gutman, S. R., Gottlieb, G. L. & Corcos, D. M. (in press). Exponential model of a reaching movement trajectory with non-linear time. *Comments on Theoretical Biology*. [SRG]

Haaland, K. Y. & Harrington, D. L. (1990) Complex movement behavior: Toward understanding cortical and subcortical interactions in regulating control processes. In: *Cerebral control of speech and limb movements*, ed. G. R. Hammond. North-Holland. [JJS]

Hagbarth, K. (1952) Excitatory and inhibitory skin areas for flexor and extensor motoneurones. *Acta Physiolgica Scandinavica* 94:1–58. [aDAM]

Hagbarth, K., Kunesch, E. J., Nordin, M., Schmidt, R. & Wallin, E. U. (1986) Gamma loop contributing to maximal voluntary contractions in man. *Journal of Physiology (London)* 380:575–91. [aSCG]

Hagbarth, K., Wallin, G., Burke, D. & Lofstedt, L. (1975) Effects of the Jendrassik manoeuvre on muscle spindle activity in man. *Journal of Neurology, Neurosurgery, and Psychiatry* 38:1143–53. [rSCG, JP]

Hahm, J.-O., Langdon, R. B. & Sur, M. (1991) Disruption of retinogeniculate afferent segregation by antagonists to NMDA receptors. *Nature (London)* 351:568–70. [aGEA]

Hahne, M., Illert, M. & Wietelmann, D. (1988) Recurrent inhibition in the cat distal forelimb. *Brain Research* 456:188–92. [TMH]

Haines, D. E., Patrick, G. W. & Satrulee, P. (1982) Organization of cerebellar corticonuclear fiber systems. In: *The cerebellum: New vistas*, ed. S. L. Palay & V. Chan-Palay. Springer-Verlag. [aJRB]

Hake, H. W. & Garner, W. R. (1951) The effect of presenting various numbers of discrete steps on scale reading accuracy. *Journal of Experimental Psychology* 42:358–66. [FJC]

Hall, L. A. & McCloskey, D. I. (1983) Detections of movements imposed on finger, elbow and shoulder joints. *Journal of Physiology (London)* 335:519–33. [aSCG]

Hallett, M., Berardelli, A., Matheson, J., Rothwell, J. & Marsden, C. D. (1991) Physiological analysis of simple rapid movements in patients with cerebellar deficits. *Journal of Neurology, Neurosurgery and Psychiatry* 54:124–33. [MH]

Hallet, M. & Khoshbin, S. (1980) A physiological mechanism of bradykinesia. *Brain* 103:301–14. [DSB]

Hallett, M., Shahani, B. T. & Young, R. R. (1975) EMG analysis of patients with cerebellar deficits. *Journal of Neurology, Neurosurgery, and Psychiatry* 38:1163–69. [MH]

Hamada, I., DeLong, M. R. & Mano, N.-I. (1990) Activity of identified wrist-related pallidal neurons during step and ramp wrist movements in the monkey. *Journal of Neurophysiology* 64:1892–1906. [aGEA]

Hamm, T. M. (1990) Recurrent inhibition to and from motoneurons innervating the flexor digitorum and flexor hallucis longus muscles of the cat. *Journal of Neurophysiology* 63:395–403. [TMH]

Hannaford, B. & Stark, L. (1985) Roles of the elements of the tri-phasic control signal. *Experimental Neurology* 90:619–34. [NL]

(1987) Late agonist activation burst (PC) required for optimal head movement: A simulation study. *Biological Cybernetics* 57:321–30. [NL]

Hansen, P. D., Woollacott, M. H. & Debu, B. (1988) Postural responses to changing task conditions. *Experimental Brain Research* 73:627–36. [VD]

Harris-Warrick, R. M. (1988) Chemical modulation of central pattern generators. In: *Neural control of rhythmic movements in vertebrates*, ed. A. H. Cohen, S. Rossignol & S. Grillner. Wiley. [CAP]

Harrison, P. J. & Jankowska, E. (1985a) Sources of input to interneurones mediating group I non-reciprocal inhibition of motoneurones in the cat. *Journal of Physiology* 361:379–401. [aDAM]

(1985b) Organization of input to the interneurones mediating group I non-reciprocal inhibition of motoneurones in the cat. *Journal of Physiology* 361:403–18. [aDAM]

Harrison, P., Jankowska, E. & Zytnicki, D. (1986) Lamina VII interneurones interposed in crossed reflex pathways in the cat. *Journal of Physiology* 371:147–66. [aDAM]

Harvey, R. J., Porter, R. & Rawson, J. A. (1977) The natural discharges of Purkinje cells in paravermal regions of lobules V and VI of the monkey's cerebellum. *Journal of Physiology (London)* 271:515–36. [JMB]

Hasan, Z. (1986) Optimized movement trajectories and joint stiffness in unperturbed, inertially loaded movements. *Biological Cybernetics* 53:373–82. [SRG, GLG, ZH, MLL, NL]

(1991) Biomechanics and the study of multijoint movements. In: *Motor control: Concepts and issues*, ed. D. R. Humphrey & H.-J. Freund. Wiley. [aGEA]

Hasan, Z. & Enoka, R. M. (1985) Isometric torque-angle relationship and movement-related activity of human elbow flexors: Implications for the equilibrium-point hypothesis. *Experimental Brain Research* 59:441–50. [AGF, ZH]

Hasan, Z. & Stuart, D. G. (1988) Animal solutions to problems of movement control: The role of proprioceptors. *Annual Review of Neuroscience* 11:199–223. [ZH]

Hawkes, R. & Gravel, C. (1991) The modular cerebellum. *Progress in Neurobiology* 36:309–27. [aJRB]

Hawkes, R. & Leclerc, N. (1987) Antigenic map of the rat cerebellar cortex: The distribution of parasagittal bands as revealed by monoclonal anti-Purkinje cell antibody mabQ113. *Journal of Comparative Neurology* 256:29–41. [NHB]

Haykin, S. (1986) *Adaptive filter theory*. Prentice-Hall. [PDN]

He, J., Levine, W. S. & Loeb, G. E. (1991) Feedback gains for correcting small perturbations to standing posture. *IEEE Transactions on Automatic Control* 36:322–32. [GEL]

Head, H. & Holmes, G. (1912) Sensory disturbances from cerebral lesions. *Brain* 34:102–254. [PM]

Hecht-Nielsen, R. (1989) Theory of the back-propagation neural network. *Proceedings of the International Joint Conference on Neural Networks*, Washington D.C., June. [HB]

Heiligenberg, W. F. (1991) *Neural nets in electric fish*. MIT Press. [MF]

Heilman, K. M., Watson, R. T. & Valenstein, E. (1985) Neglect and related disorders. In: *Clinical neuropsychology*, ed. K. M. Heilman & E. Valenstein. Oxford University Press. [aJFS]

Heister, G., Schroeder-Heister, P. & Ehrenstein, W. H. (1990) Spatial coding and spatio-anatomical mapping: Evidence for a hierarchical model of spatial stimulus-response compatibility. In: *Stimulus-response compatibility: An integrated perspective*, ed. R. W. Proctor & T. G. Reeve, North-Holland. [RWP]

Helms Tillery, S. I., Flanders, M. & Soechting, J. F. (1991) A coordinate system for the synthesis of visual and kinesthetic information. *Journal of Neuroscience* 11:770–78. [MF]

Hershberger, W. & Misceo, G. (1983) Conditioned weight illusion: Reafference learning without a correlation store. *Perception and Psychophysics* 33:391–98. [HER]

Heuer, H. (1991) Invariant relative timing in motor-program theory. In: *The development of timing control and temporal organization in coordinated action*, ed. J. Fagard & P. H. Wolff. North-Holland. [HH]

Hikosaka, K. E., Iwai, E., Saito, H. & Tanaka, K. (1988) Polysensory properties of neurons in the anterior bank of the caudal superior temporal sulcus of the macaque monkey. *Journal of Neurophysiology* 60:1615–37. [BS]

Hildreth, E. C. & Hollerbach, J. M. (1987) Artificial intelligence: Computational approach to vision and motor control. In: *Handbook of physiology*. The nervous system. High functions of the brain, vol. 5, ed. V. B. Mountcastle, F. Plum & S. R. Geiger. American Physiological Society. [aGEA]

Hill, A. V. (1938) The heat of shortening and the dynamic constants of muscles. *Proceedings of the Royal Society of London* B126:136–95. [SVA, AGF]

Hillyard, S. A., Munte, T. F. & Neville, H. J. (1985) Visuospatial attention, orienting and brain physiology. In: *Attention and performance*, ed. M. I. Posner & O. S. Martin. Erlbaum. [aJFS]

Hinton, G. (1984) Parallel computations for controlling an arm. *Journal of Motor Behavior* 16:171–94. [aGEA]

(1989) Connectionist learning procedures. *Artificial Intelligence* 40:185–234. [HB]

(1990) Special issue on connectionist symbol processing. *Artificial Intelligence* 46:1–257. [JER]

Hinton, G. & Anderson, J. A. (1981) *Parallel models of associative memory*. Erlbaum. [aEEF]

Hinton, G. E., McClelland, J. L. & Rumelhart, D. E. (1986) Distributed representations. In: *Parallel distributed processing: Explorations in the microstructure of cognition*, ed. D. E. Rumelhart, J. L. McClelland, & The PDP Research Group. MIT Press. [aGEA]

Hocherman, S. & Wise, S. P. (1991) Effects of hand movement path on motor cortical activity in awake, behaving rhesus monkeys. *Experimental Brain Research* 83:285–302. [JFK]

Hoffer, J. A. & Andreassen, S. (1981) Regulation of soleus muscle stiffness in premammillary cats: Intrinsic and reflex components. *Journal of Neurophysiology* 45:267–85. [aEB, MLL, NL]

Hoffer, J. A., Caputi, A. A., Pose, I. E. & Griffiths, R. I. (1989) Roles of muscle activity and load on the relationship between muscle spindle length and whole muscle length in the freely walking cat. *Progress in Neurobiology* 80:75–85. [aSCG]

Hoffman, W. C. (1978) The Lie transformation group approach to visual neuropsychology. In: *Formal theories of visual perception*, ed. E. L. J. Leeuwenberg & H. Buffart. Wiley. [TLC]

Hogan, N. (1982) Control and coordination of voluntary arm movements. *Proceedings of the 1982 American Control Conference* 1:552–58. [aEB]

(1984) An organising principle for a class of voluntary movements. *Journal of Neuroscience* 4:2745–54. [aEB, SVA, AGF, GLG, SRG, NL]

(1985a) The mechanics of multi-joint posture and movement. *Biological Cybernetics* 52:315–31. [aGEA, arEB, AAF]

(1985b) Impedance control: An approach to manipulation. Part I: Theory. Part II: Implementation. Part III: Application. *ASME Journal of Dynamic Systems, Measurement and Control* 107:1–24. [aEB]

(1988a) Planning and execution of multi-joint movements. *Canadian Journal of Physiology and Pharmacology* 66:508–17. [aEB, JDe]

(1988b) On the stability of manipulators performing contact tasks. *IEEE Journal of Robotics and Automation* 4:677–86. [aEB]

Hogan, N., Bizzi, E., Mussa-Ivaldi, F. A. & Flash, T. (1987) Controlling multijoint motor behavior. *Exercise and Sports Sciences Reviews* 15:153–90. [aGEA, aEB]

Hollerbach, J. M. & Atkeson, C. G. (1987) Deducing planning variables from experimental arm trajectories: Pitfalls and possibilities. *Biological Cybernetics* 56:279–92. [aEB]

Hollerbach, J. M. & Flash, T. (1982) Dynamic interactions between limb segments during planar arm movement. *Biological Cybernetics* 44:67–77. [aGEA]

Holmes, G. (1917) The symptoms of acute cerebellar injuries due to gunshot injuries. *Brain* 40:461–535. [aJRB, MH, RFT]

(1918) Disturbances of visual orientation. *British Journal of Ophthalmology* 2:449–68, 506–518. [DPC, JWG]

Holmes, G. & Horrax, G. (1919) Disturbances of spatial orientation and visual attention. *Archives of Neurological Psychiatry* 1:385–407. [JWG]

Holmqvist, B. & Lundberg, A. (1961) Differential supraspinal control of synaptic actions evoked by volleys in the flexion reflex afferents in alpha motoneurones. *Acta Physiologica Scandinavica* 54(suppl.):5–51. [aDAM]

Hongo, T., Kudo, N., Oguni, E. & Yoshida, K. (1990) Spatial patterns of reflex evoked by pressure stimulation of the foot pads in cats. *Journal of Physiology* 420:471–87. [aDAM]

Hongo, T., Kudo, N., Sasaki, S., Yamashita, M., Yoshida, K., Ishizuka, N., et al. (1987) Trajectory of group Ia and Ib fibers from the hind-limb muscles at the L3 and L4 segments of the spinal cord of the cat. *Journal of Comparative Neurology* 262:159–94. [aDAM]

Hongo, T., Lundberg, A., Phillips, C. G. & Thompson, R. F. (1984) The pattern of monosynaptic Ia connections to hindlimb motor nuclei in the baboon: A comparison with the cat. *Proceedings of the Royal Society Series B* 221:264–89. [aSCG]

Hopcroft, J. E., Joseph, D. & Whitesides, S. (1985) On the movement of robot arms in 2-dimensional bounded regions. *Society for Industrial and Applied Mathematics Journal of Computing* 14(2):315–33. [CIC]

Horak, F. (1990) Comparison of cerebellar and vestibular loss on scaling of postural responses. In: *Disorders of posture and gait*, ed. T. Brandt, W.

Paulus, W. Bless, M. Dieterich, S. Krafczyk & A. Straube. Thieme Verlag. [FBH]

Horak, F. & Anderson, M. E. (1984) Influence of globus pallidus on arm movement in monkeys. I. Effects of kainic acid lesions. *Journal of Neurophysiology* 52:290–304. [FBH, DSL]

(1984b) Influence of globus pallidus on arm movements, 2. Effects of stimulation. *Journal of Neurophysiology* 52:305–22. [DSL]

Horak, F., Nashner, L. & Diener, H. (1986) Abnormal scaling of posture responses in cerebellar patients. *Society of Neuroscience Abstracts* 12:1419. [FBH]

Horak, F. & Nashner, L. M. (1986) Central programming of postural movements: Adaptation to altered support-surface configurations. *Journal of Neurophysiology* 55:1369–81. [ADK]

Horak, F., Nutt, J. & Nashner, L. (1992) Postural inflexibility in Parkinsonian patients. *Journal of Neurological Science*, in press. [FBH]

Horcholle-Bossavit, G., Jami, L., Petit, J., Vejsada, R. & Zytnicki, D. (1988) Effects of muscle shortening on the responses of cat tendon organs to unfused contractions. *Journal of Neurophysiology* 59:1510–23. [aSCG]

(1990) Ensemble discharge from Golgi tendon organs of the cat peroneus tertius muscle. *Journal of Neurophysiology* 64:813–21. [aSCG]

Hore, J. & Flament, D. (1986) Evidenced that a disordered servo-like mechanism contributes to tremor in movements during cerebellar dysfunction. *Journal of Neurophysiology* 56:123–36. [aJRB]

(1988) Changes in motor cortex neural discharge associated with the development of cerebellar limb ataxia. *Journal of Neurophysiology* 60:1285–1302. [aJRB]

Hore, J., Preston, J. P., Durkovic, R. G. & Cheney, P. D. (1976) Responses of cortical neurons (areas 3a and 4) to ramp stretch of hindlimb muscles in the baboon. *Journal of Neurophysiology* 39:484–500. [aSCG]

Hore, J. & Vilis, T. (1984) Loss of set in muscle responses to limb perturbations during cerebellar dysfunction. *Journal of Neurophysiology* 51:1137–48. [aJRB, FBH]

(1985) A cerebellar-dependent efference copy mechanism for generating appropriate muscle responses to limb perturbations. In: *Cerebellar function*, ed. J. R. Bloedel, J. Dichgans & W. Precht. Springer-Verlag. [aJRB]

Hore, J., Wild, B. & Diener, H. C. (1991) Cerebellar dysmetria at the elbow, wrist and fingers. *Journal of Neurophysiology* 65:563–71. [MH]

Hörner, M., Illert, M. & Kümmel, H. (1991) Absence of recurrent axon collaterals in motoneurones to the extrinsic digit extensor muscles of the cat forelimb. *Neuroscience Letters* 122:183–86. [TMH, MS]

Houk, J. C. (1976) An assessment of stretch reflex function. *Progress in Brain Research* 44:303–14. [MLL]

(1979) Regulation of stiffness by skeletomotor reflexes. *Annual Review of Physiology* 41:99–114. [aDAM, TRN]

Houk, J. C., Crago, P. E. & Rymer, W. Z. (1981) Function of the dynamic response in stiffness regulation – a predictive mechanism provided by nonlinear feedback. In: *Muscle receptors and movement*, ed. A. Taylor & A. Prochazka. Macmillan. [TRN]

Houk, J. C. & Gibson, A. R. (1978) Sensorimotor processing through the cerebellum. In: *New concepts in cerebellar neurobiology*, ed. J. S. King. Liss. [JMB]

Houk, J. C. & Rymer, W. Z. (1981) Neural control of muscle length and tension. In: *Handbook of physiology*. Section 1: The nervous system, vol. 2, Motor control, ed. J. M. Brookhart, V. B. Mountcastle, V. B. Brooks & S. R. Geiger. American Physiological Society. [aGEA]

Houk, J. C., Singh, S. P., Fisher, C. & Barto, A. G. (1991) An adaptive sensorimotor network inspired by the anatomy and physiology of the cerebellum. In: *Neural networks for control*, ed. W. T. Miller, R. S. Sutton & P. J. Werbos. MIT Press. [aGEA]

Hounsgaard, J., Hultborn, H. & Kiehn, O. (1986) Transmitter-controlled properties of alpha motoneurones causing long-lasting motor discharge to brief excitatory inputs. *Progress in Brain Research* 64:39–49. [aDAM, TRN, CAP]

Hoy, M. & Zernicke, R. F. (1985) Modulation of limb dynamics in the swing phase of locomotion. *Journal of Biomechanics* 18:49–60. [CAP]

Hoy, M. G., Zernicke, R. F. & Smith, J. L. (1985) Contrasting roles of inertial and muscle movements at knee and ankle during paw-shake response. *Journal of Neurophysiology* 54:1282–94. [CAP]

Hubel, D. H. & Wiesel, T. N. (1977) Ferrier lecture: Functional architecture of Macaque monkey visual cortex. *Proceedings of the Royal Society of London* B198:1–59. [TLC]

Hughlings Jackson, J. (1931) *Selected writings of John Hughlings Jackson*, ed. J. Taylor. Hoder & Stoughton. [aSCG, AL]

Hugon, M. & Paillard, J. (1959) Depression durable du reflexe de Hoffman apres un etirement passif du muscle. *Journal de Physiologie (Paris)* 47:193–96. [JP]

Hulliger, M., Dürüller, N., Prochazka, A. & Trend, P. (1989) Flexible fusimotor control of muscle spindle feedback during a variety of natural movements. *Progress in Brain Research* 80:87–101. [aSCG]

Hulliger, M., Nordh, E., Thelin, A.-E. & Vallbo, Å B. (1979) The responses of afferent fibres from the glabrous skin of the hand during voluntary finger movements in man. *Journal of Physiology (London)* 291:233–49. [aSCG]

Hulliger, M., Nordh, E. & Vallbo, Å B. (1982) The absence of position response in spindle afferents units from human finger muscles during accurate position holding. *Journal of Physiology (London)* 322:167–79. [aSCG, AGF]

(1985) Discharge in muscle spindle afferents related to direction of slow precision movements in man. *Journal of Physiology (London)* 362:437–53. [aSCG]

Hultborn, H. & Illert, M. (1991) *How is motor behaviour reflected in the organization of spinal systems?* Motor Control: Concepts and issues, ed. D. R. Humphrey & H. J. Freund. John Wiley. [rDAM]

Hultborn, H., Illert, M. & Santini, M. (1976a) Convergence on interneurones mediating the reciprocal Ia inhibition of motoneurons I. Disynaptic Ia inhibition of Ia inhibitory interneurones. *Acta Physiologica Scandinavica* 96:193–201. [DBul]

Hultborn, H., Illert, M. & Santini, M. (1976b) Convergence on interneurones mediating the reciprocal Ia inhibition of motoneurones. *Acta Physiologica Scandinavica* 96:368–91. [TRN]

Hultborn, H., Jankowska, E. & Lindström, S. (1971a) Recurrent inhibition of interneurones monosynaptically activated from group Ia afferents. *Journal of Physiology (London)* 215:613–36. [aDAM, DBul]

(1971b) Relative contribution from different nerves to recurrent depression of Ia IPSPs in motoneurones. *Journal of Physiology (London)* 215:637–64. [TMH]

Hultborn, H., Lindström, S. & Wigström, H. (1979) On the function of recurrent inhibition in the spinal cord. *Experimental Brain Research* 37:399–403. [TMH]

Hultborn, H., Meunier, S., Pierrot-Deseilligny, E. & Shindo, M. (1986) Changes in polysynaptic Ia excitation to quadriceps motoneurones during voluntary contraction in man. *Experimental Brain Research* 63:436–38. [aSCG]

(1987) Changes in presynaptic inhibition of Ia fibres at the onset of voluntary contraction in man. *Journal of Physiology (London)* 389:757–72. [aSCG, aDAM]

Humphrey, D. R. (1986) Representation of movement and muscles within the primate precentral motor cortex: Historical and current perspectives. *Federation Proceedings* 45:2687–99. [RL]

Humphrey, D. R. & Freund, F.-J. (1991) *Motor control: Concepts and issues*. Wiley. [JT]

Humphrey, D. R. & Reed, D. J. (1983) Separate cortical systems for the control of joint movement and joint stiffness: Reciprocal activation and coactivation of antagonist muscles. In: *Motor control mechanisms in health and disease. Advances in Neurology* 39, ed. J. Desmedt. Raven Press. [aEB, DBul]

Humphrey, D. R., Schmidt, E. M. & Thompson, W. D. (1970) Predicting measures of motor performance from multiple spike trains. *Science* 170:758–62. [aEEF]

Husain, M. & Stein, J. F. (1988) Rezso Balint and his most celebrated case. *Archives of Neurology* 45:89–93. [aJFS]

Hutchins, K. D., Martino, A. M. & Strick, P. L. (1988) Corticospinal projections from the medial wall of the hemisphere. *Experimental Brain Research* 715:667–72. [aGEA]

Hutton, R. S., Kaiya, K., Suzuki, S. & Watanabe, S. (1987) Post-contraction errors in human force production are reduced by muscle stretch. *Journal of Physiology (London)* 393:247–59. [aSCG]

Hyvarinen, J. (1981) Regional distribution of functions in parietal associative area 7 of the monkey. *Brain Research* 206:287–303. [MSG]

(1982) *The parietal cortex of monkey and man*. Springer. [aJFS]

Hyvarinen, J. & Poranen, A. (1974) Function of parietal association area 7 as revealed from cellular discharges in alert monkeys. *Brain* 97:673–92. [aJFS, MSG]

Iles, J. F. (1986) Reciprocal inhibition during agonist and antagonist contraction. *Experimental Brain Research* 62:212–14. [aSCG]

Iles, J. F. & Pisini, J. V. (1986) Modulation of spinal reciprocal inhibition during postural sway in man. *Journal of Physiology (London)* 382:71P. [VD]

Iles, J. F. & Roberts, R. C. (1987) Inhibition of monosynaptic reflexes in the human lower limb. *Journal of Physiology (London)* 385:69–87. [aSCG]

Iles, J. F., Stokes, M. & Young, A. (1984) Reflex actions of knee joint receptors on quadriceps in man. *Journal of Physiology (London)* 360:48P. [aSCG]

Illert, M., Jankowska, E., Lundberg, A. & Odutola, A. (1981) Integration in descending motor pathways controlling the forelimb in the cat. 7. Effects from the reticular formation on C3-C4 propriospinal neurones. *Experimental Brain Research* 42:269–81. [DBur]

Illert, M., Lundberg, A. & Tanaka, R. (1977) Integration in descending motor pathways controlling the forelimb in the cat. 3. Convergence on propriospinal neurones transmitting disynaptic excitation from the corticospinal tract and other descending tracts. *Experimental Brain Research* 29:323–46. [DBur, AL]

Illert, M., Lundberg, A., Padel, Y. & Tanaka, R. (1978) Integration in descending motor pathways controlling the forelimb in the cat. 5. Properties of and monosynaptic excitatory convergence on C3-C4 propriospinal neurones. *Experimental Brain Research* 33:101–130. [AL]

Ingle, D. (1968) Visual releasers of prey-catching behavior in frogs and toads. *Brain, Behavior and Evolution* 1:500–18. [DI]

(1982) The organization of visuomotor behaviors in vertebrates. In: *The analysis of visual behavior*, ed. D. Ingle, M. Goodale & R. Mansfield. MIT Press. [DI]

(1983) Brain mechanisms of localization in frogs and toads. In: *Advances in vertebrate neuroethology*, ed. J.-P. Ewert, R. Capranica & K. Ingle. Plenum Press. [DI]

(1991a) Functions of subcortical visual systems in vertebrates and the evolution of higher visual mechanisms. In: *Evolution of the eye and visual system*, ed. R. Gregory & J. Cronly-Dillon. Macmillan Press. [DI]

(1991b) The striatum and spatial memory: From frog to man. In: *Visual structures and integrated functions*, ed. M. A. Arbib & J.-P. Ewert. Springer-Verlag. [DI]

Ingle, D. & Cook, J. (1977) The effect of viewing distance upon size preference of frogs for prey. *Vision Research* 17:1009–14. [DI]

Ingle, D. & Hoff, K. V. (1990) Visually elicited evasive behavior in frogs. *BioScience* 40:284–91. [DI]

Ingle, D., Jakobson, L. S., Lassonde, M. C. & Sauerwein, H. C. (1990) Deficits in interfield memory transfer in cases of callosal agenesis. *Society for Neuroscience Abstracts* 16:926. [DI]

Inhoff, A. W., Diener, H. C., Rafal, R. D. & Ivry, R. (1989) The role of cerebellar structures in the execution of serial movements. *Brain* 112:565–81. [aJRB]

Inhoff, A. W. & Wang, J. (1992) Encoding of text, manual movement planning, and eye-hand coordination during copytyping. *Journal of Experimental Psychology: Human Perception and Performance* 18:437–48. [AMG]

Ioffe, M. E. (1973) Pyramidal influences in establishment of new motor coordinations in dogs. *Physiology and Behavior* 11:145–53. [MEI]

(1991) *Mechanisms of motor learning*. Nauka. [MEI]

Ioffe, M. E., Ivanova, N. G., Frolov, A. A., Birjukova, E. V. & Kiseljova, N. V. (1988) On the role of motor cortex in the learned rearrangement of postural coordinations. In: *Stance and motion, facts and concepts*, ed. V. S. Gurfinkel, M. E. Ioffe, J. Massion & J. P. Roll. Plenum. [MEI]

Iriki, A., Keller, A., Pavlides, C. & Asanuma, H. (1990) Long-lasting facilitation of pyramidal tract input to spinal interneurons. *Neuroreport*. 1:157–160 [rDAM]

Iriki, A., Pavlides, C., Keller, A. & Asanuma, H. (1989) Long-term potentiation in the motor cortex. *Science* 245:1385–87. [aGEA]

(1991) Long-term potentiation of thalamic input to the motor cortex induced by coactivation of thalamocortical and corticocortical afferents. *Journal of Neurophysiology* 65:1435–41. [aGEA]

Ito, M. (1974) The control mechanisms of cerebellar motor system. In: *The neuroscience third study program*, ed. F. O. Schmitt & R. G. Worden. MIT Press. [RFT]

(1982) The role of the cerebellum during motor learning in the vestibulo-ocular reflex: Different mechanisms in different species. *Trends in Neuroscience* 5:416. [aJRB]

(1984) *The cerebellum and neural control*. Raven Press. [arJRB, JMB, NHB, PFCG, cJFS, RFT]

(1989) Long-term depression. *Annual Review of Neuroscience* 12:85–102. [PFCG, MI, cJFS]

(1990) A new physiological concept on cerebellum. *Revue Neurologique* (Paris) 146:564–69. [MH]

Ivkovich, D., Lavond, D., Logan, C. G. & Thompson, R. F. (1990) Measurements of reflexive reactions to different unconditioned stimulus intensities over the course of classical conditioning. *Society for Neuroscience Abstracts* 16(1):271. [RFT]

Ivkovich, D., Logan, C. G. & Thompson, R. F. (1991) Accessory abducens lesions produce performance deficits without permanently affecting conditioned responses. *Society for Neuroscience Abstracts* 17(1):869. [RFT]

Ivry, R. B. & Baldo, J. V. (1992) Is the cerebellum involved in learning and cognition? *Current Opinion in Neurobiology*, in press. [RFT]

Ivry, R. B. & Keele, S. W. (1989) Timing functions of the cerebellum. *Journal of Cognitive Neuroscience* 1:136–52. [aJRB]

Ivry, R. B., Keele, S. W. & Diener, H. C. (1988) Dissociation of the lateral and medial cerebellum in movement timing and movement execution. *Experimental Brain Research* 73:167–80. [aJRB, FBH]

Iwata, A., Thoma, T., Matsuo, H. & Suzumura, N. (1990) A large scale neural network "CombNet" and Its application to Chinese character recognition. In: *Proceedings of the International Neural Network Conference INNC*, Paris, vol. 1. Kluwer Academic. [HB]

Jacobs, K. M. & Donoghue, J. P. (1991) Reshaping the cortical motor map by unmasking latent intracortical connections. *Science* 251:944–47. [aGEA]

Jakobson, L. S., Archibald, Y. M., Carey, D. P. & Goodale, M. A. (1991) A kinematic analysis of reaching and grasping movements in a patient recovering from optic ataxia. *Neuropsychologia* 29:803–9. [DPC, MAG]

Jami, L. (1988) Propriétés fonctionnelles des organes tendineux de Golgi. *Archives Internationales de Physiologie et de Biochimie* 96:A363–78. [aSCG]

Jankowska, E. (1992) Interneuronal relay in spinal pathways from proprioceptors. *Progress in Neurobiology* 38:335–78. [aDAM]

Jankowska, E., Johanisson, T. & Lipski, J. (1981a) Common interneurones in reflex pathways from group Ia and Ib afferents of ankle extensors in the cat. *Journal of Physiology* 310:381–402. [aDAM, CAP]

Jankowska, E., Jukes, M. G. M., Lund S. & Lundberg, A. (1967a) The effect of DOPA on the spinal cord: V. Reciprocal organization of pathways transmitting excitatory action to alpha motoneurones of flexors and extensors. *Acta Physiologica Scandinavica* 70:369–88. [aDAM]

(1967b) The effect of DOPA on the spinal cord. VI. Half-centre organization of interneurones transmitting effects from the flexor reflex afferents. *Acta Physiologica Scandinavica* 70:389–402. [aDAM]

Jankowska, E. & Lundberg, A. (1981) Interneurones in the spinal cord. *Trends in Neurosciences* 4:230–33. [aSCG, rEEF]

Jankowska, E. & McCrea, D. A. (1983) Shared reflex pathways from Ib tendon organ afferents and Ia muscle spindle afferents in the cat. *Journal of Physiology* 338:99–111. [aDAM, CAP]

Jankowska, E., McCrea, D. & Mackel, R. (1981b) Oligosynaptic excitation of motoneurones by impulses in group Ia muscle spindle afferents in the cat. *Journal of Physiology* 316:411–25. [aDAM]

(1981c) Pattern of "non-reciprocal" inhibition of motoneurones by impulses in group Ia muscle spindle afferents in the cat. *Journal of Physiology* 316:393–409. [aDAM]

Jankowska, E., Padel, Y. & Tanaka, R. (1976a) Disynaptic inhibition of spinal motoneurones from the motor cortex in the monkey. *Journal of Physiology* 258:467–87. [aDAM]

Jankowska, E., Rastad, J. & Westman, J. (1976b) Intracellular application of horseradish peroxidase and its light and electron microscopical appearance in spinocervical tract cells. *Brain Research* 105:557–62. [aDAM]

Jansen, J. & Brodal, A. (1940) Experimental studies on the intrinsic fibers of the cerebellum. II. The cortico-nuclear projections. *Journal of Comparative Neurology* 73:267–321. [aJRB]

Jasmin, L. & Courville, J. (1987) Distribution of external cuneate nucleus afferents to the cerebellum II. Topographical distribution and zonal pattern. An experimental study with radioactive tracers in the cat. *Journal of Comparative Neurology* 161:497–514. [AMS]

Jeannerod, M. (1988) *The neural and behavioural organisation of goal directed movements*. Oxford University Press. [aJFS]

Jeannerod, M., Michel, F. & Prablanc, C. (1984) The control of hand movements in a case of hemianaesthesia following a parietal lesion. *Brain* 107:899–920. [aSCG]

Jenner, J. R. & Stephens, J. A. (1982) Cutaneous reflex responses and their central nervous pathways studied in man. *Journal of Physiology* (London) 333:405–19. [aSCG]

Johansson, R. S. (1991) How is grasping modified by somatosensory input? In: *Motor control: Concepts and issues*, ed. D. R. Humphrey & H.-J. Freund. Wiley. [AMG]

Johansson, R. S., Häger, C. & Bäckström, L. (1992a) Somatosensory control of precision grip during unpredictable pulling loads: III. Impairments during digital anesthesia. *Experimental Brain Research*, in press. [AMG]

Johansson, R. S., Häger, C. & Riso, R. (1992b) Somatosensory control of precision grip during unpredictable pulling loads: II. Changes in load force rate. *Experimental Brain Research*, in press. [AMG]

Johansson, R. S., Riso, R., Häger, C. & Bäckstrom, L. (1992c) Somatosensory control of precision grip during unpredictable pulling loads: I. Changes in load force amplitude. *Experimental Brain Research*, in press. [AMG]

Johansson, R. S. & Westling, G. (1984) Roles of glabrous skin receptors and sensorimotor memory in automatic control of precision grip when lifting rougher or more slippery objects. *Experimental Brain Research* 56:550–64. [aSCG, AMG]

(1987) Signals in tactile afferents from the fingers eliciting adaptive motor responses during precision grip. *Experimental Brain Research* 66:141–54. [aSCG]

(1988a) Coordinated isometric muscle commands adequately and erroneously programmed for the weight during lifting task with precision grip. *Experimental Brain Research* 71:59–71. [aSCG, AMG]

(1988b) Programmed and reflex actions to rapid load changes during precision grip. *Experimental Brain Research* 71:72–86. [aSCG, FBH]

(1990) Tactile afferent signals in the control of precision grip. In: *Attention and performance*, vol. 13, ed. M. Jeannerod. Erlbaum. [aSCG, AMG]

(1991) Afferent signals during manipulative tasks in humans. In: *Information processing in the somatosensory system*, ed. O. Franzen, & P. Westman. Macmillan. [AMG]

John, K. T., Goodwin, A. W. & Darian-Smith, I. (1989) Tactile discrimination of thickness. *Experimental Brain Research* 78:62–68. [aSCG]

Johnston, R. M. & Bekoff, A. (1989) Differential modulation of double bursting muscles in the chick. *Neuroscience Abstracts* 15:1044. [CAP]

Jones, E. G., Coulter, J. D., Burton, H. & Porter, R. (1977) Cells of origin and terminal distribution of corticostriatal fibers arising in the sensory-motor cortex of monkeys. *Journal of Comparative Neurology* 173:53–80. [aGEA]

Jones, E.G. & Powell, T. P. S. (1969) Connections of the somatic sensory cortex of the rhesus monkey. I. Ipsilateral cortical connections. *Brain* 92:477–502. [aJFS]

(1970) An anatomical study of converging sensory pathways within the cerebral cortex of the monkey. *Brain* 93:793–820. [BS]

Jones, L. A. (1986) Perception of force and weight: Theory and research. *Psychological Bulletin* 100:29–42. [HER]

(1988) Motor illusions: What do they reveal about proprioception? *Psychological Bulletin* 103:72–86. [aSCG]

Jones, L. A. & Hunter, I. W. (1983) Perceived force in fatiguing isometric contractions. *Perception and Psychophysics* 33:369–74. [aSCG]

Jordan, M. I. (1990) Motor learning and the degrees of freedom problem. In: *Attention and performance XIII. Motor representation and control*, ed. M. Jeannerod. Erlbaum. [aGEA, rEB]

Jordan, M. I. & Rumelhart, D. E. (1991) Internal world models and supervised learning. In: *Machine learning: Proceedings of the eighth international workshop*, ed. L. Birnbaum & G. Collins. Kaufmann. [PM]

Judd, J. S. (1988) On the complexity of loading shallow neural networks. *Journal of Complexity* 4:177–92. [HB]

Jurgens, R., Becker, W. & Kornhuber, H. H. (1981) Natural and drug-induced variations of velocity and duration of human saccadic eye movements: Evidence for control of the neural pule generator by local feedback. *Biological Cybernetics* 39:87–96. [PQ]

Kaas, J. H., Krubitzer, L. A., Chino, Y. M., Langston, A. L., Polley, E. H. & Blair, N. (1990) Reorganization of retinotopic cortical maps in adult mammals after lesions of the retina. *Science* 248:229–31. [aGEA]

Kalaska, J. F. (1991a) Parietal cortex area 5: A neuronal representation of movement kinematics for kinaesthetic perception and motor control? In: *Brain and space*, ed. J. Paillard. Oxford University Press. [JFK]

(1991b) Reaching movements to visual targets: Neuronal representations of sensori-motor transformations. *Seminars in the Neurosciences* 3:67–80. [JFK]

Kalaska, J. F., Caminiti, R. & Georgopoulos, A. P. (1983) Cortical mechanisms related to the direction of two-dimensional arm movements: Relations in parietal area 5 and comparison with motor cortex. *Experimental Brain Research* 51:247–60. [aEB, aEEF]

Kalaska, J. F., Cohen, D. A. D., Hyde, M. L. & Prud'homme, M. (1989) A comparison of movement direction-related versus load direction-related activity in primate motor cortex, using a two-dimensional reaching task. *Journal of Neuroscience* 9:2080–2102. [aGEA, aEEF, MH, JFK]

Kalaska, J. F., Cohen, D. A. D., Prud'homme, M. & Hyde, M. L. (1990) Parietal area 5 neuronal activity encodes movement kinematics, not movement dynamics. *Experimental Brain Research* 80:351–64. [JFK, PM]

Kalaska, J. F. & Crammond, D. J. (1992) Cerebral cortical mechanisms of reaching movements. *Science* 255:1517–23. [JFK]

Kanda, K. & Desmedt, J. E. (1983) Cutaneous facilitation of large motor units and motor control of human fingers in precision grip. In: *Motor control mechanisms in health and disease*, ed. J. E. Desmedt. Raven Press. [aSCG]

Kanda, K. & Sato, H. (1983) Reflex responses of human thigh muscles to non-noxious sural stimulation during stepping. *Brain Research* 288:378–80. [aSCG]

Kandel, E. R., Schwartz, J. H. & Jessell, T. M., eds. (1991) *Principles of neural science*. Elsevier. [CIC]

Kapur, D. & Mundy, J. L., eds. (1989) *Geometric reasoning*. MIT. [CIC]

Karanjia, P. N. & Ferguson, J. H. (1983) Passive joint position sense after total hip replacement surgery. *Annals of Neurology* 13:654–57. [aSCG]

Karnath, H. O., Schenkel, P. & Fischer, B. (1991) Trunk-orientation as the determining factor of "contralateral" deficit in neglect syndrome and as the physical anchor of our internal representation of body orientation in space. *Brain*, in press. [RAA]

Karylo, D. D. & Skavenski, A. A. (1991) Eye movements elicited by electrical stimulation of area PG in the monkey. *Journal of Neurophysiology* 65(6):1243–52. [APo]

Kasser, R. J. & Cheney, P. D. (1985) Characteristics of corticomotoneuronal postspike facilitation and reciprocal suppression of EMG activity in the monkey. *Journal of Neurophysiology* 53:959–78. [aGEA, rEEF]

Katz, P. S., Eigg, M. H. & Harris-Warrick, R. M. (1989) Serotonergic/cholinergic muscle receptor cells in the crab stomatogastric nervous system. 1. Identification and characterization of the gastropyloric receptor cells. *Journal of Neurophysiology* 62:558–70. [CAP]

Katz, P. S. & Harris-Warrick, R. M. (1989) Serotonergic/cholinergic muscle receptor cells in the crab stomatogastric nervous system. 2. Rapid nicotinic and prolonged modulatory effects on neurons in the stomatogastric ganglion. *Journal of Neurophysiology* 62:571–81. [CAP]

(1990) Actions of identified neuromodulatory neurons in a simple motor system. *Trends in Neuroscience* 13:367–73. [CAP]

Katz, R., Meunier, S. & Pierrot-Deseilligny, E. (1988) Changes in presynaptic inhibition of Ia fibres in man while standing. *Brain* 111:417–37. [aSCG]

Kawato, M. (1989) Adaptation and learning in control of voluntary movement by the central nervous system. *Advanced Robotics* 3:229–49. [ADK]

(1990) Computational schemes and neural network models for formation and control of multijoint arm trajectory. In: *Neural networks for control*, ed. W. T. Miller, R. S. Sutton & P. J. Werbos. MIT Press. [aGEA]

Kawato, M., Furukawa, K. & Suzuki, R. (1987) A hierarchical neural-network model for control and learning of voluntary movement. *Biological Cybernetics* 57:169–185. [rEB]

Kay, B. A., Saltzman, E. L., Kelso, J. A. S. & Schoner, G. (1987) Space-time behavior of single and bimanual rhythmical movements: Data and limit cycle model. *Journal of Experimental Psychology: Human Perception and Performance* 13(2):178–92. [GMa]

Keating, J. G. & Thach, W. T. (1991) The cerebellar cortical area required for adaptation of monkey's "jump" task is lateral, localized and small. *Society for Neuroscience Abstracts* 17:1381. [MH]

Keele, S. W. (1968) Movement control in skilled motor performance. *Psychological Bulletin* 70:387–403. [aGEA, JBJS, JJS]

(1981) Behavioral analysis of movement. In: *Handbook of physiology*, sect.1. *The nervous system*, vol. 2. Motor control, part 2, ed. J. M. Brookhart, V. B. Mountcastle, V. B. Brooks & S. R. Geiger. American Physiological Society. [aGEA]

Keele, S. W. & Ivry, R. (1990) Does the cerebellum provide a common computation for diverse tasks? A timing hypothesis. *Annals of the New York Academy of Sciences* 608:179–211. [MH]

Keele, S. W. & Summers, J. J. (1976) The structure of motor programs. In: *Motor control: Issues and trends*, ed. G. E. Stelmach. Academic Press. [JJS]

Keirstead, S. A. & Rose, P. K. (1988) Monosynaptic projections of single muscle spindle afferents to neck motoneurons in the cat. *Journal of Neuroscience* 8:3945–50. [GEL]

Keller, A., Arissian, K. & Asanuma, H. (1990) Formation of new synapses in the cat motor cortex following lesions of the deep cerebellar nuclei. *Experimental Brain Research* 80:23–33. [aGEA]

Kelly, T. M., Rubia, F. J., Kolb, F., McAlduff, J. D. & Bloedel, J. R. (1990a) Comparison of simple and complex spike activity in identified sagittal zones of the cat cerebellum during perturbation of the locomotor cycle using a multiunit recording technique. *Society for Neuroscience Abstracts* 16:637. [aJRB]

Kelly, T. M., Zuo, C.-C. & Bloedel, J. R. (1990b) Classical conditioning of the eyeblink reflex in the decerebrate-decerebellate rabbit. *Behavioral and Brain Research* 38:7–18. [arJRB, RFT]

Kelso, J. A. S. (1981) Contrasting perspectives on order and regulation in movement. In: *Attention and performance IX*, ed. J. Long & A. Baddeley. Erlbaum. [JJS]

Kennedy, P. R. (1990) Corticospinal, rubrospinal and rubro-olivary projections: A unifying hypothesis. *Trends in Neuroscience* 13:474–79. [GJvIS]

Khatib, O. (1987) A unified approach for motion and force control of robot manipulators: The operational space formulation. *Journal of Robotics and Automation* RA-3(1):43–53. [CIC]

Khatib, O. & LeMaitre, L.-F. (1978) Dynamic control of manipulators operating in a complex environment. *Proceedings of the Third International CISM-IFToMM Symposium*, Udine, Italy, 267–82. [rEB]

Kiendl, H., Krabs, M. & Fritsch, M. (1991) Rule-based modelling of dynamical systems. In: *Analysis and control of industrial processes*, ed. D. Popovich. Braunschweig: Langelddecke. [HH]

Kievit, J. & Kuypers, H. G. J. M. (1977) Organization of the thalamo-cortical connexions to the frontal lobe in the rhesus monkey. *Experimental Brain Research* 29:299–322. [aGEA]

Kilbreath, S. L. & Gandevia, S. C. (1992) Is voluntary control of human thumb muscles special? *Proceedings of the Australian Neuroscience Society* 3:76. [rSCG]

Kim, R., Nakano, K., Jayaraman, A. & Carpenter, M. B. (1976) Projections of the globus pallidus and adjacent structures: An autoradiographic study in the monkey. *Journal of Comparative Neurology* 169:263–90. [aGEA]

Kimura, M. (1986) The role of primate putamen neurons in the association of sensory stimuli with movement. *Neuroscience Research* 3:436–43. [aGEA]

(1990) Behaviorally contingent property of movement-related activity of the primate putamen. *Journal of Neurophysiology* 63(6):1277–96. [CIC]

Kirkwood, P. A. & Sears, T. A. (1991) Cross-correlation analyses of motoneuron inputs in a coordinated motor act. In: *Neuronal coperativity*, ed. J. Kruger. Springer-Verlag. [PAK]

Kirsch, R. F. & Rymer, W. Z. (1987) Neural compensation for muscular fatigue: Evidence for significant force regulation in man. *Journal of Neurophysiology* 57:1893–1910. [DBul]

Kitai, S. T. (1981) Electrophysiology of the corpus striatum and brain stem integrating systems. In: *Handbook of physiology: The nervous system*. Motor control, sect. 1, vol. 2, pt. 2, ed. J. M. Brookhart, V. B. Mountcastle, V. B. Brooks & S. R. Geiger. American Physiological Society. [aGEA]

Kleinschmidt, A., Bear, M. F. & Singer, W. (1987) Blockade of "NMDA". receptors disrupts experience-dependent plasticity of kitten striate cortex. *Science* 238:355–58. [aGEA]

Knapp, J. D., Taub, E. & Berman, A. J. (1963) Movements in monkeys with deafferented forelimbs. *Experimental Neurology* 7:305–15. [aSCG]

Knibestöl, M. (1975) Stimulus response functions of slowly adapting mechanoreceptors in the human glabrous skin area. *Journal of Physiology (London)* 243:63–80. [aSCG]

Koch, C. & Segev, I. (1989) *Methods in neuronal modeling: From synapses to networks*. MIT Press. [rGEA, DJ]

Koditschek, D. E. (1987) Exact robot navigation by means of potential functions: Some topological considerations. In: *Proceedings of the 1987 IEEE International Conference on Robotics and Automation* 3. Institute of Electrical and Electronics Engineers Computer Society Press. [CIC]

(1989) Robot planning and control via potential functions. In: *The robotics review*, ed. O. Khatib, J. J. Craig & T. Lozano-Perez. MIT Press. [rEB]

Kohonen, T. (1982) Self-organized formation of topologically correct feature maps. *Biological Cybernetics* 43:59–69. [aEEF, PM]

Kornhuber, H. H. & Deecke, L. (1965) Hirnpotentialänderungen bei Willkürbewegungen und passiven Bewegungen des Menschen: Bereitschaftspotential und reafferente Potentiale. *Pflügers Archiv* 284:1–17. [MEI]

Koshland, G. F., Gerilovsky, L. & Hasan, Z. (1991) Activity of wrist muscles elicited during imposed or voluntary movements about the elbow joint. *Journal of Motor Behavior* 23:91–100. [ZH]

Koshland, G. F. & Smith, J. L. (1989) Mutable and immutable features of paw-shake responses after hindlimb deafferentiation in the cat. *Journal of Neurophysiology* 62:162–73. [aSCG]

Kotlyar, B. I., Mayorov, V. I., Timofeyeva, N. O. & Shul'govsky, V. V. (1983) *Neuronal organization of conditioned behavior*. Nauka. [MEI]

Krieger, C., Shinoda, Y. & Smith, A. M. (1985) Labeling of cerebellar mossy fiber afferents with intraaxonal horseradish peroxidase. *Experimental Brain Research* 59:414–17. [AMS]

Krogh, B. H. (1984) A generalized potential field approach to obstacle avoidance control. In: *Robotics research: The next five years and beyond*, ed. Society of Manufacturing Engineers. [CIC]

Krommenhoek, K. P., Van Opstal, A. J., Gielen, C. C. A. M. & Van Gisbergen, J. A. M. (1992) Remapping of neural activity in the motor colliculus: A neural network study, submitted. [JAMVG]

Krubitzer, L. A. & Kaas, J. H. (1990) Cortical connections of MT in four species of primates: Areal, modular, and retinotopic patterns. *Visual Neuroscience* 5:165–204. [RE]

Kugler, P. N. & Turvey, M. T. (1987) *Information, natural law and the self-assembly of rhythmic movement*. Erlbaum. [JJS]

Kunesch, E., Binkofski, F. & Freund, H.-J. (1989) Invariant temporal characteristics of manipulative hand movements. *Experimental Brain Research* 78:539–46. [aSCG]

Kunzle, H. (1975) Bilateral projections from precentral motor cortex to the putamen and other parts of the basal ganglia. An autoradiographic study in *Macaca fascicularis*. *Brain Research* 88:195–209. [aGEA]

(1977) Projections from the primary somatosensory cortex to basal ganglia and thalmus in the monkey. *Experimental Brain Research* 30:481–92. [aGEA]

(1978) An autoradiographic analysis of the efferent connections from premotor and adjacent prefrontal regions (areas 6 and 9) in *Macaca fascicularis*. *Brain, Behavior and Evolution* 15:185–234. [aGEA, MSG]

Kuo, A. D. & Zajac, F. E. (1992) An analysis of the effect of muscle strength on the coordination of standing posture. *Journal of Biomechanics*, in press. [ADK]

Kuo, B. C. (1987) *Automatic control systems* (fifth ed.). Prentice-Hall. [GCA]

Kuperstein, M. (1988) Neural model of adaptive eye-hand coordination for single postures. *Science* 239:1308–11. [aGEA, aEEF]

(1991) INFANT neural controller for adaptive sensory-motor coordination. *Neural Networks* 4:131–47. [PM]

Kupfermann, I., Deodhar, D., Teyke, T., Rosen, S. C., Nagahama, T. & Weiss, K. R. (1992) *Acta Biologica Hungarica* 43:315–28. [IK]

Kupfermann, I. & Weiss, K. R. (1978) The command neuron concept. *Behavioral and Brain Sciences* 1:3–39. [IK]

Kurata, K. & Tanji, J. (1985) Contrasting neuronal activity in supplementary and precentral motor cortex of monkeys. II. Responses to movement triggering vs. nontriggering sensory signals. *Journal of Neurophysiology* 53:142–52. [aGEA]

Kurata, K. & Wise, S. P. (1988) Premotor cortex of rhesus monkeys: Set-related activity during two conditional motor tasks. *Experimental Brain Research* 69:327–43. [aGEA]

Kurylo, D. D. & Skavenski, A. A. (1991) Eye movements elicited by electrical stimulation of area PG in the monkey. *Journal of Neurophysiology* 65:1243–53. [JWG]

Kwan, H. C. (1988) Network relaxation as behavioral action. Research in Biological and Computational Vision, Departments of Computer Science and Physiology, University of Toronto. *Technical Report No. RBCV-TR-88-26*. [DSB]

Kwan, H. C., Yeap, T. H., Jiang, B. C. & Borrett, D. S. (1990) Neural network control of simple limb movements. *Canadian Journal of Physiology and Pharmacology* 68:126–30. [aGEA, DSB]

LaBella, L., Kehler, J. & McCrea, D. (1989) Differential synaptic input to the motor nuclei of triceps surae from the caudal and lateral sural nerves. *Journal of Neurophysiology* 61:291–301. [aDAM]

LaBella, L. & McCrea, D. (1990) Evidence for restricted central convergence of cutaneous afferents upon a short-latency excitatory pathway to medial gastrocnemius motoneurons. *Journal of Neurophysiology* 64:403–12. [aDAM]

Lacquaniti, F., Borghese, N. A. & Carrozzo, M. (1991) Transient reversal of the.stretch reflex in human arm muscles. *Journal of Neurophysiology* 66:939–54. [FL]

(1992) Internal models of limb geometry in the control of hand compliance. *Journal of Neuroscience* 12:1750–62. [FL]

Lacquaniti, F. & Soechting, J. F. (1986) EMG responses to load perturbations of the upper limb: Effect of dynamic coupling between shoulder and elbow motion. *Experimental Brain Research* 61:482–96. [ZH]

Lal, R. & Friedlander, M. J. (1989) Gating of retinal transmission by afferent eye position and movement signals. *Science* 243:93–96. [APo]

Laming, D. (1986) *Sensory analysis*. Academic Press. [HER]

Lan, N. & Crago, P. E. (1991) Optimal control of muscle stiffness for FNS induced arm movements. *Proceedings of IEEE/EMBS International Conference* 13(2):920–21. [NL]

Lanman, J., Bizzi, E. & Allum, J. (1978) The coordination of eye and head movement during smooth pursuit. *Brain Research* 153:39–53. [aEB]

Laporte, Y. & Lloyd, D. P. C. (1952) Nature and significance of the reflex connections established by large afferent fibers of muscular origin. *American Journal of Physiology* 169:609–21. [DBul]

Lashley, K. S. (1951) The problem of serial order in behavior. In: *Cerebral mechanisms in behavior*, ed. L. A. Jeffress. Wiley. [AMG]

Latash, M. L. & Gottlieb, G. L. (1990) Compliant characteristics single joint: Preservation of equifinality with phasic reaction. *Biological Cybernetics* 62:331–36. [SRG, MLL]

(1991a) An equilibrium-point model for fast single-joint movement. I. Emergence of strategy-dependent EMG patterns. *Journal of Motor Behavior* 23:163–78. [GLG]

(1991b) An equilibrium-point model for fast single-joint movement. II. Similarity of single-joint isometric and isotonic descending commands. *Journal of Motor Behavior* 23:179–91. [GLG]

(1991c) Reconstruction of elbow joint complaint characteristics during fast and slow voluntary movement. *Neuroscience* 43:697–712. [SVA, GLG, SRG, ZH, MLL, NL, JBJS]

Lauritis, V. P. & Robison, D. A. (1986) The vestibulo-ocular reflex during human saccadic eye movements. *Journal of Physiology (London)* 373:209–33. [RAA]

Lavond, D. G., Kanzawa, S. A., Esquenazi, V., Clark, R. E. & Zhang, A. A. (1990) Effects of cooling interpositus during acquisition of classical conditioning. *Society for Neuroscience Abstracts* 16:270. [rJRB, RFT]

Lawrence, D. G. & Kuypers, H. G. J. M. (1968) The functional organization of the motor system in the monkey. II. The effects of lesions of the descending brainstem pathways. *Brain* 91:36. [rEEF, MEI, AL]

Leclerc, N., Dore, L., Parent, A. & Hawkes, R. (1990) The compartmentalization of the monkey and rat cerebellar cortex: Zebrin I and cytochrome oxidase. *Brain Research* 506:70–78. [aJRB]

Lee, R. G., Murphy, J. T. & Tatton, W. G. (1983) Long-latency myotatic reflexes in man: Mechanisms, functional significance, and changes in patients with Parkinson's disease or hemiplegia. In: *Motor control mechanisms in health and disease*, ed. J. E. Desmedt. Raven Press. [aSCG]

Lee, D. N. & Thomson, J. A. (1982) Vision in action: The control of locomotion. In: *Analysis of visual behavior*, ed. D. J. Ingle, M. A. Goodale & R. J. W. Mansfield. MIT Press. [DPC]

Lehky, S. R. & Sejnowski, T. J. (1988) Network model of shape-from-shading: Neural function arises from both receptive and projective fields. *Nature* 333:452–54. [HB, GS]

Leiner, H. C., Leiner, A. L. & Dow, R. S. (1986) Does the cerebellum contribute to mental skills? *Behavioral Neuroscience* 100:443–54. [aJRB]

(1989) Reappraising the cerebellum. What does the hindbrain contribute to mental skills? *Behavioral Neuroscience* 103:998–1008. [cJFS]

Leinonen, L., Hyvarinen, J., Nyman, G. & Linnankoski, I. (1979) Functional properties of neurons in lateral part of associative area 7 in awake monkey. *Experimental Brain Research* 34:203–15. [MSG, JFK]

Leinonen, L. & Nyman, G. (1979) II. Functional properties of cells in anterolateral part of area 7 Associative face area of awake monkeys. *Experimental Brain Research* 34:321–33. [MSG]

Lemon, R. N. (1981) Variety of functional organization within the monkey motor cortex. *Journal of Physiology (London)* 311:521–40. [cSCG]

(1990) Mapping the output functions of the motor cortex. In: *Signal and Sense: Local and global order in perceptual maps*. eds. G. Edelman, E. Gall, & W. W. Cowan. 12:315–56. Wiley. [RL]

Lemon, R. N., Bennett, K. M. B. & Werner, W. (1991) The cortico-motor substrate for skilled movements of the primate hand. In: *Tutorials on motor neuroscience*, vol. 62, ed. G. E. Stelmach & J. Requin. Kluwer. [RL]

Lemon, R. N., Hanby, J. A. & Porter, R. (1976) Relationship between the activity of precentral neurones during active and passive movements in conscious monkeys. *Proceedings of the Royal Society London* B 194:341–73. [cSCG]

Lemon, R. N. & Mantel, G. W. H. (1989) The influence of changes in discharge frequency of corticospinal neurones on hand muscles in the monkey. *Journal of Physiology* 413:351–78. [PAK, RL]

Lemon, R. N., Mantel, G. W. H. & Muir, R. B. (1986) Corticospinal facilitation of hand muscles during voluntary movement in the conscious monkey. *Journal of Physiology (London)* 381:497–527. [aGEA, aEEF]

Lestienne, F. & Gurfinkel, V. S. (1988) In: *Vestibulospinal control of posture and locomotion*, ed. O. Pompeiano & J. Allum. *Progress in Brain Research* 76:307–13. [aJRB]

Lestienne, F., Polit, A. & Bizzi, E. (1981) Functional organization of the motor process underlying the transition from movement to posture. *Brain Research* 230:121–31. [MH]

Levine, D. S. (1991) Introduction to neural and cognitive modeling. Erlbaum. [DSL]

Levine, M. S. & Lackner, J. R. (1979) Some sensory and motor factors influencing the control and appreciation of eye and limb position. *Experimental Brain Research* 36:275–83. [aSCG]

Libet, B. (1985) Unconscious cerebral initiative and the role of conscious will in voluntary action. *Behavioral and Brain Sciences* 8:529–66. [aSCG]

Libet, B., Gleason, C. A., Wright, E. W. & Pearl, D. K. (1983) Time of conscious intention to act in relation to onset of cerebral activity (readiness-potential). The unconscious initiation of a freely voluntary act. *Brain* 106:623–42. [aSCG]

Libet, B., Wright, E. W., Finestin, B. & Pearl, D. K. (1979) Subjective referral of timing for a conscious sensory experience. A functional role for the somatosensory specific projection system in man. *Brain* 102:193–224. [aSCG]

Lichtman, J. W., Jhaveri, S. & Frank, E. (1984) Anatomical basis of specific connections between sensory axons and motor neurons in the brachial spinal cord of the bullfrog. *Journal of Neuroscience* 4:1754–63. [GEL]

Liddell, E. G. T. & Sherrington, C. S. (1924) Reflexes in response to stretch (myotatic reflexes). *Proceedings of the Royal Society* 96B:212–42. [TRN]

Liles, S. L. (1983) Activity of neurons in the putamen associated with wrist movements in the monkey. *Brain Research* 263:156–61. [aGEA]

(1985) Activity of neurons in putamen during active and passive movements of wrist. *Journal of Neurophysiology* 53:217–36. [aGEA]

Liles, S. L. & Updyke, B. (1985) Projection of the digit and wrist area of precentral gyrus to the putamen: Relation between topography and physiological properties of neurons in the putamen. *Brain Research* 339:245–55. [aGEA]

Lindsay, A. D. & Binder, M. D. (1991) Distribution of effective synaptic currents underlying recurrent inhibition in cat triceps surae motoneurons. *Journal of Neurophysiology* 65:168–77. [TMH]

Linsker, R. (1990) Perceptual neural organization: Some approaches based on network models and information theory. *Annual Review of Neuroscience* 13:257–81. [acGEA]

Lisberger, S. G. (1988a) The neural basis for learning of simple motor skills. *Science* 242:728–35. [PFCG, FBH]

(1988b) The neural basis for motor learning in the vestibulo-ocular reflex in monkeys. *Trends in Neuroscience* 11:147–52. [arJRB, cJFS, PFCG]

Lisberger, S. G., Broussard, D. M. & Bronte-Stewart, H. M. (1990) Properties of pathways that mediate motor learning in the vestibulo-ocular reflex of monkeys. *Cold Spring Harbor Symposia on Quantitative Biology* 55:813–22. [rJRB]

Lisberger, S. G. & Fuchs, A. F. (1978a) Role of primate flocculus during rapid behavioral modification of vestibulo-ocular reflex. I. Purkinje cell activity during visually guided horizontal smooth-pursuit eye movements and passive head rotation. *Journal of Neurophysiology* 41:733–63. [aJRB]

(1978b) Role of primate flocculus during rapid behavioral modification of vestibulo-ocular reflex. II. Mossy fiber firing patterns during horizontal head rotation and eye movement. *Journal of Neurophysiology* 41:764–77. [aJRB]

Lisberger, S. G. & Pavelko, T. A. (1988) Brain stem neurons in modified pathways for motor learning in the primate vestibulo-ocular reflex. *Science* 242:728–30. [rJRB]

Litvan, I., Grafman, J., Massaquoi, S., Stewart, M., Sirigu, A. & Hallett, M. (1991) Cognitive planning deficit in patients with cerebellar degeneration. *Neurology* 41 (suppl.1):225. [MH]

Llewellyn, M., Yang, J. & Prochazka, A. (1990) Human h-reflexes are smaller in difficult beam walking than in normal treadmill walking. *Experimental Brain Research* 187:321–33. [APr]

Llinas, R. (1990) *Grass lecture in neuroscience*. Conference presented at the University of Montreal. [AB]

Llinas, R. & Muhlethaler, M. (1988) Electrophysiology of guinea-pig cerebellar nuclear cells in the in vitro brain stem-cerebellar preparation. *Journal of Physiology* 404:241–58. [aJRB]

Llinas, R. & Sasaki, K. (1989) The functional organization of the olivo-cerebellar system as examined by multiple Purkinje cell recordings. *European Journal of Neuroscience* 1:587–602. [aJRB]

Lloyd, D. P. C. (1943) Conduction and synaptic transmission of the reflex response to stretch in spinal cats. *Journal of Neurophysiology* 6:317–26. [DBul]

(1946) Integrative pattern of excitation and inhibition in two-neuron reflex arcs. *Journal of Neurophysiology* 9:439–44. [TRN]

Lockery, S. R., Fang, Y. & Sejnowski, T. J. (1990) A dynamical neural network model of sensorimotor transformations in the leech. *Neural Computation* 2:274–82. [rEEF]

Loeb, E. P., Giszter, S. F., Bizzi, E. & Borghesani, P. R. (1992) Effects of dorsal root cut on the forces evoked by spinal microstimulation in the spinalized frog. *Somatosensory & Motor Research*, submitted. [rEB]

Loeb, G. E. (1983) Finding common ground between robotics and physiology (letter to the editor). *Trends in Neuroscience* 6:203–4. [aGEA]

(1984) The control and responses of mammalian muscle spindles during normally executed motor tasks. *Exercise Sport Sciences Reviews* 12:157–204. [aSCG]

(1987) Hard lessons in motor control from the mammalian spinal cord. *Trends in Neuroscience* 10:108–12. [aDAM]

Loeb, G. E., Bak, M. J. & Duysens, J. (1977) Long-term unit recording from somatosensory neurons in the spinal ganglia of the freely walking cat. *Science* 197:1192–94. [aSCG]

Loeb, G. E., He, J. & Levine, W. S. (1989) Spinal cord circuits: Are they mirrors of musculoskeletal mechanics? *Journal of Motor Behavior* 21:473–91. [GEL, CAP]

Loeb, G. E., Hoffer, J. A. & Marks, W. G. (1985) Activity of spindle afferents from cat anterior thigh muscles. 3. Effects of external stimuli. *Journal of Neurophysiology* 54:578–91. [CAP]

References

Loeb, G. E. & Levine, W. S. (1990) Linking musculoskeletal mechanics to sensorimotor neurophysiology. In: *Multiple muscle systems: Biomechanics and movement organization*, ed. J. M. Winters & S. L.-Y. Woo. Springer-Verlag. [TMH]

Logan, C. G. (1991) Cerebellar cortical involvement in excitatory and inhibitory classical conditioning. Doctoral Dissertation, Stanford University. [RFT]

Long, C. & Brown, M. E. (1962) Electromyographic kinesiology of the hand: Part III. Lumbricalis and flexor digitorum profundus to the long finger. *Archives of Physical Medicine & Rehabilitation* 43:450–60. [aSCG]

Lou, J.-S. & Bloedel, J. R. (1986) The responses of simultaneously recorded Purkinje cells to perturbations of the step cycle in the walking ferret: A study using a new analytical method – the real time postsynaptic response (RTPR). *Brain Research* 365:340–44. [arJRB]

(1988) A new conditioning paradigm: Conditioned limb movement in locomoting decerebrate ferrets. *Neuroscience Letters* 84:185–90. [aJRB]

(1992) Responses of sagittally-aligned Purkinje cells during perturbed locomotion. I. Synchronous activation of climbing fiber inputs. *Journal of Neurophysiology*, in press. [rJRB]

Low, P. A. (1984) Quantitation of muscle contraction strength. In: *Peripheral neuropathy*, ed. P. J. Dyck, P. K. Thomas, E. H. Lambert & R. Bunge. Saunders. [PRC]

Lozano-Perez, T. (1981) Automatic planning of manipulator transfer movements. *IEEE Transactions on Systems, Man, and Cybernetics*, SMC-11(10):681–98. [CIC]

Lund, S. & Broberg, C. (1983) Effects of different head positions on postural sway in man induced by a reproducible vestibular error signal. *Acta Physiologica Scandinavica* 117:307–9. [VD]

Lundberg, A. (1979a) Multisensory control of spinal reflex pathways. *Progress in Brain Research* 50:11–28. [aDAM]

(1979b) Integration in a propriospinal motor centre controlling the forelimb in the cat. In: *Integration in the nervous system*, ed. H. Asanuma & V. J. Wilson. Tokyo: Igaku-shoin. [aDAM]

(1982) Inhibitory control from the brain stem of transmission from primary afferents to motoneurons, primary afferent terminals and ascending pathways. In: *Brain stem control of spinal mechanisms*, ed. B. Sjölund & A. Björklund. Elsevier. (arDAM]

Lundberg, A. & Malmgren, K. (1988) The dynamic sensitivity of Ib inhibition. *Acta Physiologica Scandinavica* 133:123–24. [aDAM, TRN]

Lundberg, A., Malmgren, K. & Schomburg, E. D. (1977) Cutaneous facilitation of transmission in reflex pathways from Ib afferents to motoneurones. *Journal of Physiology (London)* 265:763–80. [aDAM, aSCG]

(1987a) Reflex pathways from group II muscle afferents. 2. Functional characteristics of reflex pathways to alpha-motoneurones. *Experimental Brain Research* 65:282–93. [aDAM]

(1987b) Reflex pathways from group II muscle afferents. 3. Secondary spindle afferents and the FRA: A new hypothesis. *Experimental Brain Research* 65:294–306. [aDAM, AL]

Lundberg, A. & Voorhoeve, P. E. (1961) Pyramidal activation of interneurones of various spinal reflex arcs in the cat. *Experientia* 17:46. [AL]

Luria, A. R. & Homskaya, E. D. (1964) Disturbance in the regulative role of speech with frontal lobe lesions. In: *The Frontal granular cortex and behavior*, ed. J. M. Warren & K. Akert. McGraw Hill. [JMF]

Lynch, J. C. (1980) The functional organization of posterior parietal association cortex. *Behavioral and Brain Sciences* 3:485–534. [aJFS]

Lynch, J. C., Mountcastle, V. B., Talbot, W. H. & Yin, T. C. T. (1977) Parietal lobe mechanisms for directed visual attention. *Journal of Neurophysiology* 40:362–89. [aJFS]

Lynch, J. C. & McClaren, J. W. (1989) Deficits of visual attention and saccadic eye movements after lesions of parieto-occipital cortex in monkeys. *Journal of Neurophysiology* 61:74–90. [aJFS, JWG]

Macefield, G. & Burke, D. (1990) Long-lasting depression of central synaptic transmission following prolonged high-frequency stimulation of cutaneous afferents: A mechanism for post-vibratory hypaesthesia. *Electroencephalography and Clinical Neurophysiology* 78:150–58. [aSCG]

Macefield, G., Gandevia, S. C., Bigland-Ritchie, B., Gorman, R. & Burke, D. (1991) The discharge rate of human motoneurones innervating ankle dorsiflexors in the absence of muscle afferent feedback. *Journal of Physiology* 438:219P. [rSCG]

Macefield, G., Gandevia, S. C. & Burke, D. (1990) Perceptual responses to microstimulation of single afferents innervating joints, muscles and skin of the human hand. *Journal of Physiology (London)* 429:113–29. [aSCG]

Mackay, R. P. (1954) Toward a neurology of behavior. *Neurology* 4:894–901. [JMF]

MacKay, W. A. (1982) The motor system controls what it senses. A commentary on What muscle variable(s) does the nervous system control in limb movements? by R. B. Stein. *Behavioral and Brain Sciences* 5:535–77. [PRB]

(1988a) Cerebellar nuclear activity in relation to simple movements. *Experimental Brain Research* 71:47–58. [WAM]

(1988b) Unit activity in the cerebellar nuclei related to arm reaching movements. *Brain Research* 442:240–54. [aJRB]

MacKay, W. A. & Crammond, D. J. (1987) Neuronal correlates in posterior parietal lobe of the expectation of events. *Behavioral Brain Research* 24:167–79. [JFK]

MacKay, W. A. & Murphy, J. T. (1979) Cerebellar modulation of reflex gain. *Progress in Neurobiology* 13:361–417. [aJRB, aSCG, cJFS]

MacKay, W. A. & Riehle, A. (1992) Planning a reach: Spatial analysis by area 7a neurons. In: *Tutorials in motor behavior 2*, ed. G. Stelmach & J. Requin. Elsevier. [WAM]

Mackey, M. C. & Glass, L. (1977) Oscillation and chaos in physiological control systems. *Science* 197:287–89. [DSB]

Mackey, M. C. & Milton, J. G. (1990) Feedback, delays and the origin of blood cell dynamics. *Comments in Theoretical Biology* 1(5):299–327. [AB]

Macpherson, J. M. (1988) Strategies that simplify the control of quadrupedal stance. II. Electromyographic activity. *Journal of Neurophysiology* 60:218–31. [aDAM, MLL, GMc, CAP]

Maes, P. (1991a) Learning behavior networks from experience. *Proceedings of the first European Conference on Artificial Life*. MIT Press. [SG]

(1991b) A bottom-up mechanism for behavior selection in an artificial creature. In: *From animals to animats*, ed. J. A. Meyer & S. Wilson. MIT Press. [SG]2

Magnussen, M., Enbom, H., Johansson, R. & Wiklund, J. (1990) Significance of pressor input from the human feet in lateral postural control. *Acta Otolaryngology* 110:321–27. [JDu]

Mahowald, M. & Douglas, R. (1991) A silicon neuron. *Nature (London)* 354:515–18. [rGEA]

Mai, N., Bolsinger, P., Avarello, M., Diener, H.-C. & Dichgans, J. (1988) Control of isometric finger force in patients with cerebellar disease. *Brain* 111:973–98. [aJRB]

Malamud, J. G. & Nichols, T. R. (1992) Short-range stiffness is high in slow-twitch muscle. *Biophysical Journal* 61:A294. [TRN]

Malmgren, K. & Pierrot-Deseilligny, E. (1988a) Evidence for non-monosynaptic Ia excitation of human wrist flexor motoneurones, possibly via propriospinal neurones. *Journal of Physiology (London)* 405:747–64. [DBur, AL]

(1988b) Inhibition of neurones transmitting non-monosynaptic Ia excitation to human wrist flexor motoneurones. *Journal of Physiology (London)* 405:765–83. [aSCG, AL]

Mano, N.-I., Kanazawa, I. & Yamamoto, K.-I. (1986) Complex spike activity of cerebellar Purkinje cells related to wrist-tracking movement in monkey. *Journal of Neurophysiology* 56:137–58. [aJRB]

Manor, R. S., Heilbronn, Y. D., Sherf, I. & Ben-Sira, I. (1988) Loss of accommodation produced by peristriate lesion in man? *Journal of Clinical Neuro-ophthalmology* 8(1):19–23. [JWG]

Mantel, G. W. H. & Lemon, R. N. (1987) Cross-correlation reveals facilitation of single motor units in thenar muscles by single corticospinal neurones in the concious monkey. *Neuroscience Letters* 77:113–18. [PAK]

Mao, C. C., Ashby, P., Wang, M. & McCrea, D. (1984) Synaptic connections from large muscle afferents to the motoneurones of various leg muscles in man. *Experimental Brain Research* 56:341–50. [aSCG]

Marani, E. (1981) Enzyme histochemistry. In: *Methods in neurobiology*, vol. 1, ed. R. Lahue. Plenum Press. [NHB]

Marani, E. & Voogd, J. (1977) An acetylcholinesterase band-pattern in the molecular layer of the cat cerebellum. *Journal of Anatomy* 124:335–45. [NHB]

Marchetti-Gauthier, E., Meziane, H., Devigne, C. & Soumireu-Mourat, B. (1990) Effects of bilateral lesions of the cerebellar interpositus nucleus on the conditioned forelimb flexion reflex in mice. *Neuroscience Letters* 120:34–37. [rJRB]

Marple-Horvat, D. & Stein, J. F. (1987) Cerebellar neuronal activity related to arm movements in trained rhesus monkeys. *Journal of Physiology (London)* 394:351–66. [aJRB]

Marr, D. (1969) A theory of cerebellar cortex. *Journal of Physiology* 202:437–70. [VB, PFCG, RFT]

(1982) *Vision*. Freeman. [rGEA, HH, GS, IT]

Marr, D. C. & Poggio, T. (1977) From understanding computation to understanding neural circuitry. *Neuroscience Research Progress Bulletin* 15:470–88. [PDN]

Marsden, C. D. (1982) The mysterious motor function of the basal ganglia: The Robert Wartenburg lecture. *Neurology* 32:514–39. [aGEA]

(1984) Which motor disorder in Parkinson's disease indicates the true motor function of the basal ganglia? In: *Functions of the basal ganglia*, ed. D. Evered. & M. O'Connor. Pitman. [JJS]

(1987) What do the basal ganglia tell premotor cortical areas? *Ciba Foundation Symposium* 132:282–300. [RI]

Marsden, C. D., Obeso, J. A. & Rothwell, J. C. (1983) The function of the antagonistic muscle during fast limb movements in man. *Journal of Physiology (London)* 335:1–13. [NL]

Marsden, C. D., Rothwell, J. C. & Day, B. L. (1983) Long-latency automatic responses to muscle stretch in man: Origin and function. In: *Motor control mechanisms in health and disease*, ed. J. E. Desmedt. Raven Press. [aSCG]

Marshall, K. & Xiong, H. (1991) Modulation of amino acid neurotransmitter actions by other neurotransmitters: Some examples. *Canadian Journal of Physiology and Pharmacology* 69:1115–22. [aDAM]

Martin, J. H. & Ghez, C. (1985) Task-related coding of stimulus and response in cat motor cortex. *Experimental Brain Research* 57:427–42. [aEEF]

Martin, L. & Müller, G. E. (1899) *Zur Analyse der Unterschiedsempfindlichkeit*. Barth. [AMG]

Martino, A. M. & Strick, P. L. (1987) Corticospinal projections originate from the arcuate premotor area. *Brain Research* 404:307–12. [aGEA]

Massaquoi, S. G. & Hallett, M. (1991) The kinematics of the decomposition of two joint arm movement initiation in normals and patients with cerebellar ataxia. *Society for Neuroscience Abstracts* 17:1381. [MH]

Massone, L. & Bizzi, E. (1989) A neural network model for limb trajectory formation. *Biological Cybernetics* 61:417–25. [aGEA, AAF]

Matelli, M., Luppino, G., Fogassi, L. & Rizzolatti, G. (1989) Thalamic input to inferior area 6 and area 4 in the macaque monkey. *Journal of Comparative Neurology* 280:468–88. [aGEA]

Matheson, J., Hallett, M., Berardelli, A., Weinhaus, R. & Inzucchi, S. (1985) Failure to confirm a correlation between electromyogram and final position. *Human Neurobiology* 4:257–60. [MH]

Matthews, P. B. C. (1959) The dependence of tension upon extension in the stretch reflex of the soleus of the decerebrate cat. *Journal of Physiology* 47:521–46. [MLL]

(1972) *Mammalian muscle receptors and their central actions*. Williams & Wilkins. [aEB, aSCG, RBS]

(1982) Where does Sherrington's "muscular sense" originate? Muscles, joints, corollary discharges? *Annual Review of Neuroscience* 5:189–218. [aSCG]

(1984) Evidence from the use of vibration that the human long-latency stretch reflex depends upon spindle secondary afferent. *Journal of Physiology (London)* 348:383–415. [aSCG]

(1988) Proprioceptors and their contribution to somatosensory mapping: Complex messages require complex processing. *Canadian Journal of Physiology and Pharmacology* 66:403–38. [aSCG]

(1989) Long-latency stretch reflexes of two intrinsic muscles of the human hand analysed by cooling the arm. *Journal of Physiology (London)* 419:519–38. [aSCG]

(1991) The human stretch reflex and the motor cortex. *Trends in Neurosciences* 14:87–91. [aSCG]

Matthews, P. B. C. & Stein, R. B. (1969) The sensitivity of muscle spindle afferents to small sinusoidal changes of length. *Journal of Physiology (London)* 200:723–43. [aEB]

Mauk, M. D. & Thompson, R. F. (1987) Retention of classically conditioned eyelid responses following acute decerebration. *Brain Research* 403:89–95. [rJRB, RFT]

Maunsell, J. H. R. & Newsome, W. T. (1987) Visual processing in monkey extrastriate cortex. *Annual Review of Neuroscience* 10:363–401. [cGEA]

Mayorov, V. I., Savchenko, E. I. & Kotlyar, B. I. (1977) Transformation of tactile afferent stimulus into motor command in the motor cortex in the cat. *Neurofisiologia* 9:115–23. [MEI]

Mays, L. E. & Sparks, D. L. (1980) Dissociation of visual and saccade-related responses in superior colliculus neurons. *Journal of Neurophysiology* 43:207–32. [aDAR, aJFS, JWG]

Mazzini, L. & Schieppati, M. (1992) Activation of the neck muscles from the ipsi- or contralateral hemisphere during voluntary head movements in humans. A reaction time study. *Electroencephalography and Clinical Neurophysiology*, in press. [MS]

McCloskey, D. I. (1973) Differences between the senses of movement and position shown by the effects of loading and vibration of muscles in man. *Brain Research* 61:119–31. [aSCG]

(1978) Kinesthetic sensibility. *Physiological Reviews* 58:763–820. [aSCG]

(1981) Corollary discharges: Motor commands and perception. In: *Handbook of physiology*. The nervous system, vol. 3, Motor control, ed. V. B. Brooks. American Physiological Society. [aSCG]

McCloskey, D. I., Colebatch, J. G., Potter, E. K. & Burke, D. (1983a) Judgments about onset of rapid voluntary movements in man. *Journal of Neurophysiology* 49:851–63. [aSCG]

McCloskey, D. I., Cross, M. J., Honner, R. & Potter, E. K. (1983b) Sensory effects of pulling or vibrating exposed tendons in man. *Brain* 106:21–37. [arSCG]

McCloskey, D. I., Ebeling, P. & Goodwin, G. M. (1974) Estimation of weights and tensions and apparent involvement of a "sense of effort." *Experimental Neurology* 42:220–32. [aSCG]

McCollum, G. (1992) Reciprocal inhibition, synergies and movements, submitted. [GMc]

McCormick, D. A. & Thompson, R. F. (1984) Neuronal responses of the rabbit cerebellum during acquisition and performance of a classically conditioned nictitating membrane-eyelid response. *Journal of Neuroscience* 4:2811–22. [aJRB]

McCrea, D. A. (1986) Spinal cord circuitry and motor reflexes. In: *Exercise and sports medicine*, ed. K. Pandolf. Macmillan. [aDAM, TRN]

McCrea, D. A., Pratt, C. A. & Jordan, L. M. (1980) Renshaw cell activity and recurrent effects on motoneurons during fictive locomotion. *Journal of Neurophysiology* 44:475–88. [aDAM, CAP]

McCurdy, M. L. & Hamm, T. M. (1991) Strength of recurrent inhibitory postsynaptic potentials between closely spaced homonymous and heteronymous motoneurons in the cat. *Society for Neuroscience Abstracts* 17:644. [TMH]

(1992) Recurrent collaterals of motoneurons projecting to distal muscles in the cat hindlimb. *Journal of Neurophysiology* 67:1359–67. [TMH]

McDevitt, C. J., Ebner, T. J. & Bloedel, J. R. (1987a) Changes in the responses of cerebellar nuclear neurons associated with the climbing fiber response of Purkinje cells. *Brain Research* 425:14–24. [rJRB]

(1987b) Relationships between simultaneously recorded Purkinje cells and nuclear neurons. *Brain Research* 425:1–13. [rJRB]

McIntyre, J. (1988) Reflexes and the equilibrium point control model. *Society for Neuroscience Abstracts* 14:951. [arEB, TRN]

(1990) Utilizing elastic system properties for the control of posture and movement. Doctoral Dissertation, Department of Brain and Cognitive Sciences, MIT. [arEB, AGF, GLG]

McIntyre, J. & Bizzi, E. (1992) Servo models for the biological control of movement. *Journal of Motor Behavior*, in press. [aEB]

McIntyre, A. K., Proske, U. & Rawson, J. A. (1984) Cortical projection of afferent information from tendon organs in the cat. *Journal of Physiology (London)* 354:395–406. [aSCG]

(1985) Pathway to the cerebral cortex for impulses from tendon organs in the cat's hindlimb. *Journal of Physiology (London)* 369:115–26. [aSCG]

(1989) Corticofugal action on transmission of group I input from the hindlimb to the pericruciate cortex in the cat. *Journal of Physiology (London)* 416:19–30. [aSCG]

McIntyre, A. K., Proske, U. & Tracey, D. J. (1978) Afferent fibres from muscle receptors in the posterior nerve of the cat's knee joint. *Experimental Brain Research* 33:415–24. [aSCG]

McKeon, B. & Burke, D. (1983) Muscle spindle discharge in response to contraction of single motor units. *Journal of Neurophysiology* 49:291–302. [aSCG]

McKeon, B., Hogan, N. & Bizzi, E. (1984) Effect of temporary path constraint during planar arm movement. *Society for Neuroscience Abstracts* 10:337. [aEB, GEL]

Mel, B. W. (1991) A connectionist model may shed light on neural mechanisms for visually guided reaching. *Journal of Cognitive Neuroscience* 3:273–92. [aGEA]

Menetrey, D., de Pommery, J. & Besson, J. M. (1984) Electrophysiological characteristics of lumbar spinal cord neurons backfired from lateral reticular nucleus in the rat. *Journal of Neurophysiology* 52:595–612. [JDu]

Merton, P. A. (1953) Speculations on the servo control of movement. In: *The spinal cord*, ed. G. E. W. Wolstenholme. Churchill. [aEB, aSCG, rEB, GCA, AGF, GLG, TRN]

(1964) Human position sense and sense of effort. In: Homeostasis and feedback mechanisms. *Symposia of the Society for Experimental Biology* #18. Academic Press. [PRB]

Mesulam, H.-M. (1981) A cortical network for directed attention and unilateral neglect. *Annals of Neurology* 10:309–25. [aJFS]

(1985) Patterns in behavioral neuroanatomy: Association areas, the limbic system, and hemispheric specialization. In: *Principles of behavioral neurology*, ed. M.-M. Mesulam. Davis. [BS]

(1990) Large-scale neurocognitive networks and distributed processing for attention, language, and memory. *Annals of Neurology* 28:597–613. [BS]

Meunier, S. & Morin, C. (1989) Changes in presynaptic inhibition of Ia fibres to soleus motoneurones during voluntary dorsiflexion of the foot. *Experimental Brain Research* 76:510–18. [aSCG]

Meunier, S., Penicaud, A., Pierrot-Deseilligny, E. & Rossi, A. (1990) Monosynaptic Ia excitation and recurrent inhibition from quadriceps to ankle flexors and extensors in man. *Journal of Physiology (London)* 423:661–75. [aSCG]

Meunier, S. & Pierrot-Deseilligny, E. (1989) Gating of the afferent volley of the monosynaptic stretch reflex during movement in man. *Journal of Physiology (London)* 419:753–63. [aSCG]

Mewes, K. (1988) Characteristics of rubromotoneuronal cells and their role in the control of the hand in primates. Doctoral Dissertation, University of Kansas. [aEEF]

Meyer, D. E., Yantis, S., Osman, A. M. & Smith, J. E. K. (1985) Temporal properties of human information processing: Tests of discrete versus continuous models. *Cognitive Psychology* 17:445–518. [DJ]

Meyer-Lohmann, J., Riebold, W. & Robrecht, D. (1974) Mechanical influence of the extrafusal muscle on the static behaviour of de-efferented primary muscle spindle endings in the cat. *Pflügers Archives* 352:267–78. [aSCG]

Miles, F. A., Braitman, D. J. & Dow, B. M. (1980) Long-term adaptive changes in primate vestibulo-ocular reflex. IV. Electrophysiological observations in flocculus of adapted monkeys. *Journal of Neurophysiology* 43:1477–93. [rJRB, cJFS]

Millar, J. (1973) Joint afferent fibres responding to muscle stretch, vibration and contraction. *Brain Research* 63:380–83. [aSCG]

Miller, G. A. (1956) The magic number seven, plus or minus two: Some limits on our capacity for processing information. *The Psychological Review* 63:81–97. [FJC]

Milne, R. J., Aniss, A. M., Kay, N. E. & Gandevia, S. C. (1988) Reduction in perceived intensity of cutaneous stimuli during movement: A quantitative study. *Experimental Brain Research* 70:569–76. [aSCG]

Milner, A. D. & Goodale, M. A. (in press) Visual pathways to perception and action. In: *The visually responsive neuron: From basic neurophysiology to behavior*, ed. T. P. Hicks, S. Molotchnikoff & T. Ono. Elsevier. [DPC, MAG]

Milner, T. E. (1986) Judgment and control of velocity in rapid voluntary movements. *Experimental Brain Research* 62:99–110. [aSCG]

Mink, J. W. & Thach, W. T. (1991a) Basal ganglia motor control. II. Late pallidal timing relative to movement onset and inconsistent pallidal coding of movement parameters. *Journal of Neurophysiology* 65:301–329. [DSB]

(1991b) Basal ganglia motor control, III. Pallidal ablation: Normal reaction time, muscle co-contraction and slow movement. *Journal of Neurophysiology* 65:330–51. [GMa]

Minsky, M. (1986) *The society of mind.* Simon and Schuster. [SG]

Mitchell, S. J., Richardson, R. T., Baker, F. H. & DeLong, M. R. (1987) The primate globus pallidus: Neuronal activity related to direction of movement. *Experimental Brain Research* 68:491–505.

Miyamoto, H., Kawato, M., Setoyama, T. & Suzuki, R. (1988) Feedback-error-learning neural network for trajectory control of a robotic manipulator. *Neural Networks* 1:251–65. [aGEA]

Moberg, E. (1983) The role of cutaneous afferents in position sense, kinesthesia, and motor function of the hand. *Brain* 106:1–19. [JDu]

Monoud, P., Mayer, E. & Hauert, C. A. (1979) Preparation of actions to lift objects of varying weight and texture in the adult. *Journal of Human Movement Studies* 5:209–15. [HER]

Moore, G. P., Segundo, J. P., Perkel, D. H. & Levitan, H. (1970) Statistical signs of synaptic interaction in neurons. *Biophysical Journal* 10:876–900. [PAK]

Moorhead, I. R., Haig, N. D. & Clement, R. A. (1989) An investigation of trained neural networks from a neurophysiological perspective. *Perception* 18:793–803. [RE]

Morasso, P. (1981) Spatial control of arm movements. *Experimental Brain Research* 42:223–27. [aEB]

Morasso, P. & Mussa-Ivaldi, F. A. (1982) Trajectory formation and handwriting: A computational model. *Biological Cybernetics* 45:131–42. [rEB]

Morasso, P. & Sanguineti, V. (1991) Neurocomputing concepts in motor control. In: *Brain and space*, ed. J. Paillard. Oxford University Press. [PM]

(1992) Neurocomputing aspects in modelling cursive handwriting. *Acta Psychologica*, in press. [PM]

Morel, A. & Bullier, J. (1990) Anatomical segregation of two cortical visual pathways in the macaque monkey. *Visual Neuroscience* 4:555–78. [RE]

Morgan, D. L., Prochazka, A. & Proske, U. (1984) Can fusimotor activity potentiate the responses of muscle spindles to a tendon tap? *Neuroscience Letters* 50:209–15. [aSCG]

Morgan, M. J. (1977) *Molyneux's question.* Cambridge University Press. [aJFS]

Mori, S. (1987) Integration of posture and locomotion in acute decerebrate cats and in awake, freely moving cats. *Progress in Neurobiology* 28:161–95. [GMa]

Morin, C., Katz, R., Mazières, L. & Pierrot-Deseilligny, E. (1982) Comparison of soleus H reflex facilitation at the onset of soleus contractions produced voluntarily and during the stance phase of human gait. *Neuroscience Letters* 33:47–54. [aSCG]

Morin, R. E. & Grant, D. A. (1955) Learning and performance of a key-pressing task as a function of the degree of spatial stimulus-response correspondence. *Journal of Experimental Psychology* 49:39–47. [RWP]

Moritani, R. & DeVries, H. A. (1978) Reexamination of the relationship between the surface integrated electromyogram and force of isometric contraction. *American Journal of Physical Medicine* 57:263–77. [MH]

Motter, B. C. & Mountcastle, V. B. (1981) The functional properties of the light sensitive neurons of the posterior parietal cortex studied in waking monkeys: Foveal sparing and opponent vector organization. *Journal of Neuroscience* 1:3–26. [aJFS]

Mountcastle, V. B., Andersen, R. A. & Motter, B. C. (1981) The influence of attentive fixation upon the excitability of the light sensitivity neurones of the posterior parietal cortex. *Journal of Neuroscience* 1:1218–35. [aJFS, JFK]

Mountcastle, V. B., Lynch, J. C., Georgopoulos, A., Sakata, H. & Acuna, C. (1975) Posterior parietal association cortex of the monkey: Command functions for operations within extrapersonal space. *Journal of Neurophysiology* 38:871–908. [aEEF, aJFS]

Mountcastle, V. B. & Steinmetz, M. A. (1990) The parietal visual system and some aspects of visuospatial perception. In: *From neuron to action: An appraisal of fundamental and clinical research*, ed. L. Deecke, J. C. Eccles & V. B. Mountcastle. Springer-Verlag. [DPC]

Muakkassa, K. F. & Strick, P. L. (1979) Frontal lobe inputs to primate motor cortex: Evidence for four somatotopically organized 'premotor' areas. *Brain Research* 177:176–82. [JMF]

Mugnaini, E. (1972) The histology and cytology of the cerebellar cortex. In: *The comparative anatomy and histology of the cerebellum. The human cerebellum, cerebellar connections and cerebellar cortex*, ed. O. Larsell & J. Jansen. University of Minnesota Press. [aJRB]

Muir, R. B. & Lemon, R. N. (1983) Corticospinal neurons with a special role in precision grip. *Brain Research* 261:312–16. [aEEF, cSCG, AL]

Murdoch, B., Jr. (1982) A theory for the storage and retrieval of item and associative information. *Psychological Review* 89:609–26. [BB]

Mushiake, H., Inase, M. & Tanji, J. (1990) Selective coding of motor sequence in the supplementary motor area of the monkey cerebral cortex. *Experimental Brain Research* 82:208–210. [aGEA]

(1991) Neuronal activity in the primate premotor, supplementary, and precentral motor cortex during visually guided and internally determined sequential movements. *Journal of Neurophysiology* 66:705–18. [rEEF, JT]

Mussa-Ivaldi, F. A. (1988) Do neurons in the motor cortex encode movement direction? An alternative hypothesis. *Neuroscience Letters* 91:106–111. [aEB, aEEF]

Mussa-Ivaldi, F. A., Bizzi, E. & Giszter, S. F. (1991a) Transforming plans into actions by tuning passive behavior: A field-approximation approach. In: *Proceedings of the 1991 IEEE International Symposium on Intelligent Control.* IEEE Control Systems Society. [CIC, SG]

Mussa-Ivaldi, F. A., Giszter, S. F. & Bizzi, E. (1990) Motor-space coding in the central nervous system. *Cold Spring Harbor Symposia on Quantitative Biology* 55:827–35. [arEB, SG]

(1991b) A field approximation approach to the execution of motor plans. *Fifth International Conference on Advanced Robotics*, Pisa, Italy. [SG]

Mussa-Ivaldi, F. A. & Hogan, N. (1991) Integrable solutions of kinematic redundancy via impedance control. *International Journal of Robotics Research* 10:481–91. [rEB]

Mussa-Ivaldi, F. A., Hogan, N. & Bizzi, E. (1985) Neural, mechanical and geometric factors subserving arm posture in humans. *Journal of Neuroscience* 5:2732–43. [aGEA, arEB, ZH, GEL]

(1987) The role of geometrical constraints in the control of multi-joint posture and movement. *Society for Neuroscience Abstracts* 13:347. [arEB, GEL]

Mussa-Ivaldi, F. A., Morasso, P. & Zaccaria, R. (1988) Kinematic networks. A distributed model for representing and regularizing motor redundancy. *Biological Cybernetics* 60:1–16. [aGEA, rEB]

Nagaoka, M. & Tanaka, R. (1981) Contribution of kinesthesia on human visuomotor elbow tracking movements. *Neuroscience Letters* 26:245–49. [aSCG]

Nambu, A., Yoshida, S. & Jinnai, K. (1990) Discharge patterns of pallidal neurons with input from various cortical areas during movement in the monkey. *Brain Research* 519:183–91. [aGEA]

Nardone, A., Di Francesco, G. & Schieppati, M. (1986) Reflex excitability of the soleus muscle in standing man, at rest or prior to a voluntary triceps contraction. *Neuroscience Letters* 26:S165. [MS]

Nardone, A., Romanò, C. & Schieppati, M. (1989) Selective recruitment of high threshold human motor units during voluntary isotonic lengthening of active muscles in humans. *Journal of Physiology (London)* 409:451–71. [MS]

Nardone, A. & Schieppati, M. (1988) Postural adjustments associated with voluntary contraction of leg muscles in standing man. *Experimental Brain Research* 69:469–80. [MS]

Nashner, L. M. (1976) Adapting reflexes controlling the human posture. *Experimental Brain Research* 26:59–72. [VD]

Nashner, L. M. & Grimm, R. J. (1978) Analysis of multiloop dyscontrols in standing cerebellar patients, ed. J. E. Desmedt. Karger. *Progress in Clinical Neurophysiology* 4:300–319. [aJRB, FBH]

Nashner, L. M. & Wolfson, P. (1974) Influence of head position and proprioceptive cues on short latency postural reflexes evoked by galvanic stimulation of the human labyrinth. *Brain Research* 67:255–68. [VD]

Nathan, P. W. & Sears, T. A. (1960) Effects of posterior root section on the activity of some muscles in man. *Journal of Neurology, Neurosurgery, and Psychiatry* 23:10–22. [aSCG]

Neilson, P. D. (1991) *The problem of redundancy in movement control: The adaptive model theory approach.* Workshop, Curtin University of Technology, Perth. [PDN]

Neilson, P. D., Neilson, M. D. & O'Dwyer, N. J. (1988a) Internal models and intermittency: A theoretical account of human tracking behavior. *Biological Cybernetics* 58:101–112. [PDN]

(1988b) Redundant degrees of freedom in speech control: A problem or a virtue? In: *Proceedings of the Second Australian Conference on Speech Science and Technology*, ed. M. Wagner. Australian Speech Science and Technology Association. [PDN]

(1992a) Adaptive model theory: Application to disorders of motor control. In: *Approaches to the study of motor control and learning. Advances in psychology* 84, ed. J. J. Summers. North-Holland. [PDN]

(1992b) What limits high speed tracking performance? *Human Movement Science*, in press. [PDN]

Neilson, P. D., O'Dwyer, N. J. & Neilson, M. D. (1988) Stochastic prediction in pursuit tracking: An experimental test of adaptive model theory. *Biological Cybernetics* 58:113–22. [PDN]

Neisser, U. (1976) *Cognition and reality: Principles and implications of cognitive psychology.* Freeman. [JMF]

Nelson, W. L. (1983) Physical principles for economies of skilled movements. *Biological Cybernetics* 46:135–47. [aGEA]

Newell, K. M., Carlton, L. G. & Hancock, P. A. (1984) Kinetic analysis of response variability. *Psychological Bulletin* 96:133–51. [aSCG]

Newman, W. S. & Hogan, N. (1986) High speed robot control and obstacle avoidance using dynamic potential functions. *Technical Report TR-86-042*, North American Philips. [CIC]

Newsome, W. T. & Pare, E. B. (1988) A selective impairment of motion perception following lesions of the middle temporal visual area (MT). *Journal of Neuroscience* 8:2201–11. [RE]

Newsome, W. T., Wurtz, R. H., Dursteler, M. R. & Mikami, A. (1985) Deficits in visual motion processing following visual ibotenic acid lesions of the middle temporal visual area of the macaque monkey. *Journal of Neuroscience* 5:825–40. [RE]

Nichols, T. R. (1973) Reflex and non-reflex stiffness of soleus muscle in the cat. In: *Control of posture and locomotion*, ed. R. B. Stein, K. G. Pearson, R. S. Smith & J.B. Redford. Plenum Press. [TRN]

(1987a) Coordination of muscular action in cat hindlimb by proprioceptive spinal pathways. *Neurosurgery: State of the Art Reviews* 4:303–14. [TRN]

(1987b) The regulation of muscle stiffness, implications for the control of limb stiffness. *Medicine and Sport Science* 26:36–47. [NL]

(1989) The organization of heterogenic reflexes among muscles crossing the ankle joint in the decerebrate cat. *Journal of Physiology* 410:463–77. [MLL, TRN, CAP]

Nichols, T. R. & Bonasera, S. J. (1990) Heterogenic reflex organization which mirrors specialized activation patterns of flexor digitorum longus and flexor hallucis longus muscles in the cat. *Society for Neuroscience Abstracts* 16:887. [TRN]

Nichols, T. R. & Houk, J. C. (1976) Improvement in linearity and regulation of stiffness that results from actions of stretch reflex. *Journal of Neurophysiology* 39:119–42. [NL, TRN]

Nichols, T. R. & Koffler-Smulevitz, D. (1991) A mechanical analysis of heterogenic inhibition between soleus muscle and the pretibial flexors in the cat. *Journal of Neurophysiology* 66:1139–55. [TRN]

Nichols, T. R. & Steeves, J. D. (1986) Resetting of resultant stiffness in ankle flexor and extensor muscles in the decerebrate cat. *Experimental Brain Research* 62:401–10. [MLL]

Nicoletti, R. & Umilta, C. (1989) Splitting visual space with attention. *Journal of Experimental Psychology: Human Perception and Performance* 15:164–69. [RWP]

Nielsen, J. & Pierrot-Deseilligny, E. (1991) Pattern of cutaneous inhibition of the propriospinal-like excitation to human upper limb motoneurones. *Journal of Physiology (London)* 434:169–82. [aSCG, DBur]

Niemi, P. & Näätänen, R. (1981) Foreperiod effect and simple reaction time. *Psychological Bulletin* 89:133–62. [DJ]

Niki, H. & Watanabe, M. (1976) Prefrontal unit activity and delayed response: Relation to cue location versus direction of response. *Brain Research* 105:79–88. [aEEF, aJFS]

Noga, B., Bras, H. & Jankowska, E. (1992) Transmission from group II muscle afferents is depressed by stimulation of locus coeruleus/subcoeruleus, Kölliker-Fuse and raphe nuclei in the cat. *Experimental Brain Research* 88:502–16. [aDAM]

Noga, B., Shefchyk, S. & Jordan, L. (1987) The role of Renshaw cells in locomotion: Antagonism of their excitation from motor axon collaterals with intravenous mecamylamine. *Journal of Neurophysiology* 66:99–105. [rDAM]

Nordholm, A. F., Lavond, D. A. & Thompson, R. F. (1991) Are eyeblink responses to tone in the decerebrate, decerebellate rabbit conditioned responses? *Behavioural and Brain Research* 44:27–34. [rJRB, PFCG, RFT]

Nudo, R. J., Jenkins, W. M. & Merzenich, M. M. (1990) Repetitive microstimulation alters the cortical representation of movements in adult cats. *Somatosensory and Motor Research* 7:463–83. [aGEA]

Nutt, J., Horak, F. & Frank, J. (1992) Scaling of postural responses to Parkinson's disease. In: *Posture and gait: Control mechanisms*, ed. M. Woollacott & F. Horak. University of Oregon Press. [FBH]

Ochoa, J. L. & Torebjörk, H. E. (1983) Sensations evoked by intraneural microstimulation of single mechanoreceptor units innervating the human hand. *Journal of Physiology (London)* 342:633–54. [aSCG]

O'Donovan, M. J., Pinter, M. J., Dum, R. P. & Burke, R. E. (1982) Actions of FDL and FHL muscles in intact cats: Functional dissociation between anatomical synergists. *Journal of Neurophysiology* 47(6):1126–43. [TMH]

Ogle, K. (1962) The optical space sense. In: *The eye*, vol. 4, ed. H. Davson. Academic Press. [aJFS]

Ohtsuka, K., Maekawa, H., Takeda, M. Uede, N. & Chiba, S. (1988) Accommodation and convergence insufficiency with left middle cerebral artery occlusion. *American Journal of Ophthalmology* 106:60–64. [JWG]

Ojala, J. M., Matikainen, E. & Groop, L. (1985) Body sway in diabetic neuropathy. *Journal of Neurology* 232:188. [PRC]

O'Keefe, J. & Nadel, L. (1978) *The hippocampus as a cognitive map.* Oxford University Press. [aJFS]

Optican, L. M. & Robinson, D. A. (1980) Cerebellar dependent adaptive control of the primate saccadic system. *Journal of Neurophysiology* 44:1058–76. [aJRB]

Oscarsson, O. (1979) Functional units of the cerebellum-sagittal zones and microzones. *Trends in Neuroscience* 2:143–45. [aJRB]

Ostry, D. J., Feldman, A. G. & Flanagan, J. R. (1991) Kinematics and control of frog hindlimb movements. *Journal of Neurophysiology* 65:547–62. [SG]

Ostry, D. J., Flanagan, J. R., Feldman, A. G. & Munhall, K. G. (1992) Human jaw movement kinematics and control. In: *Tutorials in motor behavior*. II, ed. G. E. Stelmach & J. Requin. North-Holland. [DJO]

Ott, E., Grebogi, C. & Yorke, J. A. (1990) Controlling chaos. *Physical Review Letters* 64:1196–99. [TLC]

Ottes, F. P., Van Gisbergen, J. A. M. & Eggermont, J. J. (1986) Visuomotor fields of the superior colliculus: A quantitative model. *Vision Research* 26:857–73. [JAMVG]

Pailhous, J. & Bonnard, M. (1992) Steady-state fluctuations of human walking. *Behavioral Brain Research* 47:181–90. [GMa]

Paillard, J. (1955) Reflexes et regulations d'origine proprioceptive chez l'Homme. *Etude neurophysiologique et psychophysiologique.* Arnette. [JP]

(1959) Functional organization of afferent innervation studied in man by monosynaptic testing. *American Journal of Physical Medicine* 38:239–47. [JP]

(1988) Posture and locomotion: Old problems and new concepts. In: *Posture and gait. Development, adaptation and modulation*, ed. B. Amblard, A. Berthoz & F. Clarac. Elsevier. [JP]

(1991) *Brain and space.* Oxford University Press. [JP]

Paillard, J. (1991) Motor and representational framing of space. In: *The brain and space*, ed. J. Paillard. Oxford University Press. [aJFS]

Paillard, J. & Brouchon, M. (1968) Active and passive movements in the calibration of position sense. In: *The neuropsychology of spatially oriented behaviour*, ed. S. J. Freedman. Dorsey Press. [aSCG]

(1974) A proprioceptive contribution to the spatial encoding of position cues for ballistic movements. *Brain Research* 71:273–84. [JP]

Palmer, S. S. & Fetz, E. E. (1985) Discharge properties of primate forearm motor units during isometric muscle activity. *Journal of Neurophysiology* 54:1178–93. [aEEF]

References

Pandya, D. N. & Kuypers, H. G. J. M. (1969) Cortico-cortical connections in the rhesus monkey. *Brain Research* 13:13–36. [aJFS]

Pandya, D. N. & Seltzer, B. (1982) Intrinsic connections and architectonics of the posterior parietal cortex in the rhesus monkey. *Journal of Comparative Neurology* 204:196–210. [aJFS, BS]

Pandya, D. N. & Vignolo, L. A. (1971) Intra- and interhemispheric projections of the precentral, premotor and arcuate areas in the rhesus monkey. *Brain Research* 26:217–33. [JMF]

Parent, A., Bouchard, C. & Smith, Y. (1984) The striatopallidal and striatonigral projections: Two distinct fiber systems in primate. *Brain Research* 303:385–90. [aGEA]

Parent, A. & De Bellefeuille, L. (1982) Organization of efferent projections from the internal segment of globus pallidus in primate as revealed by fluorescence retrograde labeling method. *Brain Research* 245:201–13. [aGEA]

Paul, R. P. (1981) *Robot manipulators: Mathematics, programming and control*. MIT Press. [aGEA]

(1987) Problems and research issues associated with the hybrid control of force and displacement. *Proceedings of the IEEE Conference on Robotics and Automation*. IEEE Computer Society Press. [aEB]

Paulin, M., Nelson, M. & Bower, J. M. (1989) Dynamics of compensatory eye movement control: An optimal estimation analysis of the vestibulo-ocular reflex. *International Journal of Neural Systems* 1:23–29. [JMB]

Pavlova, O. G. (1979) Alimentary instrumental reflex to electrical stimulation of lateral and ventro-medial hypothalamus. Doctoral Dissertation, Institute of Higher Nervous Activity & Neurophysiology. [MEI]

Pearson, K. G. (1985) Are there central pattern generators for walking and flight in insects? In: *Feedback and neural control in invertebrates and vertebrates*, ed. W. J. P. Barnes & M. H. Gladden. Croom Helm. [CAP]

Pearson, K. G. & Duysens, G. (1976) Function of segmental reflexes in the control of stepping in cockroaches and cats. In: *Neural control of locomotion*, ed. K. M. Herman, S. Grillner, P. S. G. Stein & G. Stuart. Plenum Press. [MBB]

Peck, D., Buxton, D. F. & Nitz, A. (1984) A comparison of spindle concentrations in large and small muscles acting in parallel combinations. *Journal of Morphology* 180:243–52. [aSCG]

Pellionisz, A. (1985) Tensorial brain theory in cerebellar modelling. In: *Cerebellar functions*, ed J. R. Bloedel, J. Dichgans & W. Precht. Springer Verlag. [aJRB, VB]

(1988) Tensorial aspects of the multidimensional massively parallel sensorimotor function of neuronal networks. *Progress in Brain Research* 76:341–54. [aGEA]

(1989) Tensor network model of the cerebellum and its olivary system. In: *The olivo-cerebellar system in motor control*, ed. P. Strata. Proceedings of the ENA-IBRO Symposium, Torino, Italy. Springer Verlag. [aJRB]

(1991) *The geometry of brain function: Tensor network theory*. Cambridge University Press, in press. [aJRB]

Pellionisz, A. & Llinás, R. (1979) Brain modeling by tensor network theory and computer simulation. The cerebellum: Distributed processor for predictive coordination. *Neuroscience* 4:323–48. [aJRB, PDN]

(1980) Tensor approach to the geometry of brain function: Cerebellar coordination via metric tensor. *Neuroscience* 5:1125–36. [aJRB, aDAR, TLC, PDN]

(1982) Space-time representation in the brain. The cerebellum as a predictive space-time metric tensor. *Neuroscience* 7:2949–70. [aJRB, PDN]

(1985) Tensor network theory of the metaorganization of functional geometries in the central nervous system. *Neuroscience* 16:245–73. [aJRB, PDN]

Pellionisz, A. & Peterson, B. W. (1988) A tensorial model of neck motor activation. In: *Control of head movement*, ed. B. W. Peterson & F. Richmond. Oxford University Press. [aJRB]

Péllison, D., Prablanc, C., Goodale, M. A. & Jeannerod, M. (1986) Visual control of reaching movements without vision of the limb. II. Evidence of fast unconscious process correcting the trajectory of the hand to the final position of a double-step stimulus. *Experimental Brain Research* 339:136–40. [SVA]

Penfield, W. & Boldrey, E. H. (1937) Somatic motor and sensory representation in the cerebral cortex of man as studied by electrical stimulation. *Brain* 60:389–443. [aEEF]

Perenin, M.-T. & Vighetto, A. (1988) Optic ataxia: A specific disruption in visuomotor mechanisms. I. Different aspects of the deficit in reaching for objects. *Brain* 111:643–674. [MAG]

Perret, C. & Cabelguen, J. M. (1976) Central and reflex participation in the timing of locomotor activations of a bifunctional muscle, the semi-tendinosus, in the cat. *Brain Research* 106:390–95. [aDAM]

(1980) Main characteristics of the hindlimb locomotor cycle in the decorticate cat with special reference to bifunctional muscles. *Brain Research* 187:333–52. [aDAM]

Petit, J., Filippi, G. M., Emonet-Denand, F., Hunt, C. C. & Laporte, Y. (1990) Changes in muscle stiffness produced by motor units of different types in peroneus longus muscle in the cat. *Journal of Neurophysiology* 63:190–97. [TRN]

Pew, R. W. (1984) A distributed processing view of human motor control. In: *Cognition and motor processes*, ed. W. Prinz & F. Sanders. Springer-Verlag. [JJS]

(1989) Human perceptual-motor performance. In: *Human information processing: Tutorials in performance and cognition*, ed. B. H. Kantowitz. Erlbaum. [aGEA]

Pfeiffer, F., Weidemann, H.-J. & Danowski, P. (1991) Dynamics of the walking stick insect. *Proceedings of the 1990 IEEE International Conference on Robotics and Automation*, Cincinnati, vol. 3. IEEE Computer Society Press. [JDe]

Philips, C. G. (1985) *Movements of the hand*. Liverpool University Press. [aSCG]

Phillips, C. G. & Porter, R. (1977) *Corticospinal neurones. Their role in movement*. Academic Press. [aSCG, MEI]

Piaget, J. (1963) *The origin of intelligence in children*. Norton. [PM]

Piatnitsky, E. S. (1988) Decomposition principle in mechanical system control. *Doklady AN SSSR* vol. 300 2:300–303. [AAF]

Pierrot-Deseilligny, E., Bergego, C. & Katz, R. (1982) Reversal in cutaneous control of Ib pathways during human voluntary contraction. *Brain Research* 233:400–403. [aSCG]

Pierrot-Deseilligny, E., Bergego, C., Katz, R. & Morin, C. (1981b) Cutaneous depression of Ib reflex pathways to motoneurones in man. *Experimental Brain Research* 42:351–61. [aSCG]

Pierrot-Deseilligny, E., Morin, C., Bergego, C. & Tankov, N. (1981a) Pattern of group I fibre projections from ankle flexor and extensor muscles in man. *Experimental Brain Research* 42:337–50. [aSCG]

Pinto-Hamuy, T. & Linck, P. (1965) Effect of frontal lesions on performance of sequential tasks by monkeys. *Experimental Neurology* 12:96–107. [DSL]

Pinz, A. & Bischof, H. (1990) Constructing a neural network for the interpretation of the species of trees in aerial photographs. In: *Proceedings of the 10th International Conference on Pattern Recognition*. IEEE Computer Society Press. [HB]

Pirart, J. (1978) Diabetes Mellitus and its degenerative complications: A prospective study of 4,400 patients observed between 1947 and 1973. *Diabetes Care* 1:168–88 & 252–63. [PRC]

Poggio, T. (1990) A theory of how the brain might work. In: *Cold Spring Harbor Symposia on Quantitative Biology* 55:899–910. [RAA]

Polit, A. & Bizzi, E. (1978) Processes controlling arm movements in monkeys. *Science* 201:1235–37. [aEB]

(1979) Characteristics of motor programs underlying arm movements. *Journal of Neurophysiology* 42:183–94. [aSCG]

Popov, K. E., Smetanin, B. N., Gurfinkel, V. S., Kudovina, M. P. & Shlykov, V. U. (1986) Spatial perception and vestibulomotor responses in man. *Neurophysiology (Kiev)* 18:779–87. [MBB, AP]

Porter, R. (1970) Early facilitation at corticomotoneuronal synapses. *Journal of Physiology* 207:733–45. [RL]

Posner, M. I. (1986) *Chronometric explorations of mind*. Oxford University Press. [aJFS]

Posner, M. I. & Rothbart, M. K. (1992) Attentional mechanisms and conscious experience. In: *The neuropsychology of consciousness*, ed. A. D. Milner & M.D. Rugg. Academic Press. [DPC]

Pouget, A., Fisher, S. A. & Sejnowski, S. J. (in press) Hierarchical transformation of space in the visual system. In: *Advances in neural information processing systems*, vol. 4. Kaufmann. [APo]

Poulton, E. C. (1981) Human manual control. In: *Handbook of physiology*. Section 1: The nervous system, vol. 2, Motor control, part 2, ed. J. M. Brookhart, V. B. Mountcastle, V. B. Brooks & S. R. Geiger. American Physiological Society. [aGEA]

Pratt, C. A., Chanaud, C. M. & Loeb, G. E. (1991) Functional complex muscles of the cat hindlimb. 4. Intramuscular distribution of movement command signals and cutaneous reflexes in broad, bifunctional thigh muscles. *Experimental Brain Research* 85:281–99. [JDu, CAP]

Pratt, C. A. & Jordan, L. M. (1987) Ia inhibitory interneurons and Renshaw cells as contributors to the spinal mechanisms of fictive locomotion. *Journal of Neurophysiology* 57:56–71. [CAP]

Pratt, C.A. & Loeb, G. E. (1991) Functionally complex muscles of the cat hindlimb. 1. Patterns of activation across sartorius. *Experimental Brain Research* 85:243–56. [CAP]

Precht, W. & Yoshida, M. (1971) Blockage of caudate-evoked inhibition of neurons in the substantia nigra by picrotoxin. *Brain Research* 32:229–33. [aGEA]

Prem, E., Mackinger, M., Dorffner, G., Porenta, G. & Sochor, H. (1992) Concept support as a method for programming neural networks with symbolic knowledge. *European Conference on Artificial Intelligence ECAI92*, Vienna, in press. [HB]

Pribram, K. H. (1971) *Languages of the brain.* Prentice-Hall. [BB, MEI]

Pribram, K. H., Nuwer, M. & Baron, R. (1974) The holographic hypothesis of memory structure in brain function and perception. In: *Contemporary developments in mathematical psychology 2*, ed. D. Krantz, R. C. Atkinson, R. Luce & P. Suppes. Freeman. [aEEF, BB]

Proakis, J. G. & Manolakis, D. G. (1989) *Introduction to digital signal processing.* Macmillian. [PDN]

Prochazka, A. (1986) Proprioception during voluntary movement. *Canadian Journal of Physiology and Pharmacology* 64:499–504. [aSCG]

(1989) Sensorimotor gain control: A basic strategy of motor systems? *Progress in Neurobiology* 33:281–307. [aSCG]

Prochazka, A., Hulliger, M., Trend, P., Llewellyn, M. & Durmuller, N. (1989) Muscle afferent contribution to control of paw shakes in normal cats. *Journal of Neurophysiology* 61:550–62. [CAP]

Prochazka, A., Hulliger, M., Zangger, P. & Appenteng, K. (1985) "Fusimotor set"; New evidence for α-independent control of γ-motoneurones during movement in the awake cat. *Brain Research* 339:136–40. [aSCG]

Prochazka, A., Stephens, J. A. & Wand, P. (1979) Muscle spindle discharge in normal and obstructed movements. *Journal of Physiology* 287:57–66. [aSCG]

Prochazka, A., Trend, P., Hulliger, M. & Vincent, S. (1989) Ensemble proprioceptive activity in the cat step cycle: Towards a representative look-up chart. In: *Afferent control of posture and locomotion*, ed. J. H. J. Allum & M. Hulliger. Elsevier. [APr]

Prochazka, A. & Wand, P. (1980) Tendon organ discharge during voluntary movements in cats. *Journal of Physiology* 303:385–90. [aSCG]

Prochazka, A., Westerman, R. A. & Ziccone, S. P. (1976) Discharges of single hindlimb afferents in the freely moving cat. *Journal of Neurophysiology* 39:1090–1104. [aSCG]

(1977) Ia afferent activity during a variety of voluntary movement in the cat. *Journal of Physiology* 268:423–48. [CAP]

Proctor, R. W., Lu, C.-H. & Van Zandt, T. (in press) Enhancement of the Simon effect by response precuing. *Acta Psychologica.* [RWP]

Proctor, R. W. & Reeve, T. G. (1985) Compatibility effects in the assignment of symbolic stimuli to discrete finger responses. *Journal of Experimental Psychology: Human Perception and Performance* 11:623–39. [RWP]

eds. (1990) *Stimulus-response compatibility: An integrated perspective.* North-Holland. [RWP]

Proctor, R. W., Reeve, T. G. & Weeks, D. J. (1990) A triphasic approach to the acquisition of response-selection skill. In: *The psychology of learning and motivation*, ed. G. H. Bower. Academic Press. [RWP]

Proske, U. (1981) The Golgi tendon organ. Properties of the receptor and reflex action of impulses arising from tendon organs. *International Review of Physiology* 25:127–71. [aSCG]

Proske, U., Schaible, H.-G. & Schmidt, R. F. (1988) Joint receptors and kinaesthesia. *Experimental Brain Research* 72:219–24. [aSCG]

Pylyshyn, Z. W. (1989) The role of location indexes in spatial perception: A sketch of the FINST spatial-index model. *Cognition* 32:65–97. [rJFS]

Rack, P. M. H. (1981) Limitations of somatosensory feedback in control of posture and movements. In: *Handbook of physiology:* The nervous system, vol 2, ed. V. B. Brooks. American Physiological Society. [AMG]

Rack, P. M. H. & Westbury, D. R. (1969) The effects of length and stimulus rate on tension in the isometric cat soleus muscle. *Journal of Physiology (London)* 204:443–60. [aEB]

(1974) The short range stiffness of active mammalian muscle and its effect on mechanical properties. *Journal of Physiology (London)* 240:331–50. [aEB]

Rafols, J. A. & Fox, C. A. (1976) The neurons in the primate subthalamic nucleus: A Golgi and electron microscopic study. *Journal of Comparative Neurology* 168:75–112. [aGEA]

Rakic, P. (1986) Mechanisms of ocular dominance segregation of the lateral geniculate nucleus. Competitive elimination hypothesis. *Trends in Neuroscience* 9:11–15. [aGEA]

Rall, W. (1969) Time constants and electrotonic length of membrane cylinders and neurons. *Biophysical Journal* 9:1483–1508. [DJ]

Ramos, C. F. & Stark, L. (1987) Simulation studies of descending and reflex control of fast movements. *Journal of Motor Behavior* 19(1):38–61. [NL]

Rapoport, S. (1979) Reflex connections of motoneurons of muscles involved in head movements in the cat. *Journal of Physiology (London)* 289:311–27. [GEL]

Rasnow, B., Assad, C., Nelson, M. & Bower, J. M. (1989) Simulation and measurement of the electric fields generated by weakly electric fish. In: *Advances in neural information processing systems*, ed. D. Touretzky. Kaufman. [JMB]

Ratcliff, G. (1991) Brain and space: Some deductions from the clinical evidence. In: *Brain and space*, ed. J. Paillard. Oxford University Press. [WAM]

Ratcliff, G. & Davies-Jones, G. A. (1972) Defective visual localization in focal brain wounds. *Brain* 95:46–60. [DPC]

Rauschecker, J. P. (1991) Mechanisms of visual plasticity: Hebb synapses, NMDA receptors, and beyond. *Physiological Reviews* 71:587–615. [aGEA]

Renshaw, B. (1941) Influence of discharge of motoneurones upon excitation of neighboring motoneurons. *Journal of Neurophysiology* 4:167–83. [DBul, NHB, TMH]

(1946) Central effects of centripetal impulses in axons of spinal ventral roots. *Journal of Neurophysiology* 9:191–204. [DBul]

Ribot, E., Roll, J. P. & Vedel, J.-P. (1986) Efferent discharges recorded from single skeletomotor and fusimotor fibres in man. *Journal of Physiology (London)* 375:251–68. [arSCG, JP]

Richmond, B. J. & Optican, L. M. (1987) Temporal encoding of two-dimensional patterns by single units in primate inferior temporal cortex. II. Quantification of response waveform. *Journal of Neurophysiology* 57:147–61. [IT]

(1990) Temporal encoding of two-dimensional patterns by single units in primate primary visual cortex. II. Information transmission. *Journal of Neurophysiology* 64:370–80. [IT]

Richmond, B. J., Optican, L. M., Podell, M. & Spitzer, H. (1987) Temporal encoding of two-dimensional patterns by single units in primate inferior temporal cortex. I. Response characteristics. *Journal of Neurophysiology* 57:132–46. [IT]

Richmond, B. J., Optican, L. M. & Spitzer, H. (1990) Temporal encoding of two-dimensional patterns by single units in primate primary visual cortex. I. Stimulus-response relations. *Journal of Neurophysiology* 64:351–69. [IT]

Richmond, F. J. R. & Loeb, G. E. (1992) Electromyographic studies of neck muscles in the intact cat: II. Reflexes evoked by muscle nerve stimulation. *Experimental Brain Research* 88:59–66. [GEL]

Richmond, F. J. R. & Stuart, D. G. (1985) Distribution of sensory receptors in the flexor carpi radialis muscle of the cat. *Journal of Morphology* 183:1–13. [aSCG]

Ridgway, E. B., Gordon, A. M. & Martyn, D. A. (1983) Hysteresis in the force-calcium relationship in muscle. *Science* 219:1075–77. [MH]

Riehle, A. (1991) Visually induced signal-locked neuronal activity changes in precentral motor areas of the monkey: Hierarchical progression of signal processing. *Brain Research* 540:131–37. [WAM]

Riehle, A. & Requin, J. (1989) Monkey primary motor and premotor cortex: Single-cell activity related to prior information about direction and extent of an intended movement. *Journal of Neurophysiology* 6:534–49. [aGEA]

Ritchie, L. (1976) Effects of cerebellar lesions on saccadic eye movements. *Journal of Neurophysiology* 39:1246–56. [aJRB]

Ritter, H. J., Martinetz, T. M. & Schulten, K. J. (1989) Topology-conserving maps for learning visuo-motor-coordination. *Neural Networks* 2:159–68. [aGEA, PM]

Rizzolatti, G. & Berti, A. (1990) Neglect as a neural representation deficit. *Revue Neurologique* 146:626–34. [MSG]

Rizzolatti, G. & Gentilucci, M. (1988) Motor and visual-motor functions of the premotor cortex. In: *Neurobiology of neocortex*, ed. P. Rakic & W. Singer. Wiley. [JT]

Rizzolatti, G., Gentilucci, M. & Matelli, M. (1985) Selective spatial attention: One center, one circuit, or many circuits? In: *Attention and performance XI*, ed. M. I. Posner & O. S. M. Marin. Erlbaum. [DPC]

Rizzolatti, G., Matelli, M. & Pavesi, G. (1983) Deficits in attention and movement following the removal of postarcuate (area 6) and prearcuate (area 8) cortex in macaque monkeys. *Brain* 106:655–73. [MSG]

Rizzolatti, G., Scandolara, C., Matelli, M. & Gentilucci, M. (1981a) Afferent properties of periarcuate neurons in macaque monkeys. I. Somato-sensory responses. *Behavioral Brain Research* 2:125–46. [MSG]

(1981b) Afferent properties of periarcuate neurons in macaque monkeys. II. Visual responses. *Behavioral Brain Research* 2:147–63. [MSG]

Robertson, L. T. & McCollum, G. (1991) Stimulus classification by ensembles of climbing fiber receptive fields. *Trends in Neurosciences* 14:248–54. [GMc]

Robinson, C. J. & Burton, H. (1981a) Organization of somatosensory receptive fields in cortical areas 7B, retrosinular, postauditory, and granular insula of *Macaca fascicularis. Journal of Comparative Neurology* 192:69–92. [MSG]

(1980b) Somatic submodality distribution within the second somatosensory area (S11), 7B, retroinsular, postauditory, and granular insular cortical areas of *Macaca fascicularis. Journal of Comparative Neurology* 192:93–108. [MSG]

Robinson, D. A. (1973) Models of the saccadic eye movement system. *Kybernetik* 14:77–83. [aJFS]

(1975) Oculomotor control signals. In: *Basic mechanisms of ocular motility and their clinical implications*, ed. G. Lennerstrand & P. Bach-y-Rita. Pergamon Press. [aDAR, JAMVG]

(1976) Adaptive gain control of vestibulo-ocular reflex by the cerebellum. *Journal of Neurophysiology* 39:954–69. [arJRB]

(1981) The use of control systems analysis in the neurophysiology of eye movements. *Annual Review of Neuroscience* 4:463–503. [aDAR, MF]

(1982) The use of matrices in analyzing the three-dimensional behavior of the vestibulo-ocular reflex. *Biological Cybernetics* 46:53–66. [arDAR, MF]

(1989) Integrating with neurons. *Annual Review of Neuroscience* 12:33–45. [aDAR, MF]

Robinson, D. A. & Fuchs, A. F. (1969) Eye movements evoked by stimulation of the frontal eye fields. *Journal of Neurophysiology* 32:637–48. [APo]

Roby-Brami, A. & Bussel, B. (1990) Effects of flexor reflex afferent stimulation on the soleus H reflex in patients with a complete spinal cord lesion: Evidence for presynaptic inhibition of Ia transmission. *Experimental Brain Research* 81:593–601. [arDAM]

Roe, A. W., Pallas, S. L., Hahm, J.-O. & Sur, M. (1990) A map of visual space induced in primary auditory cortex. *Science* 250:818–20. [aGEA]

Rogers, D. K., Bendrups, A. P. & Lewis, M. M. (1985) Disturbed proprioception following a period of muscle vibration in humans. *Neuroscience Letters* 57:147–52. [aSCG]

Roland, P. E. (1978) Sensory feedback to the cerebral cortex during voluntary movement in man. *Behavioral and Brain Sciences* 1:129–71. [aSCG]

(1982) Cortical regulation of selective attention in man. A regional cerebral blood flow study. *Journal of Neurophysiology* 48:1059–78. [aJFS]

(1984) Organization of motor control by the normal human brain. *Human Neurobiology* 2:205–16. [aGEA]

Roland, P. E. & Ladegaard-Pedersen, H. (1977) A quantitative analysis of sensations of tension and of kinaesthesia in man. Evidence for a peripherally originating muscular sense and for a sense of effort. *Brain* 100:671–92. [aSCG]

Roll, J. P., Gilhodes, J. C. & Tardy-Gervet, M. F. (1980) Effets perceptives et moteurs des vibrations musculaires chez l'homme normal: Mise en evidence d'une reponse des muscles antagonistes. *Archives Italiennes de Biologie* 118:51–71. [MBB, AP]

Roll, J. P. & Vedel, J. P. (1982) Kinaesthetic role of muscle afferents in man, studied by tendon vibration and microneurography. *Experimental Brain Research* 47:177–90. [aSCG]

Roll, J. P., Vedel, J. P. & Ribot, E. (1989) Alteration of proprioceptive messages induced by tendon vibration in man: A microneurographic study. *Experimental Brain Research* 76:213–22. [aSCG]

Roll, J. P., Velay, J. L. & Roll, J. R. (1991) Eye and neck proprioceptive messages contribute to the spatial coding of retinal input in visually oriented activities. *Experimental Brain Research* 85:423–31. [MBB, AP]

Rolls, E. T. & Treves, A. (1990) The relative advantages of sparse versus distributed encoding for associative neuronal networks in the brain. *Network* 1:407–21. [rDAR, JER]

Romanò, C. & Schieppati, M. (1987) Reflex excitability of soleus motoneurones during voluntary shortening or lengthening contractions. *Journal of Physiology (London)* 390:271–84. [MS]

Rosenbaum, D. A. (1980) Human movement initiation: Specification of arm, direction, and extent. *Journal of Experimental Psychology (General)* 109(4):444–74. [DJ]

(1985) Motor programming: A review and scheduling theory. In: *Motor behavior: Programming, control, and acquisition*, ed. H. Heuer, U. Kleinbeck & K.-H. Schmidt. Springer-Verlag. [aGEA]

(1991) *Human motor control*. Academic Press. [JJS]

Rosenbaum, D. A. & Saltzman, E. (1984) A motor-program editor. In: *Cognition and motor processes*, ed. W. Prinz & A. F. Sanders. Springer-Verlag. [aGEA]

Rosenberg, C. (1987) Revealing the structure of NETtalk's internal representations. *Proceedings of the ninth conference of the cognitive science society*. Erlbaum. [rDAR, JER]

Rosenbloom, P. S. & Newell, A. (1987) An integrated computational model of stimulus-response compatibility and practice. In: *The psychology of learning and motivation*, ed. G. H. Bower. Academic Press. [RWP]

Ross, H. E. (1981) How important are changes in body weight for mass perception? *Acta Astronautica* 8:1051–58. [HER]

(1991) Motor skills under varied gravitoinertial force in parabolic flight. *Acta Astronautica* 23:85–91. [HER]

Ross, H. E. & Brodie, E. E. (1987) Weber fractions for weight and mass as a function of stimulus intensity. *Quarterly Journal of Experimental Psychology* 39A:77–88. [HER]

Ross, H. E., Rejman, M. H. & Lennie, P. (1972) Adaptation to weight transformation in water. *Ergonomics* 15:387–97. [HER]

Ross, H. E. & Reschke, M. F. (1982) Mass estimation and discrimination during brief periods of zero gravity. *Perception & Psychophysics* 31:429–36. [aSCG, HER]

Ross, H. E., Schwartz, E. & Emmerson, P. (1987) The nature of sensorimotor adaptation to altered G levels: Evidence from mass discrimination. *Aviation, Space, and Environmental Medicine* 58A:148–52. [HER]

Rossi, A. & Grigg, P. (1982) Characteristics of hip joint mechanoreceptors in the cat. *Journal of Neurophysiology* 47:1029–42. [aSCG]

Rossi, A., Mazzocchio, R. & Scarpini, C. (1988) Changes in Ia reciprocal inhibition from the peroneal nerve to the soleus alpha-motoneurons with different static body positions in man. *Neuroscience Letters* 84:283–86. [VD]

Rossignol, S. & Gauthier, L. (1980) Analysis of mechanisms controlling the reversal of crossed spinal reflexes. *Brain Research* 182:31–4. [aDAM]

Rothwell, J. C. (1987) *Control of human voluntary movement*. Croom Helm Limited. [JGP]

Rothwell, J. C., Gandevia, S. C. & Burke, D. (1990) Activation of fusimotor neurones by motor cortical stimulation in human subjects. *Journal of Physiology (London)* 431:743–56. [aSCG]

Rothwell, J. C., Traub, M. M., Day, B. L., Obeso, J. A., Thomas, P. K. & Marsden, C. D. (1982a) Motor performance in a deafferented man. *Brain* 104:465–91. [PRC]

(1982b) Manual motor performance in a deafferented man. *Brain* 105:515–42. [aSCG]

Rothwell, J. C., Traub, M. M. & Marsden, C. D. (1982c) Automatic and "voluntary" responses compensating for disturbances of human thumb movements. *Brain Research* 248:33–41. [MLL]

Rowat, P. F. & Selverston, A. I. (1991) Learning algorithms for oscillatory networks with gap junctions and membrane currents. *Network* 2:17–41. [rEEF]

Rudomin, P. (1990a) Presynaptic control of synaptic effectiveness of muscle spindle and tendon organ afferents in the mammalian spinal cord. In: *The segmental motor system*, ed. M. Binder & L. Mendell. Oxford University Press. [aDAM]

(1990b) Presynaptic inhibition of muscle spindle and tendon organ afferents in the mammalian spinal cord. *Trends in Neuroscience* 13:499–505. [aDAM]

Rudomin, P., Solodkin, M. & Jiminez, I. (1987) Synaptic potentials of primary afferent fibers and motoneurons evoked by single intermediate nucleus interneurons in the cat spinal cord. *Journal of Neurophysiology* 57:1288–1313. [aDAM, PR]

Rueckl, J. G., Cave, K. R. & Kosslyn, S. M. (1989) Why are "what" and "where" processed by separate cortical visual systems? A computational investigation. *Journal of Cognitive Neuroscience* 1:171–86. [RAA]

Rumelhart, D. E., Hinton, G. E. & McClelland, J. L. (1986a) A general framework for parallel distributed processing. In: *Parallel distributed processing: Explorations in the microstructure of cognition*, ed. D. E. Rumelhart, J. L. McClelland & The PDP Research Group. MIT Press. [aGEA]

Rumelhart, D. E., Hinton, G. E. & Williams, R. J. (1986b) Learning internal representations by error propagation. In: *Parallel distributed processing: Explorations in the microstructure of cognition*, vol. 1: *Foundations*, ed. D. E. Rumelhart & J. L. McClelland. MIT Press. [aEEF, aDAR]

(1986c) Learning representations by back-propagating errors. *Nature (London)* 323:533–36. [aGEA, WAM]

Rumelhart, D. E. & McClelland, J. L. (1986) PDP models and general issues in cognitive science. In: *Parallel distributed processing: Explorations in the microstructure of cognition*, ed. D. E. Rumelhart, J. L. McClelland & The PDP Research Group. MIT Press. [aGEA]

Rumelhart, D. E., McClelland, J. L. & The PDP Research Group (1986d) *Parallel distributed processing: Explorations in the microstructure of cognition*, vols. 1 and 2. MIT Press. [aGEA, aJFS]

Rushton, D. N., Rothwell, J. C. & Craggs, M. D. (1981) Gating of somatosensory evoked potentials during different kinds of movement in man. *Brain* 104:465–91. [aSCG]

Ryall, R. W. (1970) Renshaw cell mediated inhibition of Renshaw cells: Patterns of excitation and inhibition from impulses in motor axon collaterals. *Journal of Neurophysiology* 33:257–70. [DBul]

Ryall, R. W. & Piercey, M. (1971) Excitation and inhibition of Renshaw cells by impulses in peripheral afferent nerve fibers. *Journal of Neurophysiology* 34:242–51. [DBul]

Rymer, W. Z. (1984) Spinal mechanisms for control of muscle length and tension. In: *Handbook of the spinal cord*, ed. R. A. Davidoff. Marcel Dekker. [aDAM]

Rymer, W. Z. & D'Almedia, A. (1980) Joint position sense: The effects of muscle contraction. *Brain* 103:1–22. [aSCG]

Sabin, C. & Smith, J. L. (1984) Recovery and perturbation of paw-shake responses in spinal cats. *Journal of Neurophysiology* 51:680–88. [CAP]

Sakata, H., Shibutani, H. & Kawano, K. (1983) Functional properties of visual tracking neurons in posterior parietal association cortex on the monkey. *Journal of Neurophysiology* 49:1364–80. [JWG]

Sakata, H., Shibutani, H., Kawano, K. & Harrington, T. L. (1985) Neural mechanisms of space vision in the parietal association cortex of the monkey. *Vision Research* 25:453–63. [JWG]

Sakitt, B., Lestienne, F. & Zeffiro, T. A. (1983) The information transmitted at final position in visually triggered forearm movements. *Biological Cybernetics* 46:111–18. [FJC]

Salthouse, T. A. (1984) Effects of age and skill in typing. *Journal of Experimental Psychology: General* 113:345–71. [AMG]

Saltzman, E. (1979) Levels of sensorimotor representation. *Journal of Mathematical Psychology* 20:91–163. [aGEA]

Saltzman, E. & Kelso, J. A. S. (1987) Skilled actions: A task dynamic approach. *Psychological Review* 94:84–106. [JJS]

Salzman, C. D., Britten, K. H. & Newsome, W. T. (1990) Cortical microstimulation influences perceptual judgements of motion direction. *Nature* 346:174–77. [RE]

Sanes, J. N. (1990) Motor representations in deafferented humans: A mechanism for disordered movement performance. In: *Attention and performance*, vol. 13: Motor representation and control, ed. M. Jeannerod. Erlbaum. [aSCG]

Sanes, J. N., Dimitrov, B. & Hallett, M. (1990) Motor learning in patients with cerebellar dysfunction. *Brain* 113:103–20. [aJRB, MH]

Sanes, J. N. & Evarts, E. V. (1983a) Effects of perturbations on accuracy of arm movements. *Journal of Neuroscience* 3:977:86. [aEB, aSCG, MH]

(1983b) Regulatory role of proprioceptive input in motor control of phasic or maintained voluntary contractions in man. In: *Motor control mechanisms in health and disease*, ed. J. E. Desmedt. Raven Press. [aSCG]

Sanes, J. N., Mauritz, K.-H., Dalakas, M. C. & Evarts, E. V. (1985) Motor control in humans with large fiber sensory neuropathy. *Human Neurobiology* 4:101–114. [aSCG, PRC, ZH]

Sasaki, K., Bower, J. M. & Llinas, R. (1989) Multiple Purkinje cell recording in rodent cerebellar cortex. *European Journal of Neuroscience* 1:572–86. [aJRB]

Sathian, K. & Devanandan, M. S. (1983) Receptors of the metacarpophalangeal joints: A histological study in the Bonnet monkey and man. *Journal of Anatomy* 137:601–13. [aSCG]

Sato, Y., Kawasaki, T. & Ikarashi, K. (1983) Afferent projections from the brainstem to the three floccular zones in cats. II. Mossy fiber projections. *Brain Research* 272:37–48. [NHB]

Sato, Y., Yamamoto, F., Shojaku, H. & Kawasaki, T. (1984) Neuronal pathway from floccular caudal zone contributing to vertical eye movements in cats – role of group y nucleus of vestibular nuclei. *Brain Research* 294:375–80. [NHB]

Schacher, S., Glanzman, D., Barzilai, A., Dash, P., Grant, S. G., Keller, F., Mayford, M. & Kandel, E. R. (1990) Long-term facilitation in aplysia: Persistent phosphorylation and structural changes. *Cold Spring Harbor Symposia on Quantitative Biology* 55:187–202. [MH]

Schade, J. P. & Ford, D. H. (1973) *Basic neurology*. Elsevier. [JMB]

Schady, W. J. L. & Torebjörk, H. E. (1983) Projected and receptive fields: A comparison of projected areas of sensation evoked by intraneural stimulation of mechanoreceptive units, and their innervation territories. *Acta Physiologica Scandinavica* 119:267–75. [aSCG]

Schaible, H.-G., & Schmidt, R. F. (1983) Responses of fine medial articular nerve afferents to passive movements of knee joint. *Journal of Neurophysiology* 49:1118–26. [aSCG]

Schell, G. R. & Strick, P. L. (1984) The origin of thalamic inputs to the arcuate premotor and supplementary motor areas. *Journal of Neuroscience* 4:539–60. [aGEA]

Schieber, M. (1990) How might the motor cortex individuate movements? *Trends in Neuroscience* 13:440–44. [aDAM, GMc]

Schieber, M. A. & Thach, W. T. (1985) Trained slow tracking. II. Bidirectional discharge patterns of cerebellar nuclear, motor cortex, and spindle afferent neurons. *Journal of Neurophysiology* 55:1228–70. [aJRB, aEEF]

Schieppati, M. & Crenna, P. (1984) From activity to rest: Gating of excitatory autogenetic afferences from the relaxing muscle in man. *Experimental Brain Research* 56:448–57. [aSCG]

Schieppati, M., Gritti, I. & Romanò, C. (1991) Recurrent and reciprocal inhibition of the human monosynaptic reflex show opposite changes following intravenous administration of acetylcarnitine. *Acta Physiologica Scandinavica* 142:27–32. [MS]

Schieppati, M. & Nardone, A. (1991) Free and supported stance in Parkinson's disease. The effect of posture and 'postural set' on leg muscle responses to perturbation, and its relation to the severity of the disease. *Brain* 114:1227–44. [MS]

Schieppati, M., Nardone, A. & Musazzi, M. (1986) Modulation of the Hoffmann reflex by rapid muscle contraction or release. *Human Neurobiology* 5:59–66. [aSCG]

Schieppati, M., Romanó, C. & Gritti, I. (1990) Convergence of Ia fibres from synergistic and antagonistic muscles on interneurones inhibitory onto soleus in humans. *Journal of Physiology (London)* 431:365–77. [MS]

Schieppati, M., Valenza, F. & Rezzonico, M. (1992) Motor unit recruitment in human biceps and brachioradialis muscles during lengthening contractions. *European Journal of Neuroscience* (suppl. 4) 4283:303. [MS]

Schlag, J. & Schlag-Rey, M. (1987) Evidence for a supplementary eye field. *Journal of Neurophysiology* 57:179–200. [aDAR]

Schmidt, E. M., Jost, R. G. & Davis, K. K. (1975) Reexamination of the force relationship of cortical cell discharge patterns with conditioned wrist movements. *Brain Research* 83:213–23. [aEEF]

Schmidt, R. A. (1975) A schema theory of discrete motor skill learning. *Psychological Reviews* 86:225–60. [aGEA]

Schmidt, R. A. & McGown, C. M. (1980) Terminal accuracy of unexpectedly loaded rapid movements: Evidence for a mass-spring mechanism in programming. *Journal of Motor Behavior* 12:149–61. [JBJS]

Schmidt, R. A., Zelaznik, H. N., Hawkins, B., Frank, J. S. & Quinn, J. T. (1979) Motor-output variability: A theory for the accuracy of rapid motor acts. *Psychological Reviews* 86:415–51. [aGEA]

Schomburg, E. (1990) Spinal sensorimotor systems and their supraspinal control. *Neuroscience Research* 7:265–340. [JDu]

Schomburg, E. & Steffens, H. (1986) Synaptic responses of lumbar alpha-motoneurones to selective stimulation of cutaneous nociceptors and low threshold mechanoreceptors in the spinal cat. *Experimental Brain Research* 62:335–42. [aDAM]

Schotland, J. L., Lee, W. A. & Rymer, W. Z. (1989) Wiping reflex and flexion withdrawal reflexes display different EMG patterns prior to movement onset in the spinalized frog. *Experimental Brain Research* 78:649–53. [aEB, SG]

Schouenborg, J. & Sjölund, B. (1983) Activity evoked by A- and C-afferent fibers in rat dorsal horn neurons and its relation to a flexion reflex. *Journal of Neurophysiology* 50:1108–21. [aDAM]

Schreurs, B. G., Sanchez-Andres, J. V. & Alkon, D. L. (1991) Learning-specific differences in Purkinje-cell dendrites of lobule HVI (Lobulus simplex): Intracellular recording in a rabbit cerebellar slice. *Brain Research* 548:18–22. [MH]

Schroder, K. E., Hopf, A., Lange, H. & Thorner, G. (1975) Morphemetrisch-statistsche Strukturanalysen des Striatum, Pallidum und Nucleus Subthalamicus beim Menschen. I. Striatum. *Journal für Hirnforschung* 16:333–50. [aGEA]

Schwartz, A. B., Ebner, T. J. & Bloedel, J. R. (1987) Comparison of responses in dentate and interposed nuclei to perturbations of the locomotor cycle. *Experimental Brain Research* 67:323–38. [aJRB]

Schwartz, A. B., Kettner, R. E. & Georgopoulos, A. P. (1988) Primate motor cortex and free arm movements to visual targets in three-dimensional space. I. Relations between single-cell discharge and direction of movement. *Journal of Neuroscience* 8:2913–27. [aGEA, aJRB, cSCG, DSL]

Scott, T. G. (1964) A unique pattern of localization within the cerebellum of the mouse. *Journal of Comparative Neurology* 122:1–7. [NHB]

Scudder, C. (1988) A new local feedback model of the saccadic burst generator. *Journal of Neurophysiology* 59:1455–75. [aDAR]

Seal, J., Hasbroucq, T., Mouret, I., Akamatsu, M. & Kornblum, S. (1991) Possible neural correlates for the mechanism of stimulus-response association in the monkey. In: *Tutorials in motor neuroscience*, ed. J. Requin & G. E. Stelmach. Kluwer. [RWP]

Seal, J., Riehle, A. & Requin, J. (1992) A critical reexamination of the concept of function within the neocortex. *Human Movement Science* 11:47–58. [JJS]

Searle, J. R. (1980) Minds, brains and programs. *Behavioral and Brain Sciences* 3:417–57. [DJ]

Segraves, M. A. & Goldberg, M. E. (1987) Functional properties of corticotectal neurons in the monkey's frontal eye field. *Journal of Neurophysiology* 58:1387–1419. [CLC]

Seif-Naraghi, A. H. & Winters, J. M. (1990) Optimized strategies for scaling goal–directed dynamic limb movements. In: *Multiple muscle systems: Biomechanics and movement organization*, ed. J. M. Winters & S. Y.-L. Woo. Springer-Verlag. [JMW]

Sejnowski, T. J. & Rosenberg, C. R. (1987) Parallel networks that learn to pronounce English text. *Complex Systems* 1:145–68. [aDAR]

Selemon, L. D. & Goldman-Rakic, P. S. (1985) Longitudinal topography and interdigitation of cortico-striatal projections in the rhesus monkey. *Journal of Neuroscience* 5:776–94. [aGEA]

References

Selhorst, J. B., Stark, L., Ochs, A. L. & Hoyt, W. F. (1976) Disorders in cerebellar ocular motor control. *Brain* 99:497–508. [aJRB]

Seltzer, B. & Pandya, D. N. (1978) Afferent cortical connections and architectonics of the superior temporal sulcus and surrounding cortex in the rhesus monkey. *Brain Research* 149:1–24. [BS]
 (1984) Further observations on parieto-temporal connections in the rhesus monkey. *Experimental Brain Research* 55:301–312. [BS]
 (1989) Frontal lobe connections of the superior temporal sulcus in the rhesus monkey. *Journal of Comparative Neurology* 281:97–113. [BS]
 (1991a) Post-Rolandic cortical connections of the superior temporal sulcus in the rhesus monkey. *Journal of Comparative Neurology* 312:625–40. [BS]
 (1991b) Cortical sensory and limbic projections to discrete areas of the superior temporal sulcus in the rhesus monkey. *Society for Neuroscience Abstracts* 17:1585. [BS]

Selverston, A. I. & Moulins, M. (1985) Oscillatory neural networks. *Annual Reviews of Physiology* 47:29–48. [CAP]

Severin, F., Orlovsky, G. & Shik, M. (1968) Reciprocal influences on work of single motoneurons during controlled locomotion. *Bulletin of Experimental Biology and Medicine* 66:5–9. [aDAM]

Shadmehr, R., Mussa-Ivaldi, F. A. & Bizzi, E. (1992) Postural force fields of the human arm and their role in generating multi-joint movements. *Journal of Neuroscience*, in press. [rEB]

Shambes, G. M., Gibson, J. M. & Welker, W. I. (1978) Fractured somatotopy in granule cell tactile areas of rat cerebellar hemispheres is revealed by micromapping. *Brain Behavior and Evolution* 15:94–140. [aJRB]

Shapiro, D. C., Zernicke, R. F., Gregor, R. J. & Diestel, J. D. (1981) Evidence for a generalized motor programs using gait pattern analysis. *Journal of Motor Behavior* 22(1):98–124. [GMa]

Shefchyk, S. & Jordan, L. (1985) Excitatory and inhibitory postsynaptic potentials in alpha-motoneurons produced during fictive locomotion by stimulation of the mesencephalic locomotor region. *Journal of Neurophysiology* 53:1345–55. [aDAM]

Shefchyk, S., McCrea, D., Kreillaars, D., Fortier, P. & Jordan, L. (1990) Activity of L4 interneurons during brain stem evoked fictive locomotion in the mesencephalic cat. *Experimental Brain Research* 80:290–95. [aDAM]

Shen, L. (1989) Neural integration by short term potentiation. *Biological Cybernetics* 61:319–25. [MF]

Shepard, R. N. & Cooper, L. A. (1982) *Mental images and their transformation.* MIT. [VB]

Shepherd, G. M., ed. (1990) *The synaptic organization of the brain.* Oxford University Press. [CIC]

Sherrington, C. S. (1900) The muscular sense. In: *Text-book of physiology,* vol. 2, ed. E. A. Schäfer. Young J. Pentland. [aSCG]
 (1906/1947) *The integrative action of the nervous system.* Yale University Press. [aSCG]
 (1910) Flexion-reflex of the limb, crossed extension reflex, and reflex stepping and standing. *Journal of Physiology* 40:28–121. [aDAM]

Shibutani, H., Sakata, H. & Kawano, K. (1984) Saccade and blinking evoked by microstimulation of the posterior parietalcortex of the monkey. *Experimental Brain Research* 55:1–8. [JWG]

Shindo, M., Harayama, H., Kondo, K., Yanagisawa, N. & Tanaka, R. (1984) Changes in reciprocal Ia inhibition during voluntary contraction in man. *Experimental Brain Research* 53:400–408. [aSCG]

Shinoda, Y., Yamaguchi, T. & Futami, T. (1986) Multiple axon collaterals of single corticospinal axons in the cat spinal cord. *Journal of Neurophysiology* 55:425–48. [aDAM]

Shinoda, Y., Yokota, J. & Futami, T. (1981) Divergent projection of individual corticospinal axons to motoneurons of multiple muscles in the monkey. *Neuroscience Letters* 23:7–12. [aGEA, aDAM]

Simoneau, G. G. (1992) The effects of diabetic distal symmetrical peripheral neuropathy on static posture. Doctoral Dissertation, Penn State University. [PRC]

Simoneau, G. S., Ulbrecht, J. S. & Cavanagh, P. R. (1992) Stability during standing posture in patients with distal symmetrical diabetic neuropathy, in preparation. [PRC]

Singer, I. M. (1982) Differential geometry, fiber bundles and physical theories. *Physics Today* March 1982:41–44. [TLC]

Sinkjaer, T., Toft, E., Andreassen, S. & Hornemann, B. C. (1988) Muscle stiffness in human ankle dorsiflexors: Intrinsic and reflex components. *Journal of Neurophysiology* 60;1110–21. [TRN]

Sittig, A. C., Denier van der Gon, J. J., Gielen, C. C. A. M. & Van Wijk, A. J. M. (1985) The attainment of target position during step-tracking movements despite a shift of initial position. *Experimental Brain Research* 60:407–10. [aSCG]

Skarda, C. A. & Freeman, W. J. (1987) How brains make chaos in order to make sense of the world. *Behavioral and Brain Sciences* 10:161–95. [PR, IT]

Skoglund, S. (1956) Anatomical and physiological studies of knee joint innervation in the cat. *Acta Physiologica Scandinavica* 124:1–101. [aSCG]

Slosberg, M. (1990) Spinal learning: Central modulation of pain processing and long-term alteration of interneuronal excitability as a result of nociceptive peripheral input. *Journal of Manipulative and Physiological Theraputics* 13:326–336. [rDAM]

Slotine, J.-J. E. (1985) The robust control of robot manipulators. *International Journal of Robotics Research* 4:49–64. [aEB]

Smeets, J. B. J. & Erkelens, C. J. (1991) Dependence of autogenic and heterogenic stretch reflexes on preload activity in the human arm. *Journal of Physiology* 440:455–65. [JBJS]

Smeets, J. B. J., Erkelens, C. J. & Denier van der Gon, J. J. (1990) Adjustments of fast goal-directed movements in response to an unexpected inertial load. *Experimental Brain Research* 81:303–12. [JBJS]
 (1992) Perturbations of fast goal-directed arm movements: Different behaviour of early and late EMG-responses, submitted. [JBJS]

Smetanin, B. N., Popov, K. E., Gurfinkel, V. S. & Shylkov, V. U. (1988) Effect of movement and illusion of movement on human vestibulomotor response. *Neurophysiology (Kiev)* 20:250–55. [MBB, AP]

Smith, A. M. (1990) Some cerebellar and cortical contributions to reaching and grasping. In: *Vision and action: The control of grasping,* ed. M. A. Goodale. Ablex. [DPC]

Smith, J. C., Feldmann, J. L., Schmidt, B. J. (1988) Neural mechanisms generating locomotion studies in mammalian brainstem-spinal cord in vitro. *FASEB Journal* 2:2283–88. [FBH]

Smith, J. L., Betts, B., Edgerton, V. R. & Zernicke, R. F. (1980) Rapid ankle extension during paw shakes. Selective recruitment of fast ankle extensors. *Journal of Neurophysiology* 43:612–20. [CAP]

Smith, J. L., Edgerton, V. R., Betts, B. & Collatos, T. C. (1977) EMG of slow and fast ankle extensors of cat during posture, locomotion, and jumping. *Journal of Neurophysiology* 40:503–13. [CAP]

Smith, J. L., Hoy, M. G., Koshland, G. F., Phillips, D. M. & Zernicke, R. F. (1985) Intralimb coordination of the paw-shake response: A novel mixed synergy. *Journal of Neurophysiology* 54:1271–81. [CAP]

Smith, J. L. & Zernicke, R. F. (1987) Predictions for neural control based on limb dynamics. *Trends in Neuroscience* 10:123–28. [CAP]

Smith, W. S. & Fetz, E. E. (1989) Effects of synchrony between corticomotoneuronal cells on post-spike facilitation of muscles and motor units. *Neuroscience Letters* 96:76–81. [rEEF, PAK]

Smith, Y. & Parent, A. (1986) Differential connections of caudate nucleus and putamen in the squirrel monkey. (*Saimiri sciureus*). *Neuroscience* 18:347–71. [aGEA]

Smolensky, P. (1988) On the proper treatment of connectionism. *Behavioral and Brain Sciences* 11:1–74. [GS]
 (1990) Tensor product variable binding and the representation of symbolic structures in connectionist systems. *Artificial Intelligence* 46:159–216. [GS]

Soechting, J. F. (1982) Does position sense at the elbow reflect a sense of elbow joint angle or one of limb orientation? *Brain Research* 248:392–95. [aSCG]
 (1988) Effect of load perturbations on EMG activity and trajectories of pointing movements. *Brain Research* 451:390–96. [ZH]

Soechting, J. F. & Flanders, M. (1989) Sensorimotor representations for pointing to targets in three-dimensional space. *Journal of Neurophysiology* 62:582–94. [aGEA, aSCG, aJFS, FJC]
 (1992a) Moving in three-dimensional space: Frames of reference, vectors, and coordinate systems. *Annual Review of Neuroscience* 15:167–91. [MF]
 (1992b) The organization of sequential typing movements. *Journal of Neurophysiology* 67:1275–90. [AMG]

Soechting, J. F., Helms Tillery, S. I. & Flanders, M. (1990) Transformation from head-to shoulder-centered representation of target direction in arm movements. *Journal of Cognitive Neuroscience* 2:32–43. [MF]

Soechting, J. F. & Lacquaniti, F. (1988) Quantitative evaluation of the electromyographic responses to multidirectional load perturbations of the human arm. *Journal of Neurophysiology* 59:1296–1313. [aSCG]

Solodkin, M., Jiménez, I., Collins III, W. F., Mendell, L. M. & Rudomin, P. (1991) Interaction of baseline synaptic noise and Ia EPSPs: Evidence for appreciable negative correlation under physiological conditions. *Journal of Neurophysiology* 65:927–45. [PR]

Soso, M. J. & Fetz, E. E. (1980) Responses of identified cells in post-central cortex of awake monkeys during comparable active and passive joint movements. *Journal of Neurophysiology* 43:1090–1110. [aEEF]

Sparks, D. L. (1991a) Neural encoding of location of targets for saccadic eye movements. In: *The brain and space,* ed. J. Paillard. Oxford University Press. [aJFS, DPC]

(1991b) The neural control of orienting eye and head movements. In: *Motor control: Concepts and issues*, ed. D. R. Humphrey & H.-J. Freund. Wiley. [ZH]

Stauffer, E. K. & Stephens, J. A. (1977) Responses of Golgi tendon organs to ramp-and-hold profiles of contractile force. *Journal of Neurophysiology* 40:681–91. [aSCG]

Stein, J. F. (1978) Effects of parietal lobe cooling on manipulation in the monkey. In: *Active touch*, ed. G. Gordon. Pergamon Press. [aJFS, DPC]

(1986) Role of the cerebellum in the visual guidance of movement. *Nature* 323:217–21. [GJvIS]

(1989a) Physiological differences between left and right. In: *Aphasia*, ed. F. C. Rose. Whurr. [aJFS]

(1989b) Representation of egocentric space in the posterior parietal cortex. *Quarterly Journal of Experimental Physiology* 74:583–606. [aJFS, DPC]

(1991a) Space and the parietal association areas. In: *The brain and space*, ed. J. Paillard. Oxford University Press. [aJFS]

(1991b) *Vision and language*. In: Dyslexia, ed. M. Snowling. Whurr. [aJFS]

(1991c) *Vision and visual dyslexia*. Macmillan. [rJFS]

Stein, J. F. & Glickstein, M. (1992) The role of the cerebellum in the visual guidance of movement. *Physiological Review*, in press. [cJFS]

Stein, R. B. (1974) The peripheral control of movement. *Physiological Reviews* 54:215–43. [aEB, RBS]

(1982) What muscle variable(s) does the nervous system control in limb movements? *Behavioral and Brain Sciences* 5:535–77. [JBJS]

Stein, R. B. & Capaday, C. (1988) The modulation of human reflexes during functional motor tasks. *Trends in Neuroscience* 11:328–32. [RBS]

Steinmetz, J. E., Lavond, D. G. & Thompson, R. F. (1989) Classical conditioning in rabbits using pontine stimulation as a conditioned stimulus and inferior olive stimulation as an unconditioned stimulus. *Synapse* 3(3):225–32. [RFT]

Steinmetz, M. A., Motter, B. C., Duffy, J. & Mountcastle, V. B. (1987) Functional properties of parietal visual neurones: Radial organisation of directionalisation within the visual field. *Journal of Neuroscience* 7:177–91. [aJFS]

Stelmach, G. E. & Diggles, V. A. (1982) Control theories in motor behavior. *Acta Psychologica* 50:83–105. [JGP]

Stelmach, G. E. & Hughes, B. G. (1984) Cognitivism and future theories of action: Some basic issues. In: *Cognition and motor processes*, ed. W. Prinz & A. F. Sanders. Springer-Verlag. [JJS]

Stoffer, T. H. (1991) Attentional focussing and spatial stimulus-response compatibility. *Psychological Research* 53:127–35. [RWP]

Strata, P. (1987) Inferior olive and motor control. In: *The cerebellum and neuronal plasticity*, ed. M. Glickstein & J. Stein. Plenum Press. [cJFS]

Strick, P. L. (1976) Anatomical analysis of ventrolateral thalamic input to primate motor cortex. *Journal of Neurophysiology* 39:1020–31. [aGEA]

(1983) The influence of motor preparation on the response of cerebellar neurons to limb displacements. *Journal of Neuroscience* 3:2007–20. [aJRB]

Strong, G. E. & Whitehead, B. A. (1989) A solution to the tag-assignment problem for neural networks. *Behavioral and Brain Sciences* 12:381–433. [MRWD]

Sultan, F. (1992) Vergleichende Untersuchungen uber die Flachenausdehnung der Kleinhirnrinde. Ein Beitrag zu einem Modell der Kleinhirnfunktion. Doctoral Dissertation, University of Tubingen. [VB]

Summers, J. J. (1989) Motor programs. In: *Human skills*, ed. D. H. Holding. Wiley. [JJS]

(1992) Movement behaviour: A field in crisis? In: *Approaches to the study of motor control and learning*, ed. J. J. Summers. North-Holland. [JJS]

Taga, G., Yamaguchi, Y. & Shimizu, H. (1991) Self-organized control of bipedal locomotion by neural oscillators in unpredictable environment. *Biological Cybernetics* 65:147–59. [GMa]

Taira, M. Mine, S., Georgopoulos, A. P., Murata, A. & Sakata, H. (1990) Parietal cortex neurons of the monkey related to the visual guidance of hand movement. *Experimental Brain Research* 83:29–36. [MAG]

Tanaka, R. (1983) Reciprocal Ia inhibitory pathway in normal man and in patients with motor disorders. In: *Motor control mechanisms in health and disease*, ed. J. E. Desmedt. Raven Press. [aSCG]

Tanji, J. & Evarts, E. V. (1976) Anticipatory activity in motor cortex neurons in relation to direction of intended movement. *Journal of Neurophysiology* 39:1062–68. [aGEA, aEEF]

Tanji, J. & Kurata, K. (1985) Contrasting neuronal activity in supplementary and precentral motor cortex of monkeys. I. Responses to instructions determining motor responses to forthcoming modalities. *Journal of Neurophysiology* 53:129–41. [aGEA]

(1989) Changing concepts of motor areas of the cerebral cortex. *Brain & Development* 11:374–77. [JT]

Tanji, J., Okano, K. & Sato, K. C. (1988) Neuronal activity in cortical motor areas related to ipsilateral, contralateral and bilateral digit movements of the monkey. *Journal of Neurophysiology* 60:325–43. [aGEA, cSCG, JT]

Tantisira, B. (1990) An electrophysiological and morphologcal investigation of the axonal parojection and termination of single C3-C4 propriospinal neurones in the cat. Doctoral Dissertation, University of Goteborg. [AL]

Tarassenko, L. & Blake, A. (1991) Analogue computation of collision-free paths. In: *Proceedings of the 1991 IEEE International Conference on Robotics and Automation* 1. IEEE Computer Society Press. [CIC]

Taylor, A. & Gottlieb, S. (1985) Convergence of several sensory modalities in motor control. In: *Feedback and motor control in invertebrates and vertebrates*, ed. W. J. P. Barnes & M. H. Gladden. Croom Helm. [VD, CAP]

Taylor, J. L. & McCloskey, D. I. (1990a) Ability to detect angular displacements of the fingers made at an imperceptibly slow speed. *Brain* 113:157–66. [aSCG]

(1990b) Triggering of preprogrammed movements as reactions to masked stimuli. *Journal of Neurophysiology* 63:439–46. [aSCG]

Tehovnik, E. & Lee, K. (1990) Electrical stimulation of the dorsomedial frontal cortex of the rhesus monkey. 17:372.10. [RAA]

Teyke, T., Weiss, K. R. & Kupfermann, I. (1990) Appetitive feeding behavior of aplysia: Behavioral and neural analysis of directed head turning. *Journal of Neuroscience* 10:3922–34. [IK]

Thach, W. T. (1970) Discharge of cerebellar neurons related to two maintained postures and two prompt movements. I. Nuclear cell output. *Journal of Neurophysiology* 33:527–36. [aJRB, PFCG]

(1978) Correlation of neural discharge with pattern and force of muscular activity, joint position and direction of intended next movement in motor cortex and cerebellum. *Journal of Neurophysiology* 41:654–76. [aGEA, aJRB, aEEF, cSCG, JFK]

(1980) Complex spikes, the inferior olive, and natural behavior. In: *The inferior olivary nucleus: Anatomy and physiology*, ed. J. Courville, C. de Montigny & Y. Lamarre. Raven Press. [JMB]

Thach, W. T., Goodkin, H. G. & Keating, J. G. (1992) Cerebellum and the adaptive coordination of movement. *Annual Review of Neuroscience* 15:403–42. [RFT]

Thickbrook, G. W. & Mastaglia, F. J. (1985) Cerebral events preceding saccades. *Electroencephalography and Clinical Neurophysiology* 62:277–89. [aJFS]

Thier, P. & Andersen, R. A. (1991) Electrical microstimulation delineates three distinct eye-movement related areas in the posterior parietal cortex of the rhesus monkey. *Society for Neuroscience Abstracts* 1281. [RAA, APo]

Thilmann, A. F., Schwarz, M., Töpper, R., Fellow, S. J. & Noth, J. (1991) Different mechanisms underlie the long-latency stretch reflex response of active human muscle at different joints. *Journal of Physiology (London)* 444:631–43. [VD]

Thomas, P. K. & Brown, M. J. (1987) Diabetic polyneuropathy. In: *Diabetic neuropathy*, ed. P. J. Dyck, P. K. Thomas, A. K. Asbury, A. I. Winegrad & D. Porte. W. B. Saunders. [PRC]

Thompson, R. F. (1986) The neurobiology of learning and memory. *Science* 223:941–47. [aJRB, MH]

(1989) Role of the inferior olive in classical conditioning. In: *The olivocerebellar system in motor control*, ed. P. Strata. Springer-Verlag. [rJRB]

(1990) Neural mechanisms of classical conditioning in mammals. *Philosophical Transactions. Royal Society London* B 329:161–70. [RFT]

Thompson, R. F. & Steinmetz, J. E. (1992) The essential memory trace circuit for a basic form of associative learning. In: *Learning and memory: The behavioral and biological substrates*, ed. I. Gormezano & E. Wasserman. Iowa, in press. [RFT]

Thompson, S., Gregory, J. E. & Proske, U. (1990) Errors in force estimation can be explained by tendon organ desensitization. *Experimental Brain Research* 79:365–72. [aSCG]

Thompson, W. & Tait, P. G. (1886) *Treatise on natural philosophy*. Cambridge University Press. [rEB]

Thorner, G., Lange, H. & Hopf, A. (1975) Morphometrisch-statische Strukturanalysen des Striatum, Pallidum und Nucleus Subthalamicus beim Menschen. II. Pallidum. *Journal für Hirnforschung* 16:401–13. [aGEA]

Toda, H., Takagi, M., Yoshizawa, T. & Bando, T. (1991) Disjunctive eye movement evoked by microstimulation in an extrastriate cortical area of the cat. *Neuroscience Research* 12:300–306. [JWG]

Tolbert, D. L., Bantli, H. & Bloedel, J. R. (1977) The intracerebellar nucleo-cortical projection in a primate. *Experimental Brain Research* 30:425–34. [aJRB]

Tomlinson, R. D. & Robinson, D. A. (1984) Signals in vestibular nucleus mediating vertical eye movements in the monkey. *Journal of Neurophysiology* 51:1121–36. [aDAR]

Topka, H., Massaquoi, S. G., Zeffiro, T. & Hallett, M. (1991) Learning of arm trajectory formation in patients with cerebellar deficits. *Society for Neuroscience Abstracts* 17:1381. [MH]

Topka, H., Valls-Sole, J., Massaquoi, S. G. & Hallett, M. (1992) Eyeblink conditioning in patients with cerebellar deficits. *Movement Disorders*, in press. [MH]

Torebjörk, H. E., Vallbo, Å. B. & Ochoa, J. L. (1987) Intraneural microstimulation in man: Its relation to specificity of tactile sensations. *Brain* 110:1509–29. [aSCG]

Touretzky, D., ed. (1991) Special issue on connectionist approaches to language learning. *Machine Learning* 7:105–252. [JER]

Tracey, D. J. (1979) Characteristics of wrist joint receptors in the cat. *Experimental Brain Research* 34:165–76. [aSCG]

Trotter, Y., Celebrini, S., Thorpe, S. J. & Imbert, M. (1991) Modulation of stereoscopic processing in primate visual cortex {V}1 by the distance of fixation. *Society for Neuroscience Abstracts*, 1016. [APo]

Tomlinson, R. D. (1990) Combined eye-head gaze shifts in the primate. III. Contributions to the accuracy of gaze saccades. Journal of Neurophysiology 64:1873–91. [RAA]

Tsuda, I. (1984) A Hermeneutic process of the brain. *Progress of Theoretical Physics* (suppl.) 79:241–59. [IT]

(1991) Chaotic itinerancy as a dynamical basis of hermeneutics in brain and mind. *World Futures* 32:167–84. [IT]

(1992) Dynamic link of memory-chaotic memory map in nonequilibrium neural networks. *Neural Networks* 5:313–26. [IT]

Turner, R. S. (1991) Movement and instruction related activity in the globus pallidus of the monkey. Doctoral Dissertation, University of Washington. [aEEF]

Udin, S. B. & Scherer, W. J. (1990) Restoration of the plasticity of binocular maps by NMDA after the critical period in Xenopus. *Science* 249:669–72. [aGEA]

Ueki, A., Uno, M., Anderson, A. & Yoshida, M. (1977) Monosynaptic inhibition of thalamic neurons produced by stimulation of the substantia nigra. *Experientia* 33:1480–82. [aGEA]

Umilta, C. & Nicoletti, R. (1990) Spatial stimulus-response compatibility. In: *Stimulus-response compatibility: An integrated perspective*, ed. R. W. Proctor & T. G. Reeve. North-Holland. [RWP]

Ungerleider, L. G. & Mishkin, M. (1982) Two cortical visual systems. In: *The analysis of visual behavior*, ed. D. J. Ingle, M. A. Goodale, R. J. W. Mansfield. MIT Press. [aJFS, MAG]

Uno, M., Kozlovskaya, I. B. & Brooks, V. B. (1973) Effects of cooling interposed nuclei on tracking-task performance in monkeys. *Journal of Neurophysiology* 36:996–1003. [aJRB]

Vaccaro, D. D., Agarwal, G. C. & Gottlieb, G. L. (1988) Nonlinear mechanical behavior in striated muscle and its relationship to underlying crossbridge activity. IEEE *Transactions on Biomedical Engineering* 35:426–34. [GCA]

Vallbo, Å. B. (1971) Muscle spindle response at the onset of isometric voluntary contractions in man. Time difference between fusimotor and skeletomotor effects. *Journal of Physiology (London)* 318:405–31. [aSCG]

(1973a) The significance of intramuscular receptors in load compensation during voluntary contractions in man. In: *Control of posture and locomotion*, ed. R. B. Stein, K. G. Pearson, R. S. Smith & J. B. Redford. Plenum Press. [aEB, TRN]

(1973b) Muscle spindle afferent discharge from resting and contracting muscles in normal human subjects. In: *New developments in electromyography and clinical neurophysiology*, vol. 3: Human reflexes, pathophysiology of motor systems, methodology of human reflexes, ed. J. E. Desmedt. Karger. [aSCG]

(1974a) Afferent discharge from human muscle spindles in nocontracting muscles. Steady state impulse frequency as a function of joint angle. *Acta Physiologica Scandinavica* 90:303–18. [aSCG]

(1974b) Human muscle spindle discharge during isometric voluntary contractions. Amplitude relations between spindle frequency and torque. *Acta Physiologica Scandinavica* 90:319–36. [aSCG]

(1981) Basic patterns of muscle spindle discharge in man. In: *Muscle receptors and movement*, ed. A. Taylor & A. Prochaska. Macmillan. [JP]

Vallbo, Å. B. & Al-Falahe, N. A. (1990) Human muscle spindle response in a motor learning task. *Journal of Physiology (London)* 421:553–68. [aSCG, APr]

Vallbo, Å. B. & Hagbarth, K.-E. (1968) Activity from skin mechanoreceptors recorded percutaneously in awake human subjects. *Experimental Neurology* 21:270–89. [aSCG]

Vallbo, Å. B., Hagbarth, K.-E., Torebjörk, E. & Wallin, B. G. (1979)

Somatosensory, proprioceptive, and sympathetic activity in human peripheral nerves. *Physiological Reviews* 59:919–57. [aSCG]

Vallbo, Å. B., Olsson, K. A,. Westberg, K.-G. & Clark, F. J. (1984) Microstimulation of single tactile afferents from the human hand. Sensory attributes related to unit type and properties of receptive fields. *Brain* 107:727–49. [aSCG]

Van Essen, D. C. (1985) Functional organization of primate visual cortex. In: *Cerebral cortex*, vol. 3, ed. A. Peters & E. G. Jones. Plenum Press. [JMF]

Van Gisbergen, J. A. M. & Robinson, D. A. (1977) Generation of micro- and macrosaccades by burst neurons in the monkey. In: Control and gaze by brain stem neurons, ed. R. Baker & A. Berthoz. Elsevier. [AAF]

Van Gisbergen, J.A. M., Van Opstal, A. J. & Tax, A. A. M. (1987) Collicular ensemble coding of saccades based on vector summation. *Neuroscience* 21:541–55. [JAMVG]

Van Ingen Schenau, G. J. (1989) From rotation to translation: Constraints in multi-joint movements and the unique action of bi-articular muscles. *Human Movement Science* 8:301–337. [CAP, GJvIS]

Van Ingen Schenau, G. J., Boots, P. J. M., de Groot, G., Snackers, R. J. & van Woensel, W. W. L. M. (1992) The constrained control of force and position in multi-joint movements. *Neuroscience* 46:197–207. [GJvIS]

Van Opstal, A. J. & Van Gisbergen, J. A. M. (1989a) A nonlinear model for collicular spatial interactions underlying the metrical properties of electrically elicted saccades. *Biological Cybernetics* 60:171–83. [JAMVG]

(1989b) Scatter in the metrics of saccades and properties of the collicular motor map. *Vision Research* 29:1183–96. [JAMVG]

Vartanyan, G. A. & Klementyev, B. I. (1991) *Chemical symmetry and asymmetry of the brain.* Nauka. [MEI]

Vercher, J.-L. & Gauthier, G. M. (1988) Cerebellar involvement in the coordination control of the oculo-manual tracking system: Effects of cerebellar dentate nucleus lesion. *Experimental Brain Research* 73:155–56. [aJRB]

Verfaellie, M., Bowers, D. & Heilman, K. M. (1990) Attentional processes in spatial stimulus-response compatibility. In: *Stimulus-response compatibility: An integrated perspective*, ed. R. W. Proctor & T. G. Reeve. North-Holland. [RWP]

Vilis, T. & Hore, J. (1981) Characteristics of saccadic dysmetria in monkeys during reversible lesions of medial cerebellar nuclei. *Journal of Neurophysiology* 46:828–38. [aJRB]

Volpe, B. T., LeDoux, J. E. & Gazzaniga, M. S. (1979) Spatially oriented movements in the absence of proprioception. *Neurology* 29:1309–13. [aSCG]

Von Bonin, G. & Bailey, P. (1947) *The neocortex of Macaca mulatta.* University of Illinois Press. [aJFS]

von Helmholtz, H. (1910) *Handbuch der physiologischen Optik.* Translation by Southall, J. P. C. (1925) *Helmholtz's treatise on physiological optics*, vol. 3. Dover Publications. [JWG]

Voogd, J. & Bigare, F. (1980) Topographical distribution of olivary and cortico nuclear fibers in the cerebellum: A review. In: *The inferior olivary nucleus: Anatomy and physiology*, ed. J. Courville, C. de Montigny & Y. Lamarre. Raven Press. [aJRB]

Wade, N. J. & Swanston, M. (1991) *Visual perception: An introduction.* Routledge. [PQ]

Wall, P. D. & McMahon, S. B. (1985) Microneurography and its relation to perceived sensation. A critical review. *Pain* 21:209–29. [aSCG] ⱼ

Wang, J.-J., Kim, J. H. & Ebner, T. J. (1987) Climbing fiber afferent modulation during a visually guided, multi-joint arm movement in the monkey. *Brain Research* 410:323–29. [aJRB]

Wannier, T. M. J., Maier, M. A., Hepp-Ryemond, M.-C. (1991) Contrasting properties of monkey somatosensory and motor cortex neurons activated during the control of force in precision grip. *Journal of Neurophysiology* 6:572–89. [cSCG]

Wassef, M., Sotelo, C., Thomasset, M., Granholm, A.-C., Leclerc, N., Rafrafi, J. & Hawkes, R. (1990) Expression of compartmentation antigen zebrin I in cerebellar transplants. *Journal of Comparative Neurology* 294:223–34. [NHB]

Watson, J. D. G., Colebatch, J. G. & McCloskey, D. I. (1984) Effects of externally imposed elastic loads on the ability to estimate position and force. *Behavioral Brain Research* 13:267–71. [aSCG]

Watson, R. T., Valenstein, E., Day, A. & Heilman, K. (1985) Ablation of area 7 or cortex around the superior temporal sulcus and neglect. *Neurology* 35(suppl. 1):179–80. [BS]

Webster, W. (1977) Hemispheric asymmetry in cats. In: *Lateralisation in the nervous system*, ed. S. Harnad. Academic Press. [aJFS]

Weiland, G. & Koch U. T. (1987) Sensory feedback during active movements of stick insects. *Journal of Experimental Biology* 133:137–56. [JDe]

Weiner, M. J., Hallett, M. & Funkenstein, H. H. (1983) Adaptation to lateral displacement of vision in patients with lesions of the central nervous system. *Neurology* 33:766–72. [MH]

Weinrich, M. & Wise, S. P. (1982) The premotor cortex of the monkey. *Journal of Neuroscience* 2:1329–45. [aGEA, aEEF, JFK]

Weiser, M., McElligott, J. R. & Baker, R. (1988) Adaptive gain control of the vestibulo-ocular reflex in goldfish. II. Total cerebellectomy. *Society for Neuroscience Abstracts* 14:169. [rJRB]

Weiskrantz, L. (1986) *Blindsight*. Oxford University Press. [PG]

Weissman, B. M., DiScenna, A. O. & Leigh, R. J. (1989) Maturation of the vestibulo-ocular reflex in normal infants during the first two months of life. *Neurology* 39:534–38. [aDAR]

Weizsäcker, V. V. (1950) *Der Gestaltkreis*. Thieme. [JMF]

Welsh, J. P. & Harvey, J. A. (1989a) Cerebellar lesions and the nictitating membrane reflex: Performance deficits of the conditioned and unconditioned response. *Journal of Neuroscience* 9:299–311. [aJRB, RFT]

(1989b) Intra-cerebellar lidocaine: Dissociation of learning from performance. *Society for Neuroscience Abstracts* 15:639. [aJRB, RFT]

(1991) Pavlovian conditioning in the rabbit during inactivation of the interpositus nucleus. *Journal of Physiology* 444:459–80. [rJRB]

Westheimer, G. & McKee, S. P. (1975) Visual acuity in the presence of retinal-image motion. *Journal of the Optical Society of America* 65:847–50. [JMB]

Westling, G. & Johansson, R. S. (1987) Responses in glabrous skin mechanoreceptors during precision grip in humans. *Experimental Brain Research* 66:128–40. [aSCG]

Weyand, T. G. & Malpeli, J. G. (1990) Responses of neurons in primary visual cortex are influenced by eye position. *Society for Neuroscience Abstracts* 1055. [APo]

Wickens, J. R. (1991) Corticostriatal interactions in neuromotor programming. Paper presented at the *9th International Australasian Winter Conference on Brain Research*, Queensland, New Zealand. [JJS]

Widrow, B. & Stearns, S. D. (1985) *Adaptive signal processing*. Prentice-Hall. [PDN]

Wiener, N. (1961) *Cybernetics*, 2d ed. MIT Press. [ADK]

Wiesendanger, M. & Miles, T. S. (1982) Ascending pathway of low-threshold muscle afferents to the cerebral cortex and its possible role in motor control. *Physiological Reviews* 62:1234–70. [aSCG]

Wiesendanger, R. & Wiesendanger, M. (1985) The thalamic connections with medial area 6 (supplementary motor cortex) in the monkey (*Macaca fascicularis*). *Experimental Brain Research* 59:91–104. [aGEA]

Wilkie, D. R. (1954) Facts and theories about muscle. In: *Progress in biophysics and biophysical chemistry*, vol. 4, ed. J. A. V. Butler & J. T. Randall. Academic Press. [GCA]

Williams, R. J. & Zipser, D. (1989) A learning algorithm for continually running fully recurrent neural networks. *Neural Computation* 1:270–80. [RE]

Willshaw, D. & Dayan, P. (1990) Optimal plasticity for memory matrix neurones: What goes up must come down. *Neural Computation* 2:85–93. [cJFS]

Wilson, C. J. & Groves, P. M. (1980) Fine structure and synaptic connections of the common spiny neuron of the rat neostriatum: A study employing intracellular injection of horseradish peroxidase. *Journal of Comparative Neurology* 194:599–615. [aGEA]

Wilson, V. J. & Melvill, J. G. (1979) *Mammalian vestibular physiology*. Plenum Press. [VD]

Wilson, V. J. & Peterson, B. W. (1981) Vestibulospinal and reticulospinal systems. In: *Handbook of physiology. The nervous system.* Sect. 1, vol. 2, ch. 14. American Physiology Society. [VD]

Windhorst, U. (1989) Do Renshaw cells tell spinal neurons how to interpret muscle spindle signals? In: *Progress in brain research*, vol. 80. *Afferent control of movement and posture*, ed. J. H. J. Allum & M. Hulliger. Elsevier. [TMH]

(1990) Activation of Renshaw cells. In: *Progress in neurobiology*, vol. 35. Pergamon Press. [TMH]

Windhorst, U., Hamm, T. M. & Stuart, D. G. (1989) On the function of muscle and reflex partitioning. *Behavioral and Brain Sciences* 12:629–81. [aSCG]

Wing, A. & Kristofferson, A. (1973) Response delays and the timing of discrete motor responses. *Perception & Psychophysics* 14:5–12. [aJRB]

Winograd, T. & Flores, F. (1986) *Understanding computers and cognition: A new foundation for design*. Ablex. [IT]

Winters, J. M. (1990) Hill-based muscle models: A systems engineering perspective. In: *Multiple muscle systems: Biomechanics and movement organization.* ed. J. M. Winters & S. L.-Y. Woo. Springer-Verlag. [JMW]

Winters, J. M. & Peles, J. D. (1990) Neck muscle activity and 3-D head

kinematics during quasi-static and dynamic tracking movements. In: *Multiple muscle systems: Biomechanics and movement organization*, ed. J. M. Winters & S. L.-Y. Woo. Springer-Verlag. [JMW]

Winters, J. M. & Woo, S. Y.-L., eds. (1990) *Multiple muscle systems: Biomechanics and movement organization*. Springer-Verlag. [JMW]

Wise, S. P. & Mauritz, K.-H. (1985) Set-related neuronal activity in the premotor cortex of rhesus monkeys: Effects of changes in motor set. *Proceedings of the Royal Society of London* Series B 223:331–54. [aGEA]

Wolf, S. L., Lecraw, D. E., Barton, L. A. & Jann, B. B. (1989) Forced use of hemiplegic upper extremities to reverse the effect of learned nonuse among chronic stroke and head-injured patients. *Experimental Neurology* 104:125–32. [aSCG, PRC]

Wolpaw, J. & Carp, J. (1990) Memory traces in spinal cord. *Trends in Neuroscience* 13:137–142. [rDAM]

Wood, L., Ferrell, W. R. & Baxendale, R. H. (1988) Pressures in normal and acutely distended human knee joints and effects on quadriceps maximal voluntary contractions. *Quarterly Journal of Experimental Physiology* 73:305–14. [aSCG]

Woollacott, M., Bonnet, M. & Yabe, K. (1984) Preparatory process for anticipatory postural adjustments: Modulation of leg muscles reflex pathways during preparation for arm movements in standing man. *Experimental Brain Research* 55:263–71. [VD]

Woolsey, C. N. (1958) Organization of somatic sensory and motor areas of the cerebral cortex. In: *Biological and biochemical bases of behavior*, ed. H. F. Harlow & C. N. Woolsey. University of Wisconsin Press. [aEEF]

Wurtz, R. H. & Mohler, C. W. (1976) Organisation of monkey superior colliculus: Enhanced visual responses of superficial layer cells. *Journal of Neurophysiology* 39:745–65. [aJFS]

Xu, Y., Bennett, D. J., Hollerbach, J. M. & Hunter, I. W. (1989) Wrist-airjet system for identification of joint mechanical properties of the unconstrained human arm. *Society for Neuroscience Abstracts* 15:396. [aEB]

(1990a) A wrist-mounted airjet for studying human arm joint mechanical properties. *Canadian Medical and Biomedical Engineering Society Conference*, Winnipeg, Canada. [aEB]

(1990b) An airjet perturbation device and its use in elbow posture mechanics. *IEEE EMBS Conference*, in press. [aEB]

(1991) An airjet actuator system for identification of the human arm joint mechanical properties. *IEEE Transactions on Biomedical Engineering* 38:1111–22. [aEB]

Yang, J. F., Fung, J., Edamura, M., Blunt, R., Stein, R. B. & Barbeau, H. (1991) H-reflex modulation during walking in spastic paretic patients. *Canadian Journal of Neurological Sciences* 18:443–52. [RBS]

Yang, J. F. & Stein, R. B. (1990) Phase-dependent reflex reversal in human leg muscles during walking. *Journal of Neurophysiology* 63:1109–17. [JDu]

Yao, Y. & Freeman, W. J. (1990) Model of biological pattern recognition with spatially chaotic dynamics. *Neural Networks* 3:153–70. [IT]

Yeo, C. H. (1991a) Cerebellum and classical conditioning of motor responses. *Annals of the New York Academy of Sciences* 627:292–304. [PFCG]

(1991b) Cerebellum and classical conditioning of motor responses. In: *Activity driven CNS changes in learning and development*, ed. B. Boland, J. Cullinan & K. M. Mayer. New York Academy of Sciences. [cJFS, RFT]

Yeo, C. H., Hardiman, M. J. & Glickstein, M. (1985a) Classical conditioning of the nictitating membrane response of the rabbit. II. Lesions of cerebellar cortex. *Experimental Brain Research* 60:99–113. [aJRB]

(1985b) Classical conditioning of the nictitating membrane response of the rabbit. I. Lesions of the cerebellar nuclei. *Experimental Brain Research* 60:87–98. [aJRB]

Zee, D. S., Yamazaki, Z., Butler, P. H. & Gucer, G. (1981) Effects of ablation of flocculus and paraflocculus on eye movements in primate. *Journal of Neurophysiology* 46:878–99. [aJRB]

Zeki, S. (1990) Parallelism and functional specialization in human visual cortex. *Cold Spring Harbor Symposia on Quantitative Biology* 55:651–61. [RE]

Zipser, D. (1986) Programming neural nets to do spatial computations. *ICS Report*, University of California, San Diego. [cRAA]

(1991) Recurrent network model of the neural mechanism of short-term active memory. *Neural Computation* 3:179–93. [rEEF]

Zipser, D. & Andersen, R. A. (1988) A back-propagation programmed network that simulates response properties of a subset of posterior parietal neurons. *Nature (London)* 331:679–84. [aGEA, aDAR, arJFS, RAA, CLC, MRWD, RE, APo, GS]

Index

acuity, 89
acurate premotor area, 57, 58
adaptation, 67, 72, 73
algorithm, 54–62, 94, 96, 98
allocentric, 90
alpha-gamma dissociation, 24
alpha model, 2, 4, 5, 10, 11
alpha motoneurons, 22
amorphosynthesis, 94
area 5, 91, 92, 94
area 7, 92, 93, 97
area 39, 91, 97
area 40, 91, 97
asomatognosia, 94
association, 90, 92, 95, 98
association areas, 89, 91, 92, 95, 97
astereognosis, 94
attention, 90–8
axes, 90, 92, 95

back propagation, 43, 85
Balint's syndrome, 94
basal ganglia, 54, 55, 57–9
behavioral set, 79
beta motoneurons, 22
black-box modeling, 51
Bloedel, 64, 68–70, 73
body image, 64
Brooks, 64, 65

caudate cortex, 57, 58
center of mass, 2
cerebellar cortex, 68, 69, 72, 76
cerebellar nuclei, 64, 65, 67–9, 71, 76
cerebellum, 64–8, 70, 72–6
cerebral cortex, 55, 57–9, 62
chronic unit recording, 77, 84
cingulate, 90, 91
cingulate motor area, 58
climbing fiber, 64, 68–72
connectionist models, 59–63
contact instability, 6, 10
coordinate systems, 51, 55, 58, 89, 91, 94, 95, 97, 98
coordinate transformations, 47
corollary discharge, 18
corticomotoneuronal cells, 82, 83, 84
curl of the force field, 5
cutaneous receptors, 13, 21, 25

deafferentation, 27–9
demagnification, 90, 93
denial, asomatognosia, 94
dentate nucleus, 64–8, 71, 74
displacement, 79
distortions, 89, 95, 96
distributed signals, 46–8, 51
distributed systems, 90, 96, 98
dynamic selection hypothesis, 68–71
dynamics, 55, 57, 58

Eccles, 65, 68
efference copy, 49
engineering models, 55–9, 61, 62
equilibrium point, 3, 4, 6–10

equilibrium-point hypothesis, 1, 2, 6–11
event-related potentials, 93, 94
extrinsic coordinates, 1
eye movements, 64, 65, 67, 68, 74, 75
eye muscles, 45

fastigial nucleus, 64, 67
fast movements, 6, 10, 11
feedforward computation, 2
force fields, 1, 5, 8, 9
force sensation, 26
fovea, 89, 90, 92, 93, 97
frontal eye fields, 52
fusimotor neurons, 22

GABA, 57
gain-change hypothesis, 68–70
gamma motoneurons, 22, 25
globus rallidus, 57–9
glutamate, 57, 59
goal-directed eye movements, 49
granule cell, 64, 68

hand motion, 1
hemispheres, 64, 68
hemispheric specialization, 94, 95
hidden units, 43, 46, 47, 50, 85
hippocampus, 90
holographic, 84
holographic mechanisms, 84, 85
Hughlings Jackson, 16

Ib reflex pathways, 24
inferior olive, 68
inferior parietal lobule, 91, 94, 97
initiation, 90
intention, 90, 92
interposed nucleus, 64–7, 70, 73
intraparietal sulcus, 91
intrinsic muscles, 16, 29
inverse computations, 2
inverse dynamics, 1, 2, 6, 8, 10
inverse kinematics, 2
isotonic contraction, 16

joint receptors, 13–15, 19, 21, 22

kinematics, 55, 57, 58, 61
kinesthesia, 12, 19

lambda model, 2, 4, 5, 10, 11
lateral intraparietal area (LIP), 91–3, 97
learning algorithm, 45
length-tension curve, 1, 3, 10
limbic system, 90–2
Llinas, 68, 70, 72, 75, 76
local processing, 89
localization of function, 80, 83, 87
localizing objects, 89, 90, 94, 97, 98
locomotion, 66, 68–70, 73
locomotor, 66
long loop (latency) reflexes, 22, 26

map: egocentric, 90; retinotopic, 89; topographic, 89, 90, 95, 96, 98

medial superior temporal area, 91
medial temporal area (MT), 91
memory, 72, 73
method of steepest descent, 43
microneurography, 12
microstimulation, 1, 9, 21, 27
mossy fiber, 64, 68, 69, 70
motor axons, 29
motor commands, 12, 18, 27, 29
motor control, 54–7, 60, 61
motor cortex, 17, 27, 57–9, 80
motor error, 90, 97
motor learning, 72, 73,
motor programs, 54–7, 59–62
motor system, 54–7, 59–63
movement: arm, 5, 6, 10; body, 94; execution, 2, 6, 56, 57, 59, 60; eye, 89–98; finger, 89; head, 90, 95; limb, 90, 92–4, 98; natural, 12; neck, 92; illusory, 17; planning, 2, 6; sensation, 20; sensory surfaces, 89, 96; trunk, 92; voluntary, 95
movement preparation, 57–9, 63
multijoint arm movement, 1, 4, 5, 6
muscle force, 82
muscle length-tension relationship, 1, 3
muscle spindle endings, 14–16, 18, 22; primary, 17; secondary, 17, 24
muscle synergies, 1

neglect, 94, 95, 97
neural coding, 78, 84, 88
neural integrator, 45
neural network, 42–4, 46, 49, 60, 62, 91, 95, 96, 98
neural network models, 85–8
neuronal responses, 92, 93
NMDA receptor, 59
nuclear neurons, 67–71

oculomotor system, 44
overcompleteness, 44

parallel fiber, 64, 68, 70, 72, 76
parallel processing, 56, 62, 98
parietal cortex, 50
parietal lobe, 49
perceived heaviness, 26
plastic, 72, 73
population coding, 81, 83, 87
position sense, 19
positive feedback, 45
posterior parietal cortex (PPC), 89–98
posterior parietal lesions, 94, 95
posture, 1, 2, 4, 5, 8, 11
prefrontal, 90
premotor cortex, 57, 62
premotoneuronal cells, 82–4
prepositus hypoglossi nuclei, 45
presynaptic inhibition, 23
primary cortex, 89
proprioception, 16, 17
proprioceptive feedback, 3
propriospinal system, 23
psychophysical oberservations, 1
Purkinje, 64, 66, 68–72
putamen, 57, 58

Index

reaching task, 1
reciprocal inhibition, 23
reference frames, 89, 94, 95, 98
relative recruitment time, 82
relative timing, 78, 82, 87
representation: of space, 89, 91, 93, 95, 96, 98; of coordinate transformation rules, 95, 98
retinal error, 90
retinotopic saccades, 49
robotics, 54–6
rogue cells, 46
rubromotoneuronal cells, 82–4

saccades, 46, 49, 67
sagittal zone, 64, 65, 68–71, 76
selection bias, 79, 80
self-organization, 54, 55, 60–3
semicircular canals, 45
sequential recruitment, 78, 82
servo-assistance hypothesis, 4
set-related activity, 80
Sherrington, 16
single-joint movements, 4, 6
smooth pursuit, 46
somatosensory, 80
space: action, 97; arm, 97; auditory, 89, 91, 94; egocentric, 89, 90, 94; extrapersonal, 97; limb

movement, 89, 97; ocularmotor, 89, 94, 96; perceptual, 89; peripersonal, 97; personal, 97; proprioceptive, 97; psychological, 97; real, 89, 92; retinotopic, 89, 94, 96; somaesthetic, 89, 94; stereoscopic, 89; visual, 89
spatial coordinates, 1
spatial transformation networks, 44
spinal-cord premotor circuits, 8, 9, 11
spinal frog, 8, 11
spinal reflexes, 22
springlike muscle properties, 3, 4, 10
standing, 23, 25
stiffness, 1–7, 9–11; of the human arm, 10
stiffness matrix, 5, 7
stretch reflex, 44
superior colliculi, 52, 90–3
superior parietal lobule, 91, 94
superior temporal sulcus, 91
supervised learning, 60, 61, 62
supplementary motor area, 55, 57–60
synaptic plasticity, 59–62

task dependence, 26
task-induced bias, 79
temporal transformation networks, 44
tendon organs, 16, 26
Thach, 64, 65, 67, 70, 72

thalamus, 57, 59
theorem on implicit functions, 11
timing, 82
tonotopicity, 91
top down, 85
torque-angle curves, 1
transformation; coordinate, 93, 96, 98; sensorimotor, 89, 92
trial-and-error learning, 60, 61
trunk muscles, 17
Tsukahara, 64

unimodal velocity profile, 6
unsupervised learning, 60–2

vector: direction, 96; radial opponent, 92; retinal, 96
vector hypothesis, 81
vermis, 67, 68
Vermittler hypothesis, 76
vestibular, 64, 70
vestibular nucleus, 45
vestibulo-ocular reflex (VOR), 45, 47, 67, 72, 73
virtual position, 2, 3, 6–8
virtual trajectory, 2, 3, 6, 10, 11
visuospatial, 95

wiping reflex, 8